LANDSLIDES: EXTENT AND ECONOMIC SIGNIFICANCE

PROCEEDINGS OF THE 28TH INTERNATIONAL GEOLOGICAL CONGRESS: SYMPOSIUM ON LANDSLIDES / WASHINGTON D.C. / 17 JULY 1989

Landslides: Extent and Economic Significance

Edited by
EARL E.BRABB & BETTY L.HARROD
US Geological Survey, Menlo Park, California

Convened by: Earl E.Brabb, US Geological Survey & Medardo Molina, World Meteorological Organization
Cosponsored by: US National Academy of Sciences / US Geological Survey / Association of Engineering Geologists / International Landslide Research Group

A.A.BALKEMA / ROTTERDAM / BROOKFIELD / 1989

Endpapers: Map of the world showing average slope in 5-minute quadrangles, 9.3 km on a side at the equator. The darker the area, the higher the slope and the more likely that landslides are present. Prepared from National Geophysical Data Center (Boulder, Colorado) Tape ETOPO5. Analyses and image by Robert Mark and Douglas Aitken, US Geological Survey, 1988

The texts of the various papers in this volume were set individually by typists under the supervision of each of the authors concerned.

Published by
A.A.Balkema, P.O.Box 1675, 3000 BR Rotterdam, Netherlands
A.A.Balkema Publishers, Old Post Road, Brookfield, VT 05036, USA

ISBN 90 6191 876 6
© 1989 A.A.Balkema, Rotterdam
Printed in the Netherlands

Contents

Landslides: Extent and Economic Significance, Brabb & Harrod (eds)
© 1989 Balkema, Rotterdam. ISBN 90 6191 876 6

Preface

Earl E.Brabb

The theme common to the reports in this volume is that the extent and the economic impact of landslides in the world is considerable, though not well known. Primary to this fact is that few countries have an agency responsible for gathering information about the casualties and damage caused by landslides. This is unfortunate, because the information is needed so that the impact on human and socioeconomic activities can be assessed, and cost-effective programs for reducing the impact can be developed. Part of the difficulty in recognizing the scope of the problem is that much landslide damage is masked by its association with other events—floods, earthquakes, volcanic eruptions, hurricanes, or coastal storms, even though damage from the landsliding may exceed all other costs. Moreover, the actual damage is often categorized by other terms for the landslide process, such as flash flood, avalanche, clay failure, tidal wave, cliff collapse, coastal erosion, land sinking, cave-in, mudslide, road slip, hillside slip, land slip, and debris slip.

In the San Francisco Bay region of northern California, landsliding has been a problem throughout the century, but only recently has the extent of the problem become evident. Early in this century, geologic maps covering a substantial part of the region did not show a single landslide. By 1970, approximately 1,200 landslides had been recognized. By 1980, after special programs to map landslides had been carried out by the U.S. Geological Survey and the California State Division of Mines and Geology, over 70,000 landslides had been mapped. Even this large number was not representative for all landslide processes. In 1982, an unusual rainstorm triggered more than 18,000 additional landslides, most of them in areas different from those previously mapped.

Despite the major task of assessing the extent and economic impact of landslides, even a partial assessment can be used to develop an effective program to reduce damage. In the mid-1970's, the U.S. Geological Survey in cooperation with San Mateo County, California (a 1,380 km^2 area near San Francisco), produced landslide inventory and susceptibility maps and a partial assessment of landslide costs ($4.5 million) in the county for what they considered a typical wet year. This information was used by the county staff to prepare a county regulation to reduce the density of development in landslide-prone areas to one dwelling unit per 16 ha, and to require a geologic report for any proposed development. The regulation endured substantial criticism in public hearings before the county planning commission and board of supervisors, especially from landowners, but it was adopted intact in 1973. The overall impact of that regulation was to reduce potential development of the area from more than 500,000 dwelling units to only 2,500! Landowners wanting higher density than specified in the regulation may submit a geologic report in support of higher density, and the board of supervisors will approve higher densities where warranted. This provision for exception places the burden and expense of obtaining more detailed geotechnical information on those who want to develop the land.

The experience with landslide problems of the San Francisco Bay region, San Mateo County, and other areas mentioned in this book gives rise to procedures that have been or could be helpful in raising the awareness about landslides and in eventually reducing their hazard:

o Information about landslides and their costs must be gathered by geological, engineering, geographical, and forestry professionals for towns, cities, counties, provinces, and countries so that planners, engineers, and decisionmakers can determine the scope of the problem and whether hazard-reduction techniques are available and cost effective.

o The geological, engineering, geographical, and forestry professions must pay more attention to landslide inventories that document the geographic extent of the landslide problem. Considerable emphasis should be put on developing landslide inventories at scales from 1:100,000 to 1:500,000, because decisionmakers can relate more readily to maps that encompass entire counties, provinces, and countries, and because the time required to prepare these reconnaissance maps is not substantially greater than the time required to prepare maps of small areas in substantial detail. Of course, detailed maps are needed as well.

o Geologists and other geotechnical professionals must work to translate the results of such landslide studies into maps and reports that nontechnical users can readily understand. Landslide susceptibility maps that show the location, extent, severity, and likelihood of occurrence are one example of a translated product.

o Geologists and others should ensure that this translated information is transferred to those who will or should use it. Such transfer requires delivery, assistance, and encouragement.

o Geologists, engineers, and planners must work closely with decisionmakers and their staffs to select and adopt the most appropriate landslide hazard-reduction techniques.

o Geologists and other professionals must review the effectiveness of hazard reduction techniques after they have been in use, so that good techniques can be emphasized more widely, and ineffective techniques revised or eliminated.

Solicitation of authors for the chapters comprising this volume began in late 1984. Each author was provided an outline of the kind of information needed. By June 1986, 64 authors had committed to describe 136 countries and areas, including land beneath all the oceans. Several of these reports, unfortunately, were not completed. Reports received after publication of this volume will be provided to all those that express an interest by contacting E.E. Brabb at U.S. Geological Survey, MS 975, 345 Middlefield Road, Menlo Park, CA 94025.

It is hoped that these reports will raise the awareness about the landslide problem worldwide so that more action will be taken to reduce the hazard. Nearly all of the information on the economic extent of the landslide problem has not been previously reported, and the information on the extent and character of landsliding has not been previously assembled on a country-by-country basis. Although the information in this volume will be of greatest use to geoscientists, engineers, planners, and decisionmakers, it will probably also be of interest to international investors, donors, and issuers who are initiating and funding development projects.

Landslides are ubiquitous—although this may seem self-evident, the principal obstacle to major progress in coping with them is that they are perceived to be a local problem. The fact is that they cross political, social, and economic boundaries, killing people and destroying property; what's more, unlike tornadoes and earthquakes, landslides are largely predictable and preventable using existing technology. Recognizing landslides as a worldwide issue is a primary concern, but beyond that, national and international leadership is needed to muster the resources, knowledge, and skills for reducing landslide hazards. This book is a start toward both. Recognizing a problem is the first step in solving it.

Earl E. Brabb
U.S. Geological Survey

Landslides: Extent and Economic Significance, Brabb & Harrod (eds)
© 1989 Balkema, Rotterdam. ISBN 90 6191 876 6

Foreword

Frank Press

Natural hazards, such as landslides, earthquakes, and floods, do not observe any national boundaries. On a global scale, they occur on an almost daily basis. They affect nations large and small, rich and poor. A single major natural disaster event can cause serious effects throughout an entire country or region. Such natural hazards have proven particularly devastating to developing countries.

Among natural hazards, landslides occur in virtually every country in the world. They result from heavy rains, melting snow or ice, earthquakes, volcanoes, and human activities. The most notable event in recent history was the Reventador, Ecuador, landslides of March 1987, caused by a magnitude 6.9 earthquake following a month of heavy rains. It caused 1,000 deaths and ruptured the trans-Ecuadoran oil pipeline—the nation's prime economic asset—resulting in approximately a $1.5 billion loss.

According to a recent National Research Council report, "Reducing Losses from Landsliding in the United States," on average the annual economic losses from landslides, subsidence, and other ground failures exceed those from all other natural hazards combined. They annually cause at least $1-2 billion in economic losses and 25 to 50 deaths. Despite a growing geological understanding of landslide processes and a rapidly developing engineering capability for landslide prediction and control, losses from landslides are continuing to increase. This is largely a consequence of residential and commercial development that continues to expand onto the steeply sloping terrain that is most prone to landsliding. The need to reverse the upward trend of such losses, and the belief that the technologies exist to reduce human and property losses from other hazards as well, led to the passage of a United Nations' resolution in December 1987 designating the 1990s as an International Decade for Natural Disaster Reduction (IDNDR).

The Decade will take a three-pronged approach to reduce catastrophic life loss, property damage, and the social and economic disruption caused by natural hazards. First, it will catalogue and widely disseminate what we already know about hazard mitigation and also identify gaps in that knowledge. Second, it will adapt known mitigation and preparedness techniques to each nation's unique circumstances. And, third, it will carry out a coordinated research and education program to address the gaps in knowledge and to pioneer improved mitigation practices. In essence, the Decade will be a program of shared knowledge and the shared pursuit of new knowledge. It is to be an international partnership in the truest sense, with each participating nation coming away more knowledgeable and better prepared to cope with the violent forces of nature.

The National Research Council's Division of Natural Hazard Mitigation has been focusing its efforts on addressing natural hazard reduction issues since the 1964 great Alaska earthquake. For many years, an interdisciplinary approach has been taken by the division in conducting special studies and in carrying out quick-response postdisaster investigations which are participated in by experts from the United States and abroad. This volume, which presents and discusses the extent and economic significance of landslide problems in the world, is an excellent example of how issues related to a key natural hazard can be addressed interdisciplinarily and on a global scale.

I wish to congratulate the editor and authors for their efforts in preparing this significant publication and also to express the National Research Council's commitment in continuing to address the issues in natural hazard reduction, especially in light of the International Decade for Natural Disaster Reduction program.

Frank Press, President
U.S. National Academy of Sciences

Landslides: Extent and Economic Significance, Brabb & Harrod (eds)
© *1989 Balkema, Rotterdam. ISBN 90 6191 876 6*

Foreword

Dallas L.Peck

U.S. Geological Survey interest in landslides goes back at least to the late 1800's when I.C. Russell described topographic features produced by landslides along the Columbia River and other areas in Washington and Oregon, and N.S. Shaler discussed the possibility of creating landslides when constructing roads in mountainous regions. In 1915 and 1916, Whitman Cross and G.F. Becker of the USGS participated in a committee of the National Academy of Sciences appointed by President Woodrow Wilson "to consider and report upon the possibility of controlling the slides which are seriously interfering with the use of the Panama Canal." During the first and second World Wars, Survey geologists provided information for military operations, including the stability of trenches and foundations and the trafficability of vehicles on various slopes.

Post World War II landslide investigations were largely part of a series of projects in rapidly-expanding urban areas designed to provide general background engineering geology maps to guide development. The 1964 Alaskan earthquake, with extensive loss of life and property related to landsliding, demonstrated the terrible cost of ignoring reports showing where landslides are likely to be a problem.

A surge of interest in environmental concerns in the late 1960's and early 1970's and the passage of the National Environmental Policy Act by Congress in 1969 and similar laws by state and local governments stimulated research on landslide occurrence and the mechanisms of failure. A pilot geological hazard program in cooperation with the Department of Housing and Urban Development in the San Francisco Bay region provided planners and decisionmakers with information about 70,000 landslides and predictive maps showing which areas in addition to the landslides are likely to have slope stability problems in the future. Regional and local governments have used these maps to greatly reduce potential development in the most hazardous areas and to provide guidelines for safer development in less hazardous areas.

Knowledge about recognizing and understanding landslide processes has been shared with several countries, mostly under auspices of the U.S. Agency for International Development. U.S Geological Survey landslide projects in Brazil, Costa Rica, and Bolivia have provided information for better utilization of land resources, whereas responses to earthquake- and volcano-triggered landslides, disasters or potential disasters in Nicaragua, Chile, Colombia, Yugoslavia, Guatemala, Argentina, Ecuador and New Guinea have assisted emergency-response teams and planners in coping with the problems. Exchange programs for joint research on landslides began with the People's Republic of China in 1984 and with the Italian National Research Council in 1987, and an informal exchange with the Japanese National Center for Disaster Prevention has been carried out for a number of years.

In 1980, the United States Congress authorized funding for the U.S. Geologial Survey's Landslide Hazards Program. In so doing, it established the U.S. Geological Survey as the lead federal agency for landslide research with a goal of reducing landslide hazards within the United States. Although investigations of landslides were previously supported under a variety of different U.S. Geological Survey programs, and elsewhere within the government, establishment of a separately funded landslide program focused scientific efforts and leadership and has resulted in a number of significant domestic and international accomplishments.

Major domestic accomplishments of the Landslide Program include 1) key scientific breakthroughs in the understanding of landslide mechanisms, 2) development of new techniques for landslide hazard assessment and translation of these techniques to local and state officials for hazard reduction and mitigation, and 3) increased recognition of the Nation's landslide hazard through assessments of yearly damage and institution of cooperative Federal-State landslide assessments.

Investigations of landslides triggered by the 1976 Guatemala; 1977 San Juan, Argentina; 1978 Miyagi-Ken-Oki, Japan; 1979 Montenegro, Yugosalvia; 1985 Mt. Ontake, Japan; and 1987 Ecuador earthquakes, in addition to domestic earthquakes, have added valuable information for developing and refining methods of assessing slope stability during earthquakes, which if applied in seismic microzonation could result in significant reduction in loss of life during earthquakes worldwide. Following torrential rains and catastrophic landslides in Brazil during the spring of 1988, applicable methods developed in the United States for identifying hazardous areas and for alerting those in danger were conveyed to officials facing an increasingly more serious problem daily with growing urban populations and decreasing developable safe sites.

This book, therefore, is a natural extension of our long interest in landslides and concern that the problem be recognized and considered in the decisionmaking process. Mitigation programs are most cost-effective when implemented in the early stages of development. Once the problem is fully appreciated, mitigation is usually a small part of the development budget. In calling attention to the pervasiveness of landsliding and the largely hidden costs, we hope to persuade all nations to begin mitigation programs or to support more fully those programs already underway.

Dallas L. Peck
Director, U.S. Geological Survey

Foreword

John W.Williams

Landsliding is the most natural of geologic processes. Given that the driving force for these downward movements is gravity, and the surface of the earth has and always will contain irregularities (i.e. the topography), the earth's population does and can anticipate interacting with these slope changes. The problems and challenges for the geologists, the engineering geologists in particular, is to limit the detrimental interaction. There are many complex aspects to the challenge of trying to limit the world's losses from slope instabilities which range from the mechanical aspects of the slope, to human nature and the political environment.

Our understanding of the mechanics of landsliding is ahead of our ability to use that information in the most effective fashion. One of the reasons that people are not able to utilize this geotechnical information effectively is our human response to problems, particularly disasters involving the loss of life and/or significant property damage. Memories are very short. People quickly forget the circumstances which have created past difficulties. We rebuild on the same site or in the environment which was the location for geologically induced failure just a short time earlier. Our continual reoccupation of floodplains and run-out areas for debris flows or avalanches with disastrous consequences are most depressing examples of our inability to learn from the past.

A major challenge to deal effectively with landslides is that the speed and, therefore, the perceived significance of these failures is highly variable, ranging from the dramatic collapse of mountains of rock to the imperceptible distortion of large slope areas. People generally do not get excited about things which occur so slowly that it is hard to believe something is happening. The attitude is "If I cannot see something happening, it must not be happening. If it is not happening, it probably won't happen, and, therefore, nothing needs to be fixed or avoided."

Considerations of slope instabilities often are believed to be a luxury which only concerns the more wealthy. In reality, the opposite may be more true. The poorer elements of the world's population need to be more concerned because their survival may depend on it. The poor farmer trying to feed his family in a developing country may (or may not) realize that the over-cultivation of a slope can lead to its failure or that over-collecting of wood and the resultant removal of vegetation can produce landslides. All that is in the realm of possibilities and the future. The immediate need for the farmer is food and fuel so that the family can survive. It is easy to understand the choice which will be made even though it may be a very unwise decision. These ill-considered decisions simply make survival more difficult. In those parts of the world where survival is not as challenging, and the population density and demands on the limited land surface are not as great, landuse regulation can be implemented more easily.

The problem is not that society does not have enough technical "know-how" about detecting potential slope instabilities or providinng appropriate remedial actions. The problem is getting that information appreciated and used. Education about the field of engineering geology and landslides is not enough. Engineering geologists must educate and demonstrate that it makes sense to be concerned about creating or making worse already existing slope instabilities. The individual landowner's and political authorities' points of view, particularly for the long term, must be changed. This is the challenge and until it is met, we can continue to develop better computer programs to tell us what we already know—that hills will be reduced by landsliding. We will fail to preserve and to better the environment for living for the majority of the earth's citizens. Society hears frequently about technology transfer. That's only part of the story. It is necessary to transfer the reasons for having and using the technology as well as just transferring the technology.

The most sophisticated and the latest in "high tech" research in slope stability form only a small and the final portion of the solution to the problem of the loss of life and property. The adequate understanding and mitigation of slope stability problems depends upon a comprehensive understanding of the basic geology. The latest high speed computer factor of safety analysis making use of poor geologic information does not provide good results. A false impression is given to the user of such results because of the mistaken belief that the answer must be good because of the computer's massaging of the data. These mistakes lead to exposing individuals and property to unreasonable risks or causing needless expenses associated with unnecessary mitigative measures. As is true in all scientific investigations, solutions are no better than the basic elements or premises on which they are built. The most basic element of slope stability is geology.

An adequate resolution of the problem of slope stability anywhere in the world depends upon recognizing the need and value of solutions, an understanding of the geology, and the ability to handle the appropriate data. It is towards these ends that this volume has been prepared.

John W. Williams
President, Association of Engineering Geologists

Landslides: Extent and economic significance in Canada

D.M.Cruden & S.Thomson
University of Alberta, Department of Civil Engineering, Edmonton, Alberta, Canada

B.D.Bornhold
Geological Survey of Canada, Pacific Geoscience Centre, Sidney, BC, Canada

J.-Y.Chagnon & J.Locat
Université Laval, Département de Géologie, St. Foy, Quebec, Canada

S.G.Evans & J.A.Heginbottom
Geological Survey of Canada, Terrain Sciences Division, Ottawa, Ontario, Canada

K.Moran & D.J.W.Piper
Geological Survey of Canada, Atlantic Geoscience Centre, Dartmouth, Nova Scotia, Canada

R.Powell
Geocon Inc., Missisauga, Ontario, Canada

D.Prior
Coastal Studies Institute, Louisiana State University, Baton Rouge, La., USA

R.M.Quigley
University of Western Ontario, Department of Civil Engineering, London, Ontario, Canada

ABSTRACT: Regions of Canada prone to slope movements include the offshore, the Canadian Cordillera, the Interior Plains, permafrost areas, the St. Lawrence Lowlands and the shores of the Great Lakes. The historic record of landsliding is short and localized as average population density is low, at 2.6 persons per km^2, and over 75% of the population live in cities.

Landsliding is a hazard specifically excluded from comprehensive residential insurance policies in Canada. Local and provincial governments may protect the public by zoning areas at risk as hazard land. Judicial decisions have protected these governments from liability for losses in land value arising from land zoning decisions. Recovery of losses of property in slope movements has depended on ex gratia payments by the governments.

Knowledge of the extent of landsliding in Canada is not complete. The problems with debris flows on the Pacific Coast, obvious in retrospect, were not predicted 20 years ago. Further mapping of offshore landslides is needed to determine their effects on structures and to assess seismic hazard. Other surprises may be waiting in the less well known parts of the country. It is also clear from the history of slope movements on the Interior Plains that development itself may reduce slope stability.

Independent estimates of annual loss need to be made in Canada. They may show losses of hundreds of millions of dollars per year. As a first step, a catalogue of damaging historic landslides in Canada has been set up and agencies collecting information on landslides are encouraged to contribute information both on historic and current landslide problems.

1 INTRODUCTION

Canada is the second largest country in the world. The landmass has an area of 9 970 610 km^2 of which 755 165 km^2 are covered by freshwater. It extends from 52° 37'W in Newfoundland to the Alaskan border at 141°W and spans 4634 km from 41° 41'N in Lake Erie, to 83° 05'N on Ellesmere Island. The offshore areas of Canada, including Hudson's Bay, cover over 6.5 million km^2.

We focus on regions of Canada prone to slope movements. Bornhold and Prior describe the Pacific offshore, and Piper and Moran the Atlantic and Arctic offshore. As the offshore is not well known, it has been described in more detail than the Canadian Cordillera, for instance, where recent reviews are available (Evans and Clague, 1988b). Evans discusses the Canadian Cordillera, Thomson the Interior Plains, Heginbottom contributes the section on permafrost, Chagnon and Locat review slope movements in the St. Lawrence Lowlands and Powell and Quigley survey movements along the shores of the Great Lakes (Figure 1.1). These regional contributions are summaries of papers with Tables of Historic Damaging Landslides collected in Heginbottom et al. (1989).

Each section follows a similar outline, commenting first on the geology of the region, with an emphasis on movement-prone materials and then describing movements. Finally, social and economic impacts are assessed. The Introduction reviews the population distribution, physiography, seismicity and climate of Canada before explaining the choice of the regions described and commenting on country-wide economic issues. Costs are adjusted to 1986 Canadian dollars unless otherwise specified (Statistics Canada, 1988).

The incidence of landsliding in Canada cannot be assessed reliably from existing catalogues. For instance, Spurek's (1972) encyclopedic compilation of historic destructive landslides listed 13 from Canada to give a density of 1 per 708,880 km^2 (excluding the landmass under freshwater); the 383 from Czechoslovakia gave a density of 1 per 335 km^2. Heginbottom et al. (1989) expand Spurek's Canadian list by over an order of magnitude. Among the factors responsible for this difference in density is the length of the historic record. Spurek's first landslide in Eastern Canada dates from 1823, the first in the West from 1903, the first in Czechoslovakia from 1531. The first airphoto coverage of much of Canada is less than 40 years old; before that, landslides occurred whose age is now very difficult to determine. For example, the huge prehistoric landslides described by Eisbacher (1979) from the Mackenzie Mountains are known only to be Holocene, the first historic slide of this type occurred in 1985 (Evans et al., 1987).

Not only is the historic record short but it is also localized for with a population of 25 million, average population density is low, at 2.6 persons per km^2. Canada is also highly urbanized, over 75% of the population live in cities, 58% in the St. Lawrence Lowlands. The southern Interior Plains contain five major cities, Calgary, Edmonton, Regina, Saskatoon and Winnipeg. The population of British Columbia is concentrated within the Fraser River delta and coastal plains of the Vancouver region.

1.1 Physiography

Canada is composed of three major regions: a core of Precambrian crystalline rocks forming the Shield,

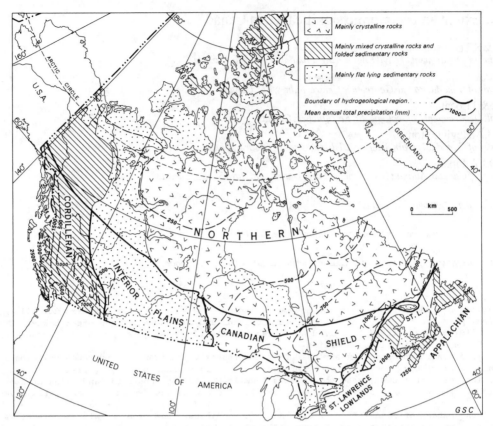

Figure 1.1 Physiographic and Hydrogeological Regions of Canada (after Geological Survey of Canada).

a surrounding crescent of younger, mainly strati-
fied rocks forming the Borderlands and the extensive
submarine areas of the Pacific, Atlantic and Arctic
shelves underlain by mainly Mesozoic and Cenozoic
strata (Figure 1.1). The surface of the Shield has
a flat, depressed centre in Hudson's Bay and an
outer rim terminated by a steep edge. The younger
rocks of the Borderlands surround the southern,
northern, and western parts of the plate as segments
of two concentric rings. The inner ring is comprised of the St. Lawrence Lowlands, Arctic Lowlands
and Interior Plains. The outer ring is formed by
mountains and plateaus, the Cordillera in the west,
the Appalachians in the southeast, and, in the
Innuitian region to the north, the mountains of the
Arctic Archipelago. The continental shelves border
the Atlantic, Pacific and Arctic coasts and have a
total area of approximately 1 354 570 km^2. The
Atlantic shelf is the largest, with an area of
582 750 km^2 and a continental slope of 321 160 km^2.
The Arctic and Labrador shelves have areas of
347 060 km^2 and 323 750 km^2 respectively. The
Pacific shelf is much smaller, with an area of
101 010 km^2 and a continental slope of 41 440 km^2
(Douglas 1970).

1.2 Seismicity

Seismic activity has induced landslides. There have
been earthquakes of greater than magnitude 7 on the
Atlantic continental margin, in the St. Lawrence
Lowlands, along the Pacific coast and in the Arctic
(Milne et al 1970). The seismic zoning map shows
the expected peak horizontal acceleration (Figure
1.2). Only the Pacific coast seismicity occurs
along an active plate margin, as the discussion of
the Pacific offshore shows. Intra-plate seismicity
may be related to rebound from loading by the large
ice sheets that covered most of Canada during the
Pleistocene.

1.3 Climate

Climate is controlled by the geography of North

America, by the movement of air from west to east
and by ocean currents. Latitude is the main factor
responsible for the westerly air flow. The
Cordillera acts as a barrier to the Pacific surface
westerlies and generates copious orographic
precipitation (Hare and Thomas 1979), west coast
precipitation often exceeds 1500 mm annually. So in
the Coast Ranges of the Cordillera, the most common
and destructive landslides are debris avalanches and
debris flows. These failures are triggered by heavy
rains that saturate the surface layer of the soil.
As this saturated mass moves down the steep slopes,
it gains both volume and velocity (Van Dine 1985).
The Great Lakes and the St. Lawrence Lowlands also
receive heavy precipitation. The Interior Plains
and the Arctic experience a more continental
climate, high Arctic regions may receive less than
100 mm of precipitation annually.

Another climatic control is shown in permafrost
areas, where many slope movements are caused by the
thawing of ground ice exposed by sea or river erosion or by the destruction of the insulating vegetation cover by forest fires. Retreats of slope
crests can be several metres per year (McRoberts and
Morgenstern, 1974). The depth of the active layer,
the layer of soil or rock that thaws in summer and
freezes in winter, is from 10 cm on Ellesmere
Island, to 15 m at high altitudes in the southern
Canadian Rockies (Figure 1.3).

1.4 Natural Materials

The distribution of movement-prone soils also
affects the distribution of landslides. The soft
soils shown in Figure 1.4 (Quigley, 1980, Figure 1)
reflect proglacial and postglacial sedimentation
during the retreat of the Wisconsin ice sheet. The
Champlain Sea, for instance, which existed between
12,500 and 10,000 B.P. in the St. Lawrence Lowlands,
deposited muds that can quickly transform from
brittle soil to viscous fluid when sufficiently
disturbed. Movements can continue into the
off-shore. Submarine mass movements can also be

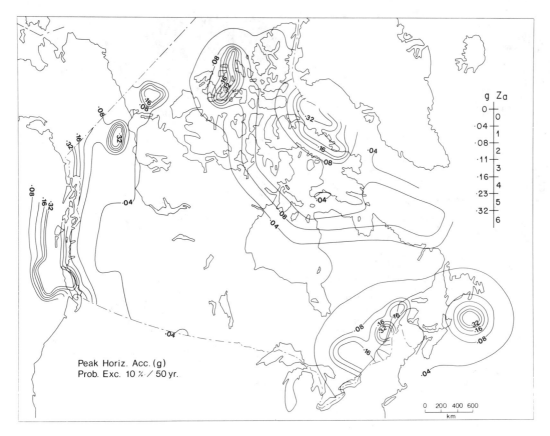

g Za
0 — 0
·04 — 1
·08 — 2
·11 — 3
·16 — 4
·23 — 5
·32 — 6

Peak Horiz. Acc. (g)
Prob. Exc. 10 % / 50 yr.

0 200 400 600
        km

Figure 1.2 Seismicity; the contours are of the peak horizontal acceleration which has a 10% chance of being exceeded in 50 years. (Geological Survey of Canada)

triggered by steep underwater slopes, large tidal ranges and earthquakes. Besides the events on fan deltas and in fjords, massive landsliding occurs on the continental margins and landslide-generated turbidity currents may transport sediments hundreds of kilometres.

On the Interior Plains, glacial sediments rest on soft rocks of Upper Cretaceous age. During deglaciation, rivers, diverted from their preglacial courses, rapidly eroded deep, steep-walled post-glacial valleys. Recent lateral erosion by these rivers of the toes of the slopes reactivates ancient landslides. Because the strata are flat-lying, the displaced masses move dominantly by translation.

Rock slope movements occur where erosion or excavation has exposed rock throughout the Canadian Cordillera, along rocky coastlines and in deeply cut river valleys in eastern Canada. Movement-prone rocks in the Cordillera were formed during or after the latest (Upper Cretaceous to Oligocene) Laramide orogeny. The products of deep Tertiary weathering form large slow-moving flows in the Intermontane Belt within the Cordillera (Bovis 1985). Here, too, Tertiary volcaniclastic rocks suffer large movements where glacial meltwater has eroded channels into late Tertiary lava capped plateaus (Evans 1984a). Quaternary volcanic centres are the most hazardous areas in the Cordillera.

1.5 Policy

The Geological Survey of Canada (GSC) is responsible for the characterization of geological, geomorphological and geotechnical properties of surface and near-surface materials and the dynamics of stability relations of terrain pertinent to the various uses of material or terrain by man and their relation to hazards or pollution. The GSC cooperates with Provincial Geological Surveys and other agencies.

Regional programs of landslide risk mapping have been undertaken by provincial agencies in British Columbia, Manitoba, Ontario, Quebec and Saskatchewan. The Geological Survey of Canada has noted major slope movements during surficial geology mapping programs and has pioneered studies of typical movements.

The information from these activities is particularily necessary to the public as landsliding is a hazard specifically excluded from comprehensive residential insurance policies in Canada. Adverse selection of risks is the reason generally given for this exclusion. Local and provincial governments may protect the public by zoning areas at risk as hazard land. Judicial decisions have protected these governments from liability for losses in land value arising from land zoning decisions (unless the zoning is carried out or administered in a negligent manner). Recovery of losses of property in slope movements has depended largely on ex gratia payments by the various levels of government.

The consequences of some landslides may extend beyond provincial boundaries. Canada is a major participant in world commodity trade. Some of its major exports are produced thousands of kilometres from its borders and carried along extensive transportation corridors. Where traffic is particularly dense and alternative routes considerably longer, parts of these corridors may be strategic. The Fraser River Canyon, for instance, holds two of Canada's four rail links to the Pacific, the Trans Canada Highway, major electrical transmission systems and a rich salmon fishery. Landslides in the Canyon have had serious economic consequences. Some crossings of the deeply incised river valleys on the Interior Plains may also be of strategic importance to the economy, as the Peace River Bridge failure illustrated. The denser transportation network of eastern Canada may still be vulnerable in

3

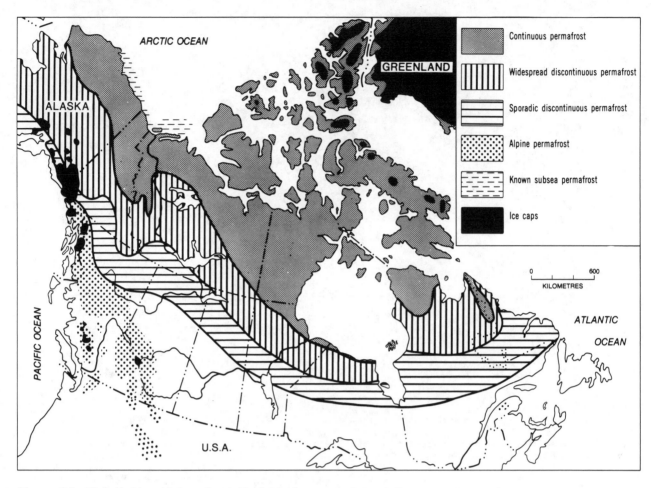

Figure 1.3 Distribution of Permafrost (Geological Survey of Canada)

the St. Lawrence Valley. The Federal Government
thus has a role to play in risk assessment.

2 PACIFIC OFFSHORE

Evidence for submarine slope movements in the offshore areas of western Canada includes damage to
buildings and wharves, distinctive bottom morphology
and sediment geometries recorded by acoustic surveying methods, and local phenomena such as unusual,
large, isolated waves in inlets. The processes
include massive landsliding (volumes up to $10^9 m^3$) on
the continental margin, debris flows in fjords
($10^7 m^3$), local avalanching of sands and boulders on
fan deltas and landslide-generated turbidity
currents which transport large volumes of sediment
up to 50 km.

Submarine mass movements have been caused by steep
underwater slopes, rapid deposition of underconsolidated sediments, large tidal ranges and earthquake
loading. In particular, the convergence and subduction of the Juan de Fuca and Explorer Plates beneath
Vancouver Island and the transform fault west of the
Queen Charlotte Islands, separating the Pacific and
America Plates, result in the high relief Coast and
Insular Mountain belts, in a steep, rugged continental margin, and in the highest seismicity in Canada
(Fig. 2.1). Glaciation, influenced by the principal
tectonic elements, has left thick accumulations of
unconsolidated sediments and carved deep fjords
along the western mainland coast, the west coasts of
the Queen Charlotte Islands and Vancouver Island.
Isostatic rebound following glaciation has exposed
sensitive marine clays in coastal areas; slope
failures in this material often continue into the
offshore. Superimposed on these geologic factors

are high rainfalls resulting in the rapid delivery
of sediments to deltas and fjords, large tides and
high waves on the exposed parts of the coast.

We discuss mapping off the Pacific Coast and then
consider submarine slope failures in fjords, the
Strait of Georgia, the open continental shelf, and
the continental slope.

2.1 Pacific Offshore Mapping

The most extensive offshore surveys are by the
Canadian Hydrographic Service for nautical charts.
Though of broad coverage and excellent quality,
these bathymetric data only suggest areas with a
history of or potential for slope failure. Echo
sounding can reveal topographic irregularities which
may be due to landsliding, but often such evidence
is inconclusive. However, hydrographic data
supported by sidescan sonar swaths and high resolution sub-bottom seismic profiles yield a three-
dimensional picture of diagnostic landslide
geometries.

The Geological Survey of Canada has been investigating the surficial geology of the British Columbia
offshore using acoustic profiling and side scan
sonar since the early 1970s but has focussed on
landslide hazards only in the southern Strait of
Georgia, Queen Charlotte Sound and selected fjords.
Similarly, surveys for various coastal and offshore
projects, such as cable and pipeline corridors,
small harbours and industrial installations, have
identified submarine slope failures only at specific
sites, such as Kitimat Arm and in the southern
Strait of Georgia (Figure 2.1). Consequently, there
are insufficient data to assess regional landslide
risk for the Pacific off shore and there is no

4

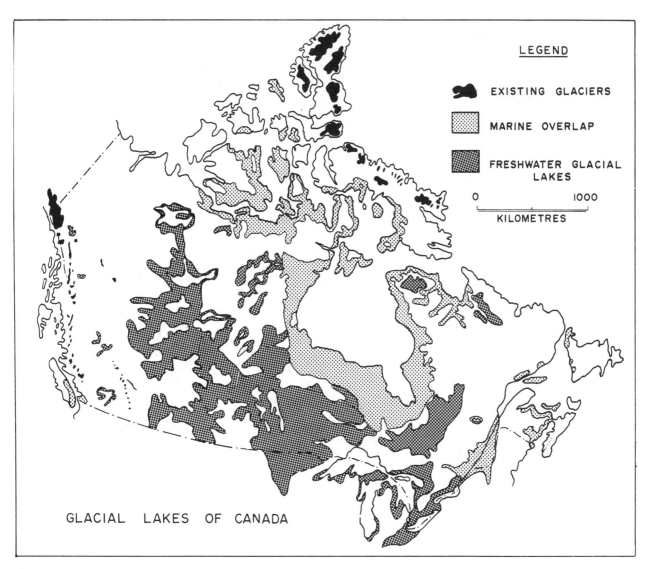

LEGEND

EXISTING GLACIERS

MARINE OVERLAP

FRESHWATER GLACIAL LAKES

0   1000
KILOMETRES

GLACIAL LAKES OF CANADA

Figure 1.4 Quaternary deposits susceptible to Slope Movements

program planned which will lead to such an assessment.

2.2 Fjords

Submarine slope movements in coastal areas are most common in fjords because of the high rates of sediment accumulation on their steep slopes. Landslides on fjord-head deltas, small fan deltas along the sides of inlets and on steep side walls near heads of fjords have resulted in either property or environmental damage.

The style and extent of failures in fjords are related to the texture of the materials; cohesive muds form debris flows, clean sands are transported long distances by turbidity currents and gravels are confined to chutes on fan deltas, not travelling beyond the base of the delta prism.

The best studied example of submarine failures of cohesive muds in a fjord is from Kitimat Arm (Bornhold, 1983; Prior et al., 1982, 1984; Johns et al., 1986). The event occurred on April 27, 1975 after an extreme low tide in Moon Bay where the failure retrogressed to the coastline and included subaerially exposed marine muds. A 75 m by 20 m pile dock disappeared and local waves, 8.2 m in height (Murty, 1979), eroded the coastline; direct damage was $ 1.32 M, the most serious of several similar events since the mid-1950s. The main

factors in the failure were rapid accumulation of weak sediments on the steep side wall near the head of the inlet, excessive loading of these sediments by construction of a crib wharf and rapid tidal drawdown (Johns et al., 1986).

Some fjord-head deltas are dominated by sands whose failure differs significantly from the cohesive muds. Bute Inlet, (Prior et al., 1986), is 80 km long and, at 660 m, one of the deepest fjords in British Columbia. At its head is a delta complex formed by two rivers which have large permanent ice fields within their drainage basins; these rivers have episodic large-scale water and sediment discharges related to moraine-dammed lake outbursts (Clague et al. 1985).

Several factors contribute to slope failures and sediment transport in Bute Inlet. Postglacial progradation of deltaic sands over less competent prodelta silts results in progressive long-term loading and sliding. Some of the largest reentrants in the delta at the fjord head may reflect periodic failure as a result of long-term effects. Infrequent high-magnitude lake outburst discharges also result in local delta slope oversteepening and excess pore pressure, initiating slope failure. Moreover, large sediment influxes lead to dense underflows, which follow preexisting slide-induced topographic lows. Similar effects appear to accompany annual discharge and sedimentation maxima.

Fan deltas along the sides of fjords are composed

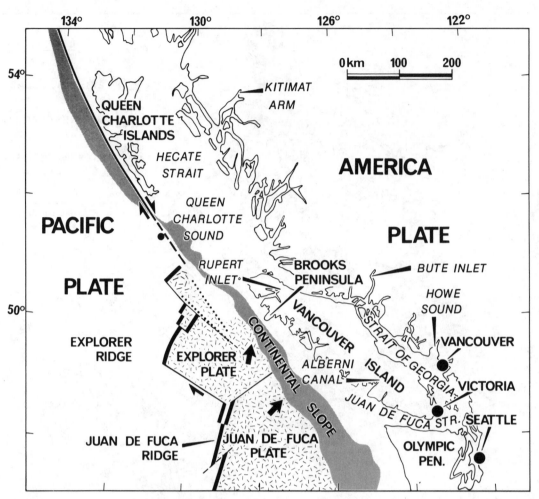

Figure 2.1   The West Coast with the locations of documented submarine landslides.

of sand, gravel and boulders and are characterized
by 15-30° underwater sediment cones cut by prominent
chutes. Terzaghi (1956) analysed the collapse of
the small fan delta at Woodfibre in Howe Sound. A
wharf and three warehouses collapsed into the inlet
forcing the closure of the pulp mill and causing
damage of between $ 2.4 M and $ 3.6 M. Soundings
after the failure indicated that the water depth had
increased by up to 13 m. The collapse occurred a
few minutes after an extreme low tide (the tidal
range is 4.3 m) in a perched body of slightly finer
sediments in a cusp-like indentation along the upper
edge of the delta front. Oversteepening of this
small zone on the delta and its relatively low
permeability led to failure. As at Kitimat, the
lower permeability of the finer grained sediment
body interfered with the tidal rise and fall of the
water table. Terzaghi (1956) stated: "As the tide
ebbs, the downward movement of the water table
behind the body is retarded, whereby the base ... is
acted upon by an excess hydrostatic pressure...
Therefore the slide occurred at extreme low tide."

Side scan sonar and shallow seismic surveys of two
similar fan-deltas in Howe Sound near Britannia
Beach (Prior and Bornhold, 1986) show delta-front
sliding, debris flows, avalanching of sediment,
shallow rotational sliding, some longitudinal
shearing, and blocky sediment accumulation at the
base of the delta slopes. These slope processes
result from a combination of factors. Bottom slopes
are steeply inclined, and rapid deltaic deposition
accompanies subaerial debris flows and torrents
during heavy rainfall and snow melt. Flood and
debris flow sedimentation on the underwater slopes
results in temporary oversteepening and rapid
loading. Shallow rotational sliding is related to

loading by splay progradation, discharging from the
delta-front chute zone onto the middle and lower
slopes. The rotational sliding, in turn, results in
the local progradation of broken and disturbed lower
delta-front sediment onto the fjord bottom. The
Britannia Beach fan-deltas develop and prograde as
subaqueous sediment transport locally reworks delta-
front sediments into a distinctive morphology but
without mass transport away from the delta.

Rogers (1980) compiled slope failures caused by
the 1946 magnitude 7.2 earthquake whose epicentre
was in central Vancouver Island. Submarine failures
in fjords buried and severed submarine cables in the
upper part of Alberni Canal on southern Vancouver
Island and eroded pilings beneath the Kildonan
Cannery located on an alluvial fan and fan-delta
system on the west coast of Vancouver Island.
Repairs to the cannery cost about $ 0.76 M. The
1949 magnitude 8.1 earthquake whose epicentre was in
the central Queen Charlotte Islands led to local
failures in fjords and inlets in the Queen Charlotte
Islands and dramatic waves (Rogers, personal
communication).

2.3   Strait of Georgia

The Strait of Georgia is both a tectonic and erosional basin which has subsided intermittently since
the Cretaceous. The morphology of the Strait has
been controlled by erosional and depositional
episodes during Pleistocene glaciations followed by
the rapid outbuilding of the Fraser River delta
(Clague et al., 1983). Thus seafloor slope failures
are common along oversteepened banks and in regions
of high sedimentation on the modern Fraser Delta

(Hamilton and Luternauer, 1983).

The most impressive submarine landslide features in the Strait of Georgia are the Foreslope Hills (Hamilton and Luternauer, 1983) off the main channel of the Fraser River at 230 to 330 m water depth in low-strength, muddy sediments on a 0.5° regional slope. The area is characterized by ridges and troughs with relief of about 20 m and crest to crest length of 400 to 600 m. Individual ridges are up to 4 km long. The remolded sediments are 35 to 100 m thick and have a total volume, including reworked local seafloor sediments, of between 5 and 14 km^3. The fauna in cores from the Foreslope Hills indicate that the sediments originated along the edge of the subtidal platform of the delta. Based on local sedimentation rate estimates, the Hills were emplaced during the last hundred years by a single large landslide lasting a few minutes to at most a few hours.

Rogers (1980) reported that after the 1946 Vancouver Island earthquake, several beaches disappeared along the east coast of Vancouver Island at Goose Spit (near Comox) and Deep Bay and at Jackson Bay and Port Neville, north of the Strait of Georgia. In Malaspina Channel, between Texada Island and the mainland, underwater telephone cables were severed.

2.4 Open Shelf

Conway and Luternauer (1984, 1985) described sediment from Queen Charlotte Sound with a blocky texture, gravelly, shelly, sandy mud layers and a rich fauna of pelecypods, gastropods, barnacles and foraminifera. They interpreted this deposit as an early post-glacial debris flow probably "generated by the collapse of oversteepened side walls from a shelf ice-stream valley, by entrainment of older material in meltwater/tidal current torrents as ice retreated across the shelf, or by seismically triggered failure of older deposits on either Goose Island Bank or the Sea Otter Shoals".

Holocence slumping has been seen in high resolution seismic records along the bank edges bordering Moresby Trough, the elongate depression which occupies southern and eastern Hecate Strait.

2.5 Continental Slope

Preliminary regional studies of mass wasting on the steep, rugged continental slope off British Columbia have recently used SeaMARC II sidescan sonar acoustic imagery supplemented by shallow seismic profiling. The transform fault margin off the Queen Charlotte Islands has abundant chutes, debris splays and small (less than 5 km^2) debris flow deposits along the landward side of Queen Charlotte Terrace. A large debris flow complex occurs on the lower slope off the southernmost tip of the Queen Charlotte Islands and covers the trace of the Queen Charlotte Fault.

The proximal parts of several large (50-100 km^2) debris flow deposits along the lower slope (water depth, > 2000 m) off Vancouver Island are frequently blocky and commonly fringed by a distal apron of smoother sediment up to 20 m thick extending to 15 km beyond the toe of the continental slope.

2.6 Conclusions

The distribution of slope instability processes is poorly known, because mapping of the offshore areas is incomplete. It is likely, based on detailed investigations at a few sites, that many similar geologic settings are also subject to underwater mass movement. The events at Kitimat, Bute Inlet, Howe Sound, Rupert Inlet and earthquake-related failures may be expected to recur. The short

history of such events, however, makes prediction of magnitudes and frequencies difficult.

Certain locations, the steeply sloping heads and flanks of inlets, (especially where accompanied by rapid sedimentation), the edges of submarine banks in the Strait of Georgia and the continental slope off both Vancouver Island and the Queen Charlotte Islands possess geologic and oceanographic conditions conducive to slope failures, especially during earthquake loading.

The construction of wharves and piers in nearshore areas, the emplacement of underwater pipelines and cables, and the choice of waste disposal sites require high resolution geophysical surveys and geotechnical analysis of sediments to evaluate the risk of slope failures.

3 THE ATLANTIC AND ARCTIC OFFSHORE

The Canadian Atlantic and Arctic offshore extends from the Gulf of Maine in the southeast to the Yukon-Alaska border in the northwest and seaward to the potential juridical limits on the continental slope and rise defined by the United Nations' Law of the Sea convention (Macnab et al., 1987). The Atlantic and Arctic continental margins are both passive margins that developed between 65 and 190 Ma. The Atlantic margin is bounded by hard rock of the Appalachian orogen and Canadian Shield, the Arctic margin by younger rocks of the Sverdrup Basin. Quaternary glaciation has had significant impact on the entire region.

3.1 Geologic Setting

On the Atlantic margin, the inner continental shelf was cut in resistant bedrock. Southern and western Newfoundland, northern Laborador and Baffin Island have deep fjords (Syvitski et al., 1986) locally with vertical walls. Nearshore troughs on the Labrador Shelf (Josenhans et al., 1986) and glacial meltwater channels on the Scotian Shelf (Piper et al., 1985a) have slopes over 30°. The outer continental shelf is underlain by Tertiary and Cretaceous sedimentary rocks, and in most places, consists of banks and intervening basins and troughs. Many of the troughs are floored by glacial till and glacial erosion has steepened their sides to over 10°. The upper continental slope has gradients between 1.5 and 3°; in places it is deeply incised by submarine canyons or smaller gullies with local vertical or undercut slopes (Josenhans et al., 1987). Many parts of the outer shelf and upper slope to water depths of 300 to 800 m are underlain by glacial till, which passes downslope into thick silt and mud deposits. Lesser gradients occur on the lower continental slope and rise. The physiographic features on the Atlantic seaboard have analogues in the ice infested waters of Hudson Bay, the Sverdrup Basin and the eastern Arctic continental shelf.

In the western Arctic, due to the dominance of the Mackenzie River, the Beaufort Sea continental shelf has a gradient of less than 0.1°. The shelf is cut by the Mackenzie Trough, whose flanks slope between 0.5 and 1.5°. On the continental shelf, up to 1200 m of Quaternary sediment overlies Tertiary deltaic sediments and Tertiary and Cretaceous bedrock. The Quaternary sediment thins to the eastern boundary (Amundsen Gulf) and to 150 m at the United States border in the west. The continental slope occurs in much shallower water than on the Atlantic margin with the shelf break at the 75 m isobath. The upper slope of the Beaufort is up to 4° where sediment failure has occurred (Hill et al., 1982). The lower slope and rise has not been mapped on the Beaufort margin due to heavy ice cover over most of the year.

The submarine landslides in the Atlantic and Arctic offshore occur in five geological environments, described in the following sections.

3.2 The continental shelf of the eastern Atlantic and Arctic

Small slides occur on steep local slopes, over 5°, on the continental shelf such as the walls of gullies and scarps of old large slides but are not easily identified using standard methods of high resolution seismic reflection profiling. Some have been identified during submersible dives (Josenhans et al., 1987: Hughes Clarke et al., 1988). Movement is generally within the upper 50 cm of sediment and may extend over the length of the local feature. Frequency is difficult to estimate but they occur easily on steep slopes where soft or loose sediment has recently been deposited. Biological activity and impacts of ice-rafted boulders are among the triggers (Josenhans et al., 1987).

3.3 The upper continental slope of the eastern Atlantic margin of Canada

Large movements which occur both on areas of smooth bathymetry and within slope canyons and gullies are most easily recognised on muddy slopes not dissected by gullies. Mass movements include rotational and bedding plane slides (Piper et al., 1985a, b; Hughes Clarke et al., 1988), leading downslope into debris flows (Mosher, 1987). Blocky debris flows occur where older sediment has moved on slopes dissected by canyons (Shor and Piper, 1988); smoother debris flows are found where only surficial sediments have failed (Piper et al., 1985a). Interstratal deformation may precede bedding plane slides (Mosher, 1987). Failure of upper slope sands is more complete, so that the occurrence of such failures can only be inferred indirectly; in the 1929 "Grand Banks" earthquake, over 100 km^3 of sand and gravel failed on the upper slope above the Laurentian Fan (Piper and Asku, 1987, Hasegawa and Kanamori, 1987).

Movements generally occur on the upper continental slope, in water depths of 300 to 2000 m. They rarely extend to the shelf break where the upper continental slope is underlain by stable glacial till (Piper et al., 1985b). Individual displaced masses range from 5 to 25 x 10^6 m^3; total displaced volume in the 1929 Grand Banks earthquake was about 100 km^3. On the Scotian slope, movement has probably been triggered by earthquakes. Movement occurred synchronously over large areas, both, in 1929, on the Laurentian Fan and St. Pierre Slope (Piper et al., 1988) and at about 12 ka on the central Scotian Slope (Piper et al., 1985a; Shor and Piper, 1988).

Movements are disguised in isolated seismic reflection profiles by erosion by Pleistocene turbidity currents on the Newfoundland and Labrador Slopes (Josenhans et al., 1987, Fig. 4). Off southern Newfoundland and Nova Scotia more detailed surveys have identified such failures (Fig. 3.1).

In general, large, detached sediment blocks do not slide into deep water; the slide in the 1929 event (Heezen and Drake, 1964) was mistakenly identified (Piper et al., 1988). Large sediment slides may occur, however, on margins with little Pleistocene glacial sediment and substantial slope erosion by submarine canyons. On the extreme southern Grand Banks and Georges Bank, toes of large sediment slides have been identified on the continental rise (Cochonat et al., 1988; Hughes-Clarke et al., 1988b).

3.4 Fjords

Due to the steep physiography of fjords (10° to 30° side wall slopes and 5° to 10° head wall slopes) and high sedimentation rates, slumps, debris flows, and turbidity currents are common. Syvitski et al. (1986) recognised sediment loading, earthquakes, and water movements as triggers. Loading of major fjord-head river deltas by long-term, high rates of sedimentation; by short-term, sediment supplies from storm floods or jokulhlaups; by the advance of delta foresets onto underconsolidated, prodelta clays; by oversteepening of the depositional slope or by the advance of tidewater glaciers can trigger movement.

In high sedimentation fjords, large slides occur in thick, rapidly accumulated muddy sequences, such as in McBeth Fiord, Baffin Island (Syvitski et al., 1986, Fig. 5.8). Retrogressive flow slides and liquefaction in silt or very fine sand, with an earthquake trigger, have been recognised in the same fjord (Syvitski et al., Fig. 5.9). Retrogressive sliding or liquefaction, several times a year, on rapidly prograding, sandy delta slopes, such as in Iterbilung Fiord, Baffin Island (Syvitski et al, 1986), may result from cyclic wave loading of over-steepened sediment. Deposition of sediment in front of tidewater glaciers, often on side wall slopes, also results in frequent failure. Frequent small failures on the sidewall slopes of fjords, as normal suspension fallout overloads steep slopes, are aided by bioturbation (Syvitski et al., 1986).

Fjords in areas of less relief have lower sedimentation rates and fewer failures. Failures in thick muddy sediments in Saguenay Fjord are related to large earthquakes, which also led to landslides on land (Schafer and Smith, 1987). Many fjords in southeastern Canada have low sedimentation rates, but accumulated thick sediments during early Holocene deglaciation. Some contain buried failures similar to those from McBeth and Saguenay fjords (Vilks et al., 1987). Former fjords along the north shore of the Gulf of St. Lawrence have been completely filled with sediment. Deltas have prograded into the Gulf, some showing failures similar to those on fjord head deltas. Larger slides in fjords, typically 10^6 to 10^8 m^3, are triggered by infrequent earthquakes. Smaller failures on prograding, sandy delta slopes may occur several times a year, and failure on steep sidewalls may also occur annually.

3.5 Upper slope of the Beaufort Sea Margin

Mass movements have been identified in 3 areas. In one, a slump scar extends at least 40 km laterally. Because the upper slope is shallow and the shelf break is at 75m, triggering by wave loading is possible. However, the Beaufort Sea is seismically active so earthquakes may also contribute to sediment instability. Hill et al. (1982) identified downslope creep in Beaufort slope sediments. As the Beaufort slope has not been surveyed in detail because ice conditions are severe even during the Arctic summer months, the extent of slope failures is not known.

3.6 Deepwater on the Atlantic Margin

On the lower Laurentian Fan, in about 4000 m of water, 3.5 kHz seismic profiles show a thin slump on the backside of a low levee (Piper et al., 1984). Schafer et al, (1985) identified mid-slope sediment slides from displaced foraminifers. Such shallow failures are isolated and poorly documented, but may be laterally extensive. The triggers are probably earthquakes.

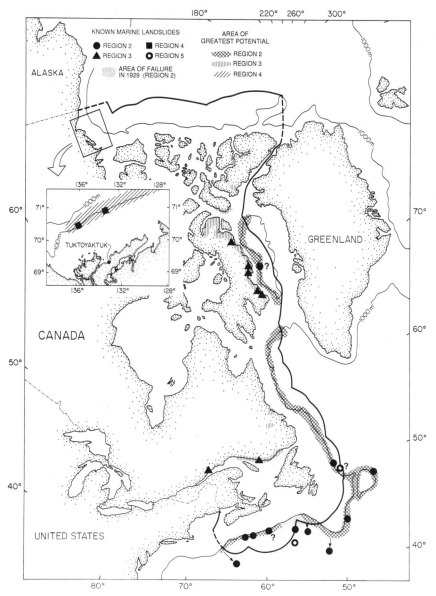

Figure 3.1 The Atlantic and Arctic Offshore with the locations of landslides.

3.7 Seismicity

The eastern Canadian continental margin has experienced large but infrequent earthquakes. The largest recorded event (M7.2), the Grand Banks earthquake, occurred in 1929 off the south coast of Newfoundland, beneath the continental slope off the Laurentian Channel within the Laurentian Slope Seismic Zone (Fig. 3.1). This 15,000 km^2 area surrounding the earthquake and the subsequent four events of M5 (Basham et al., 1983) contains half of the recorded earthquakes off the eastern Canadian margin. Earthquakes in the vicinity may be belated aftershocks of the 1929 event.

One of two models of seismic activity for the eastern Canadian margin (Basham et al., 1983) extends the onshore seismic zoning to the offshore regions with high seismic risk focused around the Laurentian Channel and Baffin Bay; the other model distributes the entire seismic risk over the eastern margin and increases the seismic risk in potential hydrocarbon production areas while decreasing the risk on the Laurentian Slope. It also implies that, although most of the margin is quiet, events equivalent to the 1929 Grand Banks event have occurred on the east coast margin about every 300 years (Adams,

1986). Reconnaisance of Holocene movements does not support such high frequencies on the continental margin.

Prior to 1981, the two seismic recording stations close to the Beaufort Sea were 900 km northeast and 150 km south of the region. During 1981 and 1982, additional stations were set up, including ocean bottom seismometers offshore. These stations were not continued, so the data for the Beaufort remains sparse (Hasegawa et al., 1979). The seismicity which has the greatest impact on the continental slope is the Beaufort Sea seismicity cluster, on the slope within an area 300 km by 200 km. The predominant recorded events within this cluster have magnitudes from 3 to 5. However, larger events occurred in 1920 (M 6.5) and in 1937 (M 5.5). But seismic and geological data are not sufficient to tie these events to failures on the slope. Consequently, the prediction of seismically induced slope instability in the Beaufort Sea is not yet possible.

In the Late Pleistocene, high sedimentation rates were common on the continental shelf and slope. Buried sediment failures on the continental slope may be the result of ice margin processes at times when continental ice sheets crossed the continental shelf (Bonifay and Piper, 1988). Sea level rises of up to 60m during the Early Holocene reduced instabilities due to wave loading and glacial retreat significantly reduced depositional rates. So, all Holocene large submarine movements on the continental slope are presumably seismically triggered.

3.8 Damage Caused by Submarine Landslides

The 1929 Grand Banks earthquake caused upper slope slumps off the Laurentian Channel and St. Pierre Bank, debris flows, and a large turbidity current (Piper et al., 1988). These submarine movements broke 28 submarine telecommunications cables owned by four different telegraph companies (Heezen and Ewing, 1952). Cable replacement cost many million dollars. Indirect damage included a major disruption of trans-Atlantic telecommunications for up to six months. Minor damage resulted from the initial seismic shock and much more severe damage from the subsequent tsunami which devastated Placentia Bay, Newfoundland, and caused 27 deaths (Doxsee, 1948).

3.9 Future Requirements

The region of potential submarine landslides on the eastern and northern Canadian margins is vast. Less than 5% of the continental slope has been adequately mapped for the evaluation of sediment stability and large areas of the coastal zones have not been charted by the Canadian Hydrographic Service. To determine the distribution, frequency, and size of movements within the Canadian juridical shelf, a significant mapping program is required. Tools for the identification of surface movements are wide-swath, sidescan sonars, and high precision bathymetric sensors; recognition of buried failures will require high-resolution, deep-water, seismic profiling systems.

Long-range, sidescan sonars such as GLORIA (Somers et al., 1978), which are capable of mapping swaths up to 20 km in deep water, recognize the largest slide masses, but are unsuited for mapping sediment failures on the continental slope. Mid and short-range, sidescan systems, such as the SeaMARC series (Ryan, 1982) and the SAR (Farcy and Voisset, 1985), recognise most submarine failures but lack adequate sub-bottom profiling, such as the deep-water boomer developed for the Seabed II program (Hutchins et al., 1985).

Precision bathymetry can use multibeam sounders such as the Seabeam system (Renard and Allenou, 1979), some swath sounding instruments that also obtain acoustic backscatter images such as SeaMARC II, or multiple, autonomous, sounding vehicles such as the Canadian Hydrographic Service Dolphin system (Dinn et al., 1987). Cost estimates for complete seabed coverage for the Atlantic continental margin (excluding the nearshore zones) ranges from $6 M for Gloria to $30 M for Seabeam. Costs for the Arctic have not yet been estimated due to the difficulties of surveying in ice-infested waters.

4   THE CORDILLERA

In the Cordillera, landslides impact directly on homes, facilities for energy production and transmission, forestry, fisheries, mining and national strategic transportation networks which access Pacific ports (Evans and Clague, 1988b).

4.1   Landslides in Rock

16 rock avalanches have occurred in the Cordillera since 1855 (Cruden, 1985, Evans, 1984b and Fig. 4.1), the largest being the 1965 Hope rockslide which had a volume of $47.5 \times 10^6 m^3$ (Mathews and

McTaggart, 1978). Quaternary volcanic rocks (Evans 1984b) and those adjacent to glaciers (Evans and Clague 1988a) are the most rockslide prone. Examples are the 1986 rock avalanche from the peak of Mount Meager (Evans 1987) in the Garibaldi Volcanic Belt and the 1959 Pandemonium Creek rock avalanche (Evans et al. 1988).

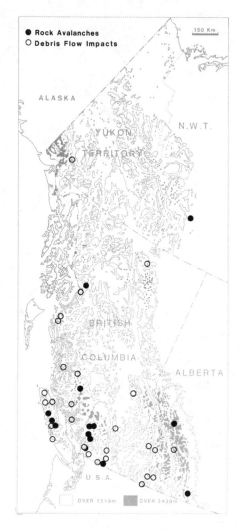

Figure 4.1   Historic rock avalanches and debris flow impacts in the Cordillera (from Evans and Clague, 1988).

Significant rock avalanches have been triggered by major seismic events such as the 1946 Vancouver Island Earthquake (Mathews, 1979; Evans, 1988), and the October 1985 North Nahanni Earthquake (Evans et al. 1987).

Very large bedrock slumps have occurred in the metamorphic rocks of the Columbia Mountains (Fig. 4.1) and have posed significant geotechnical problems in the hydroelectric development of the Columbia River. The Downie Slide, for example, has an estimated volume of $1.5 \times 10^9 m^3$ (Piteau et al. 1978) and is located within the reservoir of the Revelstoke Dam. Problematical rock slope deformations in the Coast Mountains have catastrophic potential.

Small rockfalls are common on natural and artificial rock slopes in the Cordillera (Gardner, 1970; Hungr and Evans, 1988). They are triggered by heavy rains and freeze-thaw activity (Peckover and Kerr, 1977).

4.2 Landslides in Surficial Materials

Landslides in surficial materials have been reviewed by Evans (1982a). Large slumps (Evans, 1982b) and spreads occur in Quaternary glaciolacustrine and glaciomarine sediments in the Cordillera. In 1888, a major spread in glaciolacustrine sediments blocked the Thompson River for 44 hours (Evans, 1984c). In 1880, a similar spread took place in sensitive glaciomarine clays of the Fraser Lowland (Evans, 1982a).

Debris flows are ubiquitous on steep mountain slopes in the mantles of colluvial and glacial materials (Evans, 1982b; VanDine, 1985; Hungr et al. 1987). Fatal debris flows along the Howe Sound, north of Vancouver from 1980 to 1982 were triggered by heavy rains (Evans and Lister, 1984; Church and Miles, 1987). Their frequency makes them the major landslide hazard in the Cordillera.

4.3 Damaging Landslide Events

Among 84 damaging landslide events between 1855 and 1983 (Evans and Clague, 1988b), 44% consist of debris flows, 17% involve rock slope movements of less than $10^6 m^3$ and 3.6% involve rock slope movements of more than $10^6 m^3$. 44% of the events were considered to be precipitation triggered. In 6.5% of the events, damage resulted from indirect effects to facilities beyond the limits of the landslide debris by the damming of rivers by landslide debris (Evans, 1984b, 1986) and by landslide-generated waves in natural and man-made water bodies (Fig. 4.2). A wave caused by the 1946 Mount Colonel Foster rock avalanche devastated the upper reaches of the Elk River valley on Vancouver Island for instance (Evans, 1988). Fortunately most large historical rock avalanches have taken place in remote, uninhabited areas and only 3 have impacted directly on the economic infrastructure.

365 deaths have been caused by landslide processes in the Cordillera since 1855. Between 20 and 30% of deaths resulted from debris flows and between 37 and 41% resulted from precipitation triggered events. Rockslope movements accounted for 47% of deaths with an equal percentage due to large ($>10^6 m^3$) and small ($<10^6 m^3$) events. Between 19 and 25% of the deaths were due to indirect effects.

4.4 Economic Significance

The direct and indirect costs of landslide processes in the Cordillera are difficult to estimate.

In the development of hydroelectric power in the Cordillera, B.C. Hydro have spent $61 M since 1954 on rockslopes including $30 M on the stabilization of the Downie Slide within the Revelstoke Dam reservoir, and $20M on Dutchman's Ridge upstream of the Mica Dam. To the indirect costs of structural modifications to the Cheakamus Dam must be added at least $5 M spent by the B.C. Government to buy and relocate 86 homes from the Rubble Creek Landslide fan.

Considerable sums are spent on the rockfall hazard along highways and railways in the Cordillera. An example of the impact of a large rock avalanche on a transportation corridor is the 1965 Hope Slide. The direct costs of re-establishing road access across the debris amounted to $1.05 M. In the 1983-84 financial year, $480,000 was spent on rockscaling on Highway 99 between Squamish and Vancouver. Theodore (1986) reported that Canadian National Railways (CNR) spent $28 M on rockslope stabilisation between Hope and Kamloops between 1971 and 1985. This investment dramatically reduced derailments.

A small rockfall at Hell's Gate in the Fraser Canyon during the construction of the CNR in 1914 had an effect on the Fraser River salmon fishery

from which it has not yet recovered. The debris obstructed migrating salmon in their cyclic return to their spawning grounds. In 1978 dollars, the loss to both the sockeye and the pink salmon fishery amounted to $2600 M for 1951 to 1978 (International Pacific Salmon Fisheries Commission, 1980), a loss attributable to the 1914 blockage. Fishways constructed at Hell's Gate between 1944 and 1966 cost $1.36 M to provide passage for the salmon past the obstruction.

Debris flows (Evans, 1982; VanDine, 1985; Evans and Clague, 1988b) have damaged homes in Prince Rupert, Ocean Falls, Port Alice, and the Howe Sound since 1950. Indirect costs for the construction of debris flow defences are considerable (Nasmith and Mercer, 1979; Martin et al. 1984; Hungr et al. 1987). At Port Alice direct damage caused by two debris flows in 1973 and 1975 was $700,000; the protective dykes built in 1976 cost $250,000. Direct losses due to storm-induced debris flows (Evans and Lister 1984) in the Revelstoke and Chilliwack areas totalled $2.5 M and debris flow defence measures for the new Coquihalla Highway cost $1.1 M. In the Howe Sound, large defences built to mitigate the debris flow hazard to road, railway and homes cost $20 M after the direct costs of the October 1982 rainstorm were over $1 M.

Figure 4.2 Historic landslide dams and historic landslide-generated waves in the Cordillera (from Evans and Clague, 1988).

4.5 Conclusions

The most damaging landslides both in terms of deaths and economic impact are frequent, small precipitation-triggered debris flows and rockslope movements. They formed 61% of known damaging landslides from 1855 to 1983. Landslides triggered by individual rainstorms may cause direct losses of up to $2.5 M.

Since 1951 direct costs due to landslide processes have averaged $5-10 M per year. This figure excludes the damage to the Fraser salmon fishery which between 1951 and 1978 experienced losses of $96 M per year directly attributable to the rockfall which partially blocked the Fraser in 1914. No costs have been added for the 365 historic deaths caused by landslides in the Cordillera including 76 at Frank, Alberta in 1903.

5 THE INTERIOR PLAINS

The Interior Plains are underlain by overconsolidated flat-lying Mesozoic sedimentary rocks, 900 to 1900 m of sediments have been eroded from the area since the beginning of the Tertiary (Nurkowski, 1984). The rocks are poorly indurated sandstones, siltstones and claystones or mudstones with beds of coal and bentonite which rebound and swell when load is removed and water becomes available to them (Matheson and Thomson, 1973).

In southern Alberta, Thomson and Morgenstern (1977) suggested that the strata most prone to slope movements are clay shales of marine origin, then mudstones deposited in a shallow, near shore environment. These fine grained sediments are rich in smectite that reduces ϕ'_p to about 15° for peak values and ϕ'_r to 7° or less. In general, the clay content of the poorly indurated rocks of the Interior Plains is the most important geologic factor contributing to slope instability (Locker, 1973).

5.1 Landslide Processes

The drainage of the Interior Plains was formed during the retreat of the Wisconsin continental glaciation (Gravenor and Bayrock, 1961). Post-glacial channels were rapidly eroded to produce deep, steep-walled valleys. Even preglacial channels re-occupied by rivers after deglaciation are narrow and steep-walled because the original valley had been infilled with Pleistocene deposits. The cutting of these valleys caused extensive land-sliding and while during the last 6000 years, valley deepening has been slight, lateral erosion has initiated landslides or reactivated old landslides, (Thomson and Morgenstern, 1977). The landslides are often as long as 1.5 km from scarp to toe. They are earth block slides (Varnes, 1978, Fig. 2.1) though occurring in poorly indurated rocks or slump-earth flows cut by multiple retrogressive scarps which curve downward to a common rupture surface. The earth flows at the toes are often carried away by the river.

Matheson and Thomson (1973) indicated that the rapid downcutting of valleys during deglaciation resulted in an anticlinal flexure or rebound of the strata in the valley walls and under the stream bed. This flexure caused interbed slip in the clayey strata and the weakness thus generated along the flat-lying bedding resulted in a planar rupture surface for slope movement.

5.2 Typical Landslides

Sauer (1975) discussed the influence of landslides on urban fringe development in southern Saskatchewan. Valleys have been eroded through the

Pleistocene deposits and into the Bearpaw Formation (poorly-indurated, bentonitic, Upper Cretaceous clay shale). Elsewhere, Pleistocene soft clays are overlain by sand in which there is a high water table. Both the soft clay and the clay shale are exceptionally weak, highway slopes are often designed as low as 8°. Fifteen major landslides in the City of Saskatoon (Fig. 5.1) on the South Saskatchewan River valley walls have damaged structures and interfered with park development. Clifton et al. (1981) pointed out the dominant influence of geology on slope failure and remedial measures.

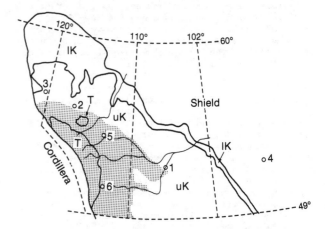

Figure 5.1 Locations of selected landslides on the Interior Plains; 1. Saskatoon, 2. Little Smoky, 3. Peace River, 4. Winnipeg, 5. Edmonton, 6. Calgary. Bedrock units are Tertiary, T, Upper Cretaceous, uK, Lower Cretaceous, lK. Stippling shows non-marine units.

A bridge 275 m long, across the Little Smoky River in north Central Alberta (Fig. 5.1), was completed in 1957. In the following year, the west abutment showed movement toward the river. Investigations revealed a landslide 700 m long on which the abutment and one pier were constructed, the scarp of the slide is 100 m above the river. At the bridge, the river flows in a preglacial channel entrenched in Upper Cretaceous, marine, thin-bedded clay shales, and is re-excavating the Pleistocene infilling, mainly till. Thomson and Hayley (1975) showed that discrete blocks were moving retrogressively, individual rupture surfaces merging downward into a common lower rupture surface. By 1980, the total slope movement was 4 m. The bridge is kept in operation by pouring a concrete pier adjacent to the existing pier every 5 years so that the slope moves the new pier under the adapted superstructure.

In October 1957, the north abutment of the Peace River Bridge (Fig. 5.1) was caught in a landslide which culminated in the collapse of the approach span and a side span (Thomson, 1958, Hardy, 1963). The loss of this bridge disrupted traffic to communities along the Alaska Highway as far north as Whitehorse, Yukon, 1475 km away. The massive gravity abutment, founded on the clay shale of the Upper Cretaceous Dunvegan Formation, acted as the anchor for the suspension cables. The approach cut allowed ingress of water to the clay shale which, in 15 years, weathered to a weak, highly plastic clay, the major cause of the landslide. The direct costs of dismantling the collapsed structure and design and construction of a new bridge (an over-deck truss 30 m longer than the original span, at 670 m), were $60 M. The indirect costs included construction and operation of the ferry crossing from 21 October to 18 November 1957 on a 24 hour per day basis; construction, maintenance and operation of the detour route, 10 km of gravel road, the crossing of the

Pine River, the decking of the Peace River railway bridge, sign posting and traffic control for two years; the cost in time and fuel to traffic of an extra 10 km; the cost to vehicles and travellers on the Alaska Highway, north of the collapsed bridge of being stranded for 5 to 7 days; the cost to businesses north of the bridge of lack of supplies, (stores, service stations, motels and gas and oil exploration near Fort Nelson, 400 km further north suffered) and the cost of the shut-down of the natural gas cleansing plant immediately north of the bridge which lost its water supply in the slide. An estimate is $20 M, yet there was no loss of life or private property and no consequent legal activity.

Slope instabilities in Winnipeg,(Fig. 5.1) along the banks of the Red and Assiniboine Rivers and their tributaries, have required legislative, legal and geotechnical measures to protect the public (Baracos and Graham, 1981). Winnipeg is on the bed of glacial Lake Agassiz which, except for riverbanks and man-made features, is flat and subject to flooding. The Red, Seine and Assiniboine Rivers have cut sinuous channels up to 15 m deep through the glacial lake sediments. Following the 1950 flood, many banks on the outsides of river curves were unstable as rapid draw-down followed the high water. Provincial legislation now requires geotechnical advice to show that any proposed work on the river bank will not result in instability or adversely affect the river flow. In the past three years about $ 10 M has been spent on riverbank improvement in greater Winnipeg.

The North Saskatchewan River is a narrow, steep-walled, postglacial river throughout Edmonton (Fig. 5.1). Thomson and Yacyshyn (1977) listed 37 landslides caused by subsidence over old coal mines, lateral erosion by the North Saskatchewan River and construction activity. Minimum landslide costs to the City of Edmonton over the last 2 decades are $8 M.

Hardy et al. (1980) discussed seven unstable slopes in Calgary (Fig. 5.1) but only one had river erosion as a contributing factor, the dominating cause was the increase in groundwater elevation due to urbanization. Drainage by horizontal or vertical drains was the major remedial technique. At some sites minor toe loading, slope reshaping and slope dressing augmented drainage installation.

5.3 Summary

Many slope movements arise with urbanization and more effort is needed on the design of set-back distances from the crests of potentially unstable river banks and on bank stabilization. Geotechnique can make important contributions to planning and legislation concerning top-of-bank construction and slope stability issues (Cruden, Tedder and Thomson, 1988).

6 PERMAFROST REGIONS

Permafrost exists where the temperature of the ground has been continuously below 0°C for two years or longer. Within permafrost, moisture in the ground is frozen, ice may occur as individual crystals or coatings on soil particles, as closely spaced lamellae (segregation ice) or as larger bodies (massive ice). The important features of ground ice are its cementing action, its sensitivity to disruption following minor changes in ground thermal regime and its ability to deform plastically. Massive ice can be treated as rock with the unusual properties of being near the melting point of the "rock" and of melting to a mobile liquid.

6.1 Extent of Permafrost in Canada

50% of Canada is underlain by permafrost (Fig. 1.3). Most of the permafrost occurs in the north but there is alpine permafrost in the Cordillera and in the Chic-Choc Mountains and other isolated localities in the Appalachians. Permafrost also occurs beneath the Beaufort Sea and the inter-island channels of the Arctic Archipelago.

Permafrost thickness is over 450 m on Melville Island and over 700 m on northern Richards Island (Judge, 1986). The uppermost layers of the ground thaw each summer and refreeze each winter. This active layer is several metres thick in the southern part of the permafrost region but only a few decimetres in the extreme northwest of the Arctic Archipelago. Within permafrost regions, not all the ground is perennially frozen. In the continuous zone, permafrost occurs beneath exposed areas of land, whereas in the discontinuous zone there are patches of unfrozen ground (Figure 1.3). The unfrozen ground increases to the south until permafrost ceases to occur.

6.2 Permafrost-Related Landslides

Landslides and slope failures in the Mackenzie Valley and western Arctic were examined during the 1970s as part of hydrocarbon exploration activities and proposals to build pipelines and highways (McRoberts, 1978, Johnston, 1981).

McRoberts (1978) recognized three forms of landslides in permafrost regions, in which permafrost and ground ice play a significant role. These forms comprise (Fig. 6.1), active layer failures (detachment failures or skin flows), thaw-slumps or retrogressive thaw-slumps (bi-modal flows, retrogressive thaw flow slides and ground ice slumps), and thermo-erosional falls. Complex landslides exhibiting features and processes associated with more than one of these three classes also occur.

Active layer failures separate the thawed active layer and its associated vegetation mat from the underlying permafrost. A typical failure is a long, narrow feature with a source high on the slope, a shallow chute down the slope and a debris lobe on the valley floor (Fig. 6.1). They occur in groups on south-facing slopes with angles of 12 to 14° at their sources; they can develop on slopes as low as 6° and can traverse slopes as low as 3°.

Active layer failures follow deep thawing of the ground, especially if it is coupled with higher than usual precipitation. Disturbance of the vegetation mat by erosion, forest fire or human activity can be triggers. Active layer failures can occur spontaneously during unusually warm thaw seasons. The permafrost immediately beneath the normal active layer is commonly ice rich (Mackay, 1971). In a warm summer, thaw may penetrate this layer raising the pore water pressure and leading to failure.

Active layer failures in the Mackenzie Valley, the Tuktoyaktuk Coastlands and the western Arctic Archipelago form largely on weak Mesozoic rocks, covered by fine grained morainal deposits and lacustrine silts and clays. Active layer failures are likely to be less common in the central and eastern Arctic, as a higher proportion of the ground surface is hard bedrock or sands and gravels which contain little ground ice, are free draining and so are less liable to fail.

Thaw-slumps are deeper and larger slope failures developing where ice-rich soil or massive ground ice is exposed at the ground surface. A thaw-slump forms a bowl in a slope, with a steep scarp in massive ice or ice-rich soil (Fig. 6.1). The floor of the bowl, during the summer thaw, is a soupy mud flowing away from the ice face. As the ice-face melts and retreats, the soil of the active layers slides down the ice-face, mixing with the meltwater

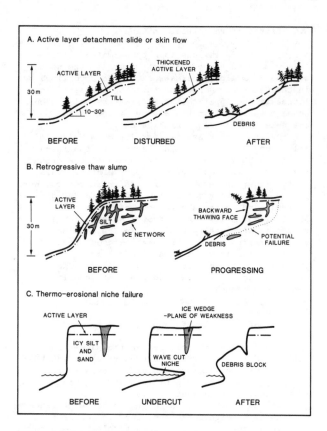

A. Active layer detachment slide or skin flow

ACTIVE LAYER   THICKENED
                ACTIVE LAYER
30 m   TILL
        10-30°   DEBRIS
BEFORE   DISTURBED   AFTER

B. Retrogressive thaw slump

ACTIVE
LAYER   BACKWARD
                             THAWING FACE
30 m   SILT
        ICE NETWORK
                        DEBRIS   POTENTIAL
                                  FAILURE
BEFORE   PROGRESSING

C. Thermo-erosional niche failure

                    ICE WEDGE
ACTIVE LAYER   -PLANE OF WEAKNESS
ICY SILT
AND
SAND   DEBRIS BLOCK
            WAVE CUT
            NICHE
BEFORE   UNDERCUT   AFTER

Figure 6.1 Stylized cross sections of landslides
related to thawing of permafrost (redrawn, in part,
after Rutter and others, 1973).

and contributing to the foot of the mud-flow.

Large, compound thaw-slumps can be tens of metres
deep and up to kilometres across. The rates of
headwall retreat range up to 0.57 m/100 hours, with
a mean of about 0.33 m/100 hours for failures on
Garry Island in 1964, 1965 and 1971 (Kerfoot and
Mackay, 1972), Banks Island in 1972 (French, 1974)
and Melville Island in 1977 to 1981 (Heginbottom,
1978, 1984a). For the Melville Island slump, both
the mean and maximum rates of headwall retreat
correlate well with the thawing index.

Thaw-slumps develop where there is a substantial
thickness of ground ice or very icy soil, generally,
in fine-grained and poorly consolidated sediments.
Thus they occur in the Mackenzie Valley, the Arctic
Coastal Plain and the north-western Arctic Islands
(Lamothe and St-Onge, 1961; Mackay, 1963b; Kerfoot
and Mackay, 1972; French and Egginton, 1973;
Heginbottom, 1978). Mackay (1963b, Fig 21)
presented a map of "ground-ice slumps" for Richards
Island, the Tuktoyaktuk Coastlands and the northern
Anderson Plain, over 100 thaw slumps occur in about
50 000 km^2.

Thaw-slumps develop by coastal retreat or riverbank erosion, active layer failures and erosion due
to melting of ice-wedges. Excavation or off-road
vehicle traffic can also initiate a thaw slump
(Heginbottom, 1973). Thaw-slumps on slopes develop
rapidly at first but then, as the mud flow drains,
dries out, and covers the ground ice, the slump
eventually stabilizes (Heginbottom, 1984a).
Thaw-slumps along coastal cliffs or river banks may
be reactivated as fluvial or wave action removes the
debris from the toe. Several large, old, multiphase
thaw slumps are known from the Beaufort Sea coast
(Mackay, 1963a, b: Rampton and Mackay, 1971; Harry
et al., 1985). The problem of coastal retreat has
received considerable attention (Mackay, 1963a,b;

McDonald and Lewis, 1973; Lewis and Forbes, 1974;
Shah, 1978; Harper et al., 1985).

Thermo-erosional falls develop where ground ice or
ice rich soil is at water level in a river bank or
coastal cliff. Thermal abrasion by the water can
lead to rapid undercutting of the cliff forming a
thermo-erosional niche (Walker and Arnborg, 1966)
(Fig. 6.1). Frozen ground above the niche can allow
an undercut as deep as 10 m to develop (Walker and
Arnborg, 1966, Czudek and Demek, 1970). The overlying frozen ground eventually fails and large
blocks of frozen ground fall to the water. The
failure plane is very steep and frequently develops
along a plane of weakness such as an ice wedge. The
falls of frozen ground protect the bank from undercutting and erosion but thermal abrasion thaws and
erodes until a fresh face is exposed. This process
is sensitive to floods and storm surges. Late thaw
season storms have the greatest effect; earlier in
the thaw season, the foot of the cliff is generally
protected by an ice-foot or snowbank.

Thermo-erosional niches develop where water is in
contact with ice-rich permafrost or massive ground
ice. They occur, therefore, along the coasts of the
western Arctic and along the banks of major rivers
and delta channels.

6.3 Economic Considerations

The permafrost region has a low population largely
concentrated in settlements. Industrial activity
has been restricted to few points, such as mines, or
to more extensive but seasonal activities associated
with mineral or hydrocarbon exploration. Air travel
and electronic communications are well developed.
Thus, the north is not vulnerable to landslides. No
deaths by landslides are known, and direct damage is
rare. Permafrost-related landslides in general do
not pose a significant hazard, but are a constraint
on development. Large, catastrophic landslides due
to failure in permafrost slopes are rare. Active
layer failures are of modest size; retrogressive
ground ice slumps expand their headwalls relatively
slowly, thermal niche failures are easy to identify
and avoid.

Given the low incidence of damage, the costs
associated with permafrost related landslides are
difficult to determine. Records of highway main-
tenace indicate failures due to thermokarst
(in-place subsidence or collapse of the ground
surface due to thawing of ground ice) are more
common, more damaging (with at least one fatality)
and more expensive. Disruption by landslides has
been considered in the routing and construction of
highways and pipelines. For example, an Environmental Working Group, responsible for assessing the
impact of construction of the Mackenzie Highway
(Heginbottom, 1975), recommended realignment of the
proposed highway in several instances.

A particular concern has been earthquake induced
landslides, though the Nahanni earthquakes of
October and December 1985, while causing several
landslides in the southern Mackenzie Mountains
(Evans et al., 1987), did not cause any damage to
the Norman Wells-Zama oil pipeline.

6.4 Implications of Climatic Change

As permafrost is a thermal phenomenon, permafrost
terrain is peculiarly sensitive to the direct
effects of climatic change. Over the next 50 to 100
years, the global mean annual temperature will
increase and this increase will be greatest towards
the poles (Harvey, 1982; MacCracken and Luther,
1985). This warmer climate will result primarily
from much warmer winters. Sea level is likely to
rise, due to expansion of the oceans and increased
melting of ice caps and glaciers.

The possible effects on permafrost terrain are well known (Carter et al., 1987). Landslides are likely to increase, the active layer will thicken, leading to more frequent failures. Some of these failures will expose massive ice or icy soil, leading to more thaw slumps. Longer periods of open-water will lead to more rapid coastal and shoreline erosion by thaw slumping or by thermo-erosional falls, enhanced by rising sea level. So, as economic development continues in the north, damage resulting from landslides will increase. The risk cannot be predicted precisely as the data-base is sparse and the population small. The risk should remain low, however, if northern development is closely regulated and controlled, and the risk of landslides is considered in the regulatory process.

7 THE ST. LAWRENCE LOWLANDS

Major rockslide and landslide areas of the St. Lawrence Lowlands are indicated in Figure 7.1. Investigation of the extent of landsliding is still limited (Lebuis et al., 1985; Chagnon, 1968; Locat et al., 1984) but detailed investigations have contributed much to understanding the slope stability of sensitive clays (Lefebvre, 1981).

7.1 The Extent of Landslides

Landslides take place mostly in Quaternary raised marine deposits. The causes of landslides are related to pore pressure changes associated with rainfall and snowmelt (Lebuis et al., 1985) to erosion (Chagnon, 1968; LaRochelle et al. 1970; Locat et al. 1984, Lefebvre 1986) or to seismic activity (Chagnon and Locat, 1987).

Postglacial sedimentary basins identified in Figure 7.1 were active between 15,000 and 6,000 years B.P. (Dionne, 1972). Their nature and relationship with the mechanical and physico-chemical properties of the sediments were reviewed by Quigley (1980) and Locat et al. (1984). Chagnon (1968) identified over 680 "sensitive clay-flow" post-glacial landslides within the St. Lawrence Lowlands and Saguenay Fjord areas. The largest is the St. Jean-Vianney landslide with an area of 20 km^2 (LaSalle and Chagnon, 1968) which has been dated back to 1663 (Legget and LaSalle, 1978).

Lebuis et al. (1985) compared the cumulative frequency of these landslides with 27 landslides, all over 1ha in area, which occurred between 1840 and 1980. Trends were similar, more than 60% of the slides exceed 1 ha. Locat et al. (1984) followed the development of landsliding in the small Chacoura River basin between 1948 and 1979. As an equilibrium profile developed in the stream by erosion of the sensitive clays, almost all the valley was affected by landslides. Erosion increased significantly in small creeks receiving large discharges from artificial drainage systems installed over the last 30 years. Although the main river is less active, tributaries become areas of landslides.

Liquefaction of soils during earthquakes is often a cause of major landslides. Numerous large landslides have been mapped in an area where major historic earthquakes have occurred and none of the original terraces were intact (Chagnon and Locat, 1987). The 1663 St. Jean-Vianney landslide was possibly caused by a Intensity X earthquake in the Charlevoix area, 130 km to the south (Legget and LaSalle, 1978). Geotechnical investigations in the Saguenay fjord have revealed numerous large landslides caused by earthquakes (Locat and Leroueil, 1988). Major landslides in the St. Boniface area, south of Shawinigan, have also been linked to the 1663 earthquake by C^{14} dating of wood found in the slide debris (Desjardins, 1980).

In James Bay, reconnaissance during hydro-electric development showed the area is underlain by extremely sensitive clays (Locat and Lefebvre, 1986) and is landslide susceptible below the marine limit (about 300 m above sea level).

Rockslides are isolated events in the Appalachians of Eastern Canada. Rockslides in Quebec City (Fig. 7.1) between 1836 and 1889 took 100 lives (35 on Sept. 19, 1889). In the Gaspe Peninsula, rockslides are common in the summer, after heavy rains, and they interrupt the road system. One major rock avalanche resulted from lightning (Dionne, 1969).

7.2 Landslide Prevention and Control

Prior to 1976, the Waters Branch of the Quebec Streams Commission and, later, the Department of Natural Resources which was responsible for riverbank maintenance, were responsible for the prevention of landslides. Many of the control works were in areas of active erosion where property was directly affected.

Since 1976 landslides in the sensitive clays of the St. Lawrence Lowlands have been mapped systematically by the Geotechnical Branch of the Department of Energy and Resources. Reports with maps (Rissmann et al., 1985, Fig. 6) were provided to the municipal authorities as input in the land management process. In 1982, the County Regional Municipalities (Municipalites Regionales de Comte) were given the responsibility of land use planning and development for which zoning maps had to be prepared. Landslide susceptibility maps on which the endangered zones were outlined were supplied by the Geotechnical Branch. Most of the St. Lawrence Lowlands were thus mapped.

In 1982 the Geotechnical Branch was disbanded and a major source of knowledge and expertise was lost to the community. The Environment Department has taken over some of the responsibilities of the Geotechnical Branch, undertaking investigations and implementing preventive measures. The Quebec Civil Protection agency may also initiate measures to insure the safety of citizens.

Few landslides have been the object of control measures before or during the slide or remedial measures after the slide. Before 1971 the Waters Branch Department of Natural Resources was responsible for control measures on rivers in Quebec, while the Federal Waters Branch was responsible for navigable rivers. In developed areas, the banks of rivers were protected from erosion whenever danger was apparent. After 1971, the Department of Natural Resources (Waters Branch and Geotechnical Branch) intervened to prevent landslides in unstable areas. The most economical interventions modified the geometry of the slope by removal of material at the top or flattening. Unstable slopes in highly sensitive clays often required different procedures such as vertical drains (Chagnon, 1975, Lefebvre et al., 1976, Chagnon et al., 1979), or counterweights at the foot of the slope (Chagnon, 1975). For failed slopes in sensitive clay a trench at the foot of the slope below the failure surface was excavated and filled with rock. An anchored counterweight was thus created which was covered with slide debris along a stable geometry (Robert and Chagnon, 1976).

A recent regional program of erosion control and landslide prevention along the shores of Lake St. Jean is a joint undertaking of the Ministry of the Environment, other concerned Departments, and the Forces motrices du Saguenay. This project should insure the stability of the shores of Lake St. Jean.

7.3 Some Major Landslides

The largest loss of life, 33, in a sensitive clay slide was at Notre Dame de la Salette on April 26,

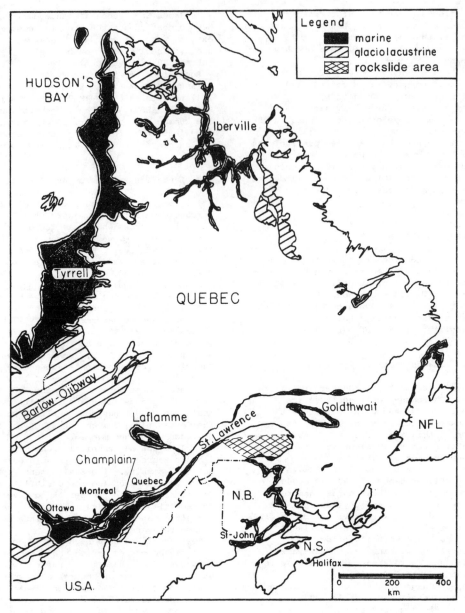

Figure 7.1 Landslide and rockslide prone areas in Quebec.

1908 (Ells, 1908). The Nicolet slide caused $5 M
damage on November 12, 1955 and 3 fatalities
(Beland, 1956).

The best known modern slide, the St. Jean Vianney
landslide (10 km west of Chicoutimi, Fig. 7.1)
destroyed 43 houses with the loss of 31 lives on May
4, 1971. $6.9 \times 10^6 \text{m}^3$ of soil was removed from
268,000 m^2, the scarp retrogressed 550 m within the
accumulation zone of a 20 km^2 slide which had
occurred in 1663 (Legget and LaSalle, 1978). A 60 x
150 m landslide on April 24, 1971 allowed the
second, larger movement to occur. The May 4th slide
moved the liquefied clay 2.8 km down to the Saguenay
River, along the Petit Bras and Aux Vases Rivers at
26 km/hr (Tavenas et al., 1971). The 42 m long
bridge crossing Aux Vases River near the Saguenay
was pushed 150 m into the Saguenay River. A 60 m
deposit of strong, highly sensitive, silty clay
overlain by 10 m of sand and slide debris is deeply
entrenched by the Aux Vases and Petit Bras whose 25
m high slopes are inclined at more than 45
degrees. The shear strength of the clay is 200 –
300 kPa, the sensitivity from 200 to infinity, and
the preconsolidation pressure is about 950 kPa.

LaRochelle (1974) indicated that this event resulted
from river erosion. In May 1971, the Quebec government decided to relocate the population, 200 houses
and families were moved to Arvida, 5 km from St.
Jean-Vianney.

Between 1836 and 1889, six rockfalls destroyed
houses on Petit Champlain street at the foot of Cap
Diamant, a 100 m cliff in Quebec City. 85 people
died. The underdip slope is made up of steeply
dipping shaly limestone and heavily fractured black
shale. Blocks are loosened and move downslope
singly or as a mass. Falls occur seemingly
independently of the climatic conditions. On
September 19th, 1889, a major rockfall destroyed 7
houses at the foot of the slope and 35 people were
killed. Correction work was undertaken by the
federal government (owner of the upper part of the
cliff and of the land behind the top of the scarp)
in 1907 when loose and overhanging rocks were
removed; in 1919 when retaining walls were erected
and in 1959 when rockbolts were installed near the
top of the slope under Dufferin Terrace (Brown and
Casey, 1960).

7.4 Social and Economic Impacts

Loss of life from landslides over the last hundred years can only be estimated from historical records. Minima are 100 deaths in sensitive clay slides, 100 in rockfalls, 10 in other types.

Records of the economic impact of landslides were published for 1971 and 1972 (Desmeules, 1972, 1973). However, there were no standard cost categories, so comparisons between different landslides are not possible. The 1971 record included the St. Jean-Vianney landslide whose cost, including relocating the local population, was later established at $ 17 M. The 1972 figures are underestimates, omitting the investigations by the Geotechnical Branch, Department of Natural Resources, Quebec. From 1972 to 1982 the Geotechnical Branch spent about $1 M/year on investigations and corrective works.

Before 1971 there were no compilations of damages or of remedial works. When a major landslide occurred, such as the Nicolet event in 1955, no money was granted for corrective works or for compensation so the financial impact on property owners was usually devastating. Landslide risk is not covered even by comprehensive insurance policies. The St. Jean-Vianney event set a precedent; the provincial government (with financial assistance from the Federal government) totally compensated private owners. But similar later interventions never established a general policy. Now citizens cannot expect compensation for losses although they may receive technical assistance and financial help for remedial works.

8   THE SHORES OF THE GREAT LAKES

2000 km of the shores of Lakes Huron, Erie and Ontario are erodible clayey or sandy soils (Figure 8.1). Shorelines eroded in glacial drift are, generally, steep bluffs that may be over 40 m high and, at Scarborough, are more than 90 m in height. Wave-induced toe erosion, especially during storms, produces thousands of landslides annually, some of them with volumes in excess of 200,000 m³. While these landslides represent a costly loss of land in urban areas such as the Scarborough Bluffs east of Toronto, they also provide littoral nourishment for southern Ontario's recreational beaches.

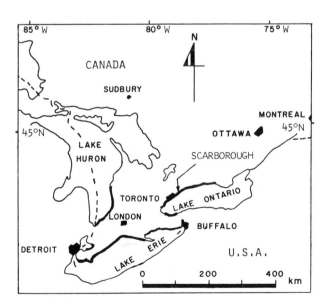

Figure 8.1 The lower Great Lakes with erodible shorelines shaded.

Here we review annual erosion rates, landslide activity, and mechanisms for public assistance. Two populous, well documented areas, the central section of the Lake Erie north shore and the Lake Ontario lakeshore along the Scarborough Bluffs are discussed in detail.

8.1 Erosion Rate/Wave Power Relationships

Toe erosion by storm-induced waves is the cause of cliff retreat along all shorelines except those which are protected. Kamphuis (1986), confirming work by Gelinas and Quigley (1973), found for clayey till bluffs on Lake Erie that

$$\overline{R} = 1.06 \; \overline{P}^{\,1.37} \tag{1}$$

where $\overline{R}$ = long-term average recession rate in m/yr and $\overline{P}$ = long-term average wave power arriving at the shoreline in kW/m. The equation demonstrates the direct dependence of erosion rate on the energy of the impacting waves. Since oversteepening results from toe erosion, the main cause of landsliding is cyclic overstressing augmented by effective strength reduction if enough time between slides allows softening to occur.

Another major factor in cliff recession rates is the shape of the offshore lake bottom. The deeper the offshore, the larger the breaking wave energy that reaches the shoreline to cause toe erosion. Boyd (1986) suggested that a deficiency of sand resulted in a concave-upwards offshore profile and more energy reaching the toe of the bluffs. These geologic factors are not included in equation (1) which is a correlation rather than a fundamental assessment.

8.2 Movement Modes

Six cyclic movement modes are typical of the clayey bluffs along the southern Great Lakes. Four were described from the north shore of Lake Erie (Quigley et al., 1977), some are similar to those in Hutchinson (1973, 1986) and Hutchinson and Bhandari (1971). Quigley and Gelinas (1976) made direct comparisons with Hutchinson's modes including a discussion of the rate of flattening of abandoned or beach protected cliffs.

Mode 1.   Linear Recession   Unvegetated clay cliffs are subject to cyclic drying and cracking followed by rainfall wetting and surface sheet sloughing to a depth of up to 10 cm. On some west central Lake Erie shorelines 27 m high bluffs retreat at 0.6m/year by sloughing in equilibrium with toe erosion caused by an average breaking wave power of 0.5 kW/m. Scallops produced by sloughing slope at 37° whereas inter-scallop ribs have vertical toe cliffs up to 20 m high and 37° tops. Toppling or block failures of these ribs seem related to joints which open on stress release. This mode of failure also occurs along sections of the Scarborough Bluffs.

Mode 2.   Slow, Circular Arc Slides   Benched and vegetated slopes up to 34 m high along the north shore of Lake Erie are cyclically oversteepened by toe erosion caused by wave power of 1 kW/m resulting in a cliff retreat of 1.4m/year. Large, slow-moving circular arc failures occur as a result of combined softening and over-stressing by toe erosion. Pre-failure and post-failure profiles repeat themselves every 10 to 25 years. Similar failures occur on Lakes Huron and Ontario.

Mode 3.   High Velocity Cyclic Slabbing   High

velocity cyclic slabs are the most spectacular failures on the Lakes creating huge, offshore debris fans during failures reported to be as loud as overhead thunder. Erosion of the sand layer at the toe causes rapid undercutting once the offshore debris fan (often over 1 ha in size) is eroded. The 42 m high cliffs near Port Stanley on Lake Erie are retreating at a rate of 3m/year under attack by 2 kW/m of breaking wave power. Smaller failures at Scarborough have been reported to generate offshore fans as far as 20 m into the lake.

Mode 4. Huge, Circular Arc Effective Stress Slides with Superimposed Cyclic Short-Term Movements One such slide, at Port Brice on Lake Erie developed during a low water cycle in 1963 when the slopes were significantly flatter and represented progressive softening due to reduction in effective stresses (Quigley and Di Nardo, 1980). Shallow movements at the cliff face continued from 1964 to 1984 resulting in cliff edge retreat at 2.2m/year. In this 20-year interval 44 m of cliff was removed and the original huge slide which had subsided only 3 m, was completely eroded.

The present sequence of shallow failures, toe erosion, sloughing and toppling seems to be in equilibrium with high water levels in Lake Erie and wave power of 2 kW/m. Another large, circular arc effective stress slide is predicted only after a long period of low water levels and reduced toe erosion.

Mode 5. Running Gullies Water-bearing surface sands and coarse silts overlying clayey bluffs are characteristically benched due to piping at the point of seepage breakout. Under conditions which are not well understood, rapid piping at the bench-top causes slabbing of sand in the upper slopes and rapid development of hanging gullies with steep lower slopes of clayey soils. In plan, some gullies are long and slender, others are nearly round. The gullies stop running when the total perimeter of the bench-top is long enough to relieve the forces of internal erosion due to seepage.

Mode 6. Internal erosion and undercutting of the slope Along the Scarborough Bluffs, large, high-backed gullies develop by piping and undercutting of the upper slope. Perched water tables occur due to permeability contrasts, most notably at the contact between the Scarborough Sand Unit and underlying Scarborough Clay Unit.

When the ground water seeps under a high hydraulic gradient towards the slope face, it forms a pipe or tunnel which expands by increased water flow and internal erosion of the sandy soil. As the pipe increases in size, the overlying soils are progressively undercut and slide in circular arcs or slabs developing large running gullies on the Bluffs. Eyles et al. (1986) described the development of such pipes and gullies which cause crest retreats from 3 m to more than 21 m/year (Geocon, 1983; Gartner Lee, 1986: MTRCA, 1980). Once such a gully starts it requires immediate remedial action.

8.3 Public Assistance

Jurisdiction over the shoreline is complex, since different aspects of development and protection are governed by various municipal, provincial and federal authorities. The Ontario Ministry of Natural Resources (MNR) has in general been the lead agency for the shoreline management of the Great Lakes. The Ministry decides annually what work will be carried out by considering risk to property and persons, property value, cost of remedial measures,

and location. Cost-benefit studies for the entire province are reviewed and the highest ranked projects are funded for the next fiscal year. If emergency action is required landowners are able to receive low interest government loans or grants.

In the largest municipality, Metropolitan Toronto, the Metropolitan Toronto and Region Conservation Authority (MTRCA) has, through MNR, jurisdiction over the management of the shoreline. Their comprehensive program of shoreline management is designed to prevent, eliminate, or reduce hazards to life and property, while preserving the natural attributes of the lakefront setting (MTRCA, 1980).

When private lakefront property is involved the Authority has three options to obtain permission for the required works. The homeowner may deed title to the land on which the coastal protective works and bank stabilization measures are built. In return, the Authority finances and constructs the necessary remedial measures. Alternatively, the homeowner can pay in full the costs of the remedial measures, or where the house is in imminent risk and it is more cost effective, all the property may be acquired by the Authority.

MTRCA has carried out remedial works at ten major sites and six minor sites along the lakefront. Measures have included toe revetments, groins, horizontal drains and placement of fill over the edge of the slope. These works have provided erosion mitigation and produced a recreational corridor along the waterfront that has increased land value. The cost of toe protection in areas with good access and available construction material is about $2700 per metre.

In other areas, the Conservation Authorities and MNR assist in the design of remedial measures for shoreline erosion. Where large shoreline management plans are required, outside consultants generally provide the necessary multidisciplinary expertise. In addition to the legal complexities of obtaining approvals for protective works, there are also technical problems because a typical waterfront landslide requires both geotechnical and coastal hydraulic studies to design the onshore and offshore protection.

9 CONCLUSIONS

9.1 The Economic Significance of Slope Movements

Landslide incidence in parts of Canada is comparable with that in other areas which are recognized as having landslide problems. There is then cause for concern with the landslide problem in Canada but the economic significance of slope movements in Canada is difficult to assess. Indirect costs, from traffic delays and economic disruption, are a substantial part of the total costs of landsliding to the economy. However, such costs are often widely distributed and represent losses of time and wages rather than direct expenditures on engineering works. Direct costs of small slope movements may be enveloped in larger, routine budget items for maintenance and repairs. It is only for large, apparently isolated events that most of the direct costs can be identified. These often represent only the costs of work carried out on the site without including the substantial overhead assumed by, say, Provincial Highway Departments in maintaining crews to respond to these emergencies. So, there are a number of causes leading to the underestimation of the effects of slope movement on the Canadian economy.

While a major synthesis of the extent of landsliding in Canada has begun, the picture is by no means complete. The problems with debris flows on the Pacific Coast, obvious in retrospect, were not predicted 20 years ago. Other surprises may be waiting in the less well known parts of the country. It is also clear from the history of slope movements on the Interior Plains, for instance, that development itself may reduce slope stability. So, to conclude that the major problems have all been identified is optimistic; however, all the major problems that have been identified have had engineering solutions at a price.

Fleming and Taylor (1980) identified losses of the order of $ 1 billion U.S. (1978 dollars) and placed damage from slope movements as among the most costly ground hazards in the U.S.A., ahead of losses from earthquakes, floods, hurricanes and tornados. Extending this model to Canada on the basis of the relative sizes of the economies and the populations of the two countries suggests losses of say $ 200 M per annum. But Canada has a larger land area at risk than the U.S.A., and a ground transportation system which is as extensive but without the density and redundancy of the U.S. network to protect it from serious disruption if one link is cut. Clearly independent estimates of annual loss need to be made in Canada. They may show losses of the order of $ 1 billion per annum.

As a first step, a catalogue of damaging historic landslides in Canada is being set up and agencies collecting information on landslides are encouraged to contribute information both on historic and current landslide problems.

9.2 The Extent of Slope Movements

The extent of slope movements is reported by the Geological Survey of Canada at scales as detailed as 1:50 000 on maps of surficial deposits (Geological Survey of Canada, 1985). Slope movements are shown by site symbols and in map units by versions of the Landform Classification System (Canada Soil Survey Committee, 1978, Chp. 17).

An elaboration of the Landform Classification System is required before its use can be extended beyond the reconnaissance studies so far undertaken by the Survey. Rib and Liang (1978) have demonstrated that it is possible to distinguish falls and topples, rotational slides, translational slides, dry flows and wet flows by skilled interpretation of suitably scaled aerial photography. An expanded classification system should be able to convey clearly these important distinctions among slope movements. Similarily, Jackson (1987) has shown that the record and activity of debris flows can be estimated from ground surface features.

More detailed studies have been made by provincial organizations across Canada (Klugman and Chung, 1976, Haughton, 1978 are examples) but the variety of styles and theoretical approaches has not allowed experience to develop to the stage when these maps can be used confidently by engineers and planners. Demonstration projects to validate and extend existing techniques in all the contrasting slope movement environments across the country would contribute to the development of codes of practice for these investigations.

As Harris (1980) has shown, municipalities and cities have mapped the stability of individual sites, requiring setbacks from hazardous slopes. While there is no requirement that jurisdictions zone sites for slope stability, information and experience gathered in servicing sites often gives special knowledge to the zoning authority which it is obliged to use reasonably. Presumably, if the authority becomes aware that sites are at risk from slope movements, it would be negligent not to inform the public of the hazard. Again, there is no common rationale to the setbacks and critical, comparative studies are needed at this scale also.

9.3 Summary and Recommendations

1. Some areas of Canada have landslide problems.
2. The cost of landslides to the Canadian economy cannot be precisely estimated with the available information. It may be hundreds of million dollars per year.
3. The major landslide problems have had engineering solutions at a price.
4. Although the outlines of the landslide problem in Canada are visible, details remain to be filled in before the picture is complete.
5. A catalogue of damaging historic landslides in Canada should be kept current.
6. The Landform Classification System should be extended to distinguish additional characteristics of landslides.
7. Planners need a rationale for the set-backs currently imposed on developments at the crests of slopes which may move.

10 REFERENCES

Adams, J. 1986. Changing assessment of seismic hazard along the southeastern Canadian margin, Proceedings, 3rd Canadian Marine Geotechnical Conference, St. John's, Newfoundland. 1: 51-54.

Baracos, A. and Graham, J. 1981. Landslide problems in Winnipeg. Canadian Geotechnical Journal. 18: 390-401.

Basham, P.W. 1983. New Seismic Zoning Maps for Canada, Geos. 12, 3: 10-12.

Basham, P.W., Adams, J. and Anglin, F.M. 1983. Earthquake source models for estimating seismic risk on the eastern Canadian margin, Proceedings, Fourth Canadian Conference on Earthquake Engineering, Vancouver, p. 495-508.

Béland, J. 1956. Nicolet landslide, Proceedings Geological Association of Canada. 8: 143-156.

Bonifay, D. and Piper, D.J.W. 1988. Late Wisconsinan ice margin on the upper continental slope off St. Pierre Bank, eastern Canada, Canadian Journal of Earth Sciences. 25: 853-865.

Bornhold, B.D. 1983. Sedimentation in Douglas Channel and Kitimat Arm: Canadian Technical Reports in Hydrography and Ocean Science. 18: 88-114.

Bovis, M. 1985. Earthflows in the Interior Plateau, Southwest British Columbia, Canadian Geotechnical Journal. 22: 313-334.

Boyd, G.L. 1986. A geomorphic model of bluff erosion on the Great Lakes. Proceedings, Symposium on Cohesive Shores, National Research Council, Ottawa. pp. 60-68.

Brown, A. and Casey, F.L. 1960. Investigation into the stability of Dufferin Terrace, Quebec, Mines Branch Investigation Report IR 60-112, Ottawa.

Canada Soil Survey Committee, Subcommittee on Soil Classification. Ottawa: 1978. The Canadian system of soil classification. Department of Agriculture. Publication 1646.

Carter, L.D., Heginbottom, J.A., Woo, M-K. 1987. The arctic lowlands. In. W.L. Graf (ed.), Geomorphic systems of North America, p. 583-628. Boulder, Colorado: Geological Society of America, Centennial Special Volume 2.

Chagnon, J.Y. 1968. Les coulées d'argile dans la province de Québec. Le naturaliste canadien. 95: 1327-1343.

Chagnon, J.Y. 1975. Stabilisation d'un glissement de terrain à Hull, Province de Québec. Proceedings 28th Canadian Geotechnical Conference, Montreal, p. 264-276.

Chagnon, J.Y., Lebuis, J., Allard, J.D. and Robert, J.M. 1979. Sensitive clays, unstable slopes, corrective works and slides in the Quebec and Shawinigan area. Guidebook, Field Trip B-11, Geological Association of Canada.

Chagnon, J.Y. and Locat, J. 1987. Geological Evaluation of the Seismicity of the Charlevoix Area, Quebec. Progress Report 2, Atomic Energy Control Board Project NO. E1103345228Y.

Church, M. and Miles, M.J. 1987. Meteorological antecedents to debris flow in southwestern British Columbia; some case histories. Geological Society of America, Reviews in Engineering Geology. 7: 63-79.

Clague, J.J., Luternauer, J.L. and Hebda, R.J. 1983. Sedimentary environments and post-glacial history of the Fraser Delta and lower Fraser Valley, British Columbia. Canadian Journal of Earth Sciences. 20: 1314-1326.

Clague, J.J., Evans, S.G. and Blown, I.G. 1985. A debris flow triggered by the breaching of a moraine-dammed lobe, Klattasine Creek, British Columbia. Canadian Journal of Earth Sciences. 22: 1492-1502.

Clifton, A.W., Krahn, J. and Fredlund, D.G. 1981. Riverbank instability and development control in Saskatoon. Canadian Geotechnical Journal. 18: 95-105.

Cochonat, P., Ollier, G. and Michel, J.L. 1988. Instability sedimentary features of the Lower Newfoundland Grand Banks slope. Geomarine Letters (in press).

Conway, K.W. and Luternauer, J.L. 1984. Longest core of Quaternary sediments from Queen Charlotte Sound: preliminary description and interpretation. Current Research, Pt. A., Geological Survey of Canada, Paper 84-A1, 647-649.

Conway, K.W. and Luternauer, J.L. 1985. Evidence of ice rafting and tractive transfer in cores from Queen Charlotte Sound, British Columbia, Current Research, Pt. A., Geological Survey of Canada, Paper 85-1A, 703-708.

Cruden, D.M. 1985. Rock slope movements in the Canadian Cordillera. Canadian Geotechnical Journal. 22: 528-540.

Cruden, D.M., Tedder, K.H., Thomson, S. 1988. Setbacks from the crests of slopes along the North Saskatchewan River Valley, Alberta. Canadian Geotechnical Journal. (in press)

Czudek, T. and Demek, J. 1970. Thermokarst in Siberia and its influence on the development of lowland relief. Quaternary Research. 1: 103-120.

Desjardins, R. 1980. Tremblements de terre et glissements de terrain: Corrélation entre des datations au ^{14}C et des données historiques à Shawinigan, Québec. Géographie physique et Quaternaire. 24: 359-362.

Desmeules, J. 1972. Les glissements de terrain au Québec en 1971. Bureau de la statistique du Québec. 11p.

Desmeules, J. 1973. Les glissements de terrain au Québec en 1972. Bureau de la statistique du Québec. 17p.

Dinn, D.F., Burke, R.G., Steeves, G.D. and Parsons, A.D. 1987. Hydrographic instrumentation and software for the remotely controlled survey vehicle 'DOLPHIN'. In Proceedings Oceans '87, Marine Technology Society and IEEE Ocean Engineering Society. Halifax. 2: 601-607.

Dionne, J.C. 1969. Note sur un éboulement à St-Fabien-sur-Mer, Côte sud du St-Laurent. Revue de Géographie de Montréal. 22: 55-64.

Dionne, J.C. 1972. La dénomination des mers postglaciaires au Québec. Cahier de géographie de Québec. 16: 483-487.

Douglas, R.J.W. 1970. Geology and Economic Minerals of Canada. Economic Geology Report,1, Geological Survey of Canada. Ottawa.

Doxsee, W.W. 1948. The Grand Banks earthquake of November 18, 1929. Publications of the Dominion Observatory, Ottawa. 7:323-335.

Eisbacher, G.H. 1979. Cliff collapse and rock avalanches (sturzstroms) in the Mackenzie Mountains, north-west Canada. Canadian Geotechnical Journal. 16: 309-334.

Eisbacher, G.H. and Clague, J.J. 1984. Destructive mass movements in high mountains; hazard and management. Geological Survey of Canada, Paper 84-16.

Ells, R.W. 1908. Rapport sur l'eboulement de Notre-Dame de la Salette, Rivière du Lièvre, Québec. Department of Mines, Canada, Publication 1031.

Evans, S.G. 1982a. The development of Big Slide, near Quesnel, British Columbia, between 1953 and 1982. Geoscience Canada. 9: 220-222.

Evans, S.G. 1982b. Landslides and surficial deposits in urban areas of British Columbia: a review. Canadian Geotechnical Journal. 19: 269-288.

Evans, S.G. 1984a. Landslides in Tertiary Basaltic Successions. Proceedings, 4th International Symposium on Landslides, Toronto. 1: 503 - 510.

Evans, S.G. 1984b. The landslide response of tectonic assemblys in the southern Canadian Cordillera. Proceedings, 4th International Symposium on Landslides, Toronto. 1: 495-502.

Evans, S.G. 1986. Landslide damming in the Cordillera of Western Canada. In Landslide Dams: Processes, Risk and Mitigation. R.L. Schuster (ed.), American Society of Civil Engineers, Geotechnical Special Publication, 3: 111-130.

Evans, S.G. 1987. A rock avalanche from the peak of Mount Meager, British Columbia. Geological Survey of Canada, Paper 87-1A, 929-934.

Evans, S.G. 1988. The 1946 Mount Colonel Foster rock avalanche and associated displacement wave, Vancouver Island, British Columbia. Proceedings, 41st Canadian Geotechnical Conference, Kitchener.

Evans, S.G. and Lister, D.R. 1984. The geomorphic effects of the July 1983 rainstorm in the southern Cordillera and their impact on transportation facilities. Geological Survey of Canada, Paper 84-1B, 223-235.

Evans, S.G. and Clague, J.J. 1988a. Catastrophic rock avalanches in glacial environments. Proceedings, 5th International Symposium on Landslides, Lausanne. 2: 1153-1158.

Evans, S.G. and Clague, J.J. 1988b. Destructive Landslides in the Canadian Cordillera; 1850-1987. Geological Survey of Canada, Paper 84-19 in press.

Evans, S.G., Clague, J.J., Woodsworth, G.J. and Hungr, O. 1988. The Pandemonium Creek rock avalanche, British Columbia. Canadian Geotechnical Journal. submitted.

Evans, S.G., Aitken, J.B., Wetmiller, R.J., Horner, R.B. 1987. A rock avalanche triggered by the October 1985, North Nahanni Earthquake, District of Mackenzie, N.W.T. Canadian Journal of Earth Sciences. 24: 176-184.

Eyles, N., Buergin, R. and Hinchenbergs, A. 1986. Sedimentological controls on piping structures and the development of scalloped slopes along an eroding shoreline; Scarborough Bluffs, Ontario. Proceedings, Symposium on Cohesive Shores. National Research Council, Associate Committee for Research on Shoreline Erosion and Sedimentation. 69-85.

Farcy, A. and Voisset, M. 1985. Acoustic imagery of the seafloor. Oceans 85, San Diego. 1005-1012.

Fleming, R.W., Taylor, F.A. 1980. Estimating the costs of landslide damage in the United States. United States Geological Survey Circular 832.

French, H.M. 1974. Geomorphological processes and terrain sensitivity, Banks Island, District of Franklins. Geological Survey of Canada, Paper 74-1A: 263-266.

French, H.M. and Egginton, P.A. 1973. Thermokarst development, Banks Island, in Permafrost, The North American Contribution to the Second International conference, National Academy of Sciences, Washington, 203-212.

Gardner, J.S. 1970. Rockfall; a geomorphic process in high mountain terrain. Alberta Geographer., 6: 15-20.

Gartner Lee Associates Limited. 1986. Engineering Geology Report, Slope Stabilization Nos. 25 and 27 Kingsbury Crescent, City of Scarborough. Metropolitan Toronto and Region Conservation Authority.

Gelinas, P.J. and Quigley, R.M. 1973. Influence of geology on erosion rates along the north shore of Lake Erie, Proceedings, 16th Conference on Great Lakes Research, International Association of Great Lakes Research.

Geocon Inc. 1983. Preliminary Stability Study, No. 1 Lakehurst Crescent, Scarborough Bluffs. Metropolitan Toronto and Region Conservation Authority.

Geological Survey of Canada. 1985. Regional systematic investigations by the Geological Survey of Canada, Map 1580A.

Gravenor, C.P. and Bayrock, L.A. 1961. Glacial deposits of Alberta. In: Soils in Canada. Ed: R.F. Legget. University of Toronto Press. Toronto, 33-50.

Hamilton, T.S. and Luternauer, J.L. 1983. Evidence of seafloor instability in the south-central Strait of Georgia, British Columbia: a preliminary compilation: Current Research, Pt. A., Geological Survey of Canada, Paper 83-1A: 417-421.

Hardy, R.M. 1963. The Peace River Highway Bridge - A failure in soft shales, Highway Research Record. 17: 29-39.

Hardy, R.M., Clark, J.I. and Stepanek, M. 1980. A summary of case histories spanning 30 years of slope stabilization in Calgary, Alberta. Proceedings, Specialty Conference on Slope Stability Problems in Urban Areas. Canadian Geotechnical Society, Toronto, Session 2, 24p.

Hare, K.F., and Thomas, M.K. 1979. Climate Canada, Wiley, Toronto.

Harper, J.R., Reimer, P.D. and Collins, A.D. 1985. Canadian Beaufort Sea physical shore zone analysis; Geological Survey of Canada, Open File Report 1689.

Harris, M.C. 1980. The state of legislation in cities across Canada. Proceedings, Symposium on slope stability problems in urban areas, Canadian Geotechnical Society, Toronto, 9p.

Harry, D.G., French, H.M. and Pollard, W.H. 1985. Ice wedges and permafrost conditions near King Point, Beaufort Sea Coast, Yukon Territory. Geological Survey of Canada, Paper 85-1A: 111-116.

Harvey, R.C. 1982. The climate of arctic Canada in a 2 x CO_2 world, Canadian Climate Centre, Report 82-5.

Hasegawa, H.S., Chou, C.W. and Basham, P.W. 1979. Seismotectonics of the Beaufort Sea. Canadian Journal of Earth Science. 16: 816-830.

Hasegawa, H.S. and Kanamori, H. 1987. Source mechanism of the magnitude 7.2 Grand Banks earthquake of November 1929: double couple or submarine landslide. Bulletin of the Seismological Society of America. 77: 1984-2004.

Haughton, D.R. 1978. Geological hazards and geology of the south Columbia River valley. Victoria Ministry of Transportation and Highways, Geotechnical and Materials Branch.

Heezen, B.C. and Drake, C.D. 1964. Newfoundland Grand Banks Slump. Bulletin American Association of Petroleum Geologists. 48:221-225.

Heezen, B.C. and Ewing, M. 1952. Turbidity currents and submarine slumps, and the 1929 Grand Banks earthquake, American Journal of Science. 250: 849-873.

Heginbottom, J.A. 1973. Effects of surface disturbance on the permafrost active layer at Inuvik, N.W.T., Canada. In Permafrost: The North American contribution to the Second International Conference, Yakutsk, p. 649-657. Washington: National Academy of Sciences.

Heginbottom, J.A. 1975. Mackenzie Highway evaluation. Geological Survey of Canada, Paper 75-1A: 509-510.

Heginbottom, J.A. 1978. An active retrogressive thaw flow slide on eastern Melville Island, District of Franklin. Geological Survey of Canada, Paper 78-1A: 525-526.

Heginbottom, J.A. 1984a. Continued headwall retreat of a retrogressive thaw flow slide, eastern Melville Islands, Northwest Territories. Geological Survey of Canada, Paper 84-1B: 363-365.

Heginbottom, J.A. 1984b. The mapping of permafrost, Canadian Geographer. 28: 78-83.

Heginbottom, J.A., Bornhold, B.D., Chagnon, J.-Y., Cruden, D.M., Evans, S., Locat, J., Moran, K., Piper, D.J.W., Powell, R., Quigley, R.W., Prior, D., Thomson, S. 1989. The extent and economic significance of landsliding in Canada, Geological Survey of Canada Paper, in preparation.

Hill, P.R., Moran, K.M., and Blasco, S.M. 1982. Creep deformation of slope sediments in the Canadian Beaufort Sea, Geo-Marine Letters. 2: 163-170.

Hughes-Clarke, J.E., Mayer, L.A., Piper, D.J.W. and Shor, A.N. 1988. Pisces IV submersible dives in the epicentral region of the 1929 Grand Banks earthquake. Geological Survey of Canada Paper, in press.

Hungr, O., Morgan, G.C., VanDine, D.F. and Lister, D.R. 1987. Debris flow defences in British Columbia. Geological Society of America, Reviews in Engineering Geology. 7: 201-222.

Hungr, O. and Evans, S.G. 1988. Engineering evaluation of fragmented rockfall hazards. Proceedings, 5th International Symposium on Landslides, Lausanne. 1: 685-690.

Hutchins, R.W., Dodds, J. and Fader, G.B. 1985. High resolution acoustic seabed surveys of the deep ocean. Proceedings, Society for Underwater Technology, London. 27p.

Hutchinson, J.N. 1973. The response of London clay cliffs to differing rates of toe erosion. Geologia Applicata e Idrogeologia. 8: 221-219.

Hutchinson, J.N. 1986. Cliffs and shores in cohesive materials: geotechnical and engineering geological aspects. Proceedings Symposium on Cohesive Shores. National Research Council, Associate Committee for Research on Shoreline Erosion and Sedimentation. 1-44.

Hutchinson, J.N. and Bhandari, R.K. 1971. Undrained loading, a fundamental mechanism of mud flows and other mass movements. Geotechnique. 21: 353-358.

International Pacific Salmon Fisheries Commission, 1980. Hell's Gate Fishways. New Westminster, B.C. 8p.

Jackson, L.E. 1987. Debris flow hazard in the Canadian Rocky Mountains. Geological Survey of Canada, Paper 86-11.

Johns, M.W., Prior, D.B., Bornhold, B.D., Coleman, J.M. and Bryant, W.R. 1986. Geotechnical aspects of a submarine slope failure, Kitimat fiord, British Columbia. Marine Geotechnology. 6: 243-279.

Johnston, G.H. 1981. Permafrost: engineering design and construction. Wiley: Toronto.

Josenhans, H.W., Barrie, J.V. and Kiely, L.A. 1987. Mass wasting along the Labrador Shelf margins submersible observations. Geomarine Letters. 7: 199-205.

Josenhans, H.W., Klassen, R.A. and Zevenhuizen, J. 1986. The Quaternary Geology of the Labrador Shelf. Canadian Journal of Earth Sciences. 23: 1190-1213.

Judge, A.S. 1986. Permafrost distribution and the Quaternary history of the Mackenzie Beaufort region: a geothermal perspective. In Correlation of Quaternary Deposits and Events around the margin of the Beaufort Sea, Geological Survey of Canada, Open File 1237: 41-45.

Kamphuis, J.W. 1986. Erosion of cohesive bluffs, a model and a formula. Proceedings Symposium on Cohesive Shores, National Research Council, Associate Committee for Research on Shoreline Erosion and Sedimentation.

Kerfoot, D.E. and Mackay, J.R. 1972. Geomorphological process studies, Garry Island, N.W.T. In Mackenzie Delta Area Monograph (D.E. Kerfoot, ed); Brock University for 22nd International Geographical Congress: 115-130.

Klugman, M.A., Chung, P. 1976. Slope-stability study of the Regional Municipality of Ottawa-Carleton, Ontario. Geological Survey, Miscellaneous Paper 68: Toronto.

Lamothe, C. and St-Onge, D.A. 1961. A note on a periglacial erosional process in the Isachsen area, N.W.T., Canada. Department of Mines and Technical Surveys, Geographical Bulletin. 15: 104-113.

LaRochelle, P. 1974. Rapport de synthèse des études de la coulée d'argile, de Saint Jean Vianney. Ministere des richesses naturelles, Québec. Rapport S-151, 75p.

LaRochelle, P., Chagnon, J.Y. and Lefebvre, G. 1970. Regional geology and landslides in the marine clay deposits of Eastern Canada. Canadian Geotechnical Journal. 7: 145-156.

LaSalle, P. and Chagnon, J.Y. 1968. An ancient landslide along the Saguenay River, Quebec. Canadian Journal of Earth Sciences. 5: 548-549.

Lebuis, J., Robert, J.-M. and Rissmann, P. 1985. Regional mapping of landslide hazards in Quebec, Symposium on slopes on soft clays. Swedish Geotechnical Institute, Report 17: 205-262.

Lefebvre, G. 1981. Fourth Canadian Geotechnical Colloquium: Strength and slope stability in Canadian soft clay deposits. Canadian Geotechnical Journal. 18: 420-442.

Lefebvre, G. 1986. Slope instability and valley formation in Canadian soft clay deposits. Canadian Geotechnical Journal. 23: 261-270.

Lefebvre, G., Lafleur, J. and Chagnon, J.Y. 1976. Evaluation of vertical drainage as a stabilizing agent in a clay slope at Hull, Quebec. Proceedings, 29th Canadian Geotechnical Conference, Vancouver. Section 5: 16-30.

Legget, R.F. and LaSalle, P. 1978. Soil studies at Shipshaw, Quebec: 1941 and 1969. Canadian Geotechnical Journal. 15: 556-564.

Lewis, C.P. and Forbes, D.L. 1974. Sediments and sedimentary processes, Yukon Beaufort Sea coast; Canada, Task Force on Northern Pipelines, Environmental Social Committee, Report 74-29.

Locat, J., Demers, D., Lebuis, J. and Rissmann, P. 1984. Prédiction des glissements de terrain: application aux argiles sensibles, rivière Chacoura, Quebec, Canada. Proceedings, 4th International Symposium on Landslides, Toronto. 2: 549-555.

Locat, J. and Lefebvre, G. 1986. The origin of structuration of the Grande Baleine marine sediments, Quebec, Canada. Quarterly Journal of Engineering Geology. 19: 365-374.

Locat, J. and Leroueil, S. 1988. Physico-chemical and geotechnical characteritics of recent Saguenay Fjord sediments. Canadian Geotechnical Journal. 25: 382-387.

Locker, J.G. 1973. The petrographic and engineering properties of fine-grained rocks of central Alberta. Alberta Research Council, Bullentin 30.

MacCracken, M.C. and Luther, F.M. 1985. Projecting the climatic effects of increasing carbon dioxide, U.S. Department of Energy, DOE/ER-0237.

Mackay, J.R. 1963a. Notes on the shoreline recession along the coast of the Yukon Territory, Arctic. 16: 195-197.

Mackay, J.R. 1963b. The Mackenzie Delta area, N.W.T., Canada. Department of Mines and Technical Surveys, Geographical Branch, Memoir 8.

Mackay, J.R. 1971. Ground ice in the active layer and the top portion of permafrost. National Research Council of Canada, Technical Memorandum. 103: 26-30.

Martin, D.C., Piteau, D.R., Pearce, R.A. and Hawley, P.M. 1984. Remedial measures for debris flows at the Agassiz Mountain Institution, British Columbia. Canadian Geotechnical Journal. 21: 505-517.

Matheson, D.S. and Thomson, S. 1973. Geological implications of valley rebound. Canadian Journal of Earth Sciences. 10: 961-978.

Mathews, W.H. 1979. Landslides of central Vancouver Island and the 1946 earthquake. Bulletin of the Seismological Society of America. 69: 445-450.

Mathews, W.H. and McTaggart, K.C. 1978. Hope rockslides, British Columbia, Canada. In B. Voight (ed.), Rockslides and avalanches. Amesterdam, Elsevier 1: 259-275.

McDonald, B.C. and Lewis, C.P. 1973. Geomorphic and sedimentologic processes of rivers and coast, Yukon coastal plain, Canada. Task Force on Northern Oil Development, Environmental Social Committee, Report 73-79.

McNab, R., Mukherjee, P.K., and Buxton, R. 1987. Canada's Continental Shelf: An Ocean Mapping Challenge, Geos, 16, 1: 18.

McRoberts, E.C. 1978. Slope stability in cold regions. In O.B. Anderstand and D.M. Anderson, (eds.), Geotechnical Engineering for Cold Regions McGraw-Hill, New York, p. 363-404.

McRoberts, E.C. and Morgenstern, N.R. 1974. The stability of thawing slopes, Canadian Geotechnical Journal. 11: 447-469.

Metropolitan Toronto and Region Conservation Authority 1980. Shoreline Management Program, Watershed Plan.

Milne, W.G., Smith, W.E.T., and Rogers, G.C. 1970. Canadian seismicity and microearthquake research in Canada. Canadian Journal of Earth Sciences. 7: 591-601.

Mosher, D.C. 1987. Late Quaternary sedimentology and sediment instability of a small area on the Scotian slope. M.Sc. thesis, Memorial University of Newfoundland, St. Johns.

Murty, T.S. 1979. Submarine slide-generated water waves in Kitimat Inlet, British Columbia. Journal of Geophysical Research. 84: 7777-7779.

Nasmith, H.A. and Mercer, A.G. 1979. Design of dykes to protect against debris flows at Port Alice, British Columbia. Canadian Geotechnical Journal. 16: 748-757.

Nurkowski, J.R. 1984. Coal quality, coal rank variation and its relation to reconstructed overburden, Upper Cretaceous and Tertiary Plains Coals, Alberta, Canada. American Association of Petroleum Geologists. Bulletin 68: 285-295.

Peckover, F.L. and Kerr, J.W.G. 1977. Treatment and maintenance of rock slopes on transportation routes. Canadian Geotechnical Journal. 14:487-507.

Piper, D.J.W. and Aksu, A.E. 1987. The source and origin of the 1929 Grand Banks turbidity current inferred from sediment budgets. Geomarine Letters. 7: 177-182.

Piper, D.J.W., Farre, J.A. and Shor, A. 1985a. Late Quaternary slumps and debris flows on the Scotian slope, Geological Society of America Bulletin. 96: 1508-1517.

Piper, D.J.W., Shor, A.N., Farre, J.A., O'Connell, S. and Jacobi, R. 1985b. Sediment slides and turbidity currents on the Laurentian Fan: sidescan sonar observations near the epicentre of the 1929 Grand Banks earthquake. Geology. 13:538-541.

Piper, D.J.W., Shor, A.N. and Hughes-Clarke, J.E. 1988. The 1929 Grand Banks earthquake, slump and turbidity current. Geological Society of America Special Paper. 229.

Piper, D.J.W., Stow, D.A.V. and Normark, W.R. 1984. The Laurentian Fan - Sohm Abyssal Plain. Geomarine Letters. 3: 141-146.

Piteau, D.R., Mylrea, F.H. and Blown, I.G. 1978. The Downie Slide, Columbia River, British Columbia. In B. Voight (ed.), Rockslides and avalanches. Amsterdam, Elsevier. 1: 365-392.

Prior, D.B., Bornhold, B.D., Coleman, J.M., and Bryant, W.R. 1982. Morphology of a submarine slide, Kitimat Arm, British Columbia. Geology. 10: 588-592.

Prior, D.B., Bornhold, B.D. and Johns, M.W. 1984. Depositional characteristics of a submarine debris flow. Journal of Geology. 92: 707-727.

Prior, D.B. and Bornhold, B.D. 1986. Sediment transport on subaqueous fan delta slopes, Britannia Beach, British Columbia. Geo-Marine Letters. 5: 217-224.

Prior, D.B., Bornhold, B.D., and Johns, M.W. 1986. Active sand transport along a fiord-bottom channel, Bute Inlet, British Columbia. Geology. 14: 581-584.

Quigley, R.M., Gelinas, P.J., Bou, W.T. and Packeer, R.W. 1977. Cyclic erosion-instability relationships: Lake Erie north shore bluffs. Canadian Geotechnical Journal. 14: 310-323.

Quigley, R.M. and DiNardo, L.R. 1980. Cyclic instability modes of eroding clay bluffs, Lake Erie Northshore Bluffs at Port Bruce, Ontario. Canada. Zeitschrift fur Geomorphologie, Supplementum 34: 39-47.

Quigley, R.M. and Gelinas, P.J. 1976. Soil Mechanics Aspects of Shoreline Erosion. Geoscience Canada. 3: 169-173.

Quigley, R.M. 1980. Geology, mineralogy and geochemistry of Canadian soft soils. Canadian Geotechnical Journal. 17: 268-285.

Rampton, V.N. and Mackay, J.R. 1971. Massive ice and icy sediments throughout the Tuktoyaktuk Peninsula, Richards Island, and nearby areas, District of Mackenzie. Geological Survey of Canada, Paper 71-21.

Renard, V. and Allenou, J.P. 1979. Sea Beam multibeam echosounding on the "Jean Charcot", description, evaluation and first results. International Hydrographic Review. 55: 35-67.

Rib, J.T., Liang, T. 1978. Recognition and Identification, In Schuster, R.L., Krizek, R.J. (eds.), Landslides, Analysis and Control, Transportation Research Board, National Academy of Sciences. Washington, Special Report 176: 34-80.

Rissman,, P., Allard, J.D. et Lebuis J. 1985. Zones exposées aux mouvements de terrain le long de la rivièr Yamaska, entre Yamaska et Saint-Hyacinthe. Ministere de l'energie et des ressources, Québec, DV 83-04.

Robert, J.M. et Chagnon, J.Y. 1976. Caracteristiques et correction d'un glissement de terrain dans les dépôts argileux de la mer de Champlain à Saint-Michel de Yamnaska. Proceedings, 29th Canadian Geotechnical Conference, Vancouver. Section 11: 1-22.

Rogers, G.C. 1980. A documentation of soil failure during the British Columbia earthquake of 23 June, 1946. Canadian Geotechnical Journal. 17: 1221-1227.

Rutter, N.W., Boydell, A.N., Savigny, K.W. and van Everdingen, R.O. 1973. Terrain evaluation with respect to pipeline construction, Mackenzie Transportation Corridor, southern part, lat. 60° to 64°N, Canada. Task Force on Northern Pipelines, Environmental-Social Committee, Report 73-36.

Ryan, W.B.F. 1982. Imaging of submarine landslides with sidescan sonar, In S. Saxov and J.K. Nieuwenhuis (eds.), Marine Slides and Other Mass movements, New York, Plenum, 175-188.

Shah, V.K. 1978. Protection of permafrost and ice rich shores, Tuktoyaktuk, N.W.T., Canada. Proceedings, 3rd International Conference on Permafrost, Ottawa, National Research Council of Canada. 1: 870-876.

Sauer, E.K. 1975. Urban fringe development and slope instability in Southern Saskatchewan. Canadian Geotechnical Journal. 12: 106-118.

Schafer, C.T., Tan, F.C., Williams, D.F. and Smith, J.N. 1985. Late glacial to recent stratigraphy, paleontology, and sedimentary processes: Newfoundland continental slope and rise. Canadian Journal of Earth Sciences. 22: 266-282.

Schafer, C.T. and Smith, J.N. 1987. Hypothesis for a submarine landslide and cohesionless sediment flows resulting from a 17th century earthquake-triggered landslide in Quebec, Canada. Geomarine Letters. 7: 31-38.

Shor, A.N. and Piper, D.J.W. 1988. A large, late Pleistocene buried blocky debris flows on the central Scotian Slope. Geomarine Letters. submitted.

Somers, M.L., Carson, R.M., Revie, J.A., Edge, R.H., Barrow, B.J. and Andrews, P.G. 1978. GLORIA II, An improved long-range sidescan sonar. Proceedings, IEEE Sub-conference on Offshore Instrumentation and Communication, Oceanology International. 16-24.

Spurek, M. 1972. Historical catalogue of slide phenomena. Studia Geographica, 19, Institute of Geography, Czechoslovak Academy of Sciences, Brno.

Statistics Canada. 1988. Construction Price Statistics, Catalogue 62-007, Supply and Services Canada, Ottawa.

Syvitski, J.P.M., Burrell, D.C. and Skei, J.M. 1986. Fjords: Processes and Products, New York, Springer-Verlag.

Tavenas, F., Chagnon, J.Y. and LaRochelle, P. 1971. The Saint-Jean-Vianney Landslide: Observations and Eyewitnesses Accounts. Canadian Geotechnical Journal. 8: 463-478.

Terzaghi, K. 1956. Varieties of submarine slope failures. Eighth Texas Conference on Soil Mechanics and Foundation Engineering, Proceedings.

Theodore, M.H. 1986. Review and case examples of rockfall protection measures in the mountain region of Canadian National Railway. In Transportation Geotechnique, Vancouver Geotechnical Society.

Thomson, S. 1958. Collapse of the Peace River Bridge. British Columbia Professional Engineer, 19, 5: 13-15.

Thomson, S. 1970. Riverbank stability study at the University of Alberta, Edmonton. Canadian Geotechnical Journal. 7: 157-168.

Thomson, S. and Hayley, D.W. 1975. The Little Smoky Landslide. Canadian Geotechnical Journal. 12: 379-392.

Thomson, S. and Morgenstern, N.R. 1977. Factors affecting the distribution of landslides along rivers in southern Alberta. Canadian Geotechnical Journal. 14: 508-523.

Thomson, S. and Tweedie, R.W. 1978. The Edgerton Landslide. Canadian Geotechnical Journal. 15: 510-521.

Thomson, S. and Yacyshyn, R. 1977. Slope instability in the City of Edmonton. Canadian Geotechnical Journal. 14: 1-16.

Van Dine, D.F. 1985. Debris flows and debris torrents in the Southern Canadian Cordillera. Canadian Geotechnical Journal. 22: 44-68.

Varnes, D.J. 1978. Slope movement and types and processes. In R.L. Schuster and R.J. Krizek (eds.), Landslides: Analysis and Control. Transportation Research Board, National Academy of Science, Washington, Special Report 176: 11-33.

Vilks, G., Deonarine, B. and Winters, G. 1987. The marine geology of Lake Melville, Labrador, Geological Survey of Canada Paper 87-22.

Walker, H.J. and Arnborg, L. 1966. Permafrost and ice-wedge effect on riverbank erosion. National Research Council, Washington, Publication 1287: 164-171.

Landslides: Extent and economic significance in the United States

E.E.Brabb
US Geological Survey, Menlo Park, Calif., USA

ABSTRACT: Landslides occur in every state of the nation and its island territories (Guam, Puerto Rico, Virgin Islands, and Samoa). They have damaged or destroyed roads, railroads, pipelines, electrical and telephone transmission lines, mining facilities, petroleum wells and production facilities, houses, commercial buildings, canals, sewers, bridges, dams, reservoirs, port facilities, airports, forests, fisheries, parks, recreation areas, and farms. Much landslide damage goes undocumented because it is considered instead with its triggering event, and thus is included in reports of floods, earthquakes, volcanic eruptions, hurricanes, or coastal storms, even though damage from the landsliding may exceed all other costs. Also, overall landslide damage tends to be underestimated because so many different terms are used for the losses, such as flash flood, avalanche, clay failure, tidal wave, cliff collapse, coastal erosion, land sinking, cave-in, mudslide, road slip, hillside slip, land slip, and debris slip. Despite uncertainty in determining damage costs, landsliding in the United States is estimated to cause an annual loss of about $1.5 billion and at least 25 fatalities.

California, West Virginia, Utah, Kentucky, Tennessee, Puerto Rico, Ohio, and Washington have the most severe landslide problem, based on landslide damage to roads, houses, and other buildings in the decade from 1973 to 1983. About 20,000,000 landslides have occurred in these and other states and outlying areas. Nearly all of these landslides were triggered by precipitation, earthquakes, volcanism, erosion, and the activities of humans.

1 INTRODUCTION

The United States has thousands of books and maps that discuss or show landsliding in small areas. These and other reports that provide an overview of the landslide type, extent, and damage in the conterminous United States are discussed in subsequent sections. None of these reports, however, provides sufficient information on a state-by-state basis so that the landslide problem can be understood, compared, and analyzed by public and private planners, engineers, and decisionmakers. Developers, lenders, and insurers at the state level, as well as elected and appointed government officials, can provide leadership, technical advice, development policies, financial incentives, and regulations to effectively reduce landslide hazards and losses. Their involvement is a prerequisite to any effective landslide hazard reduction program.

This report is primarily directed toward those public and corporate officials who have responsibility for making decisions and adopting policies for reducing landslide losses at the state level, and secondarily toward people at federal and local levels. Data are provided to show:

o Areas in each state that might be susceptible to landsliding.

o Minimum loss due to landsliding per person and per unit area.

o Rating of the severity of the hazard.

o Number of landslide publications available.

o Number of professors in each state with interest or experience in landslide problems.

o Number of members of the Association of Engineering Geologists who might be able to provide technical assistance.

o Indication of landslide inventory and susceptibility mapping available.

o Subjective rating of knowledge about landsliding in each state.

Many states, of course, already have exemplary programs to deal with landsliding hazards; for example, California and Colorado have been pioneers in developing new mapping techniques and instituting land use and grading regulations. Many other states have programs underway to determine the extent of the problem. However, all of them need additional information and encouragement and, as this report provides, a yardstick to measure progress.

The report is organized to provide landslide information on (1) geography, geology, type, process, number, and damage, and (2) the human resources available to study the subject, consistent with other reports in this volume. This information is followed by a summary of the most significant landslide reports, maps, and other information in each state selected from over 6,500 references compiled by Alger and Brabb (1985). Although the information is not consistent because of the differences in data available for each state, the extent of the problem, and the amount of work already done, the information collectively provides a useful perspective into the extent, diversity, knowledge, and cost of landslide problems in the United States.

2 GEOGRAPHY

The United States (Fig. 1) consists of the conterminous states with an area of 7,990,622 km^2, Alaska with an area of 1,520,038 km^2, and Hawaii with 16,719 km^2; it also includes countries under the sovereignty of the United States, or otherwise associated with some type of American jurisdiction or control, that extend from the Pacific Ocean to the Caribbean and have 4 million inhabitants living in a land area of about 15,500 km^2 (U.S. Geological Survey National Atlas, 1970). Of these countries outside the U.S., only Puerto Rico, American Samoa, Guam, and the Virgin Islands are mentioned in this report. The total U.S. area of 9,527,379 km^2 is fourth in the world after the Soviet Union, Canada, and China (Rand McNally Cosmopolitan World Atlas,

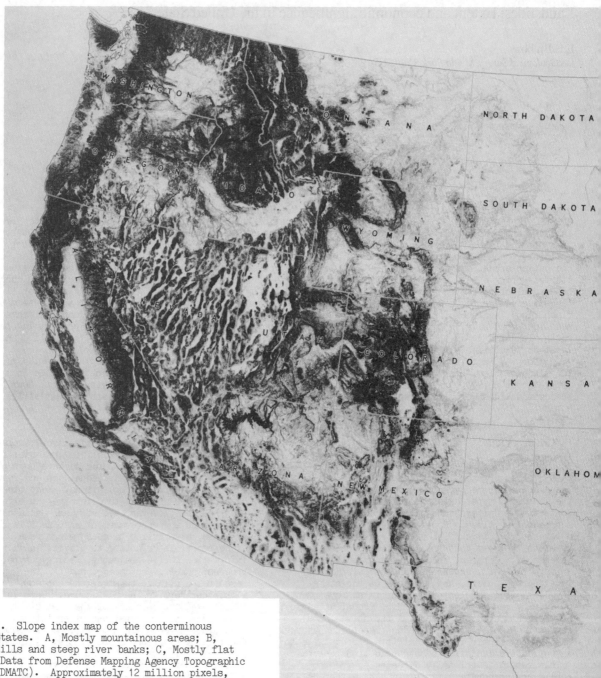

Figure 1. Slope index map of the conterminous
United States. A, Mostly mountainous areas; B,
Mostly hills and steep river banks; C, Mostly flat
areas. Data from Defense Mapping Agency Topographic
Center (DMATC). Approximately 12 million pixels,
representing an 830-meter square on the ground, were
used to construct the original map at 1:2,500,000
scale. Map prepared in 1988 by Philip J. Beilin,
Douglas S. Aitken, Joanne M. Vinton, Robert K. Mark,
and Earl E. Brabb, U.S. Geological Survey.

1980). The areas for each state and a few outlying
areas are shown in Table 1.

The average annual precipitation for the United
States is approximately 74 cm. Hawaii is the
wettest state with an average annual rainfall of 209
cm. Nevada, with an average annual rainfall of 22
cm, is the driest state (Rand McNally Cosmopolitan
World Atlas, 1980). Of the 24 high-precipitation
extremes recognized by Riordan and Bourget (1985)
that could trigger landslides, 8 are in the United
States, including the world's greatest 1-minute
rainfall, 3.1 cm in Unionville, Maryland, on 4 July
1956; the world's greatest 42-minute rainfall, 30.5
cm in Holt, Missouri, on 22 June 1947; a 24-hour
rainfall of 109 cm at Alvix, Texas, on 25 and 26
July 1979; the world's greatest average yearly
precipitation, 1,168 cm at Mount Waialeale, Hawaii;
North America's greatest snowfall in one season,
2,850 cm at Paradise Ranger Station, Mount Rainier,
Washington, 1971-1972, North America's greatest
snowfall in one storm include 480 cm at Mount Shasta
Ski Bowl, California, 13 to 19 February 1959; North
America's greatest depth of snow on the ground,
1,145.5 cm, Tamarack, California, 11 March 1911; and
North America's greatest 24-hour snowfall, 192.5 cm
at Silver Lake, Colorado, 14 and 15 April, 1921.

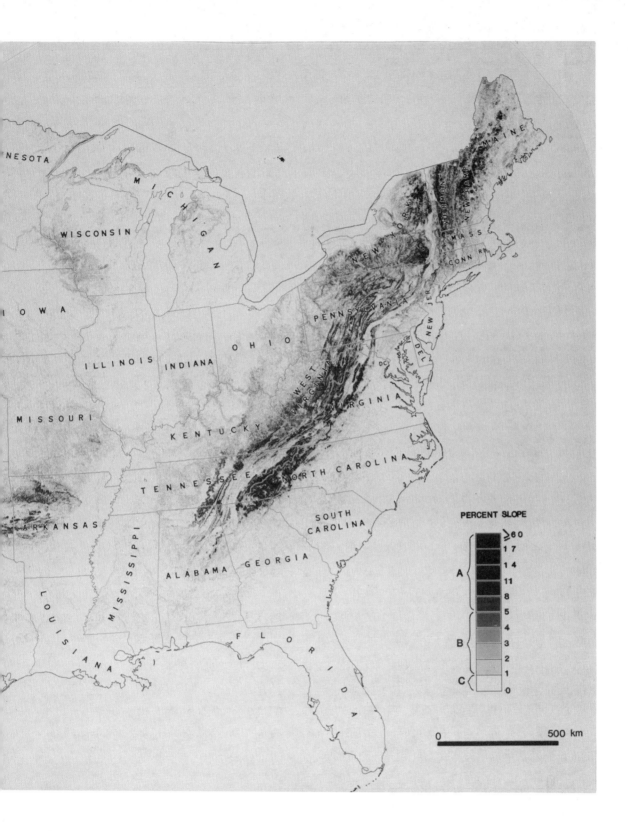

PERCENT SLOPE

A { ≥60, 17, 14, 11, 8, 5 }

B { 4, 3, 2, 1 }

C { 0 }

0   500 km

Precipitation extremes that can trigger landslides, therefore, can be expected in several parts of the United States.

Climatic regions of the eastern United States are similar to those of central and eastern Europe, the west-central Soviet Union, eastern China, Korea, and Japan. Climatic regions of the western United States are most similar to Mediterranean countries, the Middle East, large parts of China, southern Soviet Union, southern Australia, southern Argentina, and the higher parts of other Andean countries (Rand McNally New International Atlas, 1980).

The United States has been divided into physiographic provinces by various authors, but Baker and Chieruzzi (1959) and Brabb and others (in press) find difficulty in using these schemes to separate landslide-prone from landslide-free areas.

The population of the United States projected in the 1980 census to 1985 is 238,274,900 (Rand McNally Commercial Atlas and Marketing Guide, 1987). Distribution of the population by state and a few outlying areas is shown in Table 1.

The location of mountains, hills, steep valley walls, and flatlands in the conterminous United States is shown on Fig. 1. Areas for each of these

27

Table 1 — Landslide costs, extent, severity, resources available, and knowledge for the United States and some outlying areas

	Population[1]	Area km²[1]	Mountain Areas km²[3]	Hills & Steep Valleys km²[3]	People km²	Decade Damage $M[2]	$ Cost/ Person	$ Cost/ km²	$ Cost/km Mts, Hills & Steep Valleys	Landslide Hazard Rating[4]	No. of Pubs.[5]	No. of Profs.[6]	No. AEG Mbrs.[7]	Landslide Inventory Coverage[8]	Coverage Landslide Suscept.[8]	Rating of Landslide Knowledge[9]
California	26,345,300	404,815	184,274	140,480	65	1,000 ?	38	2,470	3,079	severe	960	54	749	extensive	some	fair
West Virginia	1,967,900	62,470	13,002	40,023	32	275	139	4,402	5,186	severe	77	16	12	some	some	fair
Utah	1,698,800	212,570	88,271	96,292	8	210 ?	124	988	1,137	severe	157	19	32	some	none	poor
Kentucky	3,786,200	102,740	3,026	50,729	37	190 ?	50	1,854	3,534	severe	25	8	11	some	some	poor
Tennessee	4,793,200	106,590	9,252	47,890	45	103 ?	23	1,007	1,925	severe	40	10	43	some	extensive	fair
Puerto Rico	3,196,520	8,960	na	na	357	100	30	11,160	na	severe	11	4	2	some	complete	fair
Ohio	10,836,000	106,200	79	32,307	102	100 ?	9	942	3,087	severe	119	17	29	some	some	poor
Washington	4,416,700	172,265	84,416	63,171	26	100 ?	23	581	677	severe	266	21	129	several	several	fair
Colorado	3,269,700	268,311	90,947	100,324	12	70	21	261	366	substantial	182	15	179	complete	some	fair
New York	17,966,200	122,705	17,853	72,412	146	70 ?	4	570	775	substantial	62	13	51	complete	some	fair
Pennsylvania	12,012,000	116,260	15,150	76,598	103	60 ?	5	516	653	substantial	133	8	90	extensive	some	fair
North Carolina	6,304,900	126,505	12,076	27,231	50	45.5	7	360	1,157	substantial	20	5	56	none	none	poor
Oregon	2,709,200	249,115	105,485	110,991	11	40	15	161	185	substantial	290	22	53	some	some	fair
American Samoa	32,297	197	na	na	164	20 ?	619	101,522	na	substantial	1	1	0	none	none	poor
Maryland	4,436,400	25,480	1,098	7,296	174	20	5	785	2,382	substantial	8	2	41	some	complete	fair
South Dakota	719,300	196,715	1,656	67,140	4	18	25	92	259	substantial	21	3	3	none	none	poor
Virginia	5,785,000	102,835	15,135	36,890	56	12	2	117	231	substantial	69	10	51	some	some	poor
Indiana	5,585,800	93,065	20	13,425	60	11	2	118	818	substantial	4	5	15	none	none	poor
Idaho	1,029,900	213,445	109,747	73,981	5	11 ?	11	52	60	substantial	38	11	12	complete	complete	fair
Montana	836,700	376,555	105,772	187,986	2	11 ?	13	29	37	substantial	68	7	10	complete	complete	poor
Alabama	4,079,800	131,487	2,135	45,768	31	10.5	3	80	219	moderate	12	5	19	complete	complete	fair
New Hampshire	1,000,700	23,290	7,516	12,792	43	10	10	429	492	moderate	19	3	13	none	none	poor
Alaska	537,800	1,478,458	591,939	475,543	1	10	19	7	9	moderate	85	11	37	some	some	poor
Texas	16,408,700	678,620	9,157	172,616	24	8	min.	12	44	moderate	34	10	150	none	complete	fair
Minnesota	4,224,300	206,030	494	24,648	21	7	2	34	278	moderate	4	2	10	none	none	poor
New Jersey	7,618,900	19,340	538	4,613	394	6	1	310	1,164	moderate	10	3	37	none	none	poor
Hawaii	1,071,300	16,640	11,202	2,836	64	4.5	+	270	320	moderate	18	2	7	none	none	poor
North Dakota	699,900	179,485	164	45,665	4	4	6	22	87	moderate	4	6	2	some	none	poor
Wyoming	516,400	251,200	54,584	141,691	2	4	8	16	20	moderate	77	13	5	complete	none	poor
New Mexico	1,475,700	314,255	44,278	170,732	5	4	3	13	9	moderate	17	6	23	extensive	none	poor
Mississippi	2,647,300	122,335	0	16,888	22	3.5	1	29	207	moderate	12	3	27	none	none	poor
Vermont	537,800	24,015	9,646	12,516	22	3.5	7	146	158	moderate	11	3	1	none	none	poor
Missouri	5,093,800	178,565	74	64,927	29	3 ?	min.	17	46	moderate	5	5	59	some	none	poor
Nevada	940,900	284,625	118,095	121,516	3	2.5 ?	3	11	10	low	21	6	19	some	some	poor
Louisiana	4,532,600	115,310	0	3,604	39	2.3	min.	20	638	low	5	4	10	none	none	poor
Arkansas	2,401,600	134,880	5,041	43,538	18	2	1	15	41	low	6	1	7	none	none	poor
Oklahoma	3,356,000	177,815	1,559	47,821	19	2 ?	min.	11	41	low	6	1	24	none	none	poor
Illinois	11,640,400	144,120	10	13,960	81	2 ?	min.	14	143	low	12	3	92	complete	none	fair
Arizona	3,252,200	293,985	66,988	162,184	11	2	min.	7	9	low	29	5	27	some	some	fair
Delaware	621,400	5,005	0	205	124	2	3	400	9,756	low	0	1	4	none	none	poor
Iowa	2,942,800	144,950	36	28,774	20	1.3	min.	9	45	low	5	3	5	none	none	poor
Kansas	2,470,500	211,805	0	36,557	12	1.3 ?	min.	6	36	low	2	5	45	none	none	poor
Georgia	5,990,500	150,365	3,064	31,508	40	1 ?	min.	7	29	low	4	3	35	none	none	poor
Nebraska	1,622,400	198,505	172	47,959	8	0.8	min.	4	17	very low	1	3	10	complete	none	poor
Wisconsin	4,818,000	140,965	243	39,603	34	0.7	min.	5	18	very low	11	6	11	some	some	poor
Maine	1,124,660	80,275	8,018	47,864	14	0.6	min.	7	11	very low	8	1	6	complete	none	poor
Massachusetts	5,847,400	20,265	1,613	9,293	289	0.3	min.	15	27	very low	7	3	63	none	none	poor
Dist. Columbia	621,000	163	0	79	3,810	0.2	min.	1,227	2,531	very low	6	0	12	none	none	poor
Virgin Islands	111,000	345	na	na	322	0.1	1	290	na	very low	1	1	0	none	none	poor
Michigan	9,211,900	147,510	374	30,267	62	0.1	min.	min.	3	very low	5	7	11	some	some	poor
Florida	11,435,200	140,255	0	2,322	82	0	0	0	0	very low	1	1	27	none	none	poor
South Carolina	3,396,300	78,225	532	13,481	43	0	0	0	0	very low	4	0	4	none	none	poor
Connecticut	3,175,900	12,618	550	8,113	252	0	0	0	0	very low	2	2	10	none	none	poor
Rhode Island	969,500	2,730	0	1,298	355	0	0	0	0	very low	0	0	1	none	none	poor
Guam	124,000	450	na	na	276	0	0	0	0	very low	0	?	0	none	none	poor

[1] Rand McNally, Atlas of the 50 States, 1987, and The Times, Atlas of the World, 1985. Population estimated for 1985.

[2] Minimum damage 1973-1983 to State roads and private property, in millions of dollars (Brabb, 1984a). Does not include landslide damage in 1964 Alaskan earthquake, 1969 Hurricane Camille in Virginia, or landsliding in 1980 Mount St. Helens eruption, Washington. Information for Puerto Rico, Samoa and Virgin Islands provided here for the first time (see text). A query indicates that the figure was estimated from scanty data.

[3] Data from an experimental slope index map of the United States (Fig. 1). Figures for area are preliminary and should be used only to compare one State to another.

[4] An arbitrary rating based only on the minimum damage figures provided by Brabb (1984a). Other ratings are possible, as indicated in text.

[5] Number of landslide publications listed by Alger and Brabb (1985). Only reports and maps concerned primarily with landslides are included.

[6] Number of university professors listed by Brabb and FitzSimmons (1984) as having an interest in landslide investigations and research.

[7] Number of members of the Association of Engineering Geologists members residing in the State, from 1986-1987 membership roster. Not all of these people are interested in making landslide investigations, but the figures are useful for comparing states.

[8] Comments apply primarily to landslide inventory maps of areas at least one-half of a 7 ½ -minute U.S. Geological Survey quadrangle. Although many geologic maps show landslide deposits, hardly any of these maps have a complete inventory of all landslide deposits and none of them are considered in this column. A few maps of coastal zones in Michigan and Wisconsin are considered to be adequate inventory and susceptibility maps for purposes of this report. Maps of states where the coverage is indicated as complete are at 1:500,000 scale. Most of these, such as the map for Colorado, do not adequately show debris-flow deposits.

[9] A subjective determination based on the adequacy of published information to determine the extent and kind of landslide process affecting any area in a State selected at random. Does not take into account that consultants, engineers, and geologists in highway departments, and personnel in State geological surveys have substantial knowledge about slope stability that has not been published.

features and the number of people per km[2] are given on Table 1.

3 GEOLOGY

The relation between rock type and landslide type in the conterminous United States has been described by Radbruch-Hall and others (1981); only a small part of their material is included here. The relatively youthful mountain ranges of California, Oregon, and Washington contain many poorly consolidated, clay-rich, and extensively fractured clastic units of Cenozoic age that tend to be less resistant to landsliding than the harder and older sedimentary, igneous, and metamorphic rocks of the Rocky Mountains in states like Colorado, Utah, Wyoming, Montana, and Idaho. In the Appalachian Mountains, extending from Maine to Alabama, the rocks are similar to those in the Rocky Mountains. However, they are generally more weathered and more exten-sively covered with colluvium. The full range of landslide processes operate in both areas, but earthflows seem more prevalent in the Appala-chians. Soil falls and slumps are common in unconsolidated sediment along steep river banks in the interior of the United States.

A generalized geologic map at a scale of 1:7,500,000 of the conterminous United States with overprints showing moderately and severely deformed rocks, regionally metamorphosed rocks, and rocks with mixed structural complexity has been published by Radbruch-Hall and others (1987). The same report contains engineering-geologic and environmental-geologic maps—and a map showing by colors and patterns—the distribution of various geologic conditions or processes, including landslides and steep slopes, that might cause difficulties in construction.

Figures 2 and 3 show where earthquakes could be a factor in triggering landslides. Although earth-quakes in the United States are commonly associated with California and Alaska, several other states and island areas also have earthquakes that could trigger landslides. Volcanism is a factor only in the western United States, Alaska, and Hawaii, as shown on Fig. 4. More specific geologic information is provided in descriptions for a few individual states.

4 APPROACH TO THE LANDSLIDE PROBLEM

4.1 Regional landslide information available

The first regional evaluation of the landslide problem in the conterminous United States was made by Baker and Chieruzzi (1959), who based their evaluation on questionnaires sent to state and federal agencies, companies and consultants, and a review of 267 landslide articles and texts published by 1950. They provided a list of the most troublesome types of landslides, an indication of the volume of material that moved, the names of formations that were involved, and a small-scale map at page size indicating the severity of the problem.

Radbruch-Hall and others (1982) prepared a new map of the landslide areas in the conterminous United States at a scale of 1:2,500,000 (published at a scale of 1:7,500,000). Their map was prepared by evaluating the landslide incidence and susceptibil-ity of units on a geologic map. Discussion of slope stability was organized by physiographic subdivisions.

A map showing landslide severity in the conterminous United States at a scale of about 1:11,700,000 is included in a report by Krohn and Slosson (1976). It incorporates an early version of

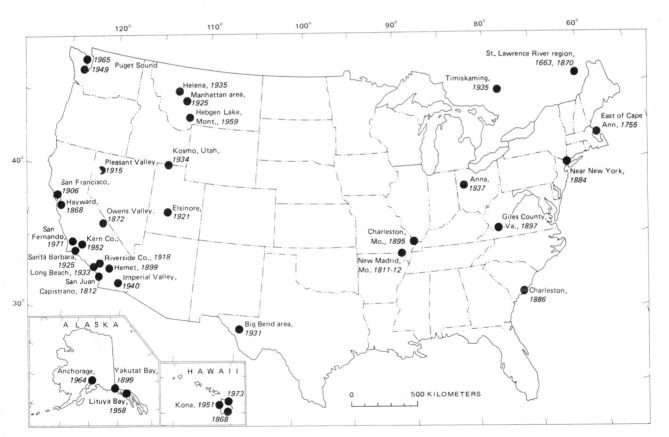

Figure 2. Location of past destructive earthquakes in the United States (from Hays, 1980). Many of these earthquakes triggered landslides (Keefer, 1984) and others could have done so.

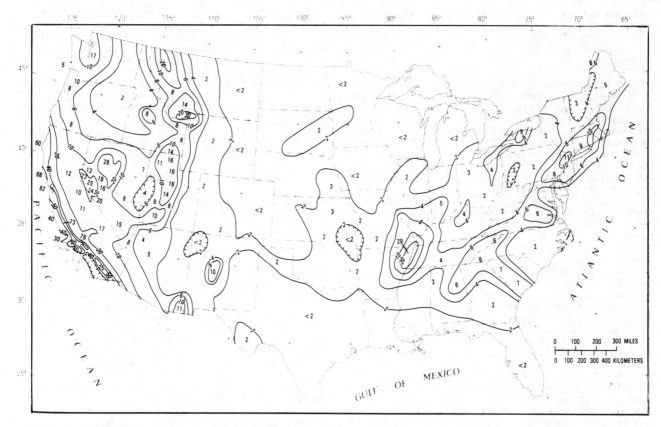

Figure 3. Earthquake probability expressed as horizontal velocity (in centimeters per second) on rock sites that have a 90-percent probability of not being exceeded in 50 years. From Algermissen and others (1982).

the map by Radbruch-Hall and others and includes many additional areas derived from several sources.

This report builds on the work by Radbruch-Hall and others (1982) and provides for the first time a state-by-state analysis based on a review of 6,500 geographically oriented references listed by Alger and Brabb (1985); a four-year reconnaissance of all states, the Virgin Islands, and Puerto Rico; and extensive conversations and cooperative programs with geologists and engineers in state geological surveys and departments of transportation. No information is provided for offshore areas, because they will be discussed in other parts of this book.

The 6,500 references to landslides in the United States listed by Alger and Brabb (1985) are divided into those that show or discuss landslides as the principal topic, and those that mention landslides incidentally; only the former are shown on Table 1. Table 1 also indicates the extent of landslide inventory and susceptibility mapping; additional information is provided in the descriptions for several states. No state has a complete inventory of landslides at a scale of 1:100,000 or larger. Many of the landslide inventories that have been prepared, even at a 1:24,000 scale, are probably incomplete. Even in California, where many different agencies have helped to make landslide inventory and susceptibility maps and where state laws require consideration of landsliding in the preparation of general plans, information about the extent and kind of landsliding is missing for at least half the state. The generally poor rating given to landslide knowledge for most states indicates partly a widespread underestimation of the debris flow problem, as mentioned, for example, by Brabb (1984b) and Baldwin and others (1987) in the San Francisco Bay region and by Clark (1987) in the Appalachian region, and results partly from a perception given by the literature and a reconnaissance of the states.

Figure 1 provides, in addition to slope, some indication of areas where landslides are most likely (darkest tones) and least likely (lightest tones) to occur. Although the amount of slope is only one of several factors that influence the regional distribution of landslides, Brabb and others (in press) indicated that in northern New Mexico, at least, a slope index map prepared from the same digital data set as the one used for Fig. 1 separated 99 percent of the areas containing deep-seated landslide deposits and 88 percent of the areas containing shallow landslides from areas where landslides are absent. The digital format also provides a means of measuring the approximate areas of mountainous, hilly, and flat terrain in each state, as indicated on Fig. 1 and in Table 1. These terrain subdivisions are arbitrary, subjective, and provisional, and they have not been tested for accuracy.

4.2 Landslide types and processes

Nearly all of the types of slope movement classified by Varnes (1958, 1978) have been reported in the United States, and many of them are illustrated in his reports. Hardly any systematic studies have been undertaken to indicate which landslide processes are dominant in various regions, although the report by Radbruch-Hall and others (1982) provided much new information. Few landslide inventory maps provide information on landslide type, but several of the inventory maps prepared in the last decade, such as the ones by Wieczorek (1984) and Guzzetti and Brabb (1987), distinguish the types of movement.

4.3 Relation of landslides to population

A comparison between the population per square kilometer on Fig. 5 and mountainous and hilly terrain where landslides are prevalent (Fig. 1)

30

extensive. Fleming and Taylor (1980a) have indicated that the average annual cost per capita of landsliding in the Cincinnati, Ohio, Appalachian Mountains area is more than four times the cost in the San Francisco Coast Ranges.

4.4 Number of landslides in the United States

Lessing and others (1983) estimate that West Virginia has 437,000 landslides. This figure is based on a sample of landslide density in thirty-one 7.5-minute quadrangles. The density is about seven landslides per square kilometer. Brabb (1985) counted 70,000 landslides on maps prepared before 1980 for nine counties in the San Francisco Bay region. In 1982, at least 18,000 new landslides were triggered, and most of these were in areas not previously recognized as landslide-prone. The total of at least 88,000 landslides comes to about five per square kilometer. In order to provide additional data, Fred Taylor of the USGS (written commun., 1988), counted the number of landslides on 80 landslide inventory maps for parts of Alabama, California, Colorado, Kentucky, Ohio, Oregon, Pennsylvania, Tennessee, Utah, and Washington. The highest state average was 10 landslides per km^2 for California; the lowest was 0.4 per km^2 for Washington and Oregon; and the average for all the states was 4.4. If the average for all states is multiplied by 4,925,820, the number of square kilometers of mountainous, hilly, and steep valley terrain in the conterminous United States and Alaska, plus the areas of Hawaii, Guam, Puerto Rico, and the Virgin Islands (Table 1), a total of 21,673,603, rounded off to 20 million is an estimate of the number of landslides in the United States and outlying areas.

4.5 Landslide costs

The first documented attempt to determine the cost of landsliding in the United States was by Smith (1958), who estimated that the yearly cost was hundreds of millions of dollars. Krohn and Slosson (1976) analyzed several reports published after the Smith report and established a figure of $20 per capita per year for damage to private dwellings in southern California. Extrapolating this figure to an estimate of 20 million people living in areas of the United States with a moderate or high landslide potential, they estimated $400 million as the yearly landslide damage to private dwellings for the nation. This figure did not include indirect costs or damage to transportation facilities, forestlands, public property, etc. Coates (1977) provided additional cost data.

Wiggins and others (1978) established for the first time an estimate of the annual landslide losses for buildings in each state. Their total of 370.3 million 1970 dollars would be approximately 986 million 1987 dollars—using Gross National Product deflator values provided by R.L. Bernkopf, USGS (verbal commun., 1988). Brabb (1984a) used largely unpublished data from State highway departments and geological surveys and came up with a much lower figure, 44 million annually for damage to private property (mostly buildings) plus about $200 million for roads. These figures for each state are used on Table 1.

Jahns (1978) indicated that landslides and subsidence caused at least $75 billion in damages from 1925 to 1975, more than triple the nearly $20 billion in damages for floods, hurricanes, tornadoes, and earthquakes during the same period. The average of $1.5 billion annually for landslide losses is nearly 3 billion in 1987 dollars.

Schuster and Fleming (1986) used the information given above, additional published data, previously unpublished data, and an estimate of indirect costs

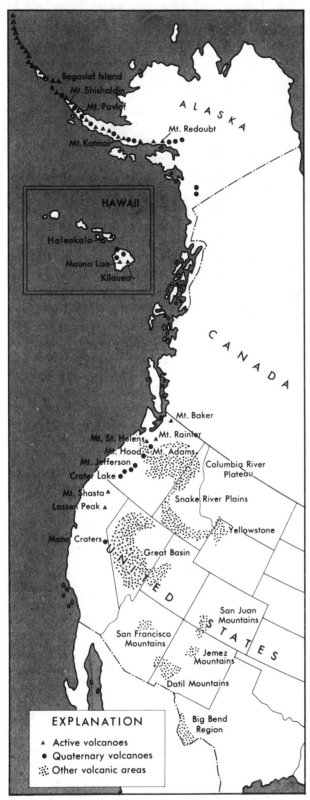

Figure 4. Volcanic areas in the western United States, Alaska, and Hawaii. From Smith and Bailey (1982).

shows the dense population in the Appalachian Mountains stretching from Alabama to Maine and the sparse population in the Rocky Mountains and other mountain systems of the American West. The western Coast Ranges extending from southern California to northwestern Washington is another region where the population density is high and where landsliding is

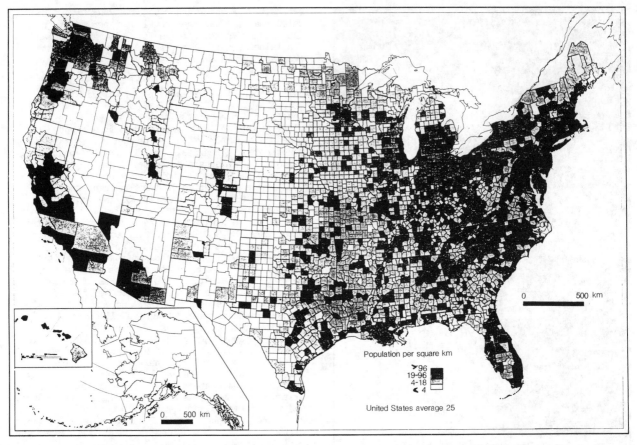

Figure 5. Population density by counties, 1980. From U.S. Bureau of the Census, PC 80-1-A1.

to determine that the annual figure for landsliding in the United States was nearly $1.5 billion, and more than 25 fatalities.

None of these reports discussed landslide damage to forests and fisheries, and the cost of this damage must be substantial. Many reports document the effect of landslides on forests, but few provide specific cost information. Megahan and others (1978) determined that the 1975 cost of repairing and stabilizing landslides in the Clearwater National Forest in Idaho was about $1 million. Almost 6 ha in forestland was damaged by 1,418 landslides that dumped more than 43,000 m³ of sediment into streams with a fishery resource—no estimate of this cost was provided. Landslide damage to roads in six U.S. National Forests in California during 1983 was $5.5 million, according to R.G. Deleissegues of the USFS (written commun., 1986). A study of Tomiki Creek in Mendocino County, California, indicates that the production of Chinook salmon and steelhead fish has been reduced 80 percent by streambank, gully, road, and landslide erosion (U.S. Soil Conservation Service, 1986). The cost of the fish lost is $843,800 annually. These figures and verbal reports from other forestry and fishery sources indicate that the total cost of landsliding is at least tens of millions, and could be hundreds of millions, of dollars annually.

The annual cost of landsliding for infrequent catastrophic events such as the 1964 Alaskan earthquake and Hurricane Camille in 1969, were not included by Brabb (1984a) because of the difficulty in establishing the initial costs and a recurrence interval. Keefer (1984, Table 9) estimated 279 million 1964 dollars (783 million in 1987 dollars) for landslide damage in the 1964 Alaskan earthquake. Plafker and Rubin (1978) indicated that the recurrence time for an earthquake like the 1964 event is on the order of 500 to 1,350 years. The average annual cost, therefore, would range from

$580,000 to $1,566,000.

Hurricane Camille in 1969 triggered debris avalanches and floods that killed 150 people and damaged or destroyed public and private property worth at least $420 million in the western part of Virginia (Williams and Guy, 1973; dollars inflated to 1987 value). The 71 cm of rain that fell in 8 hours is the amount expected once every 1,000 years, and is one of the largest quantities ever measured in the United States. If all of the costs incurred are attributed to landsliding, the average annual cost would be $420,000.

The minimum landslide costs for state roads and private property listed by Brabb (1984a) and shown

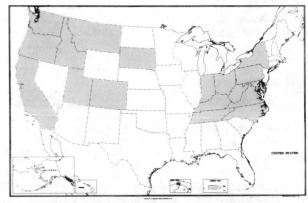

Figure 6. States with a severe or substantial landslide problem based on minimum damage, 1973-1983. American Samoa is not shown. Alaska would have been added if the earthquake of 1964 had occurred during the sample period.

on Table 1 provided a way of ranking states with the most damage. Based on these provisional and incomplete data (Fig. 6), California, West Virginia, Utah, Kentucky, Tennessee, Puerto Rico, Ohio, Washington, Colorado, New York, Pennsylvania, North Carolina, Oregon, American Samoa, Maryland, South Dakota, Virginia, Indiana, Idaho, and Montana had the highest landslide costs from 1973 to 1983. If landslide damage costs for infrequent events such as the 1964 Alaskan earthquake, 1969 Hurricane Camille, and the 1980 Mount St. Helens eruption were added as a lump sum instead of spread over a return period of hundreds of years, Washington, Alaska, and Virginia would rank near the top of the list with California.

If the landslide cost per person or the cost per unit area on Table 1 were used instead of total cost, the ranking of state landslide hazards would be substantially different. For cost per person, American Samoa has the most severe landslide hazard followed by West Virginia, Utah, Kentucky, California, Puerto Rico, South Dakota, Tennessee, Washington, Colorado, Alaska, Oregon, Montana, Idaho, New Hampshire, Ohio, Wyoming, Vermont, Minnesota, New Jersey, and North Dakota (Fig. 7).

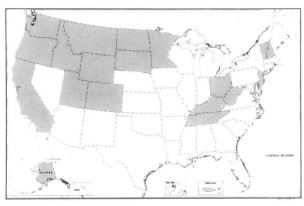

Figure 7. States with landslide costs from 1973-1983 exceeding $5 per person. American Samoa not shown.

For cost per square kilometer (Fig. 8), American Samoa is again at the top followed by Puerto Rico, West Virginia, California, Kentucky, the District of Columbia, Tennessee, Utah, Ohio, Maryland, Washington, New York, Pennsylvania, New Hampshire, Delaware, North Carolina, Hawaii, New Jersey, Virgin Islands, Colorado, Oregon, Vermont, Indiana, and Virginia.

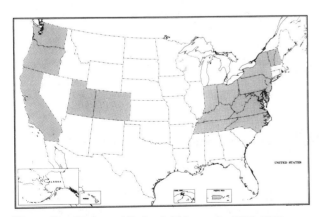

Figure 8. States with landslide costs 1973-1983 exceeding $100 per km². American Samoa, Virgin Islands, Delaware, and District of Columbia not shown.

If flat areas in a state shown on Fig. 1 are subtracted and the cost is allocated to mountains, hills, and steep valleys along rivers, some new states would be added, and the placement of several others would be changed. American Samoa and Puerto Rico would probably still be at the top (no slope data available for these areas) followed by Delaware, West Virginia, Kentucky, Ohio, California, District of Columbia, Maryland, Tennessee, New Jersey, North Carolina, Utah, Indiana, New York, Washington, Pennsylvania, Louisiana, New Hampshire, Colorado, Minnesota, South Dakota, Virginia, Alabama, Mississippi, Oregon, and Vermont (Fig. 9).

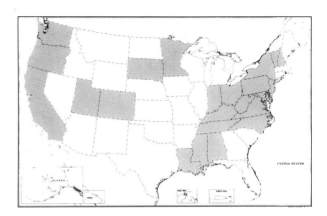

Figure 9. States with landslide costs 1973-1983 exceeding $100 per km² in mountainous and hilly terrain. Delaware and District of Columbia too small to show. No data for American Samoa, Puerto Rico, and Virgin Islands, but they would probably rank high among these states.

The quality of the information for the District of Columbia and Delaware, at least, is so questionable that these jurisdictions are probably wrongly placed on the list. Even so, California is the only western state in this list, and it is far from the top.

Figure 10 is a composite of Figs. 6, 7, 8, and 9 showing any state with a substantial or severe landslide problem according to the various criteria.

The purpose in providing these alternative rankings, despite questionable cost and slope information, is to provoke thought about what constitutes the severity of a hazardous geological process. For our purpose, total damage from 1973 to 1983 is used to determine the landslide-hazard rating for each state mentioned in Table 1; other methods might be more meaningful, if the data were accurate.

4.6 Human resources to solve landslide problems

The number of professors of geology, geography, civil engineering, and forestry with interests in and/or experience with landsliding in each state was provided by Brabb and FitzSimmons (1984) and is shown on Table 1. When compared with the damage figures for each region, the Pacific Coast States with about one half the landslide damage are underrepresented having only 97 (25 percent) of the professors; whereas the Rocky Mountain States and Central Plains States from Texas to Wisconsin are overrepresented, having 13 percent and less than 1 percent of the damage, and 21 percent and 16 percent of the professors, respectively. With respect to members of the Association of Engineering Geologists, the Pacific Coast States are similarly underrepresented, whereas the ratio of member to damages is about the same in Rocky Mountain States and is overrepresented in the Central Plains States. In other words, people with the skills to

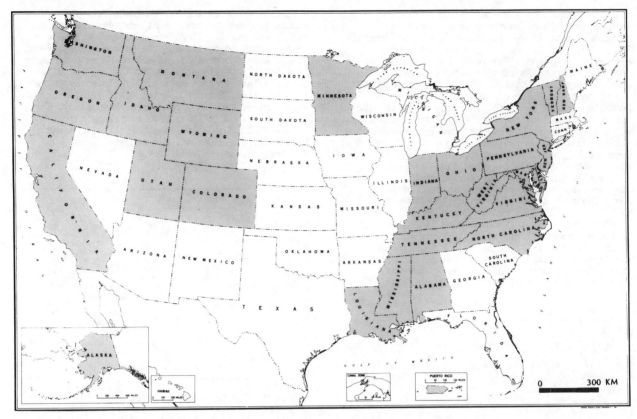

Figure 10. Composite map of Figs. 6, 7, 8, and 9 showing states with a substantial or severe land-slide problem using various criteria. American Samoa and Virgin Islands not shown. Delaware and District of Columbia too small to show at this scale.

investigate landslides are not located in places where the damage is most severe.

Alabama, Arizona, California, Colorado, Idaho, Illinois, Indiana, Maine, Maryland, Michigan, Mississippi, Missouri, Montana, Nebraska, New Mexico, New York, Oregon, Pennsylvania, Texas, Utah, Vermont, Washington, West Virginia, Wisconsin, and Wyoming have current or recent programs to map and study landslide processes in their state. The total person-years of annual effort probably does not exceed 20.

The U.S. Geological Survey devotes no more than 20 person-years per annum nationwide to landslide research. Information about landslide work conducted by other federal agencies is sparse; Sangrey and others (1985) estimated that the U.S. Geological Survey spends about one-fourth of the $10 million in research funded for landslides each year and that the Federal Emergency Management Agency, Department of Transportation, U.S. Bureau of Reclamation, National Science Foundation, U.S.D.A. Forest Service, and Federal Highway Administration are also concerned with landslides. It is estimated here at no more than 30 person-years are spent per annum, mostly by USFS personnel in the Pacific States, making assessments of potential landslide problems in areas considered for timber sales. Landslide problems on federal lands, such as in many national parks, are handled mainly on a case-by-case basis.

5 STATE-BY-STATE ANALYSIS

The 6,500 references available have been summarized in this section to provide what seemed to me to be the most significant information for each state. The unevenness in the information provided reflects differences in the landslide problem, the amount of

data available, and the amount of work being done in different states. Readers interested in a more systematic approach by physiographic province can refer to the report by Radbruch-Hall and others (1982). Unusual landslides are listed on Table 2.

Alabama

More than 300 landslides distributed throughout Alabama have been shown on maps and reports by Rheams (1982) and Rheams and others (1987). Most of them are small slumps, debris slides, and earthflows along oversteepened road cuts. Clay and sandy clay of Late Cretaceous age and shale and mudstone of Carboniferous age are the geologic units that have the most failures. Naturally occurring landslides have been recognized mainly in the northern part of the state and some of these are active or have been active in historic time. Prehistoric debris flows in the Huntsville area have been recognized by Pomeroy (1983).

The Geological Survey of Alabama has prepared several reports to assist counties in utilizing their natural resources effectively, and some of these discuss landslides. The one by Stow and Hughes (1980) is a good example.

Alaska

Landslides were responsible for a substantial part of the $1.8 billion (1987 dollars) in damages associated with the March 27, 1964, Alaskan earthquake. The landslides either destroyed buildings directly or created waves that killed at least 23 people and destroyed their homes (Hansen and others, 1966; National Board of Fire Underwriters and Pacific Fire Rating Bureau, 1964). Of the many reports describing the earthquake-triggered landslides, the one by Wilson (1967) provides a good perspective of landsliding in

Table 2. Unusual landslides in the United States

<u>Alaska</u> – 145-km-long debris flow during Pleistocene (Nichols and Yehle, 1985)
    – 90-million-m^3 debris avalanche of recent but uncertain age in Katmai area (Griggs, 1920)
    – Landslide damage of $783 million (1987 dollars) caused by 1964 earthquake was most expensive
        landslide disaster in U.S. history (Keefer, 1984)
    – Debris avalanches triggered in 1976 by blasting (Vandre and Swanston, 1977)
<u>Arizona</u> – Damming of Colorado River in Grand Canyon by slumps of uncertain age (Huntoon, 1975)
    – Many fires started by sparks from rolling and falling boulders during 1887 earthquake (DuBois and
        Smith, 1980)
    – Rockfalls in Canyon de Chelly triggered in the 1970's by sonic booms (L.E.Scott, Arizona DOT, verbal
        commun., 1982)
<u>California</u> – Mount Shasta debris avalanche with area of 450 km^2 and volume of 26 km^3 is one of the largest
        Quaternary landslides on Earth (Crandell and others, 1984)
    – Tin Mountain landslide of late Pleistocene age covers 15 km^2 and has a volume of 1.8 billion m^3
        (Burchfiel, 1966)
    – Blackhawk rockfall avalanche 7.6 km^3 in volume happened 17,400 years BP (Shreve, 1968; Stout, 1977)
    – Prehistoric Martinez Mountain rock avalanche 7.6 km long (Bock, 1977, and Baldwin, 1984)
    – 1941 debris flow 24 km long, 900,000 m^3 in volume at Wrightwood (Sharp and Nobles, 1953)
<u>Colorado</u> – Approximately 180 million m^3 of a 45 km^2 landslide complex along Muddy Creek was reactivated in
        1986 (Stover and Cannon, 1988)
<u>Hawaii</u> – Pleistocene gravel on Lanai was deposited from a giant wave generated by a submarine landslide (Moore
        and Moore, 1984)
    – Submarine landslide of unknown age 160 km long and 50 km wide has an average gradient of 2° (Moore,
        1964)
    – Alika submarine landslide formed 50,000 to 100,000 years ago has area of 4,000 km^2 and volume of
        1,500-2,000 km^3, one of the largest deposits yet described (Lipman and others, 1988)
    – August 1972 earthquakes with magnitudes from 0.8 to 1.2 cause rockfalls (Tilling and others, 1975,
        1976)
<u>Illinois</u> – Pleistocene mudflow along a glacial moraine (Hester and DuMontelle, 1971)
<u>Nevada</u> – Cumulative volume of pre-late Pleistocene landslide debris at Slide Mountain 95 million m^3 (Thompson
        and White, 1964, p. A-23)
    – May 1983 debris-flow runout area predicted (Brabb, 1984b)
<u>New Hampshire</u> – July 1885 debris flow about 11 km long (Hitchcock, 1885)
<u>New Mexico</u> – Debate over tectonic, intrusive or landslide origin of late Cenozoic folds (Foley, 1964)
    – Fold produced by landslide drilled for oil in mistaken belief that it was tectonic in origin (Reiche,
        1937, p 547)
    – 500 km^2 area of Pleistocene rock-block glides (Watson and Wright, 1963)
    – Pueblo Indians prevent rockfall nearly 1,000 years ago (Judd, 1959 and Schumm and Chorley, 1964)
    – Tenth or eleventh century rockfall used as solar marker (Newman and others, 1982)
    – July 1936 rockslide caused by lightning (Workman and Kelley, 1940)
<u>Puerto Rico</u> – 129 people killed by landslide at Ponce in October 1985, the largest number of fatalities for a
        single landslide in U.S. history (Jibson, 1986b)
<u>Utah</u> – Lateral spread with area of 9 km^2 formed 1,000 to 2,000 years ago is largest in United States (Van
        Horn, 1975)
    – June 1974 rockslide triggers movement of older, inactive, complex landslide with volume of 15 million
        m^3 (Fleming and others, 1977)
    – $200 million Thistle landslide in 1983 most expensive single landslide in U.S. history (Utah Dept.
        Public Safety, 1984)
<u>Virginia</u> – 25-km zone of giant Pleistocene rockslides with total mass of 200 to 400 million m^3 (Schultz, 1986)
    – 150 people killed by debris avalanches and floods in Hurricane Camille, August 1969 (Williams and
        Guy, 1973)
<u>Washington</u> – 20 million m^3 rockslide on glacier reformed into moraine 13,000 to 11,000 years ago (Waitt, 1979)
    – Recent 1 billion m^3 mudflow 104 km long originated at Mount Rainier (Crandell and Waldron, 1956)
    – Tree stumps transported and deposited upright by May 1980 mudflows (Fritz, 1980)
    – $500 million May 1980 Mount St. Helens rock slide, debris avalanche, and debris flows/mudflows second
        most expensive landslide disaster in U.S. history. Debris avalanche of 2.8 km^3 is world's
        largest known historic landslide. (Schuster, 1983, and this report)

Anchorage, and the map by Dobrovolny and Schmoll (1974) depicts slope stability in that area.

The many reports and maps listed by Alger and Brabb (1985) indicate that landslide processes in Alaska are as varied and extensive as those in the conterminous United States. Reports by Wahrhaftig and Black (1958), Bishop and Stevens (1964), Barr and Swanston (1970), Wu and Swanston (1980), and Carter and Galloway (1981) provide a representative sample of the literature.

American Samoa
At least 70 debris flows occurred on October 28, 1979, killing four people and causing damage to several villages (Buchanan-Banks, 1981). The amount of damage was not reported, but a hurricane on January 19, 1987, left 2,000 people homeless and caused an estimated $100 million in damages (Salt Lake Tribune, January 20, 1987, p. A3). According to Stearns (1944, p. 1310), hundreds of landslides are triggered in Samoa by each hurricane. A figure of $20 million in landslide damage per decade, therefore, is probably a reasonable estimate.

Arizona
A review of the literature and topographic maps suggests that rockfalls, rockslides, and debris flows are common in Arizona. Landslides associated with mining and road activities are a problem locally; some rockfalls have been triggered by sonic booms (L. Scott, Arizona Transportation Department, verbal commun., 1982) and by the 1887 earthquake (Dubois and Smith, 1980).

Extensive areas of rock slumps, termed Toreva blocks by Reiche (1937), occur in the northeast part of the state. Reiche indicates that most of these slumps probably originated during a wetter climate in prehistoric time, but at least two of them were triggered by historic earthquakes.

Relative slope stability maps have been prepared by Murphy and Moore (1980) and Péwé (1979) for the Tucson and Phoenix areas. Strahler (1940) has provided a good overview of landsliding in the Vermillion Cliffs area. Several debris flows in the Grand Canyon of the Colorado River, December 1966, have been described by Cooley and others (1977).

Arkansas

Thirteen landslides damaged Arkansas roads from 1971 to 1982 (J.E. Clements, Jr., verbal commun., 1982). Most of these are slumps and slump-earthflows in colluvium overlying shale in the northwestern part of the state.

California

No other state comes close to California in total landslide damage, in the number of professors and consultants concerned with landslide problems, and in the number of landslide maps and reports published. An overview and description of the landslide problem in each physiographic province has been provided by Radbruch and Crowther (1973). Landslides along the coast and on marine terraces are shown on maps and are discussed by Cleveland (1975), Pipkin and Ploessel (1974), and the California Department of Navigation and Ocean Development (1977).

California has been subdivided into 725 15-minute quadrangles by the USGS. Approximately 300 (40 percent) of these cover relatively flat areas in the Sacramento and San Joaquin Valleys, Mojave Desert, and Modoc Plateau. Of the remaining 425, about one half have landslide inventory and/or landslide susceptibility maps at scales ranging from 1:24,000 to 1:62,500 (Taylor and Brabb, 1986a).

The landslide inventory and/or susceptibility maps have been prepared by a variety of public and private organizations. Most have been prepared by the U.S. Department of Agriculture Forest Service (USFS), the California Department of Forestry, and private consultants under contract to them. The California Division of Mines and Geology (CDMG) is a major producer of landslide inventory maps and has been a pioneer in producing landslide susceptibility maps since the 1960's. Maps prepared by the USGS are mainly in the urban areas around San Francisco and Los Angeles. A few maps have been prepared by the California Department of Water Resources, by students at university earth science departments, and by consultants for a few urban areas. Thousands of maps have been prepared by consultants for site evaluation, but they cover areas too small to be useful for regional evaluation of landslide hazards.

The California Department of Transportation (CALTRANS) has a staggering amount of landslide information in their files—information that was not included in the 960 references listed by Alger and Brabb (1985). For example, nearly two-thirds of the 67 landslides in California mentioned by Taylor and Brabb (1986b) that caused fatalities, or at least $1 million in damages came from unpublished reports provided by the 11 CALTRANS district offices. Similar reports indicate that rockfalls are a problem at 511 localities along CALTRANS-maintained roads.

Highlights of USGS landslide research in California include studies of potential rockfall-avalanches (Crandell and others, 1974) and debris flows (Osterkamp and others, 1986) at volcanic centers in northern California; preparation of landslide hazard maps for land use planning in the San Francisco Bay region (Nilsen and others, 1979); the development of a regional real-time landslide warning system (Keefer and others, 1987); sediment budget studies and landslide maps in north coastal California (Nolan and others, 1976); studies in central California (Jackson, 1977) and southern

California (Scott and Williams, 1978); analysis of a high intensity rainfall event that produced 18,000 debris flows in the San Francisco Bay region (Ellen and Wieczorek, 1988); debris-flow (Campbell, 1975) and earthflow studies (Keefer and Johnson, 1983); and development of a method for predicting which areas would be likely to fail by landsliding during an earthquake (Wieczorek and others, 1985).

USFS landslide research is extensive but is difficult for geologists to find because the results are not usually published in the geologic literature. A report edited by DeGraff (1987), and another concerned with predicting landslides in clearcut patches (Rice and Pillsbury, 1982), provide a perspective on USFS research in California.

The CDMG has a program mandated by state law (AB101, 1983) to prepare landslide-hazard maps in urban and urbanizing areas. By the end of 1986, thirteen 7.5-minute quadrangles in the San Francisco, Los Angeles, and San Diego regions were either completed or were in progress. In addition, the Division since 1976 has cooperated with the California Department of Forestry in the review of timber-harvest plans and in the preparation of landslide hazard maps of forested areas along the north coast. By the end of 1985, sixty 7.5-minute quadrangles had been completed (Bedrossian, 1986). The Division also made extensive studies of landsliding in the Geyers Geothermal Resources area.

Several outstanding landslide reports have been edited, compiled, or prepared by university professors and consultants. The report compiled by Ehlig (1986), for example, describes many significant landslides in southern California and provides field-trip logs to view them. Two reports by the National Research Council (1982a, 1984) describe debris flow disasters in 1978, 1980, and 1982.

Colorado

Landslide literature for Colorado is characterized by an unusually large number of innovative and interesting reports by federal, state, university and private investigators, only a few of which can be mentioned here. A map by Colton and others (1976), one of the first for an entire state, indicates that landslide deposits cover a substantial part of the western two-thirds of Colorado. Rogers and others (1979) provide criteria for defining, recognizing, and mitigating different landslide processes, and a model ordinance for regulating land use in landslide-prone areas. A map by Mears (1977) predicting debris-flow impact pressures in the Glenwood Springs area is the first of its kind. Rogers (1987) indicates that the Colorado Geological Survey is mapping and studying 34 areas that have a variety of landslide processes; a landslide susceptibility map by Soule (1975) is typical of several that have been prepared to assist communities in mitigating or avoiding landslide hazards.

USGS Professional Paper 673 (Robinson and others, 1972) represents an unusual collaboration between geologists, geophysicists, and engineers from federal and state agencies to understand the geometry and failure mechanism of a 585,000 m^3 landslide in central Colorado. Another Professional Paper by Howe (1909) examined the distribution and causes of landslides in the San Juan Mountains, southwestern Colorado, and a report by Ives and Bovis (1978) describes the preparation of landslide hazard maps for some of the same area.

Radbruch-Hall and others (1977) indicate that large-scale gravitational spreading and displacement ("sackungen") is common in Colorado. Braddock and Eicher (1962) describe rock block slides in the north-central part of the state. Hansen (1973) mentions slumps, earthflows, mudflows, debris slides, debris flows, and rockfalls as occurring during a single storm in the Denver area.

Connecticut

Varved glacial clays in Hartford, Connecticut, have failed at two localities, according to Lane (1963). One of these was a slump-earthflow with about 75,000 m³ of material. Some rockfalls and debris flows occur on outcrops of basalt (T.L. Holzer, USGS, verbal commun., 1982).

Delaware

Slumping is a potential hazard in surficial deposits of the Piedmont Province in the northern tip of Delaware (Sundstrom and others, 1976). A few landslides have occurred along oversteepened road cuts near Wilmington (T.E. Pickett, verbal commun., 1982).

Florida

One earthflow in the northwest part of Florida has been reported by Jordan (1949).

Georgia

Many landslides occur in the northwestern part of the Georgia. Two of the 51 major debris-slide/debris-flow events in the Appalachian Highlands mentioned by Clark (1987) are in Georgia.

The Georgia Department of Transportation has an extensive file of reports on landslides that have affected highways (D. Mitchell, verbal commun., 1983).

Guam

Landslides in Guam's coral cliffs and in soils derived from volcanic rocks have been reported by Callender (1975).

Hawaii

Approximately 180 debris flows and one deep-seated landslide occurred December 31, 1987, near Honolulu when one-half the mean annual rainfall fell in one storm (S. Ellen, USGS, verbal commun., 1988).

Volcanic activity and earthquakes trigger many rockfalls and other kinds of landslides (Tilling and others, 1975, 1976). The possibility of a mudflow following an eruption in southeastern Oahu has been discussed by Crandell (1975).

An overview of landslides, mudflows, and soil avalanches in Hawaii is provided by Sterns (1966) and Wentworth (1943). A comprehensive thesis on landsliding in southeast Oahu, including landslide susceptibility maps, was written by Jellinger (1977).

Idaho

An inventory of all recognized landslides in Idaho has been nearly completed by W.C. Adams and R.M. Breckenridge of the Idaho Geological Survey.

Many reports on landsliding in U.S. National Forests in Idaho have been prepared by USFS personnel, but few of these reports have been published. (Refer to Alger and Brabb, 1985, p. 52-54 for a partial list of these reports.)

Landslides in southeastern Idaho have been mapped and described by Othberg (1984). Witkind (1972) has prepared a map showing seiche, rockfall, and earthflow hazards in the east-central part of the state, and Covington (1977) has mapped the potential rockfall hazard along the Snake River Canyon in the south-central part. Extensive landsliding in west-central Idaho has been described by Wagner (1944). Nine stereo pairs of Idaho landslides are included in a thesis by Liang (1952).

The following guidebook articles provide an opportunity for self-guided field trips to landslide areas: Clayton and others (1975), Hollenbaugh (1975), and Charboneau (1975).

A landslide susceptibility map of Idaho at a scale of approximately 1:4,000,000 was prepared by Othberg and Breckenridge (1981) from satellite images.

One of the oldest reports on landslides in the western United States is by Russell (1901). He commented on the relationship between landslides and the occurrence of sedimentary beds and volcanic ash within lava flows.

Illinois

Most landslides in Illinois are along steep banks of the Illinois and Mississippi Rivers, and along oversteepened road cuts. A report by Dumontelle and others (1971) describes typical problems. All known landslides, and the amount of damage they caused, are shown on a map by Killey and others (1984).

Indiana

Many Indiana landslide problems are related to oversteepened road cuts. Slumping of glacial lake clays caused by excessive loading or watering is described by Gray (1971). Several natural landslides have formed along steep banks of the Ohio River and its tributaries.

Sisiliano and Lovell (1971) have proposed regional and site-specific techniques for dealing with slope stability and other problems related to highway construction. Constraints on land related to slope stability have been considered in many county reports prepared by the Indiana Geological Survey; the report by Wayne (1975) is typical.

Iowa

Slumps and rockfalls along the steep bluffs of the Mississippi River and slumps along oversteepened road cuts are a recurring problem in Iowa. A rotational slump at least 150 feet deep in a coal mine area near Eldora is an example of a spectacular landslide, according to K.L. Dirk (verbal commun., 1982). Conditions under which loess in the western part of Iowa will collapse are described by Handy (1973).

Kansas

Several small landslides in shale along road cuts in northeastern Kansas have been observed by R.L. McReynolds, Kansas Highway Commission (written commun., 1982). An abstract by Katzman (1987) states that a large landslide developed in bedrock of Pennsylvanian age along bluffs of the Kansas River in 1983.

Kentucky

Landslides are common throughout Kentucky and have been a problem in the western part of the state since at least as early as 1811 (Jibson and Keefer, 1988, and Finch, 1971). A typical landslide that destroyed 17 houses and a church is described by Davies (1973). A planar block glide and other landslides in the eastern part of Kentucky are described by Froelich (1970). Many landslides have been investigated and repaired by the Kentucky Department of Transportation, which has an extensive file of landslide reports.

Preliminary landslide inventory maps have been prepared for the eastern third of the state. (Consult the bibliography by Alger and Brabb, 1985, p. 55, for the references.) A map and report by Newell (1977, 1978) provide a framework for understanding and predicting landslides and other hazards in the Appalachian Mountains.

Louisiana

Although nearly 300 landslides have been recognized in Louisiana by Professor Scott Burns and his students at Louisiana Tech University (verbal commun., 1987), most of them are small slumps along oversteepened road cuts and riverbanks. Large landslides have occurred on steep valley walls of the Mississippi River and its tributaries, and one of these severed a gas pipeline causing at least $1 million in damages (L. Guilbeau, verbal commun., 1981). Earthflows along river banks have been described by Torrey and Weaver (1984).

Maine

Maine shares with New Hampshire the distinction of

having most of the oldest reports on landsliding in the United States. A report by Hitchcock (1836), for example, describes a landslide that diverted the Presumpscot River, and a report by Morse (1869) provides additional information on this and other landslides in the Portland area. Most or all of these landslides were failures in glacial-marine deposits. Photographs of recent landslides in these deposits can be found in a report by Thompson (1979); similar landslides were described in greater detail by Amos and Sandford (1987).

Novak (1987) sent a landslide questionnaire to nearly 1,000 geologists, soil scientists, civil engineers, and public works directors in Maine. He received information on 18 earthflows or soil flows, 14 slumps, seven rockslides, four rockfalls, two block glides, one debris slide, and one lateral spread. Most of the landslides are located in glacial-marine deposits ("quick clays"), and a few are on mountain slopes and along rivers inland from the marine limit.

Maryland

A report by Balter (1970) describes several landslides in Maryland and provides design procedures for dealing with these and other landslides. Several county geotechnical maps consider cut-slope stability and other factors that could be related to landslide distribution (Piper, 1965). Computer-derived maps of slope stability and other factors that could affect land-use planning in the Baltimore-Washington urban area have been prepared by Froelich and others (1980). McMullan (1976) analyzed the relationship between landslides, coastal recession, and tropical storms. Cleaves and others (1974) have published environmental maps showing geologic factors that affect land modification, including slope stability. A landslide susceptibility map of the state at a 1:500,000 scale and a landslide susceptibility map of Prince Georges County at a 1:50,000 scale have been prepared by Pomeroy (1988, a and b). These maps and reports indicate that the landslide problem in Maryland is significant, and that it occurs mainly in clays of the Potomac Group around the northern and western margins of Chesapeake Bay.

Massachusetts

Relatively little has been written about landslides in Massachusetts since the report by Cleland (1902) describing what may have been debris flows in the northwestern part of the State. A slump/earthflow in varved lake clays in the west-central part is shown on a photograph in the report by Root (1958). A few wedge failures, rockfalls, and other landslides have affected state highways (R.P. Ierardi, verbal commun., 1982).

Michigan

No landslides have been reported from the interior part of the Lower Peninsula of Michigan, but erosion of the shoreline of the Great Lakes is extensive and has caused slumps and other kinds of landsliding in many places. Literature about the erosion is voluminous; reports by the Michigan Department of Natural Resources (1973) and the Michigan Division of Land Resources (1979) provide an excellent overview of the problem and management guidelines for dealing with the hazard. Soilfalls along Lake Michigan are shown in a report by Smith (1958, fig. 4); sand flows into Lake Michigan were reported by French (1972). Earthflows occur in "quick clays" in the Upper Peninsula (Booy, 1977; Neilson, 1965).

Minnesota

Landslides have occurred at more than 20 localities throughout Minnesota, but little of the information about them has been published (G.R. Cochran, verbal commun., 1982). Some of these are rotational slumps in glacial lake clays and others are wedge failures in a bentonite of Ordovician age. Several are

slumps in clay deposits along the Red River that forms the western border of Minnesota; one of these was described by Zejdlik (1956).

Mississippi

Mississippi, like Louisiana, has a surprising amount of landsliding considering its generally low relief. Many of the landslides involve loess, sand, and clay overlying shale bedrock along steep banks of the Mississippi River and its tributaries (Krinitzsky, 1965; Lutton, 1971; Keady and others, 1973; Childress and others, 1976; and Torrey and Weaver, 1984). Landslides away from rivers are mostly along oversteepened road and housing cuts in clay-rich formations of Late Cretaceous and Cenozoic age (W.T. Ruff, verbal commun., 1982).

Missouri

Approximately 450 landslides are shown on a landslide inventory map of Missouri by Hoffman and St. Ivany (1982). Nearly all of these are debris or earth slumps along oversteepened road cuts, mostly in the northern part of the State (see also Goodfield, 1969). Landslides in loess are common along the banks of the Mississippi and Missouri Rivers, especially where the loess overlies shale close to the surface of the ground. Loess failures in St. Louis were mentioned by Morse (1935); more recent loess failures in this city have damaged at least 15 houses, according to J.D. Vineyard (verbal commun., 1982). A program by the Missouri Geological Survey to map landslides and other hazards in St. Louis and other expanding urban areas has been described by Lutzen and Williams (1968).

Montana

Landsliding in Montana is extensive and varies in its mechanism. A magnitude 7.1 earthquake in 1959 triggered a 30-million m^3 rockslide which killed 26 people, and many rock-falls, slumps, debris avalanches, debris slides, and earthflows (Hadley, 1964). Other prehistoric, individual rockfall and rockslide/avalanches of Holocene and Pleistocene age have volumes as great as 372 million m^3 (Mudge, 1965).

Block glides, complex landslides, and other types of landslides have adversely affected many Montana highways. Most of these landslides have been described in reports by the staff of the Montana Department of Highways, but only the report by Williams and Armstrong (1970) has been published.

A map at a 1:500,000 scale showing all recognized landslides in Montana has been prepared by the Montana Bureau of Mines and Geology in cooperation with the USGS. (Daniel and others, 1985). A compilation of more than 1,000 of these landslides identifies the geologic formation involved and the types of failure where data are available.

Nebraska

Landslides in Nebraska are not a substantial problem but are more extensive than would be expected from the generally low relief. Landslides occur in loess and till of Pleistocene age and in shale of Late Cretaceous and Paleozoic age. A few rockfalls occur where resistant sandstones have formed cliffs. A landslide in Omaha has adversely affected a sewer plant, delaying a $50 million addition to upgrade the facility (Engineering News Record, 1976).

The location of all known landslides in Nebraska has been compiled in digital format by the Nebraska Geological and Natural Resources Surveys (D.A. Eversoll, written commun., 3/30/87).

Nevada

Debris flows are the principal landslide problem in Nevada; the one at Ophir Creek in the west-central part of the State (Watters, 1983) killed one person, injured four others, and destroyed several houses and other property. Innovative maps prepared by Katzer (1981), Katzer and Glancy (1978), Katzer and

Schroer (1981), and Glancy and Katzer (1977) fore-cast where this and other debris flows might impact several areas.

New Hampshire

Landslide literature in New Hampshire may date back to 1792 when Belknap described the diversion of a river (in Flaccus, 1958, p. 176). A severe storm in late August 1826 triggered debris avalanches in the northern part of the state that killed nine people and inspired poems, a ballad, a short story by Nathaniel Hawthorne, hundreds of articles, and several scientific accounts including one by Silliman and others (1829). Flaccus (1958) mentions this and many other landslides in his intriguing history of slope failures in New Hampshire from 1792 to 1957.

Debris avalanches and debris flows account for nearly all the New Hampshire landslides described in the literature, but rockslides and rockfalls are a common problem along State highways, according to F.E. Prior (verbal commun., 1982).

New Jersey

The principal landslide problems in New Jersey are slumps along steep coastal bluffs in the east-central part of the state (Minard, 1974), rockfalls from diabase ridges along the Hudson River (see photos in Liang, 1952, pl. 56a and b), and rock-slides, rock-block slides, and other landslides that have damaged state highways (H.K. Apgar, verbal commun., 1982).

New Mexico

New Mexico has a great variety of landslide processes that have damaged highways and other structures. Some of the landslide problems affect-ing highways have been described by Lovelace and others (1977) and Bennett (1974). On Sept. 12, 1988, during a thunderstorm, a boulder tumbled onto Highway 68 near Santa Fe, struck a bus, killed 5 people, and injured 14 (San Jose Mercury News, 9/13/88).

Smith (1936) indicated that sliding of till in the north-central part of the state probably occurred in two episodes of intense frost action during the Pleistocene. Reiche (1937) indicated that extensive areas of slumping, like those in Arizona, probably also occurred in prehistoric time. One of the areas of extensive slumping, termed "toreva" by Reiche, is shown in the report edited by Eckel (1958, fig. 47).

An area of rock block glides on the east flank of the Chuska Mountains in the northwestern corner of the state is at least 50 km long and averages 10 km in width (Watson and Wright, 1963).

Landslide inventory maps of the northern part of New Mexico at a 1:500,000 scale have been prepared by Guzzetti and Brabb (1987); 2,489 shallow land-slide deposits and 1,406 deep-seated landslide deposits were identified. According to Brabb and others (in press), boundaries of physiographic provinces, climatic zones, and isohyetals in New Mexico do not provide a basis for separating landslide-prone from landslide-free areas. However, a slope index map prepared from the same digital data set used for Fig. 1 separated 99 percent of the areas containing deep-seated landslide deposits and approximately 88 percent of the areas containing shallow landslide deposits from relatively landslide-free areas in the lowest slope-index category.

New York

Several hundred landslides are shown on a New York inventory at a 1:500,000 scale prepared by Fickies and others (in press). Fatalities associated with these landslides total at least 87 since 1836. Many of the fatalities were caused by slumping and flowing of glacial lake clays exposed along the Hudson River in the eastern part of the state. The literature about these deposits is extensive; the

reports by Newland (1916) and Dunn and Banino (1977) are representative. Maps showing areas vulnerable to landsliding have been prepared by Robak and Fickies (1983) and Fickies and Regan (1982).

Debris flows are numerous in at least the Adirondack Mountains in the northern part of the state (Bogucki, 1977). Rockfalls occur throughout the state and, in 1956, destroyed most of the largest hydroelectric facility serving Niagara Falls (Associated Press, 1956). Slumps and other land-slides are a conspicuous feature along bluffs exposed to waves in southeastern New York (Fuller, 1914) and along Lake Erie. Landslides in glacial deposits and other materials are shown on 10 stereo pairs in the thesis by Liang (1952). Information on 106 landslides affecting state highways was provided by V.C. McGuffey (verbal commun., 1982); data in about 800 file boxes in storage could provide additional information.

North Carolina

Rockslides, rockfalls, and slumps are a substantial problem along highways, especially in the more mountainous regions in the western part of North Carolina. Leith (1965) analyzed more than 400 of these landslides and determined that most of them occur in micaceous metasedimentary rocks and in saprolite and soil developed on these rock types. Joints and faults influence strongly the size and shape of the sliding mass.

Debris flows, debris avalanches, and debris slides are also common (Clark, 1987; Gryta and Bartholomew, 1983). Neary and Swift (1987) investigated the rainfall conditions that trigger these landslides.

North Dakota

Soil slides and slump/earthflows are common in the southwestern part of North Dakota beyond the maximum extent of Pleistocene ice sheets (Trimble, 1979). Some of these landslides have been a substantial problem in Theodore Roosevelt National Park, closing some roads and requiring others to be rerouted (Bluemle and Jacobs, 1981). Landslides occur in other parts of the state, especially along river bluffs underlain by shale of Late Cretaceous or early Tertiary age.

Ohio

Ohio has many destructive landslides and interesting and informative literature on the subject (Alger and Brabb, 1985). Landslides are prevalent in the southeastern and hilly part of the state, and on bluffs along Lake Erie. Rockfalls are a recurring problem along the steep valley walls of the Ohio River.

Three recent reports on landsliding in Ohio are especially noteworthy. A report by Fleming and others (1981) discusses several features of landslides in the Cincinnati area. An innovative probabilistic assessment of landslide susceptibility in the same area by Bernknopf and others (1988) provides a method for evaluating the benefits and costs of alternative mitigation strategies. An evaluation of slope stability in southeastern Ohio by Pomeroy (1987) illustrates well the variety of landslide problems and the adverse effects of these problems on houses and other structures; a section on advice for residents is applicable to many landslide areas.

Oklahoma

Very little published information about landsliding in Oklahoma is available, but the engineering characteristics of all geologic and soil units, including the occurrence of landslides, is described in eight reports prepared by geologists and engineers in the Oklahoma Department of Transporta-tion. A summary of some of this information was provided by Hayes (1971a and b). The areas most susceptible to landsliding appear to be in the east-central and southeastern parts of the state.

Oregon

Oregon is second only to California in the number of reports on landslides (290) mentioned by Alger and Brabb (1985), but nearly two-thirds of these are landslide inventories prepared by USFS personnel in preparation for timber sales. Several reports deal comprehensively with timber harvesting and landsliding; the ones by Swanson and Swanston (1977) and Beschta (1978) provide an overview.

Several counties and the coastal zone have been mapped for geologic hazards, including landslides. Areas where future landsliding may take place have been delineated, chiefly on the basis of steep slopes. The report and maps by Beaulieu (1977) are representative.

The coastal mountain belt, steep bluffs along the Columbia River (Palmer, 1977), and the east-central part of Oregon seem to have the largest number of landslides; but landslide deposits, including debris flows, have also been observed or reported at many other localities.

Pennsylvania

The landslide literature for Pennsylvania is impressive in its variety, comprehensiveness, and innovative approach. Particularly impressive are the reports and maps that are designed to educate decisionmakers and homeowners about landsliding and other hazardous geological processes with language and illustrations that they can easily understand. The reports by Wilshusen (1979), Briggs and others (1975), and Freedman (1977), and the map by Pomeroy and Davies (1975), are representative.

Gray and others (1979) have provided an overview of landslide processes in western Pennsylvania. They indicate that slumps, earthflows, and rockfalls are the most common landslide deposits. Hamel and Flint (1972) and Pomeroy (1982a, 1986) have provided excellent descriptions of the various landslide processes.

Debris flows and debris avalanches have had a devastating effect on several Pennsylvania communities. Reports by Pomeroy (1980, 1982b, 1984), Clark (1987), and Johnson and Rahn (1970) discuss these fast-moving landslides.

Pennsylvania landslides have been well illustrated. In addition to most of the reports mentioned previously, the thesis by Liang (1952) and the report edited by Eckel (1958) have many illustrations of landslides plus some of the techniques used to control them.

Overdip slopes are a significant factor affecting landslide susceptibility. The map by Briggs (1974), showing areas of overdip slopes around Pittsburgh, seems to be the first of its kind. The illustrations on the map showing various relations between bedding and slope are helpful in understanding the failure process.

Landslide inventory and/or susceptibility maps have been prepared for several areas in the western part of the state. The map by Pomeroy (1979) is typical; it shows areas of recent and older landslide deposits, areas most susceptible to landsliding, and steep slopes most susceptible to rockfalls.

Landslide damage to highways, houses, and other structures has been extensive. Pomeroy (1980) has indicated that landslide and flood damage in a 1977 storm was $300 million. Fleming and Taylor (1980b) determined that landslide costs in Allegheny County exceeded $4 million from 1970 to 1976.

Puerto Rico

Landslide damage to Puerto Rican roads from 1973-1983 cost at least $6 million and could have reached more than $10 million, according to Charles Gover, Puerto Rico Department of Transportation (verbal commun., 1983). The Federal Emergency Management Agency spent more than $260 million to repair flood and landslide damage in Puerto Rico between 1970 and 1985, according to Jibson (1986a, p. 18). In a storm on October 5-8, 1985, the flood and landslide damage cost at least $50 million and may have been as high as $500 million; 3,000 homes were damaged, 1,300 of them beyond repair (Jibson, 1986a). At least 129 people were killed by a landslide at Ponce (Jibson, 1986b)—the largest number of people killed by a single landslide in U.S. history.

This information suggests that $100 million in landslide damages from 1975 to 1985 is a conservative estimate for Puerto Rico.

A map at a 1:240,000 scale showing landslides and areas susceptible to landsliding was prepared by Monroe (1979). Since the map was published, several new areas of landsliding have been discovered and the debris-flow problem has proven to be much more extensive and serious than indicated on this map. A more extensive description of landslide problems in Puerto Rico is provided in a report on the Caribbean area in this volume.

Rhode Island

A small block glide along a highway cut described by Ferreira (1982) is the only landslide reported in Rhode Island.

South Carolina

The paucity of landslide reports in South Carolina is surprising in view of the extensive landsliding reported in adjacent states where geological conditions and slopes seem similar. No landslides have affected South Carolina highways in the past decade, according to R. L. Steward (verbal commun. 1982). Small soil slumps associated with gully erosion have occurred in the northern part of the State, according to Ireland and others (1939). A few rockfalls were observed on oversteepened road cuts in the northwestern part of the state on March 18, 1983. On June 26, 1987, an area approximately 70 m by 13 m in a former quarry rim collapsed and fell into the old quarry at Limestone College in Gaffney, threatening one of the college buildings (R.H. Campbell, written commun., 1987).

South Dakota

Landslides caused nearly $3 million in damages to South Dakota highways from 1978-1980, according to Bump (1979). In 1979, 43 of these slides were being monitored to determine their rate of activity. The stabilization cost of just one of these landslides is estimated to be $8 million.

Most of the landslides adversely affecting roads, houses, and other structures in South Dakota occur along the Missouri River and its tributaries where the geologic units consist of Pierre Shale (Late Cretaceous) and overlying glacial drift. According to Crandell (1952), the Pierre Shale fails mainly by slumping, whereas the glacial drift forms mainly debris slides. Earthflows and mudflows are also common. Crandell estimates that at least 75 percent of the shale exposed in the Missouri River trench and its tributaries has been moved by landsliding. Erskine (1973) has many fine photographs of these landslides and information about their stability.

Slumping in Late Cretaceous shale also occurs in Rapid City in the west-central part of the state. Slumping in rocks of early Miocene or late Oligocene age was reported by Gill (1962). Churchill (1979) indicated that mass movement and piping are the dominant processes in forming the badlands of South Dakota.

Tennessee

Nearly half of the 40 reports mentioned by Alger and Brabb (1985) describe landslides along eastern Tennessee highways where steep and inaccessible mountains and weathered and fractured rock with unpredictable stability make road building exceedingly difficult. More than 20 major landslides occurred along a 6-km road section, according to Royster (1975). Remedial measures for dealing with these landslides have been the subject of many reports, such as the one by Trolinger (1980). In 1975 and 1976, the Tennessee Department of Transportation prepared a catalog of these and 91 other major landslides in eastern Tennessee, and 43 landslides in central and western Tennessee that affected state highways from 1945 to 1973 (W.D. Trolinger, written commun., 1982). Landslides in loess and other materials were extensive in a large area in the New Madrid, Missouri, series of earthquakes beginning December 16, 1811, according to Fuller (1912); the failures have been mapped and described by Jibson (1984) and Jibson and Keefer (1984, 1988).

Landslides have also damaged houses and other structures in Knoxville, Memphis, Nashville, and Chattanooga. The Tennessee Division of Geology has prepared a few landslide inventory and susceptibility maps of quadrangles and counties in urban areas to warn people about the potential hazard, and has also prepared a partial landslide susceptibility map of the state at a 1:633,600 scale (Miller and Sitterly, 1977).

Texas

Rockfalls, earthflows, and slumps in shale and associated rocks of Cretaceous and Tertiary age have adversely affected highways and urban areas in Dallas, Waco, Austin, San Antonio, and other areas in central Texas (Abrams and Wright, 1972; Font, 1977; Allen and Flanigan, 1986). Many of these failures developed along oversteepened, human-made cuts, but some formed naturally. Naturally occurring landslides also occur in the Big Bend area and in mountains of western Texas where Trace (1942) has recognized slumped Toreva blocks as defined by Reiche (1937). Similar blocks and other kinds of landslides have been described by Rigby (1958) in rocks of Permian age.

The stability of slopes on most geological units in Texas is shown on a 1:500,000-scale map prepared by Kier and others (1977). A more detailed slope stability map of the Waco area was prepared by Font and Williamson (1970).

Utah

The May 1983 landslides in central Utah were among the most destructive, economically, in U.S. history. Damage associated with the Thistle landslide alone exceeded $200 million of the nearly $500 million in landslide and flood damage (Utah Dept. of Public Safety, 1984). Of the $200 million, $90 million was revenue lost from coal mining operations and nearly $50 million was for alternate railroad tunnels and track.

The Thistle landslide is approximately 2,440 m long and 1,220 m wide (Anderson and others, 1984). In addition to burying a railroad and highway, the landslide dammed the Spanish Fork River and created a lake that covered the town of Thistle. A tunnel was eventually cut to lower the lake and eliminate a threat of overtopping the landslide dam.

More than 2,000 debris flows were triggered in this same area in May, 1983, by abnormally high precipitation and a thick snowpack that melted quickly (Fleming and Schuster, 1985). The debris flows came down canyons into communities north and south of Salt Lake City. Wieczorek and others (1983) made the first map in Utah and one of the first maps in the United States predicting where future debris-flow activity could impact communities and how the hazard could be mitigated. Debris basins and watercourse improvements were recommended and adopted for watersheds with a high potential for debris flows. Monitoring and warning systems were installed.

A hazard mitigation plan by the Utah Department of Public Safety (1984) outlined activities and plans of federal, state, and local agencies for landslide safety, seismic safety, hazard mapping, facility siting and inspection, education and information, and funding. A conference called by Governor

Matheson led to a report (Utah Geological and Mineral Survey, 1983) discussing lessons learned, recommendations for legislative action, and the need for additional hazard research. Zeizel (1987) has provided an overview of the hazard evaluation process.

Regional overviews of landslide processes in Utah have been provided by Shroder (1971), DeGraff (1978), and Godfrey (1978). The rock type is one of the principal factors controlling landslide distribution. Most landslides occur in Tertiary, Cretaceous, and Triassic argillaceous sedimentary rocks overlain by compact well-indurated rocks such as sandstone, conglomerate and basalt. Most occur on steep slopes between elevations 1,800 and 2,400 m. North-facing slopes are generally more unstable than south-facing slopes, but this factor may be more closely related to faults and other structural conditions than to microclimate. Present mean annual precipitation does not seem to be a controlling factor, perhaps because many landslides formed when climatic conditions were different.

Two geologists have used Utah landslides to propose a landslide classification based on age. McCalpin (1984) has four categories, from active landslides to those older than 10,000 years, based on the character of the headwall and lateral scarps, drainage, internal morphology, vegetation, and toe relations. Shroder (1970) has nine categories, using similar characters but providing greater detail.

Fascinating eyewitness accounts of huge boulders being moved by debris flows (Blackwelder, 1928) provided an opportunity to interpret the origin of debris-flow deposits in Utah and elsewhere. Unfortunately, Blackwelder's emphasis on occurrences in arid regions may have led geologists to disregard the possibility of debris flows in areas with other climates, such as in the San Francisco Bay region (Brabb, 1984b).

Parry (1974) has pointed out that water-saturated quick clays are widespread in the Salt Lake City region, indicating that earthquake-generated landslides could be a real problem. Keaton and others (1987) have prepared a map showing where earthquake-induced landslides could be expected in Davis and Salt Lake Counties.

Utah landslides also impact national parks and recreation areas. Grater (1945) has discussed the major role that landslides played in shaping Zion National Park. Hansen (1961) indicated that landslides are numerous in the Flaming Gorge National Recreation Area.

Innovative landslide inventory and susceptibility maps have been prepared for a few areas in Utah. Van Horn's (1972a) inventory of part of the Salt Lake City area has block diagrams of the different types of landslides so that the terminology is easily understood. His map of susceptibility (1972b) uses slope, type of rock or surficial deposit, and the location of bedrock faults, springs and former marshes as factors in determining slope stability. A map by Fuller and others (1981) shows where different landslide processes may impact coal production.

Vermont

Surprisingly few references are available on Vermont landslides in view of the extensive areas of steep terrain, and geologic and climatic conditions that are similar to those in adjacent states where landslides are common. Ratté and Rhodes (1981) described debris avalanches in the Green Mountains and mentioned earlier accounts of similar events. Baskerville and others (1988) described a 1983 rockfall and debris slide at Smugglers Notch—a scenic tourist and recreation area in the northern part of the state. Baskerville and Ohlmacher (1988) described earthflows and other landslides in Windsor County, east-central Vermont. Newspaper accounts, abstracts, and short sections in reports prepared by the Vermont Geological Survey for environmental planning indicate that debris slides, slumps, and slump/earthflows have occurred in glacial lake clays along the Connecticut River and in other materials at a few widely scattered localities.

Information provided by the Vermont Department of Transportation (written commun., 1982) indicates that 22 landslides in silt and clay surficial deposits and eight landslides in bedrock damaged roads between 1965 and 1982. Most of the landslides in the surficial deposits seem to have been slumps. Landslides in bedrock are mainly rockslides along road cuts that have well-developed joint surfaces dipping toward the road. Rockfalls are so common that they have not been counted. The landslides are distributed throughout the state.

Virginia
Studies of debris-avalanche habitats in Virginia by Williams and Guy (1973) indicated that most of these landslides occurred on hillsides facing north, northeast, and east; on the steepest part of the hillside; on slopes greater than 35 percent; and in pre-existing hollows or depressions. Kochel (1987) determined that debris-flow deposition in central Virginia occurred at least three times during the past 11,000 years but that the frequency of debris flows in the Appalachians may be much greater on a regional scale.

Rockslides, rock-block slides, and slumps are common in the western mountainous part of Virginia (Schultz, 1986; Virginia Department of Highways, written commun., 1982), and in the northeastern part (Obermeier, 1984).

A landslide susceptibility map of Fairfax County in northeast Virginia has been prepared by Obermeier (1979). The engineering characteristics of one of the units most susceptible to landslide failure has been described by Obermeier (1984).

Virgin Islands
A report by Brabb (1984c) provides a very brief account of landslides on St. Thomas in the Virgin Islands. Additional information is provided in a report on the Caribbean area, this volume.

Washington, D.C.
Landsliding in the southeastern part of the District of Columbia has been described by Winter and Beard (1978). All of this landsliding is within clay of the Potomac Group of Late Cretaceous age. A report by Reed and Obermeier (1982) provides a good overview of landslide problems.

Washington
The eruption of Mount St. Helens, Washington on May 18, 1980, and the landslides that triggered and followed the eruption overshadow all historic landslide events in the United States. Damages in this event were at least $860 million (Schuster, 1983); perhaps half that amount or about $500 million can be attributed to landslide damage. Mullineaux and Crandell (1981) indicated that the return interval for an eruptive event is 100 to 700 years. Readers interested in details should consult Lipman and Mullineaux (1981), an overview by Voight and others (1983), and other reports listed by Alger and Brabb (1985).

The eruption was forecast by Crandell and Mullineaux (1978) as, "likely to occur within the next hundred years, and perhaps even before the end of this century" based on studies of past volcanic deposits. But even more amazing was their map showing which areas might be impacted by lava flows, pyroclastic flows, mudflows, and floods. The 1980 event followed that scenario in a remarkable fashion. Maps showing potential landslide hazards have also been prepared for Mount Baker and Mount Rainier.

Landslides along the Columbia River valley are so massive and so numerous that they have attracted the attention of geologists since the late 1800's. Hundreds of landslides were induced at one reservoir (Schuster, 1979) and several others were induced by irrigation (Schuster and Hays, 1984). A variety of landslide types also occurs in the Cascade Range (Heller, 1981), Puget Sound lowland (Thorsen, 1987), in the east-central part of the State (Garber, 1965), and elsewhere.

Landslide inventory maps at a 1:250,000 scale have been prepared for western Washington and the Columbia River basin by the Washington Geology and Earth Sciences Division. Several landslide inventory and susceptibility maps at larger scales have been prepared by that organization or the USGS in the Puget Sound region; the map by Miller (1973) is representative. Landslide inventory maps have also been prepared for several areas in National Forests by the USFS.

West Virginia
The West Virginia Department of Highways has records on 2,161 landslides that required work beyond the capability of routine maintenance crews during the period from the 1950's until 1982 (B.C. Stinnett, written commun., 9/23/82). The large number of landslides prompted the department to develop manuals for determining the severity of the problem and the funding needed to repair the damage. Department personnel have also published several reports on landslide recognition and control; the one by Long and Stinnett (1969) is representative. Most of the landslides are rock-block slides in shale of Carboniferous age, slump/earthflows in glacial lake clays of Pleistocene age, and in surficial deposits of Holocene age. Rockfalls, debris slides, debris flows and mudflows have also occurred. Several of these landslides are shown in stereopairs by Liang (1952) and in the report edited by Eckel (1958).

Reports by Lessing and others (1976, 1983) provide an excellent overview of the landslide problem in West Virginia, including an analysis of the regional factors that control landslide distribution; advice for builders, buyers, and homeowners; maps showing where recent and older landslides are located; areas where rockfalls and other landslides are likely in the future; and an unusual discussion of psychological and sociological effects suffered by victims of landslides.

Wisconsin
Landslides are a real problem along Wisconsin's shores of Lake Superior and Lake Michigan where steep bluffs underlain by glacial deposits slump into the lakes in response to wave erosion and other processes. Michelson and others (1977) provide an overview of the slope failures and bluff retreat plus detailed descriptions and maps of the areas impacted.

Rockfalls and other landslides occur on steep bedrock bluffs along the Mississippi, St. Croix, and Wisconsin Rivers and their tributaries in the western and southwestern parts of the state.

Wyoming
Landslides occur throughout Wyoming but are most numerous in the northern and northwestern parts. Bailey (1971) has prepared an excellent overview report of part of the area complete with a description of the various landslide processes, fascinating photographs of many different kinds of landslides, a discussion of forest-management practices needed in landslide areas, and a map showing landslide susceptibility. Several other short reports on landslides are contained in a book edited by Voight and Voight (1974) and in an unpublished report by Riedl (1981). An inventory of all landslides in Wyoming is nearly completed by the Wyoming Geological Survey. USGS geologists have mapped landslides and other engineering-geologic factors affecting open-pit mining of coal in the Powder

River Basin and vicinity; the report by Osterwald and others (1977) is typical.

Controversy about the origin of structural features in the Heart Mountain area, northwestern Wyoming, has been extensive and long lived. Pierce (1973, 1987) maintains that the features form a fault that moved by earthquake oscillations whereas others such as Voight (1974) and Hauge (1985) indicate that the features are landslides or fault blocks moved by volcanism, high fluid pressure, or volcanic gases.

Howard Donley Associates (1981) prepared geotechnical maps at a scale of 1:24,000 of a 1,650-km^2 area around Casper that show dormant and active landslides and areas with different landslide susceptibility.

6 REDUCING LANDSLIDE COSTS

(Several programs have been proposed to reduce landslide costs in the United States, but none has yet been adopted.) The National Research Council (Committee on Ground Failure Hazards) prepared a report (1985) recommending more effective (1) land use regulation, (2) building codes, (3) landslide mapping, (4) research on landslide initiation and landslide processes, (5) research on landslide hazard delineation, mapping, and control, (6) technology transfer, (7) national leadership, and (8) legislation to direct a governmental or private organization to lead a comprehensive national program to reduce landslide losses. Elements of such a program have been proposed by the U.S. Geological Survey (1982). Some programs already in place, such as the National Flood Insurance Program, have a landslide component that has been difficult to define and identify; this problem is discussed by the National Research Council, Committee on Methodologies for Predicting Mudflow Areas (1982b). At the request of the Federal Emergency Management Agency (FEMA), the U.S. Geological Survey prepared a feasibility study for a nationwide landslide hazard mapping effort (Campbell, 1985). A report by Olshansky and Rogers (1987) reviews the physical and legal scope of the landslide problem in the United States, presents strategies for reducing landslide damage and equitably allocating liability, and recommends a comprehensive approach for landslide-damage reduction, including an insurance program as the primary short-term strategy, and more effective grading codes and continued research as the major long-term strategies. Even though the technical capability to make significant reductions in the current rate of landslide losses is well developed in the United States, it is applied only at the local government level by a few cities and counties. (No state or U.S. national agency is assigned responsibility for the reduction of losses to landsliding.)

Even the preparation of maps and reports showing the extent, severity, kind, and expense of the landslide problem is not likely to produce a reduction in landslide losses unless additional steps are taken. As pointed out by Kockelman (1986 and written commun., 1988), an effective, comprehensive landslide hazards reduction program requires five components, each a prerequisite for its successor:

o   Conducting scientific and engineering studies of landslide processes that may be hazardous--location, size, type, likelihood of occurrence, severity, triggering mechanism, and path.

o   Translating the results of such studies into reports and onto maps so that the nature and extent of the hazards or their effects are understood by nontechnical users.

o   Transferring this translated information to those who will or are required to use it and assisting them in its use through educational, advisory, and review services.

o   Selecting and using appropriate hazard reduction techniques (see Table 3).

o   Reviewing the effectiveness of the hazard reduction techniques after they have been in use for a requisite amount of time and revising as necessary. Review of the entire program as well as the other components-- studies, translation, and transfer--may also be undertaken.

Table 3. Some techniques for reducing landslide hazards. From Kockelman (1986).

Discouraging new development in hazardous areas by:
    Disclosing the hazard to real-estate buyers
    Posting warnings of potential hazards
    Adopting utility and public-facility service-area policies
    Informing and educating the public
    Making a public record of hazards
Removing or converting existing development through:
    Acquiring or exchanging hazardous properties
    Discontinuing nonconforming uses
    Reconstructing damaged areas after landslides
    Removing unsafe structures
    Clearing and redeveloping blighted areas before landslides
Providing financial incentives or disincentives by:
    Conditioning federal and state financial assistance
    Clarifying the legal liability of property owners
    Adopting lending policies that reflect risk of loss
    Requiring insurance related to level of hazard
    Providing tax credits or lower assessments to property owners
Regulating new development in hazardous areas by:
    Enacting grading ordinances
    Adopting hillside-development regulations
    Amending land-use zoning districts and regulations
    Enacting sanitary ordinances
    Creating special hazard-reduction zones and regulations
    Enacting subdivision ordinances
    Placing moratoriums on rebuilding
Protecting existing development by:
    Controlling landslides and slumps
    Controlling mudflows and debris flows
    Controlling rockfalls
    Creating improvement districts that assess costs to beneficiaries
    Operating monitoring, warning, and evacuating systems

7 CONCLUSIONS

Landslides are a problem throughout the United States and its island territories. The western or Rocky Mountain states of California, Washington, Utah, and Colorado; Appalachian Mountain states of West Virginia, Kentucky, Tennessee, North Carolina, Ohio, and Maryland; and Samoa and Puerto Rico show up among the top 20 jurisdictions with the most landslide problems regardless of the system used on Table 1 to judge the severity of the problem. Alaska and Virginia could rank near the top based on landslide damage in the 1964 Alaskan earthquake and Hurricane Camille, respectively. New York, Pennsylvania, New Hampshire, Vermont, Indiana, Alabama, Louisiana, Mississippi, Delaware, New Jersey, and the District of Columbia in the East and Idaho, Montana, Wyoming, North Dakota, South Dakota, and Oregon in the West rank in the top 20, depending on the criteria selected to judge severity.

The lack of a National program to reduce landslide losses and the existence of landslide-prone terrain that is subjected to frequent periods of extreme rainfall, earthquakes, volcanic activity, erosion, and human activities guarantees that landslides in the United States will be a problem in the foreseeable future.

8 ACKNOWLEDGMENTS

Information for this report was provided by 122 engineers and geologists in state departments of transportation, 88 university professors, 116 geologists and engineers in state geological surveys, 68 geologists and foresters in the USFS, 10 engineers and geologists in the U.S. Bureau of Reclamation, six in the U.S. Army Corps of Engineers, and 33 in the USGS. Their names are listed by Alger and Brabb (1985), and their enormous contributions made this report possible. I am especially indebted to Christopher Alger, now with Rogers/Pacific, who assembled the file of about 6,500 references reviewed and summarized in this report. Fred Taylor, USGS, kindly interviewed 68 geologists and foresters in the USFS, copied hundreds of their reports, and provided information for my files. He also counted landslides in 80 7.5-minute quadrangles in many parts of the United States to establish a density of landslides/km^2 that could be compared with estimates in this report. Phillip Beilin, Joanne Vinton, Robert Mark, Robert Simpson, Brian Bennett, David Hooper and Thomas McCulloch, USGS, worked on the development of a provisional slope map for the United States (Fig. 1) that was used to determine the approximate area of slopes that could be a habitat for landslides.

I am also grateful to Donald Nichols and Russell Campbell, USGS, who reviewed the entire report, and William Kockelman, Robert Schuster, and David Varnes, USGS, who reviewed parts of the report.

Catherine Campbell and Jeffrey Troll, USGS, kindly provided many additional editorial comments. Lori Moore and Christopher Utter, USGS, provided photographic copies of illustrations at the correct scale for my report and many other reports in this volume.

Betty L. Harrod kindly transcribed 25 tapes of comments about landslides in nearly every state and edited the material to make it more easily understood. She also typed this and several other manuscripts in the book, formatting and reformatting the material several times to accommodate changes.

Gisela Brabb accompanied me, beginning in 1982 and ending in 1985, on most of the approximately 100,000 km of driving in a camper and automobile into every state, Puerto Rico, and the Virgin Islands to determine the extent of the landslide problem. She plotted representative landslides on 1:500,000-scale maps as we drove, and helped document with photography the kind of landslides seen. She prepared more than 500 meals and provided encouragement and support even when our camper was in second gear, buffeted by high headwinds and driving rain. How fortunate I am to have such a friend and partner!

BIBLIOGRAPHY

Abrams, T.G. & S.G.Wright 1972. A survey of earth slope failures and remedial measures in Texas. U. of Texas, Austin, Center for Highway Research. Stability of Earth Slopes Resch. Proj. 3-8-71-161, Rpt. 161-1.

Alger, C.S. & E.E.Brabb 1985. Bibliography of United States landslide maps and reports. USGS O-F Rpt. 85-585:119 p.

Algermissen, S.T. & others 1982. Probabilistic estimate of maximum acceleration and velocity in rock in the contiguous United States. USGS O-F Rpt. 82-1033.

Allen, P.M. & W.D.Flanigan 1986. Geology of Dallas,

Texas, USA. Assoc. Eng. Geol. Bull. 23:363-418.

Amos, Jeannine & T.C.Sanford 1987. Landslides in the Presumpscot Formation, southern Maine. Mn. Geol. Sur. O-F Rpt. 87-4:68 p.

Anderson, L.R. & others 1984. The Utah landslides, debris flows and floods of May and June 1983. Washington, D.C., Nat'l Acad. Press:96 p.

Associated Press 1956. Series of rockfalls topple power station into Niagara. The New York Times, N.Y., June 8, 1956:1.

Bailey, R.G. 1971. Landslide hazards related to land use planning in Teton National Forest, northwest Wyoming. USDA FS, Intermt. Region:131 p.

Baker, R.F. & Robert Chieruzzi 1959. Regional concept of landslide occurrence. Hwy Resch. Bd. Bull. 216:1-17.

Baldwin, J.E. 1984. The Martinez Mountain rock avalanche. 4th Int'l Symp. on Landslides, Toronto, Canada. Sept. 17-21, 1:447-453.

Baldwin, J.E. II, H.F.Donley & T.R.Howard 1987. On debris flow/avalanche mitigation and control, San Francisco Bay area, California. In J.C.Costa & G.F.Wieczorek (eds.). Debris flows/avalanches: Process, recognition and mitigation. Geol. Soc. Am. Rev. Eng. Geol. 7:223-236.

Balter, R.B. 1970. Landslide problems and recommended design procedures for western Maryland highways. Rpt. Md. St. Rds. Comm., Robert B. Balter Soil & Foundation Consultants, Inc., Owings Mills, Md.:62 p.

Barr, D.J. & D.N.Swanston 1970. Measurement of creep in a shallow, slide-prone till soil. Am. J. Sci 269:467-480.

Baskerville, C.A., C.A.Ratte and F.T.Lee 1988. A rockfall and debris slide at Smugglers Notch. Vermont Studies in Geology, 4.

Baskerville, C.A. and G.C.Olmacher 1988. Some slope-movement problems in Windsor County, Vermont, 1984: USGS Prof. Paper 1828:35 p.

Beaulieu, J.D. 1977. Geologic hazards of parts of northern Hood River, Wasco and Sherman Counties, Oregon. Or. Dept. Geol. & Min'l. Ind. Bull. 91:95 p., scale 1:62,500.

Bedrossian, T.L. 1986. Watersheds mapping in northern California. Calif. Geol., Feb. 1986:34-38.

Bennett, Warren 1974. Landslides on "Brazos Pass." N.Mex. Geol. Soc. 25th Ann. Fld. Conf. Guidebook:359-363.

Bernknopf, R.L. & others 1988. A probabilistic approach to landslide hazard mapping in Cincinnati, Ohio with applications for economic evaluation. Assoc. Eng. Geol. Bull. 25:39-56.

Beschta, R.L. 1978. Long term patterns of sediment production following road construction and logging in the Oregon Coast Range. Or. St. U. Water Res. Resch. Paper 8WO584. 14,6:1011-1016.

Bishop, D.M. & M.E.Stevens 1964. Landslides on logged areas in southeast Alaska. USDA FS Resch. Paper NOR-1:18 p.

Blackwelder, Eliot 1928. Mudflow as a geologic agent in semiarid mountains. Geol. Soc. Am. Bull. 39:465-484.

Bluemle, J.P. & A.F.Jacobs 1981. Auto tour guide along the South Loop road, Theodore Roosevelt Nat'l Park. No.Dak. Geol. Sur. Educ. Ser. 4:14 p.

Bock, C.G. 1977. Martinez Mountain rock avalanche. Geol. Soc. Am. Revs. in Eng. Geol. 3:155-168.

Bogucki, D.J. 1977. Debris slide hazards in the Adirondack Province of New York State. Envir. Geol. 1:317-328.

Booy, E. 1977. Instability in glacial clays; slope failures in marine and fresh-water deposits. 15th Ann. Eng. Geol. & Soils Eng. Symp. Proc.:39-55.

Brabb, E.E. 1984a. Minimum landslide damage in the United States, 1973-1983. USGS O-F Rpt.84-486:4 p.

Brabb, E.E. 1984b. Innovative approaches to landslide hazard and risk mapping. 4th Int'l. Symp. Landslides Proc., Toronto, Canada, 1:307-324.

Brabb, E.E. 1984c. Landslide potential on St. Thomas, Virgin Islands. In P.L. Gori & W.W. Hays

(eds.), A workshop on earthquake hazards in the
Virgin Islands region. USGS O-F Rpt. 84-762:97-
102.

Brabb, E.E. & Ann FitzSimmons 1984. Addresses,
topics of interest, and geographic distribution of
professors working in the United States. Natr'l.
Haz. Resch. & Appli. Info. Ctr. Spec. Pub. 8:34 p.

Brabb, E.E. 1985. On the line; losing by a
landslide. National Hazards Observer 10,2.

Brabb, E.E. & others (in press). The extent of
landsliding in northern New Mexico and similar
semi-arid and arid regions. In P.M.Sadler &
D.M.Morton (eds.). Landslides in a semi-arid
environment. Inland Geol. Soc., Riverside.

Braddock, W.A. & D.L.Eicher 1962. Block-glide
landslides in the Dakota Group of the Front Range
Foothills, Colo. Geol. Soc. Am. Bul. 73,3:317-323.

Briggs, R.P. 1974. Map of overdip slopes that can
affect landsliding in Allegheny County,
Pennsylvania. USGS Misc. Fld. Studies Map MF-543,
scale 1:125,000.

Briggs, R.P., J.S.Pomeroy & W.E.Davies 1975.
Landsliding in Allegheny County, Pennsylvania.
USGS Circ. 728:18 p.

Buchanan-Banks, J.M. 1981. The October 28, 1979,
landsliding on Tutuila, American Samoa. USGS O-F
Rpt. 81-0081:21 p.

Bump, V.L. 1979. Landslides affecting South Dakota
highways. So.Dak. Dept. Transp. Rpt.:9 p.

Burchfiel, B.C. 1966. Tin Mountain landslide,
southeastern California, and the origin of
megabreccia. Geol. Soc. Am. Bulletin 77:95-100.

California Department of Navigation and Ocean
shoreline erosion along the California coast.
Sacramento, 276 p.

Callender, G.W. Jr. 1975. Slope stability problems
on Guam. Milit. Eng. 67,439:270-271.

Campbell, R.H. 1975. Soil slips, debris flows, and
rainstorms in the Santa Monica Mountains and
vicinity, southern California. USGS Prof. Paper
851:51.

Campbell, R.H. (ed.) 1985. Feasibility of a
nationwide program for the identification and
delineation of hazards from mud flows and other
landslides. USGS O-F Rpt. 85-276.

Carter, L.David & John P.Galloway 1981. Earth flows
along Henry Creek, northern Alaska. Arctic
34,4:325-328.

Charbonneau, R.G. (ed.) 1975. Engineering geology
approaches to highway construction in central
Idaho. Geol. Soc. Am. Rocky Mt. Sec. 28th Ann.
Mtg. Fld. Trip Guidebook.

Childress, S.C., Michael Bograd & J.C.Marbel 1976.
Geology and man in Adams County, Mississippi.
Miss. Geol. Sur. Envir. Geol. Ser. 4:60-75.

Churchill, R.R. 1979. The importance of mass
movement and piping in badlands slope development.
Iowa Acad. Sci. Proc. 86,1:10-14.

Clark, G.M. 1987. Debris slide and debris flow
historical events in the Appalachians south of the
glacial border. In J.C.Costa & G.F.Wieczorek
(eds.) Debris flows/avalanches: Process,
recognition and mitigation. Geol. Soc. Am. Rev.
Eng. Geol. 7:125-138.

Clayton, J.L. & others 1975. Field trip guide; rock
alteration and slope failure, Middle Fork of the
Payette River. Boise State U., Id. Dept Geol:19 p.

Cleaves,E.T., W.P.Crowley & K.R.Kuff 1974. Towson
Quadrangle Atlas. Maryland Geol. Surv. Geol. &
Envir. Atlas Ser. 2, scale 1:24,000.

Cleland, H.F. 1902. The landslides of Mt. Greylock
and Griggsville, Massachusetts. J. of Geol.
10:513-517.

Cleveland, G.B. 1975. Landsliding in marine terrace
terrain, California. Calif. Div. Mns. & Geol.
Spec. Rpt. 119:24 p.

Coates, D.R. 1977. Landslide perspectives. Geol.
Soc. Am. Rev. Eng. Geol. 3:3-28.

Colton, R.B. & others 1976. Preliminary map of
landslide deposits in Colorado. USGS Misc. Geol.
Inv. Map I-964, scale 1:500,000.

Cooley, M.E., B.M.Aldridge & R.C.Euler 1977. Effects
of the catastrophic flood of December 1966, north
rim area, eastern Grand Canyon, Arizona. USGS
Prof. Paper 980:43 p., scale 1:62,500.

Covington, H.R. 1977. Map showing areas of potential
rockfalls in the Snake River Canyon near Twin
Falls, Idaho. USGS Misc. Fld. Studies Map MF-862,
1:24,000, 2 sheets.

Crandell, D.R. 1952. Landslides in shale at Rapid
City, South Dakota. USGS O-F Rept.:17 p.

Crandell, D.R. & H.H.Waldron 1956. A recent volcanic
mudflow of exceptional dimensions from Mt.
Rainier, Washington. Am. J. Sci. 254,6:349-362.

Crandell, D.R. & others 1974. Chaos Crags eruptions
and rockfall-avalanches, Lassen Volcanic National
Park, California. USGS J. of Resch. 2,1:49-60.

Crandell, D.R. 1975. Assessment of volcanic risk on
the island of Oahu, Hawaii. USGS O-F Rept. 75-
287:18 p.

Crandell, D.R. & D.R. Mullineaux 1978. Potential
hazards from future eruptions of Mount St. Helens
Volcano, Washington. USGS Bull. 1383-C:C1-C26,
scale 1:250,000.

Crandell, D.R. & others 1984. Catastrophic debris
avalanche from ancestral Mount Shasta volcano,
California. Geol. 12,3:143-146.

Daniel, F.E. & others 1985. Preliminary landslide
inventory map of Montana. Geol. Soc. of Am. Abst.
with Progs. 17,4:215.

Davies, W.E. 1973. The landslide at Cumberland,
Harland County, Kentucky. USGS O-F Rept. 1848:8 p.

DeGraff, J.V. (ed.) 1978. Regional landslide
evaluation; two Utah examples. Envir. Geol.
2,4:203-214.

DeGraff, J.V. 1987. Landslide activity in the Sierra
Nevada during 1982 and 1983. USDA FS, Southwest
Region, Monograph 12:74 p.

Dobrovolny, Ernest & H.R.Schmoll 1974. Slope
stability map of Anchorage and vicinity, Alaska .
USGS Misc. Geol. Inv. Map I-787-E, scale 1:24,000.

Dubois, S.M. & A.W.Smith 1980. The 1887 earthquake
in San Bernardino Valley, Sonora; historic
accounts and intensity patterns in Arizona. Ariz.
Bureau Geol. & Min. Tech. Spec. Paper 3:112 p.

DuMontelle, P.B., N.C.Hester & R.C.Cole 1971.
Landslides along the Illinois River Valley south
and west of LaSalle and Peru, Illinois. Ill. Geol.
Sur. Envir. Geol. Notes 48:16 p.

Dunn, J.R. & G.M.Banino 1977. Problems with Lake
Albany clays. Geol. Soc. Am. Rev. Eng. Geol.
3:133-136.

Eckel, E.B. (ed.) 1958. Landslides and engineering
practice. Highway Research Bd. Spec. Rept. 29.
NAS-NRC 544, Washington, D.C.:232 p.

Ehlig, P.L. 1986. Landslides and landslide
mitigation in southern California. Geol. Soc. Am.
Cordilleran Sec. 82nd Ann. Mtg. guidebook &
vol.:201 p.

Ellen, S.D. & G.W.Wieczorek (eds.) 1988. Landslides,
floods and marine effects of the storm of January
3-5, 1982, in the San Francisco Bay region. USGS
Prof. Paper 1434.

Engineering News Record 1976. Another Teton? p. 13.

Erskine, C.F. 1973. Landslides in the vicinity of
the Fort Randall Reservoir, South Dakota. USGS
Prof. Paper 675:65 p.

Ferreira, A. 1982. Rock slippage on the upper slope
of a highway cut. In O.C.Farquhar (ed.)
Geotechnology in Massachusetts. U. of Mass., Conf.
Proc., March 1980:135-140

Fickies, R.H. & P.T.Regan 1982. Geologic hazards and
thickness of overburden of the Albany, New York
15-minute quadrangle. N.Y. St. Geol. Sur. Map &
Chart Ser. 36, scale 1:24,000.

Fickies, R.H., T.J.Robak & E.E.Brabb (in press).
Landslide inventory map of New York. N.Y. St.
Geol. Sur. Map and Chart 41, scale 1:500,000

Finch, W.I. 1971. Geologic map of the Hickman
Quadrangle, Fulton County, Kentucky, and
Mississippi County, Missouri, USGS Geol. Quad. Map
GQ-874, scale 1:24,000.

Flaccus, Edward 1958. White Mountain landslides. Appalachia. 32,127:175-191.

Fleming, R.W. & others 1977. Recent movement of the Manti, Utah, landslide. 15th Ann. Eng. Geol. & Soils Eng. Symp. Proc.:161-178.

Fleming, R.W. & F.A.Taylor 1980a. Estimating the costs of landslide damage in the United States. USGS Circ. 832:21 p.

Fleming, R.W. & F.A.Taylor 1980b. Costs of landslide damage in Allegheny County, Pennsylvania. Penn. Geol. 11,5:5-9.

Fleming, R.W. & others 1981. Engineering geology of the Cincinnati area. In T.G.Roberts (ed.) Geomorphology, hydrogeology, geoarcheology, engineering geology. Geol. Soc. Am. Ann. Mtg, Fld. Trip Guidebook 18:543-570.

Fleming, R.W. and R.L.Schuster 1985. Implications of the current wet cycle to landslides in Utah, in D.S.Bowles (ed.), Delineation of landslide, flash flood, and debris flow hazards in Utah. Utah Water Res. Lab. Proc.:19-28.

Foley, E.J. 1964. The Lincoln folds, Lincoln, New Mexico. N.Mex. Geol. Sur. 15th Ann. Fld. Conf. Guidebook:134-139.

Font, R.G. & E.F.Williamson 1970. Geologic factors affecting construction in Waco. In Urban geology of greater Waco, Part IV; Engineering. Baylor Coll. Geol. Studies Bull. 12:34 p.

Font, R.G. 1977. Engineering geology of the slope instability of two overconsolidated north-central Texas shales. Geol. Soc. Amer. Revs. of Eng. Geol. 3:205-212.

Freedman, J.L. 1977. Lots of danger; property buyer's guide to land hazards of southwestern Pennsylvania. Pittsburgh Geol. Soc.: 85 p.

French, W.E. 1972. Shoreline collapse at Sleeping Bear Point, Lake Michigan (abs.). Geol. Soc. Am. Absts./Progs. 4,5:321.

Fritz, W.J. 1980. Stumps transported and deposited upright by Mount St. Helens mudflows. Geol. 8,12:586-588.

Froelich, A.J. 1970. Geologic setting of landslides along south slope of Pine Mountain, Kentucky. Hwy. Resch. Record 323:1-5.

Froelich, A.J., H.T.Hack & E.G.Otton 1980. Geologic and hydrologic map reports for land-use planning in the Baltimore-Washington area. USGS Circ. 806:26 p.

Fuller, H.K., V.S.Williams & R.B.Colton 1981. Map showing areas of landsliding in the Kaiparowits coal-basin area, Utah. USGS Misc. Inv. Ser. Map I-1033-H, scale 1:125,000.

Fuller, M.L. 1912. The New Madrid earthquake. USGS Bull. 394:59-61.

Fuller, M.L. 1914. Geology of Long Island, New York. USGS Prof Paper 82:54-56.

Garber, L.W. 1965. Relationship of soils to earthflows in the Palouse. J. of Soil & Water Conserv. 20,1:21-23.

Gill, J.R. 1962. Tertiary landslides, northwestern South Dakota and southeastern Montana. Geol. Soc. Amer. Bull. 73,6:725-735.

Glancy, P.A. & T.J. Katzer 1977. Flood and related debris flow hazards map, Washoe City Quadrangle, Nevada. Nev. Bur. Mns. & Geol. Urban Map, scale 1:24,000.

Godfrey, A.E. 1978. Land surface instability on the Wasatch Plateau. Utah Geol. 5,2:131-141.

Goodfield, A.G. 1969. Clarksville, Missouri rockfalls (abs.). Mo. Acad. of Sci. Trans. 3:105.

Grater, R.K. 1945. Landslide in Zion Canyon, Zion National Park, Utah. J. of Geol. 53,2:116-124.

Gray, H.H. 1971. Glacial lake deposits in southern Indiana—engineering problems and land use. Ind. Geol. Sur. Envir. Study 2:14 p.

Gray, R.E., H.F.Ferguson & J.V.Hamel 1979. Slope stability in the Appalachian Plateau, Pennsylvania and West Virginia, U.S.A. In B.Voight (ed.), Rockslides & Avalanches, 2. Engineering Sites. Development in Geotechnical Engineering, 14B:447-471.

Griggs, R.F. 1920. The great Mageik landslide. Ohio J. of Sci. 20:325-354.

Gryta, J.J. & M.J.Bartholomew 1983. Debris avalanche type features in Watanga County, North Carolina. In S.E.Lewis (ed.) Geologic investigations in the Blue Ridge of northwestern North Carolina. Carol. Geol. Sur. Fld. Trip Guidebook, Raleigh.

Guzzetti, Fausto & E.E.Brabb 1987. Map showing landslide deposits in northwestern New Mexico. USGS O-F Rpt. 87-70, scale 1:500,000.

Hadley, J.B. 1964. Landslides and related phenomena accompanying the Hebgen Lake earthquake of August 17, 1959. In The Hebgen Lake, Montana, earthquake of August 17, 1959. USGS Prof. Paper 435:107-138.

Hamel, J.V. & N.K.Flint 1972. Failure of colluvial slope. Am. Soc. Civ. Eng., Soil Mech. & Foundation Div. J. 98,SM2:167-180.

Handy, R.L. 1973. Collapsible loess in Iowa. Soil Sci. Soc. Am. Proc. 37,2:281-284.

Hansen, W.R. 1961. Landslides along the Uinta fault east of Flaming Gorge, Utah. USGS Prof. Paper 424-B:B306-B307.

Hansen, W.R. & others 1966. The Alaska earthquake, March 27, 1964; field investigations and reconstruction effort. USGS Prof. Paper 541:111 p.

Hansen, W.R. 1973. Effects of the May 5-6, 1973, storm in the greater Denver area, Colorado. USGS Circ. 689:20 p.

Hauge, T.A. 1985. Gravity-spreading origin of the Heart Mountain allochthon, northwestern Wyoming. Geol. Soc. Am. Bulletin 96:1440-1456.

Hayes, C.J. 1971a. Landslides and related phenomena pertaining to highway construction in Oklahoma. In Environmental aspects of geology and engineering in Oklahoma. Okla. Acad. Sci. Ann. Mtg. 2:47-57.

Hayes, C.J. 1971b. Engineering classification of highway-geology problems in Oklahoma. 22nd Ann. Hwy. Geol. Symp. Proc.:8-21.

Hays, W.W. 1980. Procedures for estimating earthquake ground motions. USGS Prof. Paper 1114:77 p.

Heller, P.L. 1981. Small landslide types and controls in glacial deposits, lower Skagit River drainage, north Cascade Range, Wash. Envir. Geol. 3:221-228.

Hester, N.C. & P.B.DuMontelle 1971. Pleistocene mudflow along the Shelbyville moraine front, Macon County, Illinois. In Till, a symposium. Ohio St. Union Press, Columbus:367-382.

Hitchcock, C.H. 1885. The recent landslide in the White Mountains. Science 6,130:84-87.

Hitchcock, Edward 1836. Sketch of the geology of Portland and its vicinity. Boston J. of Natr'l Hist. 1:306-347.

Hoffman, David & Gary St. Ivany 1982. Landslide and rockfall map of Missouri. Mo. Geol. Sur. Map, scale 1:500,000.

Hollenbaugh, K.M. 1975. The evaluation of geologic processes in the Boise foothills that may be hazardous to urban development. Geol. Soc. Am. Rocky Mt. Sec., 28th Ann. Mtg. Fld. Trip Guidebook 6:11-36.

Howard Donley Assoc., Inc. 1981. Natural and geotechnical hazards study for the urbanizing area of Natrona County, in the vicinity of Casper, Wyoming. 40 p., scale 1:48,000, Casper, Wyo.

Howe, Ernest 1909. Landslides in the San Juan Mountains, Colorado. USGS Prof. Paper 67:58 p.

Huntoon, P.W. 1975. The Surprise Valley landslide and widening of the Grand Canyon. Plateau 48:1-12.

Ireland, H.A., C.F.S.Sharpe & D.H.Eargle 1939. Principles of gully erosion in the Piedmont of South Carolina. USDA Tech. Bull. 633:143 p.

Ives, J.D. & M.J.Bovis 1978. Natural hazards maps for land-use planning, San Juan Mountains, Colorado, USA. In J.D.Ives & R.P. Zimina (eds.) Mountain geoecology and land-use implications. Arctic & Alpine Resch. 10,2:185-212.

Jackson, L.E. 1977. Dating and recurrence frequency of prehistoric mudflows near Big Sur, Monterey County, California. USGS J. Resch. 5,1:17-32.

Jahns, R.H. 1978. Landslides. In Geophysical Predictions. Nat'l Acad. Sc. Washington, DC:58-65.

Jellinger, M. 1977. Methods of detection and analysis of slope instability, southeast Oahu, Hawaii. U. of H.I. unpub. PhD. thesis:266 p.

Jibson, R.W. 1984. Earthquake-induced landslides in the New Madrid seismic zone. Seis. Soc. Am. Earthquake Notes, 55,1:19-20.

Jibson, R.W. & D.K.Keefer 1984. Earthquake-induced landslides in the central Mississippi Valley, Tennessee and Kentucky. In P.L.Gori & W.W.Hays (eds.) 1984. Proc. Symp. New Madrid Seismic Zone. USGS O-F Rpt. 84-770: 353-390.

Jibson, R.W. 1986a. Evaluation of landslide hazards resulting from the 5-8 October, 1985, storm in Puerto Rico. USGS O-F Rpt. 86-26:40.

Jibson, R.W. 1986b. Landslides resulting from the October 5-8, 1985, storm in Puerto Rico (abs.) Assoc. Eng. Geol., Abstr. & Prog. 29:52.

Jibson, R.W. & D.K.Keefer 1988. Landslides triggered by earthquakes in the central Mississippi Valley, Tennessee and Kentucky. USGS Prof. Paper 1336-C: 62 p.

Johnson, A.M. & P.H.Rahn 1970. Mobilization of debris flow. Zeitschrift fur Geomorphologie, Supplementband 9:168-186.

Jordan, R.H. 1949. A Florida landslide. J. Geol. 59,4:418-419.

Judd, N.M. 1959. The braced-up cliff at Pueblo Bonito. Smithsonian Inst. Ann. Rpt. 1958:501-511.

Katzer, T.J. 1981. Flood and related debris flow hazards map, Las Vegas SE quadrangle. Nev. Bur. Mns. & Geol. Urban map, scale 1:24,000.

Katzer, T.J. & P.A.Glancy 1978. Flood and related debris flow hazards map, South Lake Tahoe quadrangle. Nev. Bur. Mns. & Geol. Urban Map, scale 1:24,000.

Katzer, T.J. & C.V.Schroer 1981. Flood and related debris flow hazards map, Carson City quadrangle. Nev. Bur. Mns. Geol. Urban Map, scale 1:24,000.

Katzman, M.M. 1987. Metropolitan Avenue landslide, Kansas City, Kansas. Am. Geoph. Union Trans.:68,44:1285.

Keady, D.M, E.E.Russell & T.J.Laswell 1973. Acker Lake landslide, Monroe County, Mississippi. Miss. Geol. Surv. Info. Ser. MGS-73-1:24 p.

Keaton, J.R. & others 1987. Earthquake-induced landslide potential in, and development of, a seismic slope stability map of the urban corridor of Davis and Salt Lake Counties, Utah. In Proc. 23rd Symp. Eng. Geol. & Soils Eng., Idaho Dept. Transp., Boise.

Keefer, D.K. 1984. Landslides caused by earthquakes. Geol Soc. Am. Bulletin 95:406-421.

Keefer, D.K. & A.M. Johnson 1983. Earthflows: Morphology, mobilization, and movement. USGS Prof. Paper 1264:56 p.

Keefer, D.K. & others 1987. Real-time landslide warning during heavy rainfall. Science. 238:921-925.

Kier, R.S., L.E.Garner & L.F.Brown Jr. 1977. Land resources of Texas. Tex. Bur. Econ. Geol., scale 1:500,000.

Killey, M.M., Hines, J.K., DuMontelle, P.B. and Brabb, E.E. 1984. Illinois landslide inventory map. USGS Misc. Fld. Studies Map MF-1691, scale 1:500,000.

Kochel, R.C. 1987. Holocene debris flows in central Virginia. Geol. Soc. Amer. Revs. in Eng. Geol. 7:139-155.

Kockelman, W.J. 1986. Some techniques for reducing landslide hazards. Assoc. Eng. Geol. Bull. 23,1:29-52.

Krinitzsky, E.L. 1965. Geological influences on bank erosion along meanders of the lower Mississippi River. U.S. Army Eng. Waterways E.S., Vicksburg, Miss., Potamology Invest. Rpt. 12-15:30 p.

Krohn, J.P. & J.E.Slosson 1976. Landslide potential in the United States. Calif. Geol. 29, 10:224-231.

Lane, K.S. 1963. Discussion of "Experience with Canadian varved clays." Amer. Soc. Civ. Eng, J.

Soil Mech. & Founda. Div. Proc. 89,SM3,pt.1:147-156.

Leith, C.J. 1965. The influence of geological factors on the stability of highway slopes. Soc. Min. Eng. Trans. 232,2:150-153.

Lessing, P. & others 1976. West Virginia landslides and slide-prone areas. W. Vir. Geol. & Econ. Sur. Envir. Geol. Bull. 15:64 p., 23 sheets, scale 1:24,000.

Lessing, P., C.P.Messina & R.F.Fonner 1983. Landslide risk assessment. Envir. Geol. 5,2:93-99.

Liang, Ta 1952. Landslides—an aerial photographic study. Cornell U. PhD. thesis, 274 p.

Lipman, P.W. & others 1988. The Giant submarine Alika debris slide, Mauna Loa, Hawaii. J. Geogphy Res. 93:4279-4299.

Lipman, P.W. & D.R.Mullineaux (eds.) 1981. The 1980 eruptions of Mount St. Helens, Washington. USGS Prof. Paper 1250:844 p.

Long, D.G. & B.C.Stinnett 1969. Landslide recognition and control on West Virginia highways. In J.W.H.Wang & J.P.Fisher (eds.). A symposium on landslides. Ohio U. Resch. Inst., Athens, Oh., Proc. 2/26/69:98-127.

Lovelace, A.D., K. White & L. Chaturvedi 1977. Landslides in New Mexico. 15th Ann. Eng. Geol. & Soils Eng. Symp. Proc.:341-357.

Lutton, R.J. 1971. A mechanism for progressive rock mass failure as revealed by loess slumps. Intern'l J. Rock Mech. & Mining Sci.8:143-151. Mississippi.

Lutzen, E.E. & J.H.Williams 1968. Missouri's approach to engineering geology in urban areas. Assoc. Eng. Geol. Bull. 5,2:109-121.

McCalpin, James 1984. Preliminary age classification of landslides for inventory mapping. 21st Annual Eng. Geol. & Soils Eng. Symp. Proc., Moscow, Idaho:99-111.

McMullan, B.G. 1976. Agnes in Maryland; shoreline recession and landslides. In J.Davis & B. Laird, coordinators. The effects of tropic storm Agnes on the Chesapeake Bay estuarine system, p. 216-222. John Hopkins U. Press, Baltimore, Md.

Mears, A.I. 1977. Debris-flow hazard analysis and mitigation; an example from Glenwood Springs, Colorado. Colo. Geol. Sur. Info. Ser. 8:45 p.

Megahan, W.F., N.F.Day & T.M.Bliss 1978. Landslide occurrence in the western and central Northern Rocky Mountain physiographic province. Proc. 5th North Am. For. Soils Conf., Ft. Collins, Colo.:116-139.

Michelson, D.M. & others 1977. Shoreline erosion and bluff stability along Lake Michigan and Lake Superior shorelines of Wisconsin (Milwaukee County). In Shore erosion study technical report. Wisc. Geol. & Natr'l Hist. Sur., Appdx. 3:199 p.

Michigan Department of Natural Resources 1973. A plan for Michigan shorelands. Lansing, MI:135 p.

Michigan Division of Land Resources 1979. Local zoning for high risk erosion areas. Lansing, Mich.:63 p.

Miller, R.A. & P.D.Sitterly 1977. Geologic hazards map of Tennessee. Tenn. Div. Geol., Envir. Geol. Ser. 5, scale 1:633,600.

Miller, R.D. 1973. Map showing relative slope stability in part of west-central King County, Washington. USGS Misc. Geol. Inv. Map I-852A, scale 1:48,000.

Minard, J.P. 1974. Slump blocks in the Atlantic Highlands of New Jersey. USGS Prof. Pr 898:24 p.

Monroe, W.H. 1979. Map showing landslides and areas of susceptibility to landsliding in Puerto Rico. USGS Misc. Geol. Inv. Map I-1148 scale 1:24,000.

Moore, J.G. 1964. Giant submarine landslides on the Hawaiian Ridge. USGS Prof. Paper 501-D:D95-D98.

Moore, J.G. & G.W.Moore 1984. Deposit from a giant wave on the Island of Lanai, Hawaii. Science 226:1312-1315.

Morse, E.S. 1869. On the landslides in the vicinity of Portland, Maine. Boston Soc. of Natr'l. Hist. Proc. 12:235-244.

Morse, W.C. 1935. Geologic conditions governing

sites of bridges and other structures. Miss. Geol. Sur. Bull. 27:15–16.

Mudge, M.R. 1965. Rockfall-avalanche and rockslide-avalanche deposits at Sawtooth Ridge, Montana. Geol. Soc. Am. Bull. 76,9:1003–1014.

Mullineaux, D.R. & D.R.Crandell 1981. The eruptive history of Mount St. Helens. In P.W.Lipman & D.R.Mullineaux (eds.) The 1980 eruptions of Mount St. Helens, Washington. USGS Prof. Pr 1250:3–16.

Murphy, B.J. & R.T.Moore 1980. Maps showing relative slope stability and relative erodibility, central Santa Cruz River Valley, Tucson area, Arizona. USGS Misc. Geol. Inv. Map I-844-0, scale 1:125,000.

National Board of Fire Underwriters and Pacific Fire Rating Bureau 1964. The Alaska Earthquake. San Francisco:35 p.

National Research Council 1982. Storms, floods and debris flows in southern California and Arizona, 1978 and 1980. Washington, DC. Nat'l Acad. Press:487 p.

National Research Council 1982b. Selecting a methodology for delineating mudslide hazard areas for the National Flood Insurance Program. Washington, DC. Nat'l Acad. Press:35 p.

National Research Council 1984. Debris flows, landslides, and floods in the San Francisco Bay region January 1982. Nat'l Resch Conc'l. Conf. Overview & Summ., Stanford Univ., Calif. Aug. 23–26/82:83 p.

National Research Council 1985. Reducing losses from landsliding in the United States. Washington, D.C., Nat'l Acad. Press:41 p.

Neary, D.G. & L.W.Swift Jr. 1987. Rainfall thresholds for triggering a debris avalanching event in the southern Appalachian Mountains. Geol. Soc. Am. Rev. in Eng. Geol. 7:81–91.

Neilson, J.M. 1965. Landsliding and river erosion at Victoria generating station, Ontonagon County, Michigan. Geol. Soc. Am. Spec. Paper 82:140.

Newell, W.L. 1977. Map showing slope stability in the Beattyville Quadrangle, Kentucky River area development district, eastern Kentucky. USGS Misc. Fld. Studies Map MF-844, scale 1:24,000.

Newell, W.L. 1978. Understanding natural systems—a perspective for land-use planning in Appalachian Kentucky. USGS Bul. 1438:50 p.

Newland, D.H. 1916. Landslides in unconsolidated sediment; with a description of some occurrences in the Hudson Valley. N.Y. St. Museum Bull. 187:79–105.

Newman, E.B., R.K.Mark & R.G.Vivian 1982. Anasazi Solar Marker; The use of a natural rockfall. Science 217:1036–1038.

Nichols, D.R. & L.A.Yehle 1985. Volcanic debris flows, Copper River basin, Alaska. Proc. 4th Int'l Conf. & Fld. Workshop on Landslides, Tokyo:365–372.

Nilsen, T.H. & others 1979. Relative slope stability and land-use planning in the San Francisco Bay region, California. USGS Prof. Paper 944:96 p., scale 1:125,000.

Nolan, K.M., D.R.Harden & S.M.Colman 1976. Erosional landform map of the Redwood Creek drainage basin, Humboldt County, California, 1947-74. USGS Water Res. Invest. O-F Rpt. 76-42, scale 1:62,500.

Novak, I.D. 1987. Inventory and bibliography of Maine landslides. Me. Geol. Sur. O-F Rpt. 87-3:27 p., scale 1:500,000.

Obermeier, S.E. 1979. Slope stability map of Fairfax County, Virginia: USGS Misc. Fld. Studies Map MF-1072, scale 1:24,000.

Obermeier, S.E. 1984. Engineering geology of the Potomac Formation deposits in Fairfax County, Virginia, and vicinity, with emphasis on landslides. In S.E. Obermeier (ed.) Engineering geology and design of slopes for Cretaceous Potomac deposits in Fairfax County Virginia & vicinity. USGS Bull.1556:5–48.

Oshansky, R.B. & J.D.Rogers 1987. Unstable ground: Landslide policy in the United States. Ecology Law

Quaterly 13, 4:939–1006.

Osterkamp, W.R., C.R.Hupp, & J.C.Blodgett 1986. Magnitude and frequency of debris flows, and areas of hazard on Mount Shasta, northern California, USGS Prof. Paper 1396-C:21 p.

Osterwald, F.W. & others 1977. Summary report of the geology, mineral resources, engineering geology, and environmental geochemistry of the Sweetwater-Kemmerer area, Wyoming; Part B, Engineering geology. USGS O-F Rpt. 77-361:20 p.

Othberg, K.L. 1984. Geomorphology of ground failure hazards, Preston and Soda Springs 30' X 1° quadrangles, Idaho & Wyoming. ID. Geol. Sur. Tech. Rpt 84-3:30 p., scale 1:100,000.

Othberg, K.L. & R.M.Breckenridge 1981. Interpreting geologic hazards in Idaho from remotely sensed imagery. Id. Bur. Mns. & Geol. Prelim. O-F Rpt. 81-6: 65, scale 1:500,000.

Palmer, L.A. 1977. Large landslides of the Columbia River Gorge, Oregon and Washington. Geol. Soc. Am. Revs. Eng. Geol. 3:69–83.

Parry, W.T. 1974. Earthquake hazards in sensitive clays along the central Wasatch Front, Utah. Geology. 2:559–560.

Péwé, T.L., 1979, Geologic hazards in the Phoenix area. Ariz. Bur. Geol. & Min'l. Tech. Fld. Notes. 8,1-2:18–20.

Pierce, W.G. 1973. Principal features of the Heart Mountain fault and the mechanism problem. In Gravity and Tectonics. John Wiley & Sons:457–471.

Pierce, W.G. 1987. The case for tectonic denudation by the Heart Mountain fault—a response. Geol. Soc. Am. Bull. 99:552-568

Piper, H.W. 1965. Maryland engineering soil study. U. of Md. Civ. Eng. Dept., 35 p.

Pipkin, B.W. & M.Ploessel 1974. Coastal landslides in Southern California. USC Dept. of Geol. Sci., Sea Grant Pub.:19 p.

Plafker, George & Meyer Rubin 1978. Uplift history and earthquake recurrence as deduced from marine terraces on Middleton Island, Alaska. USGS O-F Rpt. 78-943:687–721.

Pomeroy, J.S. & W.E. Davies 1975. Map of susceptibility to landsliding, Allegheny County, Pennsylvania. USGS Misc. Fld. Studies Map MF-685-B, scale 1:50,000.

Pomeroy, J.S. 1979. Map showing landslides and areas most susceptible to sliding in Beaver County, Pennsylvania. USGS Misc. Geol. Inv. Map I-1160, scale 1:50,000.

Pomeroy, J.S. 1980. Storm-induced debris avalanching and related phenomena in the Johnstown area, Pennsylvania, with references to other studies in the Appalachians. USGS Prof. Paper 1191:24 p.

Pomeroy, J.S. 1982a. Landsliding in the greater Pittsburgh region, Pennsylvania. USGS Prof. Paper 1229:48 p.

Pomeroy, J.S. 1982b. Mass movement in two selected areas of western Washington County, Pennsylvania. USGS Prof. Paper 1170-B:17 p.

Pomeroy, J.S. 1983, Relict debris flows in northwestern Pennsylvania. Northeastern Geol. 5, 1:1–7.

Pomeroy, J.S. 1984. Storm-induced slope movements at East Brady, northwestern Pennsylvania. USGS Bull. 1618:16 p.

Pomeroy, J.S. 1986. Slope movements in the Warren-Allegheny Reservoir area, northwestern Pennsylvania. USGS Bull. 1650:15 p.

Pomeroy, J.S. 1987. Slope stability in the Marietta area, Washington County, southeastern Ohio. USGS Bull. 1695:47 p.

Pomeroy, J.S. 1988a. Map of susceptibility to landsliding in Maryland. USGS Misc. Fld. Studies Map MF-2048, scale 1:500,000.

Pomeroy, J.S. 1988b, Map of susceptibility to landsliding in Prince George County, Maryland. USGS Misc. Fld. Studies Map MF-2051, scale 1:50,000.

Radbruch, D.H. & K.C.Crowther 1973. Map showing areas of estimated relative amounts of landslides in California. USGS Misc. Geol. Inv. Map I-747, scale 1:1,000,000.

Radbruch-Hall, D.H., D.J.Varnes & R.B.Colton 1977. Gravitational spreading of steep-sided ridges, ("sackung") in Colorado. USGS J. Resch. 5,3:359-363.

Radbruch-Hall, D.H. & others 1982. Landslide overview map of the conterminous United States. USGS Prof. Paper 1183:25 p., scale 1:7,500,000.

Radbruch-Hall, D.H., Kathleen Edwards & R.M. Batson 1987. Experimental engineering-geologic maps of the conterminous United States. USGS Bulletin 1610:7 p., map scale 1:7,500,000.

Ratté, C.A. & D.D.Rhodes 1981. The debris avalanche in the Green Mountains of Vermont. Appalachia J.:143-145.

Reed, J.C. Jr. & S.F.Obermeier 1982. The geology beneath Washington, D.C.—The foundations of a nation's Capitol. Geol. Soc. Am. Rev. in Eng. Geol. 5:1-23.

Reiche, Parry, 1937, The toreva-block; a distinctive landslide type. J. Geol. 45,5:538-548.

Rheams, K.F. 1982. Inventory of landslides, slope failures, and unstable soil conditions in Alabama. Geol. Sur. Ala.:2 p.

Rheams, K.F., E.E.Brabb & Fred Taylor 1987. Preliminary map showing landslides in Alabama. USGS Misc. Fld. Studies Map MF-1954, scale 1:500,000.

Rice, R.M. & N.H.Pillsbury 1982. Predicting landslides in clearcut patches. In Recent developments in the explanation and prediction of erosion and sediment yield. Intn'l Assoc. Hydol. Sci. Pub. 137:303-311.

Riedl, G.W. 1981. Statewide landslide monitoring program Wyoming highway system. Wyo. Hwy. Dept. Proj. ESE 7041:51 p.

Rigby, J.K. 1958. Mass movements in Permian rocks of Trans-Pecos Texas. J. of Sedimentary Petrology 28,3:298-315.

Riordan, Pauline & P.G.Bourget 1985. World weather extremes. US Army Corps Eng. Fort Belvoir, VA, Rpt. ETL-0416:77 p.

Robak, T.J. & R.H.Fickies 1983. Landslide susceptibility within the lake clays of the Hudson Valley, New York. N.Y. St. Geol. Sur. O-F Rpt. 504.024, scale 1:100,000.

Robinson, C.S. & others 1972. Geological, geophysical, and engineering investigations of the Loveland Basin landslide, Clear Creek County, Colorado, 1963-65. USGS Prof. Paper 673:43 p.

Rogers, W.P. & others 1979. Guidelines and criteria for identification and land-use controls of geologic hazard and mineral resource areas. Colo. Geol. Sur. Spec. Pub. 6:146 p.

Rogers, W.P. 1987. Progress report for Colorado Geological Survey landslide program, 1985 through 1986. Colo. Geol. Sur. Rpt.:11 p.

Root, Arthur W. 1958. Prevention of landslides. In E.B. Eckel (ed.) Landslides and engineering practice. Nat'l Resch. Conc'l. Hwy. Resch. Bd., Washington, D.C., Spec. Rpt. 29:113-149.

Royster, D.L. 1975. Tackling major highway landslides in the Tennessee mountains. Civ. Eng. 45,9:85-87.

Russell, I.C. 1901. Geology and water resources of Nez Perce County, Idaho. USGS Water-Supply Paper 53, pt.1:75-79.

Sangrey, D.A. & others 1985. U.S. Geological Survey landslide research program; Report of the Committee for Review of the USGS Ground Failure and Construction Hazards Program. USGS Admin. Rpt. Feb. 26, 1985:19 p.

Sangrey, D.A. & A.B.Bernstein 1985. Landsliding, a hazard that can be mitigated. Nat'l Resch. Conc'l. Ground Failure Newsletter 1:6-10.

Schreve, R.L. 1968. The Blackhawk landslide. Geol. Soc. Am. Spec. Paper 108: 47 p.

Schultz, A.P. 1986. Ancient, giant rockslides, Sinking Creek Mountain, southern Appalachians, Virginia. Geology. 14:11-14.

Schumm, S.A. & R.J.Chorley 1964. The fall of threatening rock. Am. J. Sci. 262,9:1041-1054.

Schuster, R.L. 1978. Introduction, Chapter 2. In R.L.Schuster & R.J.Krizek (eds.). Landslides-Analysis and Control. Transp. Resch. Bd. Spec. Rept. 176, NAS, Washington, DC:1-10.

Schuster, R.L. 1979. Reservoir-induced landslides. Intn'l Assoc. Eng. Geol. Bull. 20:8-15.

Schuster, R.L. 1983. Engineering aspects of the 1980 Mount St. Helens eruptions. Assoc. Eng. Geol. Bull. 20:125-143.

Schuster, R.L. & W.H. Hays 1984. Irrigation-induced landslides in soft rocks and sediments along the Columbia River south-central Washington State, U.S.A. 4th Intn'l Symp. on Landslides, Toronto, Sept. 17-21:431-436.

Schuster, R.L. & R.W.Fleming 1986. Economic losses and fatalities due to landslides. Assoc. Eng. Geol. Bull. 23,1:11-28.

Scott, K.M. & R.P.Williams 1978. Erosion and sediment yields in the Transverse Ranges, southern California. USGS Prof. Paper 1030:38 p.

Sharp, R.P. & L.H. Nobles 1953. Mudflow of 1941 at Wrightwood, southern California. Geol. Soc. Am. Bull. 64,4:547-560.

Shreve, R.L. 1968. The Blackhawk landslide. Geol Soc. Amer. Spec. Paper 108:47 p.

Shroder, J.F. Jr. 1970. Landslide landforms and concept of geomorphic age applied to landslides. 21st Intern. Geogr. Cong., New Delhi, 1968. Selected Papers. 1:124-126.

Shroder, J.F. Jr. 1971. Landslides of Utah. Ut. Geol.& Min. Sur. Bull.90:51 p.

Silliman, Benjamin, C.Wilcox & T.Baldwin 1829. Miscellaneous notices of mountain scenery, and of slides and avalanches in the White and Green Mountains. Am. J. Sci. 15:217-232.

Sisiliano, W.J & C.W.Lovell 1971. A regional approach to highway soils considerations in Indiana. Purdue U. Joint Hwy. Resch. Proj. Tech. Paper 24:36 p.

Smith, H.T.U. 1936. Periglacial landslide topography of Canjilon Divide, Rio Arriba County, New Mexico. J. Geol. 44,7:836-860.

Smith, R.L. & R.A.Bailey 1982. Volcanoes in the United States. USGS pamphlet:19 p.

Smith, Rockwell 1958. Economic and legal aspects. In E.B.Eckel (ed.) Landslides and engineering practice. Nat'l Resch. Conc'l., Hwy. Resch. Bd., Washington, D.C., Spec. Pub. 29:6-19.

Soule, J.M. 1975. Geologic hazards map of Dolores, Montezuma County, Colorado. Colo. Geol. Sur. Map Ser. 4, scale 1:10,000.

Stearns, H.T. 1944. Geology of the Samoan Islands. Geol. Soc. Am. Bull. 55:1279-1332.

Stearns, H.T. 1966. Geology of the State of Hawaii, p. 32-36. Palo Alto, CA: Pacific Books.

Stout, M.L. 1977. Radiocarbon dating of landslides in southern California. Calif. Geol., May 1977:99-101.

Stover, B.K. & S.H.Cannon 1988. Reactivation of the Muddy Creek landslide, west-central Colorado, U.S.A. Nat'l Resch. Council Ground Failure Newsletter 2:8-10.

Stow, S.H. & T.H.Hughes 1980. Geology and the urban environment, Cottondale Quadrangle, Tuscaloosa County, Alabama. Ala. Geol. Sur. Atlas 9:79 p., scale 1:62,500.

Strahler, A.N. 1940, Landslides of the Vermillion and Echo Cliffs. J. of Geomorph. 3:285-300.

Sundstrom, R.W., T.E.Pickett & R.D.Varrin 1976. Hydrology, geology, and mineral resources of the coastal zone of Delaware. Del. Coastal Zone Mgmt. Prog., Tech. Rpt. 3, Sept. 1976:28 p.

Swanson, F.J. & M.M.Swanston 1977. Inventory of mass erosion in the Mapleton Ranger District, Siuslaw National Forest. USDA FS coop. study Siuslaw N.F. & the Pac. N.W. For. Range E.S.:41 p.

Taylor, Fred & E.E.Brabb 1986a. Map showing the status of landslide inventory and susceptibility mapping in California. USGS O-F Rpt. 86-100, scale 1:1,000,000.

Taylor, Fred & E.E.Brabb 1986b. Map showing landslides that have caused fatalities or at least $1,000,000 in damages from 1906 to 1984. USGS Misc. Studies Map MF-1867, scale 1:1,000,000.

Thompson, G.A. & D.E.White 1964. Regional geology of the Steamboat Springs area, Washoe County, Nevada. USGS Prof. Paper 438-A:52 p. scale 1,62,500.

Thompson, W.B. 1979. Surficial geology handbook for coastal Maine. Me. Geol. Sur.:34-37.

Thorson, G.W. 1987. Soil bluffs plus rain equals slide hazards. Wash. Geol. Newsletter 15,3:3-11.

Tilling, R.I., R.Y.Koyanagi & R.T.Holcomb 1975. Rockfall seismicity; correlation with field observations, Makaopuhi Crater, Kilauea Volcano, Hawaii. USGS J. of Resch. 3:345-361.

Tilling, R.I. & others 1976. Earthquake and related catastrophic events, Island of Hawaii, November 29, 1975; a preliminary report. USGS Circ. 740:1-29.

Torrey, V.H.III & F.J.Weaver 1984. Flow failures in Mississippi riverbanks. Proc. 4th Intn'l. Symp. Landslides Sept. 17-21, Toronto, Canada 2:355-360.

Trace, R.D. 1942. Landslide blocks along the margin of the Diablo Plateau, Texas. Fld. & Lab. 10,2:155-159.

Trimble,D.E. 1979. Unstable ground in western North Dakota. USGS Circular 798:19 p.

Trolinger, W.D. 1980. Rockwood embankment slide between stations 2001+00 and 2018+00. A horizontal drain case history. In Rock classifications and drilling and drainage. Transp. Resch. Record. 783:26-30.

U.S. Geological Survey 1982. Goals and tasks of the landslide part of a ground-failure hazards reduction program. USGS Circ. 880:48 p.

U.S. Soil Conservation Service 1986. Tomiki Creek unit, redwood empire target area, Mendocino County, California. Davis, Calif.:264 p.

Utah Geological & Mineral Survey 1983. Governor's conference on geologic hazards. Ut. Geol. & Min. Sur. Circ. 74:99 p.

Utah Dept. of Public Safety 1984. Hazard mitigation plan, Utah, 1984. Salt Lake City. Ut. Div. Emerg. Mgmnt:92 p.

Vandre, B.C. & D.N.Swanston 1977. A stability evaluation of debris avalanches caused by blasting. Assoc. Eng. Geol. Bull. 14, 4:205-223.

Van Horn, R. 1972a. Landslide and associated deposits map of the Sugar House Quadrangle, Salt Lake County, Utah. USGS Misc. Geol. Inv. Map I-766-D, scale 1:24,000.

Van Horn, R. 1972b. Relative slope stability map of the Sugar House Quadrangle, Salt Lake County, Utah. USGS Misc. Geol. Inv. Map I-766-E, scale 1:24,000.

Van Horn, R 1975. Largest known landslide of its type in the United States—a failure by lateral spreading in Davis County, Utah. Utah Geol. 2, 1:83-88

Varnes, D.J. 1958. Landslide types and processes. In E.B.Eckel (ed.) Landslides and engineering practice. Highway Resch. Bd. Spec. Rept. 29, NAS-NAC 544. Washington, DC:20-47.

Varnes, D.J. 1978. Slope movement types and processes. In R.L.Schuster & R.J.Krizek (eds.). Landslide analysis and control. Nat'l Acad. Sci. Washington, D.C.:11-33.

Voight, Barry 1974. Architecture and mechanics of the Heart Mountain and South Fork rockslide. In B.Voight & M.A.Voight (eds.) Rock mechanics, the American Northwest. Intn'l Soc. for Rock Mechs. 3rd Congress Expedition Guide, Penn. St. U. Spec. Pub.: 26-36.

Voight, Barry & Mary A.Voight (eds.) 1974. Rock mechanics; the American Northwest. Pennsylvania State Univ.:292 p.

Voight, B. & others 1983. Nature and mechanics of the Mount St. Helens rock-slide avalanche of 18 May, 1980. Geotechnique. 33,3:243-273.

Wagner, W.R.T. 1944. A landslide area in the Little Salmon River Canyon in Idaho. Economic Geol. 39,5:349-359.

Wahrhaftig, C. & R.F. Black 1958. Engineering geology along part of the Alaska railroad. USGS Prof. Paper 293-8:71-116.

Waitt, R.B. Jr. 1979. Rockslide-avalanche across distributary of Cordilleran ice in Pasayten Valley, northern Washington. Arctic & Alpine Research 11,1:33-40.

Watson, R.A. & H.E.Wright Jr. 1963. Landslides on the east flank of the Chuska Mountains, northeastern New Mexico. Am. J. Sci. 261,6:525:548.

Watters, R.J. 1983. A landslide induced waterflood-debris flow. Int'l Assoc. Eng. Geol. Bull. 28:177-182.

Wayne, W.J. 1975. Urban geology of Madison County, Indiana. Ind. Geol. Sur. Spec. Rpt. 10:21.

Wentworth, C.K. 1943. Soil avalanches on Oahu, Hawaii. Geol. Soc. Am. Bull. 54,1:53-64.

Wieczorek, G.F. 1984. Preparing a detailed landslide-inventory map for hazard evaluation and reduction. Assoc. Eng. Geol. Bulletin 21, 3:337-342.

Wieczorek, G.F. & others 1983. Potential for debris flow and debris flood along the Wasatch Front between Salt Lake City and Willard, Utah, and measures for their mitigation. USGS O-F Rpt. 83-635:25 p, scale 1:100,000.

Wieczorek, G.F., R.C.Wilson & E.L.Harp 1985. Map showing slope stability during earthquakes in San Mateo County, California. USGS Misc. Geol. Inv. Map I-1257-E, scale 1:62,500.

Wiggins, J.H., J.E.Slosson & J.R.Krohn 1978. National hazards—earthquake, landslide, expansive soil loss models. J.H.Wiggins Co. Tech. Rpt, Redondo Beach, CA:162 p.

Williams, D.A. & J.E. Armstrong 1970. Investigation of a large landslide associated with construction of I-15 near Dillon, Montana. 8th Ann. Eng. Geol. & Soils Eng. Symp. Proc.:91-108.

Williams, G.P. & H.P.Guy 1973. Erosional and depositional aspects of Hurricane Camille in Virginia, 1969. USGS Prof. Paper 804:80 p.

Wilshusen, J.P. 1979. Geologic hazards in Pennsylvania. Penn. Geol. Sur. Educ. Ser. 9:1-52.

Wilson, S.D. 1967. Landslides in the city of Anchorage. In Prince William Sound, Alaska, earthquake of 1964 and aftershocks. USDC, Coast & Geod. Sur. 2,A:253-297.

Winter, E. & B.Beard 1978. The "O" Street slide and its geologic aspects, Washington, D.C. 29th Ann. Hwy. Geol. Symp. Proc.:83-95.

Witkind, I.J. 1972. Map showing seiche, rockslide, rockfall, and earthflow hazards in the Henry's Lake Quadrangle, Idaho and Montana. USGS Misc. Geol. Inv. Map I-781-C, scale 1:62,500.

Workman, E.J. & V.C.Kelley 1940. A rock slide caused by lightning. Am. Meterol. Soc. Bull. 21,3:112-113.

Wu, T.H. & D.N.Swanston 1980. Risk of landslides in shallow soils and its relation to clearcutting in southeastern Alaska. Forest Sci. 26,3:495-510.

Zeizel, A.J. 1987. Emergency debris flow hazard evaluation during a disaster. Tokyo, Proc. Int'l. Seminar, Regional Planning for Disaster Prevention:25 p.

Zejdlik, R.C. 1956. A landslide near Oslo, Minnesota. N.Dak. Acad. Sci. Proc. 10:22-25.

Landslides: Extent and Economic Significance, Brabb & Harrod (eds)
© 1989 Balkema, Rotterdam. ISBN 90 6191 876 6

Landslides: Their extent and significance in the Caribbean

Jerome V.DeGraff
USDA Forest Service, Fresno, Calif., USA

R.Bryce
Geological Survey Division, Kingston, Jamaica

R.W.Jibson
US Geological Survey, Reston, Va., USA

S.Mora
Instituto Costarricense de Electricidad, San Jose, Costa Rica

C.T.Rogers
University of the West Indies, St.Augustine, Trinidad

ABSTRACT: Landslides are a significant process shaping the landscape of many Caribbean islands. Only on islands with low relief and limestone bedrock such as the Cayman Islands or the Bahamas can the process be considered irrelevant.

The geology and climate of the Caribbean contributes to the prevalence of landslides. The ubiquitous steep slopes resulting from tectonic and volcanic forces provide abundant locations for landslide occurrence. Other erosional processes such as wave action along island coasts and stream erosion provide a continual means for maintaining many slopes at inclinations close to the angle of repose for the materials underlying them. Tectonic and volcanic activity also creates lithologic and stratigraphic conditions favoring the occurrence of landslides.

The warm, wet climate influences the material involved with landsliding and serves as the most common mechanism for their initiation. The rapid weathering of bedrock under humid conditions creates a regolith generally weaker than the parent rock. The presence of clay either inherited from parent sedimentary bedrock or derived from weathering of metamorphic and volcanic bedrock contributes to the tendancy of soil masses to fail. The seasonal pattern of rainfall punctuated with intense storms serves as an efficient means for triggering landslides in this region.

Most landslide types are found within the Caribbean. Debris flows and slides are by far the most prevalent form. Many factors contributing to general slope instability favor development of these landslide types. Earthflows, rockslides, rockfalls, and slumps are other common but less frequently occurring landslide types.

Fatalities and injuries are a primary loss attributable to landslides in the Caribbean. In 1938, three debris flows in two days at Ravines Poisson and Ecrivisse in St. Lucia killed sixty people and injured 32 others. Estimates of the missing ranged as high as 250. More recently the catastrophic landslide in the Mameyes district of Ponce, Puerto Rico claimed the lives of at least 129 residents. Smaller landslides causing fewer individual fatalities yield a tragic toll over time. Between 1925 and 1986, twenty-five people in Dominica lost their lives to landslides in five separate events.

Destroyed or damaged structures, especially roads, are another major loss attributable to landslides. Clearance of slide debris and repair of damage are the main impacts to roads. On the narrow roads often found in the Caribbean, it does not take a large landslide to create a major impact. In St. Vincent, St. Lucia, and Dominica of the Windward Islands, the average annual cost for landslide damage to roads ranges from $115,000 to $121,000 in normal years. The average annual cost of landslide investigation, repair, and maintenance in the larger islands of Trindad and Tobago is $1.26 million and $0.96 million, respectively. On an average year, it is estimated the cost of repairing landslide damage to roads throughout the Caribbean amounts to $15 million. Other structural effects include the loss of homes. Seventeen families in Jamaica lost their homes due to a year of slow deformation by a large earthflow near Preston, St. Mary's Parish. On average, perhaps tens of houses are destroyed, and hundreds are damaged by landslides each year in Puerto Rico. In St. Vincent, water lines severed by landslides in 1981 left nearly 40 percent of the population without water for periods varying from a few days to six months. In 1986, landslides damaged pipelines to hydroelectric generating stations and reduced the total electrical generating capacity of St. Vincent by 36 percent until repairs could be made. Landslides in the vicinity of Peligre reservoir in the Dominican Republic (Hispaniola) are contributing to premature filling of the impoundment and threaten the powerhouse.

Agriculture, a major economic activity throughout the Caribbean, is impacted by interference with transportation of perishable goods to market and loss of crops. Figures for agricultural losses gathered for major storms affecting Jamaica, Hispaniola, Puerto Rico, and other islands with agrarian economies rarely distingush between losses due solely to landslides and those due to other storm effects. The magnitude of agricultural losses attributable to landslides relies on individual examples where this loss can be quantified. The Good Hope landslide in Dominica carried away or buried 3.8 hectares of crop land worked by six small farmers. An immediate economic loss of $8,000 and a delay of a few months to a few years before replanted crops are harvestable were the consequences of this event. Losses of bananas to smaller landslides on St. Lucia and St. Vincent yield losses of $4000 to $250 for individual farmers. Because agriculture is carried out mainly by individual farmers with per capita incomes ranging from hundreds to a few thousand dollars, losses of this mganitude represent a severe burden to the economies of both individual families and island nations.

1 INTRODUCTION

The small community of Good Hope, on the east coast of Dominica, West Indies, nestles along a narrow valley bottom by Grand Marigot Bay. By nightfall of November 11, 1986, the community and surrounding area had experienced several days of heavy rainfall. Earlier that day, a small landslide developed in the cutslope

of the Castle Bruce-Petit Soufriere road where it
crosses the steep hillslope above the town. Except
for this landslide, the rains had produced no other
unusual events.

Shortly after 3:00 a.m. on November 12, 1986,
nearly 17,000 cubic meters of soil and weathered rock
slid from the hillslope above the Castle Bruce-Petit
Soufriere road. Bananas, coconut, and bay trees
growing on the hillslope were swept away. The rapidly
moving slide mass destroyed citrus trees and vegetable
crops growing below the road. The slide engulfed the
health clinic near the base of the slope, seriously
injuring the nurse and killing her ten-year-old daughter in their sleeping quarters. Several meters farther downslope, the impact of the slide shoved the
primary school partly off its foundation, collapsed
the back wall, and buried the upslope side to the roof
line. As the debris came to rest, the toe of the
slide deposited material 1 to 4 meters deep along a
15-meter length of the principal street through Good
Hope. On the slope above, a 90-meter section of the
Castle Bruce-Petit Soufriere road lay blocked by the
upper slide mass (Fig. 1).

Fatalities and injuries such as the ones at Good Hope
represent a significant consequence of landslide
activity in the Caribbean. A more common economic
consequence of landslide activity is damage to roads
and structures. The destruction of the health clinic
and primary school at Good Hope represents a direct
economic impact. An indirect economic impact would be
losing use of the Castle Bruce-Petit Soufriere road.
Direct and indirect losses are associated with landslide impacts on agriculture, a primary economic
activity in many Caribbean countries.

The land areas of the Caribbean are a collection of
islands collectively known as the West Indies (Fig.
2). The northernmost islands are the Bahamas. The
other islands are grouped as the Greater and Lesser
Antilles. The Greater Antilles includes the large
islands of Cuba, Jamaica, Hispanola, and Puerto Rico.
Groups of smaller islands make up the Lesser Antilles.
The northermost group in this is the Virgin Islands.
To the south are the Leeward Islands. St. Christopher, Antigua, and Guadaloupe are among the larger
islands in this group. South of the Leeward Islands
are Dominica, Martinique, St. Lucia, St. Vincent and
the Grenadines, and Grenada. These islands form the
Leeward Islands group. At the southernmost end of the
Lesser Antilles, is the twin island nation of Trinidad
and Tobago.

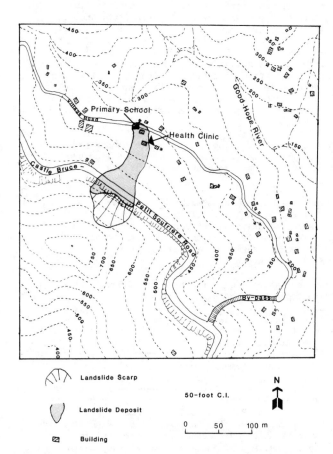

Figure 1. Map of the Good Hope landslide in Dominica.
Shading shows the area affected by the landslide
superimposed on the pre-existing topography and cultural features.

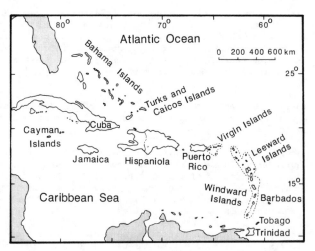

Figure 2. Location of the larger islands and major
islands groups found in the Caribbean.

2 TRINIDAD AND TOBAGO

The islands of Trinidad and Tobago are situated in the
southern Caribbean between North latitude 10° 2' and
11° 2' and West longitude 60o 30' and 61° 50'. The
islands experience a warm tropical climate with a dry
season from January to May and a wet season from June
to December. Eighty percent of the annual rainfall is
received in the wet season. Maximum intensities can
approach 50 millimeters per day. Mean annual precipitation for Trinidad varies from 1520 millimeters in
the northwest and southwest to 3300 millimeters in the
northeast (Cooper and Bacon, 1981). Average annual
rainfall for Tobago ranges from 1270 to 3050 millimeters. The maximum amount of 3050 millimeters falls on
the Main Ridge of Tobago due to topographic influence.
Both Trinidad and Tobago lies south of the main path
of hurricanes. However, Tobago has experienced two
tropical storms within this century and is frequently
affected by the southern fringes of hurricane systems
passing to the north. This can produce intense rainstorms capable of triggering catastrophic landslides.

The total land mass area of Trinidad is 4,827
square kilometers; that of Tobago is 300 square kilometers. The twin-island nation has a total population

The Good Hope landslide exemplifies both the nature
and economic significance of landslide activity
throughout the Caribbean. Landsliding is a widespread
and significant geomorphic process associated with
steep slopes and landslide-prone regolith and bedrock
common to this region. Landslides are used herein to
refer to all types of slope movements, as described by
Varnes (1978). Triggering of the Good Hope landslide
by rainfall is typical of Caribbean landslides. These
precipitation events as well as volcanic and earthquake activity are known initiators of landslides.

of 1.2 million. 1 million of which inhabit the larger
island. Eighty-eight percent of the population of
Trinidad lives on the western half of the island. The
main towns, Port of Spain and San Fernando, are situated on sheltered harbors on the west coast and are
connected to other population centers by interior and
coastal roads. In Tobago, the principal towns and
main roads are on the coast. On both islands, the
road network provides the main form of transportation.

Recurrent landslide activity, generally associated
with intense rainfall, is a common problem in both
islands. Because large sections of major roadways are
sited in landslide-prone terrain, the impact of slope
failure on the human and economic environment is
considerable.

There is a paucity of published literature on
landslides in Trinidad. No published literature on
landslides in Tobago is known. Taylor (1981b) discussed the engineering parameters of landslide-prone
clays of North, Central and South Trinidad. Brown and
others (1985) reviewed the role of geology on landslide distribution in western Trinidad. Other workers
(Rogers and Chow-Gabbadon, 1985) have described case
histories of individual landslides. Unpublished
postgraduate theses at the University of the West
Indies document a number of landslide case histories.

2.1 Geology and geomorphology of Trinidad

Trinidad is considered a physiographic extension of
eastern Venezuela (Kugler, 1959). The island can be
divided into five geologic-geomorphic units. The
first, the Northern Range, is a deeply dissected east-
west trending mountain system. It consists of low-
grade metamorphic rocks folded into a broad anticline.
The metamorphic rocks are of upper Jurassic to Cretaceous age.

The Caroni Plains is the second unit. It is a low-
lying belt of plains consisting of either Tertiary-age
terrace deposits, Recent alluvial plains, or Recent
swamps.

The third unit, the Central Range, is a ridge
trending north-northeast. A core of Cretaceous and
Eocene sediments defines an anticline flanked by
Miocene clays and reef limestones.

The Naparima and Southern lowlands, a dissected
peneplain, is the fourth unit. The Naparima lowland
is underlain by Lower Miocene calacareous marls and
sands. In contrast, the Southern lowlands are underlain by extensively folded and faulted Miocene and
Pliocene sands, silts, and clays.

The fifth unit is the Southern Range. This is a
line of low hills consisting of clays, shales, and
marls ranging in age from Oligocene to Miocene.

2.2 Landslides in Trinidad

The geographic distribution of landslides greater than
30 meters in maximum dimension is presented in Figure
3. The majority of these occur along road cut slopes.
A small number of landslides have been mapped in the
forest reserves in the east of the island. The concentration of landslides along roadways reflects both
the presence of landslide-susceptible zones and the
effects of slope-disturbing activities. Because roads
are confined largely to the populous western half of
the island, areas of recognized landslide activity are
concentrated in this same region.

Landslides in the Northern Range occur mainly in
cut slopes along coastal roads and valley roads (Taylor, 1981a). In the majority of cases, they are
initiated where roads cross slopes oversteepened by
either river or wave erosion and are triggered by
intense rainfall. Major slides have occurred where
the roadway is cut parallel to the strike of bedding,
foliation or other discontinuities causing them to
daylight into the slope face. Vibrations from blasting associated with quarrying activity in adjacent
cuts have also caused some slides.

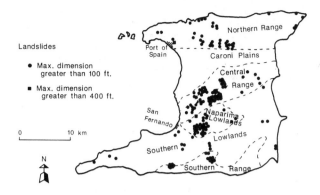

Figure 3. Distribution of landslides in Trinidad in
relation to major physiographic provinces.

Slope failures on natural slopes are also present
in the Northern Range. While these landslides occur
less frequently, they tend to be significantly larger
and deeper-seated than those associated with roads.

The phyllites of the Northern range metamorphic
series are the principal units involved with landslides in this area. This bedrock tends to underlie
the higher, steep slopes of the Northern Range. Exposures show the unit to be predominantly calcareous
phyllites interbedded with mica schist. Slates and
minor quartzites are also present. The phyllites are
inherently weak due to their high mica content, well-
developed foliation parallel to bedding and thin
bedding. Extensive jointing, faults, and shear zones
significantly reduce the strength of the rock and
decrease slope stability. Groundwater movement along
these discontinuities weathers the infilled materials
to a highly plastic clay which offers little resistance to sliding. Deep weathering profiles containing
relict structures and saprolitic zones have developed
in completely weathered sections of the phyllite.
Partially weathered outcrops are found to consist of
loosened slabs which may occur together with soil-rock
mixtures. Interbedded sequences of highly weathered
phyllite and unweathered competent quartzite are also
prone to landsliding. A thick colluvial mantle subject to landsliding and soil creep overlies in-situ
bedrock in the foothills of the range and along valley
slopes. Locally, unstable zones occur in structurally
weak zones associated with small-scale folding.

Within the phyllites, there are two main failure
modes. One mode is the development of planar slides
in partially weathered and unweathered bedrock. These
planar slides may be either slab failures moving
downslope along saturated bedding or foliation planes
and clay-filled joints, or wedge and rock block failures. The other mode is the formation of shallow
sheet slides in residual or colluvial soil. Invariably, the volume of landslide debris approaches several hundred thousands of cubic meters. Minor
rockfalls occur to a lesser extent in quartzite.

In central and southern Trinidad, landslides pose a
problem to populated areas and roads. Spoon-shaped
scars dotting the terrain seen on aerial photographs
indicate the dominance of landsliding as a geomorphic
process. Historically, large pockets of the Central
and Southern Ranges and the Naparima Lowlands are
subject to recurrent slope failures. Landslides are
prevalent along highways and secondary roads
(Diyaljee, 1984). Many secondary roads are located on
ridge crests and are flanked by slopes oversteepened
by river valley incision. As valley widening continues, the clay slopes slip into the valley bottoms
taking the roadway with them. Other common contribu-
tors to slope failures include poor drainage from
roadways, seepage of sewage from houses adjacent to
the slope, and overloading of the road shoulders by
placement of asphalt fill.

Stiff clays and clay shales of the Cipero, Karamat,

Nariva, Springvale, and Manzanilla Formations are a major influence on this landslide activity. These formations are of Oligocene to Miocene age. In the field, a represntative soil column consists of a weathered, mottled red to yellow-brown brecciated zone up to 12 meters thick. This is underlain by a less fissured transitional zone 4 to 9 meters thick. The parent clay below this soil is typically a gray to brown clay. Soil series 77, 177, 277, 377, and 278/L represent map units on the Agricultural Soil Map of Trinidad having this general profile. These clay soils are highly erodible due to their high dispersivity. This indirectly affects the stability of slopes as erosion of organic layers exposes the underlying clay to dessication.

Both the silty clays and clay shales are fissured, overconsolidated, highly plastic (PI 35%), and have a high swell potential. The dominant mineralology is montmorillonite. It makes up 40% of the clay content with illite and kaolinite together amounting to twenty percent. Calcareous or selenite gypsum fillings or both are sometimes present in fissures.

The stability of the swelling clay is influenced by its response to seasonal wetting and drying. During the dry season, dessiccation cracks, as much as 5 meters in depth, develop within the soil. Though fissured, the clays have a relatively high strength and remain stable in the dry state. With the onset of heavy rains, water readily infiltrates the ground via the tension cracks. As the clay becomes saturated, high pore-water pressures develop and swelling occurs. Calcareous cements and gypsum are dissolved and diagenetic bonds are destroyed. The ground weakens as the clay loses shear strength. Annual wetting and drying causes a gradual deterioration of the clay structure and slopes become increasingly unstable. Once the limiting condition is attained, slope failure occurs.

2.3 Geology and geomorphology of Tobago

The island of Tobago has a ridge of Cretaceous metamorphic rock passing through the central part like a backbone. The Main Ridge trends east-northeast and has abrupt slopes on the north and south. Structurally, it is an isoclinally folded anticline plunging to the northeast (Maxwell, 1948).

The Main Ridge consists of rugged and largely inaccessible terrain rising to a height of 576 meters. Active dissection by a dense network of deeply-incised streams creates long, steeply inclined slopes. The steeper gradient slopes are generally found on the northern side of the ridge. Diorite, ultramafics and basic extrusive rocks outcrop to the south and southeast of the Main Ridge.

Tertiary and Quaternary coral limestones and interbedded marls and clays form a series of low terraces in the southwest part of the island. These younger rocks represent the only coastal plain. The remainder of the island is either hilly or mountainous.

2.4 Landslides in Tobago

Because of rugged, inaccessible terrain, landslide mapping of Tobago is limited to the vicinity of existing roads. Only two roads cross the Main Ridge. Most roads in Tobago are found on the coastal plain or along the coast where the majority of the population is concentrated. The geographic distribution of active landslide areas is shown in figure 4.

Past landslide activity appears concentrated in the Mt. Dillon area. Extensive vertical landslide scarps, some exceeding 6 meters in height, lying above relatively flat sections of ground are common. The landslide debris has formed a colluvial mantle over the weathered bedrock surface. The colluvium reportedly varies in thickness from 6 to 30 meters. The colluvial soils are highly leached, have low shear strengths, and are moderately expansive. Rapid ero-

Figure 4. Location of landslide-prone areas in Tobago.

sion during heavy rainfall creates deep steep-sided gullies in the weathered mantle and often exposes bedrock. Wave erosion at the base of the slope also results in oversteepened slope angles. Under such marginally stable conditions, landslides are triggered by any mechanism which minimally disturbs the slope. These include increases in pore water pressure during heavy rains, traffic vibrations, and earthquake shaking.

Debris slides frequently develop failure planes within the weathered mantle or at the mantle-bedrock interface. Planar slides develop along unfavorably positioned foliation, joints, and bedding planes.

The Mt. Dillon and Parlatuvier Formations are most prone to landslides. These formations are exposed in the northern part of the Main Ridge from Mt. Dillon to Parlatuvier. Exposures are seen along the Roxbourough-Parlatuvier road. The thin bedded mica schists, metavolcanics, and quartzites dip steeply to the south. The rocks outcrop on steep slopes rising sharply fron the shoreline to heights above 300 meters. It appears large-scale landsliding occurred within these units in the geologic past. This activity is likely related to the same failure-producing conditions responsible for present-day landslides: unfavorably oriented joints, rock creep in highly weathered outcrops, undercutting of cliffs by wave actions, and earthquake activity.

As seen along the Roxbourough-Parlatuvier road, the mica schists and phyllites of these Main Ridge formations are completely weathered to a structureless soil with colors ranging from orange-brown to magenta. These soils deteriorate to saturated muds on wetting and cannot stand on the often high, steep slopes of the roadcut. Major rotational slides are common.

Another bedrock of the Main Ridge involved with landslides are parts of the diorite batholith. Where pockets of diorite outcrop and are deeply weathered, a fine to coarse granular residual soil forms. These locations become destabilized where high groundwater tables are present or during infiltration by intense rainfall.

The interbedded clays and marls of the coastal plain are less prone to landsliding. Groundwater flow is generally the controlling factor.

2.5 Economic impact of landslides in Trinidad and Tobago

Frequent landsliding in roadcuts in Trinidad and Tobago poses a continuing social nuisance and imposes high maintenance costs. The Ministries of Works and Agriculture conduct annual surveys of landslides along roadways in Trinidad and Tobago. Therefore, the extent of landslide activity in roadcuts is well-known while being generally unknown for natural slopes.

Two major landslides illustrate the nature of landslide activity on Trinidad. A large active landslide is located on the southern slopes of St. Barb's Hill, Laventille, in the Northern Range foothills.

The landslide was apparently initiated in 1977 by quarrying of limestone at the base of the affected slope. The limestone is in faulted contact with younger decomposed phyllites (Fig. 5). The younger rocks lost their support and moved downslope once sufficient limestone was removed. Blasting operations associated with the quarrying activity may also have contributed to triggering of the slide.

The direction of movement and configuration of the slide mass suggests movement occurred along a deep-seated southerly dipping bedding plane. The slide surface is defined by a series of subparallel scarps. These formed by repeated headward retreat of the landslide each year during the heavy rains of the wet season. The slide crown extended 400 meters upslope and engulfed 15 homes. The majority of structures in the landslide zone collapsed completely while others are damaged beyond repair. It is presently encroaching on the St. Barb's road located at the crest of the hill. The total slide area is currently 37,000 square meters.

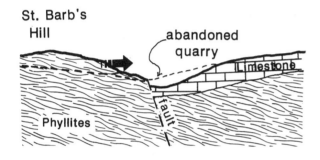

Figure 5. Cross-section of geologic conditions present at the St. Barb's Hill landslide site in Trinidad. The arrow denotes the direction of movement.

Another major landslide took place on the north-bound carriageway of the Solomon Hochoy Highway approximately 1.5 kilometers north of the Claxton Bay overpass. This highway is the main transportation artery linking the north and south parts of the island. A 21-meter section of the road shoulder was displaced during initial movement in 1984. During the rains of 1985, a larger failure developed north of, and adjacent to, the existing landslide. The later movement resulted in approximately 1-meter of vertical drop of the slow lane of the carriageway over a distance of 125 meters and necessitated closing the lane to traffic. By October 1986, the landslide encroached onto the fast lane of the highway.

The failure occurred in overconsolidated clay fill containing organic layers and sand and gravel. The initial movement took place shortly after a 1-meter trench was excavated parallel to the road shoulder to replace electricity cables. This trench served as the main access for water reaching the clay. As the clay absorbed water, it softened and lost strength. Percolating water may also have moved through the more permeable sand and gravel lenses creating perched water tables and generated higher pore pressures at the contact with the clay. Subsequent failures occurred as stormwater infiltrated the exposed scarps of the unrepaired landslide destabilizing the adjacent area. Three houses at the landslide toe suffered severe structural damage.

In Tobago, a significant landslide involving 6 hectares occurred in the Mt. Dillon area in the summer of 1964. The slide, moving toward the sea, displaced a 0.6 kilometer section of the Northside Road and blocked other parts of the roadway. The affected

section of the Northside road was rerouted to the south to preserve access between Castara and Parlatuvier. During unusually heavy rains in 1984, a series of landslides developed along the Northside Road, including the rerouted stretch. This entirely blocked off traffic. The abandoned Northside Road including the 1964 slide was upgraded and reopened to traffic as a result.

Various techniques have been applied in the repair of individual landslides. These include the use of retaining structures, particularly gabion baskets and sheet-pile walls, surface drainage, buttress fills, and benching. Wire mesh nets and fences have been utilized as methods of protection from falling rocks. Analysis for determining repairs has relied exclusively on a soil mechanics approach. This approach is not always successful and many repaired slides have failed subsequently during the next rains. A wide range of stability characteristics is exhibited over short distances for the rock and soil present in Trinidad and Tobago. Therefore, combining engineering analysis with geologic data is needed to more accurately assess the stability of slopes and improve the performance of repairs.

The development of landslide susceptibility maps to guide preliminary design and land use planning is an initial step in more accurately assessing slope stability. Information gained from site investigation permit generalizations such as those described in the proceeding sections concerning factors controlling landslide activity. Four significant parameters are slope angle, lithology and its engineering characteristics, rainfall and seismicity. The map zonation of relative landslide susceptibility is based on these parameters. Table 1 describes the parameters used in compiling a landslide susceptibility map for Trinidad. Table 2 describes the parameters used in compiling a landslide susceptibility map for Tobago.

Reliable estimates of the social and economic impact due to landslides are not available. No casualties are known to be a consequence of landsliding on Trinidad and Tobago. Although the principal damage has been to roadways, private homes, property, and schools have been affected, too. Annual damage to property and installations for both islands are not known. Also unknown is the indirect cost associated with loss of productivity due to disruption of traffic and public utilities and isolation of villages, but is likely to be very significant. The annual cost of landslide investigation, repair, and maintenance for roadways is shown in Table 3. The average annual cost was $1.26 million (US) for Trinidad from 1979 to 1986. For 1985 to 1986, these annual costs in Tobago averaged $0.96 million (US).

3 BARBADOS

The island of Barbados is located northeast of Trinidad and Tobago and due east of the Windward Islands. It is the easternmost island in the Caribbean and is situated at 13° 15' North latitude and 59° 30' West longitude (Rand McNally, 1988). This independant island nation is 430 square kilometers. Mount Hillaby, at 340 meters, is the highest point. There are about 250,000 inhabitants of Barbados.

Barbados experiences wet summers and dry winters associated with a tropical humid climate. Annual precipitation varies from 1008 millimeters on the coasts to 1650 millimeters in the interior (Peeters, 1963). Barbados lies within the main hurricane track of the Caribbean.

3.1 Geology of Barbados

Geologically, Barbados is an interesting island consisting of uplifted deep-sea facies and Pleistocene coral reefs (Fairbridge, 1975a). The higher elevations in the north are generally capped by Pleistocene coral reefs or limestones and marls of possibly Oligo-

Table 1. Factors defining relative landslide-susceptibility in Trinidad.

Factors			Landslide-Suceptibility
Landform	Relief	Geology	
Mountain Ranges			
Northern Range	Steep, mountainous, terrain	Low-grade metamorphics	High to very high in phyllites; moderate to low in limestones an quartzites.
Central Range	Moderate to steep slopes	Clays, shales, limestones and sandstones	Very high in clay pockets, high to moderate elsewhere.
Southern Range	Low hills, moderate to steep slopes	Clays with sands	Generally low, high in pockets
Dissected Lowlands			
Caroni Plains (south)	Low-lying, undulating topography; 50 to 250 feet	Sands with some clays	Generally low, moderate in pockets.
Naparima Lowlands	Undulating topography, steep slopes; 100 to 200 feet	Calcareous silty clays and marls	Very high along dissected ridges, high elsewhere.
Southern Lowlands	Low-lying undulating topography; 200 feet	Sands, silts, and clays	Generally low, moderate in pockets
Terraces			
Caroni Plains (north), Mayaro -Caroni	Low-lying topography incised by major streams, 50 to 250 feet	Sands with clays	Low to very low
Flood Plains			
Caroni Plains alluvial flats	Level ground	alluvium	Very low
Swamps			
Caroni, Nariva, and Oroupouche	Level ground	swamp deposits	Very low

Table 2. Factors defining relative landslide-susceptibility in Tobago.

Factors			Landslide-susceptibility
Landform	Relief	Geology	
Highlands			
Main Ridge (crest)	Steep, rugged mountains	Deeply weathered mica schists	Unknown, likely to be moderate to high.
Main Ridge (northern slopes)	Relatively short, precipitous slopes	Mica schists, quartzites, and metavolcanics	Very high.
Main Ridge (southern slopes)	Steep, broken by ridges	Metamorphics, diorite, and ultramafic rocks	High in metamorphics, moderate to high in diorite, and very high in shear zones.
Volcanic uplands	Undulating upland topography	Basic volcanics	Low to moderate
Lowlands			
Coastal Plain	Low-lying terraces	Coral limestone and clays/marls	Very low in limestone, moderate in clay/marl pockets.
Floodplains and swamps	Level ground	Alluvium and swamp deposits	Very low.

Table 3. Annual Cost of Landslide Damage to Roads on
Trinidad and Tobago.

| Year | Landslide Damage (in thousands of dollars)* | |
	Trinidad	Tobago
1979	360	nd
1980	520	nd
1981	790	nd
1982	1270	nd
1983	2180	nd
1984	1540	nd
1985	1820	490
1986	1630	1430

*Includes the cost of investigation, repair, and main-
tenance. Cost is given in $US as are all costs pre-
sented. The notation, nd, signifies no data.

cene or Miocene age. These limestones overlie an
oceanic series representing a deep marine facies. The
oceanic series is exposed on Mt. Hillaby and higher
elevations to the southwest. Chalks, marls, volcanic
tuff, and mudstone are the main components of the
oceanic series. A thick sequence of mudflows of
probable Eocene age called the Joes River Formation
underlie the oceanic series. The core of the island
under the Joes River Formation is the Scotland Beds.
These beds outcrop on the eastern part of the island
in the Scotland District. It is essentially a flysch
facies (Fairbridge, 1975a).

3.2 Landslides in Barbados

Small-scale landsliding is associated with coastal
escarpments where instability results from undercutting by wave-action. However, the principal landslide
activity on Barbados is confined to the Scotland
District on the northeast part of the island (Fig. 6).
Price (1958) notes reported soil movement in the
Scotland District dating from as early as the mid-
eighteen century. While these reports are very general, they describe soil sliding in large pieces with
some violence which suggests landslide movement.
These early reports establish landslide activity as a
recurring and important geomorphic process in this
part of Barbados.

Earthflows, slumps, and debris flows are the main
landslide types present in the Scotland District.
These failures occur on the upper, steep slopes within
the District and along steep slopes of deep v-shaped
valleys and sharp-crested ridges formed by stream
dissection (Prior and Ho, 1972, Peeters, 1963). These
landslides are commonly shallow failures within a few
meters to tens of meters of the ground surface. Prior
and Ho (1972) found shallow debris flows and slides on
slopes between 20 and 30 degrees. On the steeper
slopes immediately below the escarpment delineating
the Scotland District, some large slides have rafted
large blocks of the oceanic series and limestones
downslope (Peeters, 1963).

Landsliding in the Scotland District is directly
related to the response of clay present in the Joes
River Formation and Scotland Beds. Kaolinite, illite, chlorite, and montmorillinite clays are present
in the soils. The kaolinite is largely inherited from
the parent bedrock of the Joes River Formation and
Scotland Beds (Prior and Ho, 1972). Montmorillinite
content is very variable within these soils. Where
this clay is present in significant quantities, the
landslides tend to exhibit a greater tendancy for
flowage. Failures occur when the clay-rich soil
experiences a decrease in strength resulting from
increased pore-water pressure. Increased pore-water
pressure may be in direct response to intense rainfall. In other instances, pore-water pressure builds

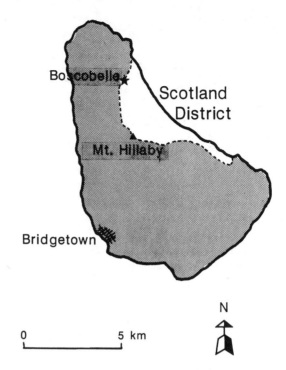

Figure 6. Location of the Scotland District in
Barbados.

on perched water tables associated with the bedding
within the bedrock or older clay-sealed slip surfaces.
Near the escarpment, percolation downward through the
limestone permits lateral seepage to produce higher
pore-water pressures during heavy rains. This mechanism apparently produced significant slope movement in
1966 after 406 millimeters of rainfall was received in
two days (Prior and Ho, 1972).

3.3 Economic impacts of landslides in Barbados

Economic impacts attributable to landslides are limited to the Scotland District. Little detailed information on the cost of landslide damage to infrastructure or agriculture is available. Cumberbatch (1966)
refers to recent catastrophic landslides devastating
villages and estates but provides no specifics. It is
inferred from a reference cited that considerable
damage occurred in the Boscobelle area on October 4,
1901.

Roads within the Scotland District incur damage
where undermined by failing slopes and blocked by
displaced slide material. Activation in early November 1987 of a small earthflow removed support from a
two-lane collector road near Welch Town. The entire
road prism along a section roughly 50 meter long was
displaced 3 to 4 meters vertically. This effectively
closed the road. Because repairs were not yet underway, the cost of restoring this road section is unknown.

Agriculture is the other primary economic impact.
Peeters (1963) notes some land is rendered untillable
by landslide activity. Ground cracking, scarps, and
gully development are some of the consequences of
landslides which renders land unuseable for farming.
Other locations are threatened by active landslides.
Concern is expressed for yet other areas where the
circumstances associated with active landslides are
present but no landsliding is evident. Soil conservation efforts include efforts to prevent pore-water
pressure increases through drainage techniques (Cumberbatch, 1966, Peeters, 1963). The cost of such
mitigation efforts are not described in published
reports.

4 WINDWARD ISLANDS

The Windward Islands extend south from 15° 45' to 11° 45' North latitude and from 60° 45' to 62° 00' West longitude (Rand McNally, 1988). They are west of Barbados and northwest of Trinidad and Tobago. From south to north, the islands are: Grenada, the Grenadines, St. Vincent, St. Lucia, Martinique, and Dominica. Martinique is an overseas department of France. The other islands are independant countries. The Grenadines are a string of small islands extending north from Grenada to St. Vincent and include Carricou, Union Island, Mustique, and Bequia. Except for Carricou which is part of Grenada, the Grenadines are part of St. Vincent and the Grenadines.

The Windward Islands lie within the trade wind belt. The rainy season occurs in summer and fall. Hurricanes and tropical storms pass over the islands during this period. However, the interior highlands of the larger islands also receive rainfall in the drier winter months and additional amounts in the summer months due to orographic uplift (Walsh, 1985). Annual rainfall in the interior highlands ranges from 10,000 millimeters in Dominica to 3750 millimeters for the lower elevation mountains in Grenada. Coastal areas receive lesser amounts ranging from 1000 millimeter on Dominica to 1600 millimeters at the southern end of St. Vincent.

4.1 Physiography and geology of the Windward Islands

Grenada is the least mountainous of the Windward Islands. Its more rugged terrain is concentrated in the central part of the island. St. Catharine at 840 meters is the highest point on Grenada (Rand McNally, 1988). Grenada and St. Vincent are nearly the same size at 344 and 345 square kilometers, respectively. In St. Vincent, the steepest terrain is found around Soufriere, a 1234-meter active volcano at the northern end of the island. The remaining steep terrain extends south through the central part of the island. On the 616-square kilometer island of St. Lucia, the central ridge hosts the steepest terrain with the north and south ends being flatter. The highest point on St. Lucia is 950-meter Mt. Gimie. At 1,100 square kilometers, Martinique is the largest of the Windward Islands. Rugged, steep terrain is found in the northern and central parts of Martinique. Mt. Pelee, at 1,397 meters, is the highest peak as well as an active volcano. A series of high peaks and connecting ridges runs the length of the 752-square kilometer island of Dominica. The highest peak is Morne Diablotin rising to 1,447 meters in the central part of the island.

The Windward Islands form a volcanic island arc. Lava flows, ash, and pyroclastic deposits ranging from Miocene to Recent in age are the principal bedrock found in these islands. Some limestones are interfingered with the volcanic layers (Fairbridge, 1975c). In general, Grenada, the Grenadines, and St. Vincent are composed of basalts and basaltic andesites. Typically, lava flows outcrop on the steeper slopes and ash underlies the gentler slopes (Walsh, 1985). St. Lucia, Martinique, and Dominica are predominately composed of acid andesite and dacitic rocks. Pyroclastic flow deposits, volcanoclastics, and lava domes, but few ash deposits, are typical for these islands.

4.2 Landsliding in the Windward Islands

Debris flows, debris slides, rockslides, rockfalls, slumps, and complex landslides are among the types of landslides found throughout the Windward Islands (Faugeres, 1966, Prior and Ho, 1972, Walsh, 1982, DeGraff, 1985, 1987a, 1988). Most landslides involve either flow or translational movement. Landslide mapping on St. Vincent, St. Lucia, and Dominica found the majority of landslides to be debris flows (DeGraff, 1985, 1987a, 1988). Table 4 characterizes the identified landslides on these three islands.

The Good Hope landslide on Dominica is one of the larger identified debris slides (DeGraff, 1987b)(Fig. 7). It involved translational movement of a soil mass. Slickensides were observed at some locations on the exposed failure plane. As it moved, the soil mass disintegrated into a disrupted mass with few discernable internal scarps or coherent blocks. Some flowage developed at the lower end of the slide mass.

From observations of exposed soil in the margins of the failure, the soil appeared residual in origin with no buried horizons, stone lines, or other indicators of significant colluvial accumulation. It is bright red and consists mainly of clay to sand-sized particles mixed with gravel-sized fragments. Some fragments could be crushed to sand by hand.

The failure plane coincided with the contact between the soil and underlying bedrock. Bedrock exposed in the failed area is a fractured, andesite. The failure plane, as defined by the surface of the bedrock, is inclined at 70 to 80 percent. The average thickness of soil over bedrock in the failed area is estimated to be 5 meters. The failed area is roughly circular in plan view and encompasses 3,630 square meters. These figures lead to an estimated volume of 17,000 cubic meters.

Debris slides and debris flows are common on mountainous slopes in the Windward Islands. Walsh (1985) noted the many debris slides and flows triggered by Hurricanes David and Frederic involved failure at depths of 2 meters or less. Similarly shallow depth to the failure plane for these landslide types is documented for Martinique (Faugeres, 1966). Prior and Ho (1972) found the majority of these slides on St. Lucia originated on slopes steeper than 35 percent.

Rockslides and rockfalls are widespread but less abundant. Bedrock escarpments on mountain slopes, steep-sided valleys, and coastal cliffs are typical localities for these landslide types. Failure usually involves a competent rock type which often has well-defined joints or similar discontinuities. Ignimbrite deposits along valleys near Rouseau, Dominica fail as rockslides and rockfalls where the prominent vertical joints create zones of weakness within the rock mass (DeGraff, 1987a). Failure of coastal cliffs occurs due to the oversteepening by wave erosion. Figure 8 shows a coastal rockslide near Dennery, St. Lucia.

Table 4. Number, size and area disturbed by past landslides on St. Vincent, St. Lucia, and Dominica.*

Island	Number of Landslides	Landslide Size (in hectares) Average	Largest	Landslide Density (per sq. km)	Terrain Disturbed (in percent)
St. Vincent	475	0.5	4.0	1.4	1
St. Lucia	430	3.0	5.0	0.7	2
Dominica	980	4.0	12.5	1.2	2

*From inventories of landslide compiled by aerial photo-interpretation and limited ground verfification (DeGraff, 1985; 1987a; 1988).

Figure 7. The Good Hope landslide in Dominica several months after failure. The Petit Soufriere-Castle Bruce road blocked by debris is visible across the mid-slope. The town is out of view in the valley to the lower left foreground.

Figure 8. Coastal rockslides in St. Lucia near the town of Dennery on the east coast. The margins of the larger, older rockslide and the smaller, more recent rockslde are highlighted by white lines.

The larger, older rockslide is visible on aerial photography taken in 1977. The younger rockslide within the older feature is seen only in 1981 aerial photography indicating the more recent failure occurred between 1977 and 1981 (DeGraff, 1985).

Slumps and complex landslides are the least common types found in the Windward Islands. Most slumps or rotational failures observed are associated with man-

disturbed slopes. In Dominica, small rotational failures triggered by Hurricanes David and Frederic were only noted on cultivated slopes (Walsh, 1982). Rotational failures seen in St. Vincent, St. Lucia, and Dominica are limited to small failures in road cuts. Prior and Ho (1972) describe complex slides involving shallow movement of the soil mantle at Moule a Chique in St. Lucia.

The steep slopes prevalent in the Windward Islands are one of the principal conditions favoring landslide development (Fig. 9). The land rises from sea level to 800 meters or more over a distance of 3 to 6 kilometers resulting in steep, rugged terrain. Examination of drainage patterns on these islands by Walsh (1985) indicates an early phase of landscape development. The resulting slopes are often at angles close to the angle of repose for the materials underlying them. Only small changes in stability conditions are required to bring such slopes close to failure.

The nature of materials underlying slopes on these islands plays a major role in landslide development. Their volcanic origin creates stratigraphic and lithologic conditions favoring landslides. Layers of alternating ash, lava, and breccia lead to locations on slopes where weaker bedrock weathers faster and undermines more competent bedrock overlying it. Location of volcanic vents in the central parts of the islands results in bedrock layers being inclined outward. The resulting presence of bedrock contacts inclined at angles less steep than the mountain slopes seems to favor landslides development. The bedrock and humid climate combine to cause deep weathering of volcanic bedrock. The resulting soil mass has a lower

Figure 9. Steep slopes in the central part of Dominica are representative of the rugged terrain typical of the Windward Islands. The margins of several recent debris flows visible on the slopes are highlighted in white.

strength than the original unweathered bedrock. Mehigan and Hartford (1985) recognized this situation for some slopes along roads in Dominica. In many instances, the soil contains significant percentages of clay. Prior and Ho (1972) found montmorillonite clay in soils tended to produce flow-type failures on St. Lucia. The clay characteristics of some soils on Dominica influence the intensity of rainfall needed to induce a landslide (Rouse et al, 1986, Rouse and Reading, 1987).

The principal triggering mechanism for landslides in the Windward Islands is rainfall. Pore-water pressure increases along discontinuities within weathered bedrock and within soil masses leads to decreased shear strength. This loss of strength coupled with the added weight of water within the saturated mass leads to failure on zones of weakness such as the soil-bedrock interface or discontinuities within the bedrock (Faugeres, 1966, Walsh, 1982). Hurricanes are one source of intense rainfall. Between September 1963 and September 1987, the following hurricanes induced landslides on one or more of the Windward Islands: Edith, Beulah, Abby, Dorothy, David, Frederic, Allen, and Emily. Storm events other than hurricanes are capable of inducing landslides. Prolonged rainfall experienced during the rainy season is capable of producing landslides as the Good Hope landslide demonstrates. In St. Vincent, Tropical Storm Danielle and associated rains in September 1986 caused more landslide damage than Hurricane Emily in September 1987.

Earthquakes and volcanic activity, recognized triggers of landslides, affect this region. All of the Windward Islands are known to have experienced earthquakes in the past. However, there is no documented instances of landslides being triggered by ground shaking. Volcanic activity is associated with landslides in the Windward Islands (Bolt, et al, 1975). Mudflows are described as part of the eruptive sequence of Mount Pelee leading up to the disasterous nuee ardante on May 8, 1902 which devastated St. Pierre, Martinique. Also in 1902, mudflows affecting the northeast and northwest parts of St. Vincent were among the events associated with the May 7th eruptive sequence of Soufriere which killed 1,500 inhabitants.

Human activities are another triggering mechanism for landslides in the Windward Islands. Roads cut into steep slopes remove support from the soil or rock mass above. The road prism may interfer with natural subsurface drainage in the slope leading to higher pore-water pressures than would occur under natural conditions. Anderson and Kneale (1985) note the terrain on these islands limits the available routes making avoidance of landslide-susceptible slopes difficult. This was the case for a major road in

Dominica constructed to improve access to the southern part of the island. The route crossed a slope segment with unfavorable bedrock lithology, groundwater conditions, and slope form. This slope, near Belvue Chopin, initially failed during road construction. It failed several times in the following years causing one death as well as impacting the road (Fig. 10). Despite stabilization efforts, it remains a potential site for future landslide activity.

Figure 10. Landslide at hairpin turn on road near Belvue Chopin in Dominica. The landslide was initially triggered by road construction. The four-foot high masonry wall visible at the toe of the landslide is part of earlier stabilization efforts.

Agricultural practices are another human activity contributing to landslides in the Windward Islands. Some farmers slash and burn the rainforest on very steep slopes to clear areas for planting bananas. The shallow-rooted banana plants are unlikely to contribute as much root reinforcement for strengthening the soil mass as the original trees of the rainforest. Rainfall reaching the ground is increased by replacing the solid rainforest canopy with the more open canopy typical of banana fields. It is suspected these differences make the slopes more susceptible to landslides. The consequences of altering the vegetation on a steep slope were demonstrated to a farmer in St. Lucia (The Weekend Voice, September 17, 1988). He maintained a banana field on a 90 to 100 percent slope cleared of forest vegetation. On September 11, 1988 following a tropical storm which drenched the island, he was an eyewitness to destruction of his field by a landslide. About 3:30 pm, the farmer observed the slope begin to move starting very slowly at the bottom and followed by the upper slope as it accelerated. Nearly 5 hectares including all of his field was carried away leaving a 1.5 kilometer long swath of exposed soil. Chief Forestry Officer Gabriel Charles and his follow rangers inspected the site and unanimously attributed the disaster to deforestation on a slope too steep to be used for banana cultivation. Even the farmer concluded, "I think it was the absence of trees with firm roots which caused the slide. I also think the squatting and the cutting of trees on the hills should be stopped..."

4.3 Economic and social impact of landslides in the Windward Islands

Landslides are known to cause fatalities and injury to the inhabitants of the Windward Islands. Both Dominica and St. Lucia provide examples of this serious landslide impact.

The unfortunate death of the nurse's daughter at Good Hope was not the first landslide-caused fatality on Dominica. Between 1925 and 1986, twenty-five Dominicans lost their lives to landslides (Fig. 11). These fatalities represent only those occurrences for

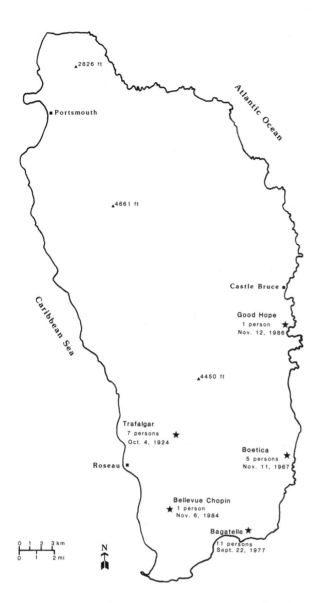

Figure 11. Location, date of occurrence, and number of fatalities due to landslides in Dominica.

which the place, time, and number of people involved could be ascertained. As such, this figure represents the minimum number of fatalities for this period. The nurse is the only injury due to a landslide which can be documented for Dominica. It is known her injuries required a recovery period of several months.

In most instances, fatalities on Dominica result from debris flows inundating homes. This was the case at the community of Bagatelle in 1977. Rain-saturated soil on a hillslope adjacent to the community mobilized into a rapidly moving mass. It engulfed four homes at the edge of the village killing eleven inhabitants. An exception to this circumstance is the death in 1984 at Belvue Chopin. Several landslides previously developed on the slope above the main road as noted in the earlier discussion of road construction triggering landslides. On November 6, 1984, another failure from this site sent a mass of debris down the road. This flow swept away a hapless pedestrian walking down the road during the storm.

A similar compilation of fatalities for St. Lucia has not been made. However, the greatest number of deaths at a single site due to landslide activity in the Windward Islands did occur on St. Lucia. The details of this event are known from newspaper and magazine accounts at the time.

In November 1938, the main road across the central highland of St. Lucia known as Barre de L'Isle, was blocked by landslides (Fig. 12). For eighteen days, landslides had prevented passage along this road connecting the capital of Castries with the southeastern part of the island. A workforce of several hundred inhabitants was present near Ravine Ecrivisse and neighboring Ravine Poisson laboring to clear the road. Normally, only a dozen families farming the local area lived near these ravines.

At 9:00 a.m. on Monday, November 21, 1938, a landslide from Ravine Ecrivisse swept into the area where the workmen were assembled (The West Indian, November 24, 1938). An hour later, a second landslide issued from neighboring Ravine Poisson. The areas engulfed by these landslides were described as "a sea of mud". The following morning at roughly 4:00 a.m., a third landslide covered an area one-half mile away. A total of sixty people are known to have died in these landslides. Another 32 persons were injured. It is unknown if all the injured survived. Estimates of missing workers ranged as high as 250 (Anonymous, 1938). An area of 10 square kilometers encompassing the ravines and vicinity was ordered evacuated and resulted in the displacement of over 500 people. In terms of loss of life, injury, and short-term disruption of people's lives due to evacuation, the landslide disaster at Ravines Ecrivisse and Poisson represents one of the worst experienced in the eastern Caribbean.

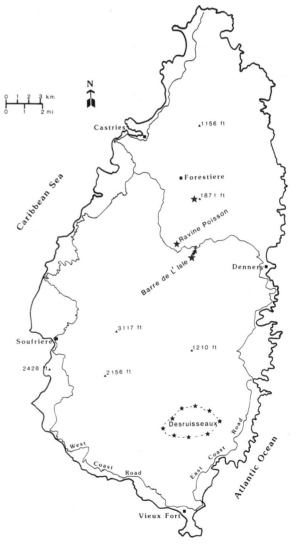

Figure 12. Location of main roads and notable landslide sites in St. Lucia.

Damage to facilities is the most common impact of landslides on the Windward Islands. Both structures and roads are destroyed or damaged by landslides. Roads blocked by landslide debris must be excavated to be returned to working order. Replacing or repairing facilities and removing debris are direct impacts. Landslides may cause indirect impacts resulting from the loss of use of damaged or destroyed facilities. Examples of damage to facilities are documented for Dominica, St. Lucia, and St. Vincent. It is assumed comparable damage has occurred on Martinique and Grenada.

Structural loss on Dominica is clearly represented by the destroyed health clinic and primary school at Good Hope. Both structures were constructed of concrete block on a poured slab foundation. The landslide completely demolished the clinic building during its passage. A few meters downslope, the primary school suffered irrepairable damage. The rear wall was shoved in by the impact of the slide mass. The entire structure was partially displaced from it foundation and the upslope side buried to the roof line with debris (Fig. 13).

Figure 13. Damaged primary school at the toe of the Good Hope landslide in Dominica. Landslide movement was from right to left across this view.

Pipelines for hydroelectric and water systems are structures suffering landslide damage on St. Vincent. Tropical Storm Danielle on September 8, 1986 triggered landslides which swept away a considerable length of the pipelines conveying water to hydroelectric stations. This affected the generating capacity at South Rivers station in the northeast part of the the island and Richmond station in the northwest. Altogether, the landslides reduced electrical generation capacity by 36 percent. The wood stave pipelines had to be fully repaired prior to restoring generating capacity some time later.

In May 1981, St. Vincent experienced a major storm during what is normally the dry season. Landslides occurring at three separate locations severed the 8-inch diameter pipeline for the Majorica water supply system. The damaged sections ranged from 5 to 20 feet in length. Nearly 40 percent of the population of St. Vincent was affected by this water system damage. Damage to the system left some inhabitants without water for a few days and others for nearly six months. Repair of the water line took six months due to the inaccessibility of the damaged sections and cost beween $87,000 and $130,000 (Central Water and Sewage Authority, Personal Comm., 1987).

The cost of clearing debris and repairing road damage to restore use is a common, serious economic impact. It is an anticipated cost when a major hurricane strikes. Figures on the cost of clearance and damage repair are avialable for Dominica, St. Lucia, and St. Vincent. It is assumed the costs for Martinque and Grenada are not dissimilar from that of their island neighbors.

In 1979, Hurricane David passed over the southern part of Dominica and was followed several days later by close passage of Hurricane Frederic. Landslide damage to roads was estimated to be $23,000 (CEPAL, 1979). Because landslides are triggered by storms other than hurricanes, slide clearance and road repair has a long-term cumulative economic impact. Between June 1983 and July 1987, over $462,000 was spent on Dominica on clearing landslide debris and associated repairs (Table 5). This represents an average annual expenditure of $121,000 (G. Elwin, Written Comm., 1987).

Table 5. Annual Cost of Landslide Damage to roads on Dominica.

Fiscal Year (June/July)	Landslide Costs* (in thousands of dollars)
1983-1984	92.8
1984-1985	269.0
1985-1986	71.7
1986-1987	63.0

*Includes only the cost of clearing landslide debris and road repairs (G. Elwin, Written Comm., 1987).

Similar costs are incurred on St. Lucia. The cost of just clearing landslide debris ranges from $38,000 to $146,000 per year which represents 2 to 6 percent of the annual road maintenance budget (J. Fevrier, Personal Comm., 1985). The range reflects differences between a normal year and a bad year with many triggering storms. In a normal year, $77,000 for repairs to roads would be added to the slide clearance figure. This represents expenditures for retaining walls, drainage, fill replacement, and similar work.

Records permitting tabulation of costs for clearing landslide debris and repairing damaged roads are not presently available for St. Vincent. Review of damage assessment reports made for Tropical Storm Danielle and the torrential rains which followed in September 1986, and for Hurricane Emily in September 1987 give some indication of the magnitude of this cost to roads. Totalling items noted as clearing of landslide debris or building of retaining walls from a district by district breakdown of road damage due to Tropical Storm Danielle yielded a total of $677,000. For Hurricane Emily, the amount was $191,000. Based on the cost of road clearance and damage from these two recent storms and the experience on neighboring islands, it is estimated the average annual cost of this impact on St. Vincent is $115,000.

Obviously, a very large landslide will result in unusually high repair costs. The Barre de L'Isle landslide, triggered by the August 3, 1980 passage of Hurricane Allen, illustrates the cost which may result from a major landslide. This landslide initially blocked the main east coast road on St. Lucia connecting Castries and Dennery. This severed the main route used to transport tourists arriving via the international airport at Vieux Fort to hotels and tourist facilities close to Castries. The landslide affected the main switchback curve on the east side of the ridgecrest. The upper part of the switchback was carried away while the slide debris came to rest on the lower switchback (Fig. 14). Clearance alone would not restore the road. A masonry retaining wall at the toe and three gabions structures within the failed area were constructed to stabilize the landslide. The gabions were placed at an average cost of $7.00 per cubic meter (Ministry of Communications, Works, and Transport, Written Comm., 1985). This figure includes excavation of the basket sites as well as assembling and filling of each basket with rock. It also includes the clearing of small landslides induced during site excavation. Work started in October 1980 and was completed by September 1982. The entire repair cost roughly $462,000. Of this total, forty-seven percent went for materials, twenty percent for labor, seven-

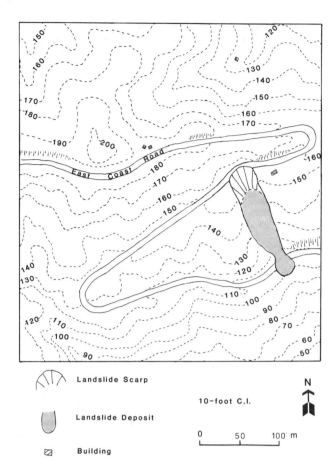

Figure 15. Several cubic yards of landslide debris on the main west coast road in St. Lucia between Castries and Soufriere. This October 22, 1985 failure shows how little material is needed to impact road use.

Landslide Scarp

Landslide Deposit

Building

10-foot C.I.

N

0   50   100 m

Figure 14. Map showing the 1980 Barre de L'Isle landslide on the main east coast road in St. Lucia. Shaded ares shows the limits of the landslide in relation to topography and cultural features.

teen percent for equipment operation, and sixteen percent for transportation. Comparable major repair sites in Dominica are found at Good Hope, Belvue Chopin and D'Leau Gommier and in St. Vincent at Wind-blow between the communities of Fancy and Owia (De Graff, 1987b, 1988).

Interference with road use does not require a major landslide like the one at Good Hope to pose a problem. Main arterial and collector roads are typically narrow and often winding on the the Windward Islands. Com-monly, the roads follow the narrow ridgetops to gain access to the more rugged interior. When a large landslide blocks a road, the transportation network may offer few alternative routes. The Castle Bruce-Petit Soufriere road blocked by the Good Hope land-slide deadends at Petit Soufriere. The people in that community have no choice but to walk around the blocked road or use a steep, temporary by-pass into Good Hope to reach Castle Bruce and other parts of Dominica. This difficult access imposes a double hardship on the local residents because the loss of the health clinic at Good Hope necessitates traveling another three miles to the nearest clinic at Castle Bruce.

It does not require a large landslide to block or interfere with traffic for short periods (Fig. 15). A landslide occurred in 1981 on St. Lucia's principal west coast road between Castries and Soufriere. About 765 cubic meters of debris from the cutslope late on Friday afternoon blocked the road. It was removed by Sunday (M. Henry, Personal Comm., 1985). During that time, people traveling to work between Castries and Soufriere or transporting perishable agricultural products for shipping from Castries were forced to wait or drive additional miles via the main east coast road to reach their destinations.

Agriculture is the main economic activity on most of the Windward Islands. Bananas are the principal export and cash crop throughout the year. Coconuts are nearly as important in terms of generating income. Other crops are grown for personal use and local markets. Landslide damage to banana plants and coco-nut trees represent an immediate loss of income for the small farmer. Replanted bananas can be bearing within a year. Coconut trees take up to twelve years before bearing fruit. This means the loss of income for the farmer persists for one to twelve years after the landslide occurs.

Catastrophic landslides carry away or bury both cash and food crops. Agricultural losses attributable wholly to landslides are difficult to determine. For example, landslides on St. Lucia in 1981 affecting crops around Desruisseaux and the neighboring villages of DeMailly and Belle Vue (Fig. 12). While their impact led to declaring this locality a disaster area, no figures on actual crop damage were gathered (G. Charles, Personal Comm., 1985). Often, figures gath-ered to show losses from a major storm represent wind, flood, and landslide-caused crop damage as a lump sum.

One of the more detailed descriptions of agricul-tural losses due to landsliding resulted from the Good Hope landslide. While these figures represent the agricultural losses for a single large landslide, they can be taken as indicative of the general losses sustained. The Good Hope landslide affected 3.8 hectares of cultivated land. This includes the land carried away by the slide and overridden by debris. Six farmers with holdings varying from 0.1 to 1.2 hectares were affected. Bananas and coconut trees represent the principal revenue-generating crops. Other tree crops lost were bayleaf, mango, breadfruit, and citrus. Root crops for personal use and sale of surplus harvest in the markets were also destroyed. Based on Ministry of Agriculture estimates, total agricultural losses were $8,000. Of this total, $5,000 represents the loss for bananas and coconuts.

Landslides smaller than the one at Good Hope occur more often, but still inflict significant losses. In November 1981, a single, moderate-sized landslide occurred on St. Lucia southeast of the community of Forestiere (Fig. 12). Starting near the top of the ridge next to Piton Flore, this debris flow carried away bananas and coconut trees. As the slide mass moved downslope, it swept away or buried banana and coconut trees owned by other farmers. This 1-hectare debris flow resulted in a loss of standing crop and near-future production valued at $4,000 (G. Charles, Personal Comm., 1985). On St. Vincent, a small debris flow occurred November 22, 1987 in a field near South Rivers. The narrow path of the flow swept five grown banana plants from the field and deposited them on the road at the base of the slope. Assuming the same value for individual banana plants used in assessing the impact of the Good Hope landslide, the loss attributable to this small debris flow is roughly $250. Because farm holdings are typically small in the Windward Islands and per capita income is between $1000 and $500, the economic hardship due to landslides illustrated by these examples is not uncommon.

5 LEEWARD ISLANDS

The Leeward Islands are found north of the Windward Islands extending to just east of the Virgin Islands between latitudes 15° 45'N to 18° 35'N and longitudes 61° 40'W and 63° 20'W (Rand McNally, 1988). St. Christopher (St. Kitts)-Nevis and Antigua-Barbuda are former British colonies which are now independent countries. Anguilla maintains an associated state status with Britian. While Montserrat remains a British colony. St. Eustace, Saba, and St. Maarten (southeastern part of St. Martin) are colonies of the Netherlands. The remainder of St. Martin and St. Barthelemny are colonies of France. Guadeloupe is an overseas department of France.

These islands lie within the trade winds belt resulting in a subtropical climate. Islands with sufficient relief receive an adequate rainfall, but those with a more subdued topography tend to be dry to semi-arid. The main hurricane track passes through these islands.

5.1 Geology of the Leeward Islands

The Leeward Islands exhibit two geologically distinct belts (Fink and Fairbridge, 1975). The island of Guadeloupe marks the southern end of the two belts and embodies characteristics of both. Guadeloupe consists of two distinct parts. Basse Terre, the western half, is dominated by Soufriere, a 1,467-meter high active volcano. To the north, the inner belt or arc of islands is also volcanic in origin. These include Montserrat, Nevis, St. Christopher, St. Eustace, and Saba. An active volcano also named Soufriere, like the volcano on Guadeloupe, is found on the island of Nevis. Andesitic flows, pyroclastic units, and volcanoclastics of recent to Eocene age dominate this belt. These volcanics are interbedded with Pliocene and Pleistocene limestones on some islands, notably on St. Eustace, St. Christopher, and Montserrat. The small cluster of volcanic islands between Guadeloupe and Dominica called the Iles d'Saints physically represents a connection of this volcanic belt to the Windward Islands.

The eastern half of Guadeloupe is called Grande Terre and is composed entirely of limestone overlying older andesitic and dacitic volcanics. Forminiferal or oolitic limestone underlies the remaining islands in the outer belt or arc of islands including Guadeloupes's offshore islands of Marie Galante and Desirade, Antigua, Barbuda, St. Barthelemy, St. Martin (St. Maarten), and Anguilla.

The inner volcanic islands tend to have higher, more rugged topography. Higher peaks include 1,467-meter Soufriere on the Basse Terre part of Guadeloupe and 1,156-meter Mt. Misery on St. Christopher. The limestone islands tend to be weathered to a topography consisting of low hills. The higher points on these islands are represented by 70-meter high Crocus Hill on Anguilla and 402-meter high Boggy Peak on Antigua.

5.2 Landslides in the Leeward Islands

Published literature on landslide occurrence in the Leeward Islands could not be found. The physical characterisitics of the islands and known landslide activity in the Windward Islands serves as a basis for inferred landslide conditions.

Small rockfall or rockslides along coastal escarpments are likely the only landslide activity present in the outer belt or limestone-dominated islands.

Their bedrock and topography are not conducive to extensive landsliding. Minor failures may result from road construction on the steeper hillslopes on some islands. A brief reconnaissance on Antigua by the senior author in 1987 failed to identify any landslides outside the coastal cliff areas. Landsliding on the outer belt of the Leeward Islands seems to be infrequent, of limited size and areal extent, and of little economic significance.

Debris flows and slides are expected on the steeper slopes underlain by volcanic bedrock on the inner islands. These conditions are comparable to those associated with landslide activity on Dominica and St. Lucia. The Basse Terre part of Guadeloupe, Montserrat, St. Christopher, and Nevis should experience a moderate amount of rainfall-triggered debris flow and debris slide activity during the wet season. In 1987, the senior author observed some fresh-appearing debris flows during aerial overflight of Basse Terre and a large rockslide on the easternmost of the Iles d'Saints. Landsliding on the inner belt of Leeward Islands is likely not unusual and may be a common occurrence. Small debris flows and slides could be locally significant on these islands. However, affected areas may be largely confined to undeveloped or little developed areas. Their effect on roads and agriculture probably represents a small, but significant impact to the inhabitants.

6 VIRGIN ISLANDS

Between latitudes 17° 40' N and 18° 50'N and longitudes 64° 75'W and 65° 10'W are a group of small islands and cays called the Virgin Islands (Weaver, 1975). These islands are divided politically into the British Virgin Islands to the north and the U.S. Virgin Islands to the south. The British Virgin Islands encompass 153 square kilometers. Totola, Virgin Gorda, and Anegada are the larger islands. At 527 meters, Mt. Sage on Tortola is the highest point in the British Virgin Islands (Rand McNally, 1988). There are 344 square kilometers in the U.S. Virgin Islands. St. Thomas, St. John, and St. Croix account for most of this area. The highest point in the U.S. Virgin Islands is Crown Mountain on St. Thomas at an elevation of 474 meters. In general, the islands are rugged with steep slopes (Weaver, 1975). Anegada is an exception being low and flat. A relatively low and flat area called the "Valley" on the southern part of Virgin Gorda and a central lowland on St. Croix represent the other significant areas of gentler terrain.

The islands are subtropical with rainfall varying between 1000 and 1300 millimeters. While dense tropical vegetation can be found in some locations, cactus and thornbush typifying drier conditions is more common.

6.1 Geology of the Virgin Islands

Weaver (1975) notes all of the Virgin Islands with the exception of St. Croix rise from the Virgin Island Bank. The relationship of this bank to the insular shelf around Puerto Rico and other structural relationships indicates the Virgin Islands are the eastern end of the Greater Antilles. Submarine lava flows, spilitic and keratophyric breccias, andesitic pyroclasitcs, andesitic breccia tuffs, and vocanic sandstones account for the majority of exposed bedrock in the Virgin Islands (Weaver, 1975). These rocks are folded homoclinally at angles averaging 40 degrees to the north in the U.S. Virgin Islands and steeper to overturned in the British. In contrast, most of the bedrock exposed on Virgin Gorda is tonalite and granodiorite from a large batholitic intrusion. Tortola and St. John have less extensive exposures of this bedrock. Limestone is found exposed in many of the islands as blocks within primarily volcanic units and as Pleistocene accumulations. On St. Croix, the Northside Range and the East End Range consist of

Cretaceous tuffaceous and volcanoclastic sediments. These rocks are both folded and faulted. A gabbro intrusion and a diorite intrusion are found in the Northside and East End Ranges, respectively. A low graben filled with Tertiary sediments separates these two mountain ranges. The bedrock geology of Anegada is markedly different from the rest of the Virgin Islands. It consists of flat-lying limestone of Pleistocene age. It is assumed this limestone platform rests on a basement of older volcanics.

6.2 Landslides in the Virgin Islands

Almost no published literature on landslides is available for the Virgin Islands. A reconnaissance of landslide potential on St. Thomas (Brabb, 1984) provides some insight on landslide activity in these islands and their social and economic impact.

Earthflows, debris slides, and individual boulders are recognized landslide types on St. Thomas. Debris flows are not documented or reported as occurring on this island. The largest landslide documented on St. Thomas is 600 meters long and 600 meters wide. It was mapped in an area about 1.5 kilometers north of Charlotte Amalie in 1979.

On April 18, 1983, a storm drenched Dorothea Bay with nearly 400 millimeters in 14 hours. It illustrates the conditions leading to landslide activity which can be found on St. Thomas. In addition to extensive flooding, this storm event produced a number of landslides. Two earthflows developed in weathered colluvium. These are small features about 30 meters long and 30 meters wide. Very small debris slides occurred in colluvium exposed at the top of some road cuts. Boulders temporarily blocked several roads. One boulder which was 6 meters in maximum diameter traveled 10 meters downslope before stoping next to and above a house (Brabb, 1984).

Assuming the landslide activity on St. Thomas is representative of landsliding in the Virgin Islands, it appears landslide activity is limited in magnitude and areal extent. From an economic standpoint, it involves an additional cost to road maintenance that is minor compared to other storm damage to roads. The threat of fatalities or injuries exists for those areas with habitations downslope from large boulder accumulations. Based on present information, it

appears even this potential would only be realized under an unusually large storm event.

7 PUERTO RICO

The Commonwealth of Puerto Rico, a possession of the United States, lies about 1700 kilometers southeast of the U.S. mainland at the eastern extremity of the Greater Antilles. Puerto Rico has an area of 9103 square kilometers, about 70 to 80 percent of which is hilly or mountainous. Approximately 60 percent of the population of 3.35 million lives in the four largest cities, San Juan, Ponce, Mayagüez, and Arecibo (Rand McNally, 1986), which are located primarily on flat or gently sloping coastal areas. Continuing growth of these urban centers, however, is pushing development onto surrounding steep slopes. The overall population density of Puerto Rico is 368 persons per square kilometer; the density outside the four major urban areas, including the mountainous interior, is probably between 100 and 200 persons per square kilometer. Average annual precipitation in Puerto Rico ranges from less than 1000 millimeters, along the southern coast, to more than 4000 millimeters, in the rain forest of the Sierra de Luquillo on the northeastern part of the island (Fig. 16). The mountainous Cordillera Central, which forms the spine of Puerto Rico, also receives considerable rainfall. Rain in Puerto Rico falls throughout the year, but about twice as much rain falls each month from May to October--the hurricane season--as falls from November to April (Monroe, 1980). Brief, intense storms are common; almost one-half of the National Weather Service's 40 weather stations record more than 13 millimeters of rainfall on 30 to 50 days each year (Calvesbert, 1970; Monroe, 1980).

Four hurricanes have passed over Puerto Rico in this century, and several others have passed close enough to produce extraordinary rainfall (Calvesbert, 1970). In October 1985, a tropical wave, which later developed into Tropical Storm Isabel, struck the south-central coast of Puerto Rico and produced extreme rainfalls. Some areas received more than 560 millimeters of rain--more than one-half the annual average--in 24 hours. Rainfall intensities reached 70 millimeters/hour for a few hours (Jibson, 1986a; 1987a; in press).

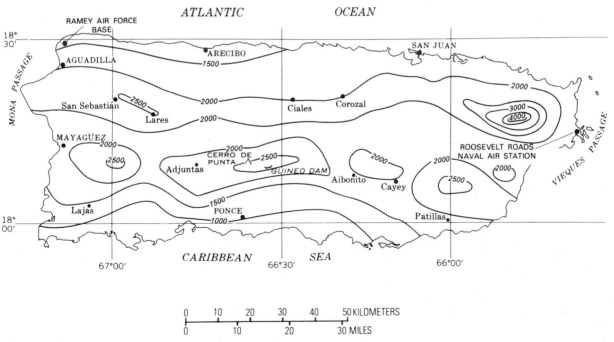

Figure 16. Annual rainfall in Puerto Rico in millimeters; data from U.S. National Weather Service (from Monroe, 1980).

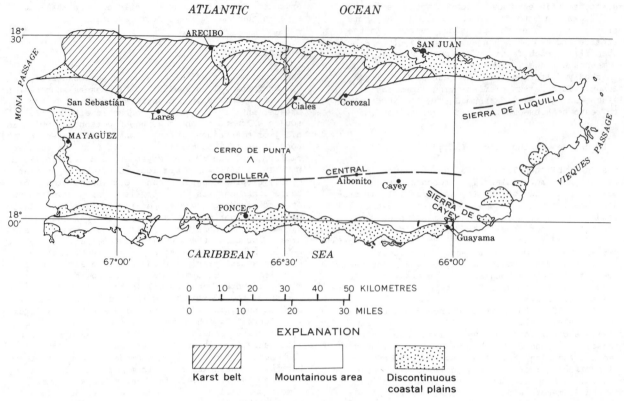

ATLANTIC   OCEAN

EXPLANATION

Karst belt   Mountainous area   Discontinuous coastal plains

Figure 17. Physiographic provinces of Puerto Rico (from Monroe, 1976).

The temperature in Puerto Rico is generally between 20 and 30°C, and the relative humidity is about 80 percent throughout the year (Monroe, 1980). The warm, moist air combined with the heavy rainfall promotes rapid, deep weathering of surficial materials, which leads to severe erosion and slope-stability problems throughout the island.

7.1 Physiography and geology of Puerto Rico

Puerto Rico can be divided into three distinct physiographic provinces (Fig. 17)--the Upland province, the Northern Karst province, and the Coastal Plains province (Monroe, 1980). The physiography and the geology of these provinces as described by Monroe (1976; 1980) are discussed below.

The Upland province, which has a maximum elevation of 1338 meters, includes the three major mountain ranges--the Cordillera Central, which forms an east-west axis through most of the southern two-thirds of the island; the Sierra de Luquillo in the northeast; and the Sierra de Cayey in the southeast. The mountainous areas are covered by dense, tropical vegetation and have been deeply incised and sculpted by streams and landslides; slopes as steep as 45° are common. Also included in the Upland province are interior lowlands surrounded by mountains and foothills to the mountains, primarily along the western, southern, and eastern flanks.

The Cordillera Central and Sierra de Luquillo consist of Lower Cretaceous to middle Eocene volcanic and sedimentary rocks (Fig. 18, TKv). The Lower Cretaceous rocks are primarily submarine volcanic ash and interlayered lava flows; reef limestones are locally present near the top of the Lower Cretaceous section. The Upper Cretaceous section includes volcanic rocks interbedded with volcaniclastic sandstone and conglomerate as well as with reef limestones. Overlying the Cretaceous rocks are Paleocene to middle Eocene tuff and sedimentary rocks. The Cretaceous and lower Tertiary rocks have been extensively folded and faulted.

The Cretaceous and Tertiary volcanic and sedimentary rocks in the interior of the island are intruded by granodiorite and diorite plutons (Fig. 18, TKi). The largest of these plutonic intrusions forms the Sierra de Cayey (Fig. 17). In west-central Puerto Rico, Cretaceous serpentine and chert are exposed (Fig. 18, Ks).

The foothills along the southern flank of the Upland province extend locally to the Caribbean Sea; they consist of Oligocene and Miocene sediments unconformably overlying the Cretaceous and lower Tertiary rocks of the mountainous interior. The Oligocene sediments (Fig. 18, To) include conglomerate, sand, and clay derived from weathering, erosion, and marine reworking of the older igneous rocks. The clastic sediments are capped by Oligocene limestone, which is overlain by several hundred meters of Miocene limestone (Fig. 18, Tm). Near-vertical limestone cliffs above steep colluvium and residuum-covered slopes are common.

The Northern Karst province (Fig. 17) includes most of north-central and northwestern Puerto Rico north of the Upland province; locally, this province extends to the Atlantic Ocean. The maximum elevation in the Northern Karst province is 530 meters along the escarpment formed by differential weathering between the igneous rocks of the Upland province and the limestone of the karst area. The Northern Karst province displays a wide variety of karst features (Monroe, 1976). The terrain, locally very rugged, ranges from steep slopes and shear cliffs to gently rolling hills. Most drainage is underground, which results in widespread collapse features.

Rocks in the Northern Karst province are Oligocene and Miocene in age. The Oligocene section (Fig. 18, To) consists of clays, sands, and gravels (all derived from the igneous rock of the mountainous areas) capped by limestone. The Miocene section (Fig. 18, Tm) consists primarily of as much as 1400 m of limestone. The weak, saturated clays of the Oligocene San Sebastian Formation underlie much of the limestone in the area, and large landslide blocks of limestone have moved downward and outward along these north-dipping

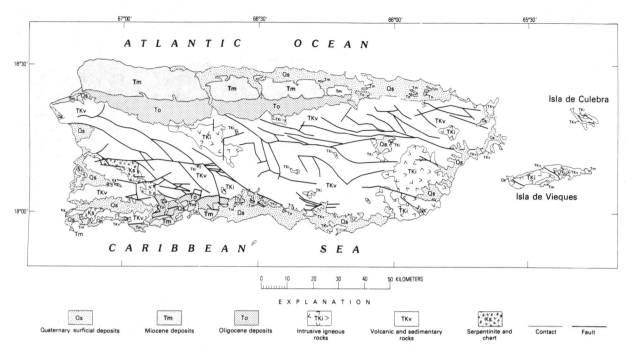

Figure 18. Generalized geologic map of Puerto Rico (from Monroe, 1980).

clays (Monroe, 1964).

The Coastal Plains province (Fig. 17) is a discontinuous, gently sloping area covered by Quaternary alluvial, estuarine, and beach deposits, which have been reworked locally by wind and wave action (Fig. 18, Qs). Inselbergs of older rocks project through the coastal plain sediments in some places. The only consolidated rock among these sediments is eolianite forming from cementation of beach sediments. Puerto Rico's major cities are built primarily in the Coastal Plain province, although population growth has pushed development onto adjacent slopes of the Upland and Northern Karst provinces.

7.2 Available literature on landslides in Puerto Rico

Geologic maps at 1:20,000 scale exist for most of Puerto Rico, and most of these maps include at least some reference to landslides. Several of the maps include landslide deposits as surficial map units.

Monroe (1979) produced a map of landslide susceptibility of Puerto Rico at 1:240,000 scale. Four categories were used to depict susceptibility--highest susceptibility was assigned to areas of past or current landslide activity, high susceptibility to areas that have slopes greater than 27° and to areas that contain "slide-prone" rock and soil, moderate susceptibility to all other sloping areas, and low susceptibility to flat-lying areas. Judgment of high and highest susceptibility was based on landslide incidence, so the map actually is a generalization of landslide incidence. Susceptibility to specific types of landslides, such as rapid debris flows, was not discussed; recent experience has shown that these types of landslides are among the most common and damaging in Puerto Rico. The map is generalized and shows large areas that have similar susceptibility to landsliding; its scale precludes the detailed quantification of susceptibility needed by the planning and engineering communities.

Monroe (1964) described large, retrogressive block slides along the northern limestone escarpment that separates the mountainous interior of the island from the northern karst area. Other landslide features associated with the San Sebastian Formation in northwestern Puerto Rico are described by Briggs and others (1970). Sowers (1971) described slope stability

problems encountered in construction projects in the weathered volcanic rocks in the rain forests of northeastern Puerto Rico. Deere and Patton (1971) discussed the relation of weathering profiles to landslide formation in Puerto Rico. Brief discussions of other types of landslide processes have been presented by Farquhar (1978), Lewis (1975), and Ortiz (1974).

A major tropical storm in October 1985 triggered thousands of debris flows as well as the disastrous rock block slide that destroyed the Mameyes district of Ponce. Landslide activity triggered by this storm was documented by Campbell and others (1985), Campbell and others (1986), and Jibson (1986a,1986b, 1986c, 1987a, 1987b); a description and analysis of the Mameyes landslide is given by Jibson (in press). Jibson (1987b) also briefly summarized landslide hazards throughout Puerto Rico.

7.3 Landslides in Puerto Rico

All major types of landslides occur in Puerto Rico, and all physiographic provinces of the island have landslides. Most of the Upland province and the Northern Karst province (Fig. 17), by virtue of high relief, steep slopes, and abundant rainfall, have continuing landslide problems. The drier southwestern part of Puerto Rico (Fig. 16) normally experiences landslides only during exceptionally heavy rainfalls; few landslides form there during periods of normal precipitation. The coastal plain has only localized landslide problems, predominantly along incised stream channels or where steeper inselbergs project through the younger surficial sediments.

Debris slides and debris flows--rapid downslope sliding or flowing of disrupted surficial rock and soil--are the most prevalent types of landslides in Puerto Rico. These landslides are particularly hazardous because they form with little or no warning and can move very rapidly down steep slopes. Structures at the base of such slopes are inundated or destroyed by the impact of the rapidly moving mixture of soil, rock, and water.

A common, but less abundant, type of landslide is rock fall--rapid movement by free fall, bounding, or rolling of bedrock detached from steep slopes. Rock falls are common on very steep natural slopes and especially on the numerous steep road cuts on the island. These landslides can be very damaging if they impact structures or passing automobiles. Recent major storms have triggered many rock falls of different sizes that closed roads and temporarily isolated parts of the island.

Block slides and slumps--masses of bedrock and overlying soil that move downslope either as intact blocks or as a collection of slightly disrupted blocks--are less common than debris slides and debris flows, but their effects can be catastrophic. Such was the case during the October 1985 storm, when the Mameyes district of Ponce was destroyed by a block slide that killed at least 129 people (Jibson, 1986a, in press). Block slides and slumps can disrupt large areas of the ground surface and thus lead to destruction of overlying structures and burial of structures downslope.

Earth flows--normally slow-moving masses of moderately disrupted earth that can move down even very gentle slopes--also occur in Puerto Rico. This movement commonly causes sufficient deformation of the ground surface to damage or destroy overlying structures or roads.

The igneous rocks in the Upland province weather rapidly to form a deep, predominantly coarse-grained saprolitic soil mantle. When saturated, this saprolite produces debris slides, debris flows, and slumps ranging from a few to several hundred meters long. In May 1985, severe storms in west-central Puerto Rico triggered hundreds of such debris slides and debris flows, which choked streams, blocked roads, and destroyed homes and other structures. Hurricanes David and Frederic in 1979 produced extreme rainfalls in northeastern Puerto Rico that triggered several debris slides as long as 750 meters and as deep as 25 meters on slopes of deeply weathered intrusive igneous rock (Fig. 19). Rock falls from natural slopes and particularly from the ubiquitous steep road cuts also are common in the igneous and sedimentary rocks of the Upland province.

Figure 19. Large debris slide in the Sierra de Luquillo triggered by heavy rainfall associated with Hurricanes David and Frederic in 1979. This slide is about 750 meters long and as deep as 25 meters.

The sedimentary rocks flanking the igneous interior of the island produce rock falls from steep cliffs and road cuts, large rock and debris slumps, and rock block slides. The deeper landslides form where limestone or sandstone overlie weaker silts and clays that act as slip surfaces. The colluvium and residuum on steep slopes composed of mudstone and particularly limestone are susceptible to failure as debris slides and debris flows; such landslides generally fail at the interface between the weathered surface material, commonly 0.5 to 2 meters deep, and the unweathered bedrock.

During the storm of October 1985, numerous debris slides and debris flows (Fig. 20) destroyed several homes and buildings between Penuelas and Coamo along the southern coast of Puerto Rico (Jibson, 1986a).

Figure 20. Debris flows at Tallaboa, between Penuelas and Ponce, triggered by the October 1985 storm.

Figure 21. Source area of debris flow near Penuelas (west of Ponce). The light-colored area is limestone bedrock; a 1-meter thick mat of colluvium and residuum on this bedrock failed and mobilized into a debris flow. The failed area is about 5 meters across on a 40o slope. A few meters upslope from the source area is a 70o bedrock face; sheet wash from this face flowed onto the top of the slope that failed and elevated pore-water pressures in the soil mat.

Figure 22. Deeply scoured debris-flow channel near Penuelas (west of Ponce).

These landslides originated as failures of thin colluvial soil mats on steep slopes (Fig. 21); the failed material scoured deep channels as it moved downslope in preexisting gullies or depressions (Fig. 22). Most of the 1985 debris slides and debris flows formed on limestone slopes underlain by a wide range of geologic structures (dip slopes and reverse-dip slopes, frac-

tured and unfractured rock, folded and homoclinal bedding, and so forth). The limestone might produce weak or permeable residuum and colluvium more susceptible to failure, and ground-water flow patterns in the limestone might favor landslide formation. Many mudstone slopes also produced debris slides and flows in 1985.

Rock falls from steep limestone slopes, in the southern flank of the Upland province and in the Northern Karst province, have repeatedly blocked major and minor roads. Most of the limestone in Puerto Rico is porous and fractured, so infiltration of rainfall can build up pore pressure within the rock and trigger rock falls. Limestone rock falls have triggered debris flows on steep colluvial slopes below near-vertical bedrock faces (Fig. 23). At one site, large limestone boulders fell from a bedrock face and impacted the head of the colluvial slope; this either disrupted the saturated colluvium enough to cause it to flow downslope, or rapidly increased the pore pressure in the colluvium and caused it to mobilize.

Large, retrogressive block slides are present in the Northern Karst province where thick limestone formations overlay clay in northwestern Puerto Rico (Monroe, 1964). These generally slow-moving landslides occur where deep river valleys expose the underlying clay beds on which sliding occurs. The block slides leave deep, vertical sided valleys between the head scarp and the upper edge of the landslide block.

Some clays in the sedimentary belt produce earth flows. The San Sebastian area, in northwestern Puerto Rico, is particularly susceptible to earth flows that form in the clayey members of the Oligocene San Sebas-

Figure 23. Rock-fall scarp above debris-flow source area west of Ponce. Several large limestone boulders fell from the bedrock face onto the top of the colluvial slope and triggered a debris flow by producing a large, instantaneous increase in pore pressures leading to flow and (or) by disaggregating the colluvium which mobilized into a flow.

tian Formation. These earth flows commonly form on gentle slopes and create a subdued, hummocky topography. Several houses and roads have been damaged severely or destroyed by these landslides (Fig. 24).

The young sediments forming the coastal plain surrounding Puerto Rico have very gentle slopes. Landsliding on the coastal plain is generally limited to small river-bank failures that choke river channels and increase erosion, which endangers homes built on the coastal plain. Steep-sided inselbergs of Tertiary and Cretaceous sedimentary and igneous rocks, which project through the young sediments, also produce landslides. If these sediments are subjected to seismic shaking, then lateral-spread landslides could form in parts of the coastal plain underlain by liquefiable sediments.

7.4 Social and economic impact of landslides in Puerto Rico

Statistics on fatalities and damage from landslides in Puerto Rico are not systematically recorded; therefore, estimating economic losses is difficult. The Mameyes landslide of October 1985 is definitely the worst landslide disaster in Puerto Rico's history (and in North American history); at least 129 people were killed, and the death toll could have been between 200 and 300. More than 100 homes were destroyed, and about as many were later condemned and removed because of continuing risk from landsliding (Fig. 25).

Confirmed reports of fatalities from other landslides are sparse. Anthony Santos (local resident, personal commun., 1985) stated that two motorists were killed by a limestone rock fall in the early 1980's near Penuelas, west of Ponce. Monroe (1979) reported that motorists have been killed by rock falls along the Rio de la Plata between Comerio and Bayamon; he also reported several instances when numerous houses were damaged or destroyed by landslides. Isolated landslides, particularly debris flows and debris

slides, reportedly have caused fatalities and destroyed property. The greatest cost to public works is road maintenance, but no estimates of total cost due to landslides are available. The frequency of serious storms in Puerto Rico suggests that a long-term average of a few (less than 5) fatalities per year could occur. On average, perhaps tens of houses are destroyed, and hundreds are damaged by landslides each year.

The storms of May and October 1985 increased public and government awareness of landslide hazards and the need to mitigate those hazards. The Puerto Rico Department of Natural Resources and the Puerto Rico Planning Board are cooperating with the U.S. Geological Survey on projects to develop landslide susceptibility maps for parts of the island. In the wake of the Mameyes landslide disaster, the Commonwealth

Figure 24. House in west-central Puerto Rico built on an intermittently active, slow-moving earthflow. Landslide movement has so deformed the house that it had to be abandoned.

Figure 25. Oblique aerial photograph showing the Mameyes landslide, on the outskirts of Ponce. The light-colored areas are severely disrupted areas; the dark, vegetated areas in the center of the landslide are the more intact slide blocks.

government, using partial funding by the U.S. Federal Emergency Management Agency, relocated hundreds of people to a new community built on stable ground. Those relocated included those whose homes had been destroyed by the landslide of October 7 and those in adjacent areas where the level of landslide hazard was judged to be unacceptably high. The ongoing concern about landslide hazards within local and Commonwealth governments could lead to implementation of policies and programs to significantly reduce landslide risk.

8 HISPANIOLA (HAITI AND DOMINICAN REPUBLIC)

The island of Hispaniola is located in the Greater Antilles between Cuba and Jamica on the West and Puerto Rico on the east. It lies between 17° and 20° North latitude and 68° and 74° West longitude. Hispaniola has an area of 76,192 square kilometers, about 75 percent of which is mountainous (Rand Mc-Nally, 1988).

Rainfall averages vary from 500 to 3000 millimeters per year. Most of the rain falls during one of the two rainy seasons, April-June or August-October. The heaviest rains are normally recorded from September to November. At least once every five years, precipitation averages as much as 400 millimeters per month and includes an intensity as high as 150 millimeters per 24 hours (Alford, 1986; Mora, 1986b). The passage of tropical storms can cause considerable variation in precipitation (Fig. 26). A notable example is the passage of Hurricanes David and Frederic during the period of August 30 and September 3, 1979. In the Dominican Republic, these two hurricanes resulted in more than 700 millimeters of rain in five days.

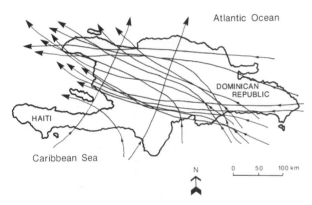

Figure 26. Paths across the island of Hispaniola taken by major tropical storms between 1886 and 1986.

8.1 Physiography and geology in Hispaniola

There are three major systems of mountains on the island of Hispaniola. To the north, there is the Cordillera Setentrional. The Cordillera Centrale-Massif Du Nord is bounded by the Cibao-Cap Haiten valley on the north and the Enriquillo-Cul-de-Sac valley to the south. The third mountain range is the Massif Du Sud in the southwestern corner of Hispaniola. The highest point on the island is found in the Cordillera Centrale. This peak, Pico Duarte in the Dominican Republic, rises to 3175 meters. The steep slopes common in these mountains are frequently subject to landslide activity.

Hispaniola has a complex geologic history tied to regional plate tectonics. Major faults generally trending northwest to southeast with some defining the margins of the three major mountain systems. The Cordillera Septentrional is composed of folded and faulted sedimentary bedrock of Paleocene to Pliocene in age (Choubert and Faure-Muret, 1976). Uplifted deposits of unconsolidated alluvium are among the units found in this mountain range. The Cordillera

Centrale-Massif Du Nord mountains are more complex with Cretaceous-age sedimentary and metamorphic bedrock and intrusives. Schist and phyllites are rock types common to the metamorphic bedrock in the Cordillera Centrale. Sedimentary bedrock includes sandstone, limestone, and siltstone. Volcanic units and Paleocene to Oligocene-age sedimentary bedrock are found in the Massif Du Sud.

8.2 Landslides in Hispaniola

Unstable slopes are a common feature on this mountainous island. While natural conditions favor landslide development, much of this activity can be attributed to human abuse of the landscape. As a result, landsliding is a major element in the very high erosion rates found on Hispanola. The detrimental consequences of landslide activity include the additional economic loss represented by long-term reduction in soil productivity and desertification in the extreme case. A significant part of the population, nearly 40 percent, live in the mountains. It is paradoxical that they should both bear some responsibility for the extreme levels of landslide activity and suffer the consequences as the victims of landslide events and erosional loss.

Published information on slope instability is virtually non-existent for Hispaniola. A number of publications provide some information of geologic conditions which can be used to identify instability prone units (CNRS, 1985, Antonini, 1986, Case and Holcomb, 1980, INDRHI, 1983, Woodring and others, 1924). Some general information on landslides and their economic and social impact is included in reports addressing more general topics such as erosion problems or regional development (OEA-DDR, 1972, NOAA-CEAS, 1979, Garvey, 1982, BDPA, 1983, Alford, 1986, and Mora, 1986a). Post-disaster reports provide some information on landslide impacts. Documented disasters with data on landslides include: December 29, 1897 Santiago earthquake, September 27, 1909 San Severo storm event, August 4, 1946 Cibao earthquake, October 12, 1954 passage of Hurricane Hazel, October 3, 1963 passage of Hurricane Flora, and the August 31-September 1, 1979 passages of Hurricanes David and Frederic (CEPAL-ONAPLAN, 1979, De La Fuente, 1976, L.G.L., 1981, Mora 1986a, 1986b, Nanita, 1982).

Landslide activity is concentrated in the mountainous areas of the islands (Fig. 27). The coincidence of steep slopes and lithologies which are landslide-prone produce this general pattern of slope instability. Most landslides occur in residual material regardless of the age of the parent material. They also occur in weathered bedrock where structures present contribute to their instability. Calcareous bedrock is the lithology least often associated with slope instability. In the central range, a slope instability develops on the schists and phyllites and intrusives. Landslides are found in areas with exposures of Tertiary-Neogene sandstone and siltstone and uplifted Pliocene-Pleistocene deposits of unconsolidated alluvium.

Land use is another significant factor in landslide development in Hispaniola. Deforestation and other misuses of steep slopes produces both landslides and severe fluvial erosion. The hydroelectric reservoir at Peligre (Morne la Yaille) illustrates the consequences of improper land use. The denuded hills around the reservoir are underlain by calcareous and clastic sedimentary rocks. Large landslides are developing in the slopes adjacent to the reservoir (Fig. 28). Landslide debris being deposited in the reservoir would aggrevate an already serous silting problem. Four kilometers downstream, the powerhouse is threatened by another landslide which could dam the river.

Fluvial erosion, earthquakes, and intense rainfall are natural processes responsible for triggering landslides in Hispaniola. The river systems in Hispaniola have reaches in disequilibrium. Downcut-

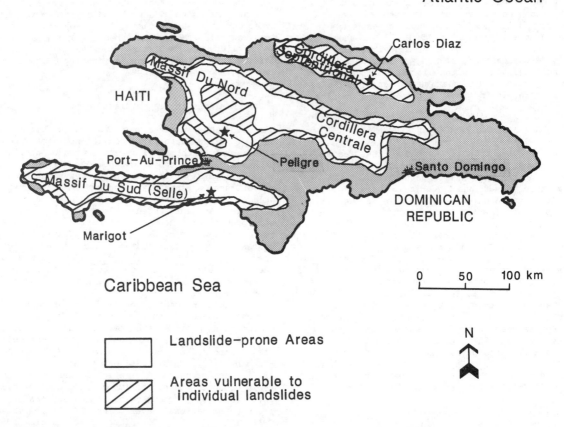

Atlantic Ocean

HAITI

Caribbean Sea

Landslide-prone Areas

Areas vulnerable to individual landslides

N

0   50   100 km

Figure 27. Distribution of landslide-prone areas and areas vulnerable to individual landslides in Hispaniola. Three landslide sites described in the text are noted.

Figure 28. A large landslide (margins highlighted in white) at the edge of the Peligre reservoir in Haiti.

ting of rivers occurs in response to base level changes due to tectonic uplift. In other reaches, the unusually heavy sediment load resulting from extensive erosion causes lateral displacement of channels. These channel changes produce oversteepened lower valley slopes favorable to landslide development (Mora, 1987). Significant landslides are present along most of the important rivers including the Artibonite, Guayamouc, Bouyaha, Fer a Cheval, and Canot. Development of high pore water pressures due to dynamic loading during earthquakes or heavy rainfall leads to landslides on the upper valley slopes.

On December 29, 1897, an earthquake with an estimated magnitude of 7.1 struck the Septentrional and Centrale Cordillera. Several significant landslides were reported induced by this tremor on both the Ciboa Valley and Caribbean sides of the range.

The Carlos Diaz, or Tamboril, landslide was likely triggered by seismic activity. It is unclear whether this feature is a rejuvenated remnant of the 1897 event or wholly a product of more recent seismic activity. The landslide developed in colluvium overlying bedrock consisting of limestone interbedded with marly sandstones and shales. Clay layers, present within the colluvium, cause perched water table conditions. Recent seismic activity affecting this area included a significant swarm in 1981 (R. Acota, Personal Comm., 1986). The site conditions and recent seismic activity produced landslide movement. This movement impacted the village of Carlos Diaz sufficiently to force relocation of about 1500 people in 1982.

A major landslide was triggered in the Haitian part of Hispaniola by heavy rainfall on June 3-4, 1985. It occurred in Chemin Dieu Moune (Marigot) in the southern part of Haiti (Prepetit and Joseph, 1986). It is reported the material mobilized was a mixture of soil and rock blocks from a residual soil.

Hurricanes David and Frederic generated heavy rainfall in the eastern (Dominican Republic) part of Hispaniola on August 31 through September 1, 1979. Landsliding was extensive with debris flows and slides being the common landslide type (Fig. 29). Secondary landslide damage resulted from these features creating short-lived landslide dams. The ponded water behind the landslide dams was released creating floods and mudflows downstream. The combined losses attributed to these hurricanes from floods, landslides, and other storm effects caused 2000 deaths and $1 billion of destroyed infrastructure and production capacity (CEPAL-ONAPLAN, 1979, Mora, 1986b).

8.3 Social and economic impact of landslides on Hispaniola

Slope instabilty is widespread and aggravates an already serious erosion problem. The normally land-slide-prone conditions of steep slopes with weathered rock subject to heavy rainfall or earthquake are made worse by land use practices which remove forest cover and disturb natural slopes (Fig. 30).

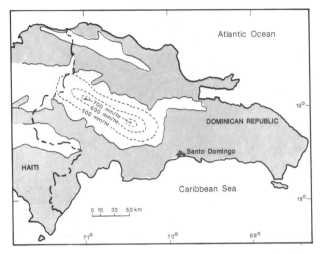

Figure 29. Main areas of landslide damage (unshaded) caused by passage of Hurricanes David and Frederic over eastern Hispaniola in August-September 1979.

Through natural and human actions, landslides cause significant losses in damaged or destroyed infrastructure. In some instances, the damage may be extensive. The portion of the $1 billion in infrastructural damage from Hurricanes David and Frederic attributed to landslides is an illustration. In other instances, landslides may cause intensive impact. The social impact of the single landslide which resulted in relocation of 1500 people from the village of Carlos Diaz illustrates an intensive impact. Landslides pose a threat to the future. Movement of existing land-slides in the vicinity of Peligre reservoir will impair the long-term function of this reservoir. More subtle is the loss of soil productivity represented by accelerated landslide erosion of productive hillslopes.

Little landslide stabilization or mitigation work is being accomplished in Hispaniola. Several studies have identified possible actions for specific projects or areas. The problem of slope instability cannot be readily separated from the severe erosion problems experienced in Hispaniola. Possible landslide mitigation or stabilization efforts are most likely to succeed as part of integrated, interdisciplinary solutions to these larger issues.

9 JAMAICA

Jamaica lies between 17° 30' and 18° 30' North latitude and 76° 00' and 78° 30' West longitude. This 10,991 square kilometer island lies south of Cuba and due west of Hispaniola (Rand McNally, 1988). Jamaica is an independant country.

Figure 30. Landslides, gullies, and other erosional features on the slopes of Massif de la Selle in Haiti. This is representative of the effects of accelerated erosion by deforestation.

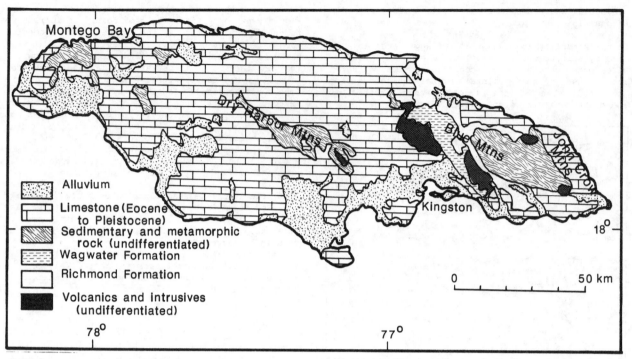

Figure 31. Generalized geologic map of Jamaica (modified from Robinson, 1975).

Like its neighboring islands, Jamaica lies within the northeast trade wind belt. The climate is subtropical with distinct rainy and dry seasons. Winter and spring coincide with the dry season with February the driest month. The rainy season is in summer and fall with the wettest month being October (Robinson, 1975). The highest average annual rainfall, 5600 millimeters, is received in the interior mountains. The southern coast has the lowest average annual rainfall at less than 1000 millimeters.

9.1 Physiography and geology of Jamaica

The interior and northern parts of Jamaica are mountainous with the mountains trending west-northwest to east-southeast. The Blue Mountains in the eastern half of the island are the highest and include 2256-meter Blue Mountain Peak, the highest point in Jamaica (Rand McNally, 1988). The Blue Mountains extend east to the John Crow Mountains on the eastern tip of the island. The western half of the island is a plateau with lower elevations than the mountainous areas. This plateau has an especially rugged central area called the Dry Harbor Mountains. The southern part of Jamaica consists of several wide coastal plains (Robinson, 1975).

More than 60 percent of Jamaica is underlain by limestone deposits of Eocene age (Fig. 31). This bedrock underlies the western plateau, the southern coastal plains, and the John Crow Mountains (Robinson, 1975, Ahmad and Jackson, 1986). Recent alluvial deposits cover the limestone of the coastal plains. The limestone capping the western plateau is breached in the central part exposing Cretaceous-age undifferentiated sedimentary and metamorphic bedrock and andesitic volcanics. Undifferentiated sedimentary and metamorphic bedrock of Cretaceous age underlie the area around Montego Bay on the northwestern end of the island and the core of the Blue Mountains in the eastern half of the islands. On its southwestern flank, the Blue Mountains are underlain by the Wagwater Formation. This formation consists of red beds, evaporites, and porphyritic andesites and some dacites (Robinson, 1975). The Richmond Formation, a flysch facies which includes local volcanics, underlies the northwestern flank of the Blue Mountains and areas on the southeast, east, and northeast.

9.2 Landslides and their impact in Jamaica

Debris flows, debris slides, earthflows, slumps, rockslides, and rockfalls are among the landslide types known to occur in Jamaica. Landslides are an important erosional process shaping watersheds in the Blue Mountains (Gupta, 1975). Shallow failures which are assumed to be mainly debris flows and slides are mapped within the Buff Bay and Yallahs river basins in the Blue Mountains. Gupta (1975) notes the persistent presence of 35 to 60 percent slopes in both river basins. His mapping shows landslides concentrated along the main streams and major tributaries, locations where slopes are maintained at steep angles by stream downcutting. The landslides involve Richmond Formation and volcanic bedrock. Landslides in these river basins occur on a frequent basis, usually in response to periodic intense rainfall. Gupta (1975) notes one failure adjacent to the Yallahs River near Mahogany Vale footbridge was triggered by Hurricane Flora in October 1963. It initiated movement of nearly 40,000 cubic meters of shattered and fractured material from the Richmond Formation.

Earthflows, debris slides, and slumps are found in many locations in the central part of Jamaica. Evidence of past landslide activity in these areas suggests the Yellow Limestone Group as well as the Richmond Formation is prone to failure (Ahmad, 1986). Slumps and rockslides are associated with cutslopes along major roads through the central and northern parts of the island. Rockfalls and rockslides occur on coastal cliffs where wave action maintains the cliff face at a steep or oversteepened angle.

Recent slope movement at Preston in St. Mary Parish illustrates some of the social and economic impact which landsliding can inflict in Jamaica. Preston is a small rural community located about 7 kilometers southwest of Port Maria. The area around Preston is on the northeastern corner of the western plateau and lies between the Port Maria and Georges rivers. The Fonthill Formation of the Yellow Limestone Group and the Richmond Formation underlie the area. The Richmond Formation is composed of mudstone which becomes weak and plastic when saturated. The mudstones of the Richmond Formation are capped by 30 to 65 meters of limestone. The limestone is horizontal to gently inclined towards the north. Some nearly vertical faults cut across the Fonthill Formation in

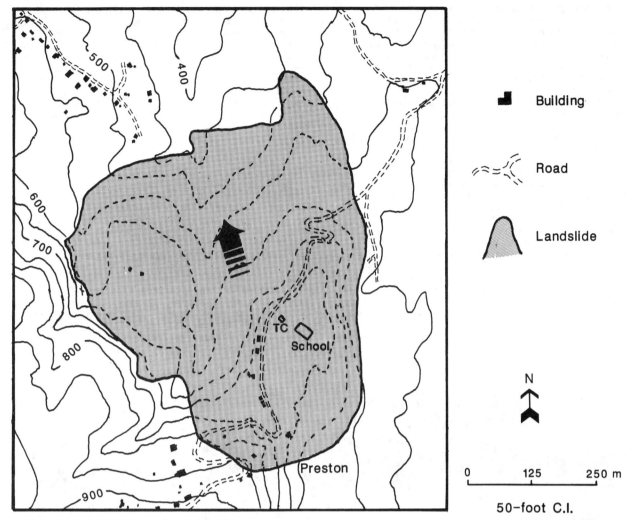

Figure 32. Map of the earthflow affecting Preston, St. Mary Parish, Jamaica. The teacher's cottage (TC) and school are noted. Dashed contour lines through the landslide (shaded area) show pre-movement topography. The large arrow indicated the general direction of movement by the earthflow.

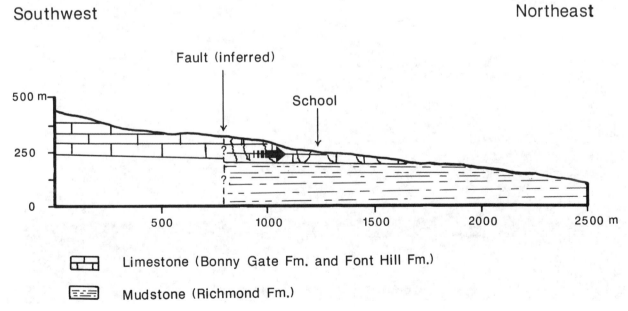

Figure 33. Simplified cross-section of the Preston earthflow showing the inferred subsurface conditions.

the vicinity of the landslide's main scarp.

The area received unusually heavy and sustained rainfall beginning in January 1986 (Fig. 32). On March 14, 1986, residents reported loud noises and ground shaking. A large tension crack developed near the fault zone damaging a number of houses. It is unclear whether the sound and vibration was due solely to landslide movement or represented movement on the fault, too. Vertical displacement was rapid between mid to late April with 2 meters accumulating in a ten day period. Ground cracking developed in other areas below the main scarp including on the school compound. These cracks damaged the teacher's cottage and the school. The residences of the 17 families in Preston were not immune to damage. Cracks and ground deformation slowly wrenched, tilted, and broke these structures. The Jamaican Geological Survey conducted photographic monitoring of these ground changes and resulting damage during this period. Measurement of the continuing movement showed an acceleration during the May and June 1986 storm events. By December 1987, the active landslide involved nearly 1 square kilometer (Ahmad, 1986). The main scarp and toe are well-developed. The main slide mass involves many ground cracks and internal scarps. Differential movement within the slide mass created a graben-like structure which is roughly 25 meters across and has nearly 12 meters of vertical displacement.

The failure plane appears to be the contact between the limestone and the underlying mudstone of the Richmond Formation (Fig. 33). The fault provided a discontinuity which isolated this overlying block and probably served as an avenue for more rapid infiltration of water to the bedrock contact. The increase in groundwater reaching this interface after the January 1986 storms caused the clayey material to become sufficiently plastic that a failure plane developed and movement of the overlying mass began. The movement is essentially translational in nature. Due to the nature of the materials involved and the movement, this feature is classified as an earthflow. The presence of four older landslides indicates movement of this type is not new to this area. Reported failures in the vicinity are documented as far back as 1946.

The current earthflow movement has inflicted considerable social and economic damage. In May 1986 upon the recommendation of the Geological Survey Division, the area was evacuated. Seventeen families were forced to abandon their homes and fields to escape the accelerated movement occuring at that time. Sections of the Preston road are totally destroyed. Many homes were slowly destroyed and others suffered significant damage. The teacher's cottage and school were damaged beyond repair and abandoned (Fig. 34). Water lines and electrical transmission lines were rendered inoperable (Ahmad, 1986). Additional expenditures will be incurred to install new utilities and roads for the area where former residents of Preston will be relocated.

Figure 34. Graben-like cracking in the Preston earthflow which caused some of the severe damage to the teacher's cottage visible in the background. Both the cottage, nearby school, and residences for 17 families were abandoned due to the damaging effects of landslide movement.

10 CAYMAN ISLANDS

The Cayman Islands are a British colony consisting of Grand Cayman, Little Cayman and Cayman Brac found south of Cuba and west of Jamaica (Rand McNally, 1988). The land area of the Cayman Islands totals 259 square kilometers.

The climate is hot and humid. The Cayman Islands are subject to hurricanes and tropical storms during May to November. The remainder of the year is generally dry.

The Cayman Islands consist entirely of limestone bedrock. Each island is a fault block of white limestone of Tertiary age. A fringe of younger limestone of Pleistocene age surrounds each island (Richards, 1975). These limestone blocks rest on an ancient submarine ridge. The islands are generally flat-lying with the highest point being 42 meters. Coastal cliffs where reefs do not fully protect the shoreline are unlikely to exceed a few meters in this area of subdued topography.

There is no reported landslide activity for the Cayman Islands. Occasional falls of limestone blocks at low sea cliffs is possible. Cayman Islands seem unaffected by landslides and likely incur no economic or social impacts from this process.

11 CUBA

Cuba is directly south of Florida and west of the island of Hispaniola. With an area of 114,524 square kilometers, Cuba is the largest of the islands in the Greater Antilles (Rand McNally, 1988). It consists of a main island about 1290 kilometers long and 70 to 200 kilometers wide with many smaller offshore islands.

The climate of Cuba is tropical. Annual rainfall varies from less than 800 to over 2200 millimeters for different parts of the island. The tradewinds tend to bring precipitation to the eastern end of the island. However, the presence of various highlands causes rainshadows where conditions are much drier. November to April is the wet season. The hurricane tracks results in most hurricanes affecting the western end of the island (Academia de Ciencias de Cuba, 1970).

11.1 Physiography and geology of Cuba

Outside of four distinctly mountainous areas, the topography is subdued with elevations less than 100 meters (Fairbridge, 1975b). The major mountain range is the Sierra Maestra in the southeastern part of Cuba. Pico Turquino, at an elevation of 1974 meters, is the highest peak in Cuba. Many other peaks higher than 1000 meters are present in this mountain range. The eastern end of Cuba is the most rugged part with the Sierra Maestra in the southeast and the Baracos Highlands in the northeast. The central part of the island includes the Santa Clara Hills rising to roughly 200 meters and the Escambray Mountains rising to nearly 700 meters. The Havana and Matanzas Highlands are found in the north-central part of the island near Havana. This is a structurally complex area. A mountainous area known as the Sierra de los Organos is found in the northwest. It is underlain by limestone producing a tropical cone karst landscape with ridgetops and peaks reaching 300 to 700 meters. Cuba is a geologically diverse island. Many of the coastal plains and interior valleys are underlain by Quaternary to Recent sediments. As noted earlier, the Sierra de los Organos represent an area of predominantly limestone. The Havana and Matanzas Highlands and the Santa Clara Hills are underlain by folded and faulted sedimentary bedrock. Sandstones, conglomerates, shales, and dolomites of Cretaceous and Tertiary age are predominant. Various metamorphic rock types make up the Escambray Mountains. In the east, the Sierra Maestra are broadly folded layers of sandstone, shale, breccia, and limestone of Paleocene age. Considerable amounts of serpentine and peridotite are exposed within folded sedimentary rock units in the neighboring Baracos Highlands.

11.2 Landslides in Cuba

Literature review found no specific references to landslide activity on Cuba even where discussions on geomorphology , geology, or soils were present (Furrazola-Bermudez et al, 1964, Academia de Ciencias de Cuba, 1970).

It seems reasonable to expect landslides to occur in the Baracos Highlands, Sierra Maestra, and Escambray Mountains in eastern and central Cuba. These areas receive the higher rainfall during the year, host many steep slopes, and include bedrock with lithologic and stratigraphic conditions which appear to favor landslide development. Failures in colluvial material leading to debris flows and debris slides are expected when intense rainfall from hurricanes and major tropical depressions affect the Sierra Maestra, Baracos Highlands, and the Havana and Matanzas Highlands.

No information on the impact of landslide activity to economic activity or inhabitants is available. However, landslides are one of several geologic hazards considered in disaster planning efforts in Cuba (D. Alonso Dominquez, Personal Comm., 1987). This suggests landslides have created sufficently significant impacts on the well-being of inhabitants to result in their consideration in governmental disaster planning.

12 BAHAMAS AND TURKS AND CAICOS ISLANDS

The Bahamas, an independent country, and Turks and Caicos Islands, a possession of the United Kingdom, are part of the archipelago that extends for about 1,000 kilometers in the Atlantic Ocean from 100 kilometers east of Miami, Florida, to 150 kilometers north of the Dominican Republic. Turks and Caicos Islands are at the southeastern extremity of the island chain. The Bahamas consist of about 700 islands, 22 of which are inhabited, and more than 2400 low, barren rock outcrops that cover an area of 13,939 square kilometers. Turks and Caicos Islands include about 30 islands covering 430 square kilometers (Rand McNally, 1986). The population of the Bahamas is 230,000, and of Turks and Caicos Islands, 8,100. The overall population density of the archipelago is thus 17 persons per square kilometer, but most of the population is concentrated in a few cities and towns.

The climate is humid subtropical; annual rainfall averages 1200 millimeters, and temperatures normally range between 20 and 30°C. Hurricanes, fairly common between May and November, have caused severe damage on several occasions.

The Bahamas and Turks and Caicos Islands occupy a large carbonate platform along the subsiding continental margin of North America (Curran, 1985). All surficial materials and bedrock are calcareous; most or all exposed materials are Quaternary in age (Curran, 1985). The topography is very subdued; most of the islands are nearly flat, and the maximum elevation of the archipelago is 63 meters. The only significant slopes or escarpments are along some coastlines where low cliffs or steep slopes, no more than several meters high, rise from the ocean (Adams, 1985).

Landslides in the Bahamas and Turks and Caicos Islands are limited to small rock falls and rock slides along some of the low sea cliffs and are in response to wave action attacking the base of the slope (Carew and Mylroie, 1985). The only other significant ground-failure phenomenon is the formation of sink holes (Mylroie, 1983), which is a fairly common occurrence. Thus, landsliding in the Bahamas and Turks and Caicos Islands is very rare, extremely limited in magnitude and areal extent, and economically insignificant.

Landslides in the Caribbean cause injury and death, damage or destroy facilities, and inflict agricultural losses. These social and economic impacts are significant for most islands in this region. In monetary terms, the losses sustained by roads is especially significant. The average annual cost of clearing and repairing roads is known only for Trinidad, Tobago, St. Lucia, and Dominica. Using these figures and general knowledge of landslide activity and roads on other islands, it is estimated the average annual cost of landslides to roads for the region is $15 million. Because this estimate used conservative assumptions, it most likely represents a minimum cost. Monetary values can be attached to crop damage to represent the short-term impact to agricultural activity. However, the long-term agricultural losses due to erosion of soil and consequent lower productivity needs to be defined in similar terms. The inability to fully quantify losses to agriculture, the primary economic activity in the Caribbean, hampers representing the total impact of landslide activity.

While considerable information on landslides exists for different Caribbean islands, it is evident a greater depth of knowledge exists for some islands than others. For a few islands, information on landslides which most assuredly occur is virtually non-existent. This illustrates the need for more comprehensive mapping of existing landslides in many Caribbean countries. The mapping should include study and documentation of specific factors influencing landslides in that locality. There is also a need to better demonstrate the social and economic impact of landslides. Information on the significance of landslides provides a clearer indication of where hazard reduction efforts should be concentrated.

A number of scientific questions exist concerning landslides in the Caribbean which merit future study. Topics having the highest priority for further research include the role of weathering and soil development in landslide formation; threshold rainfall intensities and durations needed to trigger different types of landslides in different terrains; characterization of physical properties, such as shear strength, of residuum and colluvium on steep slopes; defining debris flow runout zones; and the seismic stability of slopes. Considering the importance of agricultural activity and the deforestation on some islands, the influence of vegetation on slope stability is another important topic.

The Good Hope landslide in Dominica provides a vivid example of the nature and consequence of landslides in the Caribbean. Unfortunately, this tragedy is likely to be repeated in this region in the future. The extent to which inhabitants of these islands remain vulnerable to future landslides will depend on the emphasis placed on better defining local landslide hazard and implementing appropriate hazard reduction measures.

ACKNOWLEDGEMENTS

The senior author compiled contributions by fellow authors into the present paper. These contributions were: R. Bryce (Jamaica), R. Jibson (Bahamas, Turks and Caicos, Puerto Rico), S. Mora (Hispaniola), and C. Rodgers (Trinidad and Tobago). The senior author wrote the remaining sections and accepts responsibility for any errors in compiling the contributed material.

Mr. DeGraff and Dr. Mora gained their landslide data on the Windward Islands and Hispaniola, respectively, while assisting the Department of Regional Development, Organization of American States. The views expressed in this paper are those of the authors and do not necessarily represent those of the Organization of American States or its member states. Mr. DeGraff expresses his appreciation to the Governments of St. Lucia, the Commonwealth of Dominica, and St. Vincent and the Grenadines for permission to conduct landslide studies and for assistance from various government personnel. Mr. DeGraff was made available through the Disaster Assistance Program, a part of the Internationl Forestry Staff of the USDA Forest Service. Dr. Mora wishes to express appreciation for review of his contribution by Msc. Maria Laporte and typing by Mrs. Flory Abarca. Ms. Rodgers wishes to gratefully acknowledge the data provided by the Ministry of Works, Settlement, and Infrastructure and Works Division, Tobago House of Assembly.

REFERENCES

Adams, R.W. 1985. General guide to the geological features of San Salvador, In D.T.Gerace (ed.), Field Guide to the geology of San Salvador (3rd Edition), p.1-66. San Salvador, Bahamas, College Center of the Finger Lakes Bahamian Field Station.
Academia de Ciencias de Cuba 1970. Atlas nacional de Cuba. Havana, La Academia de Ciencias de Cuba.
Ahmad, R. 1986. Recent earth movements at Preston, St. Mary. ODIPERC News 1:1,4-5.
Ahmad, R. & T.Jackson 1986. Field guide to western margin of Wagwater fault and geology of Stilemans River quadrangle, p.1-7. Kingston, Jamaica, University of the West Indies.
Alford, D. 1986. The water and sediment balance of the upper Artibonite Basin, Haiti, p.1-33. Washington, D.C.: Organization of American States.
Anderson, M.G. & P.E. Kneale 1985. Empirical approaches to the improvement of road cut slope design, with special reference to St. Lucia, West Indies. Singapore U. Tropical Geog. 6:91-100.
Anonymous 1938. West Indian landslide kills hundreds. Engineering News-Record 121:685.
Antonini, G. 1986. Processes and patterns of landscape change in the Linea Noreste, Dominican Republic. New York: Columbia University PhD Thesis.
BDPA 1983. Cartographie thematique d'Haiti. Paris: Fonds d'Aide et de Cooperation Francaise.
Bolt, B.A., W.L.Horn, G.A.Macdonald, and R.F.Scott 1975. Geological hazards. New York: Springer-Verlag.
Brabb, E.E. 1984. Landslide potential on St. Thomas, Virgin Islands, p.97-102. U.S. Geological Survey Open-File Report 84-762.
Briggs,R.P., P.A.Gelabert, D.Jordan, E.Aquilar, R.M.Alonso & R.M.Valentine. 1970. Engineering geology in Puerto Rico. Assoc. Eng. Geol. Ann. Field Meeting Fieldtrip Guidebook 6:23.
Brown, D.,H.Romano, C.T.Rogers 1985. Landslides and flood distribution in the West Coastal area-the role of geology. Proc. 1st Trinidad Geol. Conf. Port of Spain, Trinidad.
Calversbert, R.J. 1970. Climate of Puerto Rico and U.S. Virgin Islands. Climatography of the United States, p.50-52. Washington, D.C., U.S. Environmental Science Services Administration.
Campbell, R.H., D.G.Herd, & R.M.Alonso 1985. Preliminary response activities and recommendations of the USGS landslide hazard rsearch team to the Puerto Rico landslide disaster of October 7, 1985. U.S. Geological Survey Open-File Report 85-719.
Campbell, R.H., D.G.Herd & R.W.Jibson 1986. Preliminary review of the landslide disaster of October 7, 1985, near Ponce, Puerto Rico. Geol. Soc. Amer. Abs. With Program 18:93-93.
Carew, J.L. & J.E.Mylroie 1985. The Pleistocene and Holocene stratigraphy of San Salvador Island, Bahamas, with reference to marine and terrestrial lithofacies at French Bay. In H.A.Curran (ed.), Pleistocene and Holocene carbonate environments on San Salvador Island, Bahamas. Boulder: Geological Society of America Field Trip Guidebook 2:1-10.
Case, J.E. & T.L.Holcomb 1980. Geologic-tectonic map of the Caribbean region. U.S. Geological Survey Miscellaneous Investigations Map I-1100.
CEPAL 1979. Report on effects of Hurricane David

on the island of Dominica. UN Economic and Social Council Rept. 1099.

CEPAL-ONAPLAN 1979. Repercuciones de los Huracanes David y Federico sobre la Economia y Condiciones Sociales de la Rupublica Dominicana. UN Economic and Social Council Rept. 1019.

Choubert, G. & A.Faure-Muret (Compilers) 1976. Geological World Atlas (1:10 000 000). Paris, UNESCO.

CNRS 1985. Atlas d'Haiti. Bordeaux, Universite de Bordeaux, Centre d'Etudes de Geographie Tropicale.

Cooper, St. G. & P.Bacon 1981. The natural resources of Trinidad and Tobago. London, Edward Arnold Publ.

Cumberbatch, E.R. St. J. 1966. Soil conservation in the Scotland District, Barabdos. Proc. 1st Pan. Amer. Soil Conserv. Cong., p.155-160, Brazil, Sao Paolo Ministerio da Agricultura.

Curran, H.A. 1985. Introduction to the geology of the Bahamas and San Salvador Island with an overflight guide, In H.A.Curran (ed.), Pleistocene and Holocene carbonate environments on San Salvador Island, Bahamas. Boulder: Geological Society of America Field Trip Guidebook 2:1-10.

Deere, D.U. & F.D.Patton 1971. Slope stability in residual soils. In 4th Pan. Amer. Conf. Soil Mech. and Found. Eng. Proc. 1:87-170.

DeGraff, J.V. 1985. Landslide hazard on St. Lucia, West Indies-Final Report. Washington, D.C., Organization of American States.

DeGraff, J.V. 1987a. Landslide hazard on Dominica, West Indies-Final Report. Washington, D.C., Organization of American States,

DeGraff, J.V. 1987b. Geologic reconnaissance of the 1986 landslide activity at Good Hope, D'Leau Gommier, and Belvue Slopes, Commonwealth of Dominica, West Indies. Washington, D.C., Organization of American States.

DeGraff. J.V. 1988. Landslide Hazard on St. Vincent, West Indies-Final Report. Washington, D.C., Organization of American States.

De la Fuente, S. 1976. Geografia Dominicana. S. Domingo, Ed. Quisqueyanas.

Diyaljee, V.A. 1984. roadway landslide in heavily overconsolidated Trinidad Clay. In Proc., 4th Internat. Symp. on Landslides (Toronto) 2:51-56.

Fairbridge, R.W. 1975a. Barbados. In R.W.Fairbridge (ed.), The Encyclopedia of World Regional Geology, Part 1: Western Hemisphere, p. 115. Stroudsburg, Dowden, Hutchinson & Ross.

Fairbridge, R.W. 1975b. Cuba. In R.W.Fairbridge (ed.), The Encyclopedia of World Regional Geology, Part 1: Western Hemisphere, p.252-255. Stroudsburg, Dowden, Hutchinson & Ross.

Fairbridge, R.W. 1975c. Windward Islands. In R.W.Fairbridge (ed.), The Encyclopedia of World Regional Geology, Part 1: Western Hemisphere, p.667. Stroudsburg, Dowden, Hutchinson & Ross.

Faugeres, M.L. 1966. Observations dur le modele des versants dans la region des Pitons du Carbet (Martinique). Assoc. Geographes Francais Bull. 342-343:52-63.

Farquhar,O.C. 1978. Landslides in Puerto Rico. Internat. Assoc. Eng. Geol. Bull. 16:44.

Fink, L.K., Jr. & R.W.Fairbridge 1975. Leeward Islands. In R.W.Fairbridge (ed.), The Encyclopedia of World Regional Geology, Part 1: Western Hemisphere, p.339-340. Stroudsburg, Dowden, Hutchinson & Ross.

Furrazola-Bermudez, G., C.M.Judoley, M.S.Mijailovskaya, Y.S.Miroliubov, I.P.Novojatsky, A.N.Jimenez, & J.B.Solsona. 1964. Geologia de Cuba. Cuba, Instituto Cubano de Recursos Minerales, Editora del Consejo Nacional de Universidades.

Garvey, W. 1982. Disaster relief and preparedness evaluation, Haiti. Washington, D.C., U.S. Agency for International Development Report.

Gupta, A. 1975. Stream characteristics in eastern Jamaica, and environment of seasonal flow and large floods. Amer. J. Sci. 275:825-847.

INDRHI 1983. Mapa geologico de la Republica Dominicana. Dominican Republic, PLANIACAS-BID-ATN/SF-1862.

Jibson, R.W. 1986a. Evaluation of landslide hazards resulting form the 5-8 October 1985 storm in Puerto Rico. U.S. Geological Survey Open-File Report 86-26.

Jibson, R.W. 1986b. Landslides resulting from the October 5-8, 1985, storm in Puerto Rico. Assoc. Eng. Geol. Abs. and Program. 29:52.

Jibson, R.W. 1986c. The Puerto Rico landslide disaster of October 1985. Assoc. Eng. Geol. Newsletter. 29:17-20.

Jibson, R.W. 1987a. Debris flows triggered by a tropical storm in Puerto Rico. In A.P.Schultz & C.S.Southworth (eds.), Collected papers on Eastern North American Landslides, p.9-10. U.S. Geological Survey Circular 1008.

Jibson, R.W. 1987b. Landslide hazards of Puerto Rico. In W.W.Hays & P.L. Gori (eds.), Proc. Assessment of Geologic Hazrds and Risk in Puerto Rico, p.183-188, U.S. Geological Survey Open-File Report 87-008.

Jibson, R.W. (in press) The Mameyes, Puerto Rico, landslide disaster of October 7, 1985. Engineering Geology Case Histories. Boulder, Geological Society of America.

Kugler, H. 1959. Geologic map of Trinidad. Zurich, Orelli Fossli.

Lewis, L.A. 1975. Slow slope movement in the dry tropics-La Paguera, Puerto Rico. Zeit. Fur Geomorphologie 19:334-339.

LGL 1981. Etude sedimentologique du Reservoir de Peligre, Haiti-Rapport Final. Haiti, Banque Internationale pourla Reconstrution et le Development-Electricite d'Haiti Projet PNUD/HAI/78-017.

Maxwell, J.C. 1948. Geology of Tobago, B.W.I. Geol. Soc. Amer. Bull. 59:801-854.

Mehigan, P.J. & D.N.D.Hartford 1985. Aspects of slope stability in relation to road design in the Commonwealth of Dominica. Proc. 11th Internat. Conf. Soil Mech. Found. Eng. 1:2339-2343.

Monroe, W.H. 1964. Large retrogressive landslides in north-central Puerto Rico. U.S. Geological Survey Professional Paper 501-B.

Monroe, W.H. 1976. The karst landforms of Puerto Rico. U.S. Geological Survey Professional Paper 899.

Monroe, W.H. 1979. Map showing landslides and areas of susceptibility to landsliding in Puerto Rico. U.S. Geological Survey Miscellaneous Investigations Map I-1148.

Monroe, W.H. 1980. Some tropical landforms of Puerto Rico. U.S. Geological Survey Professional Paper 1159.

Mora, S. 1986a. Etude de reconnaissance des Menaces Naturelles Dans le Basin Versant du Haut Artibonite, Haiti. Washington, D.C., Proj. Front. OEA/DDR/GOH-INGEOSA.

Mora, S. 1986b. Estudio de Reconocimiento de las Amenazas Naturales de la Region Fronteriza (occidental) de la Republica Dominicana. Washington, D.C., Proy. Des. Front OEA/DDR/GRD.

Mora, S. 1987. Analisis preliminar del fenomeno de erosion acelerada en Banica, Elias Pina, Republica Dominicana. Washington, D.C., OEA/DDR/-INGEOSA.

Mylroie, J.E. 1983. Caves and karst of San Salvador. In D.T.Gerace (ed.), Field Guide to the geology of San Salvador (3rd Edition), p.1-66. San Salvador, Bahamas, College Center of the Finger Lakes Bahamian Field Station.

Nanita, M. 1982. Disaster prevention in the Dominican Republic. Japan, Nat. Res. Center Disaster Prev. SEOPC-JICA.

NOAA-CEAS 1979. A study of the Caribbean Basin drought/food production problem-final report. Washington, D.C., U.S. Agency for International Development.

OEA-DDR 1972. Mission d'assistance technique integree; Haiti. Conseil National de Development et de Planification. Washington, D.C., Organization of American States.

Ortiz, C.A. 1974. Relationship between engineering properties and topographical expression of the rocks in Rio de la Plata's valley walls, Comerio, Puerto Rico. Intern. Assoc. Eng. Geol., 2nd

Intern. Conf. Proc. 6:1-8.

Peeters, L. 1963. Erosion et glissements de terrain a La Barbade. Rev. Belge Geographie 87:211-225.

Prepetit, C. & Y.Joseph 1986. Rapport sur le glissement de terrain produit a Dieu Moune (Marigot). Haiti, Nat. Min. Mines et Press.

Price, E.T. 1958. Notes on the geography of Barbados. Berkely, University of California MS thesis.

Prior, D.B. & C.Ho 1972. Coastal and mountain slope instability on the islands of St. Lucia and Barbados. Eng. Geol. 6:1-18.

Rand McNally 1986. The new international atlas. New York: Rand McNally.

Rand McNally 1988. World atlas of nations. New York: Rand McNally.

Richards, H.G. 1975. Cayman Islands. In R.W.Fairbridge (ed.), The Encyclopedia of World Regional Geology, Part 1: Western Hemisphere, p.226-227. Stroudsburg, Dowden, Hutchinson & Ross.

Robinson, E. 1975. Jamaica. In R.W.Fairbridge (ed.), The Encyclopedia of World Regional Geology, Part 1: Western Hemisphere, p.334-335. Stroudsburg, Dowden, Hutchinson & Ross.

Rogers, C.T. & A.Chow-Gabbadon 1985. The Parrylands DM-13 landslide. Proc. 1st Trinidad Geol. Conf. Port of Spain, Trinidad Geol. Conf.

Rouse, W.C., A.J.Reading & R.P.D.Walsh 1986. Volcanic soil properties in Dominica, West Indies. Eng. Geol. 23:1-28.

Rouse, A.J. & A.J.Reading 1986. Landslides in strong, highly permeable tropical clay soils of Dominica, West Indies. In V.Gardiner (ed.), International Geomorphology 1986, Part I, p.431-464. London: John Wiley & Sons.

Sowers, G.F. 1971. Landslides in weathered volcanicin Puerto Rico. Proc. 4th Pan. Amer. Conf. Soil Mech. and Found. Eng., 1:105-115.

Taylor, L.G. 1981a. Landslides in the Northern Range. Proc. seminar on landslides. Trinidad, Assoc. Prof. Eng. Trinidad & Tobago.

Taylor, L.G. 1981b. Parameters for the landslide-prone clays of central and south Trinidad. Proc. seminar on landslide. Trinidad, Assoc. Prof. Eng. Trinidad & Tobago.

Varnes, D.J. 1978. Slope movement types and processes. In R.L.Schuster & R.J.Krizek (eds.), Landslides-Analysis and Control, p.11-33. Washington, D.C., Transportation Research Board.

Walsh, R.P.D. 1982. A provisional survey of the effects of Hurricanes David and Frederic in 1979 on the terrestrial environment of Dominica, West Indies. Swansea Geogr. 19:28-35.

Walsh, R.P.D. 1985. The influence of climate, lithology, and time on drainage density and relief development in the volcanic terrain of the Windward Islands. In I.Douglas & T.Spencer (eds.), Environmental change and tropical geomorphology, p.93-122. London: Allen & Unwin.

Weaver, J.D. 1975. Virgin Islands. In R.W.Fairbridge (ed.), The Encyclopedia of World Regional Geology, Part 1: Western Hemisphere, p.654-655. Stroudsburg, Dowden, Hutchinson & Ross.

Woodring, W.P., J.S.Brown & W.S.Burbank 1924. Geology of the Republic of Haiti. Port-au-Prince, Dept. Public Works.

Landslides: Extent and Economic Significance, Brabb & Harrod (eds)
© 1989 Balkema, Rotterdam. ISBN 90 6191 876 6

Landslides: Extent and economic significance in Honduras

Allen P.King
USDA Forest Service, Quincy, Calif., USA

ABSTRACT: Landslides are a recognized, but little studied phenomena in Honduras, Central America. Recent investigation in the Department of Atlantida, bordering the Caribbean Sea, and the metropolitan area surrounding Tegucigalpa, the capital city, analyzed landslide frequency and factors contributing to their occurrence. Other areas in Honduras are known to be landslide-prone. No records are kept of lives lost nor value of facilities and property damaged due to landsliding. However, fatalities have occurred and roads, many homes, and other developments have suffered damage or destruction. A large slide along the Highway to the North reportedly cost over $1 million to repair. In the metropolitan area of Tegucigalpa, one landslide destroyed 200 homes and another, 50 homes. Failures in an uncompacted fill beneath a newly constructed housing project on the outskirts of Tegucigalpa damaged just-completed houses. As urban growth expands rapidly into steeper, and often less stable, terrain and development fosters industries such as logging, farming and hydroelectric projects, the risk of landslide catastrophes increases sharply.

1 GEOGRAPHY

Honduras is located within the tropics of Central America, bordered to the west by Guatemala, to the southwest by El Salvador, and to the southeast by Nicaragua (Figure 1). Most of its northern side is flanked by the Caribbean Sea, and a small portion of its southern side touches on the Pacific Ocean.

Except for the narrow Pacific coastal plain and the relatively wide Caribbean coastal plain, most of the country lies at elevations of 600 to 2,000 m. Steep, rugged landforms are typical and over 80% of the land is mountainous.

There are two general subdivisions of mountain ranges. In the north, the Central American

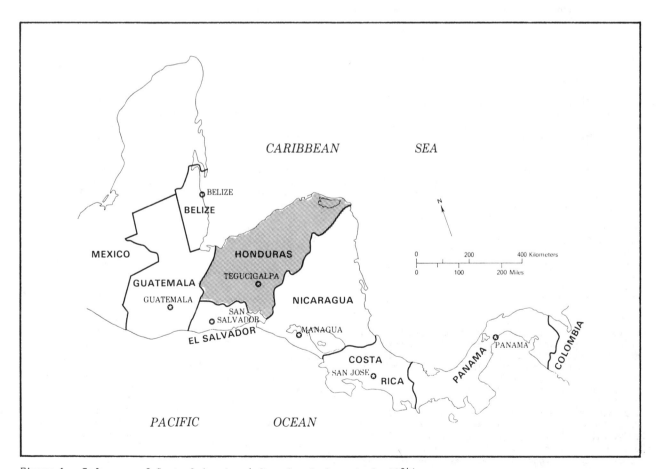

Figure 1. Index map of Central America (after Cunningham et al. 1984).

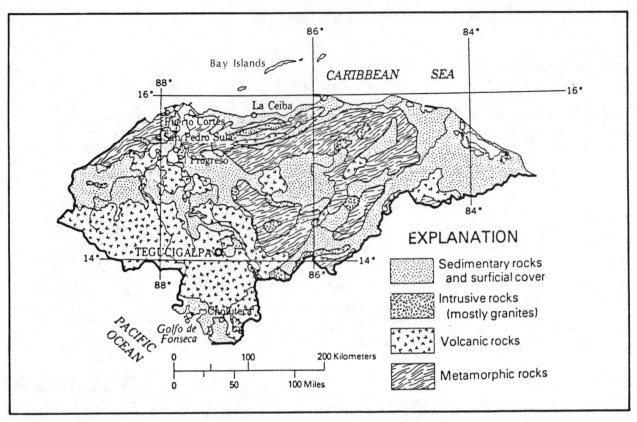

Figure 2. Major geologic divisions in Honduras, after Cunningham et al. (1984); geology simplified from Case and Holcombe (1980).

Cordillera extends from the Guatemalan border in the west to the Platano River in the east. Component ranges trend ENE-WSW, and run parallel to the north coast and to each other. The Bay Islands are summits of the undersea range of the Cordillera. In the south, the Volcanic Highlands consist primarily of highly faulted and in many areas deeply dissected Tertiary volcanics. They run from the border of El Salvador to the Nicaraguan border. Only the extreme southern part of the country forms part of the Pacific Volcanic Belt. The highest mountains are in western Honduras within the cordilleran ranges, reaching over 3,000 m.

The climate is moderate-temperate, but varies greatly with elevation. Rainfall averages 1500-2500 mm annually, most of it falling from mid-April through October. The dry season lasts from November to April. Rainfall is higher on the North Coast and northern mountain slopes, ranging from 1800 to 3000 mm annually, than on the South Coast and southern mountain slopes receiving from 1500 to 2000 mm annually. An abnormally short rainy season can result in disastrous drought in the south and west. Hurricanes have caused extensive damage and flooding on the northern coast.

The area of Honduras is 109,560 square kilometers, or roughly the size of the North American state of Ohio. The country is divided into Departments, geo-political subdivisions of which Atlantida and Francisco Morazan are two. The population of Honduras as of 1980 was 3.7 million; 36% of the population was urban at that time. The growth rate of the cities is increasing rapidly, especially in Tegucigalpa, the capital city, which has exploded from 270,000 in 1975 to an estimated 800,000 in 1987. The influx of people from rural areas has caused a dramatic housing shortage, and pushed construction of roads and new housing up into some very unstable terrain in the hills and mountains

surrounding the Tegucigalpa basin.

2 GEOLOGY AND GEOMORPHOLOGY

Although considerable work has been done since publication of the Mesozoic Stratigraphy of Honduras by Mills et al (1967), geologists are still trying to understand the stratigraphic relationships of certain rock units. According to Finch (1981), "factors that make correlation [especially of sedimentary units] difficult include: lack of access to many [densely vegetated and unroaded] areas, extensive and pervasive faulting, vegetative cover of contacts, surficial slumping, and, until recently, the lack of detailed field work over reasonably large areas." A highly generalized geologic map of Honduras is shown in Figure 2.

The Caribbean Coastal Plain (Figure 3), according to Weyl (1980), consists of a narrow belt of Tertiary and Quaternary alluvia, climbing gently through deeply decomposed hills of metamorphic and granitic rock. The easternmost portion of the country (and continuing on into Nicaragua), is a savanna called "La Mosquitia". This wide, hilly plain is a zone of tectonic subsidence which has been accumulating thousands of meters of sediments since Lower Cretaceous time, and has been greatly fragmented by countless deformational phases.

The Northern Cordillera, which is part of the Paleozoic crystalline core of Central America, is made up of gneisses, schists, phyllites and granitic intrusions. Although portions of these mountain ranges have been eroded to a gentle relief, the coast range shows extreme relief as the faulted basement rocks have been pushed up to jagged peaks overlooking the Caribbean.

The Central Cordillera contains both metamorphic

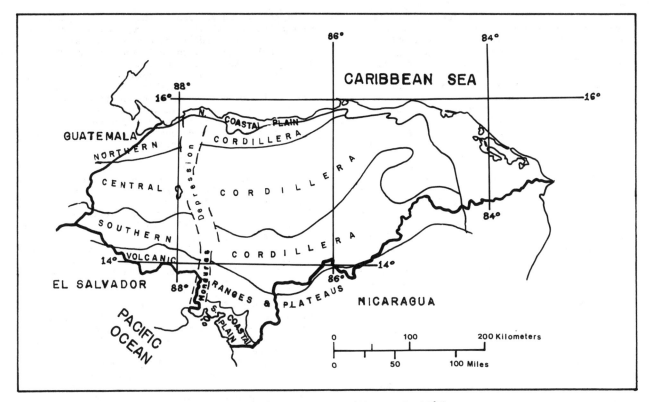

Figure 3. Morphotectonic units of Honduras (re-drawn from Mills et al. 1967).

basement rock and sedimentary Mesozoic rock, especially thick limestone deposits which often form pronounced relief from past faulting and folding, and extensive red-bed deposits. Tertiary volcanics cap portions of this area. The Northern and Central Cordillera combine to form what is known as the Central American Cordillera.

The Southern Cordillera is composed of Paleozoic metamorphic rocks, widely scattered exposures of Mesozoic and Tertiary sediments and overlying Tertiary volcanics. The volcanic deposits are built up of Oligocene to Pliocene lavas, pyroclastic rocks (mostly ignimbrites), and volcanoclastic sediments which vary in composition from rhyolites to basic andesites.

The Volcanic Ranges and Plateaus, which do not have a clearly defined boundary with the Southern Cordillera, as described by Weyl (1980), cover the southwestern part of the country and form high plateau landscapes and mountain ranges. A characteristic feature of the extensive ignimbrite deposits is the thick layered cliffs at the perimeter of the plateaus. Countless faults determine the course of the deeply incised valleys. The Pacific Volcanic Belt contributes only a few extinct volcanoes in and around the Gulf of Fonseca to the geology of southern Honduras. The Honduras Depression is a zone of more or less NS-trending faults and grabens extending through the country from the Caribbean to the Pacific, indicating a tectonic zone of tension.

Tegucigalpa and its sister city Comayaguela sit at the bottom of a mostly enclosed basin within the Southern Cordillera, carved out by the Choluteca River and its tributaries. The floor of the basin, at 900 m elevation, exposes tilted Cretaceous   red-beds of sandstone, siltstone and conglomerate, with almost no accumulation of Quaternary sediments. A few tens to hundreds of meters above the valley floor, the entire basin is surrounded by nearly flat lying Tertiary volcanic

deposits of ignimbrites and pryoclastic rock, with occasional mafic volcanic deposits. These strata form mesas and plateaus throughout and surrounding the Basin, with steep cliffs exposed around their perimeter and talus deposits at their base.

3 LANDSLIDE LITERATURE AND RELATED INFORMATION SOURCES

Very little is known about landslides in Honduras. According to the OAS (1984), "The problem of geologic and seismic hazards in Honduras is poorly recognized, and studies to prevent catastrophies caused by geologic phenomena are not conducted. It is known that there are landslides and slumps that damage highways, populated areas, agricultural areas and forests....Apparently, detailed geotechnical studies are not done in general for engineering works, and there are very few specialists is soil mechanics and geotechnology."

There are no published reports specifically describing the extent and economic significance of landslides in Honduras, and only four unpublished reports dealing with landslide activity are known to the author.

Maps and aerial photographs of high quality are generally available through official government sources. Approval to obtain these resources, however, can take weeks or months. Geotechnical professionals mostly come to Honduras from other countries to work on specific projects.

3.1 Map and aerial photo resources

As of 1980, only 40 percent of the country had been mapped topographically on 15 minute maps at the scale of 1:50,000, with 20 m contour intervals. Black and white aerial photographs of approximate scale 1:40,000, and in some areas at 1:20,000, are also avaliable. Geologic maps, produced mostly by students from the University of Texas and Wesleyan University, and by U.S. Peace Corps Volunteers are published by the Honduran National Geographic Institute, (IGN), which oversees all geologic mapping. To date only approximately 19 geologic maps at 1:50,000 scale, totaling less than 8% of the area of the country, have been completed. A geologic map of the entire country at 1:500,000 scale is available (Elvir A., 1974) but is very generalized and not suitable for site-specific work. The United Nations, other international agencies, and private companies have also produced geologic maps of certain portions of the country, at various scales. The lack of detailed geologic mapping makes landslide hazard investigations difficult and time consuming.

Although other mapping of the geology of the Tegucigalpa area exists, the most complete geologic map was compiled, but not published, by Italian geologists in the mid 80's and was obtained from the Italian consulate. (unknown author, 1986-estimated)

The report for the OAS (1984) identified a number of soils maps of varying scales and nomenclatures, but stated that in general they do not include descriptions of characteristics of soil mechanics, slope stability, nor erosion potential. Information on precipitation and surface water hydrology is gathered by various agencies, and available for a few areas on the country, but the precision and completeness of the information is questionable. There is even less information about groundwater. British geologist Brian Morris has recently been studying groundwater availability in the Department of Francisco Morazan, under the auspices of the National Autonomous Water and Sewage Service.

3.2 Geotechnical professionals

Within the entire country there are less than 10 Honduran geologists, and no known geotechnical engineers. Other than these few individuals, architects and engineers are the only professionals who normally might be concerned with slope instability, if encountered or recognized on a construction project. The Department of Geology and Hydrology of the IGN employs no geologists, nor conducts any original geologic work, but does publish geologic maps. The Department of Mines and Hydrocarbons has several geologic technicians who graciously assist geologists from other countries who come to conduct geologic studies. Most of the geologists who have contributed to an understanding of the geology and slope stability of Honduras have come from foreign governmental agencies and universities, from multi-national agencies such as the United Nations, and from international private enterprises, such as mining and petroleum companies. Some basic courses in geology are taught at the Honduran National University, directed by Dr. Marco A. Zuniga, under the civil engineering curriculum.

3.3 Regional landslide studies by the Organization of American States

The Department of Regional Development of the OAS, funded by US AID, has in recent years been developing a comprehensive disaster preparedness planning methodology for Latin America (Bender, 1984). The methodology for inventorying landslide occurrence and assessing landslide hazards was derived from work by DeGraff and Romesburg (1980). It was field tested and revised by DeGraff on various Caribbean Islands from 1985 to present, and by this author (King, 1985, and Ramirez and King, 1987) in Honduras.

The first major landslide study in Honduras (King, 1985) covered the entire Department of Atlantida (4,250 square kilometers) on the North Coast (Figure 4). It assessed location, causes and frequency of

occurrence of landslides in the low rolling hills within the coastal plain, the rugged high mountains of the Northern Cordillera, and the Bay Islands. The study included aerial photo interpretation of landslides, mapping of landslides at 1:50,000 scale, determination by isopleth mapping of which areas had a higher frequency of landslide occurrence, and compilation of a 14 page report on the methodology and findings. Due to a lack of reliable geologic mapping, a factor analysis could not be completed in order to predict the degree of hazard and zones where future slides might occur.

The second study (Ramirez and King, 1987) analyzed the 150 square kilometer metropolitan area surrounding Tegucigalpa (Figure 4) at a scale of 1:10,000.
Data from field study combined with existing information was analyzed to define which combination of factors causes the most landslide activity and hence which areas contain the greatest potential for future landslide activity. With some additional work, this information can further be translated into degree of risk to humans and developments.

The purpose of these two studies is to provide landslide hazard information which will be used in creating a master land use plan for the rich North Coastal plain and the expanding densely populated Tegucigalpa metropolitan area. The information will also be used to refine the hazard assessment methodology for application in other areas of Honduras and Latin America. Utilization of landslide hazard information is especially critical when planning major projects such as hydroelectric developments, housing developments on increasingly steeper and more unstable terrain, irrigation projects, buildings, roads, and other land disturbing activities. Both reports are on file at offices of the OAS in Tegucigalpa and Washington, D.C.

3.4 Other landslide information

Two government reports provide brief descriptions of unstable conditions around Tegucigalpa (Ramos N., and Vicencio, 1976, and Rivera, 1982). Slope stability work is known to have been conducted in the past on two major construction projects, the El Cajon hydroelectric project on the Humuya River by French geologists, and a road stabilization project on the Highway to the North, just NW of Tegucigalpa, by Japanese geologists. The latter project reportedly required extensive subsurface drainage work, and cost over one million dollars. It is assumed that geotechnical reports were written for these and possibly other major projects, however this information was unavailable for this paper.

Landslides are also mapped on a number of the published 15 minute geologic maps. Much of this sliding however is assumed to be of Quaternary age, possibly older, and may or may not have any remaining active components. These older massive slide areas could, however, indicate potential for additional instability within that general region, or within the old slides, should slope equilibrium be altered by man's activities or natural forces. Theses, dissertations or reports were prepared with most of these geologic maps and filed with the IGN. Although most were not available for this paper, they likely contain descriptions of the mapped landslide areas.

4 EXTENT AND TYPES OF LANDSLIDING

Due to the lack of studies and published reports on landsliding within Honduras, little is known about the extent and types of landslides other than along the North Coast and within the Tegucigalpa Basin.

The report by King (1985) concludes that landslides

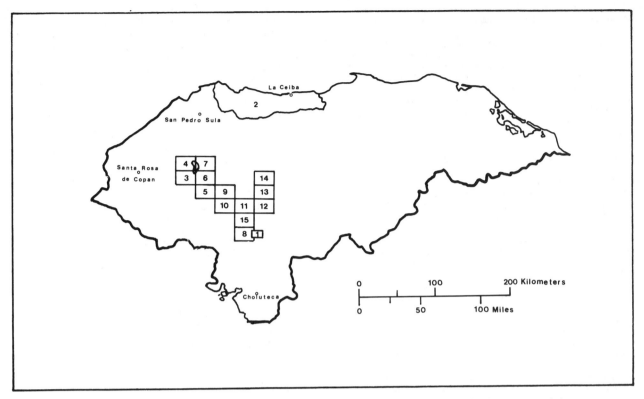

Figure 4. Map of Honduras showing areas covered by landslide hazard studies: (1) Tegucigalpa, (2) Department of Atlantida. Also shown are locations of published geologic quadrangles which contain mapped landslide areas: (3) San Pedro Zacapa, (4) Santa Barbara, (5) Siguatepeque, (6) Taulabe, (7) Santa Cruz de Yojoa, (8) Lepaterique, (9) El Rosario, (10) Comayagua, (11) Agalteca, (12) Talanga, (13) Cedros, (14) El Porvenir, (15) Zambrano.

within the Department of Atlantida along the North Coastal Plain are most often rotational slides in deeply weathered granitic or metamorphic rock. Many such slides are evident along the Progresso-La Ceiba Highway, and other roads where slopes are oversteepened and drainage is altered.

A brief reconnaisance was conducted along the highway west of La Ceiba. Road cuts up to ten meters high were frequently cut at 45 to 70 degrees. Twenty to 30 m high road cuts were equally steep, but usually terraced. Regardless, a large percentage of the cuts experienced slope movement. One section inventoried between La Ceiba and Jalimo in highly decomposed rock, showed that 78% of the cuts higher than three meters failed due either to landsliding, or in a few cases to deep gully erosion. In cuts greater than 10 m high, 85% were damaged. These slides typically contain a few tens to several hundreds of cubic meters of material. The highway was completed in approximately 1973, and slides have resulted in frequent maintenance problems ever since.

In the higher country of the Northern Cordillera mountains where slopes are steeper than in the Coastal Plain, the only development currently taking place is small-scale farming of steep slopes, timber harvesting and mining operations. These activities result in large-scale deforestation of the slopes and cause a dramatic increase in the occurrence of landslides.

Higher yet in the mountains, where no development is occurring, naturally occurring landslides are very common. Since soils are thin, shallow debris slides and rock falls are the primary modes of instability. These slides are part of the natural slope development process. The density of "fresh" slide scars in some areas is extreme, with as many as 18 individual slides identified in a single square

kilometer. These slides range in size from several hundred to several thousand cubic meters. Many of the slides traveled more than a kilometer before their energy was dissipated, leaving long narrow scars. Since soils are thin on these extremely steep slopes, many slides probably contain more rock than soil resulting in less sedimentation of the streams.

Landslides are less common on the Bay Islands, but do exist in the steepest terrain, in most cases associated with land that has been deforested. Where steep cliffs drop down to the open sea, fresh scarps are continually created by undercutting of the slopes due to wave action.

The study by Ramirez and King (1987) identifies many different types of landslides occurring within the Tegucigalpa Basin, including block falls of fractured rock, small to large complex slides with wet clayey slide planes, debris slides which cumulatively form spoon shaped depressions on some slopes, massive old deep-seated rotational features covering hundreds of square hectares, and slides in uncompacted fills and oversteepened cuts. Slides occur most frequently on, but are not limited to, steep slopes. In this study more than 300 separate landslides involving over 1300 hectares were identified and mapped, primarily from air photos but with considerable field checking. The majority of the features were less than 10 ha in size, the most common size mapped being 0.5 to 3 ha. However some slide areas were in the 40 to 60 ha range and the largest was over 100 ha.

Due to the occurrence of springs and seeps along the slopes, previous slope movements which have disrupted homes and streets, and old Spanish mine shafts beneath the city, some residents of Tegucigalpa fear hidden dangers within the nearby mountains, according to Ramos and Vicencio (1976).

In western Honduras, numerous complex slides, debris slides and slides in roadway cuts and fills were observed by this author in the steep mountainous terrain northeast of the Mayan Ruins of Copan. According to Rick Finch (personal comm., 1988), several major earth movements and many smaller slides occur along the highway between Santa Rosa de Copan and Nueva Ocotepeque. One community located just one kilometer south of Santa Rosa de Copan, where many slumping problems have disrupted the highway, is named "El Derrumbo" - The Landslide.

Fewer slides can be seen along the highways from Tegucigalpa northwest to San Pedro Sula and north to Talanga. The road which travels northeast from Tegucigalpa to San Juan de Flores is plagued with a number of massive slides in the red-beds and in deep soils derived from limestone. Displacement of the road has been an ongoing hazard and maintenance problem. The Highway to the South towards the drier climate of Choluteca passes through a few small areas of mass instability but is constructed mostly in Tertiary volcanics which appear relatively stable.

R. C. Finch (personal comm., 1988) indicated an estimated 10% of the San Pedro Zacapa quadrangle, which he mapped in 1971, is mapped as ancient/Quaternary landslides. Approximately three percent or 13.5 square kilometers of the area covered by the geologic map of the Lepaterique quadrangle (Anderson, 1987) due west of Tegucigalpa shows landslide terrain. Peace Corps Volunteer/geologist Donald Anderson (1985) states that: "Extensive areas of landslide and mudflow debris are encountered primarily in the eastern portion of the Lepaterique map area. They are most frequently associated with mass movement failures occurring along the borders of Quaternary basalt flows where resistant plateau caps have produced over-steepened flanks. Other sizable landslide features are associated with the steep slopes produced by locally upfaulted blocks. The most extensive mass of landslide debris is that of the large slide feature on the southwest slopes of El Pedregal volcano which I have informally named the 'El Aceituno' landslide. Here, over 4.5 square kilometers have been covered by slide debris..."

Slide areas are also mapped on the following geologic quadrangle maps: Santa Barbara (Finch, 1985), Siguatepeque (Curran, 1981a), Taulabe (Curran, 1981b), Santa Cruz de Yojoa (Curran, 1981c), El Rosario (Fakundiny, 1971), Comayagua (Everett, 1970), Agalteca (Emmet, et al 1970), Talanga (King, 1972), Cedros (King, 1973), El Porvenir (Simonson, 1981) and Zambrano (Dupre, 1971). A review of these maps (Figure 4) gives the following additional conclusions:

1. San Pedro Zacapa. Over 70 individual slide features are mapped southwest of Lake Yojoa, with an average size of approximately 0.7 square kilometers. Slides occur in both of the major Tertiary volcanic units, both Cretaceous limestone formations and the extensive Cretaceous/Tertiary red-bed deposits. A number of slides occur at the volcanic/red-bed contact, but many slides also occur along fault zones and adjacent to other rock unit contacts. A number of intact limestone blocks are also mapped.

2. Santa Barbara. Massive landslides of undetermined age cover 5.5% of the quadrangle in steep terrain north and west of Lake Yojoa. Some are apparently older than capping Tertiary volcanics and are traversed by faults. Most are derived from limestone bedrock which covers over 50% of the quadrangle. Average size of the slides mapped is 1.4 square kilometers.

3. Siguatepeque. Ninety-two percent of this quadrangle, which is located between Comayagua and Lake Yojoa, is mapped as various types of Tertiary volcanics. All the mapped slide areas occur in these volcanics and total 4.8 % of the quad, at an average size of 2.2 square kilometers per slide.

4. Taulabe. Although about one-half of the quadrangle is mapped as Tertiary volcanics, more than two-thirds of the 70 landslides mapped are within Mesozoic and Tertiary red-bed and limestone formations in steep country southeast of Lake Yojoa. Slides average less than one-third square kilometer each.

5. Santa Cruz de Yojoa. Most of this quadrangle, which lies just northeast of the lake, is covered by 25 different identified units of Lake Yojoa Volcanics. However only two of those units, which cover approximately one-quarter of the quad, contain landslides. These mapped slide areas cover 3.1% of the quad and average 0.6 square kilometers each.

6. El Rosario. Nine square kilometers, or 1.8% of the area, is mapped as landslides. The most slides occur in Tertiary volcanics, and most of those lie unconformably on red-beds. Several also occur in Atima Formation limestone, and one is mapped in Paleozoic schist.

7. The remaining quadrangles mentioned above have 1.5% or less of their area mapped as landslides.

Within the 13 quadrangles reviewed above a total area of 180 square kilometers is mapped as landslide terrain. It must be noted however that landslide analysis was not a primary focus of this geologic mapping. The majority of the landslides mapped are most likely ancient inactive features with no evaluation as to potential risks from future sliding.

5 CAUSES OF LANDSLIDING

A number of variables have been shown to be, or are suspected of influencing the stability of slopes in the areas which have been studied in Honduras:

1. Rock and soil types, their strength and decomposition characteristics.

2. Slope angle, which can vary greatly in different material types.

3. Amount and timing or concentration of precipitation, and corresponding surface water and groundwater flow.

4. Amount and type of vegetative cover/degree of vegetation manipulation.

5. Alteration of the landscape by humans.

6. Effects of seismic shaking.

5.1 Control exerted by rock or soil characteristics and slope

Honduran geologist Carlo Hugo Rivera (1982) was probably the first person to document instability associated with the volcanic/red-bed contact zone. Rivera describes the active rotational landsliding occurring on the slopes of El Picacho Mountain which rises over 400 m directly to the north of central Tegucigalpa. Permeable Tertiary volcanics rest unconformably on clayey impermeable Mesozoic "red-bed" sedimentary rocks. At that contact, many springs emerge, and saturated volcanic rocks detach and slide down the steep mountain slopes.

The hazard analysis by Ramirez and King (1987) verified that steep slopes near the geologic contact between red-beds sedimentary rocks and overlying Tertiary volcanics defined the extreme hazard areas where complex landslides were most likely to occur. This same phenomenon was mapped by King (1972) just south of Talanga, where Tertiary volcanics overlie Cretaceous/Tertiary red-beds of the Valle de Angeles Formation, and several large masses of volcanic material have slumped over red-beds. This relationship appears to also be significant on the geologic maps of the Taulabe, El Rosario and Comayagua quadrangles.

Higher up on the volcanic slopes above Tegucigalpa, above the influence of the saturated volcanics/red-beds contact zone, these types of slides were absent; block falls were the primary mode of failure. Highly altered mafic and andesitic volcanic flows on 45-60% slopes, tuffaceous sediments with slopes greater than 60%, and ignimbritic deposits between 30-60% slopes are identified as high hazard combinations.

Many other combinations of slope and material formed the moderate and low hazard groups. Massive tuffaceous deposits were found to stand at near-vertical cuts without failing in most areas.

5.2 Influence of precipitation and groundwater

In the foothills along the North Coast, deep soils derived from decomposed metamorphic and granitic rock units become saturated by heavy rains and subsequent excessive groundwater and surface water flow. They are especially susceptible to sliding when slopes are oversteepened, either due to stream channel cutting or man-made excavations.

Many of the numerous debris slides in the North Coast mountains, which can be readily seen on aerial photos are believed to have been caused by the heavy precipitation and powerful earth-shaking winds of the fierce hurricane Fifi in 1974.

5.3 Effects of vegetative cover and human activity

Aside from natural instability, the human factor is a primary cause of accelerated occurrence of landslides. Cut-slopes are often excessively oversteepened; fill-slopes are left uncompacted and at their natural angle of repose. Vegetation removal destroys valuable root strength and ground cover. Altered surface and subsurface drainage changes the slope equilibrium. When the rains saturate these slopes, failures occur.

Most slides which occur within inhabited areas of the North Coastal plains and foothills are initiated along road cuts or associated with farmed fields and pastures on steep slopes where the forest cover has been removed, the slash burned, and the root structure in the soil destroyed. According to a Honduran soils specialist from the community of La Ceiba in the Department of Atlantida, Vladimiro Castellanos (personal comm., 1985), the two primary causes of landslides in his region are human activity such as deforestation, road building and field burning, and steep slopes. Sr. Castellanos also stated that the campesino (poor farmer) does not respect the forest; he under-utilizes it and destroys it. People are increasingly encroaching on the forest, denuding slopes, moving into steeper terrain, and settling on potentially more hazardous land.

Aerial photos clearly show a much higher percentage of landslides in areas which have been deforested by timber harvest and farming activities than in undisturbed areas. Even on slopes where landslides are not occurring, vegetation removal is greatly increasing surface soil erosion and rapidly lowering nutrient levels and productivity of the soil.

Logging is a common activity in the foothills leading up to the mountains of the Northern Cordillera. A one-year-old logging road south of Tela was constructed in deeply weathered granitics, with steep road cuts, large uncompacted fills, and inadequate drainage. The 1985 report by King states that many rotational and translational landslides were occurring in the cuts of that road, the fill material was eroding severely, and the road would probably be impassible, without costly maintenance efforts within a year or two from that time. Construction techniques associated with timber harvest do not appear to design for slope protection and provide mitigation of drainage problems and geologically unstable terrain.

Homes built on the edge of uncompacted fills often collapse during periods of high precipitation, and have to be abandoned. There are several examples of this in new housing projects around Tegucigalpa. When poor construction techniques are used, which occurs frequently, even gentle slopes fail.

5.4 Initiation by seismic shaking

Within Honduras, the western half of the country is considered the most seismically active, although the seismic hazards are not completely understood. To date the most thorough analysis of seismic risk has been done by Kiremidgian, Sutch and Shah (1979). They concluded that since 1539 there have been numerous seismic events in Honduras which have caused a great deal of damage, including damage from earthquake-induced landslides.

Although no major earthquakes have been recorded in the country, Honduras is situated in a tectonically active area, on the western finger of the Caribbean Plate which is squeezed between the Coco's or Pacific Plate to the southwest, and the North American Plate to the north (Figure 5). In neighboring Guatemala to the north, the February 1976 Guatemalan earthquake of Richter magnitude 7.5 "generated at least 10,000 landslides that caused hundreds of fatalities as well as extensive property damage in the Guatemala City area" (Harp, Wilson and Wieczorek, 1981). The impacts on Honduras of the Guatemalan quake, and also the 1972 earthquake that destroyed Managua, Nicaragua, are unknown.

Within the last two decades, a great deal of interest has been generated in trying to understand the relationships of fault zones and plate tectonics of northern Central American countries. According to George Plafker of the U.S. Geological Survey in Menlo Park, California (Frazier,1976), "...the countries of Guatemala, El Salvador, Honduras, and probably Nicaragua are located on a continual seismic time bomb, one that has disastrously set off many times in the past and will continue to do so far into the future...earthquake damage has always been exceptionally high here."

6 SOCIAL AND ECONOMIC IMPACTS

A summary "Disaster History", on file at the U.S. Agency for International Development's (AID's) Office of Foreign Disaster Assistance (OFDA) in Washington, D.C. lists floods, hurricanes, droughts, civil strife, epidemics, and accidents, but does not single out losses from landslides (US AID, 1981). The 1974 Hurricane Fifi, the worst storm in Honduras in the 20th century, killed an estimated 8,000 people and caused $540 million in damage, including damage to all north coast primary roads and the loss of 27 primary bridges. Some of those losses are undoubtedly attributable either directly or

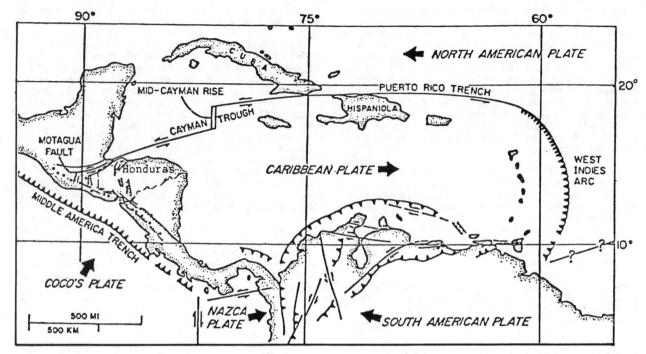

Figure 5. Plate tectonics of Central America, showing Honduras squeezed between the North American and Coco's plates (after Plafker 1977).

indirectly to landsliding.

According to a report done for the OFDA (US AID, 1981), "disaster vulnerability in Honduras is overwhelmingly related to floods like those that accompanied Hurricane Fifi....An estimated 15,000 housing units were destroyed. Most of those destroyed were rural huts of cane and thatch, though some 2,500-3,000 houses of permanent construction were also destroyed. The majority of victims lived either in the flood plains or on steep slopes which are vulnerable to mudslides. In most cases, the housing destroyed or damaged was among the poorest quality and belonged to those least able to afford repairs or replacement. Floods and mudslides in the middle of 1976 also caused substantial damage to the housing sector and contributed to a greater degree of overcrowding and the growth of uncontrolled settlements, [especially adjacent to major urban areas]. A major lesson learned from these experiences was that the site of a unit is more crucial that the materials or construction method used."

The number of people killed each year in Honduras from landslides is unknown, as is the cost of annual landslide damage. Along the road from La Ceiba southeast to Yaruca, the clearing of the vegetation has apparently caused many slides on the steep canyon slopes. Local inhabitants reported (King, 1985) that one landslide, which repeatedly slides onto the road near El Pital, has killed various people at different times in past years when it slid during rainy seasons.

In addition to the risk to human life, large investments in road construction costs are lost when roads are not planned, designed and built properly, and the soils and streams are damaged by landslides, fill and cutslope failures, surface erosion, and inadequate provision for drainage. Landslides along the North Coast Highway sometimes cause closure of that highway which is a main artery for transportation of people and goods.

A graphic example of increased landsliding due to vegetation removal is present in the Lancetilla

watershed just south of Tela. Here, a deforested area which exhibits extensive landsliding surrounds an untouched, wilderness-like "control area", where virtually no landsliding is occurring. (Figure 6). The undisturbed forest preserve, which is part of the Wilson Popenoe Botanical Garden, covers an area of approximately 12 square kilometers, and contains within its boundaries only one landslide recognizable on aerial photos. However the deforested area surrounding the preserve, with similar slope and aspect to the Lancetilla watershed, has hundreds of slides which undoubtedly occurred as a response to vegetation manipulation and other human activities.

Landslide hazards do limit land use capabilities around Tegucigalpa. The risk of further slides is high and should be taken into account as developments are planned. Rivera cautioned in 1982 that landslide occurrence would increase in the future as man continued altering the stability of the slopes with road and building construction and changes in natural drainage.

Some of the damage caused by landsliding around the Tegucigalpa area within the last 2 to 10 years includes:

1. A slide initiating on 60% slopes at the contact between red beds and overlying volcanics destroyed more than 50 homes in a poor section of Comayaguela. The residents are rebuilding their shacks on the deposited slide material with no slide stabilization measures.

2. In another area with the same geology and slope, 200 homes were destroyed and the land has been cleared and abandoned.

3. Two other large areas beneath El Picacho Mountain have suffered multiple movements which destroyed large numbers of homes more than once.

4. A recently constructed housing project in hilly terrain on the outskirts of Tegucigalpa has had many homes dismantled after settling of the large insufficiently compacted fills upon which they were constructed broke the walls apart. Cracks have occurred in the walls of many other nearby homes

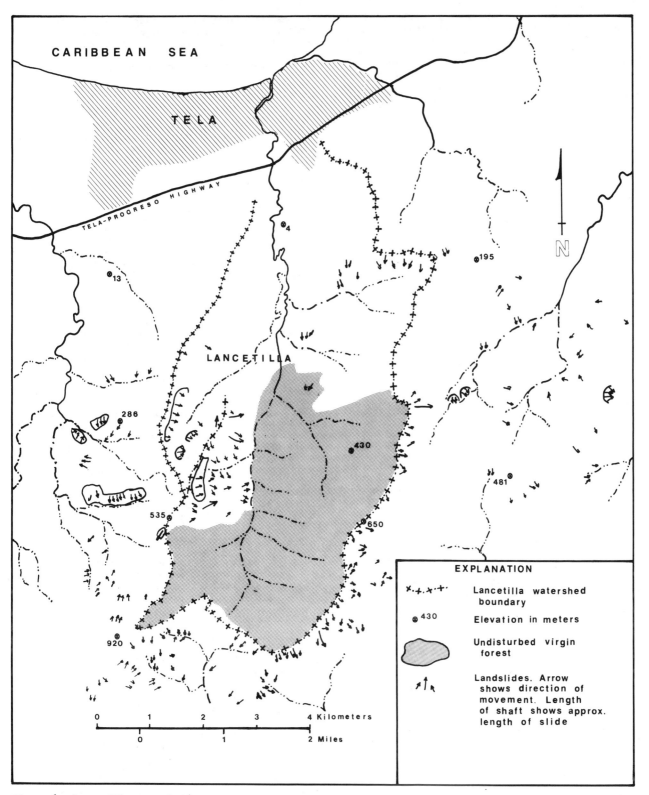

Figure 6. Lancetilla watershed, Department of Atlantida. Undisturbed forest (shaded area) of the Wilson Popenoe Botanical Garden. Surrounding watersheds (and a heavily impacted portion of the Lancetilla watershed), which have been deforested, roaded and cultivated, show numerous landslides - mostly debris slides. (Landslide identification on IGN aerial photograph no. 217, ln. 5, 1977, B&W, scale 1:40,000.)

and tenants of the housing development are extremely upset, not to mention homeless in some cases. This type of poor construction techniques was continuing in many areas as of 1987.

5. Major displacements have occurred in both the new Highway to the North and the Highway to Valle de Angeles, necessitating a major geotechnical solution costing over a million dollars in the first case, and frequent maintenance and an ongoing dangerous situation in the second.

6. A vertical wall 10 m high cut in fractured Tertiary volcanics (ignimbrite), failed and caved in the rear walls of several homes in a brand new housing project.

On the western side of the Tegucigalpa Basin, a recent (1987) invasion of poor people from rural Honduras occurred. The campesinos settled on an old massive hummocky landslide deposit covering several tens of hectares, beneath the steep cliffs of the old scarp (Ramirez and King, 1987). Not only has this been a political problem for the government, but also a problem of potential high risk to many hundreds of people who had no place else to settle and were unaware of the risks to their personal safety.

Many more examples of landslide effects and risks of future sliding exist for the metropolitan Tegucigalpa area. No formal records of loss of life from landslides were found for any area of Honduras. It should be noted, however, that even though many areas of the country are very steep and have a high incidence of landslide activity, many of those areas are uninhabited or sparsely populated, and the risks to humans are minimal.

7 MITIGATION

According to the OAS (1984), "Neither services nor offices in charge of identifying and studying natural hazards and their consequences exist in Honduras. Nor are there functionaries specifically qualified in the prevention and mitigation of natural hazards, nor others specifically charged with the protection against natural disasters...There is no legislation nor explicit regulation for prevention and mitigation of natural disasters."

The Council for National Emergencies (COPEN) is the Honduran governmental agency responsible for directing disaster relief. Many other agencies, both national and international, take part in emergency assistance and rehabilitation. However, less emphasis is placed on analyzing potential hazards and risks to people and property, and on prevention and planning for natural disasters. Landslide history is not recorded, landslide potential is not evaluated by the government, and losses due to landsliding are unknown. A search through records in various offices in the capitol city in 1987 produced no records of lost life and property damage due to landsliding. Due in part to both lack of funds and qualified personnel, COPEN has neither compiled disaster preparedness plans nor conducted technical assessments of potential hazards.

The work by the OAS, in cooperation with the Honduran planning agency (SECPLAN), is a start at providing background information about landslide and other types of hazards. The report by Ramirez and King (1987) concludes that since landslides so severely limit the capability of safe land use, landslide hazard information should be used during development of the master land use plan, as data in support of the creation of special zoning regulations, and for land disturbing projects. It also encourages creation of a governmental office with professionals in charge of landslide prevention and mitigation, and regulation of construction in unstable areas. At last report in early 1988, attempts were being made to establish that office.

Mitigation of landslides associated with roads typically involves removal of the slide debris and if necessary repair of the road surface. Occasionally slopes are re-shaped or terraced. Surface drainage may be rerouted. Subsurface drainage systems and vegetative stabilization are rarely used.

Mitigation of landlsides associated with housing developments varies greatly. One new subdivision near Tegucigalpa used grass to stabilize the surface of an uncompacted fill. The grass kept surface erosion to a minimum, but the fill settled and destroyed many homes. In another new subdivision, an attempt was made to install surface drainage along the base of a high steep cutslope, but a small landslide occurred before the project was completed due to the oversteepened cut. Mitigation of landslides does not appear to be well understood, nor part of standard operating procedures.

8 CONCLUSIONS

Little is known about the landslide situation in most of Honduras. What is presented in this paper is only a start on the vast amount of work which needs to be done. It is clear, however, that there are serious threats from landslides, which need to be considered when planning large scale developments.

The state-of-the-art of landslide identification, prevention and mitigation by Honduran geologists is assessed as being minimal to non-existent. Foreign geologists provide most of the geologic and landslide information thus far developed for Honduras. Although some reports are written and maps are published, when they leave, the foreign workers take with them most of the knowledge and experience gained.

Honduras could benefit greatly from having Honduran nationals trained in landslide identification, analysis, and mitigation. Engineering geologists could help define where and why instability problems exist. Geotechnical engineers could conduct detailed stability analyses and design appropriate engineering solutions.

Basic geologic mapping is still needed for 94% of the country. Until more Honduran geologists are trained and employed to do this type of work, foreign geologists, especially from interested universities and the Peace Corps, can fulfill an important role by mapping areas targeted for future development.

Management of landslide problems would be greatly enhanced by the creation and maintenance of a landslide management office/program with responsibility for inventory, analysis, mitigation, regulation, enforcement, documentation, coordination, and prevention activities. In addition, regulatory codes for construction in unstable areas would be very beneficial to reducing the risk to humans, structures and roads, provided they can be enforced.

9 ACKNOWLEDGEMENTS

The author would like to give special thanks to Honduran geologist Mario Casteneda and Venezuelan geomorphologist Rosa Ramirez G. for their tireless work and technical assistance during compilation of the two landslide studies, 1985 and 1987, respectively. Also, special acknowledgement and thanks to Marco Rodriquez and Pedro Mejia of the Honduran National Geographic Institute for their field assistance and research of files, and to Carlo Hugo Rivera and R.C. Finch for sharing their knowledge of Honduran geology.

The views expressed in this paper are those of the author and do not necessarily represent those of the USDA Forest Service nor of the Organization of American States or its member states.

Translation of all Spanish texts into English was done by the author who accepts responsibility for any errors. The manuscript benefitted by the review of Jerry DeGraff, Gordon Keller and Cinde King.

REFERENCES

Anderson, D.M. 1985. Geology of the Lepaterique Quadrangle, Honduras, Central America. Unpub. rpt. for Inst. Geog. Nac. & Peace Corps Honduras: 1-65.

_____ 1987. Geologic map of Honduras, Lepaterique quadrangle, scale 1:50,000. Tegucigalpa: Inst. Geog. Nac.

Bender, S.O. 1984. Project description: Natural hazard risk assessment and disaster mitigation pilot project in Latin America and the Caribbean Basin. Unpub. rpt. Dept. Reg. Devel., OAS:1-11.

Case, J.E. & T.L.Holcombe 1980. Geologic-tectonic map of the Caribbean region: U.S. Geol. Surv. Misc. Invest. Map I-1100, scale 1:2,500,000.

Cunningham, C.G., R.W.Fary Jr., M.Guffanti, D.Laura, M.P.Lee, C.D.Masters, R.L.Miller, F.Quinones, R.W.Peebles, J.A.Reinemund, & D.P.Russ 1984. Earth and Water Resources and Hazards in Central America. U.S. Geol. Surv. Cir. 925:1-40.

Curran, D.W. 1981a. Geologic map of Honduras, Siguatepeque quadrangle, scale 1:50,000. Tegucigalpa: Inst. Geog. Nac.

_____ 1981b. Geologic map of Honduras, Taulabe quadrangle, scale 1:50,000. Tegucigalpa: Inst. Geog. Nac.

_____ 1981c. Geologic map of Honduras, Santa Cruz de Yojoa quadrangle, scale 1:50,000. Tegucigalpa: Inst. Geog. Nac.

DeGraff, J.V. & H.C.Romesburg 1980. Regional landslide-susceptibility assessment for wildlands management: a matrix approach. In D.R.Coates & J.Vitek (eds.), Thresholds in Geomorphology. p.401-414. Boston, George Allen & Unwin.

Dupre, W.R. 1971. Geologic map of Honduras, Zambrano quadrangle, scale 1:50,000. Tegucigalpa: Inst. Geog. Nac.

Elvir A., R. 1974. Mapa geologico de la republica de Honduras (Geologic map of the Republic of Honduras), scale 1:500,000. Tegucigalpa: Dir. Gen. Minas e Hidrocarburos, Inst. Geog. Nac. & Banco Nac. Fomento.

Emmet, P.A., W.S.Logan & W.R.Muehlberger, 1983. Geologic map of Honduras, Agalteca quadrangle, scale 1:50,000. Tegucigalpa: Inst. Geog. Nac.

Everett, J.R. 1970. Geologic map of Honduras, Comayagua quadrangle, scale 1:50,000. Tegucigalpa: Inst. Geog. Nac.

Fakundiny, R.H. 1971. Geologic map of Honduras, El Rosario quadrangle, scale 1:50,000. Tegucigalpa: Inst. Geog. Nac.

Finch, R.C. 1979. Geologic map of Honduras, San Pedro Zacapa quadrangle, scale 1:50,000. Tegucigalpa: Inst. Geog. Nac.

_____ 1981. Mesozoic stratigraphy of Central Honduras. Am. Assoc. Petrol. Geol. Bull., 65(n 7):1320-1333.

_____ 1985. Geologic map of Honduras, Santa Barbara quadrangle, scale 1:50,000. Tegucigalpa: Inst. Geog. Nac.

Frazier, K. 1976. The shifting, stretching crust of Central America. Sci. News, 110(n 15):234-235.

Harp, E.L., R.C.Wilson, & G.F.Wieczorek 1981. Landslides from the February 4, 1976, Guatemala Earthquake. U.S. Geol. Surv. Prof. Paper 1204-A:1-35.

King, A.P. 1972. Geologic map of Honduras, Talanga quadrangle, scale 1:50,000 Tegucigalpa: Inst. Geog. Nac.

_____ 1973. Geologic map of Honduras, Cedros quadrangle, scale 1:50,000. Tegucigalpa: Inst. Geog. Nac.

_____ 1985. Landslide hazard analysis, Department of Atlantida, Honduras. Unpub. rpt. Natural Haz. Pilot Proj., Dept. Reg. Devel., OAS:1-14, 21 maps.

Kiremidgian, A.S., P.Sutch, & H.C.Shah 1979. Seismic Hazard Analysis of Honduras. The John A. Blume Earthquake Engr. Ctr., Dept. Civil Engr., Stanford U., Rpt. n. 38 (final).

Mills, R.A., K.E.Hugh, D.E.Feray & H.C.Swolfs 1967. Mesozoic stratigraphy of Honduras. Am. Assoc. Petrol. Geol. Bull., 51(n 9):1711-1786.

Organization of American States 1984. Programa de evaluacion de riesgos naturales (Natural hazards program evaluation). Leonardo Alvarez Associates. Contracted rpt. for Dept. Reg. Devel., OAS:1-42.

Plafker, J. 1977. The Guatemala earthquake and Caribbean plate tectonics. U.S. Geol. Surv. Earthquake Info. Bull., 9(n 2):1-39.

Ramirez G., R. & A.P.King 1987. Evaluacion preliminar de riesgos de derrumbes en el area de Tegucigalpa, Honduras (Preliminary evaluation of landslides in the area of Tegucigalpa, Honduras). Unpub. rpt. Natural Haz. Pilot proj, Dept. Reg. Devel., OAS:1-36, 16 maps.

Ramos N., N. & S.Vicencio 1976. Comentarios sobre la situacion del Cerro El Picacho (Comments about the situation of El Picacho Mountain). Dir. Gen. Minas e Hidrocarburos: Rpt no. 287:1-3.

Rivera, C.H. 1982. Reporte Visita Colonia "El Reparto" (Report on visit, Colonia "El Reparto"). Dir. Gen. Minas e Hidrocarburos:1-4.

Simonson, B.M. 1981. Geologic map of Honduras, El Porvenir quadrangle, scale 1:50,000. Tegucigalpa: Inst. Geog. Nac.

Unknown Author 1986 est. Geologia, proyecto aguas subterraneas y Montana El Chile para Tegucigalpa (Geology, groundwater and El Chile Mountain project for Tegucigalpa), scale: 1:25,000. Tegucigalpa: ITS Lotti & Associati, Cooperazione Tecnica Italiana.

U.S. Agency for International Development 1981. Honduras, a country profile. Evaluation Technologies, Inc. Contract No. AID/SOD/PDC-C-0283: 1-55.

Weyl, R. 1980. Geology of Central America. Berlin, Stuttgart: Gebruder Borntraeger.

Landslides: Extent and Economic Significance, Brabb & Harrod (eds)
© 1989 Balkema, Rotterdam. ISBN 90 6191 876 6

Extent and social-economic significance of slope instability in Costa Rica

Sergio Mora C.
Departamento Geologia, ICE; Escuela C.A. Geologia-CIGEFI, Universidad de Costa Rica; INGEOSA, Costa Rica

ABSTRACT: Steep topography, high intensity rainfall, volcanoes, earthquakes and weak geologic units make Costa Rica vulnerable to erosion and slope instability. Landsliding has been accelerated by road construction, deforestation, overgrazing, and exploitation of mineral resources. Slope instability is widespread and occurs as soil erosion and soil and/or rock slides. Increased population and construction in mountainous terrain exposes more people and more economic development to the landslide problem.

1 INTRODUCTION

Landslides have been a major source of social and economic loss for the population of Costa Rica at least since the 1888 Fraijanes earthquake. Earthquake and rainfall-triggered landslides were formerly considered sensational, but they have now become so widespread and commonplace that people consider them a part of their daily lives. As the population expands into hillside areas, slopes have been disturbed and the frequency of landsliding has increased.

A magnitude ML=6.1 earthquake July 3, 1983, in Pérez Zeledón focused national attention recently on landslide destruction. About 25 km² were devastated by landslide debris, and another 150 km² were affected in a direct or indirect way by landsliding. Shortly thereafter, other landslides occurred and threatened the populations close to Río Chiquito, Tapezco, San Blás, Purisil, and damaged the Braulio Carrillo Highway and other areas. Currently, slope instability has caught the attention of a large portion of the geoscientific community of Costa Rica, but little action on the part of the authorities has so far taken place.

2 LOCATION

Costa Rica is located in the Central America Isthmus, bordered on the north by Nicaragua, on the east by the Caribbean Sea, on the southeast by Panamá, and on the west and south by the Pacific Ocean (Fig. 1). The country averages about 180 km in width.

3 RAINFALL

Mean annual precipitation varies widely in Costa Rica. The driest areas get approximately 1,300 mm/year, while at the other extreme, some areas get as much as 8,000 mm/year. In the Pacific watershed, a long dry season (December through May) is followed by a high-intensity rainy season with as much as 150 mm/24 hours at least once every five years. Rainfall occurs throughout the year in the Caribbean watershed, with slightly less rain from September to November. Intensities of rainfall in this region are frequently on the order of 150 mm/24 hours every five to seven years and can reach as much as 1,500 mm per month.

4 PHYSIOGRAPHY

About 70% of Costa Rica is mountainous. The Guanacaste Volcanic Range in the northwest has five

major active or recently active volcanoes, the highest of which, Rincón de la Vieja, has an elevation of almost 3,000 m. Towards the southwest, the Tilarán Range consists of volcanic and plutonic rocks. The Central Volcanic Range has four major volcanoes, three of which have been active several times this century. The Irazú Volcano, with an altitude of 3,412 m, dominates the region. In the southeast, the Talamanca Range, comprised of volcanic and plutonic rocks, has Cerro Chirripó, the highest peak in the country (3,819 m), and Cerro

Figure 1. General location map of Costa Rica.

Kámuk (3,520 m). Both of these peaks had glaciers until about 10,000 to 12,000 years ago. A subsidiary set of mountains, the Fila Costeña Range consisting of sedimentary rocks, rises to 2,200 m south of Talamanca. All of these ranges have steep slopes averaging 40° to 50° and approaching verticality in many areas. The Central Valley, with an elevation from 800 m to 1,400 m, accommodates about 65% of the total population, but the portion of population living near steep slopes vulnerable to instability could be as high as 90% to 95%.

5 ENVIRONMENT

High rainfall, frequent earthquakes, active volcanoes, steep slopes, and weak geologic units contribute to widespread landsliding. Recent earthquakes have not occurred during periods of heaviest rainfall when the slopes would be most vulnerable to landsliding. About 22% of the country is covered by practically untouched forestland, protected by legislation, where landsliding is rare or absent.

6 PREVIOUS WORK

Before 1979, analyses of slope stability were made only to solve specific problems, mainly related to agriculture or hydropower projects. Investigations of the Rio Reventado-Taras December 4, 1963, lahar (Leeds and others, 1967; Krushensky, 1968, 1972; Instituto Costarricense de Electricidad, 1965; Umaña, 1974; and Van Ginnecken, 1978) were the exception. The lahar was triggered by the combined action of the degradation of the watershed caused by deposition of ashes from the Irazú Volcano and the activation of very intense erosion and several landslides by a rainfall storm (180 M/14 hours). Before these accounts, only newspaper descriptions and reports on landslides were available (Pittier, 1889; González, 1910).

During the 1980's, a few landslide analyses were published or released as university theses because of the extensive exposure of the population and infrastructure to hazardous conditions in many parts of the country (Granados and others, 1981; Estrada, 1987; Urena, 1985; Lezama and Leandro, 1980; Leon and others, 1981; Mora, 1986, 1988). Analyses of the probable impact of landslides on bridge sites, roads, pipelines, etc., also began to appear. These reports indicate that the geoscientific community has become aware of the importance of landslides (Mora & others, 1985; Lezama, Mora & others, 1982). However, the political situation is a different matter; little or no effective preventive work has been installed.

7 GEOLOGIC BACKGROUND

Clastic sedimentary and igneous rocks weather rapidly in Costa Rica and develop residual (regolithic) soil profiles. Hydrothermal alteration is also common. Tectonism has been intense during Tertiary and Quaternary time, producing highly fractured rocks.

8 SHEET EROSION

Large areas of the country are affected by sheet erosion (Fig. 2) which occurs where moderately steep slopes are overgrazed, deforested, or damaged by other agricultural practices. Soil on these bared slopes is removed during periods of high intensity rainfall. Organic components and other nutrients are removed from the bared soil, inhibiting renewal of vegetation and contributing to additional

Figure 2. Schematic map showing areas affected by intensive erosion in Costa Rica.

erosion. This type of erosion is widespread in the Nicoya and Santa Elena Peninsulas, the southwestern slope of the Central Valley and the El General Valley.

9 CONCENTRATED EROSION

When the flow of surface runoff water is concentrated more or less perpendicular to topographic contours along or through any kind of preferential feature such as a fault, desiccation crack, bedding surface, or scarplet along an animal path, kinetic energy is sufficient to open incisions varying in dimension with the discharge volume and slope. This process is common in Costa Rica, especially on steep slopes where vegetation has been removed and the rate of soil loss exceeds the possibility for the growth of new vegetation. The erosion forms ruts, "canadas", ravines, and badlands.

10 ADDITIONAL EROSION CAUSED BY HUMAN ACTIVITY

Road building and open-pit mineral exploitation also cause soil erosion and landsliding. The Braulio Carrillo, Interamerican, Costanera, and Osa highways, for example, have many landslides along excessively steep highway cuts installed to "save money." Quarry operations have caused the Rio Virilla, the San Blás and Rio Chiquito landslides, and placer gold mines on the Osa Peninsula have produced large quantities of sediment, have diminished the capacity of hydroelectric reservoirs, such as the Cachi Plant, and also increased ocean and river pollution.

11 SOLIFLUCTION (CREEP)

Solifluction occurs in slopes of moderate inclination (5° to 30°) at speeds on the order of a meter per year. Soil deposits involved in this type of instability are rarely thicker than 5 m but commonly have shallow water tables. Solifluction is a widespread phenomenon in Costa Rica, especially where natural or artificial drainage is poor, such as along roads where the drainage system is inadequate or on overgrazed slopes where cowpaths trap water. Deeply-rooted trees may be useful in preventing this process from expanding. Examples can be found in the areas of Fila Costeña, Tilarán,

Escazú and Esperanza on both lateritic and zonzocuitle soils (Figs. 3 and 4).

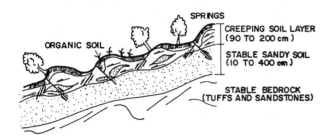

Figure 3. Example of creeping soil at the Alto de Las Palmoas and Escazú-Santa Ana Highway

Figure 4. Map showing distribution of solifluction (creep) and shallow landslides in Costa Rica

12 LANDSLIDES AND SOIL

Two kinds of soils are particularly susceptible to the occurrence of landslides:
a. Lateritic-bauxitic residual soils with large amounts of gibbsite and halloisite form deposits as much as several tens of meters thick. They are classified mostly as CH and MH (lp=30-75%) and in most places contain more natural humidity than their liquid limit (1-8%), especially throughout the long rainy season.

b. Montmorillonitic-bentonitic soils, having either residual (hydrothermal alteration of pyroclastic or old lava deposits) or transported (mangrove swamp) origin in deposits generally less than 15 m thick. They are locally called "zonzocuitle," middle-American indigenous language for "blackish-sticky soil," which describes perfectly its color and consistency. These soils are classified as CH and even OH (lp=75-150%) with large expansive behavior when water is added.

Landslides on residual soils are probably the most common type of landslide in Costa Rica (Fig. 4). Either seismic activity or heavy rainfall contribute to the mobilization of the shallower soil horizons by an increase of pore pressure. After the soil fails, intensive uphill erosion usually develops, followed by debris and mud flows of considerable dimensions. One small area of trees may be sufficient to halt the process. Landslides on residual soils have taken place recently near Cerros de Escazu, Candelaria, Savegre, and Coen.

13 MOST SIGNIFICANT LANDSLIDES

The most significant landslides economically were associated with the Tilarán (March 14, 1973) and Pérez Zeledón-División (July 3, 1983) earthquakes and the Sabalito (October 10, 1986) and Pejibaye-El Humo (July 2, 1987) storms. The Pérez Zeledón-División earthquake is unprecedented in the extent of the landslide damage (Fig. 5). Within an area of 175 km^2, the slopes affected in relation to their steepness are summarized in the following table:

TABLE 1. Areas affected by landslides during the July 3, 1983, Pérez Zeledón-División earthquake in relation to steepness (Leandro and others, 1983).

Slope degree	Area (km^2)	Percent of area affected, lithology
Less than 20°	16	12%; mostly solifluction, lateritic soils
20-35°	26	36%; altered basaltic-andesitic soils and colluvium
36-50°	71	79% colluvium; altered basaltic-andesitic rocks
51-65°	33	80%; altered granodiorite; basaltic-andesitic rocks
66-80°	21	62%; intrusive, basaltic-andesitic fresh to 56% moderately altered rocks
more than 80°	8	50%; fresh to moderately altered rocks
TOTAL	175	65%, Average

Slopes from 35° to 80° were the most vulnerable to landsliding. Many agricultural and grazing areas and about 18 km of the Inter-American Highway were destroyed. Debris avalanches and intensive river erosion provided immense quantities of silt and clay to waterways (Leandro and others, 1983).

The Pejibaye-El Humo storm of July 2, 1987, dropped more than 270 mm of rain in 4 hours (Fig 6). At least 25 km^2 of land experienced landsliding, avalanching, and flooding (Fig. 7). Losses were considerable: 14 bridges, 11 km of roads, 26 houses, 2,500 ha of coffee and sugar cane plantations, 20 days of power production at the Cachí (110 MW) hydropower plant, and at least 3 people killed (Mora, Valdés and Ramírez, 1988). The total amount of losses estimated ranges around one hundred million colones (US$12.5 million).

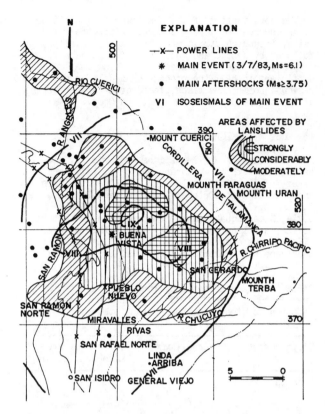

Figure 5 Isoseismals of the main event (MM:VII to IX), of the Pérez Zeledón-División earthquakes and classification of the areas affected by landslides (Leandro and Morales, 1985; Mora and Rivas, 1984; and Mora, Valdes, and Chávez, 1984)

During the October 10, 1986, Sabalito storm, a weathered, old lahar on a deforested slope slid and initiated a mudflow, killing a family of 7 and destroying 150 ha of a coffee plantation. Analysis of samples taken from the landslide two days later showed a water content of 8% more than the liquid limit of 56%.

14 SOIL SLIDES WITH CURVED OR IRREGULAR FAILURE SURFACES

Soil slides occur in thick deposits of alluvial, colluvial, and residual soil with a shallow water table (less than 15 m in depth) and on slopes from 20° to 40°. Differences in the soil composition, geology and hydrodynamic behavior of the slope are parameters defining the shape of the failure surface (semi-circular, semi-elliptical, irregular) and the total volume of the mass affected. Several small landslides of this type were generated during and shortly after the construction of the roads to the Osa Peninsula and in Valle de Talamanca, Paraíso (Cartago) and Savegre Abajo (Fig. 8).

Moderately large slides are found near Empalme, Upala and Chiripa, and larger ones (up to half million cubic meters) occurred near Tilarán, Río Chiquito and Sombrero (Paraguas). The Río Chiquito landslide (Fig. 9) was triggered during a heavy rainstorm and after over-exploitation of a quarry.

15 ROCKFALLS AND ROCK-BLOCK SLIDES

Rockfalls and rock-block slides occur on steep, rocky slopes in river gorges, canyons, narrow valleys and road cuts. Most of these landslides occurred along fractures, joints, and bedding

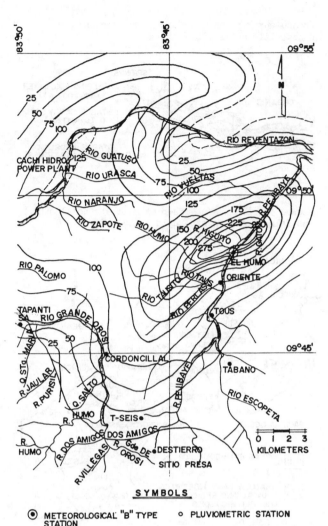

Figure 6. Isohyets from the July 2, 1987, rainstorm (between 14 and 19 hours).

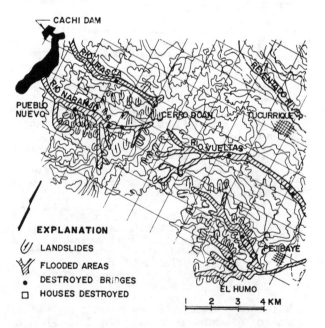

Figure 7. Landslides formed during the July 2, 1987, Pejibaye-El Humo rainstorm.

96

Figure 8. Distribution of known rock slides and curved or irregular soil slides in Costa Rica

- AREAS WITH EXAMPLES OF ROCK SLIDES.
- AREAS WITH EXAMPLES OF UPWARDLY CURVED OR IRREGULAR SOIL SLIDES

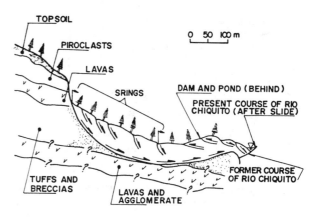

Figure 9. Schematic section of the Río Chiquito, Tres Ríos, landslide.

oriented unfavorably in relation to slope. Toppling of lava and ignimbrite columns and blocks is common in the Virilla and Río Grande River canyons (Fig. 10). Blocks and wedges have slid at a steep road cut in the microgranite of Siberia (Cerro de la Muerte, Inter-American Highway), and along bedding failures and shale in the Savegre Abajo Sandstone and Fila Cal Limestone. Undermining of river waterfalls is also common, such as in the lavas and tuffs of Prendas and in the limestones, gabbros and shales of Quebrada Canchén and Boruca (Fila Costeña) (Fig. 10).

16 MASSIVE SOIL AND ROCK SLIDES

Eighteen large landslides involving at least one million cubic meters of soil and rock masses have been identified recently in Costa Rica (Fig. 11).

A few of these with destructive potential have been studied, and are described below. Unfortunately, the lack of a centralized agency and a policy for prevention and mitigation hinders more detailed scientific and engineering research. Without such knowledge, the establishment of control, alert, and interactive preparedness systems would seem to be a waste of time and resources.

The town of Santiago de Puriscal, with 15,000 inhabitants, is slowly creeping downhill on a large landslide. Another large landslide near San Blás (Figs. 12 & 13) developed in an old lahar deposit overlying a hydrothermally-altered lava flow from the Irazú volcano. It was triggered by undermining of a quarry face. This landslide covers around 70 ha and has a volume close to 50 million cubic meters. If only a small part of this mass should dam the Reventado River and impound water, the water might all be released suddenly in a destructive wave that could drown 25,000 people in Cartago and flood an extensive agricultural area, 2 power lines, 14 bridges, a railway, 4 major aqueducts, an oil pipeline, a major hydropower reservoir (110 MW) and one of the major industrial parks in the country (ICE, 1964; Umaña, 1974; Leeds and others, 1967; Lezama and Leandro, 1980; Mora, Estrada and Delgado, 1985; Estrada, 1987; Mora, 1985, 1986) (Fig. 12).

Tapezco is another landslide with destructive potential. It formed on hydrothermally-altered sandstone and shale, and it covers about 25 ha with a volume of almost 7 million cubic meters. Movement

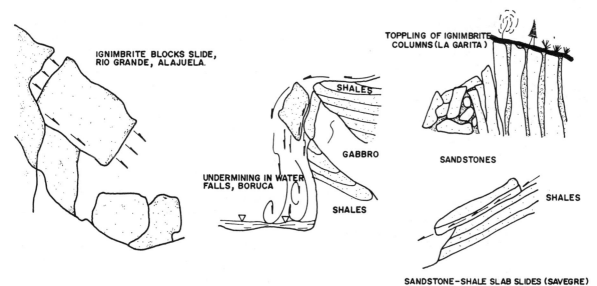

Figure 10. Schematic views of common rock slides in Costa Rica.

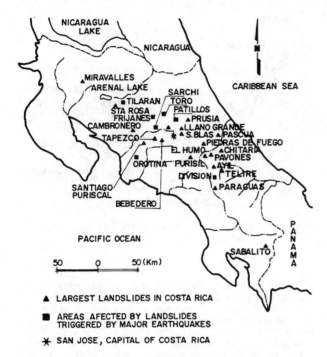

Figure 11. Map showing examples of the largest landslides in Costa Rica

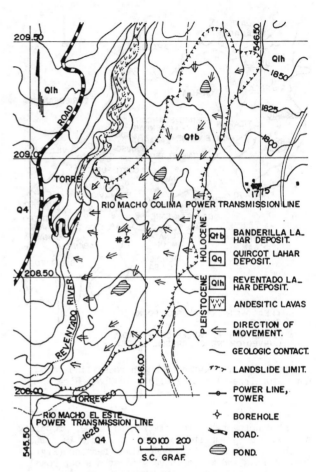

Figure 13. The San Blas landslide (After Mora, Estrado, Delgado, 1985).

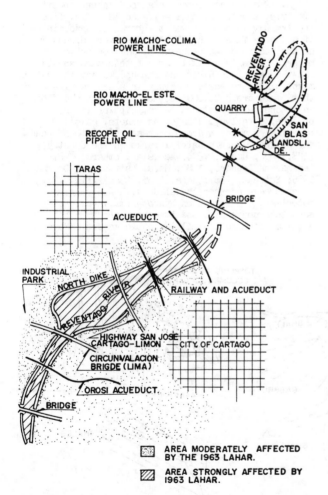

Figure 12. Most important infrastructural civil facilities associated with the hazard of the San Blas landslide.

of this landslide could dam the Uruca River. If the temporary embankment failed catastrophically, the town of Santa Ana with a population of at least 10,000, and a large agricultural area, infrastructure and lifelines, could be destroyed (Ureña, 1985; Mora, 1988) (Fig. 14).

The Reventazon River watershed contains at least four major landslides which occurred after deforestation became a common practice during the second part of this century. The Piedras de Fuego landslide (approximately 2.5 million cubic meters) formed on a highly hydrothermally-altered clastic sedimentary rock. Other landslides, such as the Pavones, Chitaría and Pascua, developed on soft, weathered and saturated sandstone, shale, and agglomerate. The Piedras de Fuego landslide continuously affects the operation of the railway to the Caribbean coast, whereas the Pavones and Chitaría affect the Turrialba-Siquirres road and power lines. The Pascua landslides endangered a town of 1,500 people during reactivation in August, 1987.

17 LANDSLIDES GENERATED BY SEISMIC ACTIVITY

Seismic activity is one of the most common triggering mechanisms in Costa Rica. The Pérez Zeledón-División earthquake of July 3, 1983 (MS=6.1; Z=13 km; MM=1X) produced the largest area of landslide destruction in Costa Rica history (Fig. 5). Another major seismic event was the March 14, 1973, Tilaran earthquake (ML=6.5; MM=1X) which generated many landslides in the Arenal Lake area (Figure 11).

Other earthquakes have induced landslides and mudflows in the Central Valley and surrounding

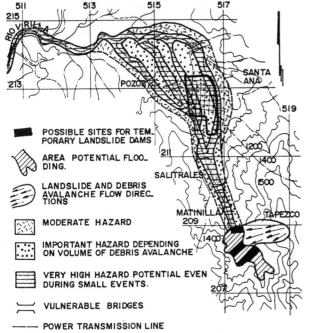

POSSIBLE SITES FOR TEM. PORARY LANDSLIDE DAMS

AREA POTENTIAL FLOO. DING.

LANDSLIDE AND DEBRIS AVALANCHE FLOW DIREC. TIONS

MODERATE HAZARD

IMPORTANT HAZARD DEPENDING ON VOLUME OF DEBRIS AVALANCHE

VERY HIGH HAZARD POTENTIAL EVEN DURING SMALL EVENTS.

VULNERABLE BRIDGES

POWER TRANSMISSION LINE

Figure 14. Areas of potential hazard from the Tapezo landslide.

areas. The December 30, 1888, Fraijanes earthquake (MM=VIII) generated a landslide that buried 6 people. Other earthquakes on August 29, 1911, Bajos del Toro I (MM=VII); June 6, 1912, Sarchí (MM=VII); March 4, 1924, Orotina (MS=7.0; MM=X, Aquacate Mountains) December 30, 1952, Patillos (MM=VII, 25 people killed); and September 1, 1955, Bajos del Toro II (MM=VIII, 6 people killed) could have generated landslides.

CONCLUSION

Slope instability and erosion are natural hazards which have resulted in losses to the Costa Rican population and economy. Most erosion is a consequence of inappropriate land use in agriculture, grazing, open-pit mining, quarrying and road construction.

Landslides develop mostly on residual (regolithic) soils derived from sedimentary or igneous rocks. Slides with curved failure-surfaces have formed on thick deposits of laterite and montmorillonite. At least eighteen large landslides of more than one million cubic meters of volume have been identified. Of those, the San Blás landslide in Cartago is by far the largest with nearly 50 million cubic meters in volume and a movement rate of 8 to 15 m per year.

Clay layers rich in montmorillonite and shallow water tables, particularly during the rainy season, facilitate slope failure. Heavy rainfall, seismic activity, and steep slopes also contribute to landsliding.

The growth in population and the development of urban areas, and life lines in mountainous terrain increase the exposure of people to erosion and landsliding.

ACKNOWLEDGMENTS

MSc. María Laport reviewed the first manuscript and provided many helpful suggestions. Mrs. Flory Abarca kindly typed it. Dr. Earl Brabb edited this paper and made many improvements.

REFERENCES

Estrada, A. 1987. Estudio geólogico-geotécnico del deslizamiento de San Blás, Rio Reventado, Costa Rica. Escuela Centroamericana de Geología, U. Costa Rica. Tesis de Licenciatura, inedita: 136.

González, C. 1910. Temblores, terremotos, inundaciones y erupciones volcánicas en Costa Rica. Imprenta Alsina, San Jose: 200.

Granados, R. & others 1981. Estudio del deslizmazamiento del Río Chiquito, Tres Ríos. Escuela Centroamericana de Geología, U. Costa Rica. Informe inédito: 19.

Instituto Costarricense de Electricidad 1965. Informe sobre el problema del Rio Reventado. Informe preliminar. San José, Costa Rica: 312.

Krushensky, R. 1968. Geology of the Río Reventado watershed, Costa Rica. USGS Rept. (1R) CR-9A: 19.

Leandro, G., C.León, F.Montalto, J.Elizondo & R.Chaves 1983. Informe geologico-geotecnico preliminar sobre el sismo de La División, Pérez-Zeledón (3 julio, 1983). Com. Reg. Emerg-Depto. Geol., ICE-RECOPE, Informe inedito: 69.

Leeds, Hills & Sewett 1967. A master plan for flood control on the Río Reventado and Río Tiribí watersheds, Costa Rica. Washington US Dept State-AID: 171.

Lezama, G. & G. Leandro 1980. Informe sobre problemas de deslizamientos en Río Reventado a la altura de la Linea de Transmision, Rio Macho-Colima. Departamento de Geologia, ICE.

Mora, S. 1985. Las laderas inestables de Costa Rica. Revista Geologica de America Central. 3:131-161.

Mora, S. 1986. Comentarios acerca de la problematica generada por las amenazas geologico-antropicas en la cuenca del Río Reventado. Cartago, Costa Rica. Informe para Ministerio de Vivienda y Asenta-mientos Humanos. Depto. de Geologica, ICE:22 p.

Mora, S. 1988. Analisis preliminar de la amenaza y vulnerabilidad generadas por el deslizamiento del Alto Tapezco, Santa Ana, Costa Rica. 4to. Seminario Nacional de Geotecnia. Asoc. Cost. Mecanica de Suelos e Ing. de Fundaciones. 14-15 abril. C.F.I.A.:16 p.

Mora, S., A.Estrada & J.Delgado 1985. Analisis del deslizamiento de San Blás, Rio Reventado, Costa Rica. Primer Simpos. Latinoamericano sobre Des-astres Naturales, noviembre. Quito, Ecuador:18 p.

Mora, S & L.D.Morales 1986. Los sismos como Fuente generadora de deslizamientos en Costa Rica y su impacto sobre las lineas vitales e infrae-structura. Primer Simposio Latinoamericano de Riesgo Sismico. San José, Costa Rica. CFIA:8 p.

Mora, S., R.Vales & C.Ramirez 1988. Los deslizamientos de Cachí-El Humo-Pejibaye, sus causas y consecuencias. 4° Seminario Nacional de Geotecnia; Asoc. Cost. Mecanica de Suelos Ing. Fundaciones, CFIA:19 p.

Perez, S. & P.VanGinnecken 1978. Mapa de Capacidad de uso del suelo de Costa Rica, escala 1:200,000. OPSA:MIDEPLAN, noticia explicativa:180 p.

Pittier, H. 1889. Observaciones y exploraciones efectuadas en Costa Rica en 1988. Tipografia Nacional:16 p.

Umaña, J. 1974. Movimientos lentos en masa hacia el Río Reventado en las cercania de la Linea de Transmision Río Macho-Colima. Depto. Geologia, ICE:38 p.

Ureña, R. 1984. Estudio del deslizamiento del Alto Tapezco, Santa Ana, Costa Rica. Escuela de Ingenieria Civil, Univ. de Costa Rica. Tesie de Licenciatura, inedita:48 p.

Waldron, H. 1967. Debris flows and erosion control problems caused by the ash eruption of Irazú Volcano, Costa Rica. USGS Bulletin 1241-I: 37.

Landslides: Extent and Economic Significance, Brabb & Harrod (eds)
© 1989 Balkema, Rotterdam. ISBN 90 6191 876 6

Slides in Panama

R.H. & J.L.Stewart
Lutz, Fla., USA

ABSTRACT: The Republic of Panama is a tropical country bounded on the north by the Caribbean Sea, on the south by the Pacific Ocean, by Colombia on the east, and by Costa Rica on the west. It has a two season tropical climate, wet and dry. Mountains extend from Colombia to Costa Rica with a lower area in the vicinity of the Panama Canal. Slides in general are found in all the mountainous and hilly regions of the country. The tropical soil profiles developed by the weathering of most rocks in the area are particulary sensitive. Many slides in Panama are the result of construction practices but many more are the result of weathering, heavy rains, and seismic activity. Little has been written about the slide problem in Panama except those resulting from the construction of the Panama Canal. Slides as the result of seismic activity are described as are a few resulting construction practices.

1 LOCATION.

The Republic of Panama is a tropical country located between the Caribbean Sea on the north and the Pacific Ocean on the south; it is bounded on the east by the country of Colombia in northwestern South America, and on the west by the country of Costa Rica. It is roughly between latitudes 7° and 10" north to 9° and 40" north, and longitudes 77° and 10" west to 83° and 3" west.

2 AREA AND POPULATION DISTRIBUTION.

Panama has a total land area of approximately 28,753 square miles with a population of slightly over 2,000,000. The greatest concentration of population is centered around Panama City, Colon and their suburbs. There are also four more large cities, Aguadulce, Chitre, Santiago, and David in the rest of the country. These cities are located in the lower, flatter areas where agriculture is the main form of employment. There are several smaller towns in the mountainous areas that have some agriculture in the form of coffee and truck gardening. The greater part of the population is located in relatively slide free areas.

3 PHYSICAL FEATURES.

The most prominent physical feature of the country is the countain chain which extends the length of the country from Colombis on the east and into Costa Rica on the west. To the east of the Panama Canal, there are two mountain chains with a broad, wide valley between them. In the central part of the country in

the vicinity of the Canal the average height of the mountain chain is very low, 200 m. The Panama Canal was constructed in the lowest part of the mountain chain. In western Panama there are broad, flat areas both to the north and to the south of the central mountain chain which reach elevations of 4,000+m. The majority of agriculture in Panama is developed in these areas to the south and north of the mountain range in the lower, flatter areas. Steep slopes are common throughout the entire mountainous areas of Panama.

4 CLIMATE.

The climate of Panama is tropical; divided into a rainy season lasting from late April to mid December, and a dry season, when little rain falls, from mid December to late April. The actual times of the ending and beginning of the different seasons can vary as much as a month. The rainfall in the country varies greatly; during the dry season the average rainfall may range from 0 to 127 mm and during the rainy season it may be as much as 630 mm per month or more. Along the southern coast the annual rainfall may vary from 762 mm to 1,270 mm, along the northern coast the average rainfall varies from 3,175 mm to more than 5,000 mm. In the mountains to the west of the Canal the average rainfall may be as much as 10,000 mm on the north side of the divide and 5,000 mm on the south side. Similar conditions exist in the mountainous areas to the east of the Canal toward the Colombia-Panama boundry.

The mean average temperature of Panama along the coast is 28° C becoming cooler as you ascend in elevation.

5 GEOLOGY.

The geology of Panama is that of a volcanic island arc with volcanic activity extending through time from the deep sea Cretaceous spreading center to the Recent. Intersperced with the volcanic activity there was a considerable amount of marine sedimentation. All of Panama was raised above sea level at the end of the Pliocene period about 3,000,000 years ago.

The rock units in the wide central valley of the Bayano-Chucunaque-Tuira rivers in eastern Panama consist of low lying, relatively flat lying, marine sediments near sea level. Steep slopes have not been

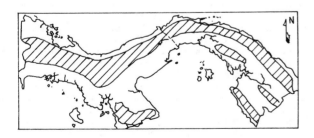

Figure 1. Mountainous areas of Panama.

developed on them and they are relatively free of sliding. The only other areas not subject to sliding are around Aguadulce, Chitre and Santiago which were developed on an old peneplane near sea level; in the lower volcanic outwash slopes of western Chiriqui province, and south of El Valle in central Panama.

The majority of natural slides in Panama are the result of the combination of deeply weathered, saprolitic, lateritic soils, and the oversaturation and overloading caused by heavy rainfall. This type of sliding occurs in all parts of Panama.

Occasionally, during the rainy season, a stalled weather front will drop unusual amounts of rain, 254 mm to 500 mm, in 24 hours. The steeper slopes will literally liquify and flow into the valleys. In general, such activity takes place in the mountains on the steeper slopes where the population is very low. There has been very little loss of life as the result of sliding.

6 TYPE OF LANDSLIDE PROCESSES AND GEOGRAPHIC EXTENT.

The types of landslide processes at work in Panama are listed as follows:
1. Slides in slopes with deep weathering profiles.
2. Slides as a result of sapping.
3. Slides as a result of seismic activity.
4. Slides as a result of manmade construction activities.
Each of these slide processes will be discussed separately.

6.1 Slides in slopes with deep weathering profiles.

A typical soil profile in Panama contains the standard A, B, and C zones but there is a difference. The A zone is usually very thin or even absent because the high bacterial count in the soil prevents little, if any, humus from forming. The A zone with high humus, and organic content can only form in the lower elevations where the soils are alkaline, with a high Ph. However, they are common at higher elevations, above 1,000 m where the temperatures are lower and the bacterial content of the soil is considerably lower.

The B zone with its low organic content and high mineral content is common and thick. The soil particles range from sand-size through clay-size particles and grade into the C zone below.

The C zone with its sand-and silt-sized particles, and weathered rock varies in thickness and grades into the weathered rock and then into the sound rock below.

A more meaningful soil profile developed from an engineering standpoint clearly portrays the size-strength characteristics that are so important in understanding the process of sliding in Panama.

Table 1. Engineering soil profile.

Grain size	Unconfined compression strength	Porosity
Clay	3-5 tons/ft^2	Low
Clay and silt	2-4 tons/ft^2	Low
Silt and clay	1-2 tons/ft^2	Moderate
Silt	.25-1 tons/ft^2	Moderate to high
Silt and sand	.25 tons/ft^2	High
Sand	.25 tons/ft^2	High
Sand and weathered rock	.25-10 tons/ft^2	Moderate to high
Sound rock	10-50+ tons/ft^2	Relatively impermible

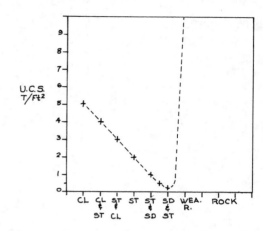

Figure 2. Typical strength curve for tropical soil profiles. U.C.S. = unconfined compression strength, CL = clay, ST = silt, SD = sand, WEA R = weathered rock, ROCK = sound, unweathered rock.

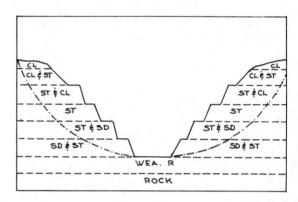

Figure 3 Typical cross-section of a road cut in a tropical soil profile. Slide plane = —·—·—·—.

This typical weathering profile forms on all but two types of rock, pure limestone, $CaCO_3$, and quartzite, SiO_2, both of which weather by solution only in the tropics at the lower elevations. Little or no soil may be found upon them. What little soil does develop is the result of impurities that were deposited at the same time as the rock. Where these rocks are found they exist at the surface as rugged hills with little soil on them.

All the other rocks, igneous, volcanic, sedimentary and metamorphic tend to have the above typical soil profile developed on them. Of course, in the wide river valleys on the flood planes the soils are different for the most part and because of their location seldom create slide problems. This is not true where a river is cutting through a deeply weathered old soil as it does in one place in the Chiriqui province of western Panama. This will be discussed further under the heading of sapping activity.

The distribution of soils that are susceptable to sliding extends over the entire country.

These saprolitic, residual soils range in thickness from 5 m to as much as 50 m thick in areas that have been subjected to tropical weathering processes throughout a greater part of Tertiary time. Some of the younger volcanoes have built up very porous pyroclastic deposits containing more deeply weathered soils than those on igneous or sedimentary rocks of the same age.

These soils, because of their strength characteristics and porosity, tend to become over loaded with water during periods of heavy rainfall. When this happens, the slopes in large areas will become practically denuded of vegetation by sliding and the stream valleys will become clogged with slide debris. This type of sliding is so common that it can be considered as a type of erosion. Luckly, in Panama such areas are not heavily populated, so there is seldom any great monitary damage caused to manmade structures. Around Panama City there have been some problems but not to any great extent at the present. As the city expands into the more hilly country surrounding it, serious problems with sliding will develop if proper precaution is not used in the construction of roads and housing areas.

The areas where sliding, as the result of heavy rains, has been most damaging are in the southern part of the Sona peninsula and the southwestern part of the Azuero peninsula. There was also a large amount of damage just east of the city of Panama in 1966.

6.2 Sliding as the result of the process of sapping

The process of sapping, in this report, is taken to be the process of erosion of a hillside around the head of a strongly flowing stream causing small landslides and resulting in the retreat of the valley head in the form of cirques formed by the sliding and removal of the material. In the province of Chiriqui in western Panama where the younger volcanics overlie a porous saprolitic soil horizon developed on still older volcanic soils, this particular type of sliding and erosion has affected an area of approximately 50 sq km. There is very little that the land can be used for, where sapping has occurred. After the original forest has been cut for timber, the land was suitable only for cattle raising and dairy farming. On the aerial photographs, this area looks like the headward erosion seen on the photographs of Mars that has been suggested to have taken place by the process of sapping. The difference is that in Panama the sapping and slide scars are covered by grass and trees.

Another area where sliding by the process of sapping has been found is in the province of Los Santos on the Azuero peninsula. In this case there is a large slide developed on a gentle high slope on a deeply weathered, saprolitic soil developed on top of a large basalt dome. The head of the slide is like a cirque with springs in it and the cirque-like feature is eroding its way up slope by slumping and sliding, and forming a true soil glacier with lateral moraines along each side of it. A soil-glacier that is slowly flowing down hill showing all the features of a normal ice glacier. This type of sliding has not been noticed in any other area in Panama.

6.3 Sliding as the result of seismic activity

Panama is part of the ring of fire that surrounds the Pacific Ocean. It is also the remains of an old volcanic arc which is still active. The land mass of Panama is the result of the interaction of the Cocos plate, the Nazca plate and the Caribbean plate. Seismic activity in Panama is still continuing because all three of these plates are still in motion. The greatest areas of seismic activity are along the Panama fracture zone between the Cocos plate and the Panama platelet which is sometimes called a part of the Nazca plate; where both plates are being subducted beneath western Panama and Costa Rica. In eastern Panama in the vicinity of the border between Colombia and Panama the Panama platelet or Nazca plate is interacting with the northwestern part of the South American plate and the Caribbean plate causing another area of abundant seismisity. Along the north coast of Panama, where the Caribbean plate is plunging under Panama, is another area of seismic activity; as is true along the southern coast of Panama where the Panama platelet or Nazca plate is being subducted. Seismic activity is much less in these last two areas as the relative motions of the plates are much slower. Needless to say Panama is in a seismically active area, and sliding as the result of seismic activity has taken place and will take place again.

Submarine sliding as the result of seismic activity took place in western Panama on July 18, 1934 when the area on which the large banana loading dock at Puerto Armuellas slid into the sea and was destroyed. A portion of the town was also destroyed at this time and the damages amounted to more than a million dollars. The earthquake was estimated to be 7.7 on the Mercali scale.

Submarine sliding also destroyed a small banana dock near the town of Tonosi as the result of an earthquake in 1914.

Portions of the Panama railroad were destroyed by sliding in 1882 by an earthquake that was centered just off the north shore of Panama in the Caribbean Sea. These slides along the railroad took place in the constructed fill areas across the deep swamps on which the railroad was constructed.

The greatest slide damage in Panama was caused by a series of earthquakes just off the shore of eastern Panama in the Pacific ocean near the Colombia-Panama border. The earthquakes began on July 11, 1976 at 16:54 hours with a shock of 6.7 on the Richter scale. The shocks continued until 20:41 hours when one of 7.0 occurred. The series of after shocks continued for several days with sporatic shocks ranging from 3.5 to 5.7 on the Richter scale. This seismic activity took place during the rainy season during a period of heavy rains when the ground was saturated and as a result the saturated soils liquified and created massive slides.

In the headwaters of the Rio Jaque and from there southward across the steep mountain sides to the Pacific Ocean the slopes are covered with deeply weathered residual saprolitic soil. These saturated slopes literally liquified and flowed down the mountain sides into the valleys. Approximately 2/3 of a 1,000 sq km area was denuded by this sliding activity. The streams and rivers were completely plugged with the jungle debris and mud that flowed from the mountain sides filling the valleys with 5 m to 15 m of slide debris. The only areas where any of the original jungle remained was laong the tops of the ridges. In many cases even the ridge tops were gone.

Along the Pacific coast southeast of the town of Jaque the area is mountainous and very steep, dropping from the mountain tops into the deep sea. In this area not only the deeply weathered saprolitic soils slid into the ocean but also a great deal of solid rock. Many of the slides along the coast were as much as a mile long and 400 m to 800 m high. In one area where the coast was 800 m high, the entire mountain slid into the ocean, rocks and all. This slide formed a circular peninsula 1 km in radius that extended 1 km from the original shore. Blocks of rock 15 m in diameter were carried a kilometer from shore and still were intact above the level of the water.

At the town of Jaque near the mouth of the Rio Jaque a large amount of damage occurred. Most of the homes in the town were of light wood frame construction built on short posts or stilts to raise them above the frequent floods caused by heavy rainfall in the headwaters of the Rio Jaque.

The town is actually built on the delta deposits of the Rio Jaque. The ground motion of the earthquake caused large parts of the delta to consolidate and in some cases slide a little. After the earthquake a large part of the town lost as much as a meter in elevation and some parts lost as much as 2.5 m in elevation. The twice daily tide submerged about one third of the town under about 1.5 m of water. Along the sea shore the beach was lowered as much as a meter in the delta area. Many areas along the shore exhibited large half moon scarps where more sliding of the beach had taken place. Some of the newly created scarps were 2 m high.

Most of the houses were thrown off their posts or stilts and were damaged to varying degrees. Most were twisted out of shape and some were totally destroyed. A monitary loss of one to two million dollars was estimated. The part of the town that became submerged at high tide was completely lost. All the structures in this area had to be removed to a more stable area out of reach of the tides.

The people in the town did what they could and adjusted to the new situation. They tore down the damaged buildings and used them to construct new one in slightly better locations. There were only two concrete block structures in the town. One was not damaged at all. The other was in the area that became submerged when the tide came in. It was badly cracked and a total loss.

With all of the described damage in the earthquake affected area there was no loss of life, and only two injuries, a wrenched shoulder and one broken leg. Luckily, the population density of the area is about one person per 1.5 sq km. The native huts, for the most part, are constructed of light wood frames with grass or leaf thatched roofs. Such structures are very light and flexible. Materials necessary for repairs or new structures are readily available in the surrounding jungle.

6.4 Slides as the result of manmade construction activities

The greatest number of slides in Panama that have affected human lives and manmade construction projects have been caused by a lack of knowledge of soil and rock characteristics and their reaction to various construction practices. The greatest number of slides are caused by making steep-sided cuts through hills and hill sides, and the improper constuction of fills where care was not taken to preserve the natural drainage. In many cases the slopes of hills are simply overloaded by the fills placed on them. When this happens it reflects a lack of sufficient study and preparation of the area under construction.

In many cases, the cuts made along hill sides and through the hills over-steepened the slopes and removed the confining pressures supporting the weaker soil materials. This practice resulted in two deaths by sliding in an area where the newly located Panama railroad was attempting to flatten slopes that had already slid once. In another case three deaths resulted from sliding during the construction of the basement structure for a bank because of a lack of understanding of the soil properties in the area. A portion of one school in Panama City was lost as the result of sliding where a large cut was made into a steep slope to increase the area for the school building. Luckily this slide took place on a weekend when there was no one in the school.

There appears to be a standard policy or gractice in Central America and Panama in highway construction which states: construct the highway with the steepest slopes that will stand for a short time, and make the cheapest possible fills to keep the original costs of highway construction low because it is easy to get money to keep the highway open once it has been made. To do the preliminary research and the construction necessary to insure a permanent, safe highway would raise the initial cost so high that it would be impossible to fund the project. Another argument is that if a highway or similar project is constructed with no sliding or other problems, then too much money was spent on the original design (over designed) and construction. Still another argument is that the construction project should be as inexpensive as possible, for the present, because less money will be spent on interest to pay the money back, as no interest would be paid on money necessary for repairs until the repairs are made. This policy seems to work. Highways are constructed and then paid for later, often at the cost of human lives.

The best and most reknown constructed slides, of course, are those along the Panama Canal. The slopes were constructed as steep as possible without taking the types of materials into account and with little or no attention paid to the geological structure of the area. Therefore, more material has been removed from the Panama Canal as the result of sliding than

was removed during the original construction. This process is still going on. In Oct. 1987 another two million yards of rock debris went into motion and completely closed the Canal for a few hours until they were able to reopen to single lane traffic. It took another two and one half months to completely reopen the second lane to traffic. Excavation of the upper part of the slide continued for several months more in an effort to further stabilize the upper slopes of the slide. Needless to say, sliding will continue along the Canal, the question is how much, and where, and what will the costs be to make the necessary repairs, what can be done to give warning when new slides are developing, and what can be done to prevent additional sliding? Steps have been taken to instrument various slopes along the Canal, to keep records of water levels, and motion of the various slopes. Further discussion of this is beyond the scope of this report.

6.5 Assessment of the state of knowledge of the landslide processes in Panama

The present state of knowledge about the extent of landslide processes in Panama is very slim. Little or nothing has been written about sliding in Panama other than the the the slides along the Canal. The only known people with any experience in the slides of Panama are those who lately worked for the Canal during the Canal Widening Studies of 1958 - 1969, and a few who are working for the Canal Commission at present. The general run of the contractors have little understanding as to the cause of landslides, or how to repair them. In some cases they welcome them because the slides add extra bonuses to their contracts for removal and repair. When large slides occur, the only recourse, usually, is to call in consultants from the United States and elsewhere to help design removal, repairs, and stabilization.

When the Canal was first constructed there was no such thing as soils engineering or engineering geology. It was the great slides along the Canal that gave the great empetus to the development of these studies. Since then, knowledge has been gained and conditions studied in order to develop stable slopes in open pit mining projects, and super highway construction. Some of this newly developed information was used with reasonable success in the Canal development projects of the 1960's.

The problem of how flat to make a slope to make it completely stable is still unknown. The effects of time on a slope are also unknown. We know that we may construct a slope that appears stable, and then to have it fail at a later date. Sometimes, this is measured in days or weeks, and sometimes it is measured in tens to hundreds of years. Where do we draw the line? Time and the changes that take place in a slope over time and the redistribution stresses within the slope are unknown and so far have not been successfully addressed.

REFERENCES

Gidigasu, A.M. 1976. Laterite soil engineering. New York, Elsevier.
MacDonald, D.F. 1915. Some engineering problems of the Panama Canal in their relation to geology and topography. Bur. Mines. 86.
MacDonald, D.F. 1932. The Panama Canal slides. Third Locks Project, Panama Canal.
Thompson, T.F. 1943 Foundations and slopes. Idem. pt. 2, sec. 5:138.
For additional references that may be applicable see: Brand, E.W. 1984. Landslides in southeast Asia: a state of the art report. ISL. 1:17-32.

Landslides: Extent and Economic Significance, Brabb & Harrod (eds)
© *1989 Balkema, Rotterdam. ISBN 90 6191 876 6*

Hazards in El Salvador from earthquake-induced landslides

Michael J.Rymer & Randall A.White
US Geological Survey, Menlo Park, Calif., USA

ABSTRACT: This report addresses the landslide situation in El Salvador, but because of limitations in available literature, is restricted to earthquake-induced landslides. The importance of earthquake-induced landslides to El Salvador was made clear by the 1986 San Salvador earthquake, which triggered several hundred landslides that killed at least 200 people. Nine other cases of earthquake-induced landslides are documented in El Salvador in the last approximately 130 years. Earthquakes that cause landslides in El Salvador include both subduction-zone events and upper-crustal events. Earthquake-caused landslides are commonly limited to a narrow zone coincident with the main volcanic chain, coastal mountains, and interior valley physiographic provinces. Widespread throughout this area is poorly consolidated volcanic tuff, a major contributing factor in the landslides caused by the 1986 San Salvador earthquake, and probably is a major contributor to the overall abundance of landslides in the region.

1 INTRODUCTION

This report gives an overview of landslide hazards in El Salvador, but because of limitations in the available literature, concentrates on landslides triggered by earthquakes. As such, this report differs from other reports in the volume in that a comprehensive summary of the landslide situation is not available. Ten cases of earthquake-induced landslides in the past 130 years are presented with as many as hundreds to thousands of slides triggered by each shock. The 1986 San Salvador earthquake, which triggered several hundred slides, is used as an example of the potential for earthquake-induced landslides in the country.

El Salvador is the smallest country in Central America with an area of 20,975 km² and a population of about 5.2 million. The capital, San Salvador, is the major city in El Salvador and has an estimated population of about 1.4 million. Other cities in the republic are Santa Ana, San Miguel, San Vicente, and Usulutan. Most cities and most of the country's population are located in the interior valley, which is approximately coincident with the main volcanic chain (Figure 1).

El Salvador is composed of five physiographic provinces (Figure 1). Of particular interest to the landslide setting of the country are the coastal ranges, interior valley, northern mountains, and main volcanic chain. Climatically induced landslides are most likely in the coastal ranges, northern mountains, and main volcanic chain, where there are locally steep slopes and heavy rainfall (Figure 2A). Rainfall in the country is concentrated in a 6 month period from May to October (Figure 2B). Physiographic provinces of interest to earthquake-induced landslides are the coastal ranges, main volcanic chain, and to a lesser extent, the interior valley, for reasons explained below.

2 GEOLOGIC AND TECTONIC SETTING

The geologic and tectonic setting of El Salvador is discussed briefly, with emphasis on potential landslide hazards. Wiesemann (1975) presented a generalized overview of the geology of El Salvador, including earlier work by various authors and geologic mapping later published by Weber et al. (1978). Schmidt–Thomé (1975) discussed the geology and tectonics of the San Salvador area in detail. This report draws from these earlier reports to provide a brief late Cenozoic context for landslides triggered by earthquakes.

El Salvador is transected by the Central American

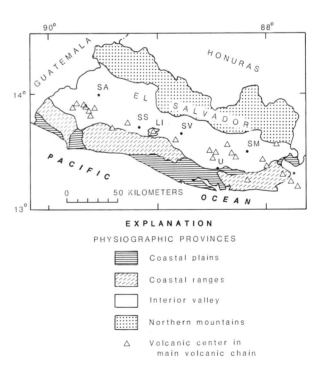

EXPLANATION

PHYSIOGRAPHIC PROVINCES

- Coastal plains
- Coastal ranges
- Interior valley
- Northern mountains
- △ Volcanic center in main volcanic chain

Figure 1. Map of El Salvador showing physiographic provinces. Major Quaternary volcanic centers (mostly in main volcanic chain) shown by triangles. SS = San Salvador, LI = Lake Ilopango, SA = Santa Ana, SV = San Vicente, SM = San Miguel, U = Usulutan.

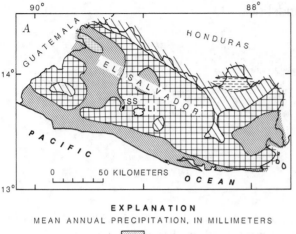

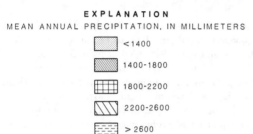

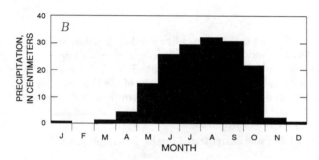

Figure 2. Rainfall distribution in El Salvador. A) Generalized isohyetal map of El Salvador (modified from Organizacion de Los Estados Americanos 1974). SS = San Salvador, LI = Lake Ilopango. B) Average monthly rainfall at San Salvador from 1966 through 1975. Data from Environmental Science Services Administration (1966–1975).

volcanic chain, which is a product of northeast-directed subduction of the Cocos plate beneath the Caribbean plate. The volcanic chain extends from Guatemala to Costa Rica. A structural depression, called a median trough by Williams and Meyer-Abich (1955) and interior valley in Figure 1, is approximately coincident with the volcanic chain.

Earthquake activity in El Salvador is of subduction-zone type, ranging in depth from 20 to 200 km, or upper-crustal type, with depths less than 20 km (Harlow et al. 1989). In spite of the greater amount of activity in the subduction zone and the greater maximum magnitudes of earthquakes therein, the shallow upper-crustal earthquakes commonly cause more damage than do the subduction-zone events (Schulz 1965; Lomnitz and Schulz 1966; Harlow et al. 1989).

The stratigraphic section of El Salvador is dominated by middle and upper Cenozoic volcanic and volcaniclastic rocks (Figure 3). Two informally named pyroclastic and epiclastic tuff units in El Salvador warrant further description because their poor consolidation and their formation of steep streambanks and riverbanks were strongly related to landslide occurrence during the 1986

San Salvador earthquake and probably earlier earthquake-induced landslides. The older of the two units is as thick as 25 m and blankets large parts of the interior valley. The younger tuff locally covers all older deposits and accounts for the flatness of much of the interior valley. Below San Salvador, the younger tuff is as thick as 25 m; west of Lake Ilopango it is greater than 50 m thick (Williams and Meyer-Abich 1955; Hart and Steen-McIntyre 1983).

3 LANDSLIDES ASSOCIATED WITH THE 1986 SAN SALVADOR EARTHQUAKE

The 1986 San Salvador earthquake, magnitude 5.4, triggered hundreds of landslides extending over an area of at least 200 km^2 (Rymer 1987). The most numerous landslides were soil falls and slides; rockfalls and rockslides, slumps, and rapid soil flows accounted for most of the remainder. Landslides discussed in this report are classified according to the system of Varnes (1978), except the term 'soil' is used to replace both 'earth' and 'debris,' and 'rapid soil flow' is used as defined by Keefer (1984); volcanic tuffs are considered as soils because of their poor consolidation and general similarity to engineering soils. A distribution map of landslides triggered by the 1986 earthquake is shown in Figure 4.

The extent of property damage and loss of life in the San Salvador area as a direct result of the 1986 earthquake-induced landsliding is not precisely known, but a conservative estimate would be approximately 100 dwellings destroyed and about 200 deaths. [Damage from the earthquake, in comparison, ranged in thousands of homes destroyed and about 1500 deaths.] Most houses damaged by landslides were within 10 m of streambanks or steep hill slopes and were generally undermined by failure of the adjacent slopes. Some dwellings at the base of slopes and streambanks were damaged by falling debris.

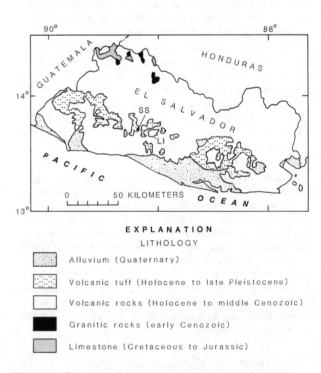

Figure 3. Generalized geologic map of El Salvador with emphasis on distribution of upper Pleistocene to Holocene volcanic tuff deposits. Modified from Weber et al. (1975). SS = San Salvador, LI = Lake Ilopango.

3.1 Falls and slides

The majority of landslides triggered by the earthquake were soil falls and slides of poorly consolidated volcanic tuff along steep valley walls, streambanks, and roadcuts. Falls and slides in the flat areas of San Salvador were almost completely restricted to streambanks and roadcuts where blocks of volcanic tuff either fell or slid toward the adjacent stream or road. In the gentle slopes immediately south of the city, soil falls and slides were most common where houses were built near steep, modified slopes that were either left unsupported or faced with a retaining wall. Farther south soil falls and soil slides were in steep natural slopes and modified slopes insufficiently supported by retaining walls.

Soil falls and slides triggered by the 1986 earthquake were generally small, less than 100 m³. The majority of soil falls and slides were either of volcanic tuff that spalled off nearly vertical slopes along streambanks or of actual soils developed on tuff that slid down local slopes. Pieces of spalled tuff were thin, generally less than 1 m thick. The small size of most soil falls and slides resulted in little damage, commonly restricted to undermining of dwellings built close to steep slopes. However, larger soil slides were much more destructive. The larger soil slides had volumes of about 300 to 1000 m³ and destroyed underfootings of homes and disrupted local pipelines. Soil slides also developed in artificial fill and locally caused damage.

Rockfalls and rockslides were the next most common types of landslide and occurred primarily in roadcuts in the hills surrounding San Salvador. Material making up these falls and slides was commonly coarse pyroclastic and epiclastic rock. Material held behind retaining walls was also locally involved in rockfalls. No rockfalls in natural slopes were noticed in the greater San Salvador area. Rockfalls were small and although they locally disrupted traffic were not significantly destructive.

3.2 Slumps

Slumps were located throughout the earthquake-affected area (Figure 4); large slumps and disrupted slumps were rather rare. The slumps were commonly larger than falls and slides, and were much more destructive.

A large disrupted slump and soil slide that resulted in an unconfirmed death toll of about 200 people is shown in Figure 5. A row of homes was located between the road shown on the right side of the figure and the ravine on the left before strong ground shaking from the earthquake caused slumping and movement, to the left in this view. The low tensile strength of the tuff here probably caused breakup of the slump mass and resulted in the destruction of houses and burial of people. This slump had an estimated volume of 4000 m³.

Slumps also occurred in fill throughout the earthquake-affected area. Most of these were in fill sections adjacent to bridges and infilled streams. Fill appeared to be one of the materials least resistant to strong ground shaking and downslope movement because of its poor compaction.

A large slump near the center of San Salvador that predated the 1986 earthquake showed renewed movement during this most recent event. This slump moved initially in association with the 1965 San Salvador earthquake. Motion of this slump was primarily vertical and was sufficient to break buried water pipes and make a road impassable. This slump had an estimated volume of 30,000 m³.

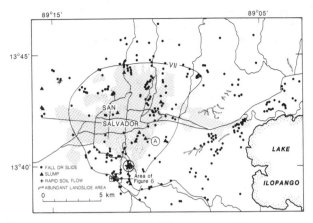

Figure 4. Location of ground failures triggered by the October 1986 San Salvador earthquake. Circled solid star denotes location of 1986 main-shock epicenter; Modified Mercalli (MM) intensity VII contour also shown. Site 'A' is location of Figure 5. San Salvador urban area and major roads shown for location (modified from Rymer 1987).

Figure 5. Remains of deadly disrupted soil slump and soil slide in the Santa Marta district of San Salvador (site 'A' in Figure 4). Jumbled area between brick road on right and ravine on left was covered with houses before earthquake. Rescue operations cleared much of the slide debris and remains of houses before photograph was taken (from Rymer 1987).

3.3 Rapid soil flows

Four mappable rapid soil flows were associated with the 1986 San Salvador earthquake (Figure 4). They were located on steep slopes with thin mantles of poorly consolidated volcanic tuff. One rapid soil flow represented in Figure 4 covered the largest area (16,500 m²) of any slope failure associated with the earthquake. This flow traveled a total horizontal distance of about 440 m, and in spite of the large area covered, had a relatively small volume of about 8000 m³. This flow is shown in Figure 6.

Figure 6 shows the geometry of the flow, indicating a complex pattern of movement. The source area for the flow was located where a local perched water table existed on an old volcanic flow. From the the source area (Figure 6) water-saturated volcanic tuff comprising the flow mass dropped about 60 to 70 m in elevation as it moved to the northwest. When the flow mass reached the bottom of the slope it ran up the opposite side of the canyon (site A, Figure 6). From here the flow moved down canyon to site B where it ran up the opposite side of the slope and also moved about 25 to 30 m upstream. A minor volume of the rapid soil flow continued farther down stream to where it filled a small reservoir, and then moved another 70 m downstream before stopping.

4 OTHER HISTORIC EARTHQUAKE-INDUCED LANDSLIDES

The 1986 San Salvador earthquake exemplifies the earthquake-induced landslide potential in El Salvador. We now present earlier examples of earthquake-induced landslides in the country. The data set we use is compiled from the literature, mostly outside of widely distributed journals, describing earthquakes and their effects in El Salvador. Table 1 lists 10 documented cases of earthquakes with associated landslides, and for each earthquake listed in Table 1, we present approximate limits of the Modified Mercalli (MM) intensity VII contour in Figure 7. In each instance the main emphasis of the documented cases was on seismological aspects of the earthquake and not on landslides. The reports, thus, do not elaborate on, nor quantify, the landslide situation; landslides were mentioned in towns and villages in the areas shown in Figure 7.

The MM VII contour is a minimum estimate of the areal limits of earthquake-induced landslides. For example, Figure 4 shows the MM VII contour includes only about half the landslides caused by the 1986 San Salvador earthquake. Careful studies of the distribution of landslides have shown that the MM VI contour best approximates the limits of landslides associated with an earthquake (Harp et al. 1981; Keefer 1984). We continue to use the MM VII contour, though, because there is a significant difference in the performance of buildings between MM VI and VII, thus allowing us to more accurately locate the limits of the MM VII contour from historical and archival records, even if it is a conservative estimate of earthquake-induced landslide limits. Also, for some of the earthquakes listed in Table 1, sufficient data do not exist to construct an MM VI contour; our mapping of the MM VII contour therefore makes for a consistent, although conservative, measure of the landslide distribution within El Salvador.

Some details of the distribution of earthquake-induced landslides for the approximately 130 year period shown in Figure 7 are worthy of note. One feature is that the events include both subduction-zone and upper-crustal earthquakes. The subduction-zone events were those in 1915, 1947, and 1982; all the others are upper-crustal events. The subduction-zone events that produce major landslides have larger magnitudes, generally ranging from 7.0 to 8.0, and are less common (White and Harlow 1989). These events also may affect a larger area, like the 1915 earthquake, but because of the greater distance from their hypocentral locations to the Earth's surface, more energy is required to produce damage. The 1982 event, for example, did not affect a significantly larger area than did the 1986 earthquake, even though the former event had a much larger magnitude (Table 1).

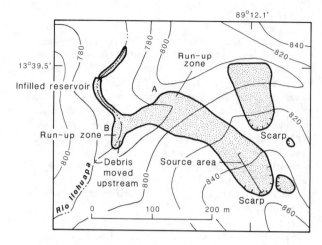

Figure 6. Map of rapid soil flow triggered in the October 1986 San Salvador earthquake showing directions of movement down initial slope and two stream canyons (modified from Rymer 1987). Sites A and B are described in the text. Contours in meters.

Table 1.—Historic earthquake–triggered landslides in El Salvador

Year	Earthquake magnitude	Brief description	Reference
1857	6 ¼	Many landslides on hills and in canyons east of Lake Ilopango	Larde 1960
1878	6 ¼–6 ½	Landslides on slopes of volcanoes near Santiago de Maria, a large slide on Cerro El Tigre buried 14 people	Larde 1952
1915	7.9	A great many landslides through-out western El Salvador	Larde 1960
1919	6.0	Many landslides on slopes of Cerro San Jacinto	Anonymous 1919
1936	6.1	Many landslides on slopes of San Vicente volcano	Levin 1940
1947	7.2	Landslides on slopes of Con-chagua volcano	Meyer-Abich 1956
1951	6.0 & 6.2*	Landslides on volcanic slopes near Santiago de Maria	Meyer-Abich 1952
1965	6.0	Landslides from San Salvador to Lake Ilopango	Lomnitz and Shultz 1966
1982	7.0	Many landslides southwest of San Salvador	Lara 1982
1986	5.4	Hundreds of landslides between San Salvador and Lake Ilo-pango with about 200 fatal-ites and at least 100 homes destroyed[†]	Rymer 1987

* Two earthquakes hours apart
[†] Death and damage estimates from landslides only, purely earthquake-related damage was much greater

Another striking feature of the distribution of earthquake-induced landslides is the general restriction of slides to the main volcanic chain, the northeast part of the coastal ranges, and the southwest part of the interior valley physiographic provinces (compare Figures 1 and 7). This is also the area in which the poorly consolidated volcanic tuffs are deposited (Figure 3).

For a more detailed look at earthquake-induced landslide hazards in El Salvador, both past and potential, we turn to the seismicity record of the country. Here, we are interested primarily in the areas of MM VII or greater that are

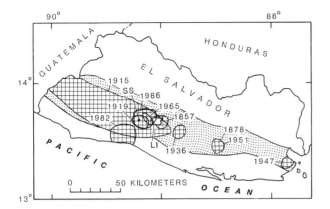

Figure 7. Location of reported areas of historic earthquake-triggered landslides in El Salvador, shown by square pattern with date. Distribution of reported landslide areas approximately coincides with main volcanic chain and coastal ranges physiographic provinces (Fig. 1) and location of poorly consolidated upper Pleistocene to Holocene volcanic tuff (Fig. 3). Doted pattern is area of MM VII for period 1955 to 1965 from Schulz (1965). SS = San Salvador, LI = Lake Ilopango.

located in areas with at least locally steep slopes. The dotted pattern in Figure 7 shows the distribution of MM VII contours for seismicity in El Salvador from 1955 to 1965 (Schulz 1965). This area may approximate the zone of major long-term earthquake-induced landslide hazards; minor earthquake-induced landslides may occur outside the dotted area. We feel justified in projecting landslide hazards throughout this area even though landslides were not mentioned in most reports on historic earthquakes. The distribution of poorly consolidated volcanic tuff, with its low tensile strength, in areas with locally steep slopes, combined with shaking intensities at least as great as MM VII strongly suggest that landslides were also triggered by most earthquakes in the square- and dotted-pattern area shown in Figure 7.

Earthquakes with MM VII or greater are common and frequent in El Salvador (Harlow et al. 1989; White and Harlow 1989). If such earthquakes cause landslides, which may be destructive and hazardous like those from the 1986 San Salvador earthquake, then the landslide hazard in El Salvador is large indeed. For example, the city of San Salvador itself has been partly to totally destroyed by earthquakes, and possibly associated landslides, 11 times since 1710 (Harlow et al. 1989). Thus, even though this report deals with only earthquake-induced landslides, we have shown that the landslide hazard is great in El Salvador, and is most commonly restricted to the coastal ranges, main volcanic chain, and interior valley areas.

REFERENCES

Anonymous 1919. El Terremoto del 28 de Abril. Diario Latino, May 13, 1919.

Environmental Science Services Administration 1966–1975. Monthly climatic data for the world. Environmental Science Services Administration. v. 19–28.

Harlow, D.H., R.A. White, M.J. Rymer, S. Alvarez, & C. Martinez 1989. The San Salvador earthquake of 10 October 1986 and its historical context. Seismological Society of America Bulletin. (in press).

Harp, E.L., R.C. Wilson & G.F. Wieczorek 1981. Landslides from the February 4, 1976, Guatemala earthquake. U.S. Geological Survey Professional Paper 1204-A. 35 p.

Hart, W.J.E. & V. Steen-McIntyre 1983. Tierra blanca tephra from the AD 260 eruption of Ilopango caldera. In P.D. Sheets (ed.), Archeology and volcanism in Central America: The Zapotitan Valley of El Salvador. University of Texas Press, Austin, Texas. 14–43.

Keefer, D.K. 1984. Landslides caused by earthquakes. Geological Society of America Bulletin. 95: 406–421.

Lara, M.A. 1983. El Salvador earthquake, June 19, 1982. Earthquake Engineering Research Institute Newsletter. 17: 87–96.

Larde, J. 1952. Geologia Salvadoreana. Biblioteca de Pueblo, San Salvador. 165 p.

Larde, J. 1960. Historia sismica y erupcio-colcanica de El Salvador. Obras Completas, Minesterio de Cultura, Departamento Editorial, El Salvador. 1: 1–576.

Levin, S.B. 1940. The Salvador earthquake of December, 1936. Seismological Society of America Bulletin. 30: 1–45.

Lomnitz, C. & R. Schultz 1966. The San Salvador earthquake of May 3, 1965. Seismological Society of America Bulletin. 56: 561–575.

Meyer-Abich, H. 1952. Terremoto en El Salvador (America Central), 6–7 de Mayo de 1951. Comunicaciones, ITIC. 1: 1–24.

Meyer-Abich, H. 1956. Los volcanes activos de Guatemala y El Salvador (America Central). Anales del Servicio Geolocico Nacional de El Salvador. 3: 1–102.

Organizacion de Los Estados Americanos 1974. El Salvador: Fase 1, zonificacion agricola. Oganizacion de Los Estados Americanos. Washington, D.C. 260 p.

Rymer, M.J. 1987. The San Salvador earthquake of October 10, 1986—Geologic aspects. Earthquake Spectra. 3: 435–463.

Schmidt-Thomé, M. 1975. The geology in the San Salvador area (El Salvador, Central America), a basis for city development and planning. Geologisches Jahrbuch. 13: 207–228.

Schulz, R. 1965. Mapa Sismico de la Republica de El Salvador. Centro de Est. Invest. Geotec. Boletin Sismologico. 10: 8.

Varnes, D.J. 1978. Slope movement types and processes. In R.L. Schuster & R.J. Krizek (eds.), Landslides—Analysis and control. National Academy of Sciences Transportation Board Special Report 176. 12–33.

Weber, H.S., G. Wiesemann, H. Lorenz, & M. Schmidt-Thomé 1978. Mapa geológico de la República de El Salvador/America Central. Bundesanstalt für Geowissenschaften und Rohstoffe: Hanover, Germany. scale 1:100,000.

White, R.A. & D.H. Harlow 1989. Significant upper-crustal earthquakes of Central America since 1900. Seismological Society of America Bulletin. (submitted).

Wiesemann, G. 1975. Remarks on the geologic structure of the Republic of El Salvador, Central America. Mitteilungen Geologisch-Paläontologischen Institut: University of Hamburg. 44: 557–574.

Williams, H. & H. Meyer-Abich 1955. Volcanism in the southern part of El Salvador. University of California Publications in Geological Sciences. 32: 64 p.

Landslides: Extent and Economic Significance, Brabb & Harrod (eds)
© 1989 Balkema, Rotterdam. ISBN 90 6191 876 6

Landslide hazards in the central and southern Andes

George E. Ericksen
US Geological Survey, Reston, Va., USA

Carlos F. Ramirez
Servicio Nacional de Geología y Minería, Santiago, Chili

Jaime Fernandez Concha
Urbanización Las Casuarinas, Monterrico, Lima, Peru

Gilberto Tisnado M.
Instituto Nacional de Investigación de Transportes, Lima, Peru

Fernando Urquidi B.
American Embassy, La Paz, Bolivia

ABSTRACT: Landslides are among the most destructive natural phenomena of the central and southern Andes, being comparable in property damage and loss of life due to vibrational destruction and collapse of buildings during earthquakes. The region considered in this report, which includes the Andean Highlands where elevations are generally above 4,000 m, the Andean front ranges, and the low-lying coastal desert region of western South America, shows great diversity of geology, topography, and climate. It is a region of active continental uplift and intense seismic activity. Most of the hundreds of landslides that occur each year are triggered by heavy rains and by earthquakes. Major earthquakes having magnitudes greater than 7, of which at least one will occur during any given 10 year period, commonly trigger hundreds to thousands of landslides, many of which have been among the largest and most destructive historic landslides in the Andes.

Landslide distributions and styles differ according to geology and topography. The most destructive tend to be those occurring in areas of great relief in the Andean Highlands and in deep valleys along the eastern and western flanks of the Andes. High-speed debris avalanches and debris flows (formed by the rupture of glacial lakes) are characteristic of glaciated mountain ranges of the Highlands. More frequent but smaller and less destructive rock falls, rock and debris slides, and soil slips also occur in this region and in deep valleys along the Andean front. Exceptionally heavy rains in the Andean Highlands also cause destructive floods, debris flows, and mudflows in these valleys, particularly along the arid western Andean front and in the coastal region. Earthquake-triggered slumps and lateral spreading in water-saturated, unconsolidated sediments have caused extensive damage to settlements and cities in the Andean Highlands and along the Pacific coast.

Statistical data about landslide frequency and destruction in the Andes are scanty, and so accurately estimating property damage and loss of life due to landslides is not possible. Nevertheless, available data suggest that in years without major landslides, destruction costs range from a few million to several tens of millions of U.S. dollars and deaths due to landslides are a few to a few tens of individuals. On the other hand, catastrophic landslides, which occur at 5- to 10-year intervals, may cause property damage in the hundreds of millions of dollars and kill hundreds or even thousands of persons.

1 INTRODUCTION

Destructive landslides are frequent and widespread in the central and southern Andes of Peru, Bolivia, Chile, and Argentina (Fig. 1), a region characterized by active plate convergence and continental uplift, intense seismic activity, and great diversity of climate. Uplift, which averages as much as a meter per thousand years, and rapid erosion have resulted in widespread unstable slopes subject to catastrophic landslide failure. Most of the hundreds of landslides that occur annually are triggered by heavy rains and earthquakes. Others are caused by man's activities such as irrigation of farmlands and construction of highways, canals, and buildings, all of which may cause changes in slope stability. Still other landslides are caused by stream or lake erosion, which undercuts former stable slopes. Some landslides occur without obvious triggering mechanisms, failure taking place when slow buildup of gravity-induced shear stress exceeds the shear strength of the landslide material.

The region of landslide hazards is large, extending over an area of about 1 million km², which includes the Andean Highlands, the lower eastern and western front ranges, and the Pacific coastal areas. The eastern lowlands of Peru, Bolivia, and Argentina, where landslides cause less destruction, are not discussed. The total population of the four countries is about 50 million; most people live in coastal cities where landslide hazards are minimal. As a consequence, landslides generally affect relatively few people, and destruction of property generally is restricted to small cities and towns, farms and farming communities, and mine settlements. Roads, railroads, and hydroelectric facilities are the works-of-man most affected by landslides in remote, sparsely populated areas.

In a typical year, rains are the most common cause of landslide failure, and rainfall patterns strongly influence incidence and distribution of landslides. Annual rainfall is seasonal and annual precipitation shows great geographic variation throughout most of the central and southern Andean region. In the Andean Highlands (altitudes above 4,000 m) in Peru

Figure 1. Index map of the central and southern Andean region, showing distribution of selected major landslides.

and Bolivia, a November-April rainy season alternates with a May-October dry season during which rainfall is sparce or absent. In this region, the maximum annual rainfall is in the eastern Andean Highlands (600-1,000 mm), increasing with decrease of altitude along the eastern Andean front to about 2,000 mm. The western Andean Highlands and coastal coastal areas of this region are are semi-arid to arid; annual rainfall in the western Highlands is 200-400 mm, decreasing with altitude along the western Andean front to less then 30 mm. The coastal desert of Peru and northern Chile lack annual rains; measurable rainfall of 1 mm or more commonly occurs only a few times in any given 10- to 20-year period. In contrast to the above northern region, the Andean region south of about lat. 27° S. has a May-October rainy season, and annual rainfall gradually increases from

about 20 mm in the north to maxima of 3,000-7,000 mm in the western Andes and and coastal lowlands of southern Chile. Rainfall in the southern areas of greatest precipitation is not seasonal, and rains may occur on more than 300 days per year. In this southern segment, rainfall decreases on the eastern side of the Andes to a minimum of about 200 mm per year. The Andean Highlands between the above northern and southern segments is a transition zone where rainfall is generally less than 100 mm per year.

Landslides tend to occur in the early to middle part of the rainy season when the ground first becomes saturated. Furthermore, annual rainfall is highly variable, and unusually heavy rains and exceptionally wet rainy seasons generally cause unusual landslide activity. Also, heavy rains in the Andean Highlands cause flash floods and debris flows in the valleys along both the eastern and western fronts and marginal lowlands. Unusual heavy rains also occur along the coast of Peru and Chile during El Niño (Goldberg and others, 1987), an interval of world-wide climatic change that occurs about every 5 years. These rains may cause catastrophic flooding, debris flows, and mudflows in valleys along the western Andean front and in coastal lowlands. Such rains also cause landslides on slopes that were stable under the prevailing desert conditions.

The greatest number of and many of the largest recorded landslides that are triggered by single events are those caused by major earthquakes. Because most earthquakes in the central and southern Andes are related to subduction, and have foci at or near the top of the subducting oceanic plate, both vibrational destruction and incidence of earthquake-triggered landslides are greatest in the coastal and western Andean regions where the subduction zone is relatively shallow. Eastward, where the subduction zone is deeper, earthquake intensities, and consequently, destruction, decrease. For example, destructive earthquakes in the eastern Bolivian Andes are far less frequent than in Peru and Chile to the west.

Published information about landslides in the central and southern Andes is scanty and, for the most part, is restricted to descriptions of major destructive landslides and characteristics and distribution of landslides related to major earthquakes. Other information about landslides, most of which is unpublished, is also found in reports by national geological institutions and governmental agencies dealing with construction of roads, hydroelectric facilities, and other public works. As a consequence, statistical information about landslide damage in the central and southern Andes is incomplete, which makes reliable estimates about extent of destruction and loss of life due to landslides impossible. Nevertheless, available information suggests that the average annual property damage is on the order of a few millions to several tens of millions of U.S. dollars. On the average, probably fewer than 10 individuals are killed by landslides during a typical year. However, major landslides, which occur at 5- to 10-year intervals, may cause property damage of hundreds of millions of dollars and kill hundreds or even thousands of persons. For example, the most destructive historic landslide of the Western

Hemisphere, the Nevados Huascaran avalanche of 1970, killed more than 20,000 individuals.

2 GEOLOGIC AND PHYSIOGRAPHIC CONTROL OF LANDSLIDES.

The central and southern Andean region is underlain chiefly by marine and continental sedimentary rocks of Palezoic and Mesozoic age that are intruded by granitic batholiths, chiefly of Cretaceous and Tertiary age, and covered by unconsolidated sediments and volcanic rocks of Tertiary and Quaternary age. The region of southern Peru, western Bolivia, northern Chile, and northwestern Argentina is a 300,000-km² volcanic field of late Tertiary to Recent age in which are several hundred recognizable stratovolcanos, only a few of which are active, and dozens of calderas that were the sources of widespread rhyolitic ash-flow tuffs. Other volcanos, several of which are active, are along the Chile-Argentina border south of Santiago, Chile. The latter are hazardous not only because of potential destruction by eruptions of ash and lava, but also because they are in a region of significant ice and snow cover where eruptions may cause destructive lahars (volcanic mudflows) in this relatively populous area. Alpine glaciers and glacial lakes are widespread in the Andean Highlands of Peru and southern Chile, and failure of morainal dams of such lakes, chiefly in the Cordillera Blanca of Peru (Fig. 2), has caused several catastrophic debris flows during the present century (Fernandez Concha, 1957).

Landslides tend to show frequencies and styles that differ according to topography and geology. The regions most prone to landslide activity are the glaciated mountain ranges and deep valleys of the Andean Highlands and deep, steep-walled canyons on the flanks of the highlands. Relatively incompetent, thin-bedded shale, siltstone, and sandstone are widespread in these regions, and extensive areas are covered with thick slope debris or glacial moraines. These are the areas where landslides are most numerous and where catastrophic landslides are most frequent. Since 1940, landslides in the central and southern Andes have caused more than a billion U.S. dollars in property damage and killed at least 25,000 persons. In this terrain of great relief, high-speed debris avalanches and debris flows (the landslide classification used in this report is that of Varnes, 1958) have been the most destructive types of landslides. Soil slips, rock and debris slides, and rock falls are also frequent and widespread in areas of steep slopes and great relief, and earthquake-triggered slides of these kinds have been particularly devastating to farmlands in some areas. Slumps and block glides on more gentle slopes and in valley bottoms have caused extensive damage to transportation routes, canal systems, and settlements.

Water-saturated, fine-grained sediments, which are widespread foundation materials of coastal cities of Peru and Chile, fail by liquefaction, slumping, and lateral spreading during earthquakes. Such types of failure caused extensive damage to buildings in Chimbote (Fig. 1) and nearby coastal settlements during the May 31, 1970, Peru earthquake, and in Concepción, Valdivia, and Puerto Montt during the southern Chile

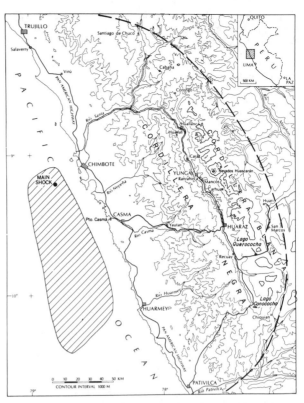

Figure 2. Map showing region affected by the Peru earthquake of May 31, 1970 (magnitude 7.7), a 30,000 km² area that was devastated by thousands of earthquake-triggered landslides, most of which were in the valley of Río Santa and the flanking Cordillera Blanca and Cordillera Negra. In addition to these features, the map shows the earthquake epicenter (main shock), the area of aftershocks (cross-hatch pattern), and the limit of earthquake damage (dashed line). From Plafker and Ericksen (1978).

earthquakes (five shocks having magnitudes of 7.5 to 8.4) of May 21 and 22, 1960 (Saint-Amand, 1961). At Puerto Montt, a large submarine silt and sand flow had the unusual effect of grounding a small ship during these earthquakes.

As previously noted, lahars are landslide hazards in the volcanic region of central and southern Chile. At the present time of limited snow and ice fields in the Andes of this region, lahars are infrequent, are small, and generally do little damage. The most destructive lahar of this century occurred during the 1964 eruption of Volcan Villarrica (a volcano in the western Andes about 60 km north of Riñihue, Fig. 1); that lahar swept down the side of the volcano and through the nearby village of Conaripe, where it destroyed several houses and killed 25 persons (Marangunić, 1974). However, during the last glacial stage of the Pleistocene, catastrophic lahars were frequent and large, as shown by remnants of lahars in Andean valleys and in the longitudinal coastal valley between Santiago and Puerto Montt (MacPhail, 1973; Marangunić and others, 1979; Abele, 1982; Fig. 1).

Catastrophic mudflows resulting from rupture of mine-tailing ponds have occured during earthquakes of the present century. Several such failures took place in central Chile during the earthquake of 1965, and one of these mudflows destroyed the mining town of El Cobre (about 100 km north of Santiago, Fig. 1) and killed more than 200 inhabitants. Tailings ponds also failed during the Peru earthquake of 1970 but are not known to have caused any significant damage or fatalities.

3 CATASTROPHIC LANDSLIDES OF THE 20TH CENTURY.

Catastrophic landslides of the 20th century include (1) Huaráz debris flow of 1941, (2) Riñihue landslides of 1960, (3) Huascarán debris avalanches of 1962 and 1970, and (4) the Mayunmarca landslide of 1974. Other less-well-known large landslides, some of which have caused extensive property damage and killed tens to hundreds of persons, have occurred, but most of these have been reported only in newspaper accounts.

3.1 Huaráz, Peru, debris flow of December 1941 (Bodenlos and Ericksen, 1955)

The Huaráz debris flow of 1941 destroyed about a quarter of the Andean city of Huaráz (Fig. 2) and killed an estimated 4,000 to 6,000 inhabitants. It was caused by sudden rupture of the morainal dam of a lake at the head of Quebrada Cohúp in the Cordillera Blanca northeast of Huaráz. A fall of ice or rock into the lake, which was the probable cause of failure of the dam, was not related to an earthquake or other known triggering mechanism. The resultant debris flow, having a volume of at least 10 million m^3, swept 23 km down the Cohúp valley, through the northern part of Huaráz, and into the Río Santa. The debris formed a temporary dam in Río Santa and the remaining water and debris swept downstream to the coast and destroyed settlements and farms in the Santa valley. The temporary dam was breached after 2 days and caused additional flooding downstream. The morainal debris remaining from the flow covered a 1-km^2 area in and near Huaráz to a depth as much as 5m. This debris contained boulders weighing as much as 700 t that had been rafted into the area by the debris flow.

This was Huaráz's first major catastrophe in its more than 300 years of existence and was only a prelude to its almost complete destruction (by collapse of buildings) during the 1970 earthquake.

3.2 Riñihue, Chile, landslides of May 1960 (Davis and Karzulovic, 1963)

The largest landslides triggered by the 1960 Chile earthquakes were three contiguous slides having a total volume of nearly 40 million m^3 that blocked Río San Pedro near the outlet of Lago Riñihue (Fig. 1), one of the large glacial lakes of southern Chile. The slides were within an 80-m-thick sequence of Pleistocene lake clays underlain by glacial till and overlain by glacial outwash. Slide movement was complex but was dominated by slumping and block gliding. The slides blocked drainage from Lago Riñihue for 63 days, while the lake level rose 26.5 m above normal. The water stored behind the dam had reached an estimated 2.5 billion m^3 when the lake began to drain through manmade canals cut into the dam to control the rate of discharge when the lake

overtopped the dam. Fortunately, these canals slowed erosion of the dam, and the lake drained to its former level without catastrophic flooding of downstream populated areas, including Valdivia, one of the major cities of southern Chile.

Other large landslides have blocked the outlet of Lago Riñihue in both historic and prehistoric times. One of these having an estimated volume of more than 100 million m^3 was triggered by an earthquake in December 1575. It blocked the outlet for nearly 4 months, and when the dam was breached, catastrophic floods swept through the downstream area.

3.3 Huascarán, Peru, debris avalanche and other landslides triggered by the Peru earthquake of May 31, 1970 (Plafker and others, 1971; Plafker and Ericksen, 1978)

Figure 3. Aerial view of the Huascarán debris avalanche of May 31, 1970, showing the source area and distribution of the Ranrahirca and Yungay lobes. Photograph by the Servicio Aerofotográfico del Perú.

The greatest number of and most destructive landslides in the Andes known to have been triggered by a single event were those associated with the Peru earthquake of May 31, 1970 (magnitude 7.7). Many thousands, perhaps tens of thousands of earthquake-triggered landslides devastated the 30,000-km^2 area affected by the earthquake (Fig. 2). These landslides were most numerous in the Andean part of the valley of

114

Río Santa (called the Callejón de Huaylas)
and the flanking high mountain ranges
Cordillera Blanca and Cordillera Negra.
The Cordillera Blanca is an ice- and
snow-covered range where many peaks exceed
6,000 m in altitude. It is cut by many
steep- to vertical-walled U-shaped glacial
valleys where earthquake-triggered rockfalls, rock debris slides, and debris
avalanches were widespread. Such landslides
blocked several streams in the Cordillera
Blanca, and one, the Huascarán debris
avalanche, described below, was the most
destructive historic landslide of the
Western Hemisphere. In addition, earthquake-triggered slumps, earth flows, and
mudflows formed in unconsolidated sediments
and lake beds on the floors of the glacial
valleys and in the Río Santa Valley. The
largest of these landslides, a 25-million-m^3
rotational slump, formed a dam on the Río
Santa at the village of Recuay (Fig. 2).
Fortunately, heavy mine equipment was
available in Recuay to cut a canal through
the dam to drain the lake before it became
large enough to cause catastrophic flooding
downstream.

In contrast to the Cordillera Blanca, the
Cordillera Negra is lower in altitude (the
undulating crest is between 4,000 and 5,000
m), lacks perennial ice and snow, has a
more rounded profile, is cut by V-shaped
stream valleys, and consists chiefly of
fractured and weathered volcanic rocks of
Tertiary age covered with thick slope debris
and morainal material. More than half the
landslides triggered by the earthquake were
in the relatively unstable weathered
bedrock and regolith of this range, and consisted chiefly of rock falls, rock and
debris slides, and soil slips, all of which
caused extensive damage to farmlands and
communities.

The 1970 Huascarán debris avalanche,
which originated on the north peak (6,655 m)
of Nevados Huascarán (Fig. 4), was preceded
by a similar, although smaller, avalanche in
1962 that was not triggered by an earthquake. This smaller avalanche, which had an
estimated volume of 13 million m^3, destroyed
most of the town of Ranrahirca and killed an
estimated 4,000 inhabitants in the town and
elsewhere in the valley of Río Shacsha (Fig.
4).

The two Huascarán peaks are the highest in
the Cordillera Blanca, and the south peak
(6,768 m) is the third highest peak in the
Western Hemisphere. Both the 1962 and 1970
avalanches began with falls of slabs of
fractured and sheeted granite from between
the altitudes of 5,400 and 6,500 m in the
near-vertical west face of the ice- and
snow-covered north peak of Huascarán (Figs.
3 and 4).

The fall causing the 1970 avalanche had an
estimated volume of between 50 and 100 million m^3, of which about 5 million m^3 consisted of ice. The slab that generated this
volume of debris is estimated to have been
between 60 and 100 m thick and had an area
of 0.6 km^2. This slab crashed onto the
glacier beneath the near-vertical face,
breaking into a chaotic mass of debris that
incorporated additional snow and ice from
the glacier, and then swept down the valley
of the Río Shacsha and into the Río Santa
(Fig. 4). The avalanche crossed the Río
Santa, flowed up the west side of the
Santa Valley a distance of several hundred
meters horizontally and about 85 m vertically, and destroyed part of the village of

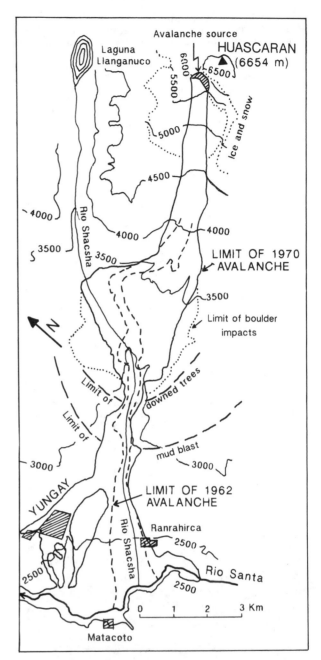

Figure 4. Sketch map showing the areas of
the 1962 and 1970 Huascarán debris
avalanches. Modified from Plafker and
Ericksen (1978).

Matacoto and killed 60 persons. In the Río
Shachsa valley, the avalanche destroyed
everything in its path, including the
northern part of the town of Ranrahirca,
which had been rebuilt after being destroyed
by the 1962 avalanche. In the lower part
of the valley, the avalanche left a mass of
debris (Fig. 3) that contains blocks of
granite estimated to weigh as much as
7,000 t.

In the 16 km from the source area to the
Río Santa (altitude 2,400 m), a vertical
drop of 3,000-4,000 m, the avalanche had an
average velocity of about 280 km/hr. On the
lower slopes of Huascarán, the avalanche may
have attained a velocity of as much as 1,000
km/hr, as indicated by boulders weighing
several tonnes that were hurled through the
air for distances as much as 4 km from mor-

ainal ridges overridden by the avalanche (Fig. 3) and by blasts of near-horizontal sheets of mud and rock debris that spattered an area extending as much as 2 km from the avalanche margin (Fig. 4). Blasts of air from beneath the avalanche in this area toppled eucalyptus trees (Fig. 4) as much as 40 cm in diameter and blew away thatch roofs of houses.

By the time it had reached the Río Santa, the avalache had become highly fluid, and it flowed both up and down the river to form a temporary dam that lasted less than half an hour. It then formed a many-kilometer-long slug of muddy water and debris consisting of trees, building materials, vehicles, and bodies of animals and humans that swept down the Río Santa some 200 km to the coast (Fig. 2) at an average velocity of 30 km/hr. This debris flow destroyed farmlands, farm buildings and communities, extensive segments of roads and a railroad, and part of a large hydroelectric installation. It killed an additional unknown number of inhabitants.

A second, much smaller lobe of the 1970 avalanche, which overtopped a 200-m-high ridge on the north side of the Río Shachsa valley (Fig. 4), obliterated the city of Yungay and killed nearly all of its 18,000 inhabitants. Destruction was awesome, bodies were torn apart, automobiles and trucks twisted together into chaotic masses (Fig. 5), and adobe blocks, the principal construction material of Yungay, were ground into the avalanche to form a final viscous pastelike material. This material now covers Yungay to a maximum depth of about 5 m; all that remains to be seen of this former city are a few of the palm trees at the site of Plaza de Armas and remnants of the cathedral (Fig. 5).

Figure 5. Site of the former Plaza de Armas of Yungay, showing remains of cathedral walls (C), and palm trees partly buried by the 5-m-thick debris now covering the Plaza de Armas. The ridge (OR) that was overtopped by the avalanche is visible in the distance. Wreckage (B) in right middle ground consists of a smashed bus and truck.

Another debris avalanche, which had a volume of less than 500,000 m^3, originated as a rock fall on the north side of the north peak of Nevados Huascarán during the 1970 earthquake. This avalanche swept into the steep-walled glacial valley to the north and formed a dam between the two Llanganuco lakes, of which only the lower is shown in figure 4. It buried a group of mountain climbers, who, unfortunately, were camped between the lakes at the time of the earthquake.

These avalanches are the only ones recorded as having originated on Nevados Huascarán since arrival of the Spanish in the 16th century. However, the Yungay area is covered with the debris of at least one enormous prehistoric avalanche that evidently also originated on the north peak of Huascarán. To judge from the debris, this avalanche was several times as large as the 1970 avalanche.

The north peak of Huascarán remains hazardous and probably will be the source of other catastrophic avalanches. In recognition of this danger, the city of Yungay was rebuilt north of its former site, and the debris-covered former city has been declared a national monument dedicated to the people killed by the avalanche.

3.4 Mayunmaraca, Peru, landslide of April 1974 (Kojan and Hutchinson, 1978)

The Mayunmarca landslide (Fig. 1) of 1974 was, in terms of a total volume of about 10^9 m^3, the largest historic landslide of the Western Hemisphere. It was a combined rock slide and debris flow that had a total length of about 8 km and a vertical drop from head to toe of 1,900 m. The duration of movement was about 3 minutes, giving an average velocity of about 130 km/hr. The slide, which was not triggered by any known event, formed a 3.8-m-long, 150-m-high (maximum) dam in the canyon of the Río Mantaro, one of the major rivers of the eastern Andes of Peru. It killed an estimated 450 persons. The dam lasted 44 days, during which time the lake behind it attained a length of 38 km, a maximum depth of 150 m, and a volume of water estimated at 6.7 X 10^8 m^3. At maximum length, the lake extended upstream nearly to the Mantaro hydroelectric plant, which is Peru's largest electrical facility. When the lake overflowed, the dam was breached and was progressively destroyed over a period of about 48 hours. The flow reached a maximum of 10^4 m^3/s and caused catastrophic flooding that destroyed villages and farms downstream. Fortunately, the area was evacuated before the dam ruptured, and no loss of life is recorded.

This sector of the Río Mantaro is underlain by a sequence of relatively incompetent, well-bedded sandstone and siltstone of the Mitu Formation (Permian) in which have occured many large landslides in the past. In addition to the Mayunmarca landslide of 1974, major landslides of the present century are known to have occurred in this area in 1930, 1945, 1960, and 1974. An active landslide that endangers the Mantaro hydroelectric facility is described in a paper by R. E. Michelena that is included in this symposium. It can be expected that other large landslides will occur in this area at relatively frequent intervals in the future.

4 CONCLUSIONS

The central and southern Andes have widespread unstable slopes that are subject to catastrophic landslide failure, caused chiefly by earthquakes and heavy rains. Many destructive landslides have occurred in the past, and it can be expected that many others will occur frequently in the future. Although much of the landslide-prone area is sparsely populated, landslides during the middle and late 20th century have killed at least 25,000 persons and caused property damage estimated to be more than a billion U.S. dollars. Engineering works aimed at mitigation of landslide hazards have been attempted in only a few places, with variable degrees of success. It can be expected that more work of this type will be performed in the future to correct landslide problems in populated areas or landslides that endanger transportation routes and hydroelectric facilities in areas of sparse population.

REFERENCES

Abele, G., 1982. El lajar Tinguiririca; su significado entre los lajares chilenos: Informe Geográfica de Chile, v. 29 p. 21-34.

Bodenlos, A. J., and Ericksen, G. E., 1955. Lead-zinc deposits of the Cordillera Blanca and Cordillera Huayhuash, Peru: U.S. Geological Survey Bull. 1017, 166 p.

Davis, S. N., and Karzulovic K., J., 1963. Landslides at Lago Riñihue, Chile: Seismological Society of America Bull., v. 53, p. 1403-1414.

Fernandez Concha, J., 1957. El problema de las lagunas de la Cordillera Blanca: Sociedad Geológica del Perú Bol., v. 32, p. 87-96.

Goldberg, R. A., Tisnado M., G., and Scofield, R. A., 1987. Characteristics of extreme rainfall events in northwestern Peru during the 1982-1983 El Nino Period: Journal of Geophysical Research, v. 92, p. 14,225-14,241.

Kojan, E., and Hutchinson, J. N., 1978. Mayunmarca rockslide and debris flow, Peru, chap. 9, in Voight, Barry, ed., Rockslides and avalanches: New York, Elsevier Scientific Publishing Co., p. 315-361.

MacPhail, D. D., 1973. The geomorphology of the Rio Teno Lajar, central Chile: The Geographical Review, v. 63, p. 517-532.

Maranguníc D., C., 1974. The lajar provoked by the eruptions of the Villarrica volcano on December 1971 [abs.]: Santiago, International Symposium on Volcanology, p. 48.

Maranguníc D., C., Moreno R., H., and Varela B., J., 1979. Observaciones sobre los depósitos de relleno de la depresión longitudinal de Chile entre los rios Tinguiririca y Maule: Instituto de Investigaciones Geológicas, II Congreso Geológico Chileno Actas, v. 3; p. 129-139.

Plafker, G., and Ericksen, G. E., 1978. Nevados Huascaran avalanches, Peru, chap. 8, in Voight, Barry, ed., Rockslides and avalanches: New York, Elsevier Scientific Publishing Co., p. 277-314.

Plafker, G., Ericksen, G. E., and Fernandez Concha, J., 1971. Geological aspects of the May 31, 1970, Peru earthquake: Seismological Society of America Bull., v. 61, p. 543-578.

Saint-Amand, P., 1961. Los terremotos de Mayo--Chile, 1960: China Lake, Michelson Laboratories, U.S. Naval Ordnance Test Station, Technical Artical 14, 39 p.

Varnes, D. J., 1958. Landslide types and processes: Highway Research Board Special Report, no. 29, p. 20-47.

Landslides: Extent and Economic Significance, Brabb & Harrod (eds)
© 1989 Balkema, Rotterdam. ISBN 90 6191 876 6

Landslides in Peru

R.E.Michelena
Michelena Repetto y Asociados S.A.

ABSTRACT: During the last 45 years a statistical record of geodynamic phenomena occurring annually in Peru has been kept. This information was probably the object of careful record by ancient civilizations flourishing in this Andean country many thousand years ago. The geological, geomorphological and hydrological conditions of this region act together giving rise to the hazard of actual and potential landslides. Additionally, the characteristics of the most important landslide of the country are reviewed. Remedial work on the latter has required an investment of over 40 million US dollars.

1 INTRODUCTION

The Peruvian territory lies on the west side of South America, with the Andes running parallel and close to the Pacific coastline, and with the Amazon Plains on the east side of the Andes. Due to strong orogenic activities, the terrain is highly vulnerable to dynamic processes such as landslides, rockfalls, mud floods, etc.

The phenomena mentioned above were widely known to the ancient cultures of Peru. In native languages, there are many words for them, like "Llapana," which is used for mud and sand flood without cobbles, and "Huaico" which is used for mud flood including cobbles and big rocks.

Chavin, an ancient culture in northern Peru which flourished 3,000 years ago, suffered the impact of the phenomena when a very large city was buried by a Huaico.

Damage resulting from these natural events is great, but details of the amount, kind, and location are difficult to obtain. Technical studies and inventories of the events are carried out mainly by the National Institute For Geology, Mining and Metallurgy (INGEMMET), but some of them are investigated by other public and private institutions.

A national inventory was initiated about 1943 and has continued to the present. The inventory indicates that a few large or recurrent events caused the death of at least 10,000 people and losses of at least 1,000 million dollars in land, crops, etc., requiring an expenditure of $100 million in remedial work. About $40 million of the remedial work was for one big event, the Tablachaca Slide No. 5.

2 THE TERRITORY

Peru's territory covers 1,285,200 km^2 and comprises three zones: the Coast, the Sierra, and the Amazon jungle. The Coast is a narrow, desertic (no rains and high evaporation) belt, without conspicuous relief. The Sierra is the Andean zone proper, with peaks above 6,000 m.o.s.l. and deep intermediate valleys. The Amazon jungle is tropical rainforest.

The Andean System has three mountainous ranges: the Coastal, Occidental, and Oriental Cordilleras. The first of these lies only along the southern part of the coastal belt, parallel to the Pacific Ocean coastline. Its northern part lies at the botton of the sea, where it sank by tectonic movements of the past.

The Occidental Cordillera is the highest of the three, whereas the Oriental is the lowest, very much exposed to the influence of the climate of the neighbouring Amazon jungle.

The deep valleys of the Cordillera and the steep sides of the hills favour hydro-geological dynamic phenomena such as landslides. The Andes' western side, facing the Pacific Ocean, is the area with Huaicos and Llapanas or mud floods. Most of these mountainous areas are exposed to active tectonic movements and show complex geological structures. There are many active faults in the northern part of the Occidental Cordillera.

The eastern side of the Cordillera is exposed to heavy rainfall (Fig. 1) and high temperatures.

Figure 1. Annual precipitation in Peru

Thick layers of residual soil overlay slopes on hills. These residual soils are, generally, clayey and become the focus for landslide problems. In recent decades, excavation for roads has disturbed the natural equilibrium of slopes. These works and water (rainfall and/or underground water) triggered slides in the residual soils. In addition, at the northern part of the Cordillera, strong earthquakes constitute a dynamic factor that increases the number of landslides. Fortunately, these phenomena occur mainly in areas with sparse population. In southern Peru, the intensity of these phenomena is relatively small.

3 HYDROLOGY

Precipitation in Peru is predominantly orographic and convective. The latter occurs in the Amazon plains where it ranges from 1500 to 4500 mm/year. The danger of landslides in this region exists only where a few hills are present.

Orographic rains are predominantly in the Andean region and where they range from a few mm/year to about 1000 mm/year in the highest areas. The rainy season lasts from November to April. During the dry season, evaporation is relatively high and the soil cracks, especially clay- and lime-rich soil. In the next rainy season, infiltration through the cracks saturates the soil and contributes to landsliding.

On mountain slopes of the Pacific side of the Andes and on coastal plains, climatic inversion by the cold Humboldt current causes dry and semi-dry weather. The extension of the dry belt fluctuates and depends on annual climate. This phenomenon is the cause for the lack of natural vegetation and extensive exposure of bare soils to weathering, because few plants can survive 3 to 5 consecutive dry years. Dry years are about 60% of all years.

The mass of detritus accumulated during dry periods is removed suddenly by the next rainfall. Depending on the type of soil and amount of slope, the consequences are Huaicos (mud, sand and rock floods), Llapanas (mud and sand floods) and landslides.

4 GEOLOGY AND GEOMORPHOLOGY

The danger for landslides is a consequence of the "young" geological characteristics of the Peruvian Andes. Steep slopes, deep valleys, strong earthquakes, and persistent orographic rains act together to create favourable conditions for actual or potential landslide hazards.

The effects of the previously mentioned phenomena are especially relevant in the case of residual soils overlaying harder deposits; contact surfaces can be lubricated by underground water resulting from rainfall, which is the case of the Tablachaca Dam slide No. 5 (Fig. 2).

In other cases, the actual or potential landslides occur in alluvial deposits, banks, or steep slopes, with little or no non-cohesive materials. There are many sites along the piedmont and steep sides of the Andean valleys where landslides are in progress.

5 TABLACHACA DAM SLIDE No. 5

Slide No. 5 is an ancient slide of about 13 million m^3 located upstream from the Tablachaca Dam and the Mantaro Hydroelectric Plant, with 50% of the country's total hydroelectric energy. In February 1982, the slide movements accelerated alarmingly, prompting the owner to undertake emergency stabilization, simultaneously with complementary studies and investigations (Figs. 3, 4, and 5). The emergency works consist of: fill buttress at the slide toe; treatment of loose sand and lime that form the foundation soil of the buttress; post-stressed anchors in a stretch adjacent to the dam where buttress size had to be reduced; underground and surface drainage; and excavation of a rock protrusion in the river curve upstream from the slide. By 1985, slide movements had decreased significantly, which indicates the effectiveness of the stabilization works. At present, information recorded with the instruments available shows little movement. However, it must be pointed out that there has recently been a period of dry years.

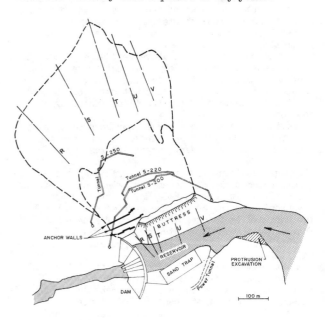

Figure 3. Emergency stabilization works at Tablachaca Dam.

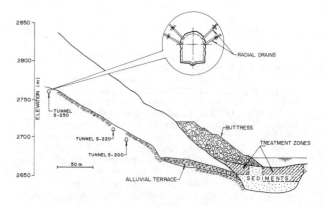

Figure 4. Detail of section T, Fig. 3, showing emergency stabilization works.

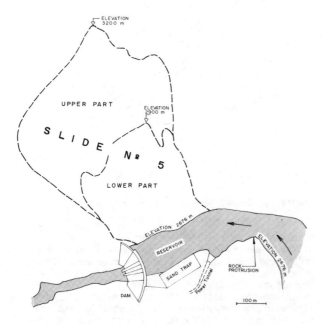

Figure 2. Tablachaca Facilities and Slide No. 5.

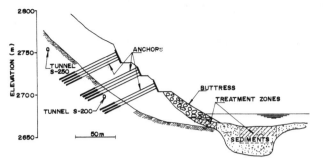

Figure 5. Detail of section R, Fig. 3, showing emergency stabilization works.

6 NATIONAL LANDSLIDE INVENTORY

INGEMMET is working on a national inventory of slides and other natural phenomena. For this purpose, the country has been divided into hydrographic basins. In special cases, important highways and railroads are being used for topographic reference.

A rough diagram (Fig.6) shows large landslide sites, and Fig. 7 showing landslides in the Rimac River basin is typical of some of the more detailed maps available. Fig.7 is a reduction of an original map at 1:100,000 scale which includes river erosion and other geodynamic phenomena. For the Central Andean Highway and Railroad, landslides are identified and defined in a technical archive, which includes suggested remedial measures in each case. The establishment of a data base with an alpha-numeric identification of each individual phenomenon is under consideration.

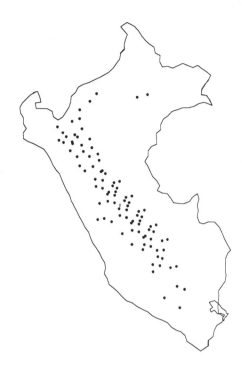

Figure 6. Large landslide sites in Peru.

Another important inventory has been made by PETROPERU (National Peruvian Oil Company) along the Northern Peruvian Oil Pipeline (800 km. long). Many landslides have been defined and plotted as part of in-depth research.

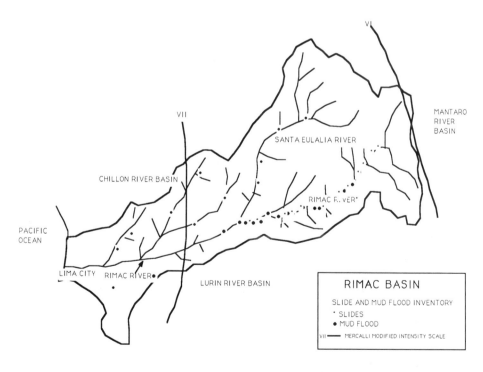

Figure 7. Rimac Basin.

Landslides: Extent and Economic Significance, Brabb & Harrod (eds)
© *1989 Balkema, Rotterdam. ISBN 90 6191 876 6*

Landslides: Extent and economic significance in Ecuador

Stalin Benitez A.
CEPE, Universidad de Guayaquil-IIEA, Guayaquil, Ecuador

ABSTRACT: Landslides are prevalent on the western slopes of the Andes where most of the 10 million people live. A landslide near Chunchi in 1983 blocked the Pan American Highway, buried vehicles, and killed at least 150 people. An even larger landslide in the same area in 1985 blocked the highway and a railroad and killed four people. In 1987, two large landslides near Cochancay buried 5 buses and killed dozens of people. The same year, landslides at the entrance to the Nambija gold mine killed 80 people.

The most disastrous landslides in Ecuador's history occurred on March 5, 1987, when earthquake-triggered landslides near Reventador killed more than one thousand people and severed 33 km of oil pipelines. The damage to the pipeline delayed oil exports for almost half a year, reducing government income by nearly 35 percent.

Landslide fatalities in Ecuador have been increasing from 10 deaths in the 1950's to 290 deaths in the 1980's, excluding the earthquake-triggered landslides in the Reventador area. Damage has been mainly to highways, railroads, vehicles, homes, irrigation channels, cattle, and cultivated fields.

1 INTRODUCTION

1.1 Geography

Ecuador, one of South America's smallest countries with an area of about 270,000 km², is near the northwest end of the continental mass, surrounded by Colombia on the north, Peru on the south and east, and the Pacific Ocean on the west (Fig. 1). Ecuador owes its name to the equinoctial line or equator, which passes through the country. Though small in size, it offers a variety of regions and climates which geographically are, from west to east:

The Galapogos Islands: world famous by its unique biota.

Territorial Sea: typified by the northern continuation of the Peru-Chilean trench.

Coast: generally lowlands and low mountains (Fig. 2).

Highlands or "Sierra": occupied by the west-central Andean Range with an average height in excess of 2,000 m. which drastically modifies the climate of the country. Historically active volcanoes occur in the northern part.

"El Oriente": comprises the sub-Andean foothills, the third eastern Andean range (though much lower than highlands) and a vast plain east of the Andean Range covered by tropical jungle that belongs to the Basin of the great Amazon River, the greatest source of ground water in the world.

1.2 Rainfall

Mean Annual Precipitation – The south coast is somewhat dry (less than 500 mm/y) (Fig. 3), affected by the same Peruvian coastal desert processes which are related to the Humboldt current. In the northern part of the coast, Equatorial tropical conditions predominate with precipitation higher than 3,000 mm per year. Rainfall is less than 1,000 mm per year in Andean valleys and substantially higher on the flanks. East of the Andes, rainfall

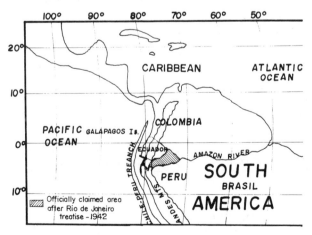

Figure 1. Map showing the location of Ecuador. Area with diagonal shading is claimed by Ecuador in accordance with the 1941 Rio de Janeiro treaty.

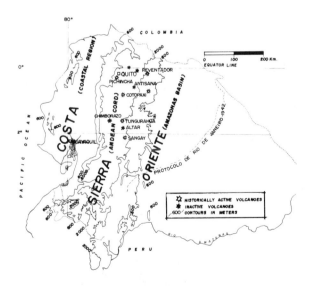

Figure 2. Physiographic map showing the main geographic subdivisions of Ecuador.

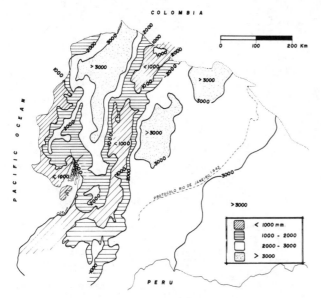

Figure 3. Mean annual rainfall from 1931 to 1960. Contours in mm.

is more homogeneous, generally from 2,000 to 3,000 mm and in places exceeding 3,000 mm per year.

Rainfall in Ecuador occurs mainly from January to June. The most intense rain occurs generally in March, when daily rainfall in Guayaquil may exceed 205 mm. Most landslides also occur during this period (Fig. 4).

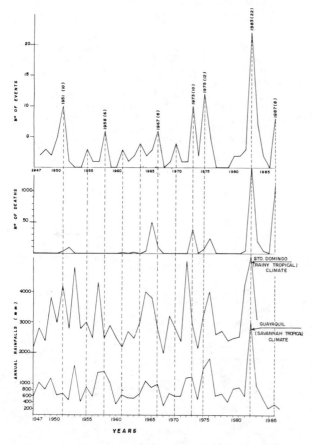

Figure 4. Relationship between rainfall and number of landslide events. Note that landslide events correlate most closely with Guayaquil maximum rainfall.

1.3 Slope

The western slopes of the Andean Range between elevations 600 m and 2,000 m and around 2°20' south latitude have the steepest slopes. This area is identified as the Pallatanga-Chunchi "alley" in this report.

The estimated population of Ecuador is 10 million inhabitants mostly evenly distributed (90%) between the coast and Sierra regions. Consequently, most of the population lives where landslides are most abundant (Fig. 5). The area most affected is the Pallatanga-Chunchi "alley" which has steep slopes and a high population density.

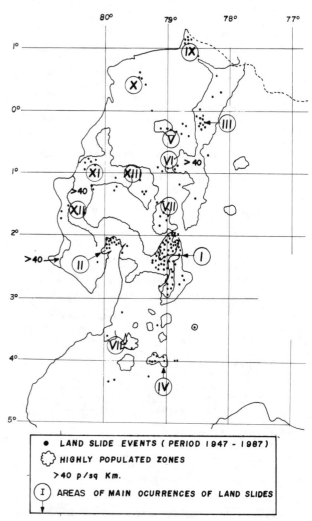

Figure 5 Landslide events and population density.

2 GEOLOGY

Continental Ecuador (Fig. 6) owes its main geological features to the subduction system between the Nazca (oceanic) and South American (continental) plates. Friction between these plates created the Andean Orogenic System with many earthquakes and important volcanic activity. The rugged Andean topography combined with seismic activity has produced the greatest natural catastrophes in Ecuador.

3 LANDSLIDE LITERATURE

There is not much literature available on landslides in Ecuador. There is no institution or government office which has statistics on landslides or is

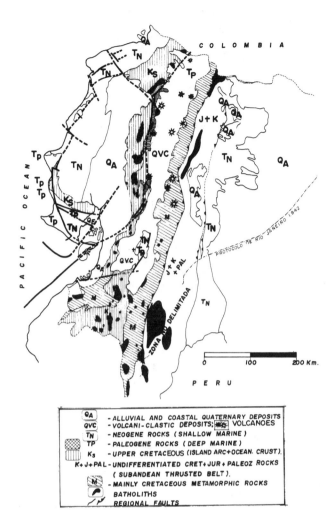

Figure 6. Geology of Ecuador.

Legend:

QA - ALLUVIAL AND COASTAL QUATERNARY DEPOSITS
QVC - VOLCANI-CLASTIC DEPOSITS; VOLCANOES
TN - NEOGENE ROCKS (SHALLOW MARINE)
TP - PALEOGENE ROCKS (DEEP MARINE)
Ks - UPPER CRETACEOUS (ISLAND ARC+OCEAN. CRUST)
K+J+PAL - UNDIFFERENTIATED CRET+ JUR + PALEOZ ROCKS
(SUBANDEAN THRUSTED BELT)
M - MAINLY CRETACEOUS METAMORPHIC ROCKS
P - BATHOLITHS
REGIONAL FAULTS

carrying out any type of systematic study. Only in 1983, when rainfall was excessive, were a few reports prepared; the one by Kojan (1983) was particularly helpful. A few unpublished reports mention landslides in Quito and Guayaquil.

4 LANDSLIDE PROCESSES

Geographic and Historic Extent – The most important landslide events reported by the national communication media throughout the 1947-1987 period are shown in Figs. 5 and 6. Areas with the largest number of landslides are designated with Roman numerals and are described below.

Area I – This area, with the largest number of landslides, is divided into two parts:

The northern part, called the Pallatanga-Chunchi alley, has the most active and continuous landslides in Ecuador. It is typified by rugged topography on the west flank of the Andes, a dense population, and geologic conditions that favor the formation of landslides. In this area, the steepest regional slopes are close to the actively-eroding Chimbo and Chanchan Rivers, which have opened deep gorges into the mountain range. These gorges were used for the Quito-Guayaquil railroad at the beginning of this century, considered at that time to be one of the greatest engineering works. A site known as "Nariz del Diablo" (devil's nose) where the railroad climbs the Andes in a zig-zag fashion to conquer the steepest slopes is world renowned. This railroad is also famous locally for various disruptions due to landsliding. Similarly, important highways in Area

I, such as the Pallatanga and Panamericana Highways, have been disrupted frequently by landslides.

The southern part contains the Canar-Cuenca section of the Panamericana Highway which had many landslides before 1975 during construction of the highway and for several years afterward.

Areas II and III – These areas include Quito and Guayaquil, the first and second most populated cities in Ecuador. Landslides in these areas are related to rapid growth of the cities during the last two decades. If this trend continues, landslides will be even worse in future years.

Areas IV to VIII – Areas with landslides along highways in the Andean foothills: the Loja-Zamora road (IV) in the east; the Santo Domingo-Aloag road (V) in the west; the Quevedo-Latacunga road (VI); the Babahoyo-Guaranda road (VII); and the Machala-Loja road (VIII). Along these routes, many landslides developed mainly after the highways were completed. In areas V, VI, and VII this was from 1950 to 1975, whereas in areas VI and VII the landslides developed from 1975 to 1987.

Areas IX to XIII – These areas along the coast also have landslides that developed soon after highway construction was completed. The relief in these areas is not as great as in the Andes, so the landslides are not as large or as numerous. The road in area X has not experienced landsliding since 1970 and the road in area XIII since 1965. Two landslides in area XIII have been associated with urban development.

5. LANDSLIDE TYPES IN RELATION TO GEOLOGY

Area I

Geology – This area is crossed by northeast-trending active fault systems (falla Pallatanga) which have strong seismic and volcanic activity and some hydrothermal alteration, such as at the Tixan sulphur deposits in the middle part. The formations are mainly volcano-clastic deposits of Tertiary and Quaternary age (Alausi and Cancagua Formations) and flysch of Late Cretaceous age (Yunguilla Formation). These formations in certain places have been extensively weathered, reducing their strength and making them more vulnerable to landsliding. They are also extensively fractured.

Landslide types – The landslides are large and complex. Many are associated with smaller mudflows and debris avalanches.

Area II

Geology – Guayaquil occupies a flood-plain of estuarine origin. Surrounding hills are underlain by the Cayo Formation (Cretaceous), a volcano-clastic flysch which decomposes easily to form expansive clays. The flysch has been quarried extensively, providing steep cuts and unstable slopes. Cutting of flysch slopes for roads and housing pods to accommodate the rapid expansion of Guayaquil has also provided ample opportunity for landsliding.

Landslide types – Rockfalls and rock mass slides in old quarry cuts.

Solifluction of residual and colluvial soils affecting inadequate foundations.

Slides of filled zones upon which structures were built.

Area III

Geology – Quito, Ecuador's capital, is on the flanks of Pichincha, a historically active volcano. Large amounts of poorly-consolidated tuff form the central cone. This material erodes easily and was formerly carried away by rivers. Constriction of the rivers by urbanization has led to flooding and landsliding.

Landslide types - Mudflows descending from the flanks of Pichincha are the principal problem, as noted by Kojan (1983).

Areas IV TO VIII
Geology - The outcropped rocks are mainly the Macuchi (volcanic) and Yunguilla (flysch) Formations of Late Cretaceous age.
Landslide types - Most of the landslides are rockfalls and rock slides triggered by highway construction.

Areas IX to XIII
Geology - Cretaceous rocks in Area XII are the same as in Area II. In the other areas, most of the rocks are brittle marine shale of Miocene age with some gypsum (Tosagua Formation). Most big events are on the cliffs near Areas XI and XII.
Landslide types - Slumps and rock-block slides along gypsum fractures.

6 STATUS OF KNOWLEDGE ABOUT EXTENT OF LANDSLIDE PROCESSES

In 1987, a commission appointed by the Junta Nacional de Defensa Civil (Civil Defense Board) of which the author was part had as its main objective the preparation of a natural hazard and risk map of Ecuador. The primary obstacle to this objective is a lack of aerial photographs, topographic maps, and geological maps. Only about 50% of the country has topographic maps at 1:50,000 scale and geological maps at 1:100,000 scale. In the Andean region, the percent of map coverage is much lower because of access problems. Consequently, the extent of the landslide problem has been estimated largely from the media, which has been concerned mainly with catastrophic events. Few field investigations have been made.

7 STATE-OF-THE-ART OF LANDSLIDE PREVENTION AND CONTROL

Landslide events in Ecuador are classified in two large groups:
   Problems along roads and railroads
   Problems in fast-growing metropolitan areas.
Landslides along road and railroads are caused mainly by a lack of government regulations and standards requiring adequate geological and geotechnical studies. Landslides in urban areas are related to the lack of urban planning that would take into account soil, geologic, and geotechnical factors. Unplanned urban growth, as in most Latin American countries, is a massive socio-economic problem practically without solution in the present conditions.

8 SHORT DESCRIPTION OF A FEW MAJOR LANDSLIDES

The landslide that caused the most fatalities from 1947 to 1987 occurred in Chunchi (Area I, Fig. 4) on March 26, 1983 during the wettest year of this century. This mass movement involved about 1,000,000 m^3 of material that slid approximately 3000 m, blocked the Panamerican Highway for 80 m, buried vehicles, and killed more than 150 people (Torres, 1983). Another even larger landslide took place on June 28, 1985 about 15 km northwest of the Chunchi landslide. Approximately 2,000,000 m^3 of material blocked a highway and railroad for approximately 300 m and killed four people. In both events, casualties could have been avoided. In the Chunchi event, for example, a Central University group of geologists warned the highway workers that a landslide had started (Geol. Guido Bonilla, personal communication, 1988), but they did not leave the place and died the next day.

In 1987, two large landslides in Area I near Cochancay buried five buses and killed dozens of people. The same year, landslides at the entrance of Nambija gold mine killed 80 people.
The most disastrous landslides in Ecuador's history occurred on March 5, 1987, when earthquake-triggered landslides near Reventador (Fig. 2) killed more than one thousand people and severed 33 km of oil pipelines. The damage to the pipeline delayed oil exports for almost half a year, reducing government income by nearly 35 percent.

9 SOCIAL AND ECONOMIC IMPACT

Figure 6 indicates that landslide fatalities in Ecuador have been increasing from 10 deaths in the 1950's to 290 deaths in the 1980's, excluding the earthquake-triggered landslides in the Reventador area. Damage has been mainly to highways, railroads, vehicles, homes, irrigation channels, cattle, and cultivated fields.
The estimated cost of physical damage, without considering life insurance or personal damage related to cattle-raising and agricultural activities, is about $4 million in the past decade, although the amount could be even higher.

10 PROFESSIONAL PERSONNEL AVAILABLE FOR LANDSLIDE RESEARCH AND/OR APPLICATION

There are dozens of photo-geologists, geomorphologists, geotechnicians, stratigraphers, tectonicists, etc., available to study landslide problems and to develop remedial measures. The success of the First Latin-American Symposium on Natural Disasters (Quito, 1985) organized by the Ecuadorian Committee on Geological Engineering and the National Politechnical College (Escuela Politecnica) shows the concern and availability of the geotechnical community if the government would provide the leadership and funding for a comprehensive program to deal with the landslide problem.

ACKNOWLEDGEMENTS

I am grateful to Marcelo Uria, who helped interpret the data, to Mrs. Nelly de Espinoza, who patiently collected data from the press, and to Mrs. Yolanda Bejar de Herdoiza, who translated the document into English, and especially to Dr. Earl Brabb who so professionally, as well as kindly reviewed and supervised this report.

BIBLIOGRAPHY

CEPE, 1987, Informe del ano 1987. Corporacion Estatal Petrolera Ecuatoriana, Quito Ecuador.
Kojan, E. 1983. Geological and engineering evaluation of disaster such as landslides, mud avalanches and floods in Ecuador. A.I.D. OFD, unpublished.
Niama, C. & M.Echeverria 1985. Inestabilidad y Riesgo en la Zona de Alausi: 106-109.
Niama, C. & M.Echeverria 1985. First Latin American Symposium on Natural Disasters. EPN-CEIG, Quito.
Torres, L. 1983. El Deslizamiento de Chunchi, III Congress of Geologic, Mines and Petroleum Engineers, CIGMIPG, Guayaquil.

Landslides: Extent and Economic Significance, Brabb & Harrod (eds)
© *1989 Balkema, Rotterdam. ISBN 90 6191 876 6*

Landslides: Extent and economic significance in Antarctica and the subantarctic

Patrick G.Quilty
Australian Antarctic Division, Hobart, Australia

ABSTRACT: The extent and economic significance of landslides in Antarctica are insignificant at present. Areas of high relief and volcanic activity do pose some potential problems but population numbers are so low that risk to humans is minor. Some glacial processes operating by forces similar to those causing landslides do pose potential future risks both in Antarctica and globally. Subantarctic islands have a greater landslide potential.

INTRODUCTION

Hazard is a way of life in Antarctica, but generally the hazards are not geological in origin except to the extent that large scale geological processes are responsible for putting Antarctica where it is.

Geological hazards become so only where population, services or structures are put at risk. The total population of Antarctica varies between some 1000 in winter to 4000 in summer (Parsons, 1987) (Fig.1). Thus, the low population, in itself, reduces the hazard to almost vanishingly small proportions.

The main geological hazards of volcanic activity, earthquakes and landslides are commonly, though by no means entirely associated with processes occurring in areas near subduction zones because it is here that elevation is high enough to generate instability and that sources of hazardous agents are abundant.

The Antarctic Plate now has only a very small proportion associated with subduction-

the Scotia Arc (Norton, 1982; Dalziel, 1982) - where the Atlantic Ocean is subduced along a length of some 400n miles below the South Sandwich Islands. Elsewhere, this large plate is bounded by constructive and conservative margins. The proportion of the plate boundary that is destructive is probably less than for any other plate. Nevertheless, there are many areas of high relief in Antarctica - the Transantarctic Mountains, Antarctic Peninsula (and its constituent blocks) and Prince Charles Mountains. Hazards identifiable in Antarctica are :

. volcanic
. those associated with fault movement
. those associated with short term glacial events

The function of this paper is to identify those areas of Antarctica which may be landslide prone, to discuss factors and processes which may influence the role of landslides as an Antarctic phenomenon and to discuss a few potentially hazardous glacial phenomena which, while not truly landslides in a normal temperate climate sense, have mass movement characteristics and some geometric aspects in common with landslides.

To date, geological research in Antarctica has been directed mainly at documenting the existence and distribution of basement rocks and relatively little effort has been directed towards gaining an understanding of modern geomorphic processes. Comments on geomorphic processes usually have been incidental to the main geological themes of any papers so far published. This may now be changing as strictly geomorphological studies are becoming more common, (eg. Pickard, 1986). Such studies have been an important aspect of offshore Antarctic studies for many years (eg. Anderson, et al., 1982; Johnson et al., 1982; Quilty, 1985).

Thus a data base for assessing the distribution of landslides and landslide processes in Antarctica is very meagre indeed. Recognition that the topic warrants consideration may lead to more relevant studies.

The area of outcrop on Antarctica is

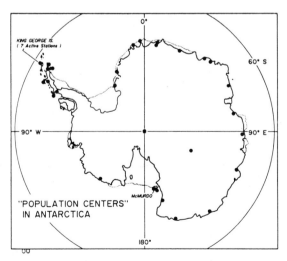

Figure 1. Outcrop and inhabited areas in Antarctica. Note the concentration in McMurdo Sound and the Antarctic Peninsula.

very small. While the total area of the continent is some $13.7 \times 10^6 km^2$, only 300 000 km^2 - about the area of Finland - is exposed. The rest is covered by an ice sheet with an average thickness of some 2.5 - 2.8km (Drewry, 1983). It is on the outcrop area, and the margins of the ice sheet, that this paper will concentrate.

As a consequence of several million years of glaciation, Antarctica has been swept almost clean of unconsolidated rock and soil, and a very large proportion of the outcrop area is of solid, fresh basement rock in which no landslide activity can be expected. There are however areas where volcanoes have produced high angle edifices of fragmental material, and others where the retreat of the ice has left major deposits of poorly consolidated or ice cored moraine. Any significant increase of temperature in these regions will generate areas of potential landslide risk.

Population density is highest on the northern part of the Antarctic Peninsula, particularly King George Island and in the McMurdo Sound area of the Ross Sea. These concentrations correspond to areas of significant outcrop, easier marine access and to higher levels of volcanic (and related potential landslide) activity.

EARTH MOVEMENTS IN ANTARCTICA

For this study, I am taking landslides to include downslope movement at a variety of scales and velocities and also to include some of these matters as they relate to ice. Gravity is always the prime cause.

Landslides and Earthquakes

Landslides occur when cohesion within an metastable mass is broken. Often, as in the case of the famous Yungay slide in Peru in 1970 and the Vaiont Dam disaster in Italy in 1963, the instability is finally triggered by an earthquake.

In Antarctica, major earthquakes are virtually unknown and any identified is a case for study (Adams, 1982). Thus Adams et al. (1985) recorded the first properly documented earthquake (magnitude 4.5) from the interior of Antarctica. It occurred in 1982. The cause of the apparent absence of Antarctic earthquakes is a subject for study in its own right.

There are many records of microseisms associated with volcanic activity (Kyle et al., 1982, Kaminuma et al., 1985) and these may be important in landslide generation, both microseisms and landslides being events (geologically) of local significance only.

While earthquakes on continental Antarctica are rare, they are well known in the oceanic area surrounding the continent (particularly the Scotia Arc- Dalziel and Elliott, 1973). In the Ross Sea there is ample evidence of intrusion and associated movement even on the present day seafloor (Cooper et al., 1987) and related seismic activity can be expected.

Small scale landslips are a common feature around lake margins bounded by sediments. Normally the sediments are cemented by permafrost in addition to any cement formed during diagenesis. The heat from the lake waters may cause thaw of permafrost allowing undercutting and collapse. This is an important mechanism of slope retreat adjacent to lakes (or the ocean in some cases) (Pickard, 1986).

Although this phenomenon is not much of a hazard to humans, it could in a sense reflect a hazard to the local environment caused by the presence of humans or by their activities elsewhere in the world. The only hazard to humans here could be dependent on changes in land form causing access tracks to disappear and a need for new ones to be developed.

In the submarine environment, turbidity currents and mass debris flows are important components of the sediment distribution system. Anderson et al., (1982) stated that "Mass movement on these margins is primarily by turbidity currents", perhaps implying a key role for submarine canyons so well known around the margins (Johnson et al., 1982). Wright et al. (1983) gave a broader list of processes, including slumping, debris flows and turbidity currents and referred to mechanisms within canyon systems and also "along the intercanyon slopes" as means of transport.

Volcanic edifices either dormant or active in Antarctica tend to have steep, snow covered slopes composed dominantly of loose, ice cemented volcanic debris. The main Antarctic localities are Mt Erebus and Mt Melbourne (McMurdo Sound) and Deception Island (northern Antarctic Peninsula). These volcanoes are, of course, a heat source leading potentially to melting of cementing permafrost with consequent downslope landslides of mixed rock and snow. The scale could be quite dramatic. All these features are close to inhabited stations and are areas of active study. Any landslide event therefore has the potential to impact human settlements or field activity and pose a hazard.

Deception Island is a locality where the combined effects of volcanic eruption and landslide (including water mass movements activity such as jokulhaups) have been documented (Roobol, 1982) and are reported to have devastated human settlements although no one was there at the time. This island constitutes the best example of volcanic/landslide activity in the vicinity of continental Antarctica.

Volcanic localities where the combined effects of volcanic activity and landslides may be expected have been documented by Gonzales - Ferran (1982, 1983). The volcanoes he listed as active (as distinct from those showing evidence of Holocene activity) are as follows (Figure 2):
1. Hallett and McMurdo Volcanic Province
    Mt Erebus
    Mt Melbourne
2. Marie Byrd Land Volcanic Province
    Whitney Peak (Mt Hampton)
    Mt Waesche
    Mt Berlin
3. South Shetland Islands Volcanic Province
    Deception Island
4. James Ross Island Volcanic Province
    James Ross Island
    Seal Nunataks
5. South Sandwich Islands Volcanic Province
    South Thule Island
    Bristol Island
    Saunders Island
    Candlemas Island
    Visokoi Island
    Leskov Island

All are composite or stratovolcanoes.

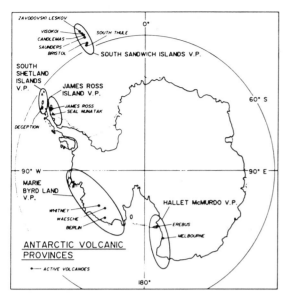

Figure 2. Active volcanism in Antarctica. Note the concentration in McMurdo Sound and the Antarctic Peninsula.

Kyle et al. (1982) referred to active and future lateral stoping and collapse as part of the expansion of the lava lake on Mt Erebus.

ICE MOVEMENT PHENOMENA

Avalanches

Avalanches occur when the grain-to-grain contact between ice particles turns to water, cohesion is lost and the mass moves under the influence of gravity. The normal trigger is an earth tremor or noise at a temperature range within which regelation can occur.

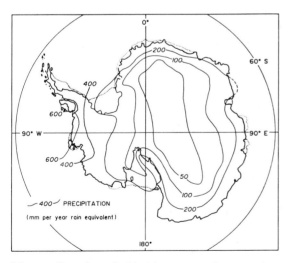

Figure 3. Precipitation over Antarctica. Isopleths are in mm rainfall equivalent per year (or 10 x gm cm^{2}yr^{-1}). Note that the highest figures are on the northern Antarctic Peninsula. Antarctica is essentially a dry, ice desert.

Antarctica is generally too cold for any regelation to occur and in addition precipitation generally is very low (Schwerdtfeger, 1983) (Fig. 3). Earthquakes are essentially absent and naturally, apart from wind, Antarctica is quiet. There are places however, where small avalanches do occur unseen by humanity. They occur on the lee side of mountain ranges which are transverse to the wind direction particularly on the Antarctic Peninsula where temperatures generally are higher than elsewhere. Certainly in this region, very steep ice slopes develop and other ice phenomena are found. Bergschrunds and cornices on lee sides of ranges all are consistent with potential ice movement.

Avalanches were observed by members of the 1907-1909 British Antarctic Expedition in the Transantarctic Mountains (David and Priestley, 1914) but, as an erosional agent, the effects were seen as insignificant compared to ice action. David and Priestley also referred to the role of the "sapping action of stormy seas" in producing sheer cliffs 60 m high.

Icebergs

Iceberg calving is a form of mass movement giving rise to hazards in two forms - to habitations established on ice shelves and to shipping.

In June 1986, the USSR station Druzhnaya was set adrift in one of three very large icebergs, each 3000-4000 km^2 in area which calved from the Filchner ice shelf. At the time it was unoccupied.

Little America, the station established and made famous by Admiral Richard E Byrd, is now lost at sea following its transport within an iceberg (but visible in the iceberg's side) in the 1960's (Lewis, 1965).

The size distribution of icebergs has been summarised by Hamley and Budd (1986) and Neshyba (1980). With modern radar techniques, these normal icebergs pose little threat to routine shipping. Should exploration in Antarctic waters develop and require long term occupation of drill sites, floating icebergs will present a hazard. Already, drilling by R.V. Joides Resolution during Ocean Drilling Program Leg 113 in the Weddell Sea and Leg 119 in Prydz Bay, required the support of a picket vessel (Maersk Master) to move icebergs and keep the drillsite safe.

If there are significant discoveries requiring establishment of production facilities, such facilities must be emplaced below depths likely to be scoured by icebergs which may be overthickened by being tilted or by having deep ice keels on them. Ample evidence exists (Barnes, 1987) of the effects of icebergs on the sea floor.

Less obvious on radar, more difficult to see and thus more dangerous are the much smaller, very dense "growlers", bodies of ice with their surface very near sea level. Their color blends in with that of sea water. Their size is up to 10-20m diameter.

A phenomenon now attracting much public interest is that of large icebergs, those which are rare (perhaps one to three around Antarctica at any one time) but immense. At the time of writing (early 1988) one exists which is approximately 160 x 45km surface area. The vertical extent has been variously reported as up to 2000m but that figure is a gross exaggeration and 200-500m probably is an upper limit. This massive iceberg was calved from the Ross Ice Shelf in October 1987. Pobeda Ice Island at 64°S, 98°E was a grounded large iceberg. It lasted at that locality until the late 1960's (Dr Ian Allison, pers. comm.) but has disappeared. In 1986, its place was taken by another and there is evidence (Allison pers. comm.) to suggest there the presence of an ice island at this locality has been a sporadic event recorded since the 1830's.

Ice Surge

Perhaps the greatest mass movement of ice noted in recent times is ice surge, described in detail by Dolgoushin and Osipova (1975) from the Pamir Mountains.

It has been postulated that there is a potential for cataclysmic ice surge from Antarctica (Wilson, 1964) with dramatic short term effects on sea level and resultant impact on settlement and installations near sea level around the world. It is now discussed in the context of results of the Greenhouse Effect.

The concept is based on the hypothesis that an ice-sheet can build up to a metastable condition and that there is either a critical change in temperature at the base of the ice or the instability reaches a critical point at which ice suffers pressure melting at the base through either geothermal or frictional heat. In either case, enormous volumes of ice move into the sea over a short period (up to several years). The ice cap eventually achieves stability. In a given ice cap, different parts seem to surge at different times.

In Antarctic terms, the phenomenon has been postulated to be a potential eventuality in two different ways.

One proposes that the East Antarctic ice sheet (the larger part, up to 4.8km thick - Drewry, 1983) containing some 26 x 10⁶km³, could surge. This assumes some instability in its present condition. It is over 4 km above sea level in places and is steeper near the margin (Drewry, 1983). In some areas it appears to be underlain by lakes and it is believed that in other areas, even though very cold, there could be basal melting and then sliding.

The other suggestion, considered more likely by some (Wilson, 1964) concerns the West Antarctic Ice Sheet which is warmer and underlain in many places by sea water (and thus floating). It is up to 2.5-3.0 km thick and contains 6 x 10⁶km³ of ice, enough to cause world sea level to rise some 6m should it all melt.

Jokulhaups

Jokulhaups occur when ice melts under an ice mass and breaks through to produce a local flood. They have been recorded on Deception Island in 1969 (Roobol, 1982), in the Casey region (Goodwin, 1988) and near Davis (Pickard, 1986). At Deception Island, it was a jokulhaup that destroyed an unoccupied Chilean station in 1969. Most jokulhaups are small but, in some circumstances, the effects could be devastating as many tonnes of water are involved.

SUBANTARCTIC ISLANDS

While Antarctica itself may be virtually earthquake and landslide free, this is not true for subantarctic areas where there are many islands and where the contrast between winter freeze and summer thaw is more pronounced. While it is fair to generalise that landslide activity is important, studies of landslide phenomena on sub-antarctic islands is, as on mainland Antarctica, virtually unknown. The following comments are not an exhaustive review but enough to indicate the relevance of the topic to subantarctic island planning.

Macquarie Island for example is a piece of seafloor raised above sea level (Varne et al., 1969; Duncan and Varne, 1988) and consists of fragmented pillow lavas and dykes with minor calcareous ooze interstitial between pillows. The east coast of the north-south oriented island is very steep and undercutting of the base has produced scree slopes which are a product of the major erosional force on the island. The area is one subject to regular and large scale earthquake shocks due to movements along the Macquarie Ridge (Jones and McCue, 1988). Landslides are a major mode of sediment transport to the ocean. Landslide prone areas on the island are not inhabited (there is a permanent station on the northern part of the island) but landslide debris makes traverse along the east coast difficult and potentially hazardous. Scott (1985) reviewed the literature appropriate to the island and referred to the link between landslips and seismic activity.

Bastien and Craddock (1976) reviewed the geology of the seldom visited Peter I Island and refer to "several avalanches observed during the 2-day visit ....", suggesting that mass movement is a very important factor in the evolution of the island.

Heard Island (Clarke et al., 1983) is an island of 2750m elevation dominated by Big Ben, an active volcano, sitting on a Paleogene carbonate basement (Quilty, et al., 1983). It is intensely glaciated and outcrop is mainly near sea level, and where rock penetrates the glacial cover at higher levels.

Although not documented, mass movement activity has had an important role in the evolution of the island and continues to do so. Mudslides up to one kilometer long have occurred from the peak of the island and apart from their importance as a transport mechanism, the redistributed material must have an effect on ice in several ways (P. Keage, pers. comm.) such as by affecting albedo and introducing a supply of liquid water.

A feature of the island is a series of ridges or buttresses radiating from the center of the island. These separate a number of cirques, the source of radially arranged short glaciers which unite in places as a piedmont glacier around the lower reaches of the volcano. Mass movement has been an important factor in the evolution of the cirques. Some of the margins of the island are composed of alluvial fans testifying to mass movement.

The Crozet Islands (Lameyre and Nougier, 1982) are very poorly known. The description of the topography would indicate that mass movement is an important element in sediment transport but no details are known.

Nougier (1982) reviewed the volcanism of St Paul and Amsterdam Islands. In both cases, he referred to the difficulty of landing because of "marine abrasion". Volcano slopes are steep, up to 30° on the outwards slope and 41° on the inward slope. In the last few thousand to few tens of thousands of years, St Paul has been reduced from 12km^2 to 6.5km^2 due to collapse of the northeastern part along a fault system.

Mass movement activity at present is minor, related to marine erosion and subsequent collapse around the island margin and perhaps in its crater.

THE FUTURE

While Antarctica remains as cold as it is now and the population as small as it is now, there is little hazard posed to humanity through landslide activity, even if landslides do occur.

There is a considerable and expanding body of opinion which suggests that Antarctica may suffer marked temperature increases over the next 50-100 years (Rind, 1984) due to heat retention by the atmosphere caused by increasing levels of the "Greenhouse gases" CO_2, CH_4 and oxides of nitrogen. The levels in the atmosphere have increased quite dramatically in the last 40 years and will continue to rise. Some of the increased heat of the atmosphere will be transferred to Antarctica. Estimates of the effects vary widely but all predict some melting of Antarctic ice and a rise in sea level.

These changes may well alter the scenario outlined earlier in this paper, so that slopes presently stable may become unstable as permafrost retreats. These matters have not, in the Antarctic context, been seriously addressed and it would be foolish to make predictions without proper analysis. At this stage, I leave the matter as one to be considered as "possible effects". Any marked increase in Antarctic population will have local effects which may exacerbate any changes caused by changes initiated else-where on earth.

If the population of Antarctica increases, even without the effects of humanity's actions discussed above, more attention may need to be given to potential landslide activity in choosing sites for installations.

CONCLUSIONS

At present, landslide hazard in Antarctica is less than for any comparable area on earth. There are very minor risks in high relief areas, particularly in areas of active volcanism which is where most human habitation exists.

In addition, there are some risks associated with glacial features such as ice surge and avalanches. All are minor.

Many subantarctic islands are volcanic in nature with high relief. Mass movement is a common factor in their erosion. Again, population density is so low that the risk to humans is very minor.

Many of these minor risks could be magnified if the earth's climate undergoes significant change over the next 50-100 years as a consequence of continued burning of large amounts of fossil fuels.

LITERATURE CITED

Adams, R.D., 1982. Source properties of the Oates Land earthquake, October 1974. In Craddock, C. (ed.) Antarctic geoscience. University of Wisconsin Press, Madison, 955-958.

Adams, R.D., Hughes, A.A., and Zhang, B.M., 1985. A confirmed earthquake in Antarctica. Geophysical Journal of the Royal Astronomical Society, 81:489-492.

Anderson, J.B., Kurtz, D., Weaver, F., and Weaver, M., 1982. Sedimentation on the West Antarctic continental margin. In Craddock, C. (ed.) Antarctic Geoscience. University of Wisconsin Press, Madison, 1003-1012.

Barnes, P.W., 1987. Morphologic studies of the Wilkes Land continental shelf, Antarctica - glacial and iceberg effects. In Eittreim S.L., and Hampton, M.A.(eds.). The Antarctic continental margin: Geology and geophysics of offshore Wilkes Land. Circum-Pacific Council for Energy and Mineral Resources, Earth Science Series, 5A:175-194.

Bastien, T.W., and Craddock, C., 1976. The geology of Peter I Island. In Hollister, C.D., Craddock, C. et al. Initial Reports of the Deep Sea Drilling Project, US Government Printing Office, Washington, 36: 341-357.

Cooper, A.K., Davey, F.J., and Behrendt, J. C., 1987. Seismic stratigraphy and structure of the Victoria Land Basin, Western Ross Sea, Antarctica. In Cooper A.K., and Davey, F.J. (eds.). The Antarctic continental margin: Geology and geophysics of the Western Ross Sea. Circum-Pacific Council for Energy and Mineral Resources, Earth Science Series, 5B: 27-76.

Dalziel, I.W.D., 1982. The early (pre-Middle Jurassic) history of the Scotia Arc region: A review and progress report. In Craddock, C. (ed.) Antarctic Geoscience. University of Wisconsin Press, Madison, 111-126.

Dalziel, I.W.D., and Elliott, D.H., 1973. The Scotia Arc and Antarctic margin. In Nairn, A.E.M. and Stehli, F.G. (eds.). The ocean basins and margins, Vol. 1. The South Atlantic. Plenum Press, New York, 171-246.

David, T.W.E., and Priestley R.E., 1914. Geology vol. 1. Glaciology, physiography, stratigraphy, and tectonic geology of South Victoria Land. Reports of the Scientific Investigations, British Antarctic Expedition 1907-1909. W. Heinemann, London, 319 pp.

Dolgoushin, L.D., and Osipova, G.B., 1975. Glacier surges and the problem of their forecasting. In International Association of Scientific Hydrology. Commission on Snow and Ice. Proceedings of the Moscow Symposium, August 1971. 292-304.

Drewry, D.J., (ed.), 1983. Antarctica: glaciological and geophysical folio. Scott Polar Research Institute, Cambridge.

Duncan, R.A., and Varne, R., 1988. The age and distribution of the igneous rocks of Macquarie Island. Papers and Proceedings, Royal Society of Tasmania 122: 45-50.

Gonzales - Ferran, O., 1982. The Antarctic Cenozoic Volcanic provinces and their implications in plate tectonic processes. In Craddock, C. (ed.) Antarctic Geoscience. University of Wisconsin Press, Madison, 687-694.

Gonzales - Ferran, O., 1983. Volcanic and tectonic evolution of the northern Antarctic Peninsula - Late Cenozoic to Recent. Tectonophysics 114: 389-409.

Goodwin, I., 1988. The nature and origin of a jokulhaup near Casey Station, Antarctica. Journal of Glaciology 34 (116): 95-101.

Hamley, T. and Budd, W.F., 1986. Antarctic iceberg distribution and dissolution. Journal of Glaciology 32 (111): 242-251.

Johnson, G.L., Vanney, J.R. and Hayes, D., 1982. The Antarctic continental shelf. In Craddock, C. (ed.) Antarctic geoscience. University of Wisconsin Press, Madison, 995-1002.

Jones, T. and McCue, K., 1988. Seismicity and tectonics of the Macquarie Ridge. Papers and Proceedings, Royal Society of Tasmania 122: 51-57.

Kaminuma, K., Ueki, S., and Kienle, J., 1985. Volcanic earthquake swarms at Mt Erebus, Antarctica. Tectonophysics 114: 357-369.

Kyle, P.R., Dibble, R.R., Giggenbach, W.F., and Keys, J., 1982. Volcanic activity associated with the anorthoclase phonolite Lava Lake, Mt Erebus, Antarctica. In Craddock, C. (ed.) Antarctic geoscience. University of Wisconsin Press, Madison, 735-745.

Lameyre, J., and Nougier, J., 1982. Geology of Ile de l'Est, Crozet Archipelago. In Craddock, C. (ed.) Antarctic geoscience. University of Wisconsin Press, Madison, 767-770.

Lewis, R.S., 1965. A continent for Science. Secker and Warburg, London, 300 pp.

Neshyba, S., 1980. On the size and distribution of Antarctic icebergs. Cold Regions Science and Technology 1: 241-248.

Norton, I.O., 1982. Paleomotion between Africa, South America, and Antarctica, and implications for the Antarctic Peninsula. In Craddock, C. (ed.) Antarctic geoscience. University of Wisconsin Press, Madison, 99-106.

Nougier, J., 1982. Volcanism of Saint Paul and Amsterdam Islands (TAAF): some aspects of volcanism along plate margins. In Craddock, C. (ed.) Antarctic geoscience. University of Wisconsin Press, Madison, 755-765.

Parsons, A., 1987. Antarctica! the next decade. Cambridge University Press, Cambridge, 164 pp.

Pickard, J., 1986. Antarctic Oasis. Academic Press, Sydney, 367p.

Quilty, P.G., 1985. Distribution of foraminiferids in sediments of Prydz Bay, Antarctica. Special Publication, South Australian Department of Mines and Energy, 5: 329-340.

Rind, D., 1984. Global climate in the 21st Century. Ambio 13: 148-151.

Roobol, M.J., 1982. The volcanic hazard at Deception Island, South Shetland Islands. British Antarctic Survey Bulletin 51: 237-245.

Schwerdtfeger, W., 1983. Weather and climate of the Antarctic. Elsevier, Amsterdam, 261pp.

Scott, J.J., 1985. Effects of feral rabbits on the revegetation of disturbed coastal slope sites, Macquarie Island. Unpub. M.A. thesis, Monash University, 196 pp.

Varne, R., Gee, R.D., and Quilty, P.G., 1969. Macquarie Island and the cause of oceanic linear magnetic anomalies. Science 166: 230-233.

Wilson, A.T., 1964. Origin of ice ages: an ice-shelf theory for Pleistocene glaciation. Nature 201: 147-149.

Wright, R., Anderson, J.B., and Fisco, P.P., 1983. Distribution and association of sediment gravity flow deposits and glacial/glacial-marine sediments around the continental margin of Antarctica. In Molnia, B. (ed.) Glacial-marine sedimentation. Plenum Press, New York, 265-300.

Landslide: Extent and economic significance in Norway

Odd Gregersen
Foundation Section, Norwegian Geotechnical Institute, Oslo, Norway

Frode Sandersen
Avalanche Section, Norwegian Geotechnical Institute, Oslo, Norway

ABSTRACT: The landslide problem in Norway is dominated by quick-clay slides. The largest of these occurred in Gauldalen in 1345 and killed 500 people. In 1983, a quick-clay landslide killed 112 people at Verdalen, and another at Rissa in 1978 destroyed 7 farms and 5 houses. On average, 15 people are killed by landslides each year, but a few catastrophes account for most fatalities.

Rockfalls and rock slides are common in western Norway where ice-eroded valleys and fjords expose bedrock on very steep slopes. Waves generated by some of these rockfalls and slides have killed as many as 73 people in a single event. Debris slides and debris flow are most common in the western coastal region.

Measures for mitigating landslides disasters in Norway including building structures to protect existing buildings, evacuating structures at risk, and preparing maps showing "safe" areas and areas at risk. The mapping program will cost 6 to 7 million US$ over a 12 year period. The annual cost of landsliding in Norway is estimated to be about 6 million US$ in average.

GEOGRAPHICAL DISTRIBUTION AND CLASSIFICATION

Norway is struck every year by a number of small and large landslides of many different categories. This article describes the categories of landslides which have greatest consequences for human life, namely those occurring in densely populated areas. These areas are southeast Norway (Oslo area), middle Norway (Trøndelag) and the coastal zone. Soil conditions in the Oslo area and Trøndelag are dominated by marine clay sediments which form quick-clay slides. The coastal zone is predominantly mountainous, with marine clays along the beach and moraine deposits on lower parts of the mountain slopes. Debris slides and rock falls/slides are most common. The geographical distribution of the different landslide categories is shown in Fig. 1. The extent of the landslide problem is illustrated in Fig. 2, which indicates the number of people killed in landslides over a hundred year period, from 1871 to 1970. While on average about 15 people are killed every year, a few large landslide catastrophes account for the majority of fatalities.

QUICK CLAY SLIDES

The phenomenon of quick clay behaviour has been discussed by Karlsrud and others (1984). The greater part of this chapter is extracted from that article.

Case histories

The largest landslide in Norway this century occurred in Rissa (Trøndelag, 1 on Fig. 1) on 29 April 1978. In the course of a few minutes, a small community with 7 farms and 5 single family houses was destroyed. The slide involved liquefaction and movement of 5 to 6 million m^3 of clay in an area of 330,000 m^2 (Fig. 3). The slide is described by Gregersen (1981).

Just before the turn of the century, in 1893, a landslide occurred in Verdalen (Trøndelag, 2 on Fig. 1). This is the largest slide during our recent history, covering an area of 3,000,000 m^2 and involving 55 million m^3 of clay (10 times the volume of the Rissa landslide), and killing 112 people.

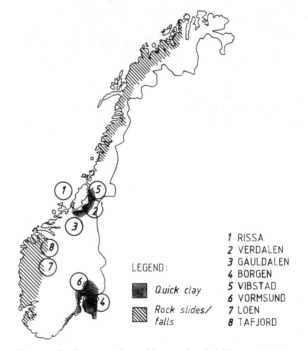

LEGEND:

▨	*Quick clay*
▧	*Rock slides/ falls*

1 RISSA
2 VERDALEN
3 GAULDALEN
4 BORGEN
5 VIBSTAD
6 VORMSUND
7 LOEN
8 TAFJORD

Figure 1. Areas vulnerable to landslide activities in Norway. Numbers refer to large landslides described in text.

The largest known quick-clay landslide occurred in Gauldalen (Trøndelag, 3 on Fig. 1) in 1345 when 500 people were killed.

Engineering geology

Quick-clay slides occur in marine sediments deposited in saline water after the last glaciation about 10,000 years ago. These clay-rich sediments cover an area of about 5,000 km^2 in Norway.

Two parallel processes form the conditions for these slides. Isostatic uplift, which amounts to nearly 200 m in southern Norway, has exposed marine sediments to surface erosion by streams and rivers,

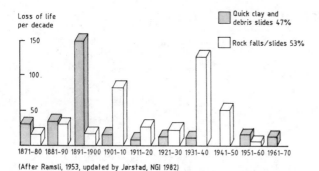

Loss of life per decade

■ Quick clay and debris slides 47%

□ Rock falls/slides 53%

(After Ramsli, 1953, updated by Jørstad, NGI 1982)

Figure 2. Loss of life due to landslide activities in Norway during the period 1871-1970. The total number of people killed in this period is approximately 700.

Figure 3. View of the landslide at Rissa in 1978. See also Fig. 9.

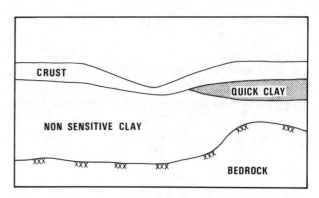

Figure 4. Location of a zone of quick clay along an eroded stream valley where bedrock is close to the surface on one side.

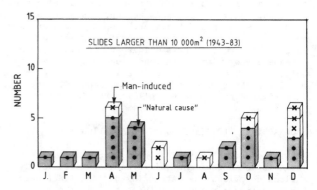

SLIDES LARGER THAN 10 000m² (1943-83)

Man-induced

"Natural cause"

Figure 5. Frequency distribution of landslides greater than 10,000 m² in area, 1943 to 1983.

resulting in increased shear stresses. Occurring along with this land upheaval is a process of leaching of salt water from the soil matrix. This weakens the structural stability of the marine clay and results in a decrease in shear strength.

The engineering geology of the Norwegian marine clays has been comprehensively discussed by Bjerrum (1954), Bjerrum and Rosenqvist (1956), and Bjerrum and others (1969). A brief review is given here because it is essential to an understanding of landslides in these deposits.

The Norwegian marine clays are rather silty, with only 20 to 40% in the clay size fraction. The clay minerals which are primarily illitic and chloritic, were deposited with a floculated structure in sea water. As shown by Rosenqvist (1946, 1955), leaching of salt water from the pore fluid changes the sensitivity of these clays from low (S_t typically 3-6) to high (S_t > 20). When the salt content drops below 1 g/l, the clay becomes a quick clay. Whereas a salty marine clay will remain plastic upon remoulding, a quick clay will transform into a liquid.

Inasmuch as the leaching process in Norwegian marine clay sediments is dependent on hydraulic gradients and time, it is virtually impossible to determine whether a leached sensitive clay occurs a a specific location without making soil borings. The arbitrary nature of quick clay formation is illustrated by the profile in Fig. 4. On the right side, a pocket of quick clay is formed by upward hydraulic gradients (artesian pressure) from the bedrock surface. On the left side, the bedrock surface lies much deeper. The hydraulic gradient is smaller, leaching is less, and little or no sensitive clay has formed.

Erosion prosesses

Quick-clay slides can be triggered by natural causes, such as erosion, or by human activity. The results of a recent study of clay slides larger than 10,000 m² in area in Norway in the last 40 years are shown in Fig. 5.

Eight of the 31 slides were triggered by small scale human activities, mostly earth works of different kinds. This chapter, however, discusses the main cause of quick-clay sides, that of erosion.

Erosion by streams and rivers, creating increasingly deeper cuts or gulleys into the marine terraces, is usually accompanied by small slips and large landslides. Changes in the originally flat, marine plateau of the Romerike district 50 km north of Oslo, after some thousand years of erosion, and landslides are shown in Fig. 6. Landslides have made a significant contribution to the overall process of denudation.

The Norwegian Geotechnical Institute (NGI) has carried out a detailed study of erosion within the area shown in Fig. 6, as described by Bjerrum and others (1969) and Foster and Heiberg (1971). Important concepts derived from this study were:

a) For a stream with a given catchment, there is a gradient below which the stream will not erode its bed (Fig. 7). On the basis of this relationship between stable gradient and catchment area, it is possible to construct equilibrium profiles for streams in the system.

b) A comparison between the equilibrium profile and the actual stream profile leads to a distinction between three zones of erosion:

(1) A zone of equilibrium in the lower range of the stream system:

(2) A zone of non-equilibrium where erosion is aggressive in the intermediate range;

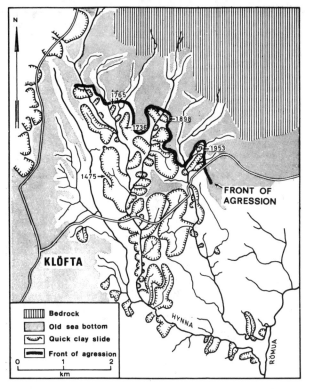

Figure 6. Erosion of the area around Kløfta, 50 km northeast of Oslo (after Bjerrum and others, 1969)

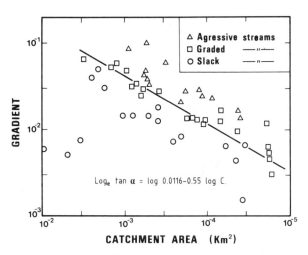

$$\text{Log}_e \tan \alpha = \log 0.0116 - 0.55 \log C.$$

Figure 7. Correlation between catchment area, stream gradient, and erosion (after Bjerrum and others, 1969)

(3) A zone nearly in equilibrium in the upper range.

The zones of erosion were found to be closely related to the geomorphology, which in turn reflects the modes by which material was brought to the stream.

The equilibrium zone (1) corresponds to areas where ancient quick clay scars are separated by low ridges and stream valleys. Mass movements arise from small rotational slides caused by meandering of the streams.

The zone of non-equilibrium (2) corresponds to the uppermost part of the slide area. It is characterised by scars of recent quick-clay slides and a steep gradient towards the intact horizontal clay plateau. The upstream limit is a "the front of aggression".

The upper equilibrium zone (3) corresponds to the

horizontal plateau of the old sea floor cut by V-shaped ravines, where slide activity is very limited. This zone is upstream of "the front of aggression".

Quick-clay slide mechanisms

The triggering of a quick-clay slide is related in many places to a local initial slide or local erosion. It may also be caused by extreme pore pressures in the ground induced by extraordinary rainfall or by thawing. Fig. 5 shows that the risk of landslides is greatest during autumn or in the thawing season.

When studying the phenomenon of quick-clay slides, two mechanisms seem to predominate: a) the retrogressive process and b) the monolithic, flake-type slide.

The retrogressive process is initiated a many localities by a local slide, leaving an unstable back-scarp in sensitive (quick) clay. The rate of retrogression can vary from slow to very rapid. Large areas may slide out, turn into liquid and flow downslope. Most of the final slide pits will have a characteristic bottle-neck shape. Such bottle-neck slides are very common; it is estimated that 30 to 40% of all mapped slides are this type. A typical bottle-neck slide, the Borgen landslide in 1955 (Oslo area, 4 on Fig. 1), is shown in Fig. 8.

The monolithic flake-type slides involve large areas which apparently slide as a monolithic unit. These slides occur within seconds and without warning. This slide mechanism can only take place where continuous layers of highly sensitive clay occur in the deposit. The authors believe that the flake-type slide mechanism is more frequent than previously appreciated; "vertical sinking", for example, is probably just a flake-type slide where the slide debris, due to topographical or other restraints, is prevented from leaving the slide pit.

Recent research at NGI on the stability of natural slopes in quick clays, by Aas (1979, 1982, 1983), indicates that the flake-type slide mechanism occurs where the overall stability of a large area is permanently low. Retrogressive sliding will occur when local instabilities are set up by initial slides, erosion and by the retrogressive sliding process itself. Factors that determine which of the mechanisms will dominate the sliding process may be:
. overall surface topograhpy
. position and thickness of the quick clay zone
. extent of the initial slide topography
. depth to firm strata
. strength characteristics of the clay
These factors are discussed in detail by Karlsrud and others (1984).

Many quick-clay landslides involve both of these two main slide mechanisms, as well as an initial slide. All the sequences of sliding were present in the huge and catastrophic landslide at Rissa (Fig. 3), as described by eye-witnesses and amateur films. The slide, shown as plan view on Fig. 9, eventually covered an area of 330,000 m^2 and involved 5 to 6 million m^3 of clay. Sliding was initiated at location A, by a minor earth fill on the shore of Lake Botnen. Retrogressive sliding took place over a 35 minute period in area B. Shortly afterwards, a huge flake-type slide occurred in area C, leading to further flake-type slides and finally a series of retrogressive failures. The flake-type sliding (areas C, D and E) took place over a very short period of time, a few minutes.

A comparison of the length to heigth ratio of slide pits (measured in the direction of sliding) versus the slide volume for all clay slides investigated by NGI during the last 15 years is shown in Fig. 10. As expected, the L/H ratio increases with slide volume. Slides involving sensitive or quick clays show an expected larger L/H-ratio than slides in low-sensitivity clay, and reach a maximum L/H-value of 15. Furthermore, the non-sensitive clay slides did not exceed a volume of about 50,000 m^3, whereas the

Figure 8. The slide at Borgen (Oslo area) in 1953.

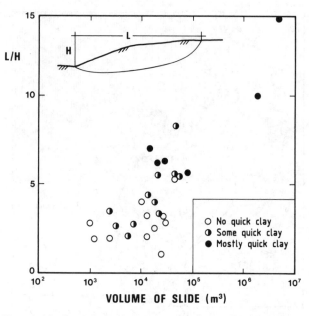

Figure 10. Length to height ratio (measured in the direction of sliding) versus slide volume for slides in clay investigated by NGI the last 15 years.

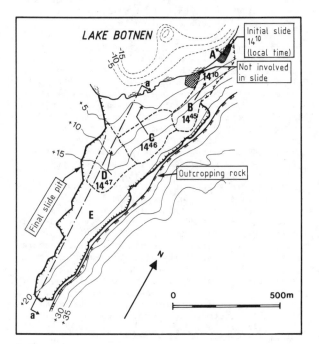

Figure 9. Plan view showing the sliding sequence in the Rissa landslide.

Figure 11. Map of principal geologic units in Norway. 1) Effusive and plutonic rocks (Permian) including Cambro-Silurian sedimentary rocks. 2) Sandstone and conglomerate (Devonian). 3) Strongly altered Cambro-Silurian sedimentary rock and igneous rocks of assumed Caledonian origin. 4) Gneissic and granitic rocks with Caledonian structure. 5)Plutonic rocks of Lofoten. 6) Gneissic and quartzitic rocks (Eocambrian). 7) Pre-Eocambrian rock complexes (Archaean rocks).

largest quick-clay slide (Rissa) was two orders of magnitude larger in size. It should also be mentioned that during the last 40 years there have been only two slides greater than 10,000 m^2 in area (or about 100,000 m^3 in volume) which did not involve predominantly quick clay. These were the monolitic flake-type slide at Vibstad along the river Namsen in 1959 (NGI, 1960), and the retrogressive slide at Ihlang near Vormsund in southern Norway in 1965 (NGI, 1967). However, in both cases there were indications of medium to high sensitivity clays in at least some parts of the slide areas. Thus one may conclude that the risk of major (> 10,000 m^2) landslides in Norway is almost entirely related to highly sensitive clay deposits.

From the study, it is evident that stability problems related to quick clays often encompass areas that are hundreds of metres away from the actual slope. It is important to bear this in mind when analysing individual building sites located on clay plateaus.

ROCK SLIDES/FALLS

The stability of rock slopes and consequences of rock slides in Norway have been discussed by Bjerrum and

Jørstad (1968). The greater part of this section is extracted from that article.

A distinction is generally made in Norway between rock falls and rock slides. Rock falls are the loosening and falling of single blocks or groups of blocks from a mountain side and are caused by processes active near the rock surface. Rock slides are a collective sliding of large rock masses along a deep-seated sliding surface.

Figure 11 shows a generalized geological map of Norway. By far the greatest number of stability problems arise in western Norway where ice-eroded deep valleys and fjords expose bedrock on very steep slopes.

Rock falls

Rock falls normally result from climate-induced processes acting at the surface. Frost shattering, chemical decomposition, temperature variations, wedging effect of roots, water pressure, and the action of snow cover are the principal factors responsible for instability. Frost shattering is the most predominant action. It occurs mainly in the spring and autumn seasons when the temperature fluctuates around the freezing point.

In comparison with frost shattering, the other factors mentioned above are of secondary importance, although they undoubtedly contribute to the loosening of blocks. Chemical decomposition of less resistant layers and zones in an otherwise hard and resistant rock mass is of some importance, especially where layers contain limestone, mica-schist and amphibolite. In heavy rainstorms, water pressure may play an important role in reducing the stability of a block. The wedging action of roots also helps move rocks, especially shists and limestones in the Oslo region.

Rock slides

In certain places, forces acting at great depths in bedrock may result in the sliding of large volumes of rock masses along deep-seated sliding surfaces. The volume of rock participating in such slides is commonly in the order of hundreds of thousands or millions of cubic metres. Nearly 350 lives have been lost in rock falls and rock slides during the period 1873 to 1940, with more than half of this number associated with three large rock slides.

A study of hillslopes with evidence for late-glacial and post-glacial landslides shows that, in the great majority of cases, slides have been detached along joint systems running almost parallel to the surface. This valley joint system, which is independent of structure and of geological boundaries, often consists of several sets of joints at various distances from the surface. The first joint is found mostly at a depth of 4 to 10 m. The distance have been formed in connection with stress changes that accompanied glacial erosion, and their development is closely associated with stored elastic energy in the rock mass originating from previous large loads of bedrock and ice.

The action of water in open rock joints probably has greater influenced stability of slopes. The water exerts an outward pressure, reduces the effective stresses and thus reduces the friction between a potential sliding body and its base. Fluctuations in precipitation, rate of snowmelt and drainage due to frost on the surface leads to great variations in water pressure throughout the year. These variations may lead to fatigue failure in the rock, bringing about a gradual extension of the open joints. In the long term, this effect is likely to cause a rock slope to fail.

The likelihood of a deep-seated slide is difficult to determine, but the large rock slides in Norway are commonly "advertised" in advance by a number of warning signs. Open joints at the surface may increase in width, perhaps several years before the occurrence of a slide. Immediately prior to a landslide, the rate of joint widening may increase substantially, and rock falls may begin. A final warning may be the observation of loud cracks from the slope.

Consequences of rock falls and rock slides

Although the volume of material that falls from mountain slopes in Norway each year amounts to millions of cubic metres, the damage caused is relatively small. From ancient times people have had sufficient understanding to place farmsteads and buildings in the least threatened areas where houses are rarely destroyed. The number of persons who perish in rock falls or slides averages 2 to 3 per annum. On the other hand, roads and buildings are exposed continuously to considerable damage by rock falls and slides. Great national disasters occur when large rock slides go into relatively narrow and deep lakes and fjords in the western part of Norway, generating waves which spread over long distances. In the present century, Norway has experienced three disasters of this kind, with a total death toll of 175 and a large loss of property. The maximum recorded heights of these waves were, chronologically, 41, 74 and 62 m.

Damage was restricted to areas 10 km from the slide, but the waves were noticed at much greater distances! (Jøstad, 1966).

The first of these disasters took place in 1905, when approximately $300,000\ m^3$ of rock fell into the lake near Loen (7 on Fig. 1) creating a wave that caused extensive damage up to 41 m above water level along the entire lake. In total, 61 people perished. In 1936, another large slide occurred at the same place, involving about 1 million m^3 of rock, a wave of 74 m high and the deaths of 73 people. The wave apparently reached a velocity of about 25 to 30 m/s, and had a wavelength many hundreds of metres. The third disaster was at Tafjord (8 in Fig. 1) in 1934 when a rock mass of more than 1 million m^3 fell from the mountain and brought along an equal volume of scree and moraine material. The resulting waves washed along the shoreline to a maximum height of 62 m. Waves 2 to 3 m in height were observed as far away as 34 km from the slide.

Landslide dams are common in Norway, but none of them apparently failed catastrophically.

Equipment and instrumentation suitable for surveying the distribution and rate of expansion of valley joints provide a way to monitor the development of deep-seated joints and the potential for rock falls and rock slides. In many places where waves generated by landslides threaten built-up areas, such instrumentation has been installed.

DEBRIS SLIDES AND DEBRIS FLOWS

Landslides on steep hillsides covered with glacial till and colluvium are quite common in Norway, especially in the western coastal region. The till can be divided in two major types: till derived from Precambrian bedrock, and till derived from Cambro-Silurian sedimentary rocks. Most till derived from Precambrian bedrock has less silt and clay (about 18.5 percent) compared to the other till (36.8 percent). The thickness of the till varies mainly with the inclination of slope; the more gentle the slope, the thicker the till. Culluvium occurs mainly on the lower part of steep slopes. The stability of the colluvium is related to parent bedrock; culluvium from fine-grained, detrial sedimentary rocks, such as shale and limestone, is slide-prone.

Structure of the soil profile is influenced strongly by the effects of frost penetration. Frost causes the upper 0.5 to 1.0 m of soil to become porous and have a high permeability. Roots and organisms in the soil enhance the porosity. Permeability of the underlying, unfrozen soil is substantially lower. This dramatic reduction in permeability is of vital importance in the stability evaluation of a slope.

In accordance with the type of movement, landslides in steep hillsides are divided in two major groups: debris slides and debris flows. The former glide on one or several sliding planes at relatively moderate velocities. The latter is a movement by plastic flow or by shear along many planes, typically with a velocity that is greater due to a higher water content. In most places, debris flows initiate as debris slides. Both debris slides and debris flow commonly follow drainage routes, which add water and cause the material to flow faster.

Initiating processes

Field measurements indicate that most debris slides are along the relatively impermeable layer constituting the lower part of the soil profile. Stability of the slope, therefore, depends largely on the shear forces acting along this sliding plane, not on the tensile strength of the soil.

Most debris slides and debris flows occur in spring and autumn associated with periods of rapid snowmelt or heavy rainfall, and often a combination of both. Excess porewater pressure increases shear stress along the potential sliding surfaces, and decreases shear resistance. Both these factors have a negative effect on slope stability.

The rate of water supply depends on rainfall intensity and meltwater production. Experience shows that slope failures are likely to occur in slide-prone areas when the water supply exceeds about 40 mm in a 6 hr period.

The inclination and aspect of soil slopes are important stability factors. The threshold angle of a slope, below which slides do not occur, is controlled by the shear strength of the slope material. In Norway, slope failures rarely occur on slopes less than bout 30°, apart from quick-clay slides. With respect to slope aspect, west-facing slopes in the maritime climate of the west coast of Norway receive greater precipitation than leeward slopes due to the orographic effect from prevailing westerly winds. In addition, the largest amounts of snowmelt water are produced in west-facing slopes where the rate of snowmelt is faster. The importance of these two factors is clearly demonstrated by the fact that most failures occur in slopes facing west.

In the continental climate of eastern Norway, snowmelt in the spring season is largely governed by sun radiation. Accordingly, slope failures in this area are most likely to occur in south- to west-facing slopes. Many landslides also take place during heavy rainshowers in late summer.

Many landslides have also been caused by human acitivity. In particular, extensive logging and construction of roads in steep hillsides have promoted slope instability.

Consequences of debris slides and flows

Only exceptionally do slope failures in steep hillsides cause damage to man and property; the most frequent consequence is closure of roads. Nevertheless, in a geomorphological context, debris slides and flows are dominant processes in the shaping of steep hillsides and the downhill transportation of slope material. Rare catastrophic events have occurred, such as in July 1789, when 68 people perished due to heavy rain and rapid snowmelt in the main valleys of southern Norway. Most of these people were probably killed by debris slides and flows.

NATIONAL LANDSLIDE POLICY

Legislation and financial support

The Norwegian Building Code prescribes an acceptable risk level for natural hazards that affect a house as 10^{-3} per year, corresponding to a 1000 year return period. For buildings that are only occasionally occupied by people, the risk level is 10^{-2} per year. Local building authorities are responsible for the approval of building licenses.

Insurance against natural hazards is included in the standard conditions for fire insurance policies in Norway. This insurance includes full compensation for all material losses due to natural hazards, such as landslides, except for loss of land. Compensation for loss of land is given by a special government office, the National Fund for Natural Disaster Assistance. The compensation is calculated as a percentage of the value, typically between 50% and 80%.

Actions for landslide danger or landslide events

A duty of the National Fund for Natural Disasters Assistance is to take precautionary measures against natural hazards. Such measures have traditionally consisted of establishing structures to protect existing buildings and the evacuation of endangered houses.

A third preventive measure, initiated some years ago, is hazard mapping. This project includes mapping of quick clay areas, rock falls and snow avalanches (Gregersen and others, 1983). The aim of this project, which will cover 80 to 90% of exposed areas, is to show "safe areas" for new construction activities and to identify buildings that are already at risk. The project is calculated to cost between 6 and 7 million US$, and will continue for a 12 year period.

When landslide events have taken place, responsibility for dealing with the disaster lies with local police. Experts are usually requested to assist the police. In most cases, the Norwegian Geotechnical Institute has provided this assistance financed by the Natural Fund.

REFERENCES

Aas, G. 1970. Vurdering av korttidsstabilitet i leire på basis av udrenert skjærfasthet. Nordiska geoteknikermøtet. Helsinki 1979. Foredrag och artiklar: 585-596. Norwegian Geotechnical Institute, Oslo, Pub. 132.

Aas, G. 1982. Stability of natural slopes in quick clays. Int. Conf. on soil Mechanics and Foundation Engineering, 10. Sthm. 1981. Proc., (3):333-338. Norwegian Geotechnical Instaitute, Oslo. Pub. 135.

Aas, G. 1983. A method of stability analysis applicable to natural slopes in sensitive and quick clays. Int. Symp. on Landslides, Linköping 1982, p. 19.

Bjerrum, L. 1954. Geotechnical properties of Norwegian marine clays. Geotechnique (4), 2, 49-60. Norwegian Geotechnical Institute, Oslo. Pub. 4.

Bjerrum, L. & I. Th. Rosenqvist 1956. Norske leirskred og deres geoteknikk. Oslo. Norwegian Geotechnical Institute, Oslo. Pub. 15, Naturen, 81, 1957, 2:108-121.

Bjerrum, L. & F. Jørstad 1968. Stability of rock slopes in Norway. NGI-publication 79:1-11.

Bjerrum, L. and others 1969. A field study of factors responsible for quick clay slides. Int. Conf. on Soil Mechanics and Foundation Engineering. Mexico 1969. Proc. 2:B31-B40. Norwegian Geotechnical Institute, Oslo. Pub. 85.

Foster, R. & S. Heiberg 1971. Erosion studies in a marine clay deposit at Romerike, Norway. Norwegian Geotechnical Institute, Oslo. Pub. 88.

Gregersen, O. 1981. The quick clay landslide in
  Rissa, Norway. Int. Conf. on Soil Mechanics and
  Foundation Engineering, 10. Sthm. 1981. Proc.
  3:431-426. Norwegian Geotechncial Institute, Oslo,
  Pub. 135.
Gregersen, O. & T. Løken 1983. Mapping of quick clay
  landslide hazard in Norway. Criteria and expenses.
  Symp. on Slopes on Soft Clays. Linköping 1982.
  Swedish Geotech.Inst., Linköping. Report 17:161-174.
Jørgensen, O. 1978. Some properties of Norwegian
  tills. NGI-Publication No. 121:9.
Jørstad, F. 1968. Waves generated by landslides in
  Norwegian fjords and lakes. NGI-publication No.79:
  13-32.
Karlsrud, K., G. Aas & O. Gregersen 1984. Can we pre-
  dict landslide hazards in soft sensitive clays?
  Summary of Norwegian Practice and Experience.
  International Symposium on Landslides. V. Proc.,
  (1):107-130.
Rosenqvist, I. Th. 1946. Om leirens kvikkaktighet.
  Statens Vegvesen. Veglaboratoriet. Meddelelse,
  4:5-12.
Rosenqvist, I. Th. 1955. Investigations in the clay-
  electrolyte-water system. Norwegian Geotechnical
  Institute, Oslo. Pub. 9.

Extent and economic significance of landslides in Sweden

Leif Viberg
Swedish Geotechnical Institute, Sweden

ABSTRACT

Landsliding is a natural hazard in many parts of Sweden. Many kinds of landslide occur, but clay slumps, in places developed into large quick-clay landslides and river and gully erosion in silt soils are the most common types of mass movement. Their occurrence coincides with many densely populated areas, where they are of great concern to authorities and individuals. In the less populated mountain areas, many kinds of mass movements occur, such as slumps, soil creep, debris and mud flows, debris avalanches, silt erosion, rock slide and rock fall. The mass movement hazard in mountain areas is more and more related to increased development. Rock slide and rock fall outside the mountain range occur locally on confined sites such as steep fault slopes and steep excavated rock slopes, e.g. for roads.

The catastrophic landslide in Tuve, Gothenburg, in 1977, increased the consciousness of landslide risk in Sweden and this has resulted in increased concern by municipalities and county administrative boards, a new national policy on landslide risk (1986) and the establishment of a Commission for research and information on landslide problems (1988).

The governmental cost of stability reinforcement works in emergency cases amounted in 1987 to about 2.5 million $US. In 1985 one site was reinforced for about 7 million $US. In the city of Gothenburg, several slopes have been stabilized during 1983-87 for about 11 million $US.

GEOLOGIC MATERIALS IN SWEDEN

The geologic materials in Sweden consist of bedrock and its overburden, in the form of unconsolidated earth materials. The earth materials were formed during and after the latest glaciation of Scandinavia. The recession of the ice from Swedish land started about 15,000 B.C. (before Christ) and the ice left Sweden about 6,000 B.C. During its melt away recession, different stages of fresh water lakes and marine seas were formed. In these waters, clay and silt were deposited close to the ice front as glacial sediments (varved clay and silt) and at later stages as postglacial sediments. The thickness of clay layers may be as much as 100 m - especially on the west coast of Sweden. The layering of the fine sediments was also influenced by wave washing during land upheaval.

The above mentioned lakes and seas covered large parts of Sweden. The highest level to which the water reached at each site is shown in Fig. 1. Below this line, fine-grained, landslide-prone sediments are found. When the load of ice was removed, the earth crust rebounded elastically and the land started to heave upwards. The water drowned parts came slowly out of water and became land. Wave washing during this land upheaval has created layers of silt and sand within clay layers. Layering of clay, silt and sand is especially found outside glaciofluvial deposits, such as eskers and deltas, where the sedimentation has been strongly influenced by seasonal changes in the sediment load. Water can easily enter the clay deposits through the permeable silt and sand layers.

Till is the most common geologic material in Sweden, covering about 75% of the land

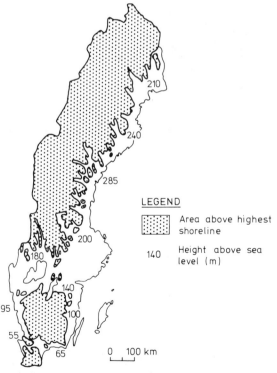

LEGEND

Area above highest shoreline

140　Height above sea level (m)

0　100 km

Figure 1.　Map showing highest shore line in Sweden.
From Fysisk riksplanering (1979)

area. The composition is normally silty-sandy till with pebbles and boulders. Till topography reflects in most cases bedrock topography. Where the bedrock is steep, the moraine surface is also steep and thus susceptible to slope movement.

The bedrock of Sweden is mainly granite and gneiss rocks of precambrium age with high strength. Weaker rock of Cambrian, Ordovician and Silurian age occurs in a few places in southern Sweden, within the mountain.

Examples of geological profiles from the southwestern and northeastern parts of Sweden are shown in Fig. 2.

LANDSLIDE FACTORS

Most Swedish landslides occur as slumps in fine-grained sediments, such as clay and fine silt. At a few localities large, catastrophic, quick-clay landslides have developed. Landslides are also found in other materials, such as till and rock.

There are many factors contributing to the development of landslides in clay sediments, as shown in Fig. 3. A short description is given here of the most important factors, such as topography, groundwater, erosion, shear strength of clay, the occurence of quick-clay, frost and thaw.

Topographic conditions

A necessary conditon for landslide is that the ground is inclined. The slopes in clay and silt sediments in Sweden are mainly of two kinds. One type is found along the sides of the bedrock valleys, where the sediments were deposited on the sloping sub-base, thereby getting an orginally sloping surface. The other type was formed in originally flat-lying sediments, into which river valleys and brook ravines were eroded down at the stage when the land was upheaved above water level.

The slope height in clay sediment may reach 30-40 m and in silt up to 50 m. The inclination in clay slopes may be as steep as 30-40° and in silt almost vertical due to false cohesion.

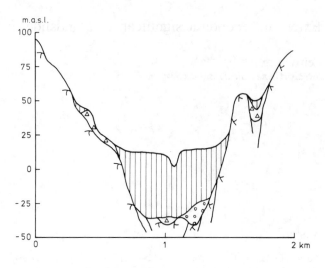

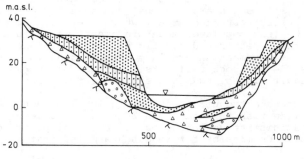

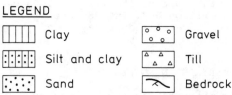

LEGEND

Figure 2.   Geological profiles from the west coast and northeastern coast of Sweden. From Fysisk riksplanering (1979)

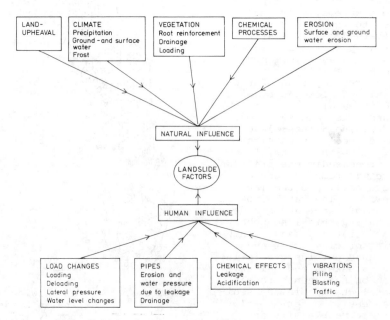

Figure 3.   Overview of landslide factors. From Viberg (1986)

142

Groundwater conditions

The groundwater pressure in both frictional and cohesive earth material is believed to play an important role as a landslide triggering factor. The frequency of landslides distributed monthly over the year, Fig. 4, shows that most landslides occur during the wet seasons of spring and autumn.

In Sweden, there are many clay slopes with artesian pressure, especially in the west coast region where high and steep rock surrounds the clay filled valleys. Most catastrophic Swedish landslides are found in areas with artesian pressure.

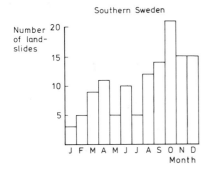

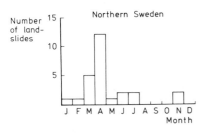

Figure 4.   Distribution of landslides per month. From Viberg (1982)

Erosion

Undermining of slopes by erosion is a common cause of landsliding. Highly erodible silt and sand slopes are abundant in valleys leading from the mountains in the northern part of Sweden. The mountain melt water causes spring floods, strong erosion and many small landslides. A few times during the last decade, heavy rains in summer have caused severe floods, erosion and, damages. In 1985, an earthdam failed completely during these summer floods.

Shear strength

Slope failure takes place either by increased stresses that exceed the shear strength or by decreased shear strength that falls below existing stresses. The shear strength and stresses in slopes are influenced by the factors shown in Fig. 3.

In stability calculations, the common practice in Sweden is to use the undrained shear strength for normally consolidated and slightly overconsolidated clays, and drained shear strength for overconsolidated clays. The undrained shear strength values received by field vane and fall cone test are reduced because the rapid tests do not reflect creep effects during permanent loading. The reduction is a function of the liquid limit. The Swedish practice in this respect is described by Larsson & al (1984).

Quick clay

Quick clay has been found in all very large (>10 ha) Swedish landslides that have been investigated. Quick clay is present where the sensitivity - the ratio of the undrained shear strength in undisturbed and remoulded state - is higher than 50 and the remoulded shear strength is lower than 0.4 kPa, Karlsson & Hansbo (1982). Quick clay is found in leached marine clays and fresh water clays. Söderblom (1969, 1974) has found that certain dispersive substances, such as phosphates and organic matter, can change clays into quick clays. A requisite for formation of quick clay seems to be the presence of low valency anions, such as Na^+. Söderblom's results has led Rosenquist (1984) to formulate a general quick clay theory, wherein the leached marine quick clay is only a special case.

The practical implications of the quick clay research are that quick clay formation is not only a natural process but also man can create quick clay by infiltration of dispersive agents into the clay.

Quick clay may occur in single or multiple layers interbedded with permeable silt and sand layers or it may form fairly homogeneous thick deposits.

Quick clay apparently influences the velocity and extension of a landslide once started. Quick clay landslides are rapid and retrogress in many cases all the way to ground where quick clays are absent.

Frost and thaw

In wintertime, the ground becomes frozen in Sweden. The average frost depth in southern Sweden is 1 to 1.5 m, and along the northern parts of the east coast the penetration depth is about 2 m. The frozen part reduces the ground water infiltration and water pressure decreases. The frost is thawed both from the surface and from below. The thawed inside soil may become very soft and fail on moderate to steep slopes. This event may initiate a larger landslide. In addition, the frost and thaw cycle creates cracks in clay, which increases the erodibility of the clay and allows ground water to penetrate more quickly.

LANDSLIDE DISTRIBTUION

Landslide inventories

A few inventories of landslides have been made in Sweden. The most comprehensive inventory of clay slides, based on a literature survey and inquiries to cities and consultants, has been made by Inganäs & Viberg (1979). Other inventories are limited to smaller areas. Examples of such inventories are mentioned here.

Frödin (1919) mapped part of Göta river valley, Wide (1972) inventoried landslides in the southwestern part of Sweden, Hjorth (1964) mapped the landslides in the Lidan river valley, Jerbo (1980) studied landslides in part of the river valley of Ångermanälven and Viberg (1982) mapped the landslides in the Göta river valley. Extract of this mapping is shown in Fig. 5.

Statens järnvägars geotekniska kommission 1914-22 (1922) made an excellent description of many landslides that were triggered in connection with the development of the Swedish railway net at the end of the 19th century and the beginning of the 20th century.

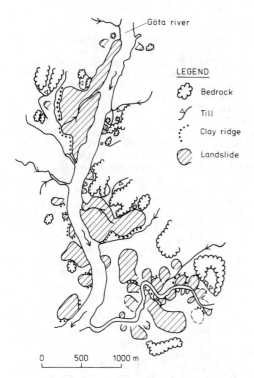

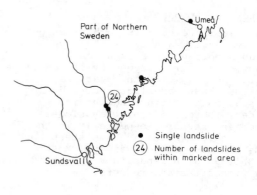

Figure 5.   Landslide map from part of
             Göta River.
             From Viberg (1982)

Landslides in clay sediments

Clay slide, slumps and quick-clay landslides
occur in many parts of Sweden. A rough subdivision of the clay areas into those with
relatively high and low landslide frequency
is shown in Fig. 6.

Figure 7.   Distribution of landslide
             in clay deposits in Sweden.
             From Inganäs & Viberg (1979)

The distribution of clay landslides up
to 1979 is shown in Fig. 7. More recent
landslides are shown in Fig. 6. During the
winter 1987/88, approximately 10 landslides
occurred, the largest of which was about
2 ha.

The distribution of clay landslides with
time, Fig. 8, reveals some interesting features, although the survey is not complete
- especially for small slides. The graph
shows distinct peaks around the turn of
the century, when the railway net was under
construction, and after World War II, when
the extensive urbanization started. It is

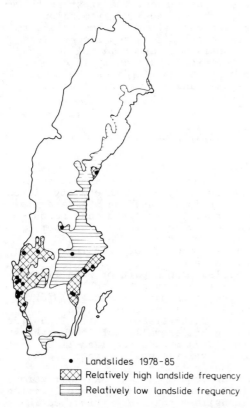

Figure 6.   Map of relative landslide
             frequency in Sweden.
             From Cato I. & Engdahl, M. (1982)

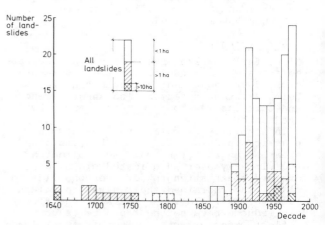

Figure 8.   Distribution of landslides
             per decade since 1650.
             From Viberg (1982)

rather evident that human actions have "assisted" nature in levelling the ground. Special attention should be focused on very large catastrophic landslides (>10 ha), the frequency of which seems to increase.

On average, one very large landslide (>10 ha) occurs every 10 years and one landslide larger than 1 ha every two years in Sweden.

Mass movements in silt deposits

In the northern part of Sweden, river valleys below the highest shore line are dominated by silt sediments, because the clay layer within the sediment deposit in general is relatively thin or missing. However, the highly erodible and frost susceptible silt sediments determine the shape of the valley landscapes. High and steep slopes may be formed and many gullies are eroded into the silt sediments. The gully erosion is normally rather slow, but may also be violent and spectacular.

The geographical distribution of the "silt valley" areas is shown in Fig. 9.

Mass movements in the Swedish mountains

Relatively little attention has been payed to landslides and other types of mass movements in the Swedish mountains because of the low population density. Snow avalanches are, of course, of great concern in ski resort areas. The rapid tourist development of the mountain areas means that the landslide risk increases and the landslide hazard is more of concern because of landslide incidents.

Many types of mass movements occur in the Swedish mountains. A comprehensive study of an area by Rapp (1960) showed that the following mass movements occurred: Rock falls, rock slides, talus creep, debris flows and debris avalanches. Examples of debris flows are shown in Fig 10.

Rock slide and rock fall

Swedish bedrock outside the mountain chain consists mainly of different types of hard rock as mentioned earlier. Granite and gneiss are the dominating types of rock. The bedrock is more or less jointed. Rock instability occur along steep fault zones, e.g. in Stockholm, at several sites at the west coast and along the motorway E4 at lake Vättern.

Rock fall is the most common rock movement. Several rock falls have occurred in excavated rock walls. Even rather small rock falls create problems for highways.

CONSEQUENCES OF LANDSLIDES

In early days, landslides were thought to be the punishment of God. The large catastrophic landslide at Intagan in the northern part of the Göta river valley in 1648 killed 128 persons, the greatest number of people in any documented landslide event in Sweden. Although a very large number of landslides have occurred, the death toll is small compared with landslide frequency. This is probably due to the fact that in most cases there have been warning signs - cracks, movements etc - so people have had time to move out of the unstable areas.

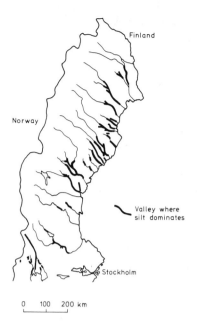

Figure 9.   Silt valleys in Sweden. From Bergquist (1986)

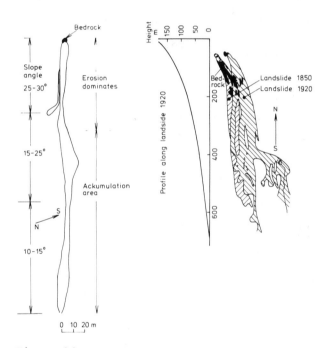

Figure 10.   Examples of debris flows in Sweden. From Rapp & Strömquist (1979) and Zenzén (1926)

The information on landslide death toll is quite reliable. The economic consequences are, however, not as easy to obtain. There is no systematic documentation of landslides in Sweden - with the exception of the work by Swedish Geotechnical Commission mentioned earlier. Some information on the consequences of some Swedish landslides is shown in Table 1.

In emergency cases, when the landslide risk is imminent or the ground has already failed, reinforcment works normally are financed by the National Rescue Service. The costs

145

Table 1. Consequences of some Swedish landslides.

PLACE	DATE	AREA ha	NUMBER killed	COST 1988 cost level US$ Mill.
Intagan	1648		128	–
Getå	1918		41	–
Surte	1950	27	1	16
Göta	1957	15	2	11
Tuve	1977	27	9	58

of such operations amounted to about 2.5
million $US for 1987. For one site in the
municipality Munkedal on the Westcoast,
reinforcement works for about 7 million
$US were carried out in 1985.

In the large city of Gothenburg, there are
many clay slopes loaded by buildings, roads
and other facilities. After an inventory
of the slopes and a classification of landslide risk based on calculated safety factors, a stabilization program started. Between 1983 and 1987 18 slopes have been
stabilized. The cost and the types of stabilization for these works are presented in
Table 2, Stadsbyggnadskontoret & Gatukontoret, Gothenburg (1987).

NATIONAL LANDSLIDE POLICY

The Swedish national policy on landslide
risk is described in detail in Ahlberg &
al (1988). Here is given the most important
features of this policy.
 The responsibility for landslide risk
is divided between the national government
and municipalities. It is regulated according to a government bill approved by Parliament in 1986. According to the bill, preventive measures against landslides in build-up
areas shall be managed in the following
ways:

1. The National Rescue Services Board
   would be provided with funds for overall surveys of landslide risk.

2. Municipalities should carry out detailed stability investigations and be
   responsible for carrying out and at
   least partially funding preventive
   measures.

3. The state would provide annually SEK
   25 million (US $ 4 million) as extra
   municipal tax equalization grants to
   municipalities with particular landslide problems for covering part orthe
   entire cost of preventive measures.

   In order to get financial support from
   the state, municipalities should have
   completed the stabilization works.
   Alternatively, they must carry through
   a landslide risk evaluation, estimate
   the cost and present for stabilization
   works, and request funds.

   For new development, the new Planning
and Building Act, enforced on 1 July 1987
regulates how landslide risk areas should
be treated. Areas with landslide risk should
be avoided   or methods should be recommended

Table 2. Data on stabilization works carried out by
Gothenburg city. From 1983-1987.

PLACE	SLIDE	TYPE OF STABILIZATION	COST 1988 price level US $ Million
Linnarhult		Relocation of brook and counterweight	5.9
Nedre Linnarhultsvägen	x	Partly filling of gully	0.04
Lärjehed koloniområde	x	Flattening of slope and counterweight	0.07
Hjällbogärdet	x	Relocation of brook and counterweight	0.29
Linnarhults industriomr.		Erosion protection	0.63
Säveån n. shore		Erosion protection	0.39
Utby Österlyckan		Lime columns	0.92
Säveån s. shore		Erosion protection	0.27
Kvibergsbäcken		Elevation of brook and counterweight	0.86
Bellevue junkyard			0.13
Angered C		Erosion protection	0.01
Lärjeholm	x	Blasted rock boulders	0.03
Bellevue Billhälls		Deloading (excavation and lightfill)	0.39
Lärjeån bridge		Bolted rock boulders and sprayed concrete	0.04
Deljöbäcken		1. Jetted concrete pillars 2. Concrete basin 3. Relocation of brook and counterweight	0.71
Nygårdsskolan		Counterweight filling	0.05
		Total cost	10.73

for improving the stability. In the latter
case, the one responsible for the preventive
measures should be identified. The municipalities have responsibility for master and
detail plans, but the County Board Administration has the right to object to and
even nullify detail plans which do not consider landslide risk properly.

REFERENCES

Ahlberg, P., Stigler, B. & Viberg, L. 1988.
 Experiences of landslide risk considerations in land use planning in Sweden.
 Proc. Vth International Symposium on Landslides, Lausanne.
Bergquist, E. 1986. Swedish terrace and
 gully landscapes (in Swedish). Uppsala
 University, Naturgeogafiska institutionen.
 UNGI Rapport Nr 63.
Cato, I. & Engdahl, M. 1982. Beskrivning
 till temakartor utvisande var särskild
 uppmärksamhet av stabilitetsförhållanden
 erfordras inom vissa bebyggda eller detalj-
 planerade områden med lerjord (in Swedish).
 Swedish Geologic Survey. Rapporter & Meddelanden: 20.
Frödin, G. 1919. Jordskreden och markförskjut-
 ningar i Göta älvs dalgång mellan Trollhät-
 tan och Lilla Edet (in Swedish). K. Vatten-
 fallsstyrelsen. Medd. 1919 Nr 19.
Fysisk riksplanering 1979. Våtmarker. Mineral-
 råvaror. Geologiska och geotekniska förhål-
 landen. (In Swedish) FRP Underlagsmatrial
 nr 3.
Hjorth, S. 1964. Slumps and ravines along
 Lidan river (in Swedish). Uppsala University. Geography.

Inganäs, J. & Viberg, L. 1979. Inventory of clay landslides in Sweden (in Swedish). Proceedings of Nordiska Geoteknikermötet 1979, Helsingfors, p. 549-556.

Jerbo, A. 1980. Ådalen in Ångermanland. An important landslide region (in Swedish).

Karlsson, R. & Hansbo, S. 1982. Classification of soils (in Swedish). Byggforskningsrådet T21:1982 (2nd revised edition).

Larsson, R. Bergdahl, U. & Eriksson, L. 1984. Evaluation of shear strength of cohesive soil (in Swedish). Swedish Geotechnical Institute. Information 3.

Rapp, A. 1960. Recent development of mountain slopes in Kärkevagge and surroundings, northern Scandinavia. Geografiska Annaler 1960:2-3 p. 71-200.

Rapp, A. & Strömquist, L. 1979. Field experiments on mass movements in the Scandinavian mountains with special reference to Kärkevagge, Swedish Lappland. Studio Geomorphologica Carpatho - Balcanica. Vol. XIII p.23-40. Krakow.

Rosenquist, I.T. 1984. Colloidal physics as basis for quick clay properties. STRiAE Vol. 19 p. 5-11: Uppsala.

Stadsbyggnadskontoret & Gatukontoret, Gothenburg 1987. Report. Stabilization works against landslides in Gothenburg 1983-87 (in Swedish).

Statens järnvägars geotekniska kommission, 1914-1922 (1922). Slutbetänkande. (in Swedish) Statens Järnvägars Geotekniska Meddelanden 2.

Söderblom, R. 1969. Salt in Swedish clays and its importance for quick clay formation. Swedish Geotechnical Institute. Proceedings No 22.

Söderblom, R. 1974. Organic matter in Swedish clays and its importance for quick clay formation. Swedish Geotechnical Institute. Proceedings No 26.

Wide, Å. 1972. Litteraturinventering av skred inträffade i Götaland t o m 1971. (in Swedish) Ex.arbete CTH Geologi. Publ. 1972:B4.

Viberg, L. 1982. Mapping and classification of the stability conditions of clay areas (in Swedish). Swedish Geotechnical Institute. Report No 15.

Viberg, L. 1986. Landslides and slope stability. Need of research and proposal for priority (in Swedish). Swedish Geotechnical Institute. Varia No 169.

Zenzén, N. 1926. Some information on the landslide on Ö Stårbetjvare in Arjeplog, Pite lappmark, Aug 3, 1920. (in Swedish) Geologiska Föreningens i Stockholm Förhandlingar, Bd 48, Häfte 2, p. 167-185.

Landslides: Extent and Economic Significance, Brabb & Harrod (eds)
© 1989 Balkema, Rotterdam. ISBN 90 6191 876 6

Landslides in Finland

Heikki Niini
Department of Economic Geology, Helsinki University of Technology, Finland

Eero Slunga
Department of Soil Mechanics and Foundation Engineering, Helsinki University of Technology, Finland

ABSTRACT: This paper reviews the Quaternary geology of Finland. It describes the geotechnical properties, the causes, and the occurrence of both natural landslides and those induced by human activities.

1 GEOLOGICAL AND GEOTECHNICAL PROPERTIES OF LANDSLIDES AREAS

In Finland, the terrain is relatively flat and differences in height are small. The most common soil type is glacial till, which was developed and preconsolidated against hard rock during the last Ice Age about 10,000 to 12,000 years ago. Older soil formations were swept out by the movements of ice. In flat depressions, the glacial soils are commonly covered by a thin layer of postglacial peat. The bearing capacity of till soils is very good, which precludes much landsliding.

Soft, fine-grained, landslide-prone sediments deposited during or after the last Ice Age are divided into four types, from youngest to oldest; Littorina, Ancylus, Yoldia, and Baltic Ice Lake sediments. The Littorina and Ancylus sediments are nearly homogeneous in structure. Yoldia sediment is homogeneous in structure or is characterized by symmictic varving. Baltic Ice Lake sediments show distinct diatactic varving and are anisotropic. Considerable sedimentation may have taken place even after the local area of water had losts its connection with the Baltic Sea and formed a separate lake basin as the land uplifted after the Ice Age.

The average thickness of the sedimentary deposits in southern and western Finland is about 10 m, but some of the clay deposits are 30 to 50 metres thick. Predominantly silty deposits in the interior of the country vary in thickness between 5 and 10 metres. Even in the valleys, the thickness of a sedimentary deposit varies considerably; a ridge of bedrock or a moraine may rise to the surface in the middle of a valley.

The humus content, which reduces the strength of soil, is highest in Littorina sediments, the average being about 4 percent. The average humus content of Ancylus and Yoldia sediments is from 1 to 1.2 percent, and the content of humus in Baltic Ice Lake sediments is 0.6 percent. The other classification properties, with the exception of the unit weight of solid particles, follow primarily the changes in clay and humus content.

For the most part, fine-grained deposits are normally consolidated. Overconsolidation occurs, however, in many places beneath a dry crust. In quite a few places, the ratio P_c/P_0 in the overconsolidated portions is as high as 2 to 4. Overconsolidation is found in all types of sediments as well as in material composed predominantly of clay and silt.

The undrained shear strength measured by the vane test may vary in fine-grained sediment beneath the dry crust in the range of about 10 to 20 kN/m^2. In a few places, substantial variations may occur in the shear strength, especially at the contact between types of sediment. The most distinct change in shear strength occurs generally in soils with organic material, where the least shear strength corresponds to the maximum value of the humus.

The sensitivity of the Finnish clay soils varies in the range of 10 to 60. Higher values are rare, and on an average the sensitivity is about 10 to 30. Thus, the Finnish clay soils may be considered rather insensitive.

2 LANDSLIDES

Large landslides are infrequent in Finland (Niini 1977). Natural landslides are most frequent in the southwestern lowlands, particularly in the Turku - Salo area, where they occur along the valleys of rivers which cut into Littorina clay deposits (Figure 1). Natural landslides also occur sometimes in the flat clay areas of southern Ostrobothnia (Hellaakoski 1920).

In Lapland, height differences are greater than elsewhere in Finland, but conditions favouring the occurrence of landslides have been of short duration. All the major landslides found there in sandy till are ancient (Kujansuu 1972).

Slow land uplift has caused landslides on the seashore in places where loose clay or gyttja has been deposited on steep, smooth-surfaced bedrock slopes. These slides are obscured by other shoreline phenomena, but traces of them are found in borings. A typical one was observed in June, 1936, at Inkoo on a small island in the Gulf of Finland (Brander 1936).

Natural landslides comprising intrusive rocks or crystalline schists have seldom occurred in Finland and only under exceptional conditions. Landslides induced by man have occurred commonly in connection with canal and other construction, for example, at railway construction (Lappalainen 1965), at the Saimaa Canal connecting Lappeenranta in SE Finland and Vyborg in the USSR (Gardemeister 1967, Slunga 1973), and at the Kimola Canal east of Lahti (Kankare 1969). The length of the biggest failures varied from 100 to 250 metres. In excavations, it is common to make slopes as steep as possible in order to avoid masses of rock or earth. Therefore, slides of varying extent have occurred at half a dozen open pit mines, in some places causing material damage. Minor landslides are common in connection with any kind of massive construction on weak ground, especially in railway, road, and waste hill constructions.

Depending on the main features of the displacement mechanism, landslides are either rockslides (mostly in open pit mining) or soil slides (the most common type in Finland). Soil slides are either turf-slab slides, rotational slides, retrogressive slump slides, or slump-earthflow slides. Movements of the first four types take place mainly on steep slopes, whereas the slump-earthflows may also occur on gently inclined slopes. Often the primary landslides are followed by smaller, secondary slides.

The causes of landslides can be classified in two ways, as: 1) either natural or due to human activities, and 2) either internal or external. The internal causes decrease the shear strength of soil or rock and external causes increase the shear stress in the ground. The basic natural cause of landsliding is in many places the presence of weak soil or layers suitable for glide surface. The surface of rupture in most of the Finnish landslides is composed of weak clay, although it may also be the surface of the bedrock.

The basic causes of the slides in Finnish open-pit mines and underground constructions are permeable joints, faults, or fissures in the slope, together with their minute shear strength, often due to slippery secondary minerals formed in ancient shearing movements or by weathering. The decrease of the shear strength of the slope material may be due to slow chemical weathering, which is seldom the case in Finland, or by an exceptional increase in the water content and pore water pressure. Usually a

decrease in shear strength is triggered by external mechanical forces.

The largest landslides occur mainly in the Littorina clay areas of the lower reaches or rivers. The high humus content of the Littorina clays acts as a basic internal cause for sliding, because it binds water and increases the shrinkage in drying and a tendency for cracks to appear in the dry crust. The cracks lower the tension strength, diminishing slope stability.

The turf-slab glides sometimes take place when the ground is still partly frozen. Freezing actually increases the strength of the ground but as the ice melts, the shear strength of soil decreases. On a slope, the surface layer may glide downwards. In Lapland, the melting has apparently caused landslides to predominate on the sunny side of slopes.

The shear stresses on slopes are in many places greatly increased due to 1) overloading of the slope, 2) removal of the frontal support, or 3) vibration. In Finland, a natural overloading of a slope resulting in a landslide is commonly due to unusual surface storage of water and abundant infiltration resulting from heavy rainfall or a rapid melting of an exceptionally thick snow cover.

The greatest number of landslides in Finland occurs in spring (May) and autumn (October – November). The frequency of landsliding in May is caused almost exclusively by the melting of the snow cover. The autumn sliding is due to high precipitation. The frequency of recent landslides has varied from year to year and shows a correlation to the mean average yearly precipitation (Aartolahti 1975).

The overloading of slopes resulting in landsliding in many places results from human activity in connection with building, storage of material, vehicle vibration, or irrigation of soil for agriculture. Natural overloading of a slope may be caused or promoted by the growth of large trees (Brander 1936).

River erosion is another common cause for landsliding in Finland. The slow removal of the frontal support of a slope causes the majority of clay landslides to form on the outer sides of river meanders. This erosion is fastest during flood season when the water is high and swift.

Water level changes may cause the removal of frontal support for earth layers susceptible to sliding in two ways: 1) When water level drops or land rises, the loss of buoyancy increases the weight of the earth mass and increases the chance of sliding; in Finland, there is a slow but regular land uplift. 2) When water level rises, the submerged land may contain material in which the increased water content causes decreased internal strength. The shear strength of the material then becomes insufficient to support the soil layers higher up. Both factors apply to many natural and artificial reservoirs with strong water-level fluctuations.

Human activities such as the building of dams, railways and road embankments, excavation and infilling of earth materials, the vibrations of machinery and blasting, and the draining of waterlogged areas also contribute to landsliding. Ditches and drains usually help to reduce the incidence of sliding, but in some places water from these drains goes into river slopes promoting erosion and sliding. Traffic vibration and wave action caused by water traffic, as well as certain agricultural and land use practices, can cause landslides in sensitive clays, such as those around Salo where slides have occurred on slopes less than 10 m high (Aartolahti 1975). The kinds of agricultural practices that tend to promote the occurrence of landsliding, acting in combination with one another (Aartolahti 1975), are: forest clearance (more landslides occur on treeless than on

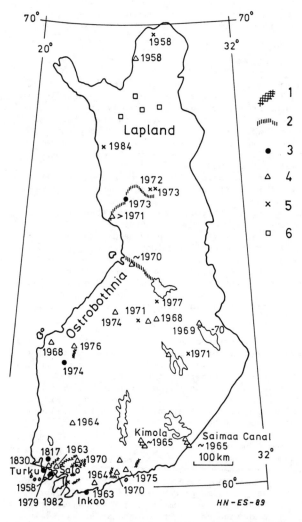

Figure 1. Occurrence of landslides in Finland. Key: 1 river-valleys with landslides, 2 river-slope failures due to breaking up of the ice, 3 considerable natural landslides, 4 considerable failures of man-made slopes (canals, earth dams, excavations, etc.), 5 considerable failures of rock-slopes (open pit mines, rock-cuttings), 6 ancient landslides.

forested land), soil tillage and grazing (more water infiltrates the ground increasing the load and decreasing the shear strength of the soil), and fencing of pastureland (stakes on the upper slopes act as a row of wedges).

Recent landslides in Finland, both natural and those caused by human activity, are mostly under 100,000 m^3. The area of activity may vary from a few m^2 up to nearly 0.1 km^2 (10 ha), and the thickness of the earth materials displaced is usually less than 10 m. The economic effect of a landslide may be considerable, depending on the location of the slide and on the nature of the work in question. For instance, on a distance of 3.7 km at the Saimaa Canal, about 80,000 m^3 soil failed causing an extra cost increase of 20 % compared with the original calculations. At the Kimola Canal also a public road was cut in addition to remarkable extra earth works. In recent years, the loss of human life in the natural landslides has been rare. The fatalities have occured in excavations and mines, ranging from one to five deaths per year.

3 NATIONAL LANDSLIDE POLICY

In accordance with Finnish legislation and codes of practice, the planning of land use and the design of roads, bridges and housing, etc., require a classification of the stability of the area in question. In some cases, however, no systematic study of risk areas is required. Because of rareness of large natural landslides one has not seen any greater reason for a consistent national landslide policy in Finland. Therefore, little has so far been done to reduce the effects of inadvertent human activities causing slides. Recently, however, even small slides and rockfalls in mines have become so important,both economically and from work-safety considerations, that the Association of Mining and Metallurgical Engineers has appointed special committees to investigate the geological causes and the design criteria for mitigating measures.

In addition to the growing volume of soil and rock excavation and removal, large-scale land and water use contribute to landsliding. Therefore, more thorough studies that estimate and reduce the risks and mitigate the damage of slides and related mass movement in Finland are needed.

REFERENCES

Aartolahti, T. 1975. The morphology and development of the river valleys in southwestern Finland. Annales Academiae Scientiarum Fennicae, A III, 116:72.
Brander, G. 1936. Jordskredet i Ingå. Deutsches Referat: Der Erdrutsch in Ingå. Terra, 48, 4: 186:192.
Gardemeister, R. 1967. On the soil and its constructional properties at the Saimaa canal construction in Finland. Eng. Geol. 2, 2: 107-115.
Hellaakoski, A. 1920. Erdschlipfe in Südwest-Finnland Fennia, 41, 5: 25.
Kankare, E. 1969. Geotechnical properties of the clays at the Kimola Canal area with special reference to the slope stability. State Inst. Tech. Research, Finland, Publ. 152. Helsinki: 134.
Kujansuu, R. 1972. On landslides in Finnish Lapland. Geol. Survey of Finland, Bulletin, 256: 22.
Lappalainen, V. 1965. Geologiska och geotekniska undersökningar vid Heinoo banuträtning, SW-Finland (in Swedish) Geologi, 17, 9-10, Helsinki: 135-136.
Niini, H. 1977. Main features and causes of landslides in Finland. Bull. Int. Ass. Eng. Geol. 16: 46-48.
Slunga, E. 1973. Om statistisk säkerhet hos jordslänt i anisotropisk lera (in Swedish). Nordisk Geoteknikermöte i Trondheim 24-26 August 1972. Föredrag. Oslo: 151-158.

Extent and economic significance of landslides in Denmark, Faroe Islands and Greenland

Stig Schack Pedersen
Geological Survey, Copenhagen, Denmark

Niels Foged
Danish Geotechnical Institute, Lyngby, Denmark

John Frederiksen
Danish State Railways, Copenhagen, Denmark

ABSTRACT: Denmark is not a country with a serious landslide problem, but several small land-slides and a few large landslides occur sporatically. Almost all the landslides are in coastal cliff areas or in embankments and cuttings in a country characterized by lowlands with little relief.

Landslides in Denmark can be classified into four types: rockfalls, rotational slides, mudslides, and mudflows. Rockfalls occur only in a very few, isolated localities. Rotational landslides and mudslides occur mainly in areas with highly to very highly plastic clays of Tertiary age lying at or close to the surface. The initial slide is rotational and subsequent movement is in the form of a mudslide or mudflow on the lower part of the original slide mass.

Few data have been collected to evaluate the expenses arising from the landslides but the expenses are probably small. The total cost of repairing and preventing slides is estimated between 1 and 3 million US$ per year. This amount does not include the loss of land from erosion and associated landsliding.

Landslides on the Faroe Islands are associated with clay within coal and lava of Tertiary age. All of the landsliding seems to have occurred in prehistoric time.

Most of the landslides in Greenland are rockfalls caused by repeated freezing and thawing of Precambrian basement rocks. Mudflows in Neogene marine clay occur when the permafrost melts. Organic-rich shale of Cretaceous and Tertiary age have slid and apparently burned by spontaneous combustion. Minor damages to human facilities have been reported.

1. INTRODUCTION

The kingdom of Denmark consists of the peninsula of Jutland and a great number of small and larger islands. The country is situated in the northern temperate zone between 54 and 57 northern latitude. Without the Faroe Islands and Greenland, Denmark covers a little less than 44,ooo km² and it has a population of just over 5 million people. The total length of the shoreline of Denmark is about 7,4oo km.

Denmark has a characteristic temperate coastal climate. The mean annual temperature is about 7 degrees C, and the warmest month, July, has a mean temperature of 16 degrees C. The winter climate is characterized by a very large number of freezing-point passages.

Annual precipitation for a normal year shows a regional variation between 45o and 8oo mm. The ground water table is at its highest in late winter and early spring, coinciding with the melting of frost in the ground. Most landslides are activated at this time of the year.

Denmark is a lowland with only small areas higher than 1oo m. The highest point is 173 m above sea level. The landscape in most areas is dominated by small, rounded hills shaped by glaciers, mostly during the last Quarternary glaciation. Some areas are almost flat. Steep gradients are rare in the natural landscape, except near the coasts.

Nearly all of Denmark is covered with glacial till and meltwater deposits reflecting the repeated Quaternary glaciation. If the glacial layers were dispersed equally over the country, they would be about 55 m thick. But this is not at all the actual situation.

In northern Jutland, they are locally more than 3oo m thick. In other areas, they are quite thin or missing. Less than lo percent of the Danish area is covered by late or postglacial deposits.

The subsurface geology of Denmark is shown on Fig. 1. The Danish area at least through the Mesozoic and the Cenozoic was a basin between the Fennoscandian craton in the northeast and the extensively eroded Hercynian range in the south. As the basin gradually filled, the shoreline moved to the west and southwest. This is the reason why uppermost subsurface deposits are younger in the west. Middle Paleocene to middle Oligocene deposits consist of highly to very-highly plastic clays, which form landslides where Quaternary deposits are thin or missing.

2. CLASSIFICATION AND MECHANISM OF LANDSLIDES

Four main types of landslides may be distinguished in Denmark, from brittle to more and more ductile deformation (Fig. 2): A) Rockfalls, B) Rotational landslides, C) Mudslides, and D) Mudflows.

2.1 Rockfall

The most brittle type of landslide is the rockfall. Rockfalls are described in many papers and textbooks concerning landslides related to the weathering of rocks, and especially to the propagation of surface-related jointing. Of course, preexisting fracture systems have an overwhelming effect on rockfall risk, but jointing may also be solely an exfoliation type of fracturing.

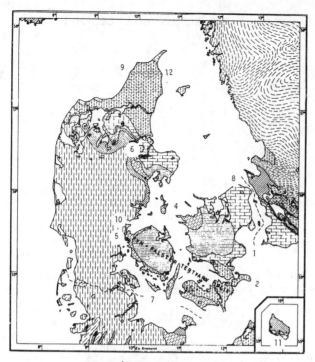

GEOLOGICAL MAP OF DENMARK
FORMATIONS AT THE BASE OF THE PLEISTOCENE

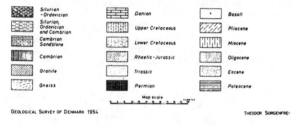

GEOLOGICAL SURVEY OF DENMARK 1954

THEODOR SORGENFREI

Figure 1. Map of Denmark showing subsurface geologic units and the most important landslide localities in the country

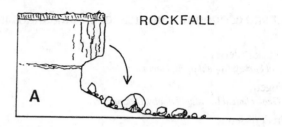

ROCKFALL

A

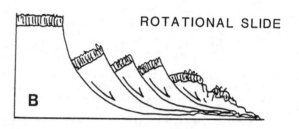

ROTATIONAL SLIDE

B

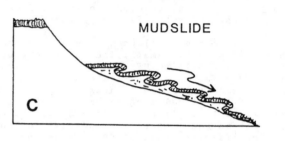

MUDSLIDE

C

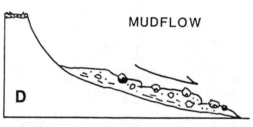

MUDFLOW

D

Figure 2. The four main types of landslides in Denmark.

In many places, rockfalls occur where a competent rock unit overlays an incompetent unit. The incompetent unit in many places is a thin bed of shale or mudstone, which when eroded, disintegrates faster than the overlying limestone. This type of erosion is typically seen in Danish coastal cliffs, where Maastrichtian and/or Danian limestone are exposed.

2.2 Rotational landslides

Rotational landslides are most common along the coastal cliffs in Denmark, creating a geomorphological feature known as the terraced coast. In general, these propagational landslides develop by successive stepping back of the curviform, decollement surface. Based on the geometry of these surfaces, the landslides can be divided into two types: A) Cylindrical slides, and B) Listric, normal-fault type. In a two-dimensional cross-section, the geometry of the cylindrical slide is approximately a circular segment. In contrast, the geometry of the listric, normal-fault slides is irregular, with the toe developing into a horizontally spreading unit.

Three main factors are important for the initiation of rotational slides: 1) rather steep, sloping surface; 2) high pore-water flow, which in general depends on the lithology; 3) erosion of the toe.

2.3 Mudslides

Mudsliding and soil creep occur on slopes dipping 1-5 degrees. A mudslide is characterized by a solid, coherent slide and a viscous matrix along the bottom of the sliding unit. The slide surface may be fractured-the transition to mudflow appears when the fracturing forms individual blocks flowing in the matrix. A buckling type of folding commonly develops in the distal part of the slide. Mudslides develop in the lower part of stabilized rotational slides at many localities (Prior, 1977). Most mudslides are related to the limit conditions of stability at a slope β, with the groundwater table at the surface:

$$\text{Tan}\,\beta = -\frac{\gamma'}{\gamma}\; \tan\ (\sim 1/2\ \tan\ \varphi')$$

where φ´is the effective residual friction angle and γ´and γ are the submerged and total density of the slide deposits in question.

The sliding is typically related to occurrences of clay, but is also very characteristic for localities in frozen ground and permafrost areas.

2.4 Mudflows

A mudflow is a viscous mass moving downslope mechanically as a liquid. In this report, mudflow includes debris flow and debris landslides activated by the melting of the snow in the spring. The main characteristic of these flows is a mass moving downslope in which the grains are held in suspension. Mudflows are commonly related to mudslides. At many localities, mudslides develop into mudflows, either during the sliding, due to loss of cohesion, or after the sliding when water in the headscarp area forms mudflows that extend over the dried-out mudslide toe.

3. LANDSLIDES IN DENMARK

Rockfalls occur only at a few localities where limestones are exposed in coastal cliffs (Fig 1). The most well known locality is the Stevns Klint coastal section (Fig. 1), which is also the type of locality for the Cretaceous-Tertiary boundary. The rocks there consist of a very hard Danian limestone underlain by a thin bed of slippery grey clay and Cretaceous chalk. The limestone has remarkable strength properties, with between 5 and 5o MPa. The Maastrictian limestone is significantly softer, so it erodes and causes rockfalls in the overhanging Danian limestone.

At Møns Klint, coastal erosion attacks soft coccolithic limestone of high porosity ($\varphi > 35\%$) and low strength ($\sigma < 5$ MPa). The natural water content w is often higher than the liquid limit of the remolded material. Landsliding at several localities has been followed by liquefaction of the slide debris, such as at Møns Klint in 1952 when a rotational slide more than loo m high from a nearly vertical cliff turned into a debris flow that extended 5oo m from the coast into the ocean.

Mudslides occur in plastic to highly plastic and fractured clays, such as those in Triassic shale on the southern part of the little island of Bornholm (locality 11). Mudslides also occur in the northern part of Jutland near the town of Frederikshavn (locality 12). There the mudslides are in "Older Yoldia Clay," a highly plastic, fractured clay, deposited in the sea during an interglacial/interstadial period in the later part of the Qqarternary epoch.

The most numerous and serious landslides occur where older Tertiary clays with a very-high to extremely-high plasticity are exposed in natural slopes, cuttings or coastal cliffs. The plasticity index in Danish Paleocene and Eocene clays is in the interval I_p 8o-3oo%, and the effective strength parameters diminish from $\varphi´$ 2o to < 12 degrees and $C´ \sim$ 4o-lo kPa. These clays are considered to have extreme swelling properties, and the intact strength degrades rapidly to effective residual parameters of $\varphi´_{res}$ 8 to 12 and $C´_{res} \sim$ 0 kPa. Sush properties caused rotational slides at Røsnæs (locality 4)

and at the Lillebælt Coast at Fredericia and Middelfart (locality 5), where the slumps rapidly degraded into mudslides and soil creep on a surface extending inland from the coast at an angle from 6-15 degrees (Prior, 1979).

Landslides also occur in very highly plastic clays at inland localities, such as at Ølst (locality 6) where the upper 1 to 2 meters of topsoil and Tertiary clay has been deformed into a sort of blanket of mudslides. Another locality is the southern coast of Ærø (locality 7) where interglacial (Eemian) marine clay deposits flowed in a glacial landscape of rather high topography. Danish clay tills rarely form landslides. Due to preconsolidation, their strength properties are very good ($\varphi´$3o to 36 and $C´\sim$ lo to 3o kPa). Even on coasts with high marine erosion, cliffs of clay till may rise to more than 3o-5o m with a β of 3o to 4o, unless the till formation has waterbearing sand layers. Typical examples are the northern coast of Zealand (locality 8) and the northwest coast of Jutland at Lønstrup (locality 9).

Inland slides in Denmark are scarce and are related mainly to human activities. A few mudflows/debris flows have taken place where water in Miocene sediments exist as springs near the foot of steep slopes, such as at Vejle Fjord (locality lo).

3.1 Economic significance

No detailed investigation has yet been made of the economic significance of landslides in Denmark, the Faroe Islands, or in Greenland. The following points are the result of information solicited from af few key persons in Denmark:

Very serious or catastrophic landslides occur in Denmark not more often than once every twenty years. Damage figures are not available for these exceptional events.

Big and small slides in cuttings and embankments occur every year, but they are by far more common in wet years. The Danish State Railway spends about $US 4oo,ooo annually on investigation, repair and protection. The Danish road system is separated into state, parish, and district roads, for which no combined picture of landslide expenses is available. A qualified guess would be to $8oo,ooo annually.

The cost of redesigning roads to avoid landslides can also be significant. When the Danish State Motorway Office built a new motorway in the eastern part of central Jutland, they carefully mapped the areas with Tertiary plastic clays by geophysical methods and borings. The motorway had to be realigned several times compared to the initial design to avoid potentially unstable cuts. No estimate of these redesign expenses is avialable, but the costs have been included generally in the overall estimate of $1 to $3 million per year.

Coastal erosion is a significant problem in Denmark, but the costs of damage and remedial action are difficult to obtain.

4. LANDSLIDES IN THE FAROE ISLANDS

The Faroe Islands consists mainly of basaltic lava flows of Tertiary age. The 3ooo m-thick lava pile can be divided into a lower, middle, and upper series. Between the lower and the middle series, there is a 15 m-thick, coal-bearing sequence.

There has never been a systematic account of landslides on the islands, but Geikie (1880) and Rasmussen (in Rasmussen and Noe-Nygaard, 1969) have provided some information. Landslides on the southern island, Suderø, have been investigated thoroughly and described by Jørgensen (1972 and 1978). Jørgensen (1978) indicates that landslides on Suderø have occurred in connection with a coal-bearing sequence and a tuff-agglomerate zone in the upper flows of a lower basalt series. "Lamellar zones," a dense network of fissures, are found especially in the lower basalt series and probably contribute to landsliding. Both the coal-bearing sequence and the tuffagglomerate zone contains clay layers which can act as a lubricant for slides.

The largest landslide on Suderø occurred in connection with a"lamellar zone." This slide is about one kilometer long and 600 m wide. All the investigated slides seem to have occurred in prehistoric time.

5. LANDSLIDES IN GREENLAND

Landslides in Greenland are influenced by permafrost, glacial ice, high topographic relief, and repeated freezing and thawing. Most of Greenland consists of Precambrian basement rocks incised by steep fjords and valleys extending from the coast to the margin of the inland ice. Repeated freezing and thawing causes exfoliation and rockfall. Mudflows occur in areas with soil cover and elevated Neogene marine clays. The flows are initiated during the summer when the uppermost half-metre of the permafrost zone has thawed. The water-saturated topsoil and disturbed marine and meltwater deposits become mobile due to the stability equation described previously. This type of mass wasting takes place on slopes dipping more than 5-15 degrees (Washburn, 1967 and Schack Pedersen, 1987).

Cretaceous and Tertiary black shales on the north coast of Nugssuaq and on the island of Disko in western Greenland have a substantial content of organic material. They are prone to landslip, and the slide masses become ignited, probably owing to spontaneous combustion (Henderson, 1969).

Rockfalls and debris slides ("ur") more than 400 m above sea level are found near the Black Angel mineralization at Marmorilik in the Umanak area, northwest Greenland. Due to the severe climate with frost activity on very steep to vertical slopes, metamorphosed limestones and dolomites continously degrade, making assessability to this mineralization very difficult.

Due to the scarcity of population in this remote land, none of the landslides seem to have damaged any human facilities.

6. LITERATURE

Danmarks Natur, Bind 1. Politikens Forlag, Copenhagen 1971.

Geikie, J. 1880. On the geology of the Faroe Islands. Trans. R. Soc. Edinb. 30:217-269.

Henderson, G. 1969. Oil and gas prospects in the Cretaceous Tertiary basin of West Greenland. Rapp Grønlands Geol. Unders. 22.

Jørgensen, G. 1972. An area of solifluction on Suderoy, The Faeroe Islands. Bull. Geol. Soc. Denmark, 21:368-373.

Jørgensen, G. 1978. Landslides and related phenomena om Suderoy, the Faeroe Islands. Bull. Geol. Soc. Denmark, 27:88-89.

Prior, D.B. 1977. Coastal mudslide morphology and processes on Eocene clays in Denmark. Geografisk Tidsskrift, bd 76:14-33.

Rasmussen, J og Noe-Nygaard A, 1970. Geology of the Faeroe Islands. Danm. geol. Unders. række 1, 25:142 p.

Schack Pedersen, S.S. 1987. Comparative studies of gravity tectonics in Quarternary sediments and sedimentary rocks related to fold belts. From Jones and Preston (eds) 1987. Deformation of sediment and sedimentary rocks. Geol Soc Special Publication 29.

Washburn, A.L. 1967. Instrumental observations of masswasting in Mesters Vig, District northeast Greenland. Medd. om Grønland, Bd 166 nr 4.

Landslides: Extent and Economic Significance, Brabb & Harrod (eds)
© 1989 Balkema, Rotterdam. ISBN 90 6191 876 6

Landslides in France: A risk reduced by recent legal provisions

J.C.Flageollet
Université Louis Pasteur, Strasbourg, France

ABSTRACT: Almost all types of landslides have occurred or might occur in France. Severall catastrophic rock-falls, slides,and debris flows have killed many people, the last one in 1970(Praz Coutant). Several large land-slides in the Alps threaten installations,dwellings, and roads; they are being closely monitored.In many com-munities,Plans d'Exposition aux Risques(PER) at 1:5,000 scale have been completed or are in progress. The PER's are legal, technical, and cartographic documents of reference with regard to landuse and building rules, as well as insurance indemnities.

RESUME: A peu près tous les types de landslides sont apparus ou sont susceptibles de se produire sur le terri-toire français; il y a eu des écroulements de roches et des glissements de grande ampleur et meurtriers dans le passé, le dernier en 1970(Praz Coutant).Environ une demi-douzaine de grands glissements menacent des instal-lations, des habitations et des routes;ils sont actuellement surveillés, principalement dans les Alpes. Les Plans d'Exposition aux Risques (PER), établis par commune à grande échelle (environ 1/ 5000 ème),sont élaborés ou en cours. Ils constituent les documents cartographiques et techniques de référence en ce qui concerne la réglementation de l'occupation du sol et des constructions, en cas de catastrophe,depuis la loi de 1982 qui a institué l'assurance "risques naturels" obligatoire.

1 INTRODUCTION

"Earth movement" is the most frequently used expres-sion in France for landslides. Earth movement also includes volcanic, hydrological, and earthquake ha-zards, and subsidence and expansion of soil. The lack of a separate category for landslide movement in go-vernment documents makes the collection of economic information difficult as best.

Most types of landslides are present in the two high mountain regions of the Pyrenees and the Alps. Other areas are much less affected, and more by slides than falls or flows.

The prevention and mitigation of landslide hazards were strongly affected in 1982 by new legislation. This legislation requires insurance against natural hazards and the mapping of hazard and risk at large scale.The mapping is carried out by French adminis-tration services under a "Plan d'Exposition aux Ris-ques (PER)". This program institutionalizes and im-improves landslide mapping known previously by the acronym "ZERMOS", and landuse mapping at 1:5,000 scale "Plan d'Occupation des Sols (POS)."

2 LANDSLIDE LOCALITIES

The number and variety of landslides in France is pro-bably not as great as in Italy and other European and Mediterranean countries, but most of the types reco-gnized by Varnes (1978) are present in sufficient number to create many problems each year. Several agencies have collected information on these landsli-des including the Bureau de Recherches Géologiques et Minières (BRGM), various public works departments, universities and private firms. Administrative archi-val documents, mainly from the 19th and 20th centuries mention more than 4,000 "earth movements," most of them landslides.

A map of mass movement-prone areas in France, pu-blished in 1980 by the BRGM and the "Commissariat" for Study and Prevention of Natural Hazards (CSPNH), indicates a number of landslides (except debris flows) which occurred more or less recently. The map shows three very large rock slides (more than 10,000,000 m3), 97 large landslides (more than 10,000 m3),466

small landslides (less than 10,000 m3),40 rockfalls (more than 100 m3),and 168 rockfalls and blockfalls (less than 100 m3).Fig.1 shows only the large and very large landslides and rockfalls.

A questionnaire sent in 1982 by the CSPNH to French communities requested additional information about landslides and other natural hazards in their terri-tory. The full extent of the landslide problem in most of France is unknown, but some generalizations are possible. Landslides are rare in massif areas underlain by hard rocks, such as the Massif Armori-cain and the Vosges. Landslides are more common but still sparse in areas underlain by Tertiary forma-tions. Landslides are abundant in the mountainous areas of the Alps and Pyrenees. For instance, 225 landslides are shown on the Saint Jean de Maurienne quadrangle at 1:50,000 scale. In the Arvan basin, 20% to 30% of the surface is covered by landslides.

Active landslides are chiefly southeast of a line between the towns of Pau and Mulhouse, in the Alps, Provence, the Pyrenees and, to a smaller extent, the Massif Central, the Jura, and Aquitaine. Slides and flows are much more numerous and frequent than falls. Landslide-prone areas where little activity is curren-tly taking place include the eastern and southern parts of the Parisian basin, the Aquitaine basin, and Auvergne.

Jagged peaks and very steep slopes in the Alps a-bove 2,000 m are habitats for frequent and large rock slides, rockfalls, and glacier-related mass movements. Lower parts of the Alps and slopes of the intermont-ane valleys have debris flows and other landslides. Habitats for landslides in the Parisian and Aquitai-ne basins are mostly restricted to ridges and cuestas underlain by limestone and other resistant rocks, and to valleys slopes deepened by Quaternary downcutting of rivers.

2.1 Paris basin

Marls and clay of Jurassic, Cretaceous, and Eocene age are units prone to landsliding in the eastern Paris basin (fig. 2). The Eocene plastic clays extend to the central part of the basin where large slumps have occurred north and south of Epernay at

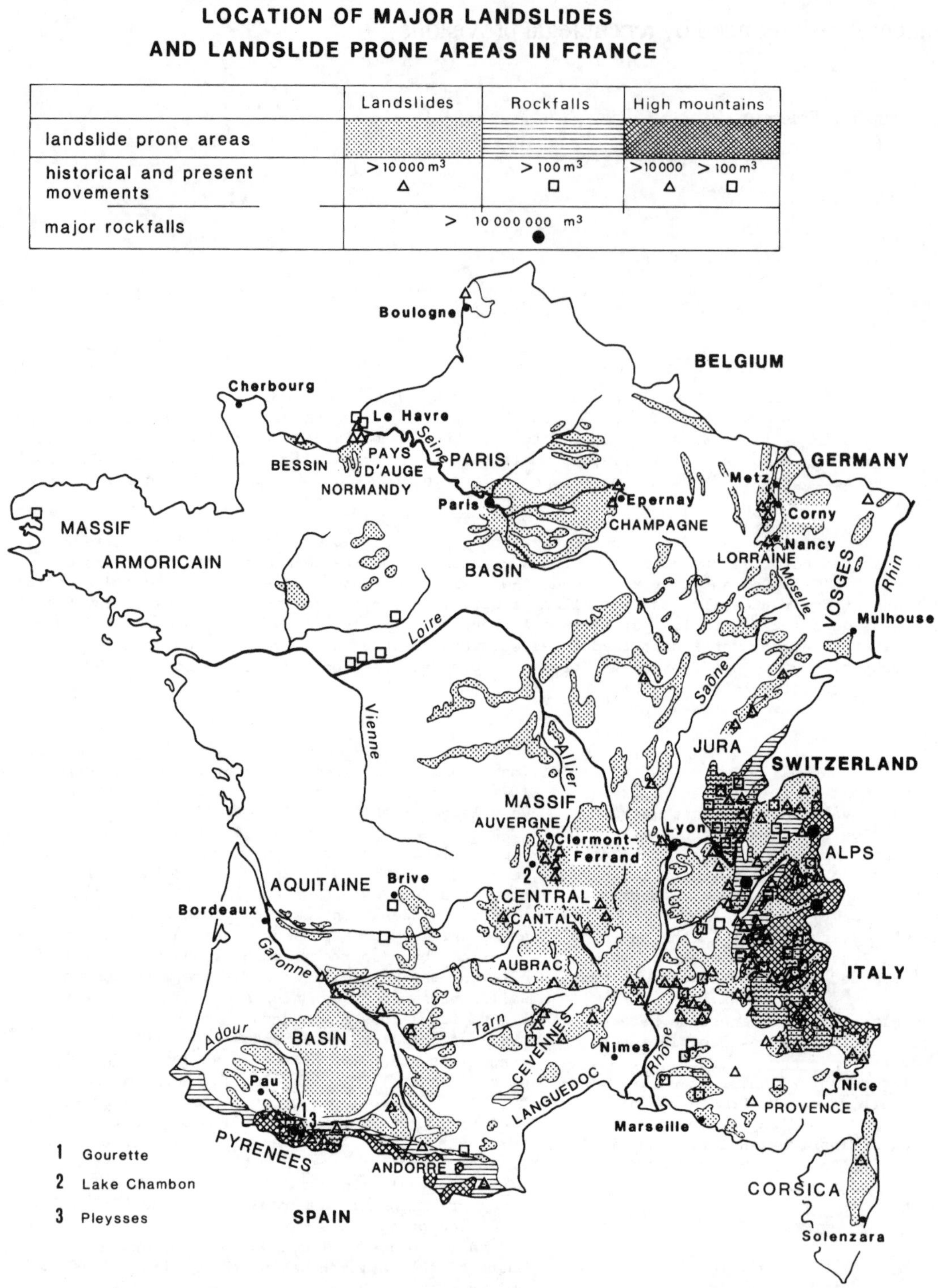

LOCATION OF MAJOR LANDSLIDES
AND LANDSLIDE PRONE AREAS IN FRANCE

	Landslides	Rockfalls	High mountains
landslide prone areas			
historical and present movements	> 10 000 m³ △	> 100 m³ □	>10000 △ > 100 m³ □
major rockfalls	> 10 000 000 m³ ●		

Figure 1. Location of major landslides in France. Does not include debris flows.

1 Gourette
2 Lake Chambon
3 Pleysses

Villers Allerand, in 1986. Some of these slumps are natural and others are human-induced, associated with slope shaping and clearing for the extention of the Champagne Vineyard, at Grauves in 1988 for example.

2.2 Lorraine

Lorraine landslides are fairly well known. Along the Seille, Meurthe, and Moselle valleys, chiefly between Nancy and Metz, hills and plateaus are underlain by Bajocian and Aalenian limestones. Beneath these limestones are marl and clay, also of Jurassic age, more or less covered by periglacial colluvium. The surface water supply is augmented by artesian flow from Jurassic sandstone. Landslides form in the marl and clay where water flow is greatest and where the bedding dips out of the slope. Most landslides (87%) have slope-angles between 14% and 21%; there are no landslides on slopes less than 10.5%. A slide at Corny is 600 m long x 600 m wide, and 40 m deep with a volume of approximately 465 million m3 . It is located in a fault zone where the beds dip as much as 8.5% toward the valley in an area where the regional dip is 2%. The landslide moved as much as 12.5 cm/h but nobody was injured.

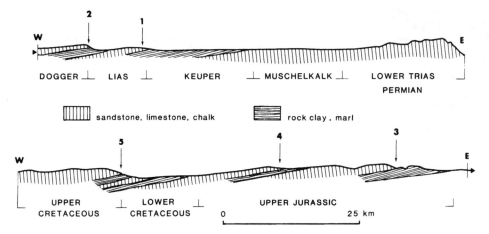

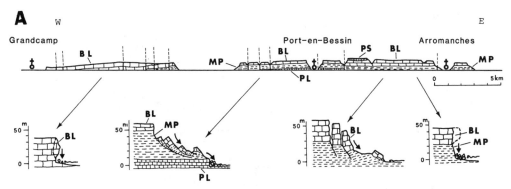

Figure 2. Morpho-structural types favorable for landsliding in the eastern Parisian basin. Slumps and planar slides are common in clay-rich beds on the escarpments. 1 lower Liasic cuesta (Sinemurian, Pliensbachian) 2 Dogger cuesta (Toarcian) 3 Oxfordian cuesta (Oxfordian s.s., Callovian) 4 Kimmeridgian cuesta 5 Coniacian-Turonian cuesta (Albian, Aptian, Barremian).

Figure 3. Landsliding prone rock clays and typical failures along the Normandy coast. A : Bessin area. PS Planet sandstone BL Bessin limestone MP Port en Bessin marl (middle and lower Bathonian) PL Porifera limestone (Bajocian)

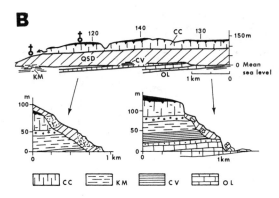

Figure 3. B : Pays d'Auge area. CC Cenomanian chalk CV Villerville clay (Oxfordian) OL Oxfordian limestone KM Kimmeridgian marl QSD Quaternary slope deposits

2.3 Normandy

In Normandy, planar slides, slumps and falls of overhanging cliffs occur mainly along the coast, where the sea erodes the slides debris and maintains the activity. On the coast east of Cherbourg between Grandcamp and Arromanches (Fig. 3 A), marl within Jurassic sedimentary rocks has an almost plastic behaviour, with a plasticity index of 23%. An entire slab of cliff, 50 m wide and 350 m long, moved towards the sea near Port

en Bessin, on August 5, 1981. The rupture surface of this rock block glide was located at the base of the marl. Almost the whole coastal slope south of Le Havre is unstable (Fig. 3 B) with rock slides, slumps, and retrogressive failures forming spectacular amphitheaters called Cirque des Graves and Fosses du Macre. In the Cirque des Graves, the last major activity was in January 1982 when the main scarp receded more than 100 m in few days. A similar landslide in the Fosses du Macre was reactivated late in 1987. Representative failures in this region are shown on Fig. 3B.

2.4 Aquitaine

The Aquitaine basin in southwestern France has numerous but generally minor landslides. Most of these slides are on valley slopes underlain by clastic sediments derived from Tertiary erosion of the Pyrenees. Erosion by rivers, such as the Garonne and Ariège, undercut these soft and clay-rich sediments promoting slumping (Fig. 4).

2.5 Central Massif

Landslides in the central Massif are small and are located on escarpments or slopes in slide-prone units, such as Permian claystone in the Brive basin ; volcanic ash, clay, and sand in the Cantal and Mont Dore areas ; Tertiary marl in the Limagne, near Clermont-Ferrand ; and Triassic and Jurassic marl in the Cevennes Mountains south of Lyons. A landslide older than

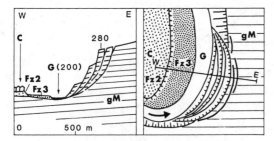

Figure 4. Slumps characteristic of failures along rivers in the Aquitaine basin, southwest France. C City of Carbonne, south of Toulouse G Garonne river Fz Alluvial terraces GM Stampian mollasse

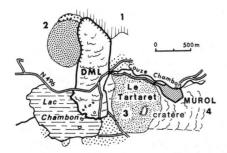

Figure 5. Landslide in late Cenozoic volcanic rocks dammed a small stream and created Lake Chambon at La Dent du Marais, in Mont Dore area, west of Clermont-Ferrand. DML Dent du Marais landslide. 1 basaltic flow 2 basaltic breccia 3 cinder cone 4 lava flow

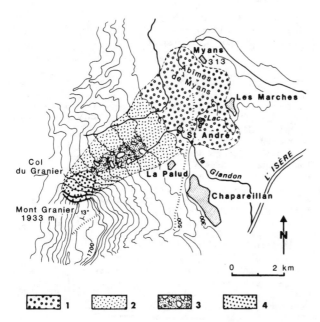

Figure 6. Mount Granier rockfall near Chartreuse, Savoie (from Flageollet, 1988). 1 slide debris with predominantly Urgonian (Lower Cretaceous) blocks and stones ; 2 slide debris with predominantly Neocomian blocks and stones ; 3 slide debris with big blocks of Hauterivian and Valanginian limestone ; 4 remobilized landslide material

6,900 BP, at La Dent du Marais south of Clermont-Ferrand, formed Lake Chambon. The material that failed is a pyroclastic basaltic breccia and cinerite of late Cenozoic age (Fig. 5).

2.6 Alps

The French Alps have many kinds of landslides from low energy and small size to high energy and large size. These mountains have suffered many landslides which are listed in Table 1. A typical rockfall is shown on Fig. 6.

Many of the landslides have developed on spectacular steep or vertical cliffs of Mesozoic limestones associated with thrust sheets (Fig. 7), in crystalline rocks of massif areas, and in Cenozoic flysch. In the Chablais and Bauges areas, slides and flows are located in thick Pleistocene glacial tills. In the Briançonnais area, Carboniferous schist, sandstone, and coal are unstable (La Ravoire debris flow).

In the Enfer ravine north of Pontamafrey, several million cubic meters of Triassic gypsum and marl failed as a large landslide which, in 1970, remobilized as an earth flow.

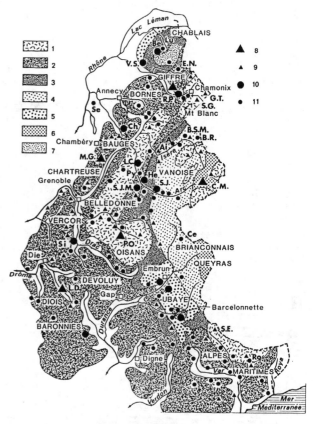

Figure 7. Landslides and geological units in the Alps. 1 Central massifs with crystalline rocks ; 2 Folded Mesozoic limestone and marl ; 3 Cenozoic flysch and thrust sheet rocks ; 4 Briançonnais thrust sheet with Mesozoic limestone and gypsum ; 5 Thrust sheet flysch and limestone covering autochtonous black marl in Embrunais and Ubaye areas ; 6 Schist thrust sheet ; 7 Chablais thrust sheet ; 8 Rock avalanche or rockfall, large size or with important morphological consequences (lake, outburst) ; 9 Small rock avalanche or rockfall ; 10 Large slide, more or less catastrophic ; 11 Small landslide.
A : Aime, Ai : Aigueblanche, Be : Bellevaux, B.R : Bec Rouge, B.S.M : Bourg Saint Maurice, Ce : Cervières, Ch : Le Chatelard, C.M : Col de la Madeleine, E.N : Entre deux Nants, G.T : Glacier du Tour, He : Hermillon, L.C : La Chapelle, L.D : Luc en Diois, Lu : Lullin, Mt. G : Mont Granier, P.O : Plaine d'Oisans, Py : Pontamafrey, R.P : Roc des Fiz, Praz Coutant, Ro : Roquebillière, Se : Serrières-en-Chautagne, S.G : Saint-Gervais, Si : Sinard, S.J : Saint Jean de Maurienne, S.M : Saint Michel de Maurienne, V.S : Viuz en Sallaz

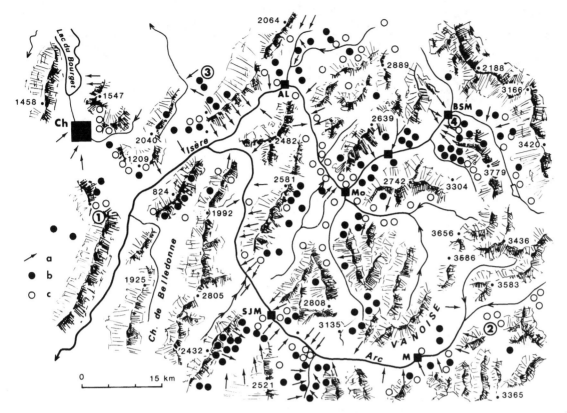

Figure 8. Landslides in Tarentaise and Maurienne regions of the Alps, Savoie Province. Located about 60 km south of Geneva (from Pachoud 1980 and Bravard 1983)
a : debris flow, b : landslide, c : rockfall. Big landslide 1 : Granier, 2 : La Madeleine,
3 : Le Chatelard, 4 : La Ravoire, Ai : Aime, Al : Albertville, B.S.M : Bourg Saint Maurice,
Ch : Chambéry, M : Modane, Mo : Moutiers, S.J.M : Saint Jean de Maurienne

In the Embrunais and Ubaye thrust sheets, autochtonous Jurassic black marl and flysch have failed. In the Isere valley, the Arc valley, and the Embrun and Barcelonnette catchment basins, landslides are abundant (Fig. 8). Metamorphic and granitic rocks in the center of the French Alps are affected by rockfalls and gravitational sagging, some of which threaten to fail catastrophically (La Clapière, Séchilienne).

Debris flows are frequent every year. Rockfalls occur less frequently, perhaps every 10 years. Complex and catastrophic landslides occur once or twice per century.

Near the town of Chamonix, on April 16, 1970, a mass of colluvial debris and trees, triggered by heavy rainfall, slid down a 30-40 percent slope and crashed into the sanitorium of Praz Coutant, killing 72 people including many children. The mass, with a volume of 30,000 to 40,000 m³, followed the path of a small, wet snow avalanche that preceded the debris slide by a few days.

2.7 Pyrenees

The Pyrenees are less affected than the Alps, but the variety of landslides is almost as large. The landslides are more spread out, occurring predominantly in the western and central Pyrenees. In the western Pyrenees near the Pic du Midi d'Ossau, rotational landslides are numerous in highly fractured Silurian and Devonian shaley rocks. Some of the slides occurred recently and others are as old as Pleistocene.

Most debris flows are located on the southern side of the Pyrenees, triggered or removed by rainfall and snowmelt at the end of spring. Earthquakes also take part in triggering or reactivating landslides such as those near Gourette and Pleysses in August 1982. In the Andorran Pyrenees in November 1982, heavy rainfall triggered earthflows and debris flows on overgrazed slopes.

3 CLIMATE

France is famous for its mild marine climate influenced by westerly wind from the Atlantic Ocean. The monthly temperature in Paris-Le Bourget varies from 5°C in January to 25°C in July ; 53 days have an average minimum 0°C, and 11 days have a daily maximum of 30°C ; annual precipitation averages 585 mm, with a maximum of 62 mm in August, and a minimum of 32 mm in March. Rainfall is generally weak in intensity from 5 to 10 mm/h at the beginning and end of a storm, and 20 to 30 mm/h for 15 to 30 minutes during the peak of the storm. The average annual duration of precipitation is 776 hours distributed in all months, with a maximum in December and January (198 h) and a minimum in July and August (86 h) (see Fig. 9). Cyclonic precipitation can, however, reach high values for this type of regime. Some winters are exceptionally rainy, producing a recrudescence of landslide activity or triggered of new landslides. The 1987-88 winter on the Normandy coasts, for example, had October to March precipitation of 680 mm, substantially more than the yearly average (452 mm) for 1980 and the previous year. Twenty-one slides occurred in the area during February, among them the big slump at Criqueboeuf with 2-1/2 million m³ of material.

The oceanic regime is modified to east of a line connecting the cities of Pau and Metz. Rainfall is greather in the eastern mountains, including the Massif Central, Vosges, Jura, Alps, and the Pyrenees. Yearly rainfall in these mountains is higher than 800 mm, reaching a maximum of 1780 mm in the Aubrac. Winters are more humid (Fig. 9).

Table 1. Major catastrophic landslides in France

1 LOCATION	2 ELEVATION		3 VOLUME (1 000 m3)	4 TYPE	5 TRIGGER	6 DATE	7 DAMAGES	8 CASUALTIES
	a	b						
PLAINE D'OISANS Oisans, Isère				4	2	1219		> 1 000
MONT GRANIER Chartreuse, Savoie	1 900	300	500	4		1248 (11,24)	H (16)	1 500 to 5 000
CERVIERES Briançonnais, Hautes Alpes	1 859	1 636		2	2	1431 (06,)	h (50)	
LA CHAPELLE Maurienne, Savoie	2 000	430		2		1431 (08,10)	h	
SAINT JEAN DE MAURIENNE Maurienne, Savoie	2 000	550		2	2 + 3	1439 (02,1)		75
LUC EN DIOIS Diois, Drôme	1 050	629	1 500	2		1442		
ENTRE DEUX NANTS Giffre, Haute Savoie	2 100	900		2 + 4	3	1602 (02,21)	H	29
LULLIN Chablais, Haute Savoie	1 600	1 000			1	1635 (04,11)	h (20)	64
BOURG SAINT MAURICE Tarentaise, Savoie	2 000	800		1 + 2	2	1636 (05,)	h (52)	
VIUZ EN SALLAZ Chablais, Haute Savoie	1 300	800	2 500	2	2	1715 (07,20)	h (20)	
HERMILLON Maurienne, Savoie	2 000	600		2	2 + 3	1740 (12,22)	h (22)	7
ROC DES FIZ Plate, Haute Savoie	1 600	1 100	20 000	4	3	1751 (08,4)	h (6)	6
AIME Tarentaise, Savoie	2 500	600		1 + 2	2	1778	n (36)	
SAINT JULIEN Maurienne, Savoie	2 000	500		2	2	1824 (07,18)	h (30)	2
BEC ROUGE Tarentaise, Savoie	2 515	1 200	3 000	2 + 4	3	1877		
SAINT GERVAIS Mont Blanc, Savoie	2 700	580	900	3	3	1892 (07,12)	h (50)	177
ROQUEBILLIERE Vésubie, Alpes maritimes	800	600	200	2	2	1926 (11,25)	h (20)	19
LE CHATELARD Bauges, Savoie	1 600	1 100	3 000	2	3	1931 (03,15)	h (15)	
SERRIERES Chautagne, Savoie	1 000	200		1	3 + 4	1936 (01,17)	h (15)	
BELLEVAUX Chablais, Haute Savoie	1 500	1 000	2 000	2	2 + 3	1943 (03,12)	h	
GLACIER DU TOUR Mont Blanc, Savoie	2 200	1 500	3 000	3	3	1949 (08,14)		6
PRAZ COUTANT Giffre, Savoie	1 500	1 200	40	2	2 + 3	1970 (04,16)	h sanitorium	72
AIGUEBLANCHE Tarentaise, Savoie				4	2	1977 (05,1)		1
BOURG SAINT MAURICE La Ravoire Tarentaise, Savoie	1 600	700	300	1	3 + 4	1981 (03,31)		

1 - Locality, area, department (French département) - 2 - a : Starting point - b : Stop - 3 - Approximately -
4 - Type of landslide - 1 : debris flow, surficial deposits ; 2 : debris flow, bedrock ; 3 : glacier-related
mass movement ; 4 : rock avalanche and rockfall - 5 - Trigger - 1 : earthquake ; 2 : rainstorm ; 3 : snowmelt
or ice melt ; 4 : human activity - 6 - Year, month, day - 7 - H : Hamlets ; h : House ; (17) : number of H or h

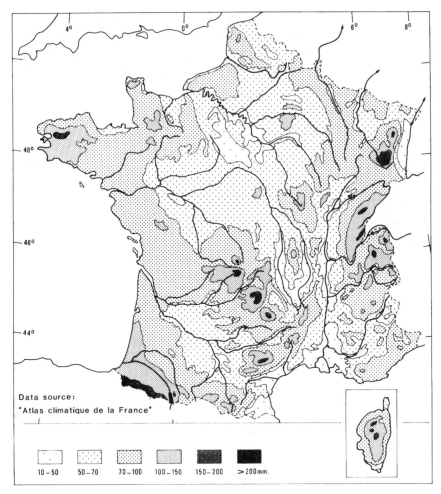

Figure 9. Mean January precipitation in France (snowfall converted to equivalent water)

The influence of the Mediterranean Sea on the climate of the southern Alps, eastern Pyrenees, Languedoc, Provence, and Corsica is well known. These areas are sheltered from western cyclonic paths, making the climate warm and dry in summer. Precipitation is abundant in the fall and can reach 1,000 mm in a few days. In Nice, the average is 862 mm in 86 days, 47 mm in July-August in 6 days, 236 mm in October-November in 18 days and 147 mm in 24 hours in November. The record is at La Clan, in the eastern Pyrenees, where 840 mm of rain fell in 24 hours on October 18, 1940. In 1982, from 2 p.m. on November 7, to 6 a.m. on November 8, more than twenty mudflows and debris-flows occurred in the mountainous area of the Solenzara basin when precipitation exceeded 400 mm in 10 hours.

On the western border of the Alps, cyclonic deluges of rain in association with uplift of warm air masses ahead of a front have totaled as much as 150 mm in 45 minutes (July 5, 1971). The rain may also be long lasting with 155 mm falling in 60 hours, such as on the Gresivaudan slopes in December, 1968. Close to the Italian frontier, rain may fall in any season, but it falls mainly in the fall and spring when warm and wet air masses come from the east, southeast, or south. Rainfalls of June 1957, which triggered numerous landslides, had this source.

4 SEISMICITY

The seismicity in France is weak compared to Turkey, Balkan countries, or North Africa. Karnik values of released energy are, for instance, 16 in the center of France, and 18 in the central Pyrenees and Rhoda-

nian Provence. Of the 624 seismic events for which epicentral intensity is known, more than onehalf are equal or less than V (no damage) on the MSK earthquake-intensity scale. Twenty-one earthquakes can be considered as catastrophic (Intensity VII - VIII) in 5 centuries, averaging 4 per century.

Earthquakes with intensities smaller than IV MSK are not usually sufficient to trigger major landslides, but they may trigger rockfalls and other landslides in very steep terrain or places where other destabilizing factors have made slopes ready to fail. The most threatened areas are the Pyrenees, and the Alps (Alpes Maritimes, Savoie, Haute-Savoie, Hautes Alpes and Alpes de Haute Provence departments).

In late August 1982, several slides in the Pyrenees were triggered by earthquakes and heavy rain (20 to 30 mm/h), especially in the areas of Eaux Bonnes and Pleysses.

5 LANDSLIDES DAMAGE

The inquiry carried out in 1985 by the CSPNH indicated that 2997 communities have earth movements problems in France, including subsidence, gullying and erosion in mountain areas, and marine and eolian erosion (Fig. 10). In 148 communities in ten departments (Indre et Loire, Haute Provence, Hautes Alpes, Corse, Gironde, Hérault, Haute Loire, Lot et Garonne, Nord and Tarn), rockfalls are cited in 32 areas, slides in 27, and flows in 4. The total surface affected is 180,447 ha. The population living in the sectors is 1,347,851.

In France, for four years, from 1983 to 1986, the damages toprivate properties have been calculated by

Figure 10. Departments in France and Corsica with ground failure problems. Dotted areas have no more than 30 communities with ground failure ; widely-spaced, vertically-lined areas have 31 to 100 communities ; and narrowly-spaced, vertically-lined areas have more than 100 communities with ground failures. Numbers refer to the number of communities with a prior right for a PER (Plan d'exposition aux risques). Information from the French Official Gazette, 1984.

using information given by insurance companies (Heuzé 1988) and by emergency public services. The earth movements including landslide damages reached $US 1,634 million in 1983, $US 640 million in 1984, $US 635 in 1985, $US 361 million in 1986. In 1983, all types of landslides were very numerous. In 1985 and 1986, they were less numerous and a lot of small landslides were not declared "natural disasters", thus their damages not taken into account by insurance companies.

6 PREVENTION AND MITIGATION

6.1 New regulations

In France from the 1970's until 1981, prevention of natural hazards had been concerned mainly with snow avalanches and floods. Landslide inventories were made by the BRGM and interdepartmental commissions for mountain and natural hazard problems. In 1981, the new Socialist government developed a more rigourous preventive policy applying to all major natural hazards on French territory, and codified this policy in a 1982 law. The policy was accompanied by better coordination of actions and financing between services of the concerned departments. To assert this policy of prevention in the face of public opinion, the Prime Minister appointed a person in charge whose rank was progressively raised from Commissioner in 1981, to Ministerial Delegate in 1983, and to Secretary of State in 1984. In 1986, when the Socialists left the Government, this person became a delegate in the Environment Ministry. Nevertheless, the law of 1982 continues to be applied. This law affirms that hazard

prevention is a duty of the State, and that protection against natural disasters is a right for all citizens. Application Decrees in May 1984 organized conditions for prevention and protection. For example, insurance companies working in France must cover natural disasters in return for a premium whose rate is fixed by a state administrator. This unique rate is 9% of the prime rate for fire and theft. A disaster is declared by a ministerial commission in which insurance representatives play an important role. Building is forbidden or controlled in dangerous zones, as defined on large scale maps called Plans d'Exposition aux Risques (PER). The 1982 law requires State agencies to make these maps.

Until January 1, 1984, the law was applied to several events, more or less catastrophic, and extended to natural hazards like wind, hail, and snow. In 1982, 1983, and 1984, the large number of costly natural disasters caused insurance companies to lose money. The cost of disasters in 1985 was not as great.

6.2 Landslide maps and plans

Although landslides are shown on most recent geological maps, hardly any of these geological maps show all of the landslides in a region. Specialized maps dealing with landslides have been prepared since 1972 under a BRGM project called ZERMOS (Zones Exposées aux Risques liés aux Mouvements du Sol et du Sous-Sol). About thirty of these maps were published by 1980. Since 1982, ZERMOS maps have been replaced by hazards maps which show probability of landslide movement, its type, and intensity. Intensity is divided into 4 ratings, very high, high, medium, and low. Probability of movement, linked to landslide activity, slope, and the strength of rocks, is shown as very probable, probable, or not probable. The hazard ("aléa") is expressed in three or four levels, high (3), medium (2), low (1), and nul (0).

PER's 'Plans d'Exposition aux Risques) are experimental risk maps for communities designated to comply with the 1982 law. Every PER includes several cartographic documents : a hazard map, a vulnerability map, a risk map, and, most important, the plan itself. The plan is at large scale, the same as landuse plans. On this plan, the zones where building is not allowed are shown in red ; those where construction is permitted if preventive measures are followed is shown in blue ; and zones without visible hazard or risk are shown in white. Additional zones can be used for areas not yet studied or for areas with hazards that cannot be forecast. Most of the communities consulted for the preparation of PER's have agreed to follow this zoning. In 1984, PER's were underway in 169 communities, among them 15 where technical studies existed before, in particular ZERMOS maps. The number was raised to 399 in 1985. In 1986, PER's were planned for 431 communities, 127 of them with landslides risk (Fig. 11). Adoption of PER's fortunately, most advanced is areas of high relief in the Alps, such as in the Isere department. Elsewhere, adoption of PER's is slower because some community officials are opposed or at least reluctant, and because official services are higher in priority. Moreover, procedures for adopting PER's are extensive and cumbersome (Fig. 12).

Even if adopted, the effectiveness of PER's in reducing landslide costs will take many years to evaluate.

6.3 Landslide prevention in mountain areas

In many mountain areas in France, surficial drainage systems have been neglected, and landslides have formed where the slopes are water-saturated. The RTM (Restauration des Terres en Montagne) Service has played an important role in improving these drainage systems. The Haute Savoie RTM, for example, restored a drainage network near Mont de Lessy where landslides had caused several problems

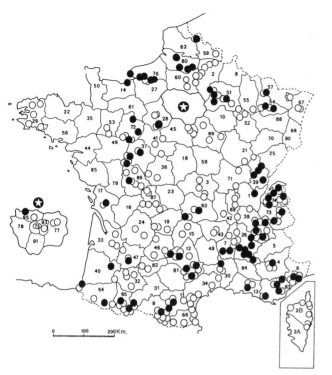

Figure 11. PER's (Plans d'Exposition aux Risques de mouvements de terrain) underway in France by 1985. Solid circles show PER's with ground failure maps, mainly landslides. Open circles show PER's with other types of hazards, mainly floods, avalanches, and earthquakes. Information from the French Official Gazette, 1985.

Reforestation is another important preventive action in the Alps and the Pyrenees. Native trees have been replaced by faster growing species from other regions. In a few areas where the soil is especially fertile, trees with high commercial value have been planted, such as Douglas fir, cedar, ash, wild cherry, and high quality walnut.

Reforestation began as nearly as 1866 in the Alps and following World War II in the Pyrenees. In the eastern Pyrenees, especially in the Haut Vallespir area, tree species have been mixed to diminish the risks of failure.

Reforestation in combination with drainage improvement has been carried out in the southern Alps, especially in a tributary of the Ubaye River in the Barcelonnette basin. In 1866, when the work started, about 65 ha was planted in a torrential basin of 2,200 ha. By 1985, 1,100 ha had been reforested, about 50% of the basin. More than 2,000 check-dams have been built. However, landslides have not been completely stopped. A slide with about 4 million m^3 of material in 1982 threatened several houses in the Valette area.

6.4 Monitoring and predicting landslides

Several unstable slopes in France are presently being closely monitored. In the Alps, for instance, movement in crystalline and calcareous rocks affects the border of Grand Maison reservoir. Cracks and signs of acceleration were observed in May 1986. The present monitoring measures deformation and hydrogeology and transmits the information by radio due to the inaccessibility of the site during the winter period. In Ville au Val, 30 km from Nancy, pore pressure is continuously recorded at the rupture surface level. In La Clapière, near Saint Etienne de Tinée, surface displacements have been used to predict that the whole slope may fail in spring 1989. A similar approach has been used for the landslide at Villerville and Cric-

queboeuf, on the coast of Normandy.

Another area monitored is the slope dominating route 91 between Grenoble and Briançon, between the villages of Vizille and Séchilienne, at a point called Les Ruines de Séchilienne. Since June 1985, crack-spacing has been watched closely by a network of 94 extensometers. A site at Sallèdes, ten kilometers from Clermont-Ferrand, was monitored from 1980 until 1986.

7 CONCLUSION

Many different kinds of landslides occur in several parts of France. The Alps and Pyrenees, which have the steepest slopes, have the greatest landslide problems. In 1983, landslide costs were at least $ US 1,634 million, they decreased in the following years. But the effect of landslides on the economy remains, most of years, moderate, approximately 0.005% of the Gross National Product in 1983, 0.001% in 1984 and 1985. Nevertheless at least 72 people were killed by landslides in 1970, and the landslide problem is now officially taken into account, like other natural hazards ; the 1982 law and the risk maps (PER) have apparently created good conditions for prevention. But the acceptation of PER's proves to be slow because of the delimitation of red zones, debated by a lot of communities.

REFERENCES

Antoine, P. & al 1987. La menace d'écroulement aux Ruines de Séchilienne (Isère). Paris, Bull. Liaison Labo. Ponts & Chaussées, 150-151: 55-64.

Antoine, P. & al 1988. Propriétés géotechniques de quelques ensembles géologiques propices aux glissements de terrain. In landslides, 5th Symposium on landslides, Lausanne, p. 1301-1306. Rotterdam, Balkema.

Arlery, R. 1979. Le climat de la France. Paris, Ministère des Transports, Direction de la Météorologie, 131 p.

Aste, J.P. 1983. Perfection et développement des méthodes inclinométriques. Bull. Assoc. Intern. Geol. Ingénieur, 26-27:5-15.

Azimi, C. et al 1988. Prévision d'éboulement en terrain gypseux. In landslides, 5th Symposium on Landslides, Lausanne, p. 531-536. Rotterdam, Balkema.

Bravard, Y. 1983. Catastrophes naturelles en Savoie. Trésors de la Savoie, 96 p.

Cazenave-Piarrot, F. et al 1984. Contrôle géologique et morphoclimatique des glissements de versants dans les Pyrénées occidentales. Documents du BRGM, 83:495-500.

Champetier de Ribes, G. 1987. La cartographie des mouvements de terrain. Des Zermos aux PER. Bull. Liaison Lab. Ponts & Chaussées, 151: 9-19.

Chardon, M. et al 1984. Géomorphologie et risques naturels dans les Alpes. 25e Congrès Intern. Géogr., In les Alpes, p. 13-41. Paris.

Cojean, R. & Gautier, P. 1984. Elaboration de cartes de sensibilité aux mouvements de terrain pour l'établissement de Plans d'Exposition aux Risques. In Mouvements de terrain, Documents du BRGM, 83: 154-161.

Combes, F. 1982. Un centenaire : le Grand Barrage Demonzey. Revue Forest. Française, 5: 80-86.

Combes, F. & Bartet, J.H. 1982. Plaidoyer pour le pin noir en Haute-Provence. Revue Forest. Française, 5: 40-49.

Courel, M.F. & Delaunay J. 1980. Les éboulements d'extension catastrophique. S.G.N. B.R.G.M. (80SGN 776GEG), 55p.

Delaunay, J. 1983. Carte des zones exposées à des glissements, écroulements, effondrements et affaissements de terrain en France. Mémoire B.R.G.M., 124, 19 p.

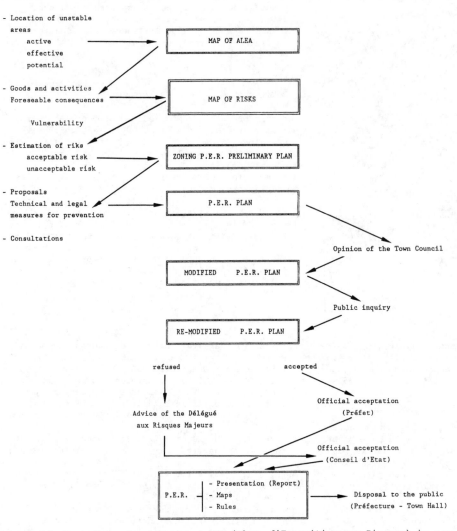

Figure 12. Procedures for adopting a PER (Plan d'Exposition aux Risques) in a community.(from Flageollet 1988).

Desfarges, J.P. & Lande, M. 1984. Mouvements de terrain en France et mutations économiques récentes en Andorre à la suite des inondations de novembre 1982. Documents du B.R.G.M., 83: 639-645.

Debelmas, J. 1979. Découverte géologique des Alpes du Nord. Editions du B.R.G.M., 84 p.

Dubie, J.Y. et al 1988. Télétransmission de l'auscultation d'un glissement : retenue de Grand'Maison, glissement du bilan. In Landslides, 5th Symposium on landslides, Lausanne, p. 399-404. Rotterdam, Balkema

Eisbacher, G.H. & Clague, J.J. 1984. Destructive mass movements in high mountains : hazard and management. Geol. Survey of Canada, Paper 84-16.

Direction de la Sécurité Civile 1980. Géographie Nationale des risques, 70 p. Ministère de l'Intérieur.

Faure, Y. et al 1988. Applications des géotextiles à la protection contre l'érosion de formations instables (Alpes). In Landslides, 5th Symposium on Landslides, Lausanne, p. 905-910. Rotterdam, Balkema.

Geze, B. & Cavaille, A. Aquitaine orientale. Guides géologiques régionaux. p. 175. Paris, Masson.

Escourrou, G. 1982. Le climat de la France. Que saisje n° 1967, 125 p. Paris, P.U.F.

Flageollet, J.C. & Helluin, E. 1984. Formations quaternaires et zonage des risques de glissements de terrain à Villerville et à Cricqueboeuf (Calvados). Documents du B.R.G.M., 83:173-183

Flageollet, J.C. & al 1987. Studies on landslides in Normandy (France) in view of their occurrence probability. Anzslide 87, Conference proceedings, p. 225-235.

Flageollet, J.C. 1988. Les mouvements de terrain et leur prévention. 224 p. Paris, Masson.

Follaci J.P. & al 1984. Crêtes doubles et perturbations de versants dans un domaine de montagne alpine (Mercantour et ses bordures). Documents du B.R.G. M., 83: 533-542

Follaci, J.P. 1987. Les mouvements du versant de La Clapière à Saint Etienne de Tinée (Alpes Maritimes). Bull. Liaison Labo. Ponts & Chaussées, 150-151:39-54

Follaci, J.P. et al 1988. Le glissement de La Clapière (Alpes Maritimes, France) dans son cadre géodynamique. In Landslides, 5th Symposium on landslides, Lausanne, p. 1323-1328. Rotterdam, Balkema.

Godefroy, T. & Humbert, M. 1983. La cartographie des risques naturels liés aux mouvements de terrain et aux séismes. Hydrogéologie et géologie de l'Ingénieur, Vol. 2: 69-90

Goguel, J. & Pachoud, A. 1981. Les mouvements de terrain du versant sud du massif de Platé, Haute-Savoie, France. 26e Congrès Géologique International, section 17. Bull. Liaison Labo. Ponts & Chaussées, spécial X: 15-26

Héraud, J. et Restituito, J. 1980. Dynamitage des ro-
chers de Coursavy, Cantal, France. 26e Congrès Géologique International, section 17, Paris. Bull.
Liaison Labo. Ponts & Chaussées, spécial X.

Heuzé, D. 1988. Vers une évaluation du coût des mouvements de terrain. Mémoire de Maîtrise de Géographie,
Université de Caen, France.

Hilly, J. & Haguenauer, B. 1979. Lorraine Champagne.
Guides géologiques régionaux, p.15. Paris, Masson.

Humbert, M., Vogt, J. & Delaunay, J. 1983. Le fichier
d'information sur les mouvements de terrain en France et ses applications. Hydrogéologie - Géologie de
l'Ingénieur, B.R.G.M., 2·

Loye Pilot, M.D. 1984. Coulées boueuses et laves tor-
rentielles en Corse : exemple de mouvements de terrain en pays méditerranéen montagnard. Documents du
B.R.G.M., 83 : 23-28

Magnan, S. 1983. L'assurance des catastrophes naturelles en France. Face au risque, 190, p. 18-24. Paris.

Margeat, H. & Michel, J.M. 1984. La loi sur les catastrophes naturelles. La Gazette du Palais, 13.11.84,
p. 500-512. Paris.

Marie, R.F. 1988. Les chutes de blocs dans le P.E.R.
de Gavarnie (Hautes-Pyrénées, France). In Landslides,
5th Symposium on landslides, Lausanne, p. 1197-1200.
Rotterdam, Balkema.

Marnezy, A. 1984. Phénomènes catastrophiques dans les
Alpes occidentales. Les Alpes, 25e Congrès Intern.
Géogr., p. 19-24. Paris.

Matichard, Y., & Pouget, P. 1988. Pluviométrie et comportement de versants instables. In Landslides, 5th
Symposium on landslides, Lausanne, p. 725-730. Rotterdam, Balkema.

Matichard, Y. & Pillard, J. 1988. Système d'auscultation en temps réel d'un versant instable. In Landslides, 5th Symposium on landslides, Lausanne, p.
459-462. Rotterdam, Balkema.

Maquaire, O. 1984. Etude du glissement du Bouffay (5
août 1981). Réflexions en vue de la prévision et de
la prévention. Documents du B.R.G.M., 83:29-40   .

Maquaire, O. & Gigot, P. 1988. La décompression à
l'arrière des falaises vives. Reconnaissance par pe-
tite sismique et rôle de l'instabilité. E xemple des
falaises du Bessin (France).Geodyn. Acta, s.p.

Mullenbach, P. 1982. Les reboisements au voisinage de
la limite altitudinale de la végétation forestière.
L'exemple de la Station du Chazelet. Revue Forest.
Française, 5 : 50-70

Pachoud, A. 1980. Influence de la disparition de l'ac-
tivité agricole traditionnelle sur la stabilité des
pentes en montagne. Risques géologiques, Mouvements
de terrain. 26e Congrès Géologique International,
section 17,p.49-54 Paris.

Perinet, F. 1982. Stations de sports d'hiver. Réflexions à propos d'un accident. Revue Forest. Française, 5 :99-111

Perrot, A. 1988. Cartographie des risques de glissement en Lorraine. In Landslides, 5th Symposium on
landslides, Lausanne, p. 1217-1222. Rotterdam, Balkema.

Peterlongo, J.M. 1978. Massif Central. Guides géologiques régionaux, p. 161. Paris, Masson.

Pincent, B. & al 1983. Mesure en place des mouvements
de versants naturels. Bull. Assoc. Intern. Géol.
Ingénieur, 26-27:107-111

Pontier, J.M. 1980. Les calamités publiques. Connais-
sances communales, p. 118 et suiv.. Paris, Berger
Levrault.

Prevoteau, M. 1982. L'indemnisation des victimes de
catastrophes naturelles. Rapport au Sénat, N° 275,
35 p. Paris.

Ramirez, A. et al 1988. Enseignements tirés de deux
écroulements par glissement couche sur couche en
terrain calcaire. In Landslides, 5th Symposium on
landslides, Lausanne, p. 1359-1362. Rotterdam, Balkema.

Richard, A. 1982. Rapport de la Commission des lois de
l'Assemblée Nationale Française, n° 718. Paris.

Rochet, L. 1987. Application des modèles numériques de
propagation à l'étude des éboulements rocheux. Bull.

Liaison Labo. Ponts & Chaussées, 150-151 : 84-95
Paris.

Rudel, C. 1982. La reforestation du bassin du Haut-
Vallespir dans les Pyrénées Orientales. Revue Forest.
Française, 5:20-31

Sauret, B. 1987. Coulées boueuses, laves torrentiel-
les. Bull. Liaison Labo. Ponts & Chaussées, 150-151,
p. 65-75.

Tazieff, H. 1983. Rapport au Président de la République. J.O. de la République Française, N° 4016, 58 p.

Tazieff, H. 1984. Rapport annuel au Président de la
République. J.O. de la République Française, N°
4030. 61 p.

Van Asch, Th.W. et al 1984. The development of landslides by retrogressive failure in varved clays.
Zeitschrift zür Geomorphologie, 4: 165-181.

Van Effenterre, C. 1982. Les barrages perméables de
sédimentation. Revue Forest. Française, 5: 87-93.

Van Genuchten, P.M.B. 1988. Intermittent sliding of a
landslide in varved clays. In Landslides, 5th Symposium on landslides, Lausanne, p. 471-476. Rotterdam, Balkema.

Varnes, D.J. 1978. Landslide types and processes. Highway Research Board Special Report 29, Nat. Acad. Sci.
544: 20-47.

Vibert, C. 1987. Apport de l'auscultation de versants
instables à l'analyse de leur comportement. Les
glissements de Lax-le-Roustit et Saint-Etienne de
Tinée. Thèse Ecole des Mines de Paris.

Vibert, C et al 1988. Essai de prévision à la rupture
d'un versant montagneux à Saint-Etienne de Tinée,
France. In Landslides, 5th Symposium on landslides,
Lausanne, p. 789-792. Rotterdam, Balkema.

Vie Le Sage, R. 1985. Rapport au Président de la République. J.O. de la République Française, 42 p.

Vogt, J. 1979. Les tremblements de terre en France.
Mémoire B.R.G.M., 96.

Weiss, E.E.J. 1988. Treering patterns and the frequency and intensity of mass movements. In Landslides, 5th Symposium on landslides, Lausanne, p. 481-
484. Rotterdam, Balkema.

Yvard, J.C. 1984. Pluviosités ét écroulements des falaises du Val-de-Loire. In "Mouvements de terrain",
Colloque de Caen - mars 1984. Documents du B.R.G.M.,
83; 453-458.

Landslides: Extent and Economic Significance, Brabb & Harrod (eds)
© 1989 Balkema, Rotterdam. ISBN 90 6191 876 6

Extent and economic significance of landslides in Spain

F.J.Ayala & M.Ferrer
Department of Environmental Geology and Geotechnics, Instituto Geológico y Minero de España (IGME), Madrid, Spain

ABSTRACT: Landslides are a serious problem in Spain, causing an estimated loss of $220,000,000 US per year. Projections indicate that the total loss for the next 30 years could exceed $6 billion. At least 117 people have been killed so far by landslides, most of them in an earthquake-triggered rockfall near Azagra (Navarra). Landslides in lignite mines, northwest Spain, required mitigation measures exceeding $30 million, and failure of coal, lead and zinc waste piles killed 17 people, put shafts out of service and polluted rivers. Submarine landslides have disrupted telephone services.

1 INTRODUCTION

After floods and erosion, landslides are the most serious geological hazard in Spain, causing an economic loss of almost 220 million US$ per year (Ayala and others, 1987a). Landslides, rockslides, and rockfalls occur mainly in mountainous zones and in eroded Tertiary basins, comprising about two thirds of Spain's land surface. These phenomena have been studied by many geologists whose contributions are summarized below (all maps and reports are in Spanish; items with an asterisk have an English legend or summary):

a) Maps and published papers

Small scale (1:1,000.000 to 1:100,000):
1:1,000.000 Map of mass movements in Spain (Ferrer and Ayala, 1987)*.
1:400,000 Instability of slopes in the Guadalquivir River Valley (Ayala and others, 1988a, in press)*.
1:200,000 General geotechnical map of Spain, issued by the Instituto Geológico y Minero (IGME); contains several sheets for each area, one of which shows landslides.
1:200,000 Slope stability in soft formations of the Madrid region (Rodríguez and others, 1986).

Medium scale (1:100,000 to 1:10,000):
For roads, several issues of the Servicio Geológico de Obras Públicas show problem areas at 1:100,000 scale.
For cities, fifteen geotechnical and geological hazards maps have been issued by the IGME. See, for example, the map of Granada by Ayala and others (1980a).

Large scale (larger than 1:10,000):
Symposium on "Instability of Slopes" (Alonso and Corominas eds., 1988).

Landslides in mining areas:
Lignite open pits of Spain (Ayala and others, 1987b)* and hard-coal open pits of Spain (Ayala and others, 1988b. In press)*.

Submarine slides:
IGME has published information about landslides on "Marine Geology sheets" at 1:200,000 scale.

Miscellaneous:
IGME "Economic and Social Impact of Geological Hazards in Spain" (Ayala and others, 1987a) discusses landslides and other geological hazards.

b) Data bases

The IGME is preparing a data base and a landslide catalog of Spain.

c) Institutional groups dealing with landslides

Instituto Geológico y Minero de España
Laboratorio de Carreteras y Geotecnia (CEDEX)
Servicio Geológico de Obras Públicas
Universities: Madrid, Barcelona, Valencia and Cantabria
Direccion General de Protección Civil.

2 GEOLOGY AND RELIEF

The geological structure of the Iberian Peninsula can be divided into four main groups: Precambrian rocks deformed before the Paleozoic and again in the Hercinian orogeny; Paleozoic rocks deformed during the Hercinian orogeny; Mesozoic and Tertiary rocks deformed during the Alpine orogeny; and Tertiary rocks not deformed in the Alpine orogeny. Spain is divided generally into the Hercinic region, formed by Paleozoic and Precambrian rocks in the western half of Spain (the Iberian Massif), and the Alpine region formed by two mountain chains, the Pyrenees in the north and the Betica Mountains in the south (figure 1). The remaining land consists of Tertiary basins and intermediate mountains. Some Paleozoic rocks also occur in the Alpine intermediate mountains.
Stratigraphy is of great importance in considering the landslide potential of

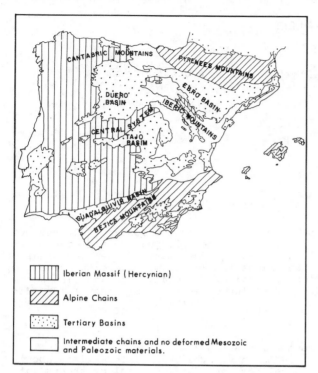

Figure 1. Main structural units of the Iberian Peninsula.

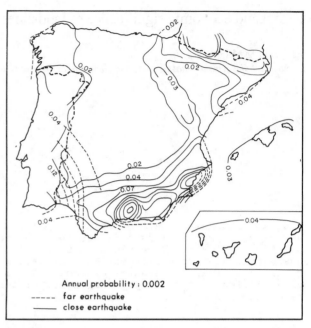

Figure 2. Map of Spain showing seismic acceleration in gravity acceleration units (Seismoresistent Rules Commission, 1987).

different geological materials. Variations in the composition, cementation, strength, hardness, porosity and permeability, and the rock structure help determine whether geologic units will fail or remain stable under various triggering conditions. Most of the mountainous areas in Spain consist of ancient materials which have undergone deformation and diagenesis; the large basins consist of Tertiary materials that generally have a horizontal or subhorizontal structure.

In general, rotational and translational landslides occur in soft formations, such as clay-rich and sand-rich sedimentary deposits of Tertiary age. Rockfalls and rockslides occur in hard, fractured, and compacted igneous and volcanic materials, such as those which constitute the mountain chains.

Materials in the Iberian Massif are mainly granite and other Paleozoic rocks, in places covered with thick sedimentary sequences and volcanic materials.

3 SEISMICITY

Most of the seismic zones are located in the Betica Mountains (Málaga, Granada, Almería, Murcia) and in the Pyrenees Mountains (Fig. 2). An earthquake triggered the Tivissa landslide (Tarragona), and a study in Granada City showed a strong relationship between neotectonic faults and other landslides (Ayala and others, 1980a).

4 CLIMATE

Two main characteristics define the climate of Spain: 1) influence of sub-tropical, maritime polar, continental, and Mediterrenean air masses; 2) a high average elevation, one of the highest of

any country in Europe. The socalled "green zone" of Atlantic climate is located in the northern part of Spain (Fig. 3), and the "brown zone" with continental and Mediterranean climate characterizes the rest of the country. Each of these zones is divided into different regions according to temperature, rainfall, influence of the sea, and orography as shown in Fig. 4. The maximum rainfall is along the northern coast in the "green zone", where it rains 120 to 180 days per year. Areas with continental climate have 40 to 120 days of rain per year, depending on orography; the Mediterranean region in the east and northeast, including Las Baleares Islands, have between 40 and 60 rainy days per year; and the southeastern part of Spain, constituting the more arid

Figure 3. Climatic Division of Spain (Font, 1983.)

area, has less than 20 days of rain per year.

Mediterranean climate is characterized by torrential rains. During the floods of November, 1987, the daily rainfall in some areas of the southeast surpassed the average annual rainfall. Rainfall is directly related to the frequency of landslides in Spain. The rainy winter--spring of 1988 was accompanied by a significant increase in the number of landslides; many of these landslides were reactivations of ancient movements.

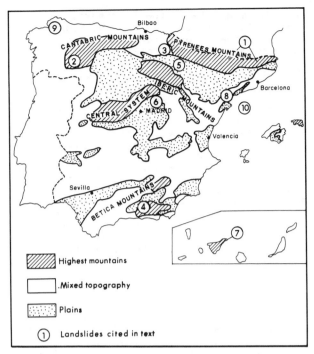

Figure 5. Relief of Spain and location of landslides cited.

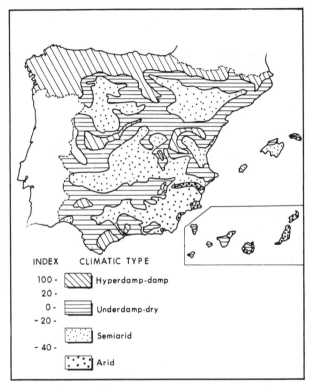

Figure 4. Simplified climatic map of Spain including the Thornthwaite index (modified from Justo and Cuellar, 1973, and Rodríguez, 1975).

5 TYPES AND DISTRIBUTION OF LANDSLIDES

The distribution of landslides cited in the text is shown on Fig. 5. The largest landslides with millions of cubic meters of rock or debris occur in the mountains in the northern, north-central and southern part of Spain. The maximum heights of these mountains is 2,500 to 3,500 meters. Rockslides, rock avalanches, debris flows, earth flows, and rock-block glides occur during heavy rainfall in these areas. Mud flows and earth flows associated with heavy rainfall periods have been recognized and studied in the Pyrenees (Corominas and Alonso, 1984). Large ancient landslides have been recognized there by Garcia Yague and Garcia Alvarez (1988). The largest mass movements have occurred in largely uninhabited areas so the economical and social impacts are limited to roads and railways.

On the relatively flat plains, landslides are restricted to areas cut by rivers or to the gentle slopes of valleys where differential erosion of limestone and

shale or limestone, shale and gypsum has produced rock ledges. Rockfalls and rock-block slides develop from these ledges. Rotational landslides are also common. Human-induced landslides can occur anywhere, even in relatively flat terrain if steep slopes are produced in the cut.

Heavy rainfall, weathering, decompression, ice or water pressure, and differential erosion contribute to rockfalls, rotational and translational landslides, mudflows, and debris flows involving moraine deposits in the Pyrenees (Corominas and Alonso, 1984). Mass movements in this area are commonly mudflows and debris flows, which is places involved tens of millions of cubic meters of material. Debris flows affect mainly moraine deposits and in several places

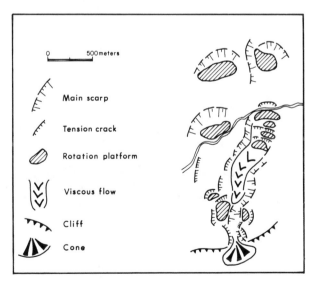

Figure 6. Sketch of a mudflow originating from rotational landslides (Corominas, 1984).

have diverted the course of rivers.

Most landslides with flow mechanisms in the Pyrenees originated as rotational landslides (Fig. 6). Water pressure at the toe of the slide mass increased, causing the material to flow downslope at speeds of tens of meters per hour. The clay materials involved in these flows have a plasticity index of less than 20 percent, allowing residual friction angles of about 18 degrees (Corominas, 1984).

Huge rockslides involving Paleozoic rocks and moraine deposits are also common in the Pyrenees. One example is a large ancient landslide more than 3 km long in the El Forn area (Andorra) between France and Spain (number 1 in Fig. 5). The slide occurred after a glacier retreated (Fig. 7), 20,000 to 10,000 year ago, when landslide material from faulted Paleozoic schist and from moraine deposits above the schists failed again, partly due to glacial erosion at the toe of the slope (Soutadé 1988).

Torre del Bierzo, is a rockslide that menaces a railway in the Iberian Massif, northwest Spain (number 2 in Fig. 5). The landslide involves folded and faulted Paleozoic shale, quartzite, and sandstone (Fig. 8) which had failed at some time in the past. The toe of this ancient landslide was reactivated recently when material around the train tunnel was removed, as well as the tunnel. Weight from coal waste in the upper part of the landslide probably contributed to the recent failure. The failure surfaces are mainly in intensely fissured shale 8 to 10 meters thick. The water table during the failure process is shown on Fig. 8. The part of the landslide which was reactivated covered approximately 250,000 square meters, had a volume of 3.8 million cubic meters and moved as much as 10 cm/month.

Remedial measures proposed (Ayala, Ferrer and Aparicio, 1988) include retrieval of mining water and anchorage. Anchorage would increase the safety factor from 1 (actual) to 1.05 at a cost of $8.5 million. The geotechnical parameters, in situ, calculated from back-analysis, are the residuals, $C' = 0$, $\emptyset' = 14°$. It may be necessary to move the railway out of the slide area.

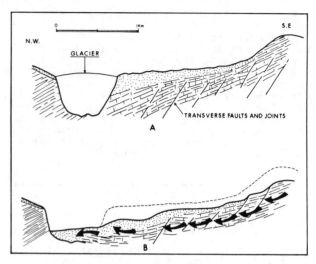

Figure 7. The landslide of "El Forn" in the Andorra region, Pyrenees (Soutade, 1988). A, Geologic cross-section of the El Forn area during maximum glaciation. Landslide deposits shown by dotted pattern was derived from bedrock and moraines. B, Recession of the glacier and formation of a series of slumps and slump/earthflows.

Inza is a complex landslide in Cretaceous flysch in the Cantabric Mountains (no. 3 on Fig. 5). The landslide occurred in 1714 when a heavy rainfall triggered a rockslide measuring one million cubic meters in carbonate rocks and weathered shale at the top of a hill in Inza. The rockslide turned into an earthflow which descended at an average speed of 20 m/day beyond the old village (Fig. 9). A back analysis with an infinite slope model (flownet parallel to the surface) showed that the geotechnical parameters controlling the slide were residual $C'= 0$, $\emptyset' = 25°$. Some of the weathered shale 4 to 10 meters deep, with plastic clay (LL = 59.1%) having high activity (A = IP/%, 2 m = 2.42) was also mobilized (Ayala, Aparicio, and Sanz, 1987). Many other areas of shale-rich flysch, such as at Azpiroz Mountain Pass in the same area, have similar instability problems.

In Olivares (Granada), a slump/earthflow

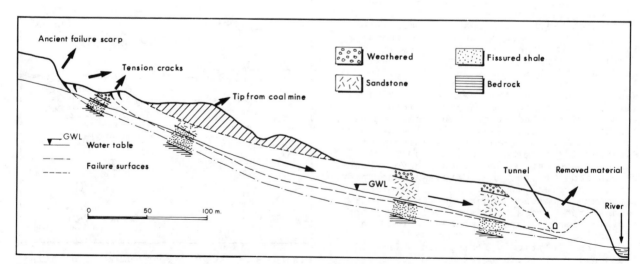

Figure 8. Torre del Bierzo landslide in sedimentary rocks, northwest Spain (Ayala, Ferrer and Aparicio, 1988).

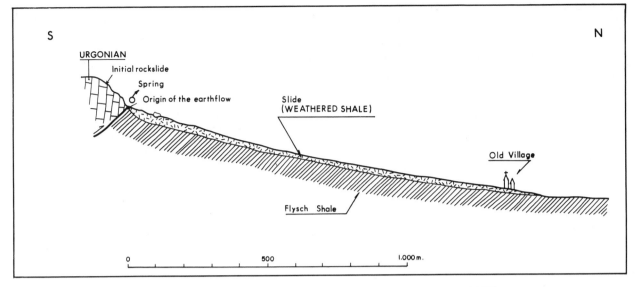

Figure 9.   Sketch of the Inza landslide (Ayala, Aparicio and Sanz, 1987).

in Miocene clay in April, 1986, alarmed
the population because of the possibility
of flooding if a natural dam were formed
at the toe of the slide. The landslide is
shown in Fig. 10 (locality 4, Fig. 5). The
clay has medium-high plasticity (LL =
33-60%), medium-high expansivity,
non-active (A = 0.61-0.81), and a moisture
content between 10% and 66% (Rodríguez and
Duran 1988). A linear relation C_u - LI (LI
= liquidity index) explains the flow due
to high moisture content. During the flow,
C_u was near 0.5 t/m², it was 6 t/m² after
dessication. This strength increase is
reached with a decrease of LI from 0.6 to
0.0 in altered specimens. The slide began
with a speed of 2 meters/hour and 13 days
later stabilized when the water drained

naturally.
  Rockslides in Paleocene gypsum in the
Ebro River basin occurred at Azagra
(Navarra, locality 5, Fig. 5), in 1874,
when one hundred people died. In the last
few years, rockslides have dammed the Arga
(Carcar) and Aragon Rivers (Milagro,
Falces) causing floods. The landslides are
related to the transformation of anhydrite
to gypsum with an increase in volume of
about 40%. The original stratification is
changed and the rock is converted into a
homogeneous mass. At other localities
where the gypsum is primary, rock-block
slides occur. In those areas, the
geotechnical parameters are about C' = 0
and ∅' = 20° corresponding to residual
values of thin clay layers interbedded
with gypsum (Ayala, Aparicio, and Conconi,
1988). Instability occurs in the form of
wedges or rotational slides with tension
cracks (Fig. 11). The cliffs, as much as
90 meters high, may have originated by the
hydration process (Del Valle, 1987). Toe
scour by rivers plays an important role
in the instability processes.
  Landslides in Guadalajara province in
the Tajo River area (locality 6, Fig. 5)
occur in sand, clay, gypsum, and

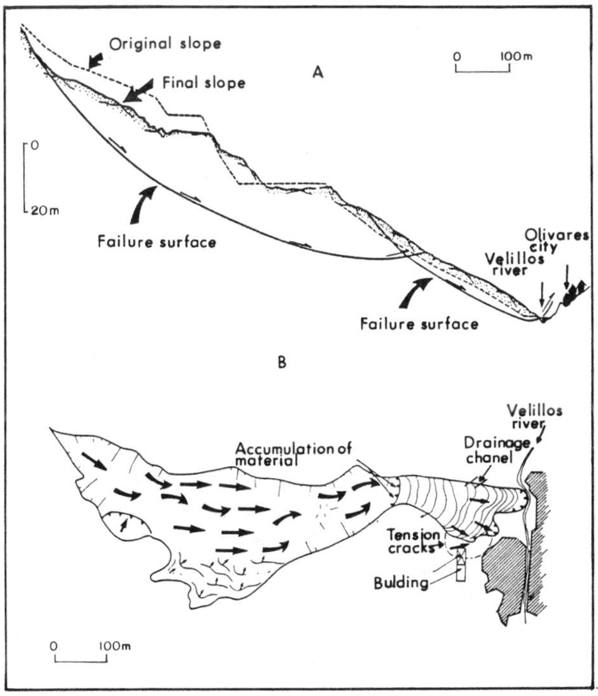

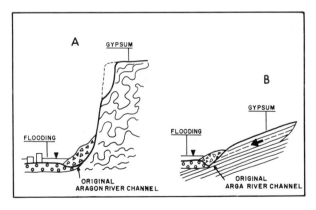

Figure 11. Landslides in gypsum in
Navarra. A, Milagro Area. Gypsum
transformed from anhydrite. B, Carcar
area. Slide in primary gypsum (Ayala,
Aparicio, and Conconi, 1988).

Figure 10. Sketch of the Olivares
landslide (Duran, 1986).

calcareous sediments of Tertiary age and fluvial origin. Most of the slumps are landslides on slopes underlain by clay-rich sediments in areas of active stream erosion (Martinez and Garcia Yague, 1988). The average slopes are between 17° and 27°, and the mean precipitation is about 500 mm/year. Fig. 12 shows a typical slump in this area. Nine large landslides with areas up to 2 km² and a volume greater than 10 million m³ have been identified in this area (Garcia Yague and Garcia Alvarez, 1988).

In Burgos City in the Duero River basin, many similar slides were reactivated during spring rains in 1988. One of these threatened a residence for the elderly. On the same slope several years ago, a slide damaged a road (Ayala and others, 1988c).

In Tenerife, Canary Islands (locality 7, Fig. 5), rockfalls and landslides occur mainly on the islands which have a thick layer of soil and volcanic formations formed in successive eruptions as late as 1971. The volcanic formations consist of lava and pyroclastic ash which form cliffs and steep slopes. Landslides also occur as rockfalls where the lava has extensive columnar jointing and easily-eroded ash beds, as shown on Figure 13. In December,

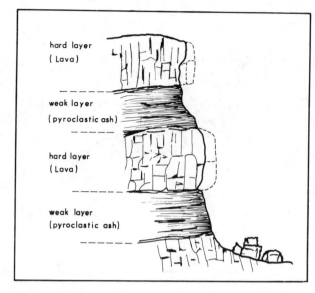

Figure 13. Rockfalls in jointed lava undercut by erosion of weak ash beds in the Canary Islands (Ferrer and Ayala, 1988).

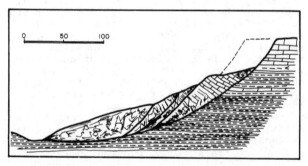

Figure 12. Representative landslide in Tertiary deposits of the Tajo River area, Guadalajara (García Yague and García Alvarez, 1988).

1987, a rockfall on the motorway of Santa Cruz de Tenerife - Las Teresitas Beach blocked traffic and required the evacuation of some buildings (Ferrer and Ayala, 1988).

In October, 1845, several earthquakes occurred in the region of Tivissa, Tarragona (locality 8, Fig. 5); the strongest had an intensity of 6 (MKS) and a unified magnitude of 5.2. Several landslides were triggered, the largest of which was a rock-block slide in Manou (Saenz, 1988). A limestone plate 20 m thick, 120 m long, and 60 m wide, with a volume of 120,000 m³ moved downslope (Fig. 14). The failure surface was in a layer of clay interbedded in the limestone. The strength of the clay could have been reduced by heavy rains in the region before the earthquakes. Back-analysis of the rockslide assuming a horizontal acceleration of 0.066g at the time of failure is C = 0, Ø = 25° to 30°.

6 LANDSLIDES RESULTING FROM MINING

The most important landslides are those related to brown lignite exploitations in

northwest Spain, Galicia region (locality 9, Fig. 5), where rainfall is about 1650 mm/year. In the Puentes de García Rodríguez Mine, with an annual production of 14 million tons, a slide of 450 million tons occurred in 1983 (Ayala and others, 1987b). Deep pumping and unloading of the head area to stabilize the slide cost more than 30 million US$. The rocks involved are Paleozoic phyllites, C' = 0, Ø' = 17°, Miocene clays, C' = 0, Ø' = 14°, and lignites. Similar landslides problems occur in the Meirama Mine in the same region.

Several failures of mining waste piles have killed people. Landslides in zinc waste piles at the Reocin Mine, Cantabria, in 1965, for example, killed 12 people. Similar failures at Ortuella, Basque country, killed 5 people. In the Mosquitera coal mine, Asturias, a slide of waste put the main shaft out of service. In Goizueta, Navarra, during a 1983 flood, a tailing dam broke and allowed lead-zinc waste to pollute the Urumea River that passes through San Sebastian City (Ayala and Del Valle, 1984).

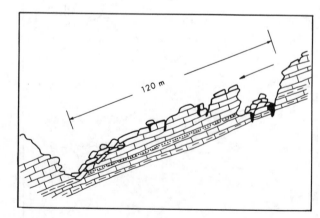

Figure 14. Tivissa rock-block slide, Tarragona, triggered by an earthquake (Saenz, 1988).

7 SUBMARINE SLIDES

Large mudflows (turbidites) occur in the Mediterranean Sea (-900 m) at the front of the Ebro delta (Fig. 15 and locality 10, Fig. 5). Five big mudflow complexes (lobes) with an area between 150 and 300 km^2 are the largest slides in Spain. Movement of sediments on the continental slope has produced serious disruptions of telephone cables to the Canary Islands.

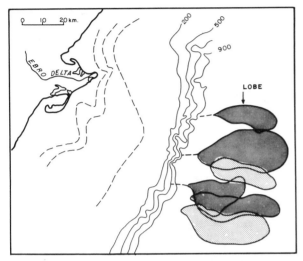

Figure 15. Submarine landslides offshore from the Ebro delta. Depth contours in meters. (Lamont-Doherty & IGME, 1985).

8 ECONOMIC SIGNIFICANCE OF LANDSLIDES

In order to determine the economic significance of geological hazards in Spain, the Geological and Mining Institute has carried out a project, (Ayala and others, 1987a) using the methodology of the "Master plan for California" (Alfors, Burnett, and Gray, 1973) with several of its criteria modified and adapted to Spanish conditions. The main goal of the project has been to estimate total expected losses for the period 1986-2016 (30 years). For landslides, the expected losses are 6.6 billon US$ for this 30 year period. This estimate is for landslide damage to urban areas, roads, and railways including direct damages and stabilization costs. The damages were estimated in the following manner:

1. The landslide hazard and loss in urban areas was estimated for all of the 1:50,000 quadrangles in Spain (Fig. 16). The degree of hazard is expressed numerically:

        1. None
        2. Low
        3. Medium
        4. High

2. Losses per event PS (pérdidas por suceso, US$/event) are determined:
PS = CG x PA x CP x FC where CG = Geological cost coefficient, PA = Persons in the area, CP = Proximity coefficient and FC = Catastrophe factor. CG measures the losses predicted per person and event (US$/person x event).
The values are:

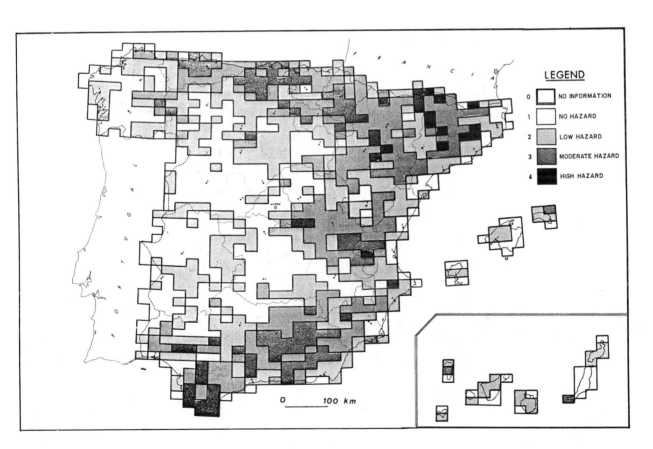

Figure 16. Landslide hazard in Spain (Ayala and others, 1987a).

Hazard level	CG (US$/person x event)
1	0
2	1.38
3	48.34
4	73.66

To determine PA, all populations of more than 1,000 persons have been considered. CP measures the impact of an event on the 1:50,000 scale maps for each population. The values are:

0:	Population out of risk
0.5:	Partially in risk
1:	Inside the risk

The total population exposed to landslides risk are:

Hazard level	Population exposed	% Spanish Population
Low	3,560,741	9.25
Medium	2,129,256	5.53
High	1,827,508	4.75
Total	7,517,505	19.5

FC reflects the costs due to loss of human life and the effects of catastrophes on a regional scale. The final values have been determined from those considerated by Alfors, Burnette, and Gray (1973) fitting with real values of human losses in historic events in Spain. These final values are, for the different levels of hazards considered:

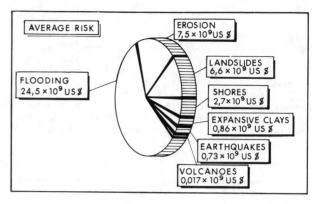

Figure 17. Expected losses from geological hazards in Spain from 1986 to 2016 (Ayala and others, 1987a).

Hazard level	FC
1: None	1
2: Low	1
3: Medium	1.1
4: High	1.5

3. The total losses PT (pérdidas totales, US$) for 30 years are then determined:

PT = PS x FS where FS = Frequency of events. For landslide risk, FS = 1/500.

PT has been computed for each 1:50,000 quadrangle. PT is 72.2 million US$.

Landslide losses for railways and roads

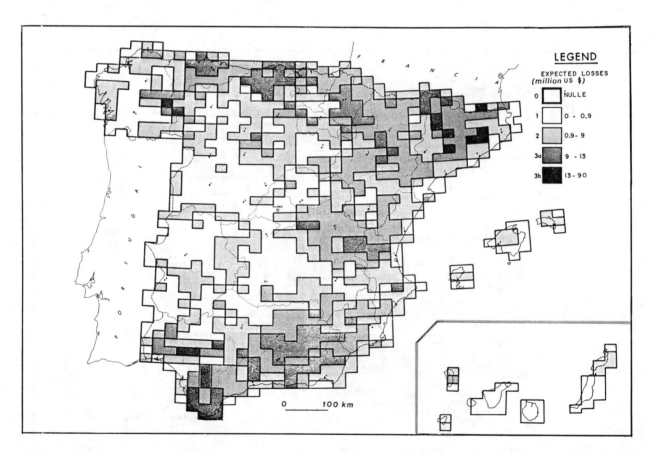

Figure 18. Expected economic losses for landslides in Spain for the period 1986 to 2016 (except for motorways and railways) (Ayala and others, 1987a).

176

are much more important. For railways, the National Company RENFE, has reported an annual landslide loss of 43.5 US$. The losses in roads have been estimated as one-third this amount per kilometer, but the length is almost seven times greater (80,000 km versus 12,710 km). The total expected losses are 4.9 billion US$ for 30 years without economic growth and 6.6 billion US$ with 2% economic growth. The annual loss per person is 5.7 US$; the annual loss is 0.08% of the Spanish Gross National Product.

The importance of landslides compared to other geological hazards is shown on Fig. 17. Landslides are 15.4% of the expected losses, third in rank after floodings and erosion.

The geographical distribution of expected landslides losses is shown on Fig. 18.

Landslides (especially rockfalls) also kill people. In 1955, a rockfall in Huelva city killed 9 people (Ayala and others, 1980b). The Azagra rockfall mentioned previously killed 100 people. Based on information compiled by Kates (1978) and an Insurance Company, Munchener Ruck (1978), about 70 people are expected to be killed by landslides in Spain in the next 30 years.

AKNOWLEDGMENTS

The authors wish to acknowledge and thank Dr. José Medialdea, Head of Marine Geology, for his information about submarine landslides, Carmen Carabias and María Teresa Navajo for typing, and Luis Sainz Trápaga and Esteban de la Cruz for making the figures. All of these people work for the IGME.

REFERENCES

Alfors, J., J.L. Burnett & T.E. Gray 1973. Urban geology master plan for California. Bull. Calif. Div. Mines and Geology, n. 198.

Alonso, E. & J. Corominas (eds.) 1988. II Symposium on unstable slopes, Andorra. In Spanish.

Ayala, F.J. & others 1980a. Geotechnical map of Granada city, scale 1:25,000. IGME. In Spanish.

Ayala, F.J. & others 1980b. Geotechnical map of Huelva city, scale 1:25,000. IGME. In Spanish.

Ayala, F.J. & J. del Valle 1984. Analysis of the failure of Goizueta tailing dam and the contamination produced in the Urumea river. International Simposium on Mining and Metallurgy. Barcelona. In Spanish.

Ayala, F.J., V. Aparicio & E. Sanz 1987. Analysis of the Inza landslide (Navarra) in 1714-15. Bull. Mining and Geological Institute of Spain. T. 98. IGME. In Spanish.

Ayala, F.J., & others 1987a. Economical and social impact of geological hazards in Spain. IGME. In Spanish (*)

Ayala, F.J., & others 1987b. Slope stability in lignite open pits of Spain. IGME. In Spanish (*).

Ayala, F.J. & others 1988a. Slope stability in the Guadalquivir River Valley. IGME. In press. In Spanish (*)

Ayala, F.J. ,& others 1988b. Slope stability in hard-coal open pits of Spain. IGME. In Spanish (*).

Ayala F.J. and others, 1988c. Study and stabilization of a landslide affecting a residence for the elderly in Fuentesblancas (Burgos). Unpublished. In Spanish.

Ayala, F.J., V. Aparicio & G. Conconi 1988. Study of the unstable slopes in the gypsum cliffs of Navarra. II Symposium on unstable slopes. Andorra. In Spanish.

Ayala, F.J., M. Ferrer & V. Aparicio 1988. Report on the landslide affecting the railway Palencia-La Coruña, in Torre del Bierzo (León). Unpublished. In Spanish.

Corominas, J. 1984. On the formation of mud-flows in the Catalan Pyrenees. I Symposium on slope instability in the Pyrenees. U.P. Barcelona. In Spanish.

Corominas, J. & E. Alonso 1984. Slope instability on the Catalan Pyrenees: causes and typologies. I Symposium on slope stability. U.P. Barcelona. In Spanish.

Del Valle, J. 1987. Halocinetic structures and geological hazards. I Course on Geological Hazards. IGME. In Spanish.

Duran, J.J. 1986. The Olivares landslide. The geologist, number 20. In Spanish.

Ferrer, M. & F.J. Ayala 1987. Map of mass movements in Spain, scale 1:1,000.000. IGME. In Spanish (*).

Ferrer, M. & F.J. Ayala 1988. Study and stabilization of rockfalls affecting a highway in Tenerife (Canary Islands). IGME. Unpublished. In Spanish.

Font, I. 1983. Climatology of Spain and Portugal. National Institute of Meteorology. Chap. 1 and 5. In Spanish.

García Yague, A. & J. García Alvarez 1988. Big landslides in Spain. II Symposium on unstable slopes. Andorra. Ed. by Corominas and Alonso. In Spanish.

IGME. Collection of Geotechnical General Maps of Spain, scale 1:200,000. In Spanish.

IGME. Collection of Marine Geology Maps of Spain, scale 1:200,000. In Spanish.

IGME 1986. Map of risks from expansive clay in Spain. IGME. In Spanish*.

Justo, J.L. & V. Cuellar 1973. Thornthwaite index map of Spain. Soil mech and transport lab. Bull. 89. In Spanish.

Kates, R.W. 1978. Risk assessment of environmental hazard. Scope 8. John Wiley & Sons, New York.

Lamont-Doherty Geological Observatory & Instituto Geológico y Minero de España, 1985. Submarine fans of the Ebro River Delta. Unpublished Report.

Martinez, J.M. & A. García Yague 1988. Slope instability in the continental Miocene: Matayeguas river, Tajo River Tertiary basin. II Symposium on unstable slopes. Andorra. In Spanish.

Munchener Ruck (Insurance Company) 1978. World Map of Natural Hazard. Munchener Ruckversicherungs-Gesellschaft, Munich (R.F.A.).

Rodríguez, J.M. 1975. Study of the expansive clays. Soil Mech. & Transport Lab. Bull. 108. In Spanish.

Rodríguez, J.M. & others 1986. Slope stability in soft formations in the Madrid region. IGME. In Spanish.

Rodríguez, J.M. & J.J. Duran 1988. The Olivares landslide (Granada), April

1986. II Symposium on unstable slopes. Andorra. Ed. by Alonso and Corominas. In Spanish.

Saenz, C. 1988. A landslide provoked by the Tivissa earthquake. II Symposium on unstable slopes. Andorra. Ed. by Alonso and Corominas. In Spanish.

Seismoresistant Rules Commission 1987. Provisional unpublished report. In Spanish.

SGOP. Collection of maps for studies of roads and highways, scale 1:100,000. In Spanish.

Soutade, G. 1988. El Forn landslide, Andorra. II Symposium on unstable slopes. Andorra. Ed. by Alonso and Corominas. In French.

Landslides: Extent and Economic Significance, Brabb & Harrod (eds)
© 1989 Balkema, Rotterdam. ISBN 90 6191 876 6

Landslides in Portugal – Extent and economic significance

L. Fialho Rodrigues & A. Gomes Coelho
Laboratório Nacional de Engenharia Civil, Lisbon, Portugal

ABSTRACT: This paper examines the landslide situation in Portugal including the Atlantic islands of Azores and Madeira. In spite of its small area (89 000 sq.km) Portugal presents considerable contrasts concerning both geology, physiography, rainfall and population, which strongly influence the geography distribution, frequency and severity of landslides. In Portugal landslides rarely reach imposing dimensions and only very few cases can be regarded as purely natural geological hazards. The majority of landslides have been triggered by human activity although always related to periods of continued rainfall. For these reasons landslides are of higher significance in the western mesozoic sedimentary borderland, a region of lower relief energy but where the majority of important construction activities for housing, roads and highways have been developed. Procedures for hazard zonation and prevention of landslides have been developed only for few particular areas of increasing urban growth where the majority of landslides are reactivated and controlled by the residual strength of quaternary clayey solifluction deposits over Jurassic marls. Although losses due to landslides are very dispersed and difficult to quantify it can be said that annual costs are significant if direct and indirect costs are considered.

1 LOCATION

Portugal is located at the southwestern end of Europe on the West side of the Iberian Peninsula (figure 1). The territory with a total land area of 88 607 sq.km has a rectangular shape extending for a maximum of 561 km in length and varying in width between 112 and 219 km. Besides the continental land, the Portuguese territory also comprises the Atlantic archipelagos of Azores and Madeira.

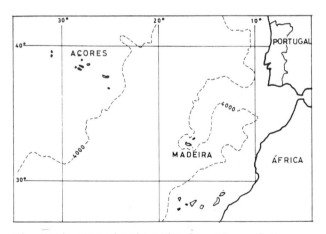

Figure 1. Map showing the location of Portugal and Portuguese Atlantic Islands.

In spite of its small area the continental territory shows considerable contrasts concerning both geology, physiography climate and population distribution which strongly influence the occurrence, spatial distribution, frequency and severity of different types of instability in natural slopes.

The 1986 census gave the population as 10 240 000 including Madeira (270 000) and Azores (253 500). The density of population averages 109,2 per sq. km. There are however great irregularities of distribution. The area of higher demographic concentration is the northwestern and central coastal zone within which lie the metropolitanean areas of Lisbon and Porto presenting densities from 600 to 750 per sq. km.

2 PHYSIOGRAPHY

Portugal has not extensive mountainous areas of rugged and high relief, the elevation being less than 400 m above sea level over 72% of its entire area (figure 2). There is a natural contrast from north to south stressed in the Hercynian Massif by higher altitudes in the north. The mountains of the northwest (Gerez-1507 m, Larouco-1525m, Marão-1415 m) and the Central Cordillera (Estrela-1991 m, Açor-1418 m, Lousã-1204 m) where granites predominate intruding metamorphic rocks are the highest peaks of a mountainous plateau dissected by deep valleys.

In the south, occupying nearly a third of the total land area of Portugal, lies a gently undulating open surface of low altitude (< 200 m) scarcely interrupted by minor rises such as Serra do Caldeirão (575 m) and Serra de Monchique (902 m), near the Algarve border.

Steep-sided valleys prevail in the northern and central mountainous regions in the Precambrian and Paleozoic schistous metamorphic rock masses crowned by quartzitic crests. Granitic terrain generally form high plateaux often rugged in places especially near rivers where scarps may be found.

The western and southern sedimentary borderlands have generally a low relief of gently rolling hills with some steep natural slopes often related to small escarpments of cuestas formed by the outcrop of resistant layers in monoclinal structures. Here, high steep slopes and scarps are only common in the naked karst topography of limestone regions.

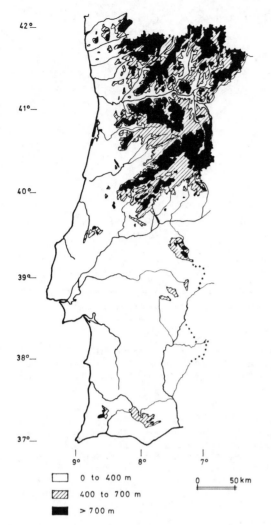

Figure 2.Map of Portugal showing physiography.

0 to 400 m
400 to 700 m
> 700 m

0   50 km

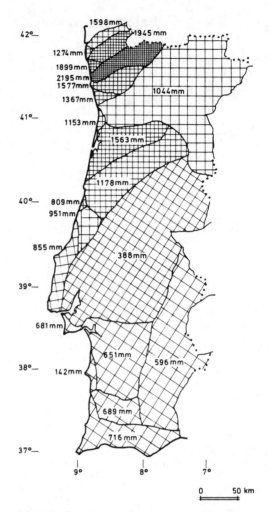

Figure 3. Mean annual precipitation in the Portuguese hydrographic basins (after Len-castre, 1984).

0   50 km

3 RAINFALL

Portugal has a Mediterranean regime strongly influenced by its Atlantic position. The Mediterranean influence is more remarkable in the south characterised by mild winters with low precipitation and dry and hot summers. North of Tejo river the Atlantic exerts considerable influence increasing humidity, moderating temperature and reducing the period of summer drougth.

Mean annual rainfall presents variations as a function of latitude, altitude and distance from the sea. The mountainous regions of the northwest and the Central Cordillera with mean annual rainfall above 1000 mm contrast with the rest of the country. Maximum va-lues have been recorded in the central zone (1991 m) with 2500 mm and in the northwest (1508 m) with 3500 mm. On the contrary, the northeast region presents low precipitation showing the barrier efect of the north-western and central mountains between the Atlantic influence and the interior.

Figure 3 shows mean annual precipitation in the Portuguese hydrographic basins. Annual rainfall averages 959 mm for the whole country ranging from 596 mm in the South to 2195 mm in the northwest.

Most of the rain falls in Autumn and Winter (figure 4). Precipitation increases after September reaching maximum values from No-vember to March.

Short term intensities above 110 mm in a

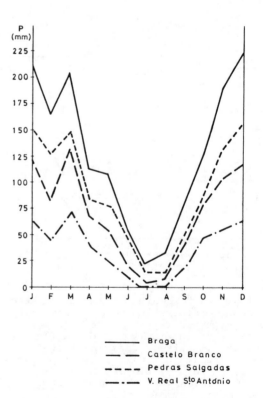

Figure 4. Monthly distribution of rainfall.

Braga
Castelo Branco
Pedras Salgadas
V. Real Sto António

24 hour - period are rare. The maximum intensities may be found in three regions (Daveau, 1972): the mountains of the Nort-West and their extern foreland, the Central Cordillera, the southern uplands of Algarve and the lowlands of eastern Algarve (figure 5). In the northwest heavy daily rains may fall throughout the whole year although they are more frequent during the winter. In the southern region of Portugal, they only appear in winter and the day of heaviest rainfall in the year may represent a high proportion of the annual precipitation.

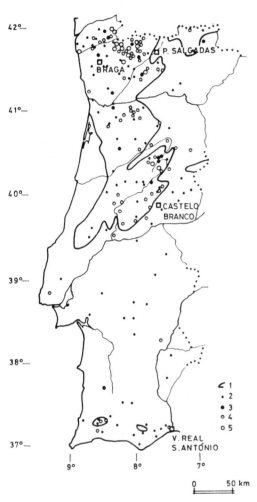

Figure 5. Location of places which have recorded precipitation values above 110 mm per day in the period 1931-60 (after Daveau, S., 1972) (1 - annual isohyet of 1000 mm; 2 and 4 - stations having recorded 110 to 180 mm per day, 3 and 5 - stations having recorded more than 180 mm per day).

Computed values of maximum intensities recorded in Lisbon area for a period of 108 years (Lencastre, 1984) have shown that every 10 years may fall 24,2 mm in 30 min. and 33,4 mm in 1 hour; every 50 years these values reach 31,5 mm and 49,7 mm respectively. However, in November 1967 castastrophic floods in Lisbon area were caused by an excepcionally heavy rainfall of 60 mm in a 60 min-period and 129,7 mm in a 6 hour-period.

Slope failures generally occur in winter from December to February after heavy rains and there is a clear relationship between rainfall and frequency of landslides. Slope failures tend to occur whenever the rains continue to increase gradually saturating larger an deeper areas.

4 GEOLOGY AND LANDSLIDE-PRONE UNITS

The territory may be divided into three major structural units as shown in figure 6:

- Hercynian Massif
- Western and Southern sedimentary borders
- Lower Tejo and Sado Basins

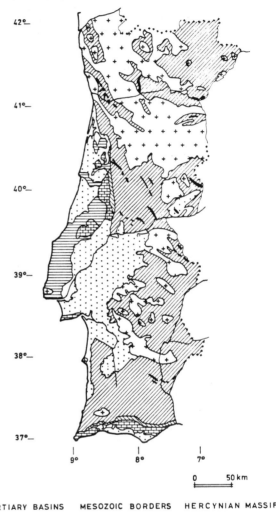

Figure 6. Map of Portugal showing Geology
1 - mesozoic sediments, 2 - limestones, 3 - - sandstones and marls 4 - granitic rocks, 5 - Metamorphic rocks, 6 - quartzitic crests.

4.1 Hercynian Massif

This area represents the Precambrian-Paleozoic stable craton of the Iberian Peninsula. All the rocks within this area were deformed and metamorphosed to varying degrees by the effects of the Hercynian orogeny which ended in Late Paleozoic time. Main geological formations consist of metamorphic rocks, intruded by large irregularly shaped masses of post - -orogenic granitoids.

Within this area steepness is a key factor controlling the location and extent of landslides which affect both granitic and metamorphic rocks.

4.1.1 Granitic rocks and residual soil

Granitic rocks are only modestly involved in slope failure with the exception of falls sometimes affecting steeper slopes in fractured zones. Steep slopes and scarps along valleys in sound granite present generally paralel jointing and exfoliation linked with pressure reliese and temperature changes. These slopes may be affected by rockfalls due to unfavourable orientation and density of discontinuities and weathering along them.

As a rule, depth of residual soil does not exceed 15 to 20 m. The zones of weathered granite typically occur along the surface of upland plateaux and under alluvial terraces. Deep residual soil may also occur in slopes along tectonized zones as well as in zones affected by deuteric alteration (kaolinization) owing to hydrothermal processes.

Generally the base of slopes in residual soil is covered by colluvial deposits con - sisting of fine-grained weathering products and rock fragments accumulated by rainwash.

Slope instability in residual soil and colluvium is frequently related to gully erosion caused by intense run -off mainly in zones affected by desforestation or earth - works for construction. Shallow debris slides have also occurred in slope undercut by river erosion and human activity. Although infrequent, at least one case of earthflow occurred in the north of Portugal. Heavy rainfall triggered the movement of a loose mass of colluvium which as a narrow flow reached the foot of the slope destroying one house and causing three casualties.

4.1.2 Metamorphic rocks

Precambrian and Paleozoic metamorphic rocks (schist, phyllite, metagraywake and quartzite) are strongly deformed and fractured forming near the surface a closely jointed rock mass. A common feature of steeper slopes in these anisotropic rock masses is the terminal bending or downward curvature of layers caused by creep and solifluction, particularly when the surfaces of layering, shaly cleavage or foliation have high angles of dip. These slopes are frequently disturbed in excavation works for roads and dams because there is a proneness to slide on the surfaces of overturned layers.

The steep slopes of large valleys have a regular morphology as they are almost always covered by colluvial deposits extending downslope and levelling out any depression in bedrock. These colluvial deposits consist of little weathered heterometric platy rock fragments embedded in a clayey matrix.

The schist mountains of the Central Cordillera and North of Portugal are frequently disturbed by slope movements involving both surface deposits and the disrupted and sheltered zones of the bedrock.

Another type of slope instability affecting these metamorphic formations, very frequently triggered by excavation works for road construction, is related to rock and debris slides along predisposed surfaces of layer - ing, cleavage or schistosity, in the case of downslope dipping structures. This type of planar translational slides occur predominantly in long slopes presenting low to medium dip angles of the underlying rocks.

4.2 Western and Southern sedimentary borderlands

Geological formations in the western border represent the remnants of a larger basin formed and deformed during Mesozoic and Cenozoic time. Deposition in the basin continued until the end of Cretaceous when basic intrusives were emplaced in the areas of Sintra, Sines and Monchique. In the South, the Algarve border consists primarily of Mesozoic rocks which are deformed by east- -west trending structures.

Within these units, landslides are tipi - cally controlled by lithology, geological structure and hydrogeological conditions, steepness being a secondary factor. On the other hand, it must be observed that coastal sedimentary borderlands coincide with zones of higher concentration of people and economic activity where the majority of important construction activities for housing, roads and highways have been developed. For these reasons, landslides and other mass movements have been relatively more frequent in these areas of lower relief energy than in the mountainous regions of the Hercynian Massif.

In the mesozoic borders slope instability can be related basically to the following landslide-prone units:

- Upper Cretaceous sequences of marls, clays, sands and sandstones;

- Middle Cretaceous sequences of lime - stones and marls;

- Upper Jurassic marly-calcareous sequen - ces

4.2.1 Upper Cretaceous sequences of marls, clays, sands and sandstones

These formations constitute a relevant part of the Late-Cretaceous sequences of the Pombal - Condeixa - Soure and Aveiro areas in the western sedimentary border. Litholo - gical complexes gathered into this unit consist of marine-marginal and continental terrigenous detrital deposits (silty sands, sandstones, marls and clays) often present- ing trough-cross bedding and flaser cross bedding arrangements. These formations may be affected by rotational and complex land - slides closely connected with high pore- -water pressures within sandy layers or lenticles lying between or alternating with clayey beds. An example of slope failure in these formations is the Condeixa landslide caused by excavations carried out for the construction of the Lisboa - Porto Highway (figure 7).

4.2.2 Middle Cretaceous sequences of limestones and marls

Middle Cretaceous formations ranging from Cenomanian to Turonian are widespread in the areas of Lisbon - Cascais - Ericeira and Nazaré - Leiria - V.N. de Ourém, in the western sedimentary border. These formations are composed of limestones (bioclastic li - mestones, cherty limestones, reef limestones marly limestones) with interbedded layers of marls and clays. Their conditions differ according to their lithological characteristics, geological structure (bedding,jointing and faulting) and occurrence of interbedded clayey layers along which failure and sliding can take place. The majority of landslides are generated along clayey bedding planes dipping downslope. These planar slides have been generally set in motion as a result of slope undercutting by excavation.

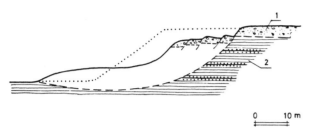

1 _ SAND AND GRAVEL

2 _ CRETACEOUS CLAYEY SANDS AND CLAYS

Figure 7. Condeixa landslide.

This type of slope failure is exemplified by the Monsanto landslide which occurred after the construction of the highway to the Tejo Bridge in Lisbon (figure 8). The Monsanto landslides occurred in a slope formed by a complex of Cenomanian marly limestones, marls and clays, capped by Turonian reef limestones, dipping about 7° downslope. The slipped rock mass of 300 000 m^3 volume severely damaged and interrupted the new road over a distance of 250 m. The slide was generated along a gypseous clay layer interbedded in dolomitic and marly limestones. Undercutting for road construction was carried out in 1964. First signs of movement were noticed in the begining of 1967. The final abrupt slide occurred one year latter in February 1968. The slide is thought to have been caused by long-term deformation, strain softening and progressive failure along the gypseous clay layer.

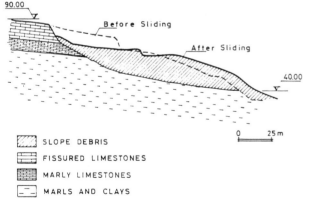

SLOPE DEBRIS

FISSURED LIMESTONES

MARLY LIMESTONES

MARLS AND CLAYS

Figure 8. Monsanto landslide.

4.2.3 Upper Jurassic marly-calcareous sequences

The most important zone within the western sedimentary borderland which is prone to landslides is the Alhandra - Vila Franca de Xira area, 15 to 20 km north of Lisbon, along the rigth bank of Tejo river, consisting of

Upper Jurassic formations ranging from Lusitanian to Kimmeridgian. Within this unit four lithological complexes have been recognized (figure 9):

- Freixial Complex - composed of interbedded marly sandstones and marls;

- Pterocerian Complex - composed of limestones interbedded with marly sandstones and marls;

- Amaral Complex - composed of corallic limestones and poorly consolidated marly limestones interbedded with marls;

- Abadia Complex - composed of a thick recurrent sequence of marly and clays with sparse interbedded micaceous sandstones.

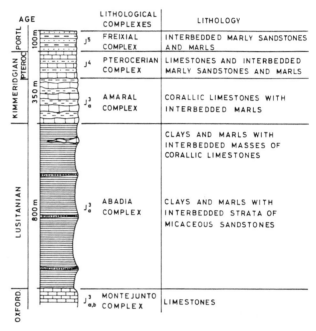

Figure 9. Lithostratigraphic complexes of upper Jurassic formations.

Morphological features of the area are a consequence of the contrast between upper limestone beds and the soft underlying argillaceous layers of the Abadia complex. Hard limestones form a caprock occurring close to the crest of scarps, the lower part of slopes being concave or rectilinear and formed by soft clays and marls covered by a mantle of colluvial deposits. These slope debris are composed of small and coarser fragments of hard limestones embedded in a clayey matrix produced by Pleistocene solifluction.

Perched groundwater may occur in the upper limestones after periods of continued precipitation. Confined groundwater is generally present in jointed permeable sandstones and limestones interbedded in impermeable clays and marls.

In this area the most frequent slope failures affect mainly the clayey quaternary solifluction deposits where landslides may occur even in gentle slopes. Evidence of creep in these deposits may be seen everywhere mainly as a consequence of wetting and drying of clay. Generally, continuous creep gives rise to microscarps and a rapid landslide may occur especially if unloading of the toe of the slope takes place or after a period of continued rainfall. These land-slides are generally shallow, characterized by low depth/length ratios and controlled by the thickness of colluvium which varies between 2 and 8 m. Very often the movement develops as a slide in the head area passing downslope into a mudflow (figures 10, 11 and 12). The groundwater issuing from the under-lying water bearing layers of the bedrock is an important contributing factor towards in-creasing pore pressures in colluvium and flowing of the saturated clayey soil.

In this area there is also much morpholo-gical evidence of translational slides of limestones and marls along bedding planes which sometimes evolve in elongated flows. They leave peculiar arched scars with flat bottoms in the dipslope side of valleys.

These translational slides are a con-sequence of seepage along contacts between limestones and interbedded marls and under-cutting of the slope by stream erosion as occurred in the Roucas landslide (figure 12).

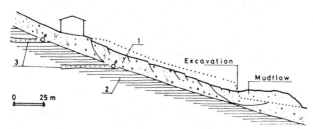

Figure 10. Bom Retiro landslide.

1 _ COLLUVIUM OF CLAY AND LIMESTONE DEBRIS
2 _ JURASSIC SHALY MARLS
3 _ INTERBEDDED MICACEOUS SANDSTONES
♂ SPRING

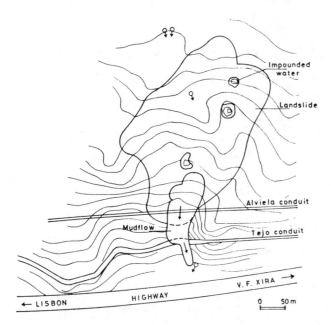

Figure 11. Roque - Anés landslide.

1 _ CLAYEY COLLUVIAL DEPOSITS
2 _ MIOCENE CLAYEY SANDS
3 _ JURASSIC LIMESTONES
4 _ JURASSIC SHALY MARLS
♀ _ SPRING

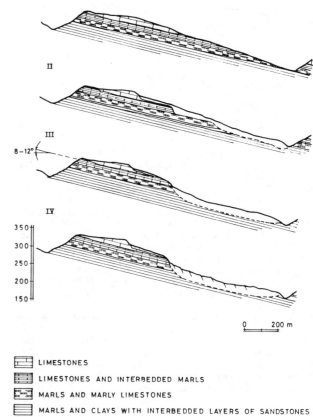

Figure 12. Roucas landslide.

LIMESTONES
LIMESTONES AND INTERBEDDED MARLS
MARLS AND MARLY LIMESTONES
MARLS AND CLAYS WITH INTERBEDDED LAYERS OF SANDSTONES
SLOPE DEBRIS

There is also much evidence of slope move-ment caused by slow plastic deformation of soft marls and clays. In some zones, close to the edge of limestone caprock it may be seen that blocks of limestone separated from the scarp moved downwards sinking into the underlying soft marls (figure 13). This slow process of block sliding is important since it causes the disturbance of marls in slopes to a certain depth. It must be observed that the majority of landslides in this area are caused by reactivation of ancient landslides. In February 1979, following a period of con-tinued rainfall three extensive landslides and innumerable other small mass movements occurred in the area. In every case they were located in zones where ancient landsli-des had taken place as was clearly identi-fied in aerial photography.

4.3 Lower Tejo and Sado basins

This unit makes up much of the western--central Portugal and is a Tertiary basin

184

which started being formed during the Oli -
gocene. The Tertiary formations consist of
thick sequences of continental and marine
deposits of Oligocene, Miocene and Pliocene
age. Within this unit landslide distribution
is restricted to a few singular areas and
appears to be influenced by slope steepness
and lithology. Two main prone-units may be
considered:

- Miocene overconsolidated clays of Lisbon
  region

- Miocene continental deposits of Santarém

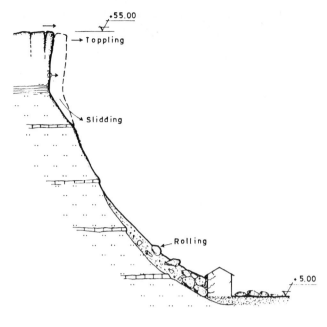

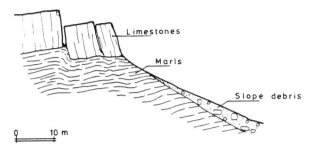

Figure 13. Block sliding caused by slow plastic deformation of Jurassic marls

4.3.1 Miocene overconsolidated clays of Lisbon region

In the area of Lisbon the rocks of Miocene
age dip generally southwards and eastwards,
at an average angle of 8 - 10°, towards the
subsident Tejo basin, forming monoclinal
structures. These monoclinal structures affect a considerable thickness of rocks of
differing lithological composition (clays,
sands, calcarenites and limestones) and
resistance to erosion. The drainage pat -
tern developed by means of differential erosion along lines of weakness, such as those
afforded by clays, incoherent sands and
poorly cemented calcarenites. The resulting
landforms are characterized by assymmetric
valleys comprising a steeper slope in which
the intervening harder and more permeable
well-cemented sandstones and limestones are
left upstanding forming a cuesta with a
steep scarp-face. The stability of these
slopes is controlled by the behaviour of the
Miocene overconsolidated clays which often
form the base of the slopes. Stiff overconsolidated clays generally present non-syste-
matic fissures and slikensides caused by
unloading. Besides they are sensitive to
humidity variations which cause expansion -
-contraction effects and desintegration. The
evolution of these slopes proceeds by shallow
slides and mudslides in the lower part of
slopes affecting the decompressed weathered
clays and undermining the resistant limestone and sandstone beds causing rock falls
and scarp recession. Rock falls occur on
steep (generally steeper than 50°) slopes and
scarp-faces and generally originate in the
upper part of slopes (figure 14).
This is the case of the southern bank of
the Tejo estuary which presents a final
stretch consisting of a narrow straight
"channel" with a marked assymmetry of the
slopes in the river banks, due to the gentle
southward dipping of the Miocene beds. The
resulting southern river bank has been occupied by tank farms and other port facilities
which are often threatened by shallow slides
and rock falls.

CALCARENITE

SILTY FINE SAND

MARLY CLAY

CALCAREOUS SANDSTONE

Figure 14. Caparica rock fall.

4.3.2 Miocene continental deposits of San-tarém

The old town of Santarém is located 70 km N
of Lisbon, on the right bank of Tejo river.
The valley side is formed by hill-slopes
about 2 km long and 80 to 100 m high, slopping 35 to 40° towards the river throughout
most of its extent. Geological formations
consist of continental deposits of Miocene
age and the geological structure as a
whole is horizontal or dipping downslope at
a very gentle angle (2 to 3°).
The dominant geomorphological feature is
a cover of upper resistant Santarém lacus -
trine limestone beds resting on the weaker
underlying sandy and silty clays (figure 15).
This cover or cap-rock is approximately
constant in elevation forming a plateau and
some separated flat-topped interfluves.
The town developed in this platean on the
top of slopes and the railway runs along the
base of slopes between the hill side and
the river. Deep dissection by rill and gully
erosion has penetrated the upper limestone
beds and the underlying sandy and clayey
formations.
It is apparent that past evolution of these
slopes was controlled by bank erosion and
undercutting by the river, mainly at Portas
do Sol, the slopes of which are located in
the point of maximum curvature of a concave
stretch of the river. Thus it seems that
the construction of the embankment for the
railway in the middle of the 19th century,
had a stabilizing effect since it impeded
toe erosion and basal removal by the river
stream.
Hydrogeological conditions are characterized by perched water tables in the sandy
layers causing seeps and spring lines along
the contact with clayey impermeable beds.
Many active processes acting on these slo -
pes are run-off erosion, weathering, seep -

age in water bearing soils and undermin - ing of more resistant beds by differential erosion. Run-off in these steep slopes cau - ses a downslope increase of erosion and steepening in the upper part of slope once lowering and recession of the top is impeded by the most resistant limestones. Weathering by rainwash and wetting and drying, causes a fall in cohesion of silty and clayey soils and the development of physical desintegra- tion by expansion, fissuring and cracking.

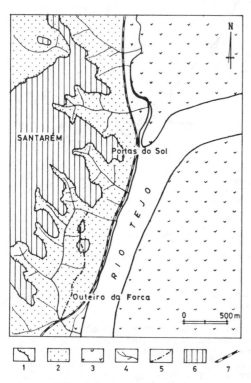

Figure 15. Geomorphological sketch of Santa- rém 1 - edge of limestone caprock, 2 - sandy and clayey soils, 3 - alluvial plain, 4 - - gullies, 5 - interfluve, 6 - urbanized area, 7 - railway.

Slope instability in the area, mainly along the railway line, is well documented since the 19th century, the first study on the stability of the area having been made in 1862 a few years after the inauguration of the railway (Ribeiro, 1862). The northern zone of these slopes at Portas do Sol and Quebradas was affected by shallow land- slides concerning the weathering profiles of slopes and debris of older slides and colluvial deposits, as well as falls and topples of narrow spans of soil and/or rock masses from the upper steeper part of slopes. The main cause of these slides and falls was the loss of cohesion of soils owing to weathering. The final factor of sliding was the rainfall to which they were always related. In some places the sliding and falling soil hit repeatedly the railway line causing traffic disruption. The selection of measures, design and construction of stabilizing works were particularly dif - ficult on account of the steepness and im - possibility to lower the slope profile owing to the location of the ancient Castle of Santarém at the top of the slope, and the location of the railway in the base.

A different type of landslide affected the same geological unit in the southern zone. The Outeiro da Forca landslide occurred in 1968 and was investigated by Oliveira (1972). The movement consisted of a deep landslide of the translational type which took place along a thin interbedded clay layer of high plasticity and organic matter content (fi - gure 16). The first signs of movement, cracks and small settlements were detected two weeks before. The final landslide last- ed four hours during which the soil was displaced about 30 m. The landslide created a scarp of about 20 to 30 m and the displa- ced soil mass slided over the existing ground surface and reached the railway line.

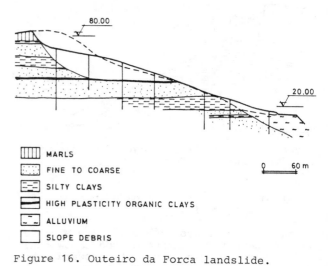

▥ MARLS	
FINE TO COARSE	
SILTY CLAYS	
HIGH PLASTICITY ORGANIC CLAYS	
ALLUVIUM	
SLOPE DEBRIS	

Figure 16. Outeiro da Forca landslide.

5 ATLANTIC ISLANDS

5.1 Madeira archipelago

The Madeira archipelago comprising Madeira, Porto Santo and Desertas lies in the North Atlantic at latitude 17° 0', longitude 32° 45', 900 km southwest of Lisbon (figure 1). The islands are predominantly volcanic al - though no historical activity is recorded other than earthquakes, the last major seism occurred in 1918. Madeira, the largest island of the group is densely populated (270 000) and highly cultivated whereas Por- to Santo is scarcely populated and the De - sertas are uninhabited.

Mean annual rainfall in Madeira presents variations as a function of altitude (figu- re 17) ranging from 500 mm in the southern coast (Funchal) to more than 3000 mm in the central mountains (Pico Ruivo-1861 m). The majority of the rain occurs in winter from November to Marsh and short term intensities may be high.

Madeira is arcuate (figure 17) being 62 km E-W and 23 km N-S at its broadest. It rises rapidly to 1861 m at Pico Ruivo only 7 km from the north coast and its shape is typi- cal of a shield volcano with intense modi- fication of the original constructional sur- face by deep fluvial dissection. The volca- nic rocks of Madeira can be divided into four eruptive complexes the oldest being at least Middle Miocene while most of the capping la- vas are younger than 4 million years. A se - ries of ashcones are scattered over the shield surface and rejuvenescent valley erup- tion have occurred. The oldest complex is predominantly composed of coarse piroclastics and is cut by a plexus of dykes the principal focus of which being the mountainous center of the island. The younger complexes consisting predominantly of basaltic lavas with interbedded tuff but few dykes, form a lava cap dipping towards the sea. Deep steep-sided ravine-like valleys trend northwards and southwards from the central mountains forming a centrifuge drainage pattern of dipslope torrential streams. High steep cliffs dominate the coastline almost all around the island.

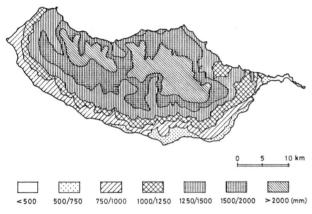

0   5   10 km

| <500 | 500/750 | 750/1000 | 1000/1250 | 1250/1500 | 1500/2000 | >2000 (mm) |

Figure 17. Mean annual precipitation (mm) in Madeira island as a function of altitude.

The predominant types of landslides are rock falls and debris slides which frequently occur both in coastal cliffs and in the headwall scarps of steep-sided valleys. Rock falls generally occur as spalling failures mainly from upper parts of steep walls and marine cliffs, commonly near the slope crest. Particularly prone to rock falls and debris slides is the oldest volcanic complex composed of a cahotic accumulation of coarse pyroclasts, owing to its relatively low tensile strength as compared with most of other volcanic rocks. In this older pyroclastic complex failure takes place mainly through intact rock and not along preexisting fractures as it is common in basaltic lava. In the upper complexes of basaltic lava interbedded with tuffite, weathered and highly fractured slopes are most susceptible to rock falls which are controlled by the network of fractures or joints.

Shallow debris slides are also common in gentler slopes involving the movement of shallow slope debris on the surface of the bedrock.

In Madeira, landslides in valley slopes frequently block the main river drainage holding back the water and endangering the villages and farms downstream owing to the breaching of debris dam by impounded water and subsequent flooding. Another type of special hazard is related to rock falls in marine cliffs giving rise to high swells which may invade the villages located near the sea. In Madeira the majority of casualties due to landslides have been caused in this way.

5.2 Azores archipelago

The Azores archipelago lies in the North Atlantic at latitudes from 36° 55' to 39° 43'N and longitude from 25° to 31° W.

The archipelago consists of 9 islands with a total land area of 2247 sq.km occupied by 253500 people, the largest island being S. Miguel (797 sq.km) and the smallest Corvo (17 sq.km).

All the islands are of volcanic origin and present high relief and rugged mountains with the exception of Santa Maria and Graciosa which have low relief energy and moderate altitudes. The islands present typical volcanic landforms related to subrecent and recent volcanism. The Pico island with a volcanic cone reaching 2345 m is the highest mountain of the Portuguese territory. Almost all the islands are crossed by east-west trending ranges sloping northwards and southwards, dissected by deep and narrow valleys. The coast is generally high with steep cliffs.

The Azores archipelago have the same pluviometric regime as the continent and Madeira with a more regular distribution of precipitation along the year and a shorter period of drought in summer.

The Azores archipelago is located on the crest of the Mid-Atlantic Ridge (triple junction) in a seismically active zone. Therefore earthquakes constitute one of the most important landslide inducing agents. The predominant type of landslides are rock falls and debris slides. Debris slides generally involve slopes of volcanic pumiceous tuff and palaesols. As an example is often cited the disastrous landslide of Vila Franca do Campo (S. Miguel) occurred in 1522. Vila Franca do Campo is located in the foothill of the Serra de Água de Pau (Fogo volcano) the slopes of which are made up of fine trachytic pyroclastics consisting of pumice deposits and palaeosols. The landslide was preceded by a period of heavy rainfall and triggered by an earthquake. The seismic shock transformed the saturated pumice into a fluid mass producing an extensive earthflow that buried the village causing hundreds of deaths.

Rock falls commonly affect cliffs and steep slopes of clinker and tuff overlain by weathered and fractured basaltic layers. The clinker is easily eroded in the lower zone of the slope causing the overhanging of upper layers of basaltic lava.

The only tracts of land favouring the settlement of people and agriculture along the shore consist of slope debris deposited in the foot of marine cliffs originated by ancient rock falls and debris slides. Owing to its unfavourable location under steep cliffs these places called "fajãs" are exposed to hazardous landslides. The last disastrous event over a "fajã" occurred in December 1987 in Ponta da Fajã (Flores island) affecting a headwall scarp between 500 and 550 m high in basalt, tuff and pumiceous soils. The slide involved about 150 000 m^3 of debris and buried several houses. After the slide, fracture-bounded blocks were still visible in the scarp as well as extension cracks behind the headwall. The site was considered hazardous to anyone occupying Ponta da Fajã and people were compelled to leave the place.

6 LANDSLIDE PREVENTION AND CONTROL

In Portugal, geologic formations where landslides are common and broad landslide problem areas where instability have taken place in the past or actually, as well as the environmental conditions under which they occur, are well known. In some of these areas, landslides are attracting an increasing concern owing to urban growth and construction. An example is the northern region of Lisbon along the right bank of the Tejo river where alluvial flat areas are already densely occupied by housing and industrial plants. In the sixties excavation cuts for the construction of the Lisbon-Porto highway, which runs along the foothills of the river bank, have caused instability and sucessive landslides in the adjacent hillside zone. In the same slopes, running paralel to the highway, are located three important conduits of the Lisbon water-supply system. Since then they became constantly threatned by slope instability. In addition, the construction of the highway triggered an increasing urban growth which has progressively invaded the hillslopes surrounding the villages bordering the river and the town of Vila Franca de Xira. All over this area, landslides associated with hillside development and with sites of high vulnerability became a serious local problem (Coelho, 1979).

Thus, the need to control urban growth and to protect people and structures have stressed the neccessity of landslides hazard analysis and risk assessment mainly for the following situations:

- in landslides prone areas under urban development, to estimate unstable zones where movements can occur or be triggered by construction activities; this calls for the preparation of large scale maps (1/10 000) that meet the needs of urban planning and construction;

- in zones of recognized high vulnerability in terms of damage potential to people and existing structures, to identify hazardous sites and to plan and design of warning and protective measures.

The first step in hazard zonation and prevention of landslides has been the inventory and classification of landslides present in the area by means of photointerpretation of 1/15 000 black and white infrared aerial photography and field survey. This approach generally allows an objective identification and mapping of ancient landslides as well as the analysis of their relations with lithology, morphology, geological structure, hydrogeological conditions and engineering properties of soil and rock. On the basis of a synoptic evaluation of the different factors interfering with stability, zones of different degree of risk of instability are mapped as a second step. This evaluation depending as it does on the intuition and expertise of the interpreter is liable to introduce some subjective judgement. In several cases it must be verified as a further stage in the field and by means of stability analysis. At last derivative maps may be prepared aiming at establishing the basis of hazard zoning maps at 1/25 000 to 1/10 000 scale, each zone being characterized by specific limitations or constraints: i) definition of "non aedificandi" (non building) zones; ii) zones where construction should depend on preconstruction investigation and design followed by careful construction procedures; iii) stable zones where landslide hazards are not to be expected. The experience shows that such maps supply planners and engineers with data permiting them to foresee the consequences of slope evolution and to plan accordingly. Besides in areas of high susceptibility they are an essential base for the enforcement of legal restraints on land use.

7 SOCIAL AND ECONOMIC IMPACTS

In Portugal landslides seldom reach imposing dimensions and only very few cases can be regarded as purely natural geological hazards. The majority of landslides are reactivated and have been triggered by human activity although always related to periods of continued rainfall. For these reasons landslides are of higher significance in the western sedimentary mesozoic borderlan, a region of lower relief energy but where the majority of important construction activities for housing, roads and highways have been developed. Even in this region losses due to landslides are very dispersed and individual slope failures are not generally so catastrophic as other geological hazards such as floods or earthquakes the damages of which are concentrated in a flood plain or in an urban area. Direct and indirect costs of a great number of small landslides are difficult to evaluate as the respective data are very sparse and incomplete and difficult to gather. On the other hand, in the case of severe landslides, a significative fraction of costs is related to maintenance, field instrumentation and warning as well as to remedial measures costs, including site investigation analysis and design, which are also difficult to determine. Nevertheless, a crude estimation shows that since 1958 at least 62 events have occurred causing 31 casualties and it is believed that total annual costs may be significant if direct and indirect costs are considered.

8 LANDSLIDE LITERATURE AVAILABLE

The majority of the worthwhile studies concerning important landslides have been carried out by the research officers of the LNEC (Laboratório Nacional de Engenharia Civil) and published as internal reports.

An overall survey of the landslide situation has been attempted by the same institution for the Serviço Nacional de Protecção Civil (Coelho, 1985) through a national inquiry in order to gather data about damaged facilities and economic and social losses. Other few studies have been carried out by students and researchers of Universities (Brum Ferreira, 1984; Costa, 1985, 1986; Carvalho & Lamas, 1987).

REFERENCES

Coelho, A.G. (1979). Engineering Geological Evaluation of Slope Stability for Urban Planning and Construction. Bull. Int. Ass. Eng. Geology, 19, 75-78, Krefeld.
Coelho, A.G. (1985). Estudo dos Fenómenos de Instabilidade dos Taludes Naturais. Relatório Interno, LNEC, Lisboa.
Carvalho, A.R. & Lamas, C. (1987). Carta de Movimentos de Terrenos nos Taludes da Margem Sul do Tejo. Geotécnico, Boletim da Secção Autónoma de Geotecnia da Universidade Nova de Lisboa, nº 3, Lisboa.
Costa, C. (1985). Fenómenos de Instabilidade nas Escarpas da Margem Sul do Tejo. Geotécnico, Boletim da Secção Autónoma de Geotecnia da Universidade Nova de Lisboa, nº 1, Lisboa.
Costa, C. (1986) - Nota Preliminar sobre a Evolução Geomorfológica das Escarpas de Almada nos Últimos Cem Anos. Geotécnico, Boletim da Secção Autónoma de Geotecnia da Universidade Nova de Lisboa nº 2, Lisboa.
Daveau, S. (1972). Repartition Géographique

   des Pluies Exceptionnellement Fortes   au
   Portugal. Finisterra, Rev. Portuguesa de
   Geografia, vol. VII-13, Lisboa.

Ferreira, A.B. (1984) - Mouvements de Terrain
   dans la Région au Nord de Lisbonne. Conditions morphostructurals et Climatiques.
   Comm. du Colloque Mouvements de Terrains.
   Sèrie Documents du BRGM nº 83.

Lencastre, A. (1984). Lições de Hidrologia.
   Faculdade de Ciências e Tecnologia da Universidade Nova de Lisboa.

Oliveira, R. (1972). An Example of the In -
   fluence of Lithology on Slope Stability.
   24th Geol. Congr., Montreal.

Ribeiro, C. (1862). Relatório àcerca da ques-
   tão que foi proposta à Comissão Geológica
   do Reino em ofício do Ministério das Obras
   Públicas MOP, unpubl. report, Lisboa.

Landslides: Extent and Economic Significance, Brabb & Harrod (eds)
© 1989 Balkema, Rotterdam. ISBN 90 6191 876 6

Landslides: Extent and economic significance in Central Europe

Adam Kotarba
Institute of Geography, Polish Academy of Sciences, Krakow, Poland

ABSTRACT: Landslides are a common and serious problem in the mountainous regions of Poland, Czechoslovakia, Hungary, Romania and Bulgaria, and along the Danube River and its tributaries. At least 1,000 landslides have damaged roads in Poland; 548 villages and towns in Czechoslovakia are at risk. Nine villages in Bulgaria have been abandoned and 12 others are in danger from landsliding. Hungary's largest iron and steel center along the Danube has been adversely affected by a landslide 1,300 m long. Sporadically, people have been killed by landslides since 1960, at least 11 of them in Bulgaria.

1 INTRODUCTION

Central Europe includes Poland, Czechoslovakia, Hungary, Romania and Bulgaria covering an area of 882,000 km² on which 94.2 million people live (in 1983). The natural conditions of these countries are greatly different, from the plains of the western part of the Russian platform (northern Poland) to the central European mountains of the Alpine system (Slovakia, Hungary, Romania and Bulgaria).

The Hercynian massifs of western Europe extend to the middle Polish uplands and the western part of Czechoslovakia (Bohemian massif) as a part of the young, west European platform. The northeastern rampart of the Bohemian massif has developed as a horst-like mountain system (the Sudetes) on the border between Poland and Czechoslovakia. The Ore Mountains (Fig. 1) form the northwestern area between Czechoslovakia and the German Democratic Republic, elevated above the Bohemian massif (upland).

The Carpathian orogenic system is about 1,500 km long and 2,500 m high with the highest peak (2,663 m) in Czechoslovakia. This epigeosynclinal mountain arc covers the territory of Slovakia, south Poland, the USSR and Romania. Two large tectonic depressions, the Pannonian in Hungary and the Transylvanian Basin in Romania, are located within the Carpathian arc. A third flat area, south of the Carpathians, is the lower Danube lowland in the territories of south Romania and north Bulgaria. The Stara Planina (Balkan) mountain range in Bulgaria, elevation 2,000 m, corresponds to the Alpine System. Rila-Rhodope, an old mountain massif in Bulgaria, rises to 2,925 m.

In Central Europe, both erosional and depositional landforms are present. When an ice sheet covered the areas of north and central Poland during the Pleistocene, glacial and glaciofluvial sediments were deposited. Erosion-depositional landforms developed within these deposits on the Polish lowlands and uplands. Inter-Carpathian depressions related to neotectonic subsidence, in contrast, were filled by thick Tertiary and Quaternary deposits. In the lowland areas, large European rivers, such as the Danube with tributaries in Czechoslovakia, Hungary, Romania and Bulgaria, and the Vistula and Oder (Odra) in the Polish territory, were incised. During the cold periods of the Pleistocene, most of the mountainous and upland areas were affected by periglacial denudation and accumulation processes. Mountain glaciation occurred in the uppermost parts of the Carpathians, Stara Planina and Rila-Rhodope Mountains, and on a lesser scale, in the Sudetes.

The climate of Central Europe is considered humid-temperate, transitional between maritime and continental influences. Mean annual precipitation on the Polish Lowland (Warszawa), on the Hungarian Plain (Bratislava, Budapest) and on the Danube Plain in Romania (Bucuresti) is similar, 580-640 mm. On mountain summits, annual rainfall totals are substantially higher, reaching 1,800 mm in the Carpathians. Maximum monthly totals are measured in July in Warsaw (Warszawa) and in June in Bratislava, Budapest, Bucuresti and Sofia, and are as high as 76-92 mm (World Weather Records, 1979). In the Carpathians, summer rainfall (July) is very high (200-250 mm) and mainly associated with thunderstorms. Long duration precipitation and extensive cloudiness are related to warm atmospheric fronts, while short, high intensity rainfall is related to cold fronts. A single rainstorm can drop 200-300 mm of water during a 24-hour period. Such values were measured at different places in the Carpathians (Balteanu, 1976; Cebulak, 1983), but the probability of their occurrence is only about 1 percent.

Geological, geomorphological and climatic conditions suggest that in different regions of Central Europe, a great variety of landslides and other mass move-ments can occur. These phenomena are conditioned by lithology, relief energy, neotectonic movements and earthquakes, local climate and human impact. The scope of this paper is to describe landslide processes in Central Europe and to show their general impact on the economy.

Slope mass movements in the Central European mountains have been described in scientific literature from the beginning of the 20th century. The incidence of natural catastrophes connected with landslides is highest for those living in the Alps.

2 TYPE OF LANDSLIDE PROCESSES IN CENTRAL EUROPE

Dominant mass movement processes are grouped into four categories based on the mechanics of displacement and the factors controlling and contributing to unstable conditions on slopes and following the universal classification of slope movement in Czechoslovakia of Nemčok and others (1972). The elementary categories of movements that are most common in the European climatic conditions are creep, sliding, flow and fall. These categories are generally accepted in all countries. This classification summarizes earlier classifications by Kleczkowski (1955) in Poland and Nemčok and Rybar (1968) and Zaruba and Mencl (1961) in Czechoslovakia. The movement is further classified by the scheme of Varnes (1958, 1978) who considers the type of material and the velocity. Detailed field studies show that additional categories of movement,

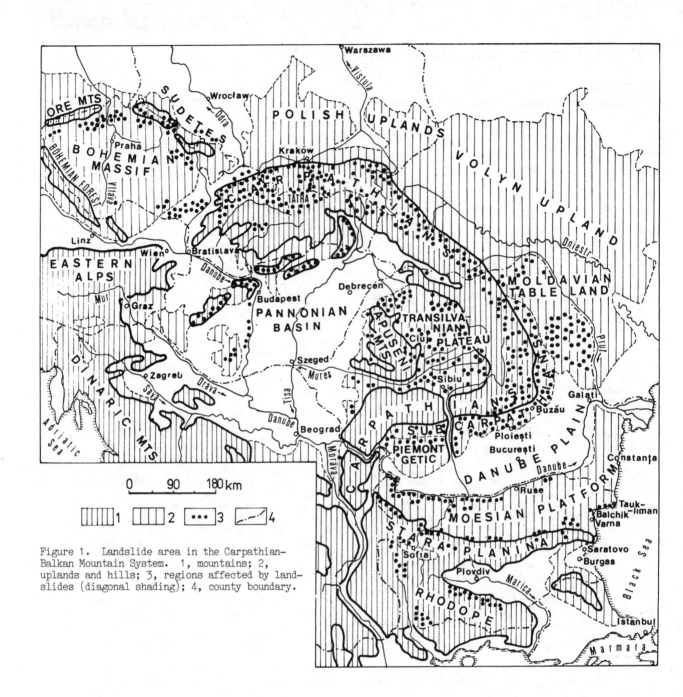

Figure 1. Landslide area in the Carpathian-Balkan Mountain System. 1, mountains; 2, uplands and hills; 3, regions affected by landslides (diagonal shading); 4, county boundary.

topples and lateral spreads, can also be easily recognized in Central Europe. In the Carpathian studies, Nemčok (1982) added rock glaciers, known from the high-mountain massif of the Tatra, as a part of surficial creep. In high-mountain regions, debris flows were included as landslides.

In Hungary, Pecsi (1971) distinguished between a landslide in the strict sense (a slip plane) and landslides in the broader sense (lack of a definite slip plane). He includes rockslides and slope slides, slips of stratum, slice-slides, slumps, slump-earthflows and block slides as "true" landslides.

Bulgarian researchers paid special attention to the geological structure of the sliding mass displace-ments. All landslides were divided into four types: block, packet, compact and consistent (Avramova-Tacheva and Voutkov, 1974). Most important from an engineering-geological point of view are those belonging to the first type, i.e., block-type landslides having as a mechanical model a two-layer medium with the upper layer of brittle, hard rocks and the lower layer of plastically deformable rocks (Kamenov and others, 1977).

The Romanian point of view on landslides is summarized by Tufescu (1964). Topographic expression observable in the field is important in distinguishing four essential landslide types. Geomorphological features are related to the kind of material in which the landslide occurs and to the rate and direction of motion.

3 LANDSLIDES IN POLAND

In Poland, 54 percent of the territory is characterized as lowlands elevated to 150 m above sea level; 42 percent consists of upland areas rising to 500 m, and only 3 percent of southern Poland has mountainous relief. In this area, both lithology and structure favor frequent occurrence of landsliding. The area is divided into two main tectonic units, the Outer and Inner Carpathians. The Outer Carpathians consists of flysch rock complexes. The areas particularly affected by landslides are those where large synclines or monoclines of rigid, porous sandstone layers rest on impermeable plastic clay shale and shale-sandstone strata. These sedimentary rocks of Cretaceous and Paleocene age have been strongly folded and thrust-faulted. The Beskidy Mountains consisting largely of flysch were uplifted more than 1,500 m (Babia Gora, 1,725 m) and dissected by valleys 200–300 m deep. Precipitation of 800–1,000 mm/yr. promotes structural landslides, comprised of flysch rocks and Quaternary periglacial deposits (Fig. 2). The Carpathian foothills (Pogórze) are also formed of the flysch series, but are less resistant to erosion and are characterized by rounded, broad ridges leveled at an altitude of 350 to 600 m.

Large complex landslides in the Carpathians have a

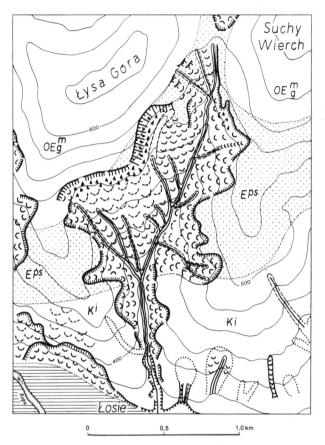

Figure 2. Landslides in the Lysa Gora area, Beskid Niski, Poland. Example of a landslide valley. Landslide location closely related to lithological facies of the Carpathian flysch. OE m/g, Magura sandstone; E_{ps}, variegated shale; k_i, Inoceramus-bearing flagstone. Stream incision gives impulse for landslide rejuvenation (Kotarba, 1974).

long history of movement. Large colluvial masses from which the landslides are derived formed under periglacial conditions and in the postglacial period (Gil and others, 1974). Slides developed infrequently during the Holocene, coinciding with humid periods (Starkel, 1966). Detailed investigations indicate that most active landslides today developed in the areas affected by mass movement in the past. This relationship is the same within the entire Carpathian arc in Poland, Czechoslovakia, Hungary and Romania.

The best known rock slump developed in the Beskid Wysoki (high Beskid Mts.) on the north slope of the Babia Gora, 1,725 m a.s.l. This failure was described as a glacial landscape because of the huge niches and tongues. More recent studies (Zietara and Zietara, 1958; Starkel, 1960; Alexandrowicz, 1978) reveal that this gigantic failure is a landslide formed by deep-seated creep (Nemčok, 1982). The total amount of displaced rock material is probably a few km^3 and was probably one of the greatest natural catastrophes in the history of the

Carpathians.

Many smaller landslides were also triggered in the past. In 1907, a rock mass of 10 million m^3 nearly 2 km long was displaced in the village of Duszatyn in the eastern Beskidy Mountains. Another landslide occurred in the wet summer of 1913 near the village of Szklarka (Beskid Niski, low Beskid Mts.) where about 3.5 million m^3 of waste and rocks were displaced in an area of nearly 50 ha (Sawicki, 1917).

Deep, structural sliding occurs when the seasons are characterized by prolonged humid weather, sometimes continuing for years. Substantial slope failures were triggered during such occurrences in the Flysch Carpathians.

The most common slope movement types in the Beskid Mountains are slump earthflows and multiple rotational slides. Each individual large landslide is distinguished by simultaneous creep, sliding, flow and fall movements, resulting in specific failures and high rates of displacement (Gil and Kotarba, 1977). According to Pavlov (1903), the landslides are divided into two categories, delapsive and detrusive. Delapsive slope movements are initiated at the base of the slope by erosion. The displacement spreads upslope to cover a higher and higher position on the slope. As described by Ahnert, the landslide surface in this category has external equilibrium, which is due to the relationship between slope and fluvial processes. Detrusive slope movement starts on the upper segment of the slope. The lower slope segment moves in reaction to pressure from the upper mass of water and material. Of the active landslides in the Flysch Carpathians, 91 percent are delapsive, caused by lateral erosion and downcutting of rivers and streams. Only 5 percent of landsliding is detrusive, triggered by overloading from rain or meltwater infiltrating into colluvial masses. In the Sudetes, large landslides are related to a small volcanic massif of the Kamienne Mountain Range, 935 m a.s.l. The largest landslide, located at Grzmiaca, is one-half kilometer long with an area of 30 ha (Pulinowa and Mazur, 1970). The rocks involved are melaphyre, porphyry and tuff interbedded with schist and sandstone. Much smaller landslides occur in Precambrian gneiss and schists, such as those in the Snieznik Massif, 1,425 m a.s.l.

Unique, large-scale, block-type movements are well known in the Stolowe Mountains, 919 m a.s.l. This tourist area (Szczeliniec Wielki Massif) is a large platform of Cretaceous sedimentary rocks consisting of flat-lying, sandstones overlying plastic clay marl. The flat-topped massif is affected by slow (6–8 mm/yr.) movements in the marginal zone. Separate blocks and rock columns move away from the centre of the massif (Pašek and Pulinowa, 1976; Pulinowa, 1972). The blocks change their position, lean over, break, and fall down the slope. Other toppling failures are common in the area (Fig. 3), and in the Bohemian Massif of Czechoslovakia.

The slopes of the Polish lowlands are rarely affected by landslides. Most failures are concentrated along the biggest rivers where relatively high banks are eroded laterally. Catastrophic mass displacements have occurred in Warsaw (Warszawa), Plock, Dobrzyń, Wloclawek and Toruń along the Vistula (Wisla) River (Fig. 4). In other areas of the Polish lowlands, sandy Miocene sediments and variegated, plastic Pliocene clays overlain by Quaternary glacial drift deposits have been deformed by the Scandinavian Ice Sheet. Large landslides developed during interglacial periods when 80 to 100 m deep erosional valleys were cut in these sediments. Today, substantial rejuvenation of these fossil landslides takes place when lateral erosion or wave action and water level changes occur (Banach, 1977), such as along the Vistula Reservoir near Dobrzyń (Fig. 4). The most hazardous stretch is located on high cliffs on the Baltic coast (Subotowicz, 1982).

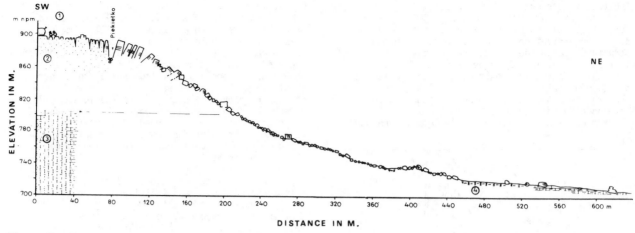

Figure 3. Toppling failures in Szczeliniec Wielki, Stolowe Mountains, Poland. 1, sandstone of the upper sedimentary stage; 2, sandstone of the lower sedimentary stage; 3, marl; 4, peat of boreal age (Pasek and Pulinowa, 1976).

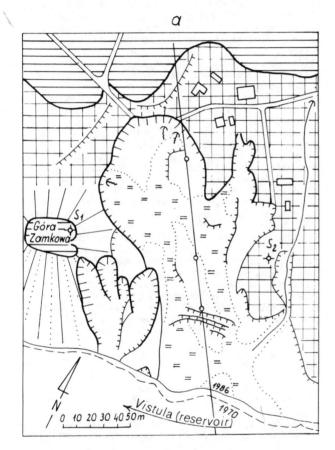

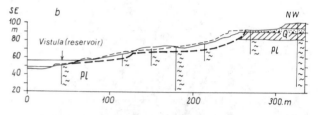

Figure 4. Geomorphological sketch and cross-section of the landslide at Dobrzyn on the Vistula River bank, north Poland. 1, extent of moraine plateau; 2, moraine plateau not affected by landslide (stable); 3, moraine residual hill; 4, slopes affected by shallow mass movement, fall, and creep; 5, slopes not subject to mass movement; 6, landslide niches; 7, secondary scarp within the landslide; 8, swamp; 9, spring and creek from ground water; 10, survey station and measurement point; Q, Quaternary sediment; Pl, Pliocene sediments (Banach, 1977).

4 LANDSLIDES IN CZECHOSLOVAKIA

About 85 percent of Czechoslovakia has upland and mountainous terrain; only 13 percent can be classified as lowland. High mountains cover 0.8 percent of the country. As expected, the landslide hazard is a serious problem, especially with the high density of population (120 persons per km^2). Also the geological features are characterized by a great variety of lithological complexes susceptible to mass movement and tectonic activities. The geology, along with the humid climate of Central Europe, helps to explain why more than 9,000 landslides were recognized, registered and plotted on a map, scale 1:25,000, by Nemcok and Rybar (1968). At least 343 of the 1984 landslide catastrophes in Europe described during the last sixteen centuries were located in Czechoslovakia (Spurek, 1972).

There are two basic morphostructural units: the West Carpathian mountain system formed by the Alpine orogeny and the Bohemian Massif developed as a platform on an old, peneplanized Variscian mountain system rejuvenated by later block tectonics. The west Carpathian region is divided into several geomorphological units (Matula, 1968; Mazur and Luknis, 1980): a core of high mountains and highlands, Flysch highlands and uplands, volcanic highlands and uplands, intra-mountain depressions and intra-Carpathian lowlands. According to Nemcok (1982), slope deformations on the central ridges of the Flysch highlands (middle mountain type of relief) occur rarely and are deep-seated failures, such as in the Babia Gora area located on the frontier between Poland and Czechoslovakia. Areas of Flysch uplands are elevated to 1,200 m a.s.l. and are characterized by many slope failures in areas where rhythmic rock complexes consist of sandstone and shale (Novosad, 1966). Various types of mass movements develop within periglacial slope formations up to 20 m thick. Slump-slides and earthflows as long as 1-3 km are common. In some localities, the volume of displaced mass is as much as one million m^3. In one area of 774 km^2, 4,392 landslides have been recorded (Nemcok, 1982).

Tatry, Nizke Tatry and other smaller mountain groups belonging to the core of high mountains and uplands within the Inner Carpathians are areas where Precambrian metamorphic rocks and Variscian granites and limestone-dolomite formations are affected by deep-seated creep (Nemcok and Pasek, 1969; Mahr

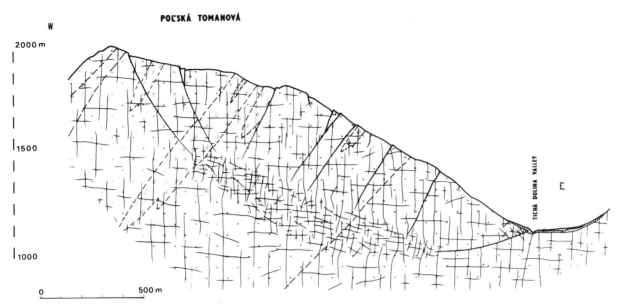

Figure 5. Cross-section of the Polska Tomanova on the Czechoslovak-Polish boundary, Tatra Mountains, showing deep-seated deformation within grandiorite, granitoids, paragneiss and amphibolite (Mahr and Nemcok, 1977).

and Baliak, 1974). Most of these failures developed on high mountain ridges and slopes, glaciated during the Pleistocene and remodelled into glacial cirques and trough valleys. The rate of mass displacement was not measured geodetically, so only substantial ridge-trench topography gives an indication of deep-seated creep (Fig. 5).

In the areas of the Carpathian volcanic highlands in central Slovakia, various types of andesites, basalts and their tuffs of Tortonian and Sarmatian age occur. Where volcanic rocks overlie Tertiary clay, silt, and marl beds dissected by valleys, sliding of blocks occurs (Malgot, 1977; see Fig. 6). The most extensive and disastrous example of a block slide was triggered in 1960–61 at the town of Handlova where about 180 houses, a highway, and a railway were destroyed. This failure was 1,630 m long, 80–110 m wide and 25–30 m thick, with a total volume of 20 million m^3 (Zaruba and Mencl, 1969). Maximum rate of displacement was 6.3 m per 24 hours. This catastrophic failure was triggered by long-lasting precipitation from June to December, 1960.

Intramontane depressions and lowlands in Czechoslovakia are rarely affected by landslides. Most of the failures are closely related to lateral stream erosion or wave action in the vicinity of artificial

dams. Shallow sliding and surficial creep is mainly within Quaternary fluvial, lacustrine and periglacial slope sediments (Nemcok, 1982).

In the Bohemian Massif, typical mass movements are related to nearly horizontally bedded, weakly-consolidated Permo-Carboniferous and Upper Cretaceous sedimentary and volcanic rocks. Block-type slope movements occur within solid sandstone complexes lying on clay or marl, weakly-consolidated plastic sediments (Nemcok, Pasek and Rybar, 1982; Rybar and Zvelebil, 1980). A typical example of big block displacements developed along the deep Labe River canyon and its tributaries through the Ceskie Stredohori. Toppling failures are also common in these areas (Fig. 7).

5 LANDSLIDES IN HUNGARY

Hungary is a typical lowland country with 65 percent of the area below 200 m a.s.l. The Great and Little Hungarian Plains dominate the landscape and are divided by uplands and low mountains 400–600 m above sea level (Transdanubian region). Only about 2 percent of the Hungarian territory belongs to the Alpine and Carpathian systems. In spite of this fact, almost one thousand significant landslides

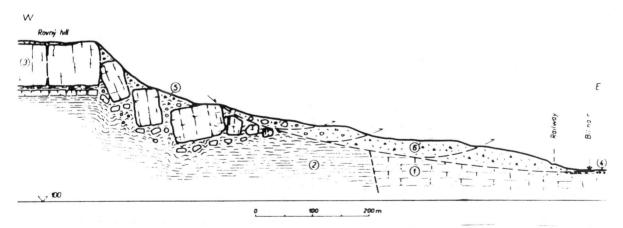

Figure 6. Lateral-spread landslide near the village of Kostov, northwestern Bohemia, Czechoslovakia. 1, Cretaceous sandstone; 2, tuff and tuffite; 3, nepheline basanite; 4, terrace gravel; 5, diluvial deposits (loam and debris); 6, slide masses (Pasek and Demek, 1969).

Figure 7. Schematic cross-section through a sandstone rock slope in the Labe Canyon area, Decin Highland, Czechoslovakia. 1-4, rock massif (lower Turonian quartz sandstone; 5-8, surface deposits, Quaternary; 9, rockwall blocks drawn to scale; 10, blocky and stony debris; 11, newly-formed slope surface (Zvelebil, 1985).

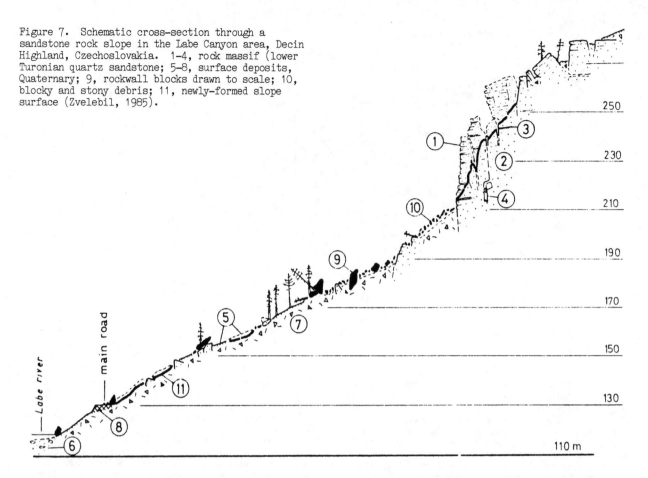

have been mapped in Hungary (Pecsi and others, 1976). The main areas of landsliding cover the Transdanubian hill and highland regions in the western part of the country, and the mountain-foreland and intermontane basins of the north Hungarian Highland Range (the Carpathian system). The Transdanubian region is composed of Oligocene and Miocene clay, sand, and marl overlain by Pleistocene loess deposits (Fig. 8). The largest landslides occur in the vicinity of Esztergom, Gerecse and Komlo and are related to hilly areas dissected by Pleistocene valleys. The north Hungarian Highland Range consists of young volcanic rocks, mainly pyroclastics, which fail characteristically as block slides.

The last important region where landslides occur frequently is on the west bank of the Danube between Budapest and the state frontier with Romania. The river is bordered by steep bluffs 50 m high near Budapest and about 15 m high close to the frontier. The most hazardous landslides have developed on these bluffs where upper Pannonian silty clay and marsh deposits, together with a thick Quaternary loess series, have been cut by the Danube. Other reaches that consist entirely of Quaternary loess deposits or upper Pannonian strata are less affected by sliding (Pecsi and others, 1979).

The most serious economic problem in Hungary related to landslides happened when Dunaujváros, Hungary's largest iron and steel centre, was built in the 1950's. Housing in the new town and the iron and steel plant were located close to the Danube on the loess plateau, 40-50 m elevation above river level (Pecsi and Scheuer, 1979). To protect this area from landslides, expensive mitigation devices including drain holes, pumped wells, drainage galleries and trenches were installed. In spite of these defense works along a 3 km long reach of the Danube bluff, mass movements occurred. The greatest mass displacement occurred in February, 1964, along a 1,300 m stretch of the bluff. A mass of clay

having a volume of about 10 million m^3 slipped rapidly toward the Danube. No one was killed or injured, but the water supply to the iron works was disrupted and production held up (Pecsi and Scheuer, 1979). The series of slides had a nearly vertical shear plane in the area where the rupture developed, and was almost horizontal at the base of the bluff. This well-known large landslide area was carefully studied and an artificial terracing system was installed along a 4 km stretch of the bluff. Now the river does not directly undercut the bank, so the landslide risk is markedly diminished.

Failures of the Dunaujváros type are classified as imbricated slice-slides (slumps). This peculiar type of landslide, which differs from other slump-slide types distinguished by Pecsi (1971), is also described as occurring in other localities, such as the Dunaföldvar (Fig. 9). In September, 1970, a new landslide was triggered as a system of slices 45-50 m thick on a Dunaföldvar loess bluff. This land-slide was activated as a result of heavy filtration of ground water on impermeable clays within loess after a very wet season. The volume of the land-slide mass was about one million m^3 (Pecsi, 1979).

6 LANDSLIDES IN ROMANIA

The Carpathian Mountains dominate the Romanian territory and 70 percent of the area is classified as mountainous and upland with 31 percent of that area elevated above 800 m a.s.l. Intermontane lowlands and basins cover 30 percent of the total territory. Substantial landslide areas are related to topographic relief and climatic conditions, and frequent and intense earthquakes trigger additional landslide processes.

The density of population in the mountains reaches 40-70 persons per km^2. Human impact on the Romanian environment is an additional factor in mass movement in large areas, particularly in the Subcarpathians (especially Buzau region) situated southeast of the

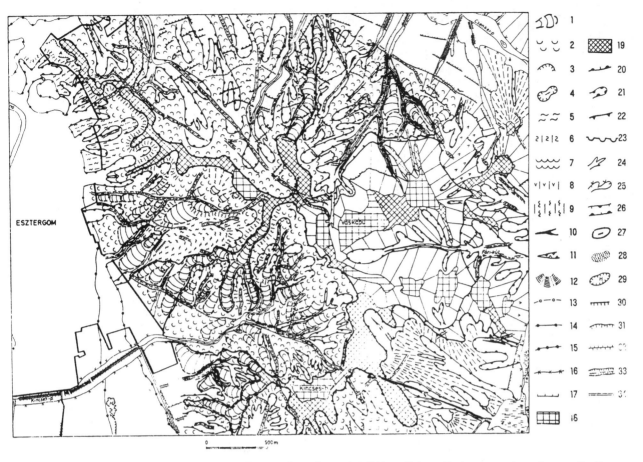

Figure 8. Detail from the geomorphological map of surface stability of town Esztergom and environs, North Hungary. I. Forms due to mass movement: Fossil forms: 1, stabilized tongue or mound of former slide or slump; 2, inactive fossil landslide slopes. Recent forms: 3, rupture front (main scarp) of landslide; 4, depression enclosed by disturbed material; 5, slopes temporarily stable; 6, unstable slopes prone to sliding; 7, active mobile sliding slopes. Related phenomena: 8, slopes threatened by rill erosion, 9, slope wash; 10, gullies, 11, big gullies, 12, alluvial fan of gully erosion. II. Other genetic forms of relief: 13, terrace number I of the Danube River; 14-16, river terraces of the Danube no. II/a, II/b and III; 17, stable low river bank; 18, plateau; 19, derasional ridge; 20, derasional valley remodelled by linear erosion; 21, derasional valley; 22, horst; 23, derasional step; 24, aggraded flat valleys on flood-plain; 25, valley bottom margin; 26, valley creek; 27, blow hole; 28, sand dunes. III. Man made forms: 29, open-pit mining; 30, man made terraces; 31, cuts; 32, dykes; 33, loess gully; 34, roads (surveyed and compiled by A.Juhesz) (Pecsi, Juhesz and Schweitzer, 1976).

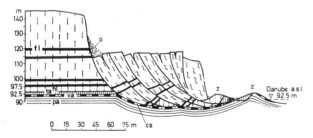

Figure 9. The Dunafoldvar River bank landslide on the Great Hungarian Plain. 1_1, loess displaced by sliding; 1_2 waste mound of earlier slides; hl, pale pink sandy loess; o, talus; z, earth mound and Pannonian clay unwarped from the Danube stream bed; fl, fossil soils; ta, dark grey clayey loam soil; pa, Pannonian clay; Va, red clay; cs, sliding plane (Pecsi, 1979).

Carpathian chain (Balteanu, 1974, 1976). Other areas of active mass movement are the Moldavian Tableland, the Translyvania Plateau, the Paleogene Flysch Carpathians and peri-Carpathian piedmont (Balteanu, 1974, 1976; Posea and Ielenicz, 1976; Posea and Popescu, 1976; Morariu, 1974).

In the Subcarpathian region, numerous catastrophic mass movements have been influenced by extreme meteorological events, and earthquakes (Grumazescu, 1973). Active tectonic forces and the weak character of the Neogene molasse deposits (clay, marl, conglomerate, gravel, sand, and sandstone) are responsible for triggering mass movements on an unprecedented scale. In the Buzau Subcarpathian area, "solifluction, mudflows, earthflows, land-slumps, landslides, and rockfalls cover 60-70 percent of the slope surface" (Balteanu, 1974). Earthflows have a downslope surface transfer rate from a few meters to 20-30 m in a month. Mudflow speeds are as high as 100-120 m per day. The major period of mass movement coincides with snow thaw in the first half of March, and the maximum rainfall at the end of June and the beginning of July (Balteanu, 1974). Complex valley landslides up to 1.5 km long, 80 m wide and 4 m deep are typical geomorphic features, and both sliding and flowing occur within the moving masses (Fig. 10).

The climatic and lithological conditions are supplemented by recent earthquakes. According to Enescu and others (1974), the recurrence interval of earthquakes of magnitude 7, Richter scale, is 50 years for the period 1471-1974. Catastrophic effects of the earthquake of March, 1977 (magnitude 7.2) in the Buzau Carpathians and Subcarpathians were carefully studied by Balteanu (1979). At that time, block-glide type mass movements affected

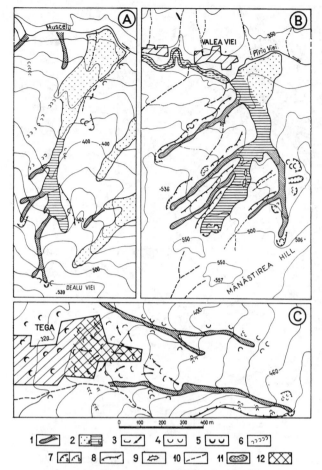

Figure 10. Examples of slopes affected by mass movements in the Buzau Subcarpathians, Romania. 1, mudflow; 2, complex valley landslide: fixed (a), active (b); 3, complex mass movement (deep landsliding, slump, and mudflow); 4, active landslide (deeper than 2 m); 5, old landslide; 6, sheet slide; 7, main (a) and secondary (b) scarp; 8, steep slope; 9, outlier; 10, active gully erosion; 11, temporary lake in river channel; 12, part of village destroyed by mass movement (Balteanu, 1976).

parent rock to a depth of 7-15 m and were accompanied by intensive rockfalls, debris avalanches and accelerated debris creep. The surface lowering rate of 0.6 mm per event for a small Paleogene flysch catchment was calculated on the basis of direct field measurements after the event. The corresponding value for another catchment, composed of poorly-cemented Villafranchian gravels, was 1.8 mm (Balteanu, 1979).

In the montaneous regions of the eastern Carpathians consisting of Paleogene flysch, landslide processes predominate in slope morphology. Deep mass dis-placements affecting parent rocks are divided into deep valley slides and slope slides (Posea and Ielenicz, 1976). The deep valley slides are 0.2-5.0 km long, 50-60 m wide in the central section and more than 500 m wide at the toe. The surface area is 2-3 km² with an approximate depth of moving mass on the order of 10-12 m. Slope slides are 2 km long and cover a similar surface area. Many of the slope slides develop in the mountain forest belt, about 1,000 m a.s.l. Their lower sections, or toe zone, are under plantations (Fig. 11).

Landslides on the Moldavian Tableland and the Transylvanian Plateau develop within sandy-clay monoclinal rock complexes. According to Posea and Popescu (1976), mass movements are followed by

infiltration and piping. Non-resistant soils and rocks, when soaked with water, are affected by soilfall and rotational slumping. The lowest section of these detrusive landslides has a translational character. Other deep mass movements are related to sandstone, tuff and sandy complexes. Delapsive rotational landslides termed "glimee" (Morariu and Garbacea, 1968) are common features in these regions. The "glimee" landslides were formed at the end of the Pleistocene and are rejuvenated only occasionally during exceptionally wet seasons (Morariu, 1974).

7 LANDSLIDES IN BULGARIA

About 66 percent of Bulgarian terrain is typically upland, 13 percent is mountainous, and flat areas of the Danube Plain cover about 20 percent. Mean density of population amounts to 76 persons per km², with higher densities on lowlands and in the Black Sea coastal area.

Large landslides occur on the northern part of the Black Sea coast within nearly horizontally-bedded Miocene and Quaternary sediments, mainly marl, sand, clay, and loess (Moesian Platform). The largest block-landslide area related to neotectonic movements and eustatic variation of the Black Sea basin during the Quaternary are up to 30 km long and 100 m deep (Kamenov and others, 1973). The best known block-type landslides are located in the Taukliman (the Bay of Birds) and near the towns of Balchik and Varna. The last one has a sliding cirque 5-6 km in diameter and a depth of up to 100 m. Rejuvenation of an old landslide during the 1971 spring season caused displacement of about 5 million m³ of sand-rich rock masses overlying Miocene clay (Kamenov and others, 1973).

On the northern Black Sea coast and along the Danube River banks, numerous block-type landslides

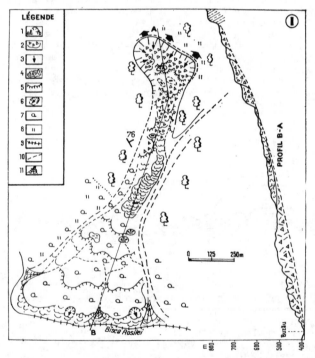

Figure 11. Example of valley landslide at Valea Oii, Romanian Carpathians. 1, landslide niche, a) less than 10 m high, b) higher than 10 m; 2, displaced material; 3, rockfall; 4, youngest mudflow tongue; 5, toe of landslide triggered in 1941; 6, active landslide; 7, landslide surface stabilized by vegetation; 8, slope surface occasionaly affected by mass movement; 9, stream; 10, marginal drainage line; 11, alluvial fan (Posea and Ielenicz, 1976).

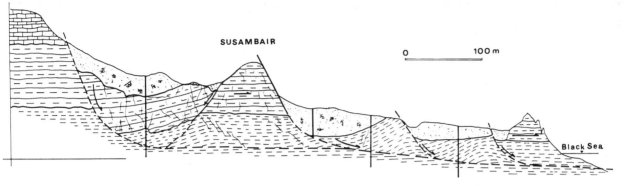

Figure 12. Cross section of the block-slide at Balchik, Bulgarian Black Sea coast (Evstatiev & Rizzo, 1984).

were studied by Kostak and others (1981). Annual measurements at Taukliman slide show that continuous movement of about 0.26 mm per year is influenced by seismic activity (Vrancea earthquake 1977) and rainfall in 1977. At least two, often opposite, movements exist with creep and slide alternating in time (Rizzo and Tzvetkov, 1985).

At the Balchik landslide system (Fig. 12), old photographs and paintings, as well as direct measurements, were used to calculate a remarkable movement on the order of 20-50 cm per year. Such phenomena are especially hazardous in the densely populated coastal zone (Rizzo and Tzvetkov, 1985). Mass movements along the Danube River bank are ancient; most were triggered during the Pliocene and Pleistocene. Now they are reactivated by lateral erosion related to intensive antrophic changes of the water level and relatively frequent earthquakes. The thickness of the failures is as much as 60-80 m and they fall into two categories of lateral spread and translational slides. At some localities, such as near the village of Botevo, Vidin district, an average value of lateral retreat of high river banks due to erosion reaches 1 m per year (Simenova, 1979). Earth slumps, topples and collapsing occur along the entire right bank of the Danube (Fig. 13).

In the vast pre-Balkan (Stara Planina) massif, folded flysch structures of Upper Cretaceous and Paleocene age are affected by deep-seated block landslides. Within these structures dipping toward the Black Sea, large landslides with a complex mechanism (such as Sarafovo landslide, Cap Emine) occur on the coast.

Large landslides also occur in mountainous areas. One of the largest Bulgarian failures occurred in the Rhodope Mountains near the town of Pestera, where the volume of displaced rock masses

was more than 1 million m^3. In the Visocica valley in Stara Planina, a gigantic earthflow was triggered in 1963. Rapid flow of 4 million m^3 of material dammed the valley up to 36 m and created a 0.5 km long lake. The town of Nis, located in the valley, was in danger so the lake was artifically drained.

8 IDENTIFICATION, CONTROL, AND SOCIAL AND ECONOMIC IMPACT OF LANDSLIDE PROCESSES IN CENTRAL EUROPE

Systematic identification and registration of landslide processes in Central Europe started after World War II. Early studies concentrated on the mechanisms and main factors causing the most spectacular landslides. The catastrophic landslides of 1960-61 in Czechoslovakia when a part of the town of Handlová in Central Slovakia was destroyed became the primary motivation for systematic investigation and registration of all landslide areas. From 1961 to 1979, engineering-geological works were installed on slopes. The approximate cost of these protective measures at Handlova was as high as 320 million Kčs, while the total cost of damaged facilities was much higher (Nemčok, 1982). Similar problems were noted in other countries, especially in towns located on high terraces of the Vistula River in Poland and the Danube River in Hungary, Romania and Bulgaria. Facilities on the Black Sea coast in Romania and Bulgaria and part of the Baltic coast in Poland are also greatly affected by landsliding.

Identification of unstable areas and a quantitative rating of the degree of instability were essential parts of the program accomplished in Central Europe. Geomorphological and geological mapping was done at scales of 1:10,000 in Hungary, and 1:25,000 in Czechoslovakia, Poland, part of Romania and Bulgaria. Data collected were used to construct engineering geological maps of Bulgaria at 1:500,000 scale (Kamenov and Iliev, 1963), a synoptical map of landslide areas in Czechoslovakia at 1:1,000,000 (Nemčok and Rybař, 1968), and a general geomorpho-logical map of Poland and Hungary at 1:500,000 (Starkel and others, 1980; Pecsi, 1972). A geomorphological map of the Danubian countries at map scale 1:2,000,000 prepared by Pecsi with the help of special consultants from all the countries concerned contains qualification of current geo-morphic processes within the relief types of the Carpatho-Balkan mountain system. Landslide topography is marked on this map published in Vienna in 1978.

All data collected during the landslide registration are deposited in the archives of the Geological Survey in Poland and Czechoslovakia, in the Geographical Research Institute of the Hungarian and Romanian Academies of Sciences, and in the Geological Institute of the Bulgarian Academy of Sciences.

In the Slovak Carpathians, the extent of landsliding was complied by Nemčok (1982) and is

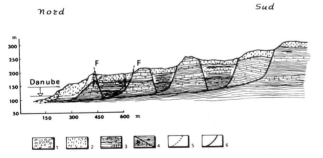

Figure 13. Cross-section of the block slide in loess on the Danube River bank in Bulgaria. 1, fluvial deposits; 2, Holocene loess; 3, Pleistocene-Holocene interlayered clay and loess; 4, Cretaceous clay unit; 5, recent landslide; 6, slip plane of the block landslide; F, tension cracks formed in 1981 (Rizzo, Tzvetkov and Slavov, 1984).

Table 1. Extent of landsliding in the Slovak Carpathians (Nemcock, 1982)

Geomorphological-geological unit	Area affected by mass movements km²	Number of failures	Percent of total area
Core high mountains	160.6	691	5.7
Core highlands	30.0	40	0.36
Flysch highlands	51.4	147	1.17
Flysch uplands	774.0	4,516	11.6
Volcanic highlands	366.9	2,500	12.0
Volcanic uplands	42.0	75	1.6
Intra-mountain depressions	80.7	984	0.86
Intra-Carpathian lowlands	5.0	51	0.036
Total	1,483.6	9,194	3.01

shown on Table 1. The total impact of landslides on the Slovakian economy is difficult to determine precisely, although Nemcok indicates that 548 villages and small towns are at risk. In addition, landslide costs include both direct and indirect losses from mass movements on 127 sections (46 km) of railway tracks and 800 sections (270 km) of state roads. These data from Slovakia are similar to those from the Polish Flysch Carpathians where Bober and Dziewański (1985) reported about 8,500 landslides in an area of 25,000 km². Nearly 1,700 km² or about 6 percent of the Polish Carpathians is mobilized periodically. In Bulgaria during the last few years, 9 villages were abandoned by the inhabitants and 12 towns are in danger of damage from landsliding.

The greatest landslide damage to Polish roads before World War II occurred when the main road from Kraków to Zakopane, a winter sport centre in Poland, was constructed. Five large landslides were stabilized by protective works estimated to cost from $10,000 to $350,000 for each landslide (Królikiewicz, 1978). From 1970 to 1974, this road was totally reconstructed for a distance of 18.4 km. The cost of one kilometer of road construction was 26.2 million zloty, making a total cost of nearly a half billion zloty (about $5 million in 1974).

In the Polish Carpathians, more than one thousand landslides affect state roads. The problem is much more serious in the Romanian Carpathians, especially in the Subcarpathian region which is subject to earthquake activity. Therefore, in many cases it is difficult to distinguish between losses from landslides and losses from earthquakes.

In addition to the economic losses due to slope movement, the loss of human life has occurred sporadically, mainly in Romania and Bulgaria. Deaths due to slope-failure disasters in Bulgaria are listed in the following table (Frangov and Avramova-Tacheva):

Site		Year	Number of people killed
Towns:	Oriachovo	1960	3
	Varna	1968	3
	Smolian	1971	4
Villages:	Cibor (Michailovgrad District)	1982	1

The above data on the social and economic impact of landslide processes indicate that the landslide problem in central Europe is substantial, even in comparison with other parts of Europe. The following historical catalogue of major landslide catastrophes from 328 to 1970 on Table 2 published by Spurek (1972) shows the problem in central Europe compared to several other countries.

Table 2. Catastrophic landslides in Europe (Spurek, 1972)

Country	Type of mass movement		
	Rockfall (A)	Rockslide (B)	A + B
Italy	148	553	701
Switzerland	153	162	383
Austria	77	162	239
Germany	46	92	137
France	41	55	96
Czechoslovakia	40	343	383
Poland	0	25	25
The rest of Europe (Including Hungary and Romania)	4	38	42
Total	603	1,984	2,586

ACKNOWLEGMENTS

This paper is based on the results of research in
Poland, Czechoslovakia, Hungary, Romania and
Bulgaria published in different journals and
books. I particularly wish to thank Jan Rybař,
Institute of Geology and Geotechnics, Czechoslovak
Academy of Sciences, Prague; Adam Kertesz,
Geographical Research Institute, Hungarian Academy
of Sciences, Budapest; and Gheorgi Franov and Elka
Avramova-Tacheva, Laboratory of Geotectonics,
Bulgarian Academy of Sciences, Sofia, for their help
in data collection. I am sure that not all has been
resolved to their satisfaction, but the paper is
greatly improved by their efforts.

REFERENCES

Alexandrowicz, S.W. 1978. The northern slope of
Babia Gora Mt as high rock slump. Studia Geomorph.
Carpatho-Balc. XII:133-148.
Avramova-Tacheva, E. and V.Voutkov 1974. On the
typification of landsliding phenomena (in
Bulargian) Bull. Geol. Inst. Eng. Geol. and
Hydrol. Bulgarian Acad. Sci., Ministry of Heavy
Industry XXIII:133-147.
Balteanu, D. 1974. Some investigations on present-
day slope processes in the Romanian
Subcarpathians. Rev. Roum. Geol. Geophys. et
Geogr. Geographies 18, 1:25-30.
Balteanu, D. 1976. Some investigations on the
present-day mass movements in the Buzău
Subcarpathians. Rev. Roum. Geol. Geophys. et
Geogr. Geographie 20:53-61.
Balteanu, D. 1979. Effects of the March 4, 1977
earthquake on slope modelling in the surroundings
of the Pătîrlagele Research Station. (The Bazău
Carpathians and the Subcarpathians). Studia
Geomorph. Carpatho-Balc. XIII:175-189.
Banach, M. 1977. Rozwoj osuwisk na prawym zboczu
doliny Wisly miedzy Dobrzyniem a Wloclawkiem (The
growth of landslides on the right-bank slope of
the Vistula valley between Dobrzyn and
Wloclawek). Geographical Studies 124,
Warszawa:101 p.
Bazynski, J. and A.Kuhn 1970. Rejestracja osuwisk w
Polsce (Landslide registration in Poland). Przegl.
Geol. 3.
Bober, L. 1984. Rejony osuwiskowe w polskich
Karpatach fliszowych i ich zwiazek z budowa
geologiczna regionu (Landslide areas in the Polish
Flysch Carpathians and their connection with the
geological structure of the region). Biul. Inst.
Geol. 340. Z badan geol. w Karpatach XXIII:115-
162.
Bober, L. and J.Dziewanski 1985. Engineering-
geological problems and water-engineering
structures. In Problems of Quaternary Geology of
the Polish Carpathians, Guide to excursion 5,
Carpatho Balcan Geological Association XIII
Congress, Cracow, Poland 1985:13-15.
Cebulak, E. 1983. Maximum daily rainfalls in the
Tatra Mountains and Podhale Basin. Zesz. Nauk. UJ,
Prace geogr. 57:337-343.
Evstatiev, D. and V.Risso 1984. Sull'origine ed
evoluzione delle frane nella zona di Balchik, sul
Mar Nero (Bulgaria). Geol. Applicata e
Idrogeologia XIX:289-301.
Gil, E., E.Gilot, A.Kotarba, L.Starkel, and
K.Szczepanek 1974. An Early Holocene landslide in
the Niski Beskid and its significance for paleo-
geographical reconstructions. Studia Geomorph.
Carpatho-Balc. VIII:69-83.
Gil, E. and A.Kotarba 1977. Model of slide slope
evolution in flysch mountains, an example drawn
from the Polish Carpathians. Catena 4:233-248.
Gil, E. and L.Starkel 1979. Long-term extreme
rainfalls and their role in the modelling of
flysch slopes. Studia Geomorph. Carpatho-Balc.
XIII:207-220.

Grumazescu, H. 1973. Subcarpati dintre Cilnău si
Susita. Studiu geomorfologic. Ed. Academiei,
Bucuresti.
Kamenov, B., I.Ilviev, and E.Avramova-Tacheva 1977.
Conditions for the origin, mechanism and dynamics
of block landslides in Bulgaria. Bull. Intern.
Assoc. Eng. Geol. 16, Krefeld 1977:98-101.
Kamenov, B., I.Iliev, St.Tsvetkov, E.Avramova, and
G.Simeonova 1973. Influence of the geological
structure on the occurrence of different types of
landslides along the Bulgarian Black Sea coast.
Geol. Applicata e Idrogeol. VIII, 1:209-220.
Kleczkowski, A. 1955. Osuwiska i zjawiska pokrewne
(Landslides and related phenomena). Warszawa, Wyd.
Geol.
Koštak, B. and E.Avramova-Tacheva 1981. Propagation
of coastal slope deformations at Taukliman,
Bulgaria. Bull. Intern. Assoc. Eng. Geol. 23,
Krefeld:67-73.
Kotarba, A. 1974. Modelling of flysch slopes by
landslips illustrated by examples chosen from the
Polish Carpathians. Abhandl. der Akad.
Wissenschaften in Göttingen, Math.-Physikal.
Klasse III, 29:102-110.
Królikiewicz, A. 1978. Osuwiska na obejsciu Mogilan
(Landslides in the vicinity of Mogilany). In
Oswiska i sposoby zapobiegania im. Wyd.
Komunikacji i Lacznosci, Warszawa:19-37.
Mahr, T. and F.Baliak 1974. Regional investigation
of slope deformations in the high mountain area of
the West Carpathians. Proc. Xth Congress CBGA,
V:169-178, Bratislava.
Mahr, T. and A.Nemčok 1977. Deep-seated creep
deformations in the crystalline cores of the Tatry
Mts. Bull. Intern. Assoc. Eng. Geol. 16:104-106.
Malgot, J. 1977. Deep-seated gravitational slope
deformations in neovolcanic mountain ranges of
Slovakia. Bull. IAEG, 16: 106-109, Krefeld.
Morariu, T. 1974. Le système des glissements de
terrain en Roumanie. Rev. Roum. Géol. Géophys. et
Géogr. Géographie 18, 1:9-17.
Nemčok, A. 1982. Zosuvy v Slovenskych Karpatoch
(Landslides in the Slovak Carpathians). Ed. VEDA,
Bratislava.
Nemčok, A. and J.Pašek 1969. Deformacie horskych
svahov (Deformation of mountain slopes). Geol.
Prace, Spravy 50:5-28.
Nemčok, A., J.Pašek and J.Rybař 1972. Classification
of landslides and other mass movements. Rock
Mechanics 4:71-78.
Nemčok, A., J.Pašek and J.Rybař 1982. Mass movements
on the slopes in Czechoslovakia. Problems of
Geomechanics 8, Yerevan:30-44.
Nemčok, A. and J.Rybař 1968. Landslide
investigations in Czechoslovakia. Proc. of the 1st
Session of the Intern. Assoc. Eng. Geol.,
Prague:183-198.
Novosad, S. 1966. Slope disturbances in the Godula
Group of the Moravskoslezske Beskydy Mountains.
Sbor. geol. Ved, R.HIG, 5:71-86, Praha.
Pašek, J. and J.Demek 1969. Mass movements near the
community of Stadice in North Western Bohemia.
Studia Geogr. 3, Brno:17 p.
Pašek, J. and M.Z.Pulinowa 1976. Block movements of
Cretaceous sandstones in the Stolowe Gory Mts.
Poland. Bull. Intern. Assoc. Eng. Geol. 13:79-82.
Pécsi, M. 1979. Landslides at Dunafőldvár in 1970
and 1974. Geogr. Polonica 41:7-12.
Pécsi, M., A.Juhász and F.Schweitzer 1976. A
magyarországi felszinmozgasos terůletek
terképezése (The mapping of areas affected by
landsliding in Hungary). Fůldrajzi Értesitő 2-
4:223-235.
Pécsi, M. and G.Scheuer 1979. Engineering geological
problems of the Dunaujváros loess bluff. Acta
Geol. Acad. Sci. Hungaricae 22, 1-4:345-353.
Pécsi, M., F.Schwietzer and G.Scheuer 1979.
Engineering geological and geomorphological
investigation of landslides in the loess bluffs
along the Danube in the Great Hungarian Plain.
Acta Geol. Acad. Sci. Hungaricae 22, 1-4:327-343.

Posea, Gh. and M.Ielenicz 1976. Types de glissements dans les Carpates de la courbe (Bassin du Buzău). Rev. Roum. Géol. Géophys. et Géogr. Géographie 20:63-72.

Posea, Gh. and N.Popescu 1976. Les glissments massifs dans les Piemonts Pericarpatiques. Rev. Roum. Géol. Géophys. et Géogr. Géographie 20:45-52.

Pulinowa, M.Z. 1972. Poscesy osuwiskowe w srodowisku sztucznym i naturalnym (Landslide processes in natural and artificial environments). Dokumentacja geogr. 4. Warszawa:112 p.

Pulinowa, M.Z. and R.Mazur 1970. Stare osuwisko w Grzmiacej w Sudetach (An old landslide at Grzmiaca in Sudety Mountains). Wszechswiat:7-8.

Rizzo, V. and S.Tzvetkov 1985. Complex movements in deep block-type landslides. In Progress in mass movement and sediment transport studies--Problem of recognition and prediction. CNR-PAN Meeting, Torino, Dec. 5-7, 1984, Torino:95-106.

Rizzo, V., S.Tzvetkov and P.Slavov 1984. Aspetti e micromovimenti di un graben in una frana a blocchi in Bulgaria. Geol. Applicata e Idrogeol. XIX:259-268.

Rybar, J and J.Zvelebil 1980. Felssturz bei Hrensko im Elb-sandsteingebirge. Z angew. Geol, 26, 3:153-155, Berlin.

Sawicki, L. 1913. Krajobrazy lodowcowe Zachodniego Beskidu (Glacial landscapes of the Western Beskid). Rozpr. AU, III, 13:1-22.

Sawicki, L. 1917. Osuwisko ziemne w Szymbarku i inne zsuwy powstale w 1913 roku w Gaicyi Zachodniej (Earthslide at Szymbark and other slides triggered in 1913 in Western Galizia). Rozpr. Wydz. Mat.-Przyr. PAU, A, 56:227-313.

Simeonova, G.A. 1979. Rock erodibility as an element in the change of erosion stability and assessment of the geodynamic activity of the right slope regions of the hydrotechnical complex of Nikopol-Turnu-Magurele. Bull. Intern. Assoc. Eng. Geol. 20:156-158.

Starkel, L. 1960. Rozwoj rzezby Karpat fliszowych w holocenie (The development of the Flysch Carpathians relief during the Holocene). Prace geogr. 22, Warszawa:239 p.

Starkel, L. 1966. Post-glacial climate and the moulding of European relief. In World Climate from 8000-0 B.C. Royal Met. Soc. London:15-33.

Subotowicz, W. 1982. Litodynamika brzegów klifowych wybrzeza Polski (Lithodynamics of the Baltic cliffs). Ed. Ossolineum:152 p.

Špurek, M. 1972. Historical catalogue of slide phenomena. Studia geogr. 19, Brno:178 p.

Tufescu, V. 1964. Typologie des glissements de Roumaine. Rev. Roum. Géol. Géophys. et Géogr. Géographie 8:141-147.

Varnes, D.J. 1958. Landslide types and processes. In Landslides and Engineering Practice, H.R.B. Spec. Report 29:20-47.

Varnes, D.J. 1978. Slope movement types and processes. In Landslides Analysis and Control. Spec. Report 176. Nat'l Acad. Sci.:12-33.

World Weather Records 1961-1970, vol.2, Europe. U.S. Dept. of Commerce, Asheville, N.C. 1979.

Záruba, Q. and V.Mencl 1961. Ingenieurgeologie. Berlin,-Prag, Verlag der Tsch. Akad. Wiss.

Zietara, T. 1988. Landslide areas in the Polish Flysch Carpathians. Folia Geogr. ser. geogr.-phys. XX:21-31.

Zietara, K. and T.Zietara 1958. O rzekomo glacialnej rzeźbie Babiej Góry (On supposed glacial relief of the Babia Góra Mt.). Roczn. Nauk. Dydakt. WSP 8:55-77.

Zvelebil, J. 1985. Failure of the sandstone rock slope in the Labe Canyon, Děčin Highland, Czechoslovakia. Proc. Intern. Symp. on Landslides III, Univ. Toronto Press:591-596.

Landslides: Extent and Economic Significance, Brabb & Harrod (eds)
© 1989 Balkema, Rotterdam. ISBN 90 6191 876 6

Coping with landslide problems in Czechoslovakia

J.Rybář
Institute of Geology and Geotechnics, Czechoslovak Academy of Sciences, Prague, Czechoslovakia

S.Novosad
Geotest, Ostrava, Czechoslovakia

ABSTRACT: The paper presents the characterization of the principal stages of studying and coping with the landslide problem in Czechoslovakia, including the whole-state registration of slope deformations. Theoretical results of investigation and examples of landslide control are added.

RESUME: Ce travail contient la caractéristique des phases principales d'etude et solution du probléme des glissements de terrain en Tchécoslovaquie, l'enregistrement des pentes instable inclus. Les résultats théorétiques d'investigation et exemples des mesures de stabilisation sont présentés.

Czechoslovakia, a Central European country occupying more than 128,000 km², is an area of high landslide incidence. In the geologically younger mountainous part – Slovakia – to the East, about 3 % of the area are affected by slope movements. In the western Bohemian part, where the terrain is less diversified, only 1 % of the total area is exposed to slide hazards.

In Czechoslovakia, the study of slope movements has a long tradition. Špůrek (1983) attempted to represent graphically to what extent the landslides have been investigated during last hundred years on the basis of publications, archive reports and experts reports (Fig.1). In the first stage the slope movements were only studied from the geological and morphological points of view and with a few exceptions, the papers have a descriptive character. The engineering-geological approach to the problem of slope movements appeared in the 1920' when professor Quido Záruba, the founder of the Czechoslovak engineering-geology began to investigate the landslides. The number of papers dealing with slope movements was gradually increasing, until World War II brought a certain stagnancy and reduction of the number of publications. In the 1950', however, there was an upsurge of interest in slope movements of various types, in connection with the post war development of capital construction. An important stimulating event was the landslide, which in 1960/1961 destroyed part of the mining town Handlova in central Slovakia (Záruba and Mencl 1969). The attitude of public authorities towards the relevance and solution of this problem changed essentially. The situation resembled the circumstances that developed in Sweden during the years 1914-1922, when the disastrous landslides near the Lake Aspen acted as an impulse to systematic investigation of landslides along the Swedish railway lines.

Czechoslovakia was among the first world countries which carried out the whole-state registration of all hazardous sliding prone areas of economic significance. the areas mapped in 1962 – 1963 accounted for 62% of the territory. On the government decision about 30 engineering geologists took part in the work and slide areas were documented and plotted in maps on a scale of 1 : 25,000 (Rybář et al. 1965). The card index and maps were deposited in the Central geological archives GEOFOND in Prague and Bratislava. Subsequently the maps of slide areas on a scale of 1 :200,000 have been compiled, in which the economic importance of the individual areas has been emphasized. Together with the characterization of structures seriously threatened by sliding, these documents were given to the relevant Departments and regional administrative

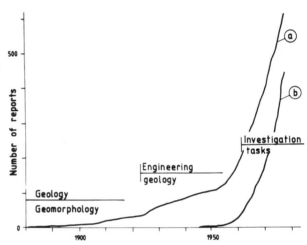

Fig.1 Mass curves of published studies (a) and archive records (b) on slope movements in Czechoslovakia (according to Špůrek).

organs.

The central geological archives introduced a duty to consult the documentation on every locality before engineering- geological investigations are proposed. They cannot be opened unless this rule is obeyed. This arrangement has promptly come into use, although such regulations are usually not so readily accepted.

The collection of data on slope failures in the archives is steadily increasing. Thus, for example, the documentation was complemented by mapping of several areas where at the time of registration a wider occurrence of lanslides was not presumed, and of areas of minor economic values (e.g. scarcely inhabited wooded highlands). The archive of landslides is currently complemented by data obtained by engineering-geological investigations carried out for individual investing actions, and other reports whose delivery into the Geofond archive is compulsory. As the original registration cards did not provide data that could be used in fully automated computer technique, as early as 10 years after the registration works were closed it was urgent to consider a modernization of the filling system. At the elaboration of a new registration system (Pašek et al. 1977) much attention has been devoted to the transfer of the primary data (Špůrek 1987).

At the present time, the examination of archive records on slope movements is taken for granted not only by engineering geologists but also by many construction, forest and agricultural engineers. The contact with the visitors, whose number is constantly increasing, is advanced by the disposal of computer graphics, geological charts included. Moreover, a direct examination of original maps and field documentation is available to every visitor, if desired. The computer outputs are advantageous, in particular, in preparing design of a wider regional extent.

The access of Geofond records to a user is also mediated by 'regional geologists', i.e. geologists employed in regional administrative offices. They are in a close contact with the Geofond and they can influence the designs of construction and mining works in case the occurrence of slope deformations were underrated. With a few exceptions, it has been achieved that records on slope movements are considered in preparation of every housing and industrial estate, in planning routes of motorways, railway lines and pipelines.

In systematic registration carried out in 1962-1963 about 9,000 landslide prone areas were recorded. After 25 years their number rose to 15,000. It has been confirmed that natural slope movements developed almost exclusively on slopes that were subject to sliding already in the past. Also landslides triggered by human activity originate usually by reactivation of ancient slope failures or in geological structures susceptible to sliding. The registration of all cases of slope failures has conditioned the limitation of the risk of new landslide development to a minimum. The national -economic contribution is appreciable, although for the lack of appropriate bases it cannot be enumerated.

The interest in slope movements did not fade out with their registration but has remained in the limelight of Czechoslovak engineering geologists. They have continued the tradition founded by professor Quido Záruba, who solved a number of intricate problems concerning the construction practice, and beginning with the year 1922 he published tens of papers on this topic.
In co-operation with professor Vojtech Mencl, expert in geomechanics, he wrote the book 'Landslide and their control', which was published in several languages. The last up-dated and enlarged edition in English appeared in 1982 (Záruba and Mencl 1982). Professor Záruba educated a number of engineering geologists concerned for landslide problems who have applied the experience gained during the whole-state documentation, to the co-ordinated study of selected aspects of the origin and development of slope movements.

To begin with, sliding areas immediately endangering economically important structures were selected on the basis of the systematic registrations. At engineering geological investigation and monitoring of slopes new research and monitoring methods have been examined and verified. Thus, e.g. new methods of the assessment of the depth of slide surface have been developed, using a slope extensometer (Rybář 1968), the method of brittle conductors (Hickl 1977) and other unusual procedures (Fussgänger and Rybář 1981). The application of geoacoustic determination of rock noise depth extent has proved successful (Novosad 1979). In order to established the relationship between the changes in pore pressure at different depths of the landslide and the rate of movement, the pore pressure recorders were placed in a landslide artificially induced in Tertiary claystones (Škopek et al. 1972).

Of general importance was the proposal of a new classification of slope movements (Nemčok et al. 1972). On the basis of geomechanical character and rate of movement the known types of slope processes were divided into four principal groups. creep,

sliding, flow and fall. The classification became obligatory for all Czechoslovak working places and was taken over by engineering geologists in both West and East Germany. It was also taken into consideration in the classification used dominantly in North America, proposed by D. J. Varnes.

The summary elaboration of registration results provided information on the character and distribution of slope deformations in individual engineering geological regions (Nemčok et al. 1982), and map of sliding areas of Czechoslovakia on a scale of 1 : 1,000,000 was constructed (Nemčok and Rybář). In the map the occurrences of natural and anthropogenic slope movements are plotted and the principal engineering geological regions prone to sliding are delimited.

The results of systematic registration of landslides in Czechoslovakia and chiefly the many years' and worldwide collecting of data on slope failures formed the basis for compiling a historic catalogue of landslides and rockfalls in Europe (Špůrek 1972); the supplement covering the period of 1970 - 1987 is prepared for print.

Špůrek (1982) has analysed the effects of climatic and cosmic factors on the origin of slope movements in detail, which enabled him to raise a long-term prognosis of the oscillation of their frequency.

Much attention has been devoted to the mapping methods of sliding events. The principles of their graphic illustration in maps of different scales have been developed and verified on maps of scales 1 . 1 : 25,000, 1 : 10,000; 1 : 5,000 and 1 : 200 (Rybář 1973).

The mapping methodology has been significantly enriched by extensive mapping of large areas in mountainous regions of Slovakia (Mahr and Malgot 1978, Malgot and Mahr 1979). Attention was concentrated on

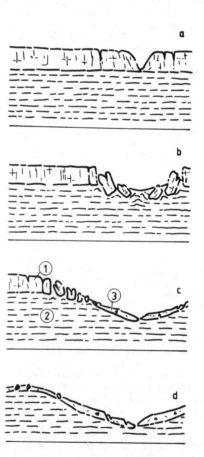

Fig.2 An example of development stages (a – d) of the deformation of a slope built up of two different rock complexes. 1 - rigid rocks, 2 - plastic rocks, 3 - waste

SW

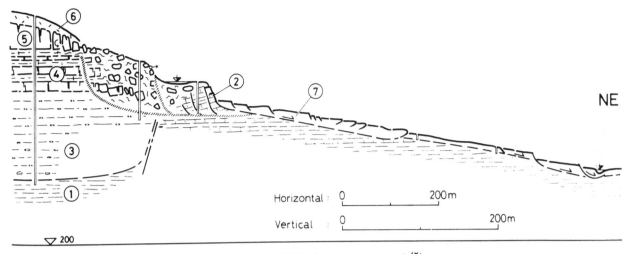

Fig.3 Geological profile of the landslide near Děčín (according to Rybář).
Upper Creataceous: 1 - marly claystone (Coniacian), 2 - sands and sandstones intercalated with claystones and sandy clays (Coniacian - Santonian). Tertiary: 3 - fresh-water sediments, 4 - pyroclastic rocks, 5 - basalts, 6 - debris, 7 - slid mass.

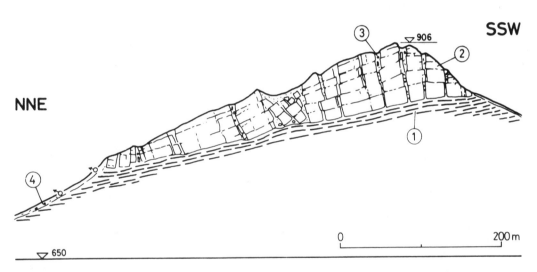

Fig.4 Profile of the crest of Lukšinec Hill in the Moravskoslezské Beskydy Mts. (according to Novosad 1966). 1 - pelitic beds, 2 - sandstones, 3 - gullies, 4 - slope debris.

maps of a scale of 1 : 10,000, in response to the demands of practice. They have become an indispensable basis for planning the development of housing estates in constricted valleys where slope failures are a very frequent phenomenon.

Special investigation has been directed to the evolution of the activity of very slow slope movements. From the geological aspect they are represented by long-term movements, in which creep acts as predominant mechanism. Their velocity is so small that the movement is not observable in the field and they are mostly regarded as fossil events. If the creep deformations are not revealed in time, being neglected by construction or mining works, the slightest interference may trigger uncontrollable movements which can result in sliding or falls of disastrous character.

A very sensitive method of dilatometric measurement in fractures between rock blocks was developed and verified at tens of localities in

Czechoslovakia and abroad (Košťák 1988). Direct measurement using a TM 71 device complemented with the prototype of automatic registration for whole-year observation in inaccessible areas has provided definite information on very slow movements on slopes where movements are not apparent morphologically nor demonstrable by geodetic measurement (Košťák and Avramova-Tacheva 1981, Košťák and Rybář 1978, Košťák and Cacon 1988). The optic-mechanical device TM 71 having a long-term high stability has been proved useful recently in scanning the slow-movement activity in the head scarp of the historic landslide near the town of Frank (Alberta, Canada).

Linked-up with the pioneer studies of professor Záruba (Záruba 1956,1958) are the investigations of block movements of the creep type, which are in Czechoslovakia unusually frequent. In most cases we are concerned with sinking of solid rigid rock blocks into plastic substratum (Fencl 1966, Pašek 1974, Pašek and Košťák 1977, Malgot 1977, Nemčok and Baliak

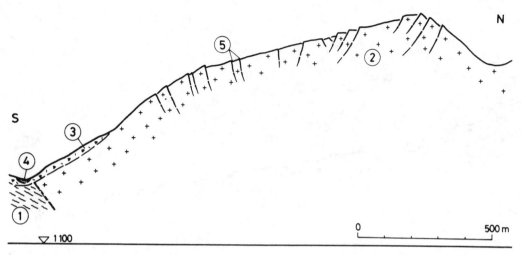

Fig.5 Profile of the crest of Chabenec Hill (according to Nemčok and Mahr).
1 - migmatitic orthogneiss, Proterozoic, 2 - biotit-quartz diorite and granodiorite,
(Paleozoic), 3 - boulders and gravel of glacial deposits, 4 - gravel and sand of fluvial
deposits, 5 - sliding surfaces.

1977). The mechanism of movement is shown schematically in Fig.2 (Rybář and Nemčok 1968) and the slope event from northern Bohemia is represented in Fig.3. Less frequent is sliding of rigid blocks along a predetermined surface or zone (Fig.4, Novosad 1966, 1978b).

The understanding of the mechanism of creep movements of the block type was facilitated by physical models constructed of optically sensitive photoplastic materials (Košťák 1977).

In the high crystalline mountains of the Slovakian Carpathians still another type of very slow slope movements has been studied, i.e. sagging of mountain slopes (Fig.5 and Fig.6), extending to a depth of about 400 m (Nemčok 1972, Mahr 1977, Mahr and Nemčok 1977). Almost 700 localities have been assessed (Nemčok 1982) and the measurements have shown that some apparently fossil disturbances are active even under present geological conditions. The surface creep events, including the rock glaciers, have also been studied in detail (Fig.7).

Our knowledge gained from the primary registration of landslides has been greatly amplified also in other regions. In the eastern part of the country the region of young volcanites, the Carpathian flysch region and the region of tertiary tectonic basins were systematically investigated (in Nemčok 1982). In the Bohemian regions to the west, slope

deformations were studied in the Permo-Carboniferous area at the foot of the Krkonoše Mts. and in the sandstone facies of the Bohemian Cretaceous Basin (Kalvoda and Zvelebil 1983, Zvelebil 1988).

An object of summary analysis was the slope stability in Tertiary brown-coal basins in relation to the structural-geological conditions in Tertiary (Rybář 1978); attention was also given to the affects of the abnormal horizontal stress release (Rybář 1971).

No less successful have been proved the results of systematic geological survey. Appropriate geoelectrical, seismic, magnetometric, thermal and geoacoustic methods have been elaborated for investigations of creep deformations of the block type and of landslides (Müller 1977, Novosad 1979). The continuous close contact of engineering geologists and geomechanics with geophysicists has added much to the store of knowledge. Of great value are, e.g. the results obtained in determination of the kneadded zone in plastic sediments underlying the deformations of block type.

The progress of knowledge in engineering geology and related fields is given to a high degree by the difficulties and urgency of problems to be solved. It also holds true for the studies of slope movements in Czechoslovakia. One of the impulses which provoked their intensifications was the destruction of a part of the town of Handlová in 1960 and 1961, as is emphasized in the introduction of this paper.

Systematic investigation of slope movements, particularly on high mountain slopes was instigated by the construction of pumped storage power plants in late 1960'. The studies draw from experiences gathered at the construction of dams and operation of reservoirs (Záruba 1978) both in Czechoslovakia and abroad. Serious problems of slope stability had to be solved at the selection of the damsite or during construction of hydraulic works Orlík, Dobšiná, Žermanice, Krpelany-Sučany-Lipovec (Záruba and Mencl 1954, 1958) and Morávka (Novosad 1978b).

The problems of slope stability connected with the filling and operation of reservoirs are still more serious. In Czechoslovakia they were encountered practically in all reservoirs impounded by high dams. The engineering-geological survey was aimed at a timely prognosis of landslide hazards, and appreciation of contingent landslide risk. It also should provide data needed for the assessment of feasible preventive stabilization and its cost as a basis on which the effectiveness of preventive measures would be evaluated. Extensive slides were

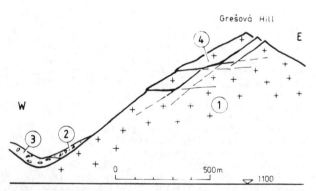

Fig.6 Profile of the Grešová ridge (after Nemčok).
1 - granitoids (Paleozoic), 2 - slope debris, 3 -
glaciofluvial deposits, 4 - blocks shifted along
faults.

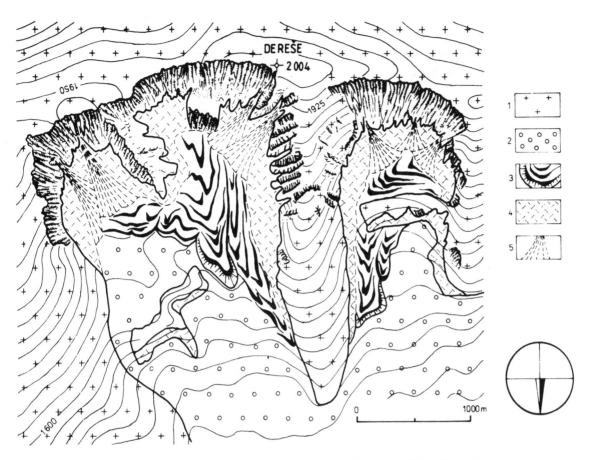

Fig.7 Map showing rock glaciers in the basin below Mt. Dereše in the Nízké Tatry Mts.
(according to Nemčok and Mahr). 1 - biotite-quartz diorite and granodiorite (Paleozoic),
2 - boulders and gravel of the glacial moraine deposits, 3 - moraines of fossil rock glaciers,
4 - loamy-stony debris, 5 - debris cones (Quaternary).

reactivated chiefly on the banks of the reservoirs
Orava (Horský and Müller 1972), Nechranice (Rybář 1977,
Záruba et al. 1986), Šance (Novosad 1979, 1984) and
Liptovská Mara.

On the basis of information derived from mapping
of dormant landslides according to their
morphological features, the areas most seriously
threatened by sliding have been delimited, and in
individual cases the value of possible damage has
been estimated and appropriate corrective measures
have been proposed. In the reservoir of the
Liptovská Mara Dam, e.g., earth works and drainage
boreholes were carried out. In the Šance reservoir,
where the cost of preventive stabilization of the
Řečice rockslide (volume 8 to 10 million m³) was
estimated at c.treble the amount of possible damage,
it was not performed. In 1970 before the first
filling of the reservoir, monitoring of slope
movement and of adverse factors was introduced. The
results confirmed the presumed reactivation of the
rockslide; they also provided valuable information
usable for the assessment of the effects of various
factors - precipitation, flooding of slopes,
alternation of annual seasons - on the development
of slope deformation, mainly on the transition from
the primary and secondary creep to the stage of rapid
slide. The experience thus gained is used for
estimation of the behaviour of analogous rockslides
originated either under extreme natural conditions
(intensive precipitation, etc.) or by changes in
natural conditions due to man's activity. In the
Moravskoslezské Beskydy Mts. itself there are tens
of similar slides. The continuously controlled Šance
rockslide is moreover used for testing new methods of
surface movements monitoring (Novosad 1978a).

The Czechoslovak geologists used the results of

long-term investigation of landslides on the
reservoir banks gained in their home country also
abroad. The most significant of their tasks was the
investigation of Slope failure No.5, which was
reactivated by filling the Tablachaca reservoir on
the Mantaro river in Peru. In case of a abrupt slope
failure the function of the system producing c. 45%
of electric power of the country would has been
endangered (Novosad et al. 1979). Thanks to
extensive remedial measures, the movement has been
slowed down appreciably, and the landslide is
regarded as the largest in the world where expert
knowledge succeeded in averting disastrous risk
(in P.P.Repetto, proc. of 11th Int.Conf. SMFE, San
Francisco).

The passing and consistent enforcement of the law
of the agricultural soil protection in the 1970'
limited the use of arable plains for building up and
recommended to settle the areas of lower soil
quality, including hilly lands. The decision led to
the necessity of building up even the landslide-prone
slopes, particularly in the densely peopled
mountainous areas in Slovakia. Of 70 Slovakian towns
with more than 5,000 inhabitants, 52 can spread only
onto the slopes (Nemčok 1982), most of which are
disturbed by slope movements. For this reason
a number of towns, despite a higher risk, have
proceeded to hillside development after extensive
preventive measures. It can be demostrated with the
development of the towns Košice, Prievidza, Banská
Bystrica, Zvolen or Považská Bystrica.

Unusully high demands on the extent of remedial
measures have led to industrialization of these
works. The classic procedures requiring hard manual
labor have almost disappeared. Recently, drainage
boring has been most widely used in Czechoslovakia;

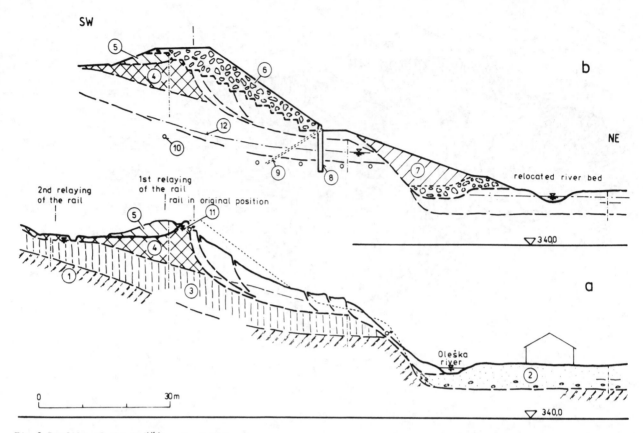

Fig.8 Profile of the Košťálov landslide.
a – situation in January 1975, b – situation in August 1977. Permian: 1 – claystones, marlstones and sandstones. Quaternary: 2 – sandy gravel and sands, 3 – slope loams, in places with debris, 4 – loamy-stony made-up-ground of primary embankment. Remedial measures: 5 – provisional embankment for the 1st rail relaying, 6 – material removed from the primary embankment, 7 – new part of the embankment and support of quarry stone, 8 – pile wall, 9 – anchors, 10 – drainage boreholes, 11 – original ground-water table, 12 – drained ground-water table.

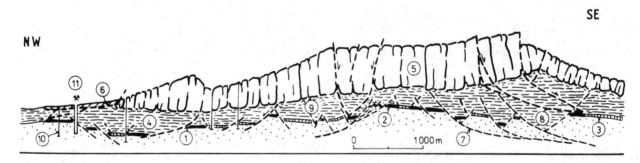

Fig.9 A profile of brown-coal deposits at the foot of the Vtáčnik Mts.
1 – underlying conglomerates, 2 – coal seam, 3 – coal seam partly affected by surface mining, 4 – overlying clays and tuffites, 5 – andesite and agglomerate tuff, 6 – displaced material, 7 – surface of rupture in prevolcanic sediments of gravitional-tectonics origin – sliding surfaces of recent natural landslides, 8 – sliding surfaces activated by underground coal mining, 9 – kneaded zone, 10 – boreholes, 11 – air shaft

in the last 20 years almost 90% of localities have been drained in this way. The drainage is usually combined with additional stabilization methods, most frequently earth works, piling or anchored pile walls. An example of the combination of methods employed most frequently in Czechoslovakia is shown in Fig.8 (Rybář 1979). In January 1975 a landslide disturbed a 120 years old railway embankment in north-eastern Bohemia. The traffic on the main railway line was interrupted, the river bed at the slope foot was buried and for a short time also the settlement on the fluvial plain was flooded. After provisional arrangement (laying of side-tracks and

removal of slipped material from the river bed) two fans of horizontal drainage boreholes and an anchored pile wall were constructed and appropriate earth works were made. Cast-in-place piles 1020 mm in diameter were used.

In the last years, the development of engineering geology has been greatly stimulated by the stability problems connected with the exploitation of mineral raw materials. In Czechoslovakia, a country densely populated (120 inhabitants per 1 km²), the mining has many hundred years history. Most of the deposits easily workable were exhausted already in the past so that at the present time many deposits are opened

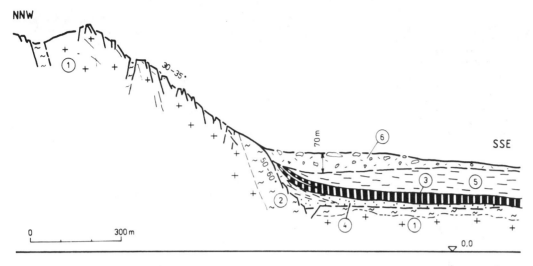

Fig.10 Profile at the area of the foot of the Jezerka slope, in front of the Čs. armády brown-coal mine. 1 – crystaline complex, 2 – weekended zones. Miocene 3 – coal seam, 4 – sandy sediments, 5 – clayey sediments. Quarternary 6 – debris (accumulation of rockfall material).

under extremely difficult engineering-geological conditions.

A wide range of problems is connected with the development of underground mining of coal under the the slopes and ridges of the volcanic Vtáčnik Mts. in central Slovakia. The mining works provoke activation of dormant slope movements. Andesite blocks up to several hundred metres thick set into movement endangering the safety of workers and structures at depth and at the surface of the massif as well. The common mining approach cannot be applied to predicting possible deformations at the head of mining. This can trigger slope deformations extending even several kilometres beyond the established protection zones (Malgot and Otepka 1978, Rybář and Malgot 1982). Figure 9 shows one profile of the Handlová coal deposit.

Problems of another type have arisen in brown coal opencast mines of the foot at the Krušné Hory Mts. in northern Bohemia. Sedimentary filling of a Tertiary basin should be there extracted over a length of more than 50 km. In critical sectors the fault walls should be relieved by cuts at first 250 m and then after 600m high. From engineering geological point of view it cannot be ruled out that

the danger of deep slope movements might increase producing deformations of the adjacent mountain slopes formed of crystalline rocks. A careful analysis served as a basis for a set of maps of 1.200,000 scale showing the stability conditions in the outcrop area of the basin (Rybář 1987).

The giant Čs.armády Mine was the first that has approached to the mountain slopes. One of the critical profiles below the Jezerka slope is shown in Fig.10. The height of the slope edge above the opencast bottom will be at the beginning 550 m attaining gradually up to 700 m. Should the prospective Kohinoor Mine be realized, the height will reach 1000 m. It must be noticed that the slope below Jezerka Hill was in the Pleistocene and partly still in the Holocene modelled by recurrent rockslides and rockfalls. Tertiary deposits at the foot of the slope are overlain by an accumulation of 27 million m^3 of crystalline rock materials, locally more then 70 m thick.

The stability of high slopes in front of the Čs.armády Mine has occupied much attention of engineering geologists. Detailed investigation provided input data for mathematical and physical modelling of high slope stability. Among other methods , the solution of the state of stress was developed and tested on models made of optically sensitive materials by the freezing method of stress state determination in a centrifuge – Fig. 11 (Málek 1987). For the assessment of critical inclinations of high slopes in opencasts the engineering geological empirical approach was proposed (Zika et al.1988). The method of probability evaluation of the strength of a rock massif was elaborated using the theory of percolation (Košťák 1982). This method that considers a wide range of mechanical properties of the rock massif, eliminates the disadvantages of deterministic methods by introducing a qualitatively new concept, i.e. the degree of high-slope disturbance risk.

In hazardous areas the monitoring of changes in front of mine workings has become an important procedure. The monitoring involved the 'warning criteria' and the emergency design of the opencast, which establishes the procedures to be used in individual stages.

The results of monitoring combined with the documentation of exposures opened during operation have provided a basis on which the original results of mathematical modelling (Doležalová and Rozsypal 1987, Rozsypal 1988) are verified in the Čs.armády Mine. The observational method of design is here

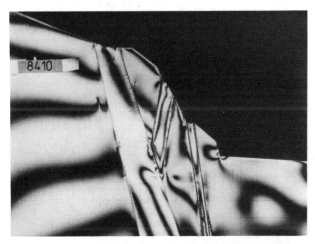

Fig.11 Example of photoelastic assessment of the stress state of a rock slope using the freezing method in a centrifuge (according to Málek).

concerned.

Theoretical principles of spatial and chronological prognoses of the development of slope
disturbance and of the mechanism and extent of slope
failures have been elaborated in order to obtain
information needed in solving the stability problems
of high and extremely high slopes which are expected
to arise in the near future. The program also
benefited from the results of Zvelebil (1985), which
he gained in preparing the prognosis of disastrous
rockfalls in the sandstone complex of the Bohemian
Cretaceous Basin.

CONCLUSIONS

The paper summarizes the characteristics of the
principal stages of the studies and solution of the
landslide problems in Czechoslovakia. An important
landmark in their investigation became the landslide
that in 1960-1961 destroyed part of the mining town
Handlová. This event instigated a systematic
registration of all hazardous slope movements in the
country. The application of the results gathered in
practice is enabled by the service of the Central
Geological Archive Geofond.

The registration work brought a lot of valuable
knowledge which was used with advantage in
co-ordinated investigation of several selected
problems concerning the origin and development of
slope movements. An overview of the results obtained
is here presented. The selection of the problems to
be studied was to a high degree given by the demands
of practice. The examples cited relate, in
particular, to stability problems caused by both
surface and underground coal mining under difficult
geological conditions.

REFERENCES

Doležalová, M. and A.Rozsypal 1987.Field measurements
and computational models for solving an open pit
mine stability problem. Proc.2nd Int.Symp. Field
measurements in geomechanics.Rotterdam: Balkema.
Fencl J. 1966. Types of landslides in the Cretaceous
Basin of Bohemia (in Czech). Sbor. geol.
Věd, Ř.HIG 5:23-41. Praha:ÚÚG.
Fussgänger, E. and J.Rybář 1981. Experience in
determination of the depth of an active slide
surface (in Czech). Geol.Průzk. 23:50-53. Praha.
Hickl, J. 1977. Equipment developed for measuring
landslides and for indicating the plane of sliding.
Bull. IAEG 16:217-218. Krefeld.
Horský, O. and K.Müller 1972. The inundation areas of
the Orava reservoir (in Czech). Sbor.geol.Věd,Ř.
HIG 10:59-71. Praha:ÚÚG.
Kalvoda, J. and J.Zvelebil 1983. The dynamics and types
of slope failures during the development of the Labe
valley in the Děčín Highlands (in Czech). Acta
Montana 63:5-74. Praha.
Košťák, B. 1977. Photoplastic slope deformation
models. Bull. IAEG 16:221-223. Krefeld.
Košťák, B. 1982. Probabilistic strength predictions in
a rock massif (in Czech). Acta Montana 58:3-93.
Praha: ÚGG ČSAV.
Košťák, B. 1988. Instability detection in rock. Proc.
Environmental Geotechnics and Problematic Soils and
Rocks, Bangkok 1987, p. 555-564. Rotterdam,
Brookfield: Balkema.
Košťák, B. and E.Avramova-Tacheva 1981. Propagation of
coastal slope deformations of Taukliman, Bulgaria.
Bull. IAEG 23:67-73. Krefeld.
Košťák, B. and S.Cacon 1988. Monitoring and
interpretation of sandstone block movements on a
table hill margin. Proc. 5th Int. Symp. on
Landslides, Lausanne. 1:439-442. Rotterdam: Balkema.
Košťák, B. and J.Rybář 1978. Measurements of the
activity of very slow slope movemnents. Grundlagen
u. Anwendung d.Felsmechanik, p. 191-205. Clausthal:
Trans Tech Publications.

Mahr, T. 1977. Deep reaching gravitation deformations
of high mountaion slopes. Bull. IAEG 16.121-127.
Krefeld.
Mahr, T. and J.Malgot 1978. Zoning maps for regional a
urban development based on slope stability. 3rd int.
Congr.IAEG, Sec.I. 1:124-137. Madrid.
Mahr, T. and A.Nemčok 1977. Deep-seated creep
deformations in the crystalline cores of the Tatry
Mts. Bull. IAEG 16:104-106. Krefeld.
Málek, J. 1987 Solution of the state of stress of the
rocky slopes by centrifugal modelling. Proc. 1st
Conf.on Mechanics, 6:118-121. Praha: ÚTAM ČSAV.
Malgot, J. 1977. Deep-seated gravitational slope
deformations in neovolcanic mountain ranges of
Slovakia. Bull. IAEG 16:106-109. Krefeld.
Malgot, J. and T.Mahr 1979. Engineering geological
mapping of the West Carpathian landslides areas.
Bull. IAEG 19:116-121. Krefeld.
Malgot, J. and J.Otepka 1977. Gravitational slope
deformations near Handlová. Bull. IAEG 15:63-65.
Krefeld.
Müller, K. 1977. Geophysical methods in the
investigation of slope failures. Bull. IAEG
16:227-229. Krefeld.
Nemčok, A. 1972. Gravitational slope deformations in
the high mountains of the Slovak Carpathians (in
Slovak). Sbor geol.Věd, Ř. HIG 11:77-97. Praha:ÚÚG.
Nemčok, A. 1982. Landslides in the Slovak Carpathians
(in Slovak). Bratislava:Veda.
Nemčok, A. and F.Baliak 1977. Gravitational deformatior
in mesozoic rocks of the Carpathian mountain ranges.
Bull. IAEG 16:109-111. Krefeld.
Nemčok, A. and J.Pašek and J.Rybář 1972. Classificatic
of landslides and other mass movements. Rock
Mechanics, 4, 2:71-78. Wien-New York.
Nemčok, A. and J.Pašek and J.Rybář 1982. Mass movements
on the slopes in Czechoslovakia. Problems of
Geomechanics 8:30-44. Yerevan.
Novosad, S. 1966. Slope disturbances in the Godula
Group of the Moravskoslezske Beskydy Mountains (in
Czech). Sbor.geol. Věd, Ř.HIG 5:71-86. Praha:ÚÚG.
Novosad, S. 1978a. The use of modern methods in
investigating slope deformations. Bull. IAEG
17:71-73. Krefeld.
Novosad, S. 1978b. Depth creep in a slightly inclined
coarse flysch. 3rd .Int.Congr. IAEG, Vol.2:70-75.
Madrid.
Novosad, S. 1979. Establishing conditions of
equilibrium of landslides in dam reservoirs by means
of the geoacoustic (rock noise) method. Bull IAEG
20:138-144. Krefeld.
Novosad, S. 1984. The quantitative prediction of the
development of the Řečice landslide as a result of
long-term monitoring of the movement. Proc. 27th
IGC, vol.8:112-114. Moscow.
Novosad, S. and R.Barvínek and M.S.de La Torre 1979.
Estudio de estabilidad del derrumbe No.5 en el
reservorio de Tablachaca de la central
hidroelectrica del Mantaro. Proc. 6th Panamerican
Congr. ISSMFE, vol.1:331-344. Lima.
Pašek, J. 1974. Gravitational block-type slope
movements. Proc. 2nd int.Congr. IAEG, V-PC-1.1. Sao
Paulo.
Pašek, J. and B.Košťák 1977. Block-type slope movements
(in Czech). Rozpravy ČSAV 87:33-58. Praha:Academia.
Pašek, J. and J.Rybář and M.Špůrek 1977. Systematic
registration of slope deformations in
Czechoslovakia. Bull. IAEG 16:48-51. Krefeld.
Rozsypal, A. 1988. Problems of prognosis of slope
deformation of open pit mines. Proc. 5th Int.Symp.
on Landslides, Lausanne, 2:1233-1236.
Rotterdam:Balkema.
Rybář, J. 1971. Tektonisch beeinflusste
Hangdeformationen in Braunkohlenbecken. Rock
Mechanics, 3, 3:139-158. Wien - New York.
Rybář, J. 1973. Representation of landslides in
engineering geological maps. Landslide - The Slope
Stability Review, 1, 1:15-21. Eureka, Cal.
Rybář, J. 1977. Prediction of slope failures on water
reservoir banks. Bull. IAEG 16:64-67. Krefeld.
Rybář, J. 1978. Influence of discontinuities on the

stability of slopes in the Miocene. Proc. 3rd Int.
Congr. IAEG, Sec.II, Vol: 2.82-87. Madrid.
Rybář, J. 1979. Slope deformations in north-eastern
Bohemia (in Czech). Sbor. 22.konf. ČSMG, 287-299.
Trutnov.
Rybář, J. 1987. The engineering-geological zoning of
the outcrop part of the North-Bohemian brown-coal
basin at the footing of Krušné Hory Mts. Acta
Montana 77:3-64. Praha:ÚGG ČSAV.
Rybář, J. and J.Malgot 1982. Engineering geology in the
exploitation of mineral deposits. Proc. 4th Congr.
IAEG, Vol. II, Th.1:333-342. New Delhi.
Rybář, J. and A.Nemčok 1968. Landslide investigations in
Czechoslovakia. Proc. 1st Session of the IAEG,
183-198. Prague.
Rybář, J. and J.Pašek and L.Řepka 1965. Dokumentation
der systematischen Untersuchung der Rutschungsgebiete
in der Tschechoslowakei. Eng.Geol. 1:21-29.
Amsterdam:Elsevier.
Škopek, J. and J.Rybář and J.Dobr 1972. Pore-water
Pressure Observations in a Landslide. 24th
Int.Geol.Congress, Sec. 13:150-159. Montreal.
Špůrek, M. 1972. Historical catalogue of slide
phenomena. Stud.geogr. 19:1-180. Brno.
Špůrek, M. 1982. Long-range prognosis of fluctuation
of the sliding events frequency. Čas.Min.Geol.,27,
3:261-267. Praha.
Špůrek, M. 1983. Hundred years of landslide
investigations in Czechoslovakia (in Czech). Geol.
Průzk. 29:256-258. Praha.
Záruba, Q. 1956. Superficial quasi-plastic
deformations of rocks (in Czech). Rozpravy ČSAV, Ř.
MPV 66:15-135. Praha.
Záruba, Q. 1958. Bulged valleys and their importance
for foundation of dams. 6th Congres des grands
barrages, C.30:509-525. New York.
Záruba, Q. and V.Mencl 1954. Engineering geology (in
Czech). Praha:Nakl.ČSAV.
Záruba, Q. and V.Mencl 1958. Analysis of a landslide
near Klačany on the river Váh (in Czech). Rozpr.
ČSAV 68:31 pp. Praha:Nakl.ČSAV.
Záruba, Q. and V.Mencl 1969. Landslides and their
control. Amsterdam, London, New York:Elsevier.
Prague:Academia.
Záruba, Q. and V.Mencl 1982. Landslides and their
control. Praha:Academia.
Záruba, Q. and J.Rybář and Z.Kudrna 1986. Analysis of a
deep landslide on the bank of the reservoir.
Inž.geol. a chidrogeol., BAN, 15/16:9-17. Sofia.
Zika, P. and J.Rybář and Z.Kudrna 1988. Empirical
approach to the evaluation of the stability of high
slopes. Proc. 5th Int.Symp. on Landslides, Lausanne,
Vol. 2:1273-1275. Rotterdam, Brookfield:Balkema.
Zvelebil, J. 1985. Time prediction of a rockfall from
a sandstone rock slope. Proc. 4th Int.Symp. on
Landslides, 3:93-95. Toronto:University of Toronto
Press.

Landslides: Extent and Economic Significance, Brabb & Harrod (eds)
© 1989 Balkema, Rotterdam. ISBN 90 6191 876 6

A review of landslide processes on the territory of the Soviet Union

L.K.Ginzburg
Ukrspetsstroiproekt, Dnepropetrovsk, USSR

ABSTRACT: The article briefly reviews some problems associated with the landslide phenomena on the territory of the USSR. It deals with the landslide events in various regions of the Soviet Union; examples of some large-scale soil shifts and their causes are given, and organizational schemes for landslide observation and fighting are shown. A good deal of the information contained in the article is taken from the literature referenced below.

INTRODUCTION

The territory of the Soviet Union equals 22.4 mln sq.km, with a population exceeding 270 mln people /1/. Obviously, the relief and the natural conditions on such a large territory are very diversified. The relief in the USSR is very non-uniform and contrasting /1/. The average surface elevation of the country is about 430 m above sea level, while its highest point - the Kommunism peak in the Pamir reaches 7495 m and the bottom of the Karagie depression in Western Kazakhstan lies 132 m below the ocean level.

The USSR is mostly a country of vast plains which totally comprise 68 percent of its territory. The regions with flat terrain are confined to platforms - tectonically quiet stable formations of the Earth's crust, where the folding movements were over long ago.

The mountainous regions feature a complex geological structure and a heavily broken and contrasting relief. Territories with a mountainous relief totally occupy less than 1/3 of the territory of the Soviet Union. They are mainly found along the southern and eastern outskirts of the USSR, forming an enormous amphitheatre framing the plains of the western half of the country. In the south this mountainous belt includes the Carpathians, the Crimean mountains, the Caucasus, Kopetdag, the Pamirs, Tien Shan, the mountains in the South of Siberia,with the mountainous systems of the North-Eastern Siberia and the Far East in the East. They mostly compise the so-called medium-high mountains ridges and massifs of 1500-2000 m height. However, the summits of some mountain ranges in the South of the country raise above 4-5 thous.m, while in the Pamirs and Tien Shan their height reaches 6-7 thous. m. The mountain massifs are cut throughout by a network of deep, often canyon-like valleys through which violent streams flow.

Since mountains cover about 7 mln sq.km of the territory of this country, the problem of landslide stabilization is very acute in the Soviet Union. In non-developed mountainous regions where the surface of the mountains remained untouched, landslides and downfalls occur very seldom, and if they occur, they usually cause no harm, while in the regions of human activity where under-cutting of slopes, removal of topsoil, development of deep foundation pits begin to take place, the landslide processes are activized and they often bring serious harm to the national economy of the country.

The Soviet Government adopted a detailed decision"On the Improvement of Nature Protection and Utilization of Natural Resources" which establishes reliable control for the proper utilization of land, water, forests, bowels of the earth and other natural resources, control of activities to prevent erosion of soils and to establish regulations for their recultivation,prevention of atmospheric air, soils, surface and underground water pollution, preservation of water protecting forests, vegetable and animal life. The same government decision allows to start any activities in mountainous regions only after an adequate checking of stability of slopes with a simultaneous provision of the necessary anti-landslide operations.

Nevertheless, landslides occur not only in the mountainous regions of the USSR. Large territories of plain areas are cut with ravines whose edges often become a cause of landslide movements. Such ravines are rather numerous on the territory of the Ukraine, Moldavia, Central Asia, etc.Even in the capital of the USSR -Moscow which is generally situated on a flat terrain, landslides occur from time to time.

River banks prone to water abrasion are also places of landslide occurence.Landslides and soil downfall on steep banks occur on the such rivers as the Dnieper, Volga, Yenisei and many others.

The stability of the open pit edges often deteriorates when minerals are mined opencast. Thus, in the Soviet Union a great deal of attention is paid to the stabilization of open pit edges and prevention of downfalls of the surrounding soil masses. Underwater landslides occur in the USSR as well. One example of underwater soil sliding down from the sea shore into the sea is Pitsunda peninsula in the Caucasus, where great efforts are being taken now to preserve the benches.

Such landslides are characteristic for many areas of this country. The main of such areas are shown on the map/2/,Fig.1.

Figure 1. Schematic map of the Soviet Union

Activization of the landslide processes is often connected with the periods of rainfalls. Showers saturate the soil masses being in their ultimate condition, thus making the slopes unstable. Therefore it is important to take into consideration the amount of rainfalls in various regions of the country. The average annual precipitation in the Soviet Union is shown on the map /1/, Fig.2.

Seismic influence is another important factor which causes instability of landslide-prone slopes. In the USSR earthquakes occur in the western, southern and eastern regions of the country.

The problem of landslide prevention is rather urgent. Displacements of large masses of soil on unstable slopes often inflict substantial damage to the national economy of the country, interferes with the construction activities and normal functioning of industrial enterprises. Unstable slopes occupy large territories of Russia, the Ukraine, Moldavia, Central Asian Republics /4/. During the construction of the Baikal-Amur railway line there also occurred problems with designing the roadbeds on landslide slopes and with installing antilandslide facilities /4/. Hence, aspects of protecting territories against landslides and of their monitoring, as well as provision of the most effective antilandslide measures are of substantial importance nowadays.

ORGANIZATION OF ANTILANDSLIDE SERVICE, DEVELOPMENTS AND PUBLICATIONS

The founder of the soil mechanics as a science, prof. K.Terzagi said quite clearly at the IV International Congress (London, 1957):"Those of you who had little or no experience with landslides on natural slopes may be rather optimistic to think that the problem of estimating the degree of stability of such slopes is dependent mostly on the degree of taking samples from the mass and on the methods of analysis and computations soils. In many cases it proves to be impossible to estimate the degree of slope stability from the landslide viewpoint before the landslide proper occurs".

Hence, regular monitoring of landslide slopes is of great importance for timely prevention of the possibility of landslide occurence. In the Soviet Union such duties are performed by landslide stations located in numerous regions of the country /5/. These organizations in the form of engineering-geological parties, carry out studies of the landslide process regime in various regions of the USSR.

Such studies include geodetic observation of the dynamics of shifing of the countersunk and surface marks arranged on a slope, and of the level of the underground water, as well as periodic visual observation of unstable slopes and drilling of holes to study the surfaces of slide when such necessity occurs. The results of such studies are plotted on diagrams of landslide movements which allow to forecast the moment of catastrophic shifts /5/.

In addition to landslide stations, in many dangerous regions of the country there are antilandslide departments which, first, act as customers for constructing antilandslide facilities, and, second, give permission for any construction projects within a landslide-prone region. This enables to prevent disruption of slope stability by the activity of man and ensures provision of adequate protective facilities.

Numerous other organizations belonging to different administrative and industrial departments are also involved into the problems of landslide studies, forecasting and prevention /5/. All enterprises involved in

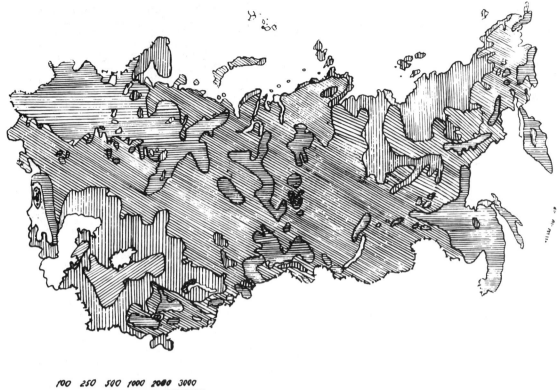

100 250 500 1000 2000 3000

Figure 2. Average annual precipitation, mm

designing or construction on slopes or in their vicinity, face the necessity of solving landslide problems to a certain extent. The number of publications on that subject is also growing, mostly in the form of short articles /5/. However, the number and volume of publications dealing with landslides is not yet adequate to give a detailed characteristic of the numerous regions in the USSR. Many of the important mechanisms of the landslide process development, especially the time-dependent ones, have not been revealed; the problems of its development and dynamics have not been adequately studied, forecasting of landslides has not yet been adequately developed /5/.

As stated in the above quotation by K.Terzagi, an estimate of the slope stability degree by computation still remains quite tiresome. There exist dozens of methods for computing the slope stability coefficient, which give totally differing results. Numerous research, design and surveying organizations in the Soviet Union are engaged in the studies of such computation methods, yet the problem is still far from being solved, both in this country and abroad.

The methods of designing antilandslide means and protective facilities used worldwide are far from being perfect as well.Taking into consideration the fact that the damage caused by the landslides costs millions of roubles,it is a must throughout the world not to save money on the development of reliable methods for studying stabilization of landslides.

LANDSLIDES IN THE BASIC REGIONS OF THE EUROPEAN TERRITORY OF THE USSR

One of the most developed parts of our coun-

try /5/ is the European territory of the USSR; various types and classes of landslides have grown there threatening numerous facilities pertaining to national economy (towns and other settlements, roads and pipelines, hydrotechnical facilities, agricultural, etc).

Within the territory of the Northern Caucasus /5/ slipping landslides are widespread /Fig.3/ which are mostly developed in the Paleogene and Neogene, often weathered clayey rocks.

The main portion of the slipping landslides within that region have a surface of shift which is usually confined to weathering fractures or other weakened zones.

Within the Mountainous Crimea region /5/ the slipping landslides are confined to clayey states of Medium Jura and Taurida series which includes a complex set of flish deposits of the Upper Triassic and Lower Jurassic age. The landslides are most often connected with the lithological differences, which are presented by a finely-rhythmical alteration of the argillites, aleurolites and sandstones. Of a more sofisticated nature is the mechanism of gigantic landslides which occur as a result of the separation of limestone massifs from the Crimean Yaila. They form large tornaway blocks which are observed in numerous places of the Southern seacoast of the Crimea. The movement of the blocks of rock occurs along large technical fractures which are especially numerous in the folded zones of Yaila (Fig.4).

Shift landslides /5/ are prevalent on the larger part of the European territory of the USSR.They are widespread along the banks of numerous water reservoirs and large rivers of the Russian plain (the Volga, Kama, Oka, Dnieper, Dniester, Bug, Don, etc.) and

215

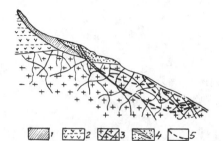

Figure 3. Geological section of a landslide in the igneous rocks of the Caucasus:
1 - slide rocks; 2 - alternating layers of andesitobasalts and limnic clays; 3 - fractured granites; 4 - fractures filled with a clayey material; 5 - supposed surfaces of slide.

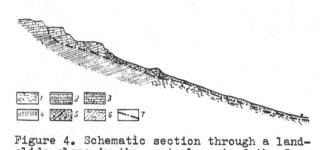

Figure 4. Schematic section through a landslide slope in the central part of the Southern coast of the Crimea:
1 - clayey slates of the Tavrida series;
2 - alternation of clayey slates, aleurolites and sandstones of the Middle Jura;
3 - fractured claystones of the Upper Jura;
4 - sea clays of the Quarternary age; 5 - shifted large blocks of the Jura rocks;
6 - landslide accumulations; 7 - the surface of slide.

their tributaries. Besides, such landslides of rather substantial size, occur in Moldavia, on the coasts of the Black and Azov seas, on the Stavropol height, the Nothern Caucasus, etc. Such landslides sometimes catch large thickness of soil. They are initiated on slopes with a height of 20-120 m and above. The largest landslides are of a frontal shape, the extent of some landslides reaching 1100 m. The volume of some of the landslide bodies amounts to several and dozens of millions of cubic metres.

The complexity of natural conditions, difficulty of access and a great depth of the rock entrainement by the landslides interfere with carrying out a detailed exploration in order to find out some characteristic features of the mechanism of that class of landslides. Substantial difficulties also occur when studying their dynamics due to slow shifts of rock blocks to the stage of landslide preparation, which may last more than a millenium. Such mode of the process development is sometimes interrupted by landslide movements of the second and higher order when noticeable deformations occur on some sections of the landslide slope, with the velocities of shifts sharply growing.

Landslides in cretaceous clayey rocks are also frequent in the European part of the USSR /5/. Such landslides were mostly detailed studied on the banks of the Kuibishev water reservoir and within the Saratov re-

gion. As examples of large landslides confined to the clays of Barremian and, partly, of the Aptian layer, the landslides of the Sokolovaya Gora and Uvek-gora("gora" means mountain) in Saratov may be taken. In spite of the long period of studying them, many important details of the mechanism of these landslides still remain unclear in some aspects. The landslides feature a frontal shape with a distinctly observable block-type structure. The width of some landslide stages reaches 50-80 m. The length of the whole landslide slope along the axis of shift is 400 m on Sokolovaya gora and 300 m on Uvek-gora. The width of these landslide areas is 1500 and 800 m respectively, while the height of the slope from the shore line to the edge of the plata is about 137 and 128 m. The position of the main surface or shift zone was not specified exactly, and according to the majority of researchers it passes through the clays of the Barremian layer. However, it is supposed that there are several surfaces of shift within the mass of the Barremian clays on the slope of Sokolovaya gora, thus leading to the conclusion about a multilayered motion of this landslide. The last catastrophic shift of the landslide on Sokolovaya Gora which occured in July 1968, caught almost the whole landslide slope. In the middle portion of the slope the horizontal shifts reached several metres. Similar catastrophic movements also occured on the Uvek-mountain.

Generally, landslides along the Volga banks are very widespread /6/, and a great deal of attention is being paid to fight them.

Among the most active landslide regions of the European part of the USSR we should also mention: the Black Sea Coast in the region of Odessa where large-scale shore-reinforcement activities are carried out to fight the landslides; the banks of the Dnieper in the region of Kiev, Kremenchug, etc; ravine areas in Dnepropetrovsk, Kharkov, Rostov; the slopes of the Carpathians where very large sums are spent to stabilize the landslides, etc.

Landslides in Moldavia should be noted in particular as being developed in horizontal clays of the Sarmatian layer /5,7/. Exploration of some of the landslide areas revealed their complex multilayered structure. As an example, the Bykovets regions may be taken where the landslide slope has a length along the axis of shift up to 460 m, the width being about 1500 m. The average slope inclinations vary from 10° in the lower part to 20-25° in the upper one. From 4 to 7 slide stages of 50-70 m wide are distinguished on the slope. The depth of the rocks caught by the landslide usually does not exceed 20 m. As a result of one of the largest movements on the slope a landslide occured at night from the 11-th on the 12-th of March, 1967, which caught the western part of the slope. The volume of the shifted masses amounted to about 3 mln m3 (Fig.5). Abundant precipitations during the autumn, winter and spring seasons resulted in a sharp increase of the hydrodynamic pressure and deterioration in the strength characteristics of the soil, and gave an impulse for the landslide.

It is impossible to describe all the large scale landslides of the European part of the Soviet Union in a single article, but even from the stated above it is quite cle-

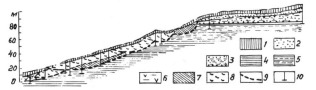

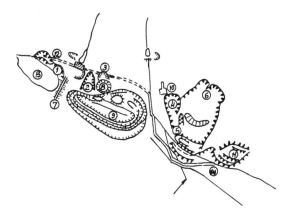

Figure 5. Geological section through a land-slide slope in the region of Bykovets (Moldavia):
1 - loams; 2 - finegrained sands; 3 - fine sands with clay inter layers; 4 - layered clays with aleurite interlayers; 5 - aleurites; 6 - coarse grained sands with interlayers of humusized clays; 7 - fresh-slipping clays; 8 - severely kneaded structureless clayey rocks; 9 - surface of the current landslide shifts; 10- geological survey well.

Figure 6. General chart of the Angren region:
1- dam-adjacent landslide (20 mln m3), occured in 1954; 2 - Turksky (15 mln m3), 1954; 3 - Dzhigiristansky, 1969; 4 - Zagasansky (20 mln m3), 1958; 5 - Bagaransky (400 thou m3), 1950; 6 - Atchinsky (800 mln m3), 1971; 7 - dam; 8 - Dzhigiristansky open pit; 9 - open pit colliery; 10 - underground mine; 11 - disposal areas; 12 - water drain tunnel; 13 - Ahangaransky water reservoir; 14 - Ahangaran-river excurrent canal.

ar that this problem is very important and requires constant attention.

LANDSLIDES OF CENTRAL ASIA

In Central Asia landslides most often occur in the contact zone of loess soils with other rocks, through "weak" clay bands.The volume of the shifted masses reaches from 10 thou.m3 to 40 mln m3. In such cases a landslide is,as a rule, characterized by the durable preparation period and relatively fast and catastrophic movement of masses.

In 1981, about two thousand of large-scale current landslide processes were registered in the South-East of Central Asia /8/. Landslides with a volume from tens thousands to hundreds of millions cubic metres are referred to large scale landslides.

The folded mountainous region of Central Asia which occupies 56 mln hectares has a mountainous landscape in which grandiose rises of Earth's crust are connected /8,9/. It includes knots of mountain systems where ridges of Tien Shan and the Pamirs meet. The mountain structures of Tien Shan within the borders of the Soviet Union extend from west to east 1200 km with a maximum width in the west equal to 400 km.

Central Asia is a classical region of occurence of loess rocks of deluvial, deluvial-pluvial and propluvial origin which occur mostly in foothill regions with absolute elevations from 500 to 1500 m and in mountainous regions - from 1500 to 3000 m. The majority of the genetic types of the loess rocks in Central Asia was formed by the action of water. The finegrained material which the loess bed is comprised of is initially produced in mountainous and Alpine zones as a result of weathering of bed rocks. The initial material for the production of loess is actually prepared in the mountainous zone /8,9,10/.

In Central Asia the loess rocks in the landslide slopes most often move on sand-clay deposits, claystones and tuf sandstones. In each case the weakened zone (the surface of slide) has its own characteristics features expressed in the form, power and character of the influence on the modification of the physical, chemical and strength properties of the rocks.

One of the typical regions of development of landslides in clayey rocks is the Angren

region. It contains /8/ eight landslide areas (Fig.6); the main areas are the following: Bagarnsky, 1950 (volume 400 thou. m3), Turksky, 1954 (25.0 mln m3), Verhne-turksky, 1954 (20.0 mln m3), Zagasansky, 1958 (about 20 mln m3),Dzhigiristansky,1969 (100 thou m3) and over the open pit of burnt rocks, 1969 (800 thou m3). In 1972 on the left-bank slope of the Ahangaran river a huge moving landslide started to develop - the Atchinsky landslide, which is a vivid example of the influence of the activity of man on the changing in geological environment.

The formation of the Atchinsky landslide is connected with the extension of the dis-location trough which has been formed during 17 years /8/. As a result of gas release from the coal seam 5-15 m thick at a depth 100-130 m in the middle part of the slope of an area 1.05 km2 a cavity has been formed with a subsequent sinking of the earth surface by 5.0 m. All that led to the loss of the lateral support for the mass of the Cretaceous and Paleogene rocks located above along the slope, to the change of the underground water regime and formation of the Atchinsky landslide with a volume of up to 700-800 mln m3 on an area about 8.0 km2. On the right bank in the area of Teshiktash settlement within the 1 st terrace there appeared bulge-out swells which caused destruction of houses, deformation of the vehicle road, of the river bed, water excurrent canal and the bridge across the Ahangaran river /8/. Further activization of the land-slide could lead to a critical situation with catastrophic consequences. First signs of the landslide formation were noticed in 1972, in the upper part of the slope. Gaps formed in the loess rocks, while on the right bank two bulge out swells were observed situated opposite one another at a distance of 1.5-1.8 km, with the development of fractures inclined to the Ahangaran river bed at an angle of 25-30°. The fracture on the bulge-out swell 1 (the eastern part of the landslide) extended for 175 m and de-

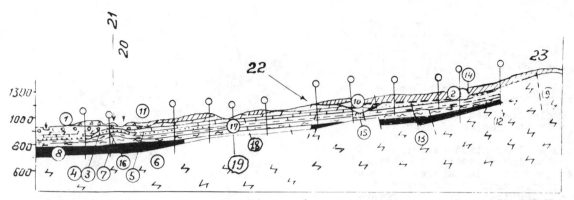

Figure 7. Geological section through the Atchinsky landslide:
1 - alluvial gravels; 2 - loess rocks; 3 - paleogene marls; 4 - limestones; 5 - sandstones, sands, clays; 6 - cretaceous gravelites, sandstones, clays; 7 - Juraceous clays and sandstones; 8 - coal; 9 - quartz porphyrys; 10- deposits of ancient landslide movements; 11 - downfalls; 12 - Shaugazsky thrust; 13 - line of faults; 14 - break-off wall; 15 - well dump cut depth; 16 - flexure folds; 17 - surface of slide; 18 - bandslide movement inclination angle according to a surface mark; 19 - gassed-out space; 20 - the Ahangaran river; 21 - bulge-out swell; 22 - dislocation trough; 23 - surface marks.

formed a road and houses. On the bulge-out swell 2(north-western part of the landslide) the bulge-out fracture was observed for 200-250 m and deformed the bridge across the river and houses.

An orientation section through the axis of the Atchinsky landslide is shown in Fig.7.

Extensive studies of the Atchinsky landslide according to a unified programme were started in 1974 /8,9/. The work was performed systematically, the results were discussed twice a year. Additional consultants were invited when it was necessary. All that ensured high quality of research carried out and allowed to find timely solutions for the problem of providing protective measures against the landslide of the "century" /11/.

The danger of such landslide is quite frequent in Central Asia.

Heavy showers in Tajikistan in spring 1987 /12/ caused much trouble in mountainous regions-heavy landslides, powerful mudflows, downfalls. In the town of Nurek houses were destroyed and accidents took place on the mountainous highway Nurek-Dushanbe. Calamity destroyed a large section of the road, several irrigation facilities, bridges across the mountainous rivers. A settlement Sari-Guzar suffered heavy losses. The resulting material damage amounted to millions of roubles /12/. The situation was worsened by a simultaneous earthquake with a force 3-4. Downfalls occured simultaneously in twenty places.

Landslides of enormous volume became active in Kirgizia in March 1988 /13/. After sleep people first of all cast alarmed glances at the steep slopes which like a stone necklace surrounded the village of Katurgan. The huge tongue of the landslide seemed to get frozen in its place. But every time it was noticed by an attentive eye that the enormous mass of clay (as specialists calculated later, its volume equalled 1.5 million m3) slowly but continuously moved to the settlement, about 15-20 cm per day. Only in the last days before the houses were caught by the many meters thick layer of mud, the landslide quickened its pace. There have been several such landslides this year. There were no casualties

only due to the fact that people were removed in time from three settlements. Machinery, aviation, foodstaffs and clothing were urgently provided. The damage suffered by such regions of Kirgizia amounted to over ten million roubles /13/.

At present special geological teams were formed in Kirgizia which observe the landslide conditions and issue annual bulletins with the landslide forecasts in the Republic, and send warnings about the current situation.

Such activities must obviously be extended, not only in Central Asia, but all over the Soviet Union, and throughout the world.

LANDSLIDES CONFINED TO THE ACTION OF WATER

Water abrasion was always the cause of numerous landslides. Steep river banks often fall down as a result of their intensive undermining by water. It is worth to mention, for instance /3/, the fall-down-landslide of 25 mln m3 volume of rock on the Zeravshan river (Central Asia, April,1964), when numerous enterprises and the town of Pedzhikent were endangered.

A landslide unexpectedly started in February, 1974 on one of the newly built mainroads in the city of Gorky, on the Oka river /3/. The volume of the soil masses which were shifted to a depth of 15 m amounted to about 150 thou m3.

The landslide phenomena along the banks of the newly formed water reservoirs are quite common since the soil masses which earlier had been above the level of the underground water, get moistened. These phenomena become complicated almost in all cases when the water level in the reservoirs fluctuates. Sometimes in such cases thinks occur due to the efflux of sand masses from the strata of the bank slopes. In one of the settlements along the bank of the Dneprovsky water reservoir an unexpected subsidence with an amplitude of shift up to 15 m was observed /3/. It should also be noted that the products of downfalls and landslides decrease by themselves the useful volume of the water reservoirs and in many instances accelerate the process of their silting. Slumping of the loamy slide rock sheets which cover the steep fractured rocky edges of ravines often leads to an increase in filtration losses in the roots of dams.

Coastal landslides (pelagic and fluvial) are always connected with the water area, and that is usually noticeable in their behaviour. For instance, a huge landslide near the settlement Zolotoy Plyazh (Golden Beach) in the Crimea sometimes terminates in bulge-out swells in the sea bottom . Similar patterns have numerous caucasian landslides as well.

However, an inverted picture is also rather frequent, viz."dragging away" of the sea beach material by the sea, slitting of the material along the steep underwater slope. Thus the so-called underwater landslides are generated.

A similar picture is observed nowadays at one of the best health resorts - on the Pitzunda Cape in the Caucasus /14/.

With every gale the south-western shore of Pitzunda began to retreat, with the relict pines falling down. As far as twenty years ago the pine forest on the left bank of the Bzyb river (which falls into the sea here) was hundred metres off the sea. Since that time the sea approached and most of the trees perished.

This, to our mind, occurs because the submarine slope is very steep here, and the waves at the shore line are as high as in the open seas. All the changes take place over a short distance and at the very last moment. Just by the shore a wall of water rises and falls straight on the beach,thus a powerful surf flow is formed which runs up the beach tens of metres. The waves with an average height of three-four metres on deep water, increase in height within the surf zone to five-six metres, i.e. almost half as high. The work performed by such waves in the shore zone is enormous, therefore they change the shore severely and destroy strong concrete walls. Oceanologists calculated that the strikes of the waves about 7 m high per kilometer of the shore line have an energy of about one million h.p.

Various structures of wave suppressors have been proposed, but their effectiveness raised great doubts. Traditional groins and breakwaters proved unacceptable for Pitzunda because the submarine slope is very steep there. Specialists have thought over the problem of building a gigantic pneumatic breakwater around the cape, in order to suppress the waves at a far distance off the beach by means of compressed air supplied through pipelines. Such breakwaters proved feasible in some sea ports. But it was too difficult and expensive to realize such a project at Pitzunda.

An extensive research program which was undertaken, proved that the main reason of such negative phenomena at Pitzunda is the dragging away of the beach material and absence of its restoration by the Bzyb river which had changed its bed. The main problem in Pitzunds is the protection of the shore. To this end it is necessary, first of all, to return the mouth of the Bzyb river to its original place and make the river "to work for the shore". If it does not give the required result, then it would be necessary to fill the beach with pebble brought from other places. The required quantity may be defined only by experience. It should be done without delay since each passing year makes it more difficult to

protect the shore, and a moment may come when nothing will help and enormous material expenses would be necessary /14/.Therefore the work is being done there very actively now.

CONCLUSION

It is impossible to describe in a single article all the landslides which occur on such vast areas as the territory of the Soviet Union. We have not paid enough attention, for instance, to the frequent landslides which occur on the vehicle and rail roads of the country. Thus, in April 1985 a landslide in one day cut the railway line /15/ near Sochi. By the way, Sochi (a health resort on the Black Sea Coast of the Caucasus) in far from being a safe place from that point of view /15/. 60 percent of the territory of that town (it covers 145 km along the sea shore) are prone to landslides.

Similar defects are noted for the coastal areas of the Azov and Caspian seas. Various landslides have developed on the slopes of the valleys of the largest rivers of Siberia,the Yenisei, Ob and their tributaries, as well as on the shores of the unique lake Baikal. These landslides are presented by numerous regional types and differ from the landslides in the European part of the USSR /16/. The formation of downfalls and landslides on the Baikal shores is also fostered by the seismic processes which determine their inherent features.

About 47 percent of the USSR territory is in the zone of permafrost. Landslides are also widespread in this zone. They are closely connected with the thermal conditions of the soils and have a number of their own features.

Thus, the problem of landslides includes many various aspects and requires close attention since the damage inflicted by the landslides is enormous. It was estimated[17], for instance, that during less than two centuries only on one Soviet city-Odessa the landslides took away almost seven hundred hectares of land, often together with the houses and other buildings. This area exceeds the territory of the historical centre of Odessa with dozens of its picturesque blocks of buildings.

When preparing territories for construction on inclined sites the hardest conditions among all the slope-inherent processes are created by landslides. The development of landslides on a construction site increases the costs /18/ of the initial operations by 20-100 percent.

All that dictates that we have to constantly study the landslides and develop effective antilandslide means.

It seems reasonable to conclude this article with the words of the Soviet writer Chingiz Aitmatov. A character of one of his novels says:"what is the cause of landslides, these inevitable motions, when huge hillsides and even mountains move and collapse, widely opening the hidden abiss of the Earth. And people get horrified seeing what chasm is under their feet. The perils of landslides are in the fact that the catastrophe ripens unnoticeable, from one day to another, because the underground waters continuously undermine the basis of the rocks and only a minor vibration of the

earth, a thunder or a shower is enough for
a mountain to start slowly but inevitably
sliding down. An ordinary collapse occurs
unexpectedly and at once. But a landslide
moves formidably and there is no power or
force which is able to stop it..."

REFERENCES

1. Countries and peoples. Popular scientific
   publication on geography and etnography
   in 20 volumes.Volume-Soviet Union.General
   review. Russian Federation. Moscow, Mysl,
   1983.
2. Maly atlas mira. Glavnoye Upravleniye geo-
   dezii i Kartographii(under the Council of
   Ministers of the USSR), M.,1981.
3. Maslov N.N.Mechanics of rocks in the practice of construction(Landslides and their
   elimination).Moscow,Stroiizdat, 1977.
4. Ginzburg L.K. Antilandslide stabilizing
   structures. Moscow, Stroiizdat, 1979.
5. Küntzel V.V. Features of the landslide
   process on the European territory of the
   USSR, Moscow, Nedra, 1980.
6. Cheprasov A.F.Landslides and experience
   in fighting them. Nizhnevolzhskoye knizh-
   noye izdatelstvo. Volgograd, 1972.
7. Landslides of Moldavia and environment
   control.Theses of reports, conference in
   Kishinev, June, 27-28, 1983.,Kishinev,
   1983.
8. Niyazov R.A. Formation of large landslides in Central Asia. Fan Publishers,Uzbek
   SSR.Tashkent, 1982.
9. Niyazov R.A. Landslides in loess rocks of
   the south-eastern part of Central Asia.
   Fan Publishers, Uzbek SSR., Tashkent,1974
10. Presnuhin V.I., Reiman V.M.,Markov A.B.
    On classification of landslide deforma-
    tions in Tadzhikistan. Collection of ar-
    ticles. Hydrogeology and engineering geo-
    logy.Donish Publishers, Dushanbe, 1975.
11. Dimov G."Landslide of the century" is
    stopped. Izvestiya.February,11,1987.
12. Karpov A.,Orlov V. Echo of the terrible
    natural calamity. Izvestiya, May,6,1987.
13. Shipitko G. Kak sneg na golovu? (From
    the place of the accident). Izvestiya,
    April,8, 1988.
14. Menshikov V.L., Peshkov V.M. The shore
    of Pitzunda: facts and hypotheses. Moscow,
    Mysl Publishers, 1980.
15. Arsenyev V. Landslide in broad daylight.
    Izvestiya, March,18,1985.
16. Yemelyanova E.P.,Zolotarev G.S. Land-
    slides in the USSR and problems of their
    study. Materials of a conference on the
    problems of studying and fighting land-
    slides.Kiev,1964.
17. Knop A., Shmyganovsky V. Really a golden
    shore. Izvestiya, July, 17, 1985.
18. Bileush A.I.,Seredian Y.I.,Marchenko A.G.
    Shtekal A.S. Engineering preparation of
    territories in complex conditions. Budi-
    velnik Publishers, Kiev, 1981.

Landslides: Extent and Economic Significance, Brabb & Harrod (eds)
© *1989 Balkema, Rotterdam. ISBN 90 6191 876 6*

The extent and economic significance of the debris flow and landslide problem in Kazakhstan, in the Soviet Union

A.Yu.Khegai & N.V.Popov
Kazselezashchita, Alma-Ata, USSR

ABSTRACT: This paper discusses landslide problems, especially debris flows in the mountainous regions of Kazakhstan in the northern and western Tien Shan Mountains, the Dzungarian Alatau Range, and the Altai Mountains. The extent and economic significance of debris flows and landslides in these areas are different and depend on a variety of factors. Protective measures to mitigate and reduce the hazard have been developed mainly in the northern Tien Shan Mountains. The cost of these structures and landslide damage exceeds $500 million.

1 INTRODUCTION

This report on the debris flow and landslide problems in the mountains of Kazakhstan (Fig. 1) is based on data obtained by Kazglavselezashchita

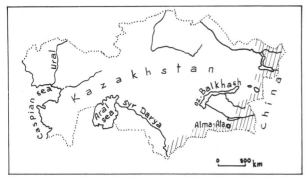

Figure 1. Mountainous regions of Kazakhstan (diagonal lines).

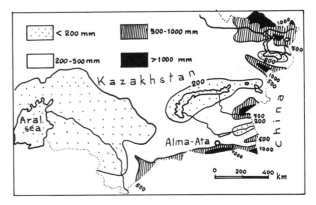

Figure 2. Average annual precipitation distribution.

researchers. Kazakhstan (or the Kazakh SSR) occupies 2.17 million sq.km² in the southern part of the Soviet Union. Debris flows and landslides occur widely in the southeastern part of this republic. Historically, these areas have had the most development. Some of these debris flows have impacted Alma-Ata, the capital of Kazakhstan, with approximately 5 million people, and other towns and settlements near and in the mountains. Debris flows and landslide activity in the last few decades increased substantially, causing damage estimated to exceed 700 million rubles (about $US 500 million).

Precipitation in the mountainous regions of Kazakhstan varies with elevation; the highest peaks receive the most precipitation. Annual precipitation ranges from 250 to 1,000 mm, and even more in some areas (Fig. 2).

Southeastern Kazakhstan is seismically active. Several active faults traverse the region.

Glaciation also contributes to landsliding. More than 2,270 glaciers have formed enormous reserves of loose material, high, steep ridges and reservoirs of water that trigger debris flows and landslides, particularly in the Trans-Ili and the Dzungarian Alatau Ranges.

2 LITERATURE ON DEBRIS FLOWS AND LANDSLIDES

Published literature on debris flows and landslides

in the northern Tien Shan Mountains region is ample. Mushketov (1890) and Bogdanovich and others (1914) reported on debris flows and landslides after disastrous earthquakes. Medoyev (1938), Gorbunov (1939), and Palgov (1947) also described landslides. More recent reports on landslides include those by Kolotilin (1963), Fleishman (1964, 1970), Raushenbach (1967), Gorbunov (1974), Makarevich (1970), Zems (1974), Vinogradov (1977), Mochalov and Stepanov (1977), Plekhanov (1984), Popov (1985), Keremkulov (1986), and Medeuov (1986). Many other published reports on landslides and debris flows are available.

An engineering-geological survey has been prepared for a large part of Kazakhstan, especially the most developed areas. In 1986, a map at 1:1,000,000 scale showing areas of debris flow hazard in Kazakhstan was issued jointly with the Institute of Geology, Kazakh SSR Academy of Sciences. More detailed surveys have been carried out in many of the most hazardous areas, such as along rivers in the Trans-Ili and Dzungarian Alatau Ranges.

3 DEBRIS FLOW AND LANDSLIDE MITIGATION MEASURES

General schemes have been developed to protect populated areas against debris flows, landslides, and rockfalls. These include design and construction of engineering works to slow down, deflect, and trap debris flows; avoidance of the most hazardous areas; development of landslide and debris flow warning systems; mapping of areas vulnerable to debris flows; and study of dangerous slope processes. Preventive works envisioned by this general scheme in the Trans-Ili Alatau Range are nearing completion. Preventive measures against landslides and debris flows in other mountainous regions of Kazakhstan are being adjusted to reflect changes in glacial erosion and more intensive development than originally planned.

4 DEBRIS FLOWS AND LANDSLIDES IN THE NORTHERN TIEN SHAN MOUNTAINS.

The Trans-Ili and Jungei Alatau Ranges are in the northern Tien Shan Mountains region of Kazakhstan. The ranges were formed by Caledonian folding (Abdulin, 1981). The most elevated axial parts of these ranges are composed of granitic rocks. Volcano-sedimentary and sedimentary rocks of various composition and age comprise the lower beds. These rocks play a leading role in the formation of solid material in debris flows and landslides.

The large size and frequent occurrence of debris flows and landslides in Kazakhstan make this republic the most hazardous in the Soviet Union. One of the most important factors in the formation of these landslides, according to Medoyev (1938), is frequent seismic activity associated with several tectonic faults (Fig. 3). The largest landslides

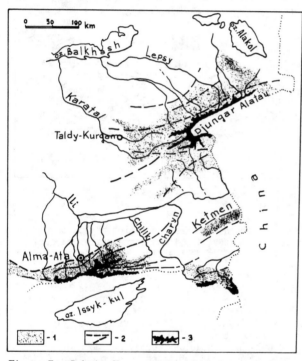

Figure 3. Debris flows and landslides in northern Tien Shan Mountains. 1, Areas of moderate relief where landslides are common; 2, major faults; 3, areas of high relief where large debris flows originate.

are associated with fractured rocks along the so-called "thermal line," a fault with thermal and mineralized water discharge along the northern slope of the Trans-Ili Alatau Range. A spectacular rockfall in the Issyk River valley and the formation of the Issyk rock-dammed lake 8 thousand years ago are related to this fault.

Landslides are also common in the basins of the Talgar, Malaya, and Bolshaya Almaatinka Rivers. A tectonic fault in this area has generated many strong earthquakes, including a disastrous one in 1887 (M=7.3). According to Mushketov (1890), this earthquake caused many landslides in the mountains from 1,500 to 1,800 m above sea level. The total

area subject to rockfalls and landslides amount to about 2,240 sq.km. The volume of some earthflows was estimated at 40-100 million cu m. Due to high moisture content, many landslides and earthflows moved almost like debris flows. Many landslides blocked rivers, forming temporary lakes. When the landslide dams failed some days after the earthquake, debris flows formed (Mushketov, 1890).

The Chilik-Kemin fault (Fig. 4) separating

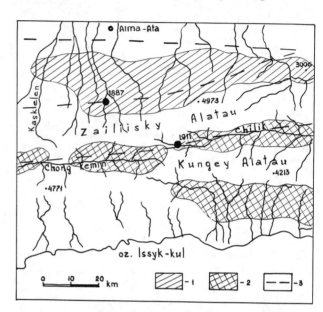

Figure 4. Zones of mass-scale debris flows and landslides associated with 1887 and 1911 earthquakes in the Trans-Ili Alatau Range near Alma-Ata. 1, Area of landsliding associated with the 1887 earthquake; 2, area of landsliding associated with the 1911 earthquake; 3, fault zone.

the Trans-Ili and Kungei Alatau Ranges is one of the regional tectonic faults. Disastrous earthquakes occurred along this fault in 1885, 1887, and 1911. The earthquake in 1911 (M=8.2) occurred during a dry, cold period when little moisture was in the ground. Nevertheless, many large landslides and earthflows caused considerable damage and heavy casualties, especially in the Jungei Alatau Range where the zone of fractures extended 200 km. A number of rock-dammed lakes formed in the northern slope of the Kungei Alatau Range in the basins of the Kaindy, Kolsai, and Uryukty Rivers.

No earthquake-triggered landslides have formed in recent years. Most of the recent landslides are smaller and are related to road building, improper drainage, and excessive grazing.

Debris flows continue to be a problem, however. The most vigorious debris flows are confined to the central parts of the ridges in the Turgen-Uzunkargaly interfluve, Trans-Ili Alatau Range. In this region, the piedmont plain is composed essentially of alluvial fans formed from debris flows. Large boulders transported by these debris flows have been carried as much as 30 km from their origin. The debris flows are triggered by high-intensity rainfall (showers), the breaking of ice and moraine dams, and the melting of snow. The most significant debris flows are confined to zones with Recent and Late Quaternary moraines at elevations between 2,500 and 3,500 m.

Three different debris flow mechanisms have been distinguished by Vinogradov (1977): erosion transport, erosion displacement, and displacement types. Most disastrous debris flows have been highly destructive due to displacement and erosion processes on frontal escarpments of ancient moraines. The significant debris flow foci (downcuts) up to 70 to 100 m deep are in the basins of the Issyk, Bolshaya and Malaya Almaatinka, Talgar, Chemolgan, and Chilik Rivers. Data on reliably documented disastrous debris flows are summarized in Table 1.

country. Vast alluvial fans bear witness to vigorous debris flow activity in the past, the fans manifesting themselves particularly in the piedmont plains on the northern slope of the ridge in the Bien-Lepsy interstream area. Taking account of the

TABLE 1. The main disastrous debris flow occurrences in the Trans-Ili Alatau Range

River Basin	Date	Trigger	Volume million cu m
Aksai	8.06.1887	Earthquake	40
B.Almaatinka	8.06.1887	Earthquake	70
M.Almaatinka	7.07.1921	Shower	1.0 - 3.0
B.Almaatinka	8.07.1950	Shower	1.2
M.Almaatinka	20.08.1951	Glacial lake outburst	0.5
M.Almaatinka	7.08.1956	Glacial lake outburst	1.0
Issyk	6.07.1958	-	4.0
Issyk	7.07.1963	-	5.8
M.Almaatinka	15.07.1973	-	3.8
B.Almaatinka	3.08.1977	-	5.0
Talgar	21.06.1979	-	0.1
Kaskelen	23.07.1980	-	2.0
Issyk	30.07.1982	Shower	1.0
B.Almaatinka	28.06.1988	Shower	1.0

Of special concern are dynamic landslide lake systems occurring at many places in intermontane valleys. Lateral and bottom erosion, the suffosion processes and filtration cause gradual, in some cases rapid, slaking and damage of the landslide dam. Such dammed lakes are justly referred to as temporary (Kolotilin, 1961). Their location in the middle-mountain zone makes them a potential debris flow threat. When the dam breaks, destructive floods and debris flows result. These outbursts were observed in 1983, 1984, and 1988 on the northern slope of the Kungei Alatau Range in the basins of the Kaindy, Kolsai, and Uryukty Rivers where lakes with volumes from 0.7 to 5.7 million cubic meters burst through landslide dams and carried debris flows and floods into the valleys below. All of the landslide dams had formed during the earthquake of 1911.

Landslide and debris flow processes are also active in the eastern periphery of the Terskei Alatau and Sarydjaz Ranges in the basins of the Karkara and Bayankol Rivers within the boundaries of Kazakhstan. Landslides and rockfalls there are also confined to zones of active tectonic faults (Blagoveshchensky, 1979). Their economic effect is low because the territory has little development.

5 DEBRIS FLOWS AND LANDSLIDES IN THE DZUNGARIAN ALATAU RANGE

The Dzungarian Alatau Range in Kazakhstan is a system of several parallel ridges divided by depressions. In the north, the range is surrounded by plains of the Balkhash-Alakol Lake basin and in the south by plains of the Ili depression.

Judging from debris flow activity, the Dzungarian Alatau Range is the most hazardous region of the

fact that, as a whole, the region has been intensively developed only since the 20th century, reliable historical knowledge about debris flow occurrence in the past is lacking. However, the deposits indicate debris flows with volume exceeding several millions of cubic meters.

A disaster in April, 1959 in Tekeli, situated in a depression, can be distinguished among the well documented debris flow occurrences. The disaster was triggered by a vigorous shower at a time when snow was melting. Many landslides, earthflows, and debris flows destroyed dozens of houses, industrial buildings, and other projects, causing heavy causualties.

Most debris flows in this region occur during heavy rainfall, from 30 to 50 mm. The most vulnerable areas are underlain by loess and loess-like loam in low-mountain areas like the territory on the southern slopes of the range in the Khorghos-Borokhudzir interstream area. Debris flows occur at regular intervals, namely once or twice every two years, although they are usually not very vigorous.

In the last twenty years, the Dzungarian Alatau Range has turned into a highly hazardous debris flow region. One of the debris flows in September 8 and 9, 1982 in the basin of the Sarkand River, formed from an outburst of a glacial lake. The debris flow with a volume of more than 2 million cu. m. extended more than 50 km causing considerable destruction in the valley. However, the people were warned about the danger and there were no serious causualties. Additional debris flows related to similar glacial lake outbursts are expected in the region of the Toksanbai Range intersecting the Dzungarian Alatau Range.

Landslide activity in the Dzungarian Alatau Range northwest of the Lepsy River (basins of the Zhananty

and Rgaity Rivers) is quite different. The most
significant landslides, similar to those in other
mountainous regions of Kazakhstan, are confined to
zones of seismicity. According to researchers at
the Institute of Geology, Kazakh SSR Academy of
Sciences, the most significant landslides occurred
in the Kolpakov and Tunkuruz depressions. Many old
landslide deposits are widespread on the valley
slopes of the Zhenishke, Tentek, Orta-Tentek, and
other rivers. Some of these landslides are still
active.

The debris flow at Tekeli, mentioned previously,
is only a small part of the landslide problem in
that area. Slopes of mountains near Tekeli are
covered extensively by landslides. However, all
these landslides are small in size.

Large tectonic faults are present in the
Dzungarian Alatau Range. The epicenters of strong
earthquakes are confined to places where these
faults intersect secondary fractures. Earthquake-
triggered landslides block many intermontane valleys
and form lakes such as Nizhny and Verkhny Zhasylkol
Lakes in the basin of the Lepsy River. The volume
of the landslide dams is 35 and 48 million cu m,
respectively. If the land-slide dams contain blocks
and coarse waste, they are likely to remain in
place.

Several reservoirs formed from a combination of a
landslide and a moraine in the northern and southern
slopes of the Dzungarian Alatau Range in the basins
of the Khorghos, Bolshoy Usek, Baskan, and other
rivers. As a whole, landslides throughout the
Dzungarian Alatau Range are widespread, but they
have had little effect on the regional economics
because of the sparce development. In the last few
years, however, the emergence of new settlements,
industrial projects, and communication networks in
the mountains has raised concern about the potential
impact of these landslides. Engineering and
geological conditions are now taken into account in
designing and constructing buildings and other
structures in the mountains. Special landslide
mitigation measures are also envisioned.

Considering the increased activity of debris flows
and the high potential debris flow hazard on slopes
of the Dzungarian Alatau Ranges, the
Kazglavselezashchita has worked out complex schemes
for protecting populated areas against hazardous
exogenetic processes. Plans have been made to
considerably accelerate mitigation measures,
including the construction of debris flow traps,
reforestation, drainage, and mechancial
stabilization of hillsides.

6 DEBRIS FLOWS AND LANDSLIDES IN THE WESTERN TIEN
SHAN MOUNTAINS

The Talas Alatau, Kirghiz, Ugam, Maidantal,
Korzhentau, and Karatau Ranges in the Tien Shan
Mountains are represented in Kazakhstan mainly by
their northern offshoots. The region is complicated
structurally. The maximum elevation of the Kirghiz
Alatau Range is 4,337 m near the town of Karumbash;
4,488 m near the town of Manas along the Talas
Alatau Range; and 2,176 m near the town of Bestau in
the Karatau Range. The substantial height of these
ranges and their geographical position on the route
of Atlantic air masses favors heavy precipitation
ranging from 900 to 1,000 mm annually. This
precipitation and the high elevation also contribute
to the current glaciation in the high-mountain zones
of the ranges (Fig. 5).

Debris flows are widespread in the western Tien
Shan Mountains, especially in the Talas Alatau
Mountains and on the Kirghiz Range where debris flow
deposits form confluent alluvial fans. In some
places, the deposits are 50 to 100 m thick.
However, there is not much information about debris
flow occurrences in historical time. People
probably avoided these hazardous areas.

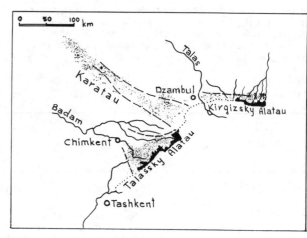

Figure 5. Debris flows and landslides in the
western Tien Shan Mountains.

Widespread debris flows did occur in April, 1959,
when practically all the mountainous regions of
southeastern Kazakhstan were subjected to heavy
rainfall and rapid melting of snow. Debris flows
were most destructive in the basins of the Keles,
Ugam, Sairam, Aksu, and other rivers. The material
loss was estimated in the tens of millions of rubles
and there were heavy casualties.

After 1959, the intensity and extent of debris
flow activity have decreased, although local debris
flows in separate regions of the Kirghiz Range and
the Talas Alatau Range have occurred almost every
year.

The Karatau Range, extending more than 420 km from
the southeast to the northwest, is a peculiar debris
flow region. Despite its insignificant height,
debris flows triggered by heavy rainfall are
prevalent, especially in the center of the basins of
the Bugun and Karashik Rivers. Damage from the
debris flows is negligible because the region is
sparsely populated. Nevertheless, considering
potential development of industry and transportation
networks, the debris flows must be studied to
develop suitable mitigation measures.

The western Tien Shan Mountains are very active
seismically. The largest faults have a northwest
strike. Landslides and rockfalls in the valleys of
the Ugam River and its tributaries are related to
earthquakes along these faults. The earthquake-
triggered landslides dammed several drainages and
formed lakes. Similar lakes occur in the Maidantal
Range and in the highest part of the Talas Range.
Landslides have also been caused by the removal of
natural vegetative cover, especially in regions
underlain by loess-like loams such as the low-
mountain part of the Talas Alatau Range in the
Tyulkubas region.

The plan for protecting populated areas in the
mountainous regions of southern Kazakhstan
envisioned a variety of protective measures in the
Kirghiz and Talas Alatau Ranges.

7 DEBRIS FLOWS AND LANDSLIDES IN EASTERN KAZAKHSTAN

Two large areas with an extensive debris flow
hazard, the Altai and Saur-Tarbagatai Mountains, are
in eastern Kazakhstan. The Altai region is
characterized by highly dissected mountain
topography. Drainage divides in these mountains are
commonly at 2,000 to 3,000 m elevation. The highest
elevation, at the town of Belukha, is 4,506 m.

The main ranges of the region are the Katun, the
Tigerets, the Kholzun, the Listvyaga, the Narym, the
Sarym-Sakty, the Southern Altai, and the
Tarbagatai. The most elevated high mountain areas
have active glaciation. Annual precipitation in

these high mountain areas is as much as 1,500 m. Most of the precipitation falls in the warm period of the year.

Information about debris flow activity in the Altai Mountains is spotty and incomplete. Basins of the Uba, Ulba, Kurchum, Koldzhir, and Naryn Rivers, tributaries of the Irtysh River, seem to have the most debris flows (Fig. 6). Many debris flows occur

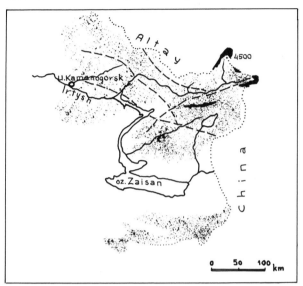

Figure 6. Debris flows and landslides in the Altai Mountains.

on the northern slope of the Sarym-Sakty Range in basins of the Sarym-Sakty and Ush-Kungy Rivers, as well as on southern slopes of the Katun Range in the basin of the Belaya Berel River and the southern Altai Range in the basin of the Bukhtarma River. Debris flows occur in those localities every two or three years. In 1953, disastrous debris flows occurred in the Belaya Berel River when a moraine-dammed lake broke through the dam. Other debris flows triggered by high-intensity rainfall developed in 1963, 1969, and 1975 along the Sarym-Sakty River. The material loss by debris flows in the Kazakhstan Altai Mountains has been estimated by Selezashchita specialists as several millions of rubles. A number of settlements and road networks are within areas affected by debris flows.

Other kinds of landslide processes are not frequent in the Altai Mountains. Their activation is most often related to activities of man than with natural factors. Several natural landslides in surficial materials were triggered by rainfall along the Bukhtarma River in 1958 to 1960, and other landslides have been triggered recently by river erosion. In spite of the small number of these natural landslides, they have destroyed buildings, structures, and land used in agriculture.

Block-type landslides have been observed in the basin of the Bukhtarma River near the Chinghiztal settlement. Many small local landslides with volumes less than 1,000 m^3 occur in the basins of the Kara-Koba and Kurchum Rivers. Landslides are rare in the mountainous areas of the Zaisan depression and the Saur-Tarbagatai Range where rainfall is scarce.

Despite the small number and small size of landslides other than debris flows, they cause considerable damage, mostly to roads. Landslide protection for automobile roads, railways, and individual settlements is envisioned in the scheme for protecting populated areas in eastern Kazakhstan against destructive exogenetic processes.

REFERENCES

Abdulin, A.A. 1981. Geology of Kazakhstan, Nauka. Alma-Ata.:312.
Kolotilin, N.F. 1961. Deformations of hillsides and coastal slopes in the conditions of seismic and debris flow regions in the South-Eastern Kazakhstan. Alma-Ata. AN KazSSR publishers:155.
Medoyev, G.Ts. 1938. Geological conditions of the formation of mud- and stone streams in the basin of the Malaya Almaatinka River. Proc. of the KGMI. Alma Ata. 1:5-22.
Mushketov, N.V. 1890. Verny earthquake in May (June) 28, 1887. Proc. of Geolkom. V.X. No.1. SPb.
Vinogradov, Yu.B. 1977. Glacial breaking-through floods and debris flows. Gidrometeoizdat. L.:154.
Zhandayev, M.Zh. 1972. Geomorphology of the Trans-Ili Alatau and problems of formation of river valleys. Nauka. Alma-Ata:160.

Landslides: Extent and Economic Significance, Brabb & Harrod (eds)
© 1989 Balkema, Rotterdam. ISBN 90 6191 876 6

Landslides: Forms, processes and economic significance in Israel

Moshe Inbar
Department of Geography, University of Haifa, Haifa, Israel

ABSTRACT: Landslide activity in Israel is not very extensive; rockfalls are common along steep slopes of gorges in the Northern basaltic areas and slumps on steep slopes underlain by chalk or marl in the Mediterranean areas. Landslides are a major problem along the cliffs of the densely populated Mediterranean coast. No major landslides have been reported from the extremely arid regions of the country. Major landslides are connected with exceptionnel rainfall and earthquakes. No significant economic damage by landsliding has occurred in the last forty years, but there has been considerable investment in expensive engineering works in protection of roads and buildings. A review is made of the published literature and problems to be studied are suggested.

1 INTRODUCTION

And his feet shall stand upon the mount of Olives, which is opposite Jerusalem on the East, and the mount of Olives shall split in two, half toward the east and the other half toward the west, and here shall be in it a great valley; and half of the mountain shall be left toward the north, and half of it toward the south.

And you shall flee to the valley of the mountains; for the valley of the mountains shall reach the place of disaster and you shall flee as you fled from the earthquake in the days of Uzziah King of Judah.   ZECHARIAH XIV: 4.

This description by the prophet Zechariah is the oldest reference of a probable landslide in the Holy Land (Wachs and Levitte, 1983). The event occurred presumably during the Uzziah earthquake in the period 790-745 B.C. (Schalem, 1949). However, Israel is not a region with extensive landslide activity, nor is it characterized by intense seismic activity (Kafri, 1988).

Areas with steep slopes underlain by chalk or marl are susceptible to landslide activity. Rockfalls are common in deep river canyons and along cliffs. The development of new areas and expansion of already-built areas in the last

century have increased landslide activity by several fold and the risk of damage to human lives and property.

2 PHYSIOGRAPHY AND GEOLOGY

The physiography of Israel is characterized by several climates and slope forms which determine a large number of regions and subregions. The country is about 400 km long and 40 to 80 km wide with a total area of 28,000 sq. km. (Fig.1).

It can be divided broadly into 3 main physiographic units: 1) the Mediterranean coastal plain, 2) the central mountainous backbone, and 3) the Jordan Rift valley, which is part of a transform plate boundary. Each unit comprises several morphotectonic and climatic provinces, but they can be subdivided into three main units: northern, central, and southern. The northern and central have Mediterranean climate and the southern has a hot, desert climate (Fig. 2).

The elevation of the narrow coastal plain rises gradually from 10-20 m at coastal cliffs to about 100m. The central mountainous area has elevations generally between 500 and 1,000 m, reaching 2000 m in the north. The Jordan rift is a linear depression as much as 400 m below sea level in the Dead Sea area. It extends from the northern Jordan valley in the north to the Dead Sea and Arava depression in the south, trending north-south along the eastern border of Israel a distance of 420 km. Fresh fault scarps several hundred meters long, occur in the rift along with echelon strike-slip faults, pull-apart basins, and compressional structures (Garfunkel, 1981).

The geology of Israel is outlined in Fig.3. Detailed geological maps at 1: 20,000 and 1: 50,000 scale are available for most of the country. The coastal plain is underlain by Quaternary marine and continental deposits. The main rock types are partly-consolidated sandstone of Pleistocene age interbedded with red-brown sandy loam (Yaalon, 1967).

The central mountainous area is underlain by several hundred meters of carbonate marine rocks of Mesozoic and Cenozoic age. Marl and soft chalk rocks in several formations of the Cretaceous and Paleogene Mount Scopus group are the most susceptible one to mass movements. The thickness of the group reaches over 250 m, filling synclinal basins, a common structural configuration in the

Figure 1. Map showing the location of Israel.

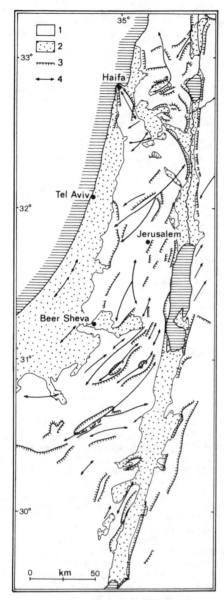

Figure 2. Physiography of Israel. 1. Mountains
and hills; 2. Valleys and plains; 3. Escarpments;
4. Anticlinal axis.

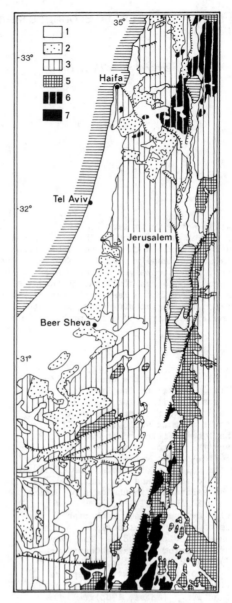

Figure 3. Geology of Israel
1. Neogene-Quaternary; 2. Paleogene, mainly marine
carbonates; 3. Upper Cretaceous, marine; 4.
Triassic-Jurassic, mainly marine; 5.
Paleo-Messozoic, mainly continental sandstones; 6 .
Meso-Cenozoic volcanics; 7. Precambrian, mainly
acid intrusives.

Middle East (Arkin, 1988).
   The Jordan valley is filled with alluvial and
lacustrine clastic sediments.

3 CLIMATE

Israel's climate is influenced by three main
factors: altitude, latitude and proximity to the
sea. Rainfall is Mediterranean in character with
a summer dry season and a winter rainy season
(Fig.4). Mean annual rainfall ranges from 1,500 m
in the upper elevations of the northern
mountainous area to 30 mm in the southern part of
the country. About half of the total area
receives less than 200 mm annual rainfall (Fig.5).
Rainfall intensities are about 50-100 mm/day, and
they decrease from the coastal plain area to the
inland areas. Intensities of 20 mm/hour have a
50% annual frequency.
   A breeze from the Mediterranean Sea is common
during the summer months. Easterly, dry desert
winds occur in the transition seasons from October
to November and March to April.

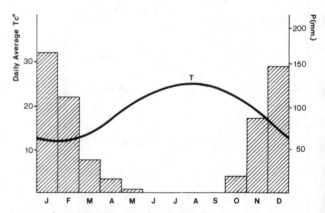

Figure 4. Mt. Carmel climograph showing
temperature and mean monthly precipitation.

228

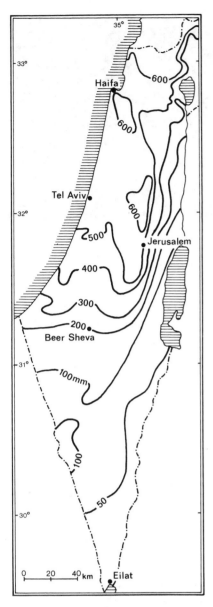

Figure 5. Map showing mean annual precipitation, 1931-1960. Isohyets in mm. (Sharon and Kutiel, 1986).

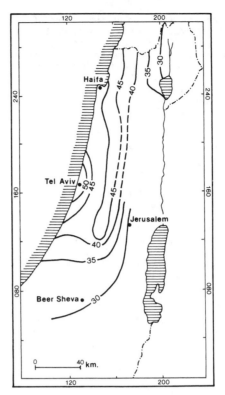

Figure 6. Isopluvial map indicating the amount of rain, in mmm. that could fall in 60 minutes with a return period of 20 years (5% proability) (Schein and Buras, 1973).

The mean annual temperature for the Mediterranean area is 18 C.

4 DISTRIBUTION OF LANDSLIDES

Two types of landslides are common in the mountainous area: rockfall and slump, as used by Nilsen and Brabb (1975). Rockfalls occur along steep slopes of basaltic gorges in the Golan Heights, the Jordan River gorge and the Lower Galilee. They are promoted by heavy rains and occasionally triggered by earthquakes. The falling material consists mainly of basaltic columns which overlie weathered lava flows. Scars left by the rockfalls are mostly ten meters high and about 30 m wide with a volume in hundreds of cubic meters (Inbar and Evenir, 1987).

In the extremely arid region, no major landslides have been reported. The process of escarpment retreat in Sharm el Sheikh cliffs, southern Sinai, was described as: a) disintegration of a lower clastic layer and, b) collapse of a cap-rock resistant layer by rockfall

process. Rate of retreat estimates are low: 0.1-0.2 m/1000 years on average for a granite escarpment, with 0.1 m as a minimum value, and 2.0 m/1000 years as a maximum rate (Yair and Gerson, 1974). No collapse of blocks was observed during a major earthquake (VIII-IX, amended Mercalli scale) which affected the area in March 1969.

Landslides and rockfalls are a major problem along the Mediterranean coast of Israel because of the dense population and the most occupancy of the area for recreation during most of the year. The coast is characterized by a narrow beach of 10 to 20 m in width and bordered by a steep cliff up to 45 m high. The cliff consists of alternate layers of sandy loam, sandy silt and eolianite (calcareous sandstone) at different stages of cementation. The cliff is undergoing erosion by waves and by piping processes from rainwater percolating through the soft, sandy loam layer (Arkin and Michaeli, 1985). Protection against marine wave erosion is not enough to prevent sliding of the cliff due to sharp angles (Wiseman and Hayati, 1971). Rate of recession of the cliff in one area for the period of 1949-1969 was 1.5 m per year, based on air photo analysis (Almaliah, 1971).

Submarine slumping occurs on the continental margin of Israel. The slumps were probably triggered by seismic activity (Almagor and Wiseman, 1980, and Frydman and Talesnick, 1988). Data from bathymetric charts show scars as wide as 3 km and as long as 4 km.

5 TYPES OF LANDSLIDES

5.1 Slumps

Slumps are the most common landslide in the

229

Galilee and the Mediterranean mountainous area of Israel. Several of these have been described by Wiseman and others (1970) and Wachs and Levitte (1978). In populated areas, the potential risk is high and slumps are of major concern. During two major earthquakes in 1837 and 1927 in the Galilee, eyewitnesses indicate that most of the damage and human casualties were caused by seismically-triggered landslides. These were located on chalk and marl dip slopes greater than 15% (Wachs and Levitte, 1980). Landslides on steep marl slopes also develop when rainfall is heavy, such as the exceptional 1968/69 winter (Inbar, 1987). Two major landslides that occurred in that event were described by Wiseman and others (1970) and Safra (1973). In M'ar village, one of the landslides caused damage to several houses and reactivated an old landslide. Minor slide movements continued in following years. The other landslide, in the Saar River area, moved several tens of thousands of cubic meters of rock. It was probably the largest slide recorded in modern times in that area (Safra, 1973).

The weakening of the marl is due to a process of summer chemical disintegration and winter mechanical disintegration. The seasonal repetition of these processes leads to increasing porosity, water percolation and further weakening of the slope (Arkin, 1988). Under these conditions and on steep slopes, heavy rains and seismic activities promote slumps.

The largest number of people killed in a landslide in this century was in the southern Dead Sea badland areas where 20 people were killed Dec. 20, 1970 when a small cliff of marl collapsed. The collapse happened a few days after 6 mm rain had fallen in the area (Ma'ariv, 31/12/70). No seismic movement was recorded previous to the event.

Old slumps were identified in several areas of marl and chalk overlain by conglomerate in Canada Park and the Mt. Scopus area. They were studied by D. Wachs and others (1986) in the Canada Park area and by Wachs and Levitte (1983) in the Mt. Scopus (Jerusalem) area. Archaeological structures located at the toe of the slumps in Canada Park indicate that the slides are older than 2,800 years. Decrease of slope angle from about 25 to 9 increased stability, although creep processes are still active (Wachs and others, 1986). A similar recent slump was described in the foothills of the Hebron Mountains by Shalem (1947). In the Jerusalem area, the Mount of Olives slide may be the one described vividly in the book of Zacharia (XIV) and related to the earthquake described in the Bible as the Uzziah event (790-745 B.C.). The major earthquake probably triggered the slide in soft chalk areas of the Mount of Olives and did not affect the main area of the old city of Jerusalem, which is built on hard limestone and dolomite (Wachs and Levitte, 1983). Old landslides are also found in the Lower Galilee where the basaltic caprock is underlain by soft sedimentary rocks (Yair, 1971). In the southern Golan area, which has similar stratigraphy, a series of fossil and active landslides were found recently. The slides are triggered mainly by heavy rainfall on steep slopes.

5.2 Rockfalls

Rockfalls and other landslides are frequent in the Jordan gorge and on the Golan Plateau where rivers flowing toward the Jordan rift valley have deep, entrenched valleys in basalt. Difference in the strength of the basalt layers and steep slopes are the main factors promoting these landslide processes (Evenir, 1988).

Rockfalls in the Jordan gorge were mapped and studied by Inbar and Evenir (1987). Erosion by

Figure 7. Rockfall in the Jordan gorge.

the river promotes rockfalls and keeps the base of the slopes clean, so the slopes remain steep (Fig. 7).

The largest boulders in the rockfalls are 3.5 m x 2 m x 1.5 m, about 20 tons. Rockfalls are also common in roads cut into hard limestone or dolomite. Terraces, wire nets, and other engineering devices are used to reduce rockfall risk.

6 ECONOMIC ASPECTS

Engineering works to mitigate landslides, such as terraces, walls, and metallic nets are expensive. The estimated expense for slope protection in the city of Haifa, population approximately 250,000, is about $30,000 US annually. No figures are available for the entire country, but a rough estiamte is about one-half million dollars annually, mainly for protection of roads and buildings.

No major economic damage by landsliding has been reported in the last 40 years, since the establishment of Israel. The tragic event which killed 20 people was discussed in the first part of this report. Several other people have been killed by rockfalls and collapsing of the coastal cliffs.

7 AGENCY RESPONSIBILITY

The Geological Survey of Israel is the main government agency dealing with landslide problems. Two studies with slope stability maps have been completed by them for the Galilee and the Jerusalem areas. Research work is also carried out by university faculty in the earth sciences and engineering. Several of their theses deal with landslides (see bibliography).

8 CONCLUSION

The critical questions which should be studied are, 1) the spatial distribution of landslides in relation to physiogrpahic factors, lithology, tectonics, and slope morphology; 2) the role humans play in promoting landslides, and 3) the age and recurrence of landslides. The major aim of these studies should be to determine which areas are most vulnerable to landsliding so that these areas can be avoided or studied carefully before development takes place.

REFERENCES

Almagor, G. & G. Wiseman 1980. Submarine slumping and mass movements on the continental slope of Israel. In S. Sayod and J.E. Nieuwenhuis (eds.) Marine slides and other mass movements. Plenum Press, NY: 95-128.

Almaliah, M. 1971. Present day processes in the Natanya-Hofit cliff. Dept. of Geography, Univ. of Haifa. Unpublished manuscript. 70p. (in Hebrew).

Arkin, Y. and Michaeli, L., 1985. Short and long-term erosional processes affecting the stability of the Mediterranean coastal cliffs of Israel. Engineering Geology: 153-174.

Arkin, Y. 1986. Geotechnical factors influencing marl slopes in Israel. Report 27/86 Geological Survey of Israel. Jerusalem. 386p.

Arkin, Y. 1988. Disintegration of marl slopes in Israel. Environ. Geol. Water Sci.: 5-14.

Evenir, M. 1988. Morphology and dynamic processes of landslides on the basaltic slopes of the Jordan canyon. M.A. Thesis, Univ. of Haifa, 102p. (in Hebrew).

Frydman, S. and M. Talesnick 1988. Analysis of seismically triggered slides off Israel. Environ. Geol. Water Sci.: 21-26.

Garfunkel, Z. 1981. Internal structure of the Dead Sea leaky transform (Rift) in relation to plate kinematics. Tectonophysics: 81-108.

Inbar. M. 1987. Effects of a high magnitude flood in a Mediterranean climate: A case study in the Jordan River basin. In L. Mayer and D. Nash (eds.) Catastrophic Flooding. Allen & Unwin: 333-353.

Inbar, M. & M. Evenir 1987. Landslides in the Jordan gorge between the Hulah and the Kinneret. Dept. of Geography Report, Univ. of Haifa. Israel. 24 p. (in Hebrew).

Kafri, U. 1988. Environmental geology in Israel. Environ. Geol. Water Sci.: 3-4.

Ma'ariv 1970. Daily newspaper 31/12/70. Ma'ariv, Tel Aviv, Israel (in Hebrew).

Nilsen, T.H. and E.E. Brabb 1975. Landslides in studies of seismic zonation of the San Francisco Bay region. In R.D. Borcherdt (ed.) Studies for seismic zonation of the San Francisco Bay Region. US Geol. Survey Prof. Paper 941- A: 75-86.

Safra, D. 1973. The Saar River landslide. Teva Va'aretz: 16-21 (in Hebrew).

Schein, Z. and N. Buras 1973. Rainfall intensities in Israel. Israel J. of Earth Sc.: 15-30.

Shalem, N. 1947. A landslide in the Hebron area. In N. Shalem 1973 Collection of Papers, Qiryat Sefer. Jerusalem: 130-141 (in Hebrew).

Shalem, N. 1949. The seismicity of Jerusalem. In N. Shalem, 1973 Collection of Papers, Qiryat Sefer, Jerusalem: 270-308 (in Hebrew).

Sharon, D. and H. Kutiel 1986. The distribution of rainfall intensity in Israel, its regional and seasonal variations and its climatological evaluation. J. of Climatology: 277-291.

Wachs, D. and D. Levitte 1978. Damage caused by landslides during the earthquakes of 1837 and 1927 in the Galilee region. Report Hydro 5/78, Geol. Survey of Isarel. Jerusalem.

Wachs, D. and D. Levitte 1981. Earthquake-induced landslides in the Galilee. Israel J. of Earth Sc.: 39-43.

Wachs, D. and D. Levitte 1983. Earthquake risk and slope stability in Jerusalem. Report EG/2/83. Geol. Survey of Israel, Jerusalem: 8p.

Wachs, D., B. Buchbinder, and A. Sneh 1986. Old landslides in the Canada Park area: Geological and environmental implications. Israel J. of Earth Sc.: 158-165.

Wiseman, G., G. Hayati, S. Frydman, B. Aisenstein, D. David and A. Flexer 1970. A study of a landslide in the Galilee. Pub. 146. Technion Faculty of Civil Eng. Haifa.

Wiseman, G., G. Hayati 1971. Stability study of a coastal cliff. Progress Report 1970-71. Technion-Israel Institute of Technology, Haifa (in Hebrew).

Yaalon, D.H. 1967. Factors affecting the lithification of Eolianite and interpretation of its environmental significance in the coastal plain of Israel. J. Sediment Petrol: 1189-1199.

Yair, A. 1971. Geomorphological processes on marl slopes. Jerusalem Studies in Geography, Hebrew Univ., Jerusalem: 156-190.

Yair, A. and R. Gerson 1974. Escarpment retreat in an extremely arid environment (Sharm el Sheikh, Southern Sinai Peninsula) Z. Geomorph. Suppl. Bd. 21: 202-215.

Landslides: Extent and Economic Significance, Brabb & Harrod (eds)
© 1989 Balkema, Rotterdam. ISBN 90 6191 876 6

Landslide vulnerability in the Arabian Peninsula and Egypt

H.A.El-Etr & M.S.Yousif
Ain Shams University, Cairo, Egypt

M.M.Hamza
Suez Canal University, Ismailia, Egypt

A.A.Hussein
The Geological Survey of Egypt, Cairo, Egypt

ABSTRACT: Using basic maps depicting relief, physiography, drainage, rainfall, geology, roads, seismicity and earthquakes an interpretative landslide vulnerability map was prepared. Areas of higher risk are related to tectonic sutures of divergent (Red Sea), convergent (Zagros) and transform fault (Gulf of Aqaba – Dead Sea) types.

Physiography

The study area involves the eastern part of the Great Sahara. Figure (1) generalizes the prominent relief. Highest topography characterizes the region around the Red Sea and the arc of Oman. This is followed, further in land by plateau and plain areas. The central part of the Arabian Peninsula is demarcated by a series of arcuate ridges and cuestas mainly facing west. Tertiary lava flows (Harrats) are common in the Red Sea region. Substantial parts of the plain areas are occupied by sand dunes and other desert surface lag deposits.

Drainage is illustrated in Figure (2) and is mainly disintegrated and internal. Well-defined water divides bound the Red Sea on the east, west and north. The size of the drainage tributaries is rather small and of low furcation orders. To the east of the water divide of the eastern margin of the Red Sea, drainage ends in the cuestas and plains of central Arabia. Because of evident structural and lithological controls, the drainage network in this part is more conspicuous, more integrated and in which larger angles of juncture other than the acute gravity controlled type are rather common.

Most of the drainage net in south Arabia drain north towards Rub El khali (the Empty Quarter). Drainage in the arc of Oman (southern Persian Gulf) is rather radial.

- The drainage network is a relict geomorphic feature formed during the Quaternary

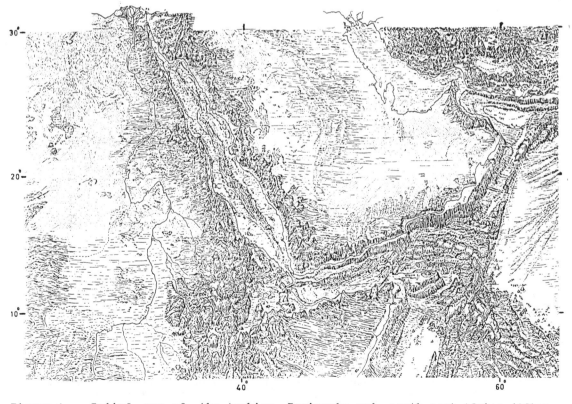

Figure 1 . Relief map of the Arabian Peninsula and north-east Africa (After Heezen and Tharp, 1964)

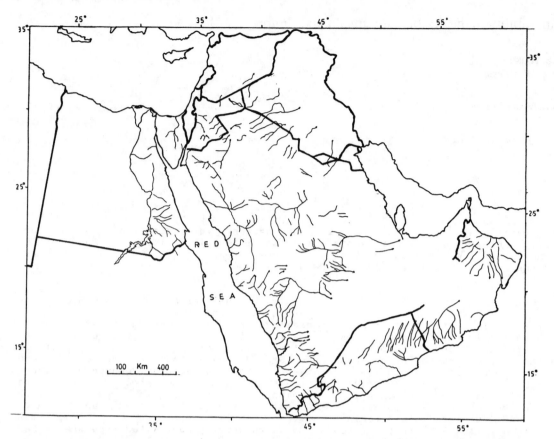

Figure 2 . Drainage pattern. Arrows point downstream towards debouching sites. Designation : Thin lines, main water divides; Heavy solid lines, international political boundaries (After the New Oxford Atlas,1978).

times mainly by the action of water under a wetter climate. At present, the channels are basically dry and frequently are filled by drifting loose sand and dunes. However, desert torrential storms and concomitant flash floods are not infrequent. In such cases, substantial amounts of water collects in short times and locally may cause serious damages.

Although the study area is known by its extreme aridity some rain may fall particularly in the mountainous coastal fringes including the Hijaz Mountains and Red Sea Hills as well as the southwestern (Yemen) and southeastern (Oman) corners of the Peninsula. Figure (3) shows the rainfall distribution. Maximum recorded amounts are up to 1000 mm per year. A small amount of rain falls on the Egyptian Mediterranean coastal belt.

Geology

The study area includes a wide variety of rock types representing a span of time that extends for more than 2000 Ma (Fig. 4).
The oldest known (Archean) rocks are those occurring at Gebel Oweinat, in the southwest corner of Egypt and northwestern Sudan (>2500 Ma) and in the eastern-most reaches of the Arabian Shield in Afif and Ar Rayn (>2000 Ma). Inbetween these old continental masses, and on both sides of the Red Sea, the Arabian-Nubian Shield, mostly of Upper Proterozoic age, crops out (Al-shanti,1980). In the arc of Oman, on the Persian Gulf, rocks of Paleozoic and older ages incorpora-

ting slices of Mesozoic ophiolites are exposed. A prolific Paleozoic to Tertiary sedimentary succession overlaps the shield on both sides, extending to the Persian Gulf to the east and into the Nile Valley to the west. Also, the Red Sea is bounded by narrow coastal plains dominantly of Tertiary to Quaternary age. The sedimentry cover is

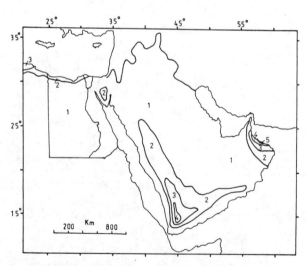

Figure 3 . Mean annual precipitation. Number designations : 1,<100 mm; 2,100-<250 mm; 3, 250 -<500 mm; 4,500-<750 mm; 5, 750-<1000 mm (After the New Oxford Atlas, 1978).

234

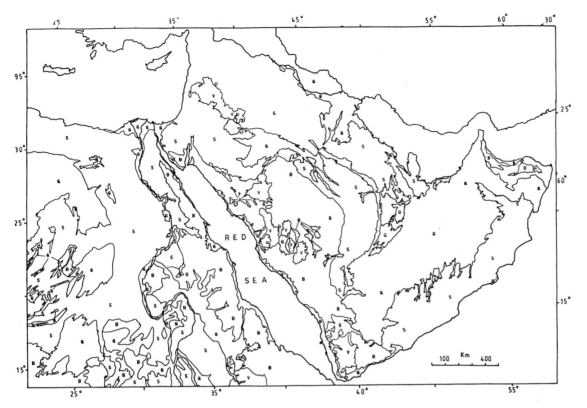

Figure 4 . Simplified geological map. Letter designations : B, Basement complex, predominantly precamb rian, S, Paleozoic through Tertiary sedimentary succession; V, Tertiary volcanics; Q, Quaternary desert surface deposits and sand dunes, river alluvium, and coastal sediments (After Unesco,1976).

petroleum bearing where conditions are favorable, in eastern Arabia and the Gulf of Suez. Flood basalts of Tertiary and younger age form scattered patchy areas (Harrats) on both sides of the Red Sea are evidently related to the Red Sea rifting episode.
Quaternary sediments are represented by desert lag deposits and sand dunes, wadi alluvium and coastal sediments.
Tectonically, the Arabian peninsula is a microplate surrounded by divergent (Red Sea), convergent (Zagros) and transform fault (Gulf of Aqaba-Dead Sea) types plate margins. Such areas were tectonically active for a long period of time, particularly during the Tertiary. At present they are in a waning phase.

Communication

The highway network is well developed and many new ones have been constructed particularly in the last two decades (Fig.5). It connects the main centers of human activities and settlements. Least inhabited is the southeastern part of the Peninsula west of Oman. The central part of Arabia is traversed east-west by several high-quality highways. Also, highways parallel the eastern and western sides of the Red Sea and an important highway extends along the Nile in a northsouth direction. Few connection roads join the western Red Sea highway with that along the Nile. The construction of some of these highways especially those connecting the mountain front with the Red Sea coast in Arabia required extensive tunnelling, bridging and slope stabilization works. Nevertheless these are man-made

features that accentuate landslide vulnerability.

Seismicity

Figure (6) illustrates the epicenters of earthquake events with magnitude Ms 3.9 or greater occurring from 1900 through 1983 (Riad et al., 1985). It is evident that the lands of Arabia and Egypt are only slightly affected by earthquakes. Major centers of activity, however, are found in the adjacent seas particularly the southern part of the Red Sea and the western part of the Gulf of Aden and the Arab Sea bounding the Peninsula from the south. Moreover, the eastern side of the Persian Gulf is highly seismic; being a part of the tectonically-active zagros mountains fold belt (subduction-collision plate margin). Figure (7) outlines the damage by earthquakes in the study area, during the last 2000 years. The Red Sea margins and southern Iran were the most affected.
The south eastern border of the Red Sea witnesses high level of microseismic activity in association with active faults. Some major events may take place, such as that which destroyed Dhamar,Yemen on 13-12-1982 (Shehata et al.,1983). Due to the reactivation of spreading in the Red Sea trough during the last 2 m.y readjustments are taking place in the coastal plain and mountain escarpment. Moreover, a substantial part of the tectonism can be related to salt intrusion especially around Jizan, in southwest Saudi Arabia. Besides, minor earthquakes are taking place away from plate margins, in intraplate environments. One

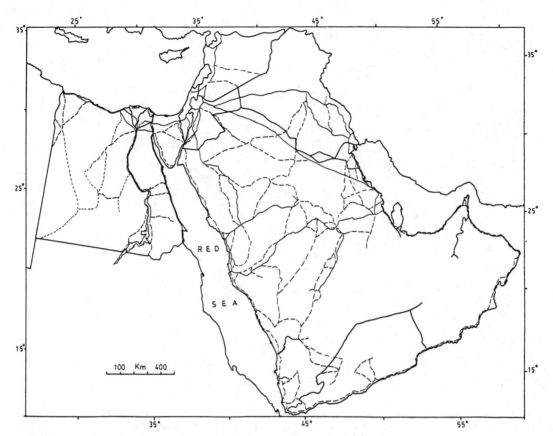

Figure 5 . Highway network (After the New Oxford Atlas,1978).

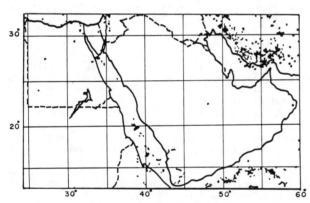

Figure 6 . Seismicity of the middle east., Symbols indicate events with magnitude Rs 3.9 or greater occurring from 1900 through 1983 (After Riad et al.,1985)

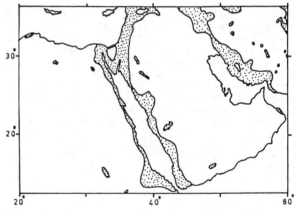

Figure 7. Areas damaged by earthquakes in the Middle East during the last 2000 years (After Cidensky and Rouhban,1983).

such event took place in Tabah village, Saudi Arabia on 6-9-1984 (Roobol et al., 1985). The site is related to an ancient volcanic center with a wide deep vent filled with sand and gravel eroded from tuffs. This vent fill acts as an excellent aquifer.

Rapid depletion of the aquifer (for agriculture and road construction purposes) caused drying out and shrinkage of the vent fill material with the development of deep vertical fractures that form a concentric ring around the vent. With these the seismic events are associated (4 at least in the period 1982 - 1984).

Landslide Risk

Based on the aforementioned maps of Figure (1) through (7), Figure (8) was constructed to show areas most prone to landsliding. Generally, these are areas of high relief, well developed incised drainage, rainy climate, on or adjacent to seismically active belts and traversed by highway plexus. Due to the generalities of the basic data available, the map, however, is of a preliminary nature.

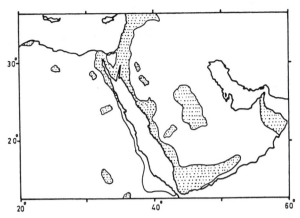

Figure 8 . Map showing areas of relatively high landslide vulnerability (sti-
ppled).

References

Al-Shanti, A. 1980. Evalution and mineral-
  ization of the Arabian-Nubian shield.Inst.
  App. Geol (Jeddah), Bull.3.
Cidinsky,K & B. M. Rouhban 1983. Asssessment
  and mitigation of earthquake risk in the
  Arab Region. Unesco.
Heezen, B.C. & R. Tharp 1964 . Physiographic
  diaqram of the Indian Oceean. The geologi-
  cal Society of America.
Riad, S., R.H. Smith, H. Meyers, W. Rineh-
  art, M. Rockwell,and T.A. Delaney 1985.
  Seismicity of the Middle East (1900 -
  1983).
  National Geophysical Data center, Colo.
Roobol, M.J., S.A. Shouman & A.M. Al-Solami
  1985. Earth tremors, ground fractures and
  damage to buildings at Tabah. Saudi Arabia
  Deputy Ministry for Mineral Resourcs. Ope-
  n File Report DGMR-OF-05-13
Said,R. 1962. The geology of Egypt, Elsevier
  pub.
Shehata, W.M.,A. Kazi, F.A. Zakir, A. M. A-
  llam 1983. Preliminary investigation on
  Dhamar earthquake, North Yemen, of
  December 13,1982. Bull. Fac. Earth Sci,
  King Abdulaz iz Univ., V.5, P. 23-52.
The New Oxford Atlas 1978. Oxford University
  Press.
Unesco, 1976. Geological world atlas
  (1:10,00 0,000).

Landslides: Extent and Economic Significance, Brabb & Harrod (eds)
© 1989 Balkema, Rotterdam. ISBN 90 6191 876 6

Landslides in Ghana and their economic significance

J.K.Ayetey
Geotechnical Division, Building and Road Research Institute (CSIR), U.S.T., Kumasi, Ghana

ABSTRACT: Landslides occur throughout Ghana but the largest ones are in the Voltaian Formation of Paleozoic age. Fortunately most of these landslides are in sparsely settled areas. Small landslides are common in Birrimian metamorphic rocks of Precambrian age.
   Joints, foliation, differential weathering and vegetation are important factors affecting the distribution of landslides. The type of soil is also important; lateritic soils hold up well on slopes but saprolitic soils fail easily.
   Landslides have devastated forests, farms, and roads. The total cost is estimated between half a million and a million dollars depending on the annual rainfall intensity.

1 INTRODUCTION

Landslides first seriously caught the attention of the highway authorities when in 1969 a major trunk road was cut by a landslide for about a week. In order to clear the road, a study was conducted at the site of the landslide and it became apparent that the Voltaian rocks which formed scarps would continue to cause landslides from time to time. This prompted a research project to be initiated which with time was extended to cover the whole of Ghana.

In this paper, major landslide-prone areas have been delineated and minor but frequent landslides documented.

2 DEFINITION OF TERMS

Several classifications of mass wasting processes are available (Sharpe, 1938; Varnes, 1958) but most of these are modifications of the original Sharpe classification. No attempt will be made to present another classification, but it is important that the terms used in this paper be defined. Mass wasting is a general term covering all natural processes by which large masses of earth material are moved by gravity from one place to another. These processes may be slow or rapid. Muller (1964) estimated the speed of creeping movements before the Vajont slide as 20cm to 30cm a day. Rock slides and avalanches may attain speeds of several meters per second.

The nature of movement is also important in defining any process of mass wasting. Movements with a vertical as well as a horizontal component which take place on definite slide planes are termed slides. Those with vertical and little or no horizontal translation are termed falls. Subsidence is, therefore, a special type of fall in which all parts of the body except the upper part are enclosed. When a movement has a vertical as well as a horizontal component, but does not take place on a definite slip plane, it is termed a flow. The term "landslide" will not be used, as Sharpe does, to describe all types of rapid movement but will be restricted to rapid movement in which soil or unconsolidated earth material moves as one or several units on a single or on many slip planes. Slump is a special type of landslide in which the slip plane is more or less circular and the material moved may be rotated backward on a horizontal axis to the slope on which it descends. The material moved characterizes the type of movement.

Therefore, there can be rock slide or a landslide, a rock creep or a soil creep.

It has become the practice to classify slope failures in jointed rocks according to the mode of failure. Four main modes of failures are: a) circular failures, b) plane failures, c) wedge failures, and d) toppling failures. The failure types are directly dependent on the attitudes of structural discontinuities in the rock. Whenever a movement is made up of two or more types, it is termed a complex movement.

3 GEOLOGY OF GHANA

Ghana is underlain mainly by Precambrian metamorphic and granitic rocks (Fig. 1), all the metamorphics of which have been folded and intruded by mafic rocks. Most of the remainder of the country consists of Paleozoic sedimentary rocks. Small areas on the coast are underlain by flat or gently dipping sedimentary rocks of late Paleozoic and Jurassic-Cretaceous age.

The Precambrian rocks include the Dahomeyan system which underlies the Accra Plains and the southern Volta region. It consists mainly of massive crystalline granulite, gneiss, migmatite with subordinate quartz schist, biotite schist and remnants of sedimentary rocks. The Birrimian system, which covers about one-sixth of the total area of the country, is subdivided into a lower series of slate, phyllite, greywacke, tuff and lava, together with schist and gneiss derived from these rocks, and an upper series of greenstone, mainly metamorphosed basic and intermediate lava and pyroclastic rocks with some hypabyssal igneous rocks and intercalated bands of phyllite and greywacke. Persistent bands of manganiferous phyllite and gondite (manganese-garnet and quartz-rock) occur in the Upper Birrimian in various parts of the country.

The Birrimian rocks are isoclinally folded with dips generally steeper than 60°. The average trend of the fold axis is northeast-southwest, but locally the strike could swing towards the north or the east. Near the Black Volta in the north of the country, the regional strike is slightly west of north.

The Tarkwaian system is made up of of schist, quartzite, phyllite and meta-conglomerate derived from the Birrimian. It reaches a thickness of 3,300 m in the Tarkwa area. Areas of Ghana underlain by the Tarkwaian are far smaller than those underlain

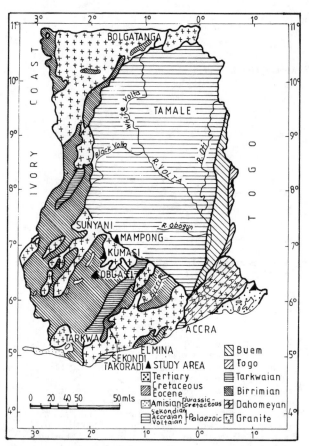

Figure 1. Map showing general geology of Ghana and landslide study areas.

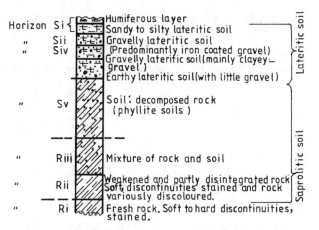

Figure 2. Typical point profile showing all possible residual horizons.

by the Birrimian; the largest lies between Konongo and Tarkwa in a forested region.

The Togo series consists of metamorphosed arenaceous and argillaceous sedimentary rocks and some sandstone, shale and limestone. These rocks underlie the Akwapim and Togo Ranges where the harder rocks such as quartzites form the hills and the shale and phyllites form the intervening valleys.

The Buem series underlies a large area of the country west of the Akwapim Togo Range. The series consists of shale, sandstone, quartzite and volcanic rocks with subordinate limestone, tillite, grit and conglomerate.

The Voltaian Formation of Paleozoic age, which covers about two-fifths of Ghana, consists chiefly of sandstone, shale and mudstone. These rocks are essentially flat-lying. More resistant beds form escarpments at many localities.

4 JOINTS AND FOLIATION

Joints and foliation are an important factor in slope failure. A study of 200 joints and foliation planes along 21 road cuts in the vicinity of Obuasi, Kumasi, and Sunyani indicates that the discontinuities average 2 m in length and are spaced an average of .283 per m.

5 WEATHERING

Weathering is the most important factor affecting a slope in the tropics. A typical weathering profile is shown in Figure 2. Lateritic horizons in the profiles perform well as slope material. However, saprolitic soils are unsatisfactory as slope material. Engineering soils maps (figures 3a and

3b) were prepared separately for saprolites and laterites.

Differential weathering is important in undermining resistant sandstones. Sandstones in the Voltaian Formation are undercut as much as 6 m at some localities.

6 SLOPE FAILURES

The Birrimian system underlies the most developed and populated areas in Ghana. A few major slope failures occur in surface mines but most of the landslides in this rock system are minor and not economically significant.

The Voltaian Formation has many large landslides but, fortunately, in places with little development. Most of the landslides occur where scarps have formed, in areas where the Voltaian overlies the Birrimian.

Major landslide areas in Ghana are shown on Fig. 4. Most of these are underlain by the Voltaian Formation. One of the major slides in this unit occurred July, 1968, when 1,500 m³ of rock, soil, and vegetation blocked the Kumasai-Mampong road for 10 days.

Minor slides are listed in Table 1. All of these took place along oversteepened road cuts.

7 CAUSES OF FAILURES

Failure of the Voltaian Formation along the Kumasai-Mampong road was related to weathering and jointing. Most of the failures in the Birrimian system were plane and wedge-type. A slump in weathered gneiss had a circular slip surface. More specifically: (a) Structural discontinuities – The Voltaian Formation along the Kumasai-Mampong road consists of massive, extensively-jointed sandstone with interbedded shale, and mudstone. Differential weathering of the softer shale and mudstone undercut the sandstone as much as 4m. Tension crack developed in the massive sandstone which eventually failed. (b) Texture of soils – The particle-size distribution of the debris was important in subsequent failures of the decomposed Voltaian material. After the rockfall, huge boulders formed high stable slopes. As the boulders decomposed into smaller particles, the strength changed considerably and the slope became unstable and failed again. (c) Vegetation – The landslide in the Kumasai-Mampong road area is in the rain forest area of Ghana. Large roots of trees in this area penetrate deeply into the rock through joints (Plate 1), acting as wedges and, in time, pushing over the shattered rocks. During windstorms, tree and root movement may also cause the widening of rock joints.

A

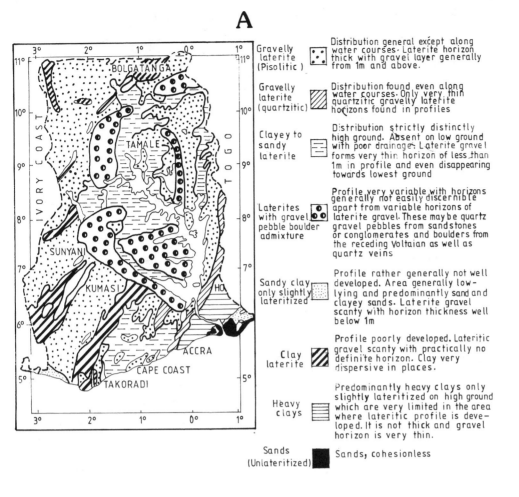

Gravelly laterite (Pisolitic) — Distribution general except along water courses. Laterite horizon thick with gravel layer generally from 1m and above.

Gravelly laterite (quartzitic) — Distribution found even along water courses. Only very thin quartzitic gravelly laterite horizons found in profiles

Clayey to sandy laterite — Distribution strictly distinctly high ground. Absent on low ground with poor drainage. Laterite gravel forms very thin horizon of less than 1m in profile and even disappearing towards lowest ground

Laterites with gravel pebble boulder admixture — Profile very variable with horizons generally not easily discernible apart from variable horizons of laterite gravel. These may be quartz gravel pebbles from sandstones or conglomerates and boulders from the receding Voltaian as well as quartz veins

Sandy clay only slightly lateritized — Profile rather generally not well developed. Area generally low-lying and predominantly sand and clayey sands. Laterite gravel scanty with horizon thickness well below 1m

Clay laterite — Profile poorly developed. Lateritic gravel scanty with practically no definite horizon. Clay very dispersive in places.

Heavy clays — Predominantly heavy clays only slightly lateritized on high ground which are very limited in the area where lateritic profile is developed. It is not thick and gravel horizon is very thin.

Sands (Unlateritized) — Sands, cohesionless

B

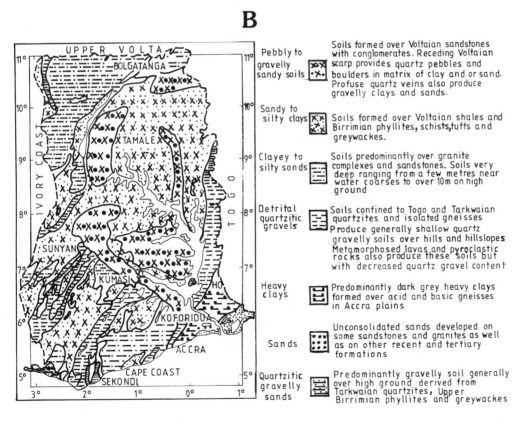

Pebbly to gravelly sandy soils — Soils formed over Voltaian sandstones with conglomerates. Receding Voltaian scarp provides quartz pebbles and boulders in matrix of clay and or sand. Profuse quartz veins also produce gravelly clays and sands.

Sandy to silty clays — Soils formed over Voltaian shales and Birrimian phyllites, schists, tuffs and greywackes.

Clayey to silty sands — Soils predominantly over granite complexes and sandstones. Soils very deep ranging from a few metres near water coarses to over 10m on high ground

Detrital quartzitic gravels — Soils confined to Togo and Tarkwaian quartzites and isolated gneisses Produce generally shallow quartz gravelly soils over hills and hillslopes Metamorphosed lavas and pyroclastic rocks also produce these soils but with decreased quartz gravel content

Heavy clays — Predominantly dark grey heavy clays formed over acid and basic gneisses in Accra plains

Sands — Unconsolidated sands developed on some sandstones and granites as well as on other recent and tertiary formations

Quartzitic gravelly sands — Predominantly gravelly soil generally over high ground derived from Tarkwaian quartzites, Upper Birrimian phyllites and greywackes

Figure 3. Map showing engineering soil units in Ghana. A, Laterite and lateritic soils; B, Saprolitic soils.

Table 1. Conditions of Cuttings and Natural Slopes of Roads in Some Formations of Ghana

LOCATION	Cutting Depth in feet	Slope Angle	Parent Rock	Nature of Soil	Atterbergs Limits L.L	P.L	P.I	Shear Parameters c'	0	Approx. date of Cutting	REMARKS
1. Saltpond by-pass	27'	64°	Lower Birrimian mica schist	Red micaceous silty clay	42	23	19	3. PSI	17°	1959	Local failure taking place. Present slope is 55°.
2. Cape Coast by-pass chainage 10 + 00	16'	54°	"	Light Red highly micaceous silts	47	27	20	4. "	15°	1960	Local failure due to soil erosion taking place: by and large stable.
3. Cape Coast by-pass chainage 57 + 00	13'	63°	"	Reddish brown micaceous silt	39	30	9	3. "	18°	1960	No failure noticed so far.
4. Cape Coast by-pass 179 + 00	23'	58°	"	Reddish brown micaceous silt with patches of kaolin	47	21	66	2.5-5"	17°-21°	1960	Failed after 1961 heavy rains. Present slope 53°.
5. Sekondi College face	48'	49°	Sandstone, clay shale	Light grey and red mottled clays.	65	31	34	3. "	18°	Not known (Natural)	Failed in 1962. Remedial measures include drainage, growing of grass, etc. Stable since then.
6. Sekondi Residency face	142'	53° & 61°	Clay shale and and sandstone	Heavy mottled clays.	78	29	49	0-3 "	13°	Natural	Lower slopes started failing and a crack developed on the top of the slope.
7. Agona Akim Road Mile 7.	14'	67°	Upper Birrimian Phyllite	Clay of high plasticity	62	25	37	2.5 "	19°	1961	Failed soon after a heavy rain. Failed again in 1962. Present slope 47°.
8. Agona Akim Road Mile 10.	16'	57°	"	"	-	-	-	-	-	1962	Surface failure, not deep. Final dressing at 54° is standing very well.
9. Agona Akim Road Mile 9.	19'	68°	"	"	67	31	36	1. "	17°	1962	Failed soon after construction due to heavy rains.
10. Agona Akim Road Mile 13.	18'	61°	"	"	54	29	25	0	16°	1962	Surface failure after heavy rainfall. Present angle of slope 54° standing very well.
11. Mpataba Elubu Road Mile 17.	17'	63°	Sandstone, shale	Dark grey heavy clay	89	36	53	0	11°	1963	Deep failure, 4 months after cutting during the rainy season. Remedial works carried out. Occurrence of a slip again after 8 months. Present slope of 43° is overgrown by vegetation.
12. Mpataba Elubu Road Mile 23.	13'	85°	Sandstone & shale	Dark grey heavy clay	78	32	46	2 PSI	140°	1963	Failed after heavy rainfall and construction. Deep failure. Failed twice again after a couple of months. The natural stable angle 45°.
13. Kumasi Obuasi Road Mile 18.	17'	68°	Upper Birrimian phyllite	Micaceous silty clay	19	23	16	5 "	21°	1961	Standing without any failure. Quart veins in the cutting and surface are hard due to laterization.
14. Kumasi Obuasi Road Mile 21.	14'	64°	"	"	41	26	15	3 "	19°	1961	Standing without any failure.
15. Kumasi Obuasi Road Mile 22.	15'	57°	"	"	-	-	-	-	-	1961	No failure.
16. Kumasi Accra Road Mile 5.	17'	64°	Muscovite granite	Silty clay	41	26	15	4	21°	Not known	Standing without showing any signs of distress. Surface hard due to laterization. No signs of surface erosion.
17. Kumasi Accra Road Mile 6.5.	14'	68°	"	"	37	23	14	3	25°	"	"
18. Kumasi Accra Road Mile 3.5 behind AGIP Petrol Dump.	17'	63°	"	"	31	27	14	1	23°	"	"
19. Kumasi Mampong Road Mile 23.	14'	54°	Sandstone	Silty clay	34	21	13	4 "	25°	"	Stable.
20. Accra Coast at the back of Usher Fort.	210'	51°	Interbedded sandstone & shale.	Heavy clays (5 ft) overlying interbedded sandstone & shale	58	31	27	3 "	210°	"	Except for local failures due to slumping & weathering in the shale, the performance of the slope is satisfactory.
21. Accra Coast near Black Star Square	32'	47°	Clay shale	Heavy clays weathered product of shale, overlying horizontally bedded shale. Dipping downwards towards the sea at an angle of 13°.	47	21	25	0	23°	"	As a result of weathering along bedding in the shale, which causes local surface failure, the slopes are performing satisfactorily.

8 SCREE SLOPES

In the dry parts of Ghana, erosion of scarps generally leads to the formation of scree slopes with thin, cohesionless material. As the spaces within the blocky deposits become filled with fine material, the slope develops some cohesion. Gerber and Scheidegger (1974) stated that the development of cohesion is particularly rapid if the scree contains clayey components. During infrequent heavy rainfall, many minor slides occur.

Gerber and Scheidegger (1974) used a mechanical model to study the mechanics of debris slides on a scree slopes. This model seems to explain the existence of tongue-shaped deposits on scree slopes in Ghana (Plate 2).

CONCLUSION

Major landslides in Ghana are mainly associated with undercut massive sandstones of the Paleozoic Voltaian Formation where this unit is exposed along escarpments. Joints, heavy rainfall, tropical

242

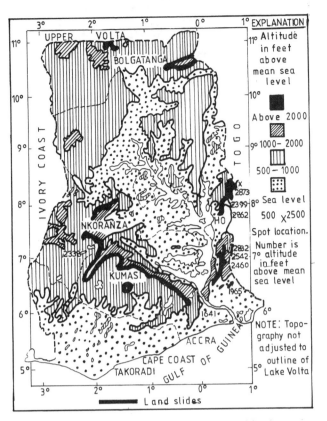

Figure 4. Relief map showing major slide-prone areas.

Plate 2. Jamasi landslide with headwall about 60m high in background. The slide extended 250m, bull-dozing timber, devastating farms, and damaging a road.

Plate 1. Destabilization of rock by roots penetrating into joints and crevices.

weathering, and extensive tree cover may all con-tribute to slope failure. Landslides in this formation have devastated forests, farms and roads.

Minor landslides are common along oversteepened road cuts and on scree slopes.

Hardly any landslide maps have been prepared for Ghana. Documentation of past landslide activity in the Voltaian Formation is especially needed because this unit has produced the most economically serious landslides.

ACKNOWLEDGEMENTS

I had very useful discussions with Dr. K. Amonoo-Neizer, Deputy Director of the Institute on some results of research data analysis relating to the Slopes Project, for which I am grateful. The services of Mr. T.K. Adubuah, the draughtsman, and Mrs. Elizabeth Anaba, the typist, are also thankfully acknowledged.

REFERENCES

Gerber E. & A.E.Scheidegger 1974. On dynamics of scree slopes. Rock Mechanics 6:25-38. (c) by Springer-Verlag.
Muller, L. 1964. The rockslide in the Vajont valley. Rock Mech. & Eng. Geol. 2:148-212.
Sharpe, C.F.S. 1938. Landslides and related phenomena. Columbia U. Press, New York: 137.
Varnes, D.J. 1958. Relation of landslides to sedimentary features. Applied sedimentation, John Wiley & Sons, New York.

Landslides: Extent and Economic Significance, Brabb & Harrod (eds)
© 1989 Balkema, Rotterdam. ISBN 90 6191 876 6

Landslides in Nigeria – Extent, awareness, social and economic impacts

C.O.Okagbue
Department of Geology, University of Nigeria, Nsukka, Nigeria

ABSTRACT: Landslides occur in severely gullied areas, in road cuts, embankments and in steep residual or colluvial terrains. Heavy rainfall is the main causative factor for most slides.
  Both the state of knowledge about extent of landslide processes and state-of-the-art landslide prevention and control are rather low in the country.

1 INTRODUCTION

Nigeria, the largest and most populous country in Black Africa, is geographically located in the tropics within approximately longitudes 2°30' and 15°00.E and latitudes 4°00' and 14°00'N (Fig. 1). The area is about 923,768 sq km, while the population is estimated at 100 million (last census in 1963 showed total population as 55.63 million). Although only a few catastrophic (destroying much property) landslides have been recorded, slope movements are quite common and located mostly in gully areas, road cuttings, river banks, and in a few places on very steep hills and highway embankments.

2 RAINFALL

Nigeria experiences high seasonal rainfall; the amount depends on the relative location. Rainfall decreases both in duration and amount from the coast to the interior, except where altitudinal effects create islands of higher rainfall, for instance, in the Jos Plateau. The coastal areas receive more than 4,000 mm in the rainy season from April to October, whereas the extreme north receives less than 250 mm spread over 3 to 4 months. Fig. 2 shows the mean annual rainfall, and Table 1 gives the monthly distributions in selected cities.

  Rainfall intensity, rather than the total annual rainfall, is a critical factor for landslide generation. For instance, 45 mm of rain has been known to fall within 1 hour in Aguata near Nanka in southeastern Nigeria (Grove, 1951). Violent local storms, which are common at the beginning (April) and end (October) of the rainy season, are powerful

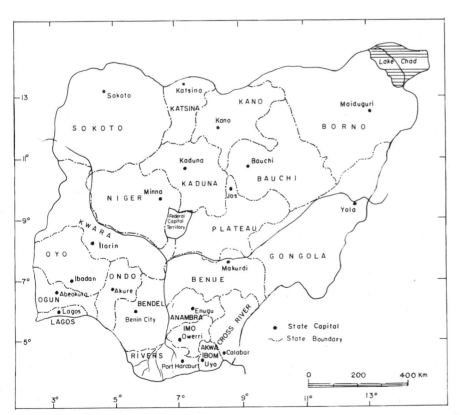

Figure 1. Map of Nigeria showing states and their capitols.

Table 1. Monthly distributions (mm) of rainfall in selected cities across Nigeria

Location	No of Years	J	F	M	A	M	J	J	A	S	O	N	D	Year	Relief m. above s.l.
Sokoto	39	0	0	0.8	10	51	101	154	231	139	13	0	0	700	4.68
Hadejja	10	0	0	0	2	23	58	155	231	76	2	0	0	547	0.41
Kano	54	0	0.2	2	10	66	110	204	302	132	15	0	0.2	841	5.99
Yelwa	12	5	3	3	13	124	114	190	211	262	58	2	0	985	5.17
Kaduna	36	2	3	13	64	150	180	216	302	269	74	5	2	1278	11.30
Bauchi	12	0	2	5	36	89	150	231	366	180	38	2	2	1095	9.75
Jos	24	3	5	25	99	194	218	323	287	211	38	5	2	1408	21.22
Zungeru	14	0	5	20	59	122	153	169	246	296	68	7	0	1145	4.31
Minna	28	2	7	19	60	153	183	207	280	304	134	12	2	1363	7.48
Ilorin	28	8	20	54	116	177	192	148	124	259	148	31	10	1287	13.71
Offa	10	8	18	51	104	166	186	138	116	252	139	29	11	1218	20.93
Ogbomosho	15	9	18	52	104	167	188	139	117	255	140	29	11	1229	18.46
Lokoja	38	10	13	43	112	152	163	175	180	236	135	18	5	1242	3.40
Makurdi	12	5	10	20	124	226	193	185	193	279	150	18	2	1403	3.36
Ibadan	26	15	31	90	151	160	214	172	104	184	164	48	12	1345	18.52
Ondo	24	13	46	112	153	173	226	223	145	220	188	49	11	1589	12.71
Lagos	69	28	47	102	144	254	437	248	62	144	200	68	27	1761	15.00
Enugu	28	16	31	85	149	255	259	182	174	300	242	44	17	1754	19.44
Benin	25	24	32	83	170	215	298	286	217	303	246	76	16	1966	12.15
Warri	41	56	146	227	279	378	444	307	432	318	-	119	37	2784	0.56
Forcados	54	72	188	297	362	362	492	579	398	561	413	156	47	3619	0.56
Calabar	42	55	73	154	228	314	414	460	423	415	308	188	50	3070	4.00
P.H.	15	33	61	127	180	226	335	323	335	384	272	147	46	2471	3.85
Ibi	57	3	7	32	91	172	181	152	147	234	127	9	2	1157	4.38
Birnin Kirbi	35	0	0	5	13	69	109	185	267	155	18	0	0	821	4.08
Maiduguri	45	0.3	0.1	0.1	7	38	66	186	221	102	19	0.2	0	640	3.82
Yola	31	0.8	0.2	8	58	118	158	169	203	194	83	4	0	996	3.87
Potiskum	-	0	0	0	10	51	91	216	259	137	23	0	0	787	5.39
Ogoja	-	13	20	58	124	229	282	208	259	292	300	54	13	1852	11.55

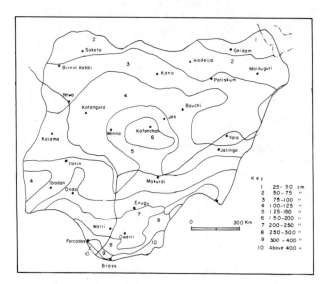

Figure 2. Mean annual rainfall (source Oguntoyinbo, 1983).

agents of runoff and sheet wash, causing landslides at the heads of gullies and in storm-water channels. Continued and intense precipitation from May to October in the southern parts of the country is also an effective landslide agent. Thorough saturation of the soil induces mass earth movements by slumping, caving, slow downhill creep, and sometimes fast debris flows.

3 LITERATURE

Regrettably, very little published literature deals primarily with landslide and slope stability problems in Nigeria. Most of the work done on mass movement has been on gully erosion. Although most of the gully advances are accompanied by land-sliding, very few authors have treated them as such.

Nevertheless, many of the writers have used the words "slumps" and "mass wasting" in their descriptions of events that lead to gully advancement. Notable among such writers are Ologe (1973, 1974), Ogbukagu (1976), Egboka and Okpoko (1985), and Egboka and Nwankwor (1986). The few reports that have dealt with landsliding and (or) debris flows are by Jeje (1979), Okagbue and Abam (1986), and Okagbue (1988).

Internationally, landslide that accompany gully processes have received little attention. Worthy of noting, however, are the works of Taylor and Johnson (1973), Bradford and Piest (1977), Bradford et al. (1978), and Bradford and Piest (1980).

4 GEOMORPHOLOGY AND GEOLOGY

4.1 Prevalence and location of steep slopes

In most of Nigeria, there is a close correspondence between landforms and underlying rocks at both regional and local levels. Of the area covered by sedimentary formations, sandstones form positive relief elements such as hills and escarpments, whereas argillaceous sediments form lowlands and valleys.

In areas where plateaus, mesas, or buttes are formed from the erosion of sedimentary formations, they are invariably terminated by steep scarps, e.g. Sokoto Plains on the Gwandu Formation in the northwestern part of the country and the Bida Plains on Nupe Sandstones in the west-central part of the country. Cuestas occurring on the Benue and Gongola valleys (east middle belt area) are associated with ridges and hills. Cuestas are also found in the southeastern part of the country at the Enugu/Okigwe escarpment, the Udi/Nsukka escarpment, and the Awka-Orlu Uplands. Ezechi (1987) reports that the gradient in some places is about 12% to 30%, while at other localities, it ranges from 20% to 30%. Steep slopes also occur in the Ekiti area of western Nigeria.

In areas underlain by igneous and metamorphic (locally referred to as Basement Complex) rocks, the

246

variation in lithology seems to reflect the variety of landforms. For example, gneisses, migmatites, and schists at many localities underlie the more or less undulating to flat plains, although such plains are for the most part dotted by isolated rocky hills. Metamorphosed sedimentary rocks, especially quartzites and mica schists, form impressive north-south trending ridges in the western half of the country from southwestern Nigeria to eastern Sokoto state in the northwestern part of the country. Older granites and other granitic rocks are associated with hilly and highland terrains, especially steepsided inselbergs, rock massifs, and mountain ranges. The more resistant younger granites have given rise to massive hills on the Jos Plateau and massifs associated with the Jemaa Platform in the neighborhood of Kafanchan in the north-central part of the country. The Mandara Hills (in the northeast), the Adamawa Plateau (also in the northeast), the Akovolwo hills (in the middle belt), the Bambariko (Ikom) hills, and the Obudu and Oban massifs (all in the southeast) are landform regions that also have steep slopes. In general, therefore, steep slopes are not restricted but can be found in almost all parts of the country. A few of these slopes, e.g. the Mambilla Plateau, the Akovolwo hills, the Enugu/Okigwe escarpment, and the Bambariko hills in recent years have been the sites of complex landslides.

4.2 Geology

The geology of Nigeria is rather complex so that only a brief, simplified account can be given here. The country is underlain by Basement Complex and sedimentary rocks, each occupying about 50% of the country's land mass. The Basement Complex rocks (comprising granites, gneisses, schists, migmatites, etc.) are found in the western, northern, and in the Oban and Obudu areas of southeastern Nigeria. The sedimentary rocks are found in the coastal areas of eastern and western Nigeria, the Niger and Benue Valleys, and the Sokoto, Gongola, and Chad basins. Fig. 3 shows a generalized geologic map.

Of the sedimentary units, only the sandstones (especially the Ajali Sandstone of Upper Maestrichtian age, the Nsukka Formation of Upper Maestrichtian-Danian age, and the Nanka Sands of Lower Eocene age (shown in Fig. 4) are known to slide frequently. These formations are also ravaged by gully erosion, which is the causative factor of

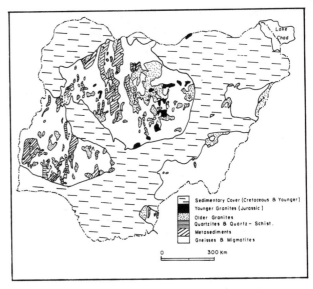

Figure 3. Generalized geologic map of Nigeria.

sliding in most places.

The Nsukka Formation, which consists of an alternating succession of sandstone, dark shale, and sandy shale with thin coal seams at various horizons, is not naturally landslide prone. However, the basal sandstone, about 15 m thick, often suffers gully erosion which ultimately leads to minor landslides. The Ajali Sandstone is also affected by severe gullying and landsliding. The formation consists of thick, friable, poorly-sorted sandstones, typically white in color but iron stained at a few localities. Where it caps the Enugu escarpment (e.g., north of Oji River), the formation is incised by deep canyons and gullies along the higher slopes of the scarp. The Nanka Sand is probably the most affected by gullying and landslides. The formation consists of thick sequences of loose and friable sandbeds with little or no cement, and a few interbeds of thin shale, claystone, and ferruginous sandstone horizons.

Landslides in the igneous and metamorphic rocks are uncommon, except where residual soils derived from deep weathering of the rocks suffer minor

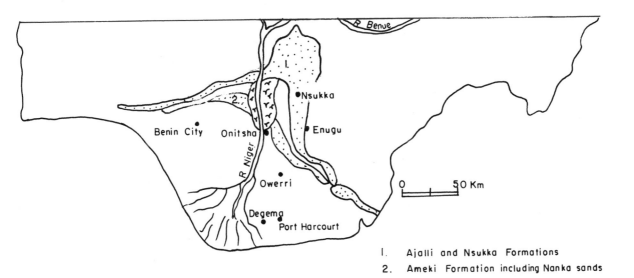

1. Ajalli and Nsukka Formations
2. Ameki Formation including Nanka sands

Figure 4. Geologic formations that experience frequent landslides.

slumps after heavy rains. Such slumps have mostly been inconsequential, except in a few steep hills where they have triggered massive debris flows.

5 GEOGRAPHIC EXTENT OF LANDSLIDES

Of the 21 states that make up the country (Fig. 1), only about five experience frequent and notable landslides. Four of these (Anambra, Imo, Cross River and Bendel) are states ravaged by serious gully erosion; the fifth (Gongola) experiences landslides in the Mambilla Plateau area, mostly as a result of steep slopes. Geographically, these states make up only about 1/8 of the country's total landmass. However, this is not to say that some landslides do not occur in other states. For example, gullies have been observed in the watershed areas of Kaduna State and Shemanker Valley on Jos Plateau (Grove, 1952, 1956, 1957); parts of Sokoto State (Prothero, 1962); parts of Gongola Basin (Dowling, 1965); Kano close-settled zone (Mortimore and Wilson, 1965); Zaria area (Ologe, 1972, 1973), parts of Benue State, especially Ankpa; and in Kwara State, especially in the basin of the Kampe River. Ologe (1986) notes that gullying has affected a broad east-west belt across the Nigerian savanna from the eastern highlands to the Sokoto basin. These gullies invariably lead to several small slides and slumps. A few landslides unconnected with gullying have been reported near Ile-Ife in Oyo State (Jeje, 1979), near Umuahia in Imo State (Okagbue, 1988), near Jato Aka in Benue State (Nwajide et al., in press), near Ikom in Cross River State (Abu, 1987), near Isuawa in Anambra State (Abbey, 1987), and near Umomi and Ugwalawo in Benue State (Akor, 1987). Most of the landslides caused by heavy rainstorms occurred along steep slopes and involved residual/colluvial materials.

5.1 Population distribution relative to landslides

Factors of gullying in Nigeria have been examined under two major subheadings: human, or anthropological, and physical. The most important human factor involves heavy population pressure on the available land, the consequent over cultivation and (or) overgrazing, and the usual cultivation methods often involving mound making on steep slopes and alignment of ridges parallel to the slope direction. Population pressure is also suspected to

result in numerous engineering works accompanied by excavations and disruption of natural water courses. It is interesting to note that the intensely gullied areas are also densely populated (compare Figs 4 and 5). This perhaps lends some credence to the anthropological factor, although such relationship can hardly be on a one-to-one basis. For example, almost all the landslides associated with hills have occurred in forest lands that are either uninhabited or very sparsely populated. Therefore, some slides are virtually unconnected with population pressure.

5.2 Types and distribution of landslide processes

In most of the severely gullied areas of southeastern Nigeria, the lateritic soil overburden, which in some places is as much as 20 to 30 m in thickness, has been dissected exposing very thick deposits of soil composed of an upper, homogenous and cohesive, reddish-brown unit and a lower, cohesionless, white unit having thin shale-siltstone interbeds. The walls of the complex gullies stand almost like cliffs.

The landslide processes which lead to gully erosion occur by several modes such as slumping (rotational sliding), planar gliding (translational sliding), and in some areas, toppling (rotational falling). Slumps are the most common type of slide. The large ones measure from 70 m to 90 m from crown to toe. During the rainy season, saturated sandy soil at the toe of slumps can be remolded to the consistency of a viscous fluid and flow tens of meters. Slumping is commonly retrogressive, i.e., new portions of the surface of rupture form progressively upslope in previously unfailed soils. The maximum depth of the surface of rupture ranges from about 20 to 30 m for most slumps, the deepest recorded is about 45 m (Okagbue, 1986). Slumps are concentrated in the homogenous reddish soil overburden.

Planar glide blocks develop mostly where clay dips in the direction of the slope and where the slope of the gully wall is relatively gentle. Some of the slides have pressure-induced ridges and hummocks near the toes. At most localities, the impermeable clay beds separating the sand are a locus of concentration for both slumps and planar glide blocks. The clay is constantly lubricated by seeping water and, therefore, becomes an easy slip plane. A few topples occur where relatively stronger interbeds, mostly iron-cemented, are present. Subaerial erosion, combined with sloughing and caving of uncemented sands underlying the iron-cemented layer, lead to undermining and the development of overhangs and tension cracks. Okagbue (1986) has given detailed descriptions of these failure processes.

Along non-gullying river banks, the sliding processes are mainly by slumping and toppling, depending upon the relative thickness of the cohesive upper stratum of the banks' stratigraphy (Okagbue and Abam, 1986). In the Mambilla Plateau area, landslides occur in the rainy season mostly in the form of slumps and are most prominent in the eastern and southwestern parts of the plateau. Undermining of the slope toe by running water is a major factor in the sliding process.

Landslides in residual soils (in Basement Complex areas) are mostly slumps. However, a few debris flows have been reported. One occurred on a ridge near Ile Ife, Oyo State (Jeje, 1979); two others along the slopes of the Akovolwo hill ranges near Jato Aka, Benue State (Nwajide et al., in press); one along the Bambariko hills near Ikom, Cross River State (Abu, 1987); and another along the Umomi and Ugwalawo hill ranges near Idah Benue State (Akor, 1987).

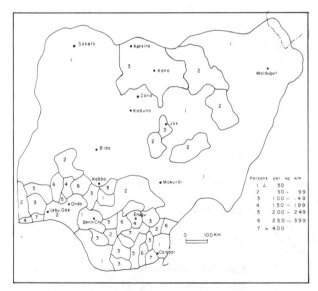

Figure 5. Population density, 1963 (adapted from Afolayan, 1983).

5.3 Status of knowledge about extent of landslide processes

The majority of people in Nigeria are unaware of the extent of landslide processes. Understandably, this stems from the fact that many people are not affected by the processes. Those whose lands are ravaged by gully erosion, however, are aware on a non-technical level. The Federal government, those state governments whose areas of jurisdiction are affected, and several scientists (geologists, soil scientists, foresters and agricultural engineers) are generally aware about landslide processes and continue to battle the problem. A few years ago, for example, the Federal government awarded a multi-million dollar contract to engineering firms to help combat gully erosion. State governments have tried from time to time to educate the public about techniques to prevent or minimize the effects of erosion. These efforts are hampered by lack of an overall terrain evaluation of the country to guide land use planning or to highlight areas of potential instability. Some people have been forced to learn about this ecological problem if they live near the few sites of debris flows. These people have on a few occasions fled their homes thinking that an earthquake had occurred.

5.4 State-of-the-art landslide prevention and control

Gully erosion which leads to landsliding has been seriously tackled in Nigeria. Because the generation of the gullies has been attributed to the material properties of the geologic formations affected, the human aspect of forest clearing and land misuse, and the large amount of surface runoff usually involved, control has been based largely on techniques capable of reducing the erosive capacity of the flood waters. Techniques capable of reducing either the quantity of the flood water flowing in the drainage network or reducing their velocity and techniques for increasing the resistance of the soil relative to the erosive capacity of the flood waters are employed.

To reduce the erosive capacity of the flood water, two types of measures have generally been adopted. One type has involved the construction of hydraulic regulation works that integrate a drainage network with storage ponds to cut off flood crest and lower hydraulic loads of interceptor canals. The interceptor canals, which are commonly located at the head of the advancing gully-channels, drain runoffs from areas adjacent to the gullies and discharge them in artificial reservoirs (ponds) constructed where deep infiltration can occur. The other type has involved the installation of channel structures to lead surface water away from the gully areas.

Within the hydraulic regulation structures, stabilization works such as check dams (mostly timber type) have been constructed on the main channels of gullies, revetments and hedges at the inner gully slopes.

The revetments have been in the form of wicker-work fences used to reduce surface flow and to stabilize the gully slopes. The fences are formed by stakes about 10 cm thick driven into the ground close together and interwoven with braces. The wicker-work fences in some places are accompanied by tree planting to help strengthen the soil.

Tree planting may be the oldest form of control measure; the government advocates tree planting both for control and prevention. Its use is based on the premise that the tree roots help to strengthen the soil by binding the soil particles together. Also, gully-slope flattening has been carried out by contracting firms involved in erosion control.

The control measures so far adopted in the gullied areas appear to have been successful in the shallow (less than 15 m deep) gullies cut mainly into red clayey earth. They have failed in deep gullies cut into very permeable and cohesionless sand where the gully walls are indented with seepages at various horizons. These failures have raised the question as to whether the remedial measures which take care of only the surface runoff and fail to accommodate the disastrous effects of groundwater are really effective.

Tree planting as a stabilizing measure has been effective only on minor and shallow gullies (less than 5 m deep). The reason appears to be that the disturbances (toe cutting by runoff) in deeper gullies originate several meters below the zone of influence of tree roots. In deep gullies where the slopes are constantly undermined by flood and internal erosion, the trees close to the gully edge usually slide down onto the gully floor as soon as the slope is undermined.

Therefore, although serious efforts are made to control gully erosion, little attention has been paid to such widely recommended landslide control measures as groundwater drainage. Moreover, construction of retaining walls has seldom been attempted either because of financial constraints or lack of know-how.

In a few cases where minor landslides have involved highway cuts, repairs have been carried out by the government on an individual basis with no specific rules. Because a majority of these cut slope failures do not involve large volumes of materials and because their consequences are usually minor (temporary closure of road, at worst), no serious attempts have been made to prevent or control them. Failed slopes are usually cut back and surface drainage provided. Subsurface drainage, retaining walls, and ground anchors have virtually never been used in such circumstances. The author has also observed a landslide involving a highway where the downdropped road section was wrongly corrected by simply filling to the original grade level. The result of this loading was a recurrence of the slide. The landslide needed subsurface drainage, along with ground anchor, to be corrected (Okagbue, 1988).

Attempts have been made at a few locations in the Niger delta to control riverbank failures by the use of filter mattresses on flattened back slopes (Abam and Okagbue, 1986). So far, the attempts have proved successful, but because the project is capital intensive, very few locations have been covered, and the likelihood of extending the measure to other affected areas seems remote.

5.5 Short description of a few major landslides

Many landslides occur in gully areas but none has been studied on an individual basis, probably because none has been catastrophic. The few major landsides studied are of non-gully origin. One occurred on a ridge near Ile-Ife (southwestern Nigeria) on October 21, 1976. It look the form of a debris flow whose scar occupied 0.31 ha with a debris yield of 2,500 m^3 and debris deposits covering about 0.30 ha. The flow, according to Jeje (1979), seemed to have resulted from three major factors: the persistent, moderately intense rainfall for several days before the event; the poorly permeable soil and substratum on the hillslope; and the inefficiency of subsurface drainage on the hillslope.

Although the scar of the flow covered only 0.30 ha, about 450 cocoa trees, 35 kola nut trees, and 10 orange trees were destroyed near Ile-Ife (Jeje, 1979) by uprooting or were broken in the middle. Since the flow occurred on a hillside forestland, no corrective measures were undertaken.

Another major landslide on November 7, 1985 involved the only paved road linking the city of Umuahia and the towns of Bende and Ohafia in the southeastern part of the country. After a day of heavy rain, the slide broke the highway into three segments. The middle section dropped to such a depth that it was nearly impossible to use the road again. The landslide measured about 200 meters from crown to toe with a width approaching 190 m in the middle section area. Vertical displacement along the crown scarp was about 1.8 m. At one boundary of the slide, vertical drop measured about 1.7 m, while at the other boundary, it measured about 1.4 m. The movement had severed the road into three segments with the unstable middle section moving away relative to the other two sections (Fig. 6). Detailed studies by Okagbue (1988) revealed that the slide occurred in clay shale where the mineralogy

Figure 6. A landslide along the Umahia-Bende Road in southeastern Nigeria.

was predominantly kaolinite and illite, with the possible presence of vermiculite, chlorite, and traces of montmorillonite. The slide was triggered by a sudden rise in ground water level and it occurred along a previous slip surface.

The major storm-induced debris flows of September 24 and 29, 1979, (Nwajide et al., in press) occurred along the Akovolwo ranges in Benue State. Two movements developed on bedrock slopes covered by a veneer of colluvium and (or) residuum no more than 1 1/2 to 2 m thick. Each started as a slip (Fig. 7) and evolved into a major debris flow. At Mbaav, one of the flow sites, the move-ment exploited an existing fracture-controlled mountain stream channel, branched, and rejoined farther downslope. The mid-channel width was about 70 m before branching and rejoining toward the foot. The length of flow was as much as 1 1/2 km. At Mberev, the other flow site, the mass moved along a minor mountain flood path which was subsequently widened,

Figure 7. A landslide that turned into a major debris flow in the Akovolwo hills of Benue state.

deepened, and then formed into a new stream channel. The width of the debris flow was about 50 m at the mid-channel, widening to over 100 m at the foot, with the length about 2 km. Boulders of all shapes and sizes were dislodged and transported varying distances in both debris flows. Trees were broken, uprooted, and buried by the rubble. No remedial measures were necessary because the move-ments occurred in forested hillsides, although not too far (less than 4 km) from rural villages.

6 SOCIAL AND ECONOMIC IMPACTS

No statistics are available for assessing the casualties or damage caused by landslides in Nigeria. However, the impact of the gully systems and the associated landslides in southeastern Nigeria has been devastating on human resources and ecology.

Farming constitutes the main livelihood of people in the affected areas. With farming jeopardized at a constantly increasing rate by gully advancement through landsliding, there is an obvious threat to human survival. It is estimated that about 20 ha of land is lost annually and this translates, in agricultural income, to millions of dollars annually. The cost of remedial measures such as gullies in Anambra State alone is estimated at $11.3 million (Anambra State Government Report, 1985). About the same amounts are probably needed for Imo and Cross River States.

Besides the destruction near Ile-Ife of the cocoa, kola nut, and orange trees (Jeje, 1979), the footslope normally used for cultivation of food crops was covered by sterile, stony debris rendering an approximate area of 0.30 ha useless for cultivation.

In the Bambariko hills debris flow of August 23, 1987, long stretches of farmlands were completely buried. Crops such as yams, plantains, and cassava, some already harvested and stored, were destroyed. Immediate losses were estimated at more than 0.33 million dollars. Long term losses are obviously greater because the valleys, which had hitherto provided the rich agricultural land that served as the pivot of the people's livelihood, are now covered by sterile stony debris.

Other social impacts include the perpetual fear in which the people who live in the affected areas are subjected. Many have had to abandon their homes to the menacing landslides. In 1979, the people of Mbaav and Mberev, (two small villages in the vicinity of Jato-Aka, Benue State) vacated their homes when the debris flows reached avalanche speeds in the Akovolwo ranges. They had mistaken the deafening noise for earthquakes. The Bambariko people (particularly in the eastern Boki village) have recently (November, 1987) sent a delegation to the Nigerian Federal government asking for relocation. They fear that the Bambariko hills which have witnessed two major debris flows, one in 1983 and the most recent in 1987, might "revolt" again.

Table 2 shows a few selected gullies and the impact on the local communities. Apart from the social impact, the landslide debris is quickly carried into streams, resulting in silting and a threat to the water supply and aquatic life.

7 SUMMARY AND CONCLUDING REMARKS

Landslides in Nigeria, although very scantily reported in the literature, occur mostly in sedimentary terrains that are severely gullied. They take the form of slumps, block translational sliding, and, in some places, topples. In igneous and metamorphic (locally referred to as Basement Complex) areas, landslides occur as slumps in highly weathered residual soils and (or) colluvial deposits. A few high velocity debris flows in

Table 2. A few selected gully/landslide sites in southeastern Nigeria and their impact on local communities

Gully/landslide site	Facilities damaged or threatened
Ekwulobia	Aguata High School buildings threatened; residential areas and farmlands dangerously being encroached upon; people in Ula village vacating their residences
Uga	Water intake pipes exposed; villagers living along the approach route fast being driven away.
Akpo	Akpo Boys' High School threatened.
Onitsha-Nsugbe Road	Bridge situated about 300m behind Progress Hotel complex, Onitsha and near Offia-Oyibo threatened
Mbaukwu	Old Mbaukwu-Nibo road now abandoned; parts of central school, Mbaukwu, carried away.
Nnobi-Ideani-Uke	Nnobi community secondary school, Ideani and Uke Girls' secondary school threatened.
Awuda-Nnobi	A decked building lost, many other buildings threatened.
Ududonka-Agulu	Many buildings lost in 1974 landslide.
Amichi-Ekwulumili	A customary court lost. Bridge across Nsasa River very threatened.
Achi	Achi-Maku Road undermined.
Nkisi	Greater Onitsha water scheme threatened, water pipe lines exposed and broken in places; inhabitants around the area abandoning homes.
Ngwo	Enugu-9th mile corner old road cut off and abandoned.
Eke	Dormitory of the girls' secondary school and a building of the central school at Afo Ojebe as good as lost.

regions of high relief and steep slopes have been recorded in Basement Complex areas. Landslides also occur along oversteepened road cuts and in a few places along highway embankments. Most of these slides are water induced because they occur most often during the rainy season (April-October).

At present, there is little notable interest in Nigeria, either by the Government or by scientists, in this problem except, perhaps, gully erosion. Three reasons for this seem apparent: (1) landslides are not a common occurrence throughout the country, (2) no catastrophic landslides claiming lives and huge amounts of economic losses have occurred, and (3) very few personnel seem to have an interest or technical training in landslide problems. The result is that no terrain evaluation is available to guide land use planning or to highlight areas of potential instability. It is hoped that nature will continue to keep the country free from catastrophic landslides as it has so far, but population increase and the resultant pressure on land may well interfere in future with the present equilibrium.

8 ACKNOWLEDGMENTS

The author benefited much from discussions with colleagues in the Department of Geology, University of Nigeria. Dr. K.O. Uma of that department helped to proofread the manuscript. The drawings were done by Mr. A. Dibia of the Geography Department. The author thanks all of them.

REFERENCES

Abam, T.K.S., and Okagbue, C.O., 1986. Construction and performance of river bank erosion protection structure in the Niger Delta Bull. Assoc. Eng. Geol. 23(4): 499-506.
Abu, B.D., 1987, Living in fear. Newswatch Magazine, Nigeria, Oct. 26: 32.
Akor, J., 1987, Volcanic eruption in Benue State. Sunday Concord, Nigeria, Nov. 8: 16.

Afolayam, A.A., 1983. Population. In Oguntoyimbo, J.S., Areda, O.O., and Filani, M. (eds.), A geography of Nigerian development, Heinemann Educational Books (Nig.) Ltd.: 113-123.
Anabram State Government, 1985. Gully units in Anambra State with cost estimates of remedial measures. Interministerial comm. rept. on soil erosion, Anambra State, Government House, Enugu.
Bradford, J.M., and Piest, R.F., 1977. Gully wall stability in loess derived alluvium. Soil Science Society of America Journal. 1 (1): 115-122.
Bradford, J.M., Piest, R.F., and Spomer, R.G., 1978. Failure sequence of gully headwalls in western Iowa: Soil Science Society of Am. J. 42: 323-328.
Bradford, J.M., and Piest, R.F., 1980. Erosional development of valley-bottom gullies in the upper midwestern United States. In Coates, D.R., and Vitek, J.R. (eds.), Thresholds in geomorphology: 75-101.
Dowling, H.W.F., 1965. The mode of occurrence of laterites in northern Nigeria and their appearance on aerial photography. Eng. Geol. 1: 221-233.
Egboka, B.C.E., and Okpoko, E.I., 1985. Gully erosion in the Agulu-Nanka region of Anambra State, Nigeria, in Challenges in African hydrology and water resources. Proc. Harare Symposium, IAHS Publ. 144: 335-347.
Egboka, B.C.E., and Nwankwor, G.I., 1985. The hydrogeological and geotechnical parameters as causative agents in the generation of erosion in the rain forest belt of Nigeria. J. of African Earth Sciences. 3 (4): 417-425.
Ezechi, J.I., 1987. An engineering geologic study of selected gully sites in Anambra and Imo States, Nigeria. University of Nigeria, Nsukka, Department of Geology, M.S. thesis.
Grove, A.T., 1951. Land use and soil conservation in parts of Onitsha and Owerri provinces. Nigeria Geol. Survey Bull. 21. Zaria, Gaskiya Corp.
Grove, A.T., 1952. Land use and soil conservation on the Jos Plateau. Geological Survey of Nigeria Bull. 21.
Grove, A.T., 1956. Soil erosion in Nigeria. In Geographical essays on British tropical lands,

Steel, R.W., and Fisher, C.A., (eds): George
Philip & Sons, Ltd., London: 79-111.

Grove, A.T., 1957. The Benue Valley: Government
Printer, Kaduna.

Jeje, L.K., 1979. Debris flow on a ridge near Ile-
Ife, western Nigeria: Journal of Mining &
Geology, 16 (1): 9-16.

Mortimore, N.J., and Wilson, J., 1965. Land and
people in the Kano close settled zone: Ahmadu
Bello University, Dept. of Geography, Occasional
Paper No. 1.

Nwajide, C.S., Okagbue, C.O., and Umeji, A.C., (in
press). Slump-debris flows in the Akovolwo
Mountains area of Benue State, Nigeria: Natural
Hazards.

Ogbukagu, IK. N., 1976, Soil erosion in the northern
parts of the Awka-Orlu Uplands, Nigeria: J. of
Mining & Geol. 13: 6-19.

Oguntoyinbo, J.S., 1978, Climate. In Oguntoyinbo,
J.S., Areda, O.O., and Filani, M. (eds.), A
geography of Nigerian development, p.83-94.
Heinemann Educational Books, Ibadan, Nigeria.

Okagbue, C.O., 1986. Gully development and advance
in a rain forest of Nigeria. Proc. 5th Int.
Congress Int. Assoc. Engr. Geol., Buenos Aires:
1999-2010.

Okagbue, C.O., and Abam, T.K.S., 1986. An analysis
of stratigraphic control on river bank failure:
Eng. Geol. 22: 231-245.

Okagbue, C.O., 1988 (in press). A landslide in a
quasi-stable slope: Eng. Geol.

Ologe, K.O., 1972. Some aspects of the problem of
modern gully erosion in the northern states of
Nigeria: Paper presented at the Hydrological
Technical Committee Symposium, A.B.U. Zaria,
Sept., 1972.

Ologe, K.O., 1973. Gullies in the Zaria area: a
preliminary study of headscarp recession:
Savanna, 1: 55-66.

Ologe, K.O., 1974. Kudingi gully: headscarp
recession, 1969-1973: Savanna, 3: 87-90.

Ologe, K.O., 1986. Soil erosion characteristics,
processes and extent in the Nigerian savanna.
Proc. National Workshop on Ecological Disasters.
In Erosion, Fed. Univ. of Sci. and Tech.,
Owerri. Fed. Ministry Sci. & Tech., publ.

Prothero, R.M., 1962. Some observations on
desiccation in northwestern Nigeria: Erdkunde,
6: 111-119.

Taylor, J.C., and Johnson, H.P., 1973. Gully bank
erosion: mechanics and simulation by digital
computer: Iowa State Water Resources Research
Institute Project No. 4 - 034-1A.

Landslides: Extent and Economic Significance, Brabb & Harrod (eds)
© 1989 Balkema, Rotterdam. ISBN 90 6191 876 6

Landslides in Kenya: A geographical appraisal

K.M.Rowntree
Department of Geography, Rhodes University, Grahamstown, RSA

ABSTRACT: The landslide problem in Kenya has been little researched and there is a lack of information on both the distribution of landslide phenomena and their economic and social significance. The landslide risk in Kenya is assessed therefore in relation to geology, rainfall and slope gradient. Areas identified as being most at risk are those affected by major tectonic activity associated with the formation of the rift valley, with deeply weathered volcanic soils and a high to moderate mean annual rainfall. Many of the areas thus identified also carry a high and increasing population density due to the limited areas of land in Kenya with sufficient rainfall to support agriculture. The economic cost of landsliding to the rural community is considered therefore to be significant.

1 INTRODUCTION

The landslide problem in Kenya has attracted little attention from administrators and researchers alike. The focus has been more on the effects of surface water erosion as a source of the rivers' high sediment yield (Dunne, 1979) and as a cause of land degradation (Barber, 1983). However, as this paper demonstrates, the potential for landslides is locally significant and is concentrated in areas of moderate to high population density so that it does represent a hazard to land development. In the absence of well documented research on the distribution of landslides and related phenomena in Kenya, this paper attempts to assess the risk of landsliding by examining the magnitude and distribution of causative factors: namely geology, rainfall and topography. These factors will then be related to the social and economic consequences of landslides in terms of population distribution and land usage. No attempt will be made to put a cost to landslide damage as the relevant data is not available to the author.

For a comprehensive review of landslide phenomena, their causes and consequences the reader is referred to the work of Crozier (1986). Literature relating to tropical areas is reviewed by Thomas (1974).

Landslides are favoured in the tropics by deep weathering processes (Thomas, 1974) which produce thick zones of incoherent materials of low shear strength, high void ratio and high clay content. Basic rocks in particular give rise to landslide prone materials. Frequent associations occur between landslide activity and the presence of duricrusts.

Landslides in the tropics are of greatest importance in tectonically active areas of high relief and high rainfall; they are generaly absent from the vast shield areas of the intertropical zone. According to Thomas (1974) they are confined to slopes of between 30 and 60 degrees; on steeper slopes the regolith is too thin to support slides as such, on gentler slopes creep dominates over slides. The form of the landslide depends on the depth of weathering. Deep weathering favours rotational slips whereas slides and avalanches occur in less deeply weathered material in areas of high drainage density and deep dissection.

Thomas (1974) gives the main trigger factors in the tropics as high rainfall concentrated into a short wet season, earthquakes and tree fall. The effect of tree fall is to allow increased infiltration of water into the regolith.

Neither Crozier (1986) nor Thomas (1974) make mention of Kenyan landslides, though examples are cited from neighbouring Tanzania (Temple and Rapp, 1972), Uganda (Pallister, 1956) and Sudan (Ruxton and Berry, 1961). Given the occurrence of landslide phenomena in these three countries the existence of significant landslide activity within Kenya itself would appear likely. Direct evidence does in fact come from the work of Kamau (1981), the only researcher to make a systematic study of Kenyan landslides. He describes extensive mass movements on the eastern dipslope of the Aberdares Mountains. The results of his study are described in more detail later in this paper.

2 KENYA: LOCATION AND PHYSIOGRAPHY

Kenya is located on the east coast of Africa and is bisected by the Equator (Fig. 1). A major

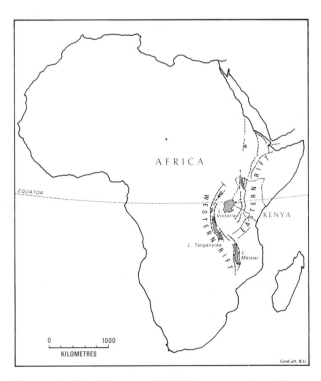

Figure 1. Location map of Kenya

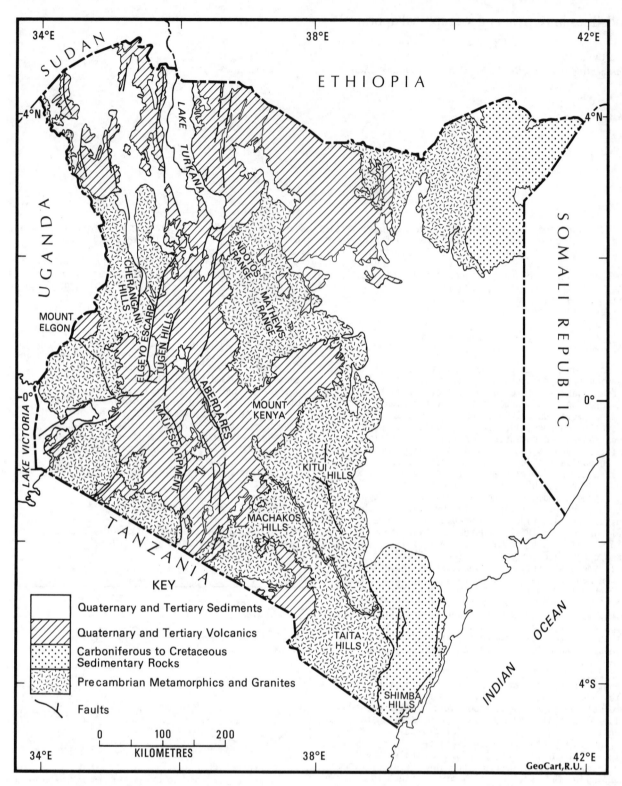

Figure 2. Simplified geology of Kenya

topographic feature is the north-south trending rift
valley which forms a linear trough between adjacent
high plateaus to the east and west. Vulcanism
associated with the tectonic activity which formed
the rift system is a widespread source of rock
material and was responsible for the mountain massifs
of Mount Kenya, the Aberdares and Mount Elgon (Fig.
2). The rift system is therefore an important
feature in determining the distribution of landslide
potential in Kenya through its control on topography,
rock materials and climate.

3 GEOLOGY

The geological map of Kenya (Fig. 2) shows four
major geological systems, the metamorphic rocks of
Precambrian age, sedimentary rocks of Carboniferous
to Cretaceous age, Tertiary and Quaternary volcanics
and unconsolidated Tertiary and Quaternary
sediments. Also evident from Fig. 2 is the
widespread distribution of fault lines associated
with tectonic activity along the rift system.

Precambrian rocks of the basement and associated systems outcrop over large areas of the eastern plateau and in the west near Lake Victoria. These rocks are composed of metamorphosed sedimentary rocks such as gneisses, schists and quartzites (Cahen et al., 1984) with local granite intrusions. The characteristic topography of the eastern plateau is that of undulating plateau or peneplain broken by inselberg type uplands. Major hill masses include the Ndoto and Mathews ranges to the north, the Machakos and Kitui hills in central Kenya and the Taita Hills in the south east.

Sedimentary rocks of Carboniferous to Cretaceous age outcrop near the south coast and in north-eastern Kenya. Hilly topography, for example that of the Shimba hills, is confined to Triassic rocks, composed of sandstones and grits.

Volcanic rocks of Tertiary and Quaternary age are widespread throughout central and northern Kenya, being associated with the rift system. The earliest lavas are Miocene in age and are predominately phonolitic. Major tectonic activity in the late Pliocene was accompanied by a change to trachyte volcanism (Baker et al., 1972). Other lava types include basalts, nephelinites and alkali rhyolites. Flood lavas are widespread on the plateaus adjacent to the rift valley and within the graben of the rift valley floor, whereas central vent eruptions form major mountain masses: Mount Kenya, Mount Elgon, and the Aberdares range. The lavas are predominantly of alkaline type and under a humid climate have weathered to give deep, clay rich, well structured soils, which are subject to landsliding on steep slopes.

Lastly, Tertiary and Quaternary sediments, composed mostly of lacustrine and fluviatile deposits, are widespread within the rift valley floor and in eastern Kenya. These deposits generally occur in low energy environments and are therefore not prone to landsliding. There are, however, exceptions as when the sediments are disrupted by faulting as has happened in the Lake Baringo basin (Rowntree, 1985).

The rift system of east Africa is a notable feature of the Kenyan landscape and provides an obvious focus of steep slopes, weak seismic activity and potential landslide activity. The geology of the rift system is described in detail by Baker et al. (1972) and is summarised below.

The most impressive section of the rift system in Kenya is the Gregory rift which extends northwards from the southern border with Tanzania to the Baringo and Suguta grabens south of Lake Turkana. The main scarps range from 300 to 1,600 m in height and are en echelon in plan, forming a complex graben 60 to 70 km. wide. North of latitude 1° N., the rift valley gradually widens towards Lake Turkana so that the escarpment slopes are less marked. An arm of the rift system, the Kavirondo rift, extends south west into Lake Victoria with a width of between 15 and 25 km.

The present day rift valley is the result of major tectonic activity at the end of the Pliocene (about 2.5 m.y.). Central Kenya was uplifted by around 1,500 m; at the same time the margins of the rift trough were warped down and strongly faulted, to produce a true graben for the first time. Smaller scale movements continued throughout the Quaternary when Plio-Pleistocene volcanics and Pleistocene sediments of the graben floor were cut by swarms of closely spaced faults with throws of up to 150 m.

The rift system in Kenya is comparatively aseismic at present and although earth tremors occur no major earthquake has yet been recorded.

4 RAINFALL

Prolonged or intense rainfall, or more particularly a combination of the two, are among the most important triggers of landslides (Nilsen, 1986; Crozier, 1986). Humid tropical climates are also

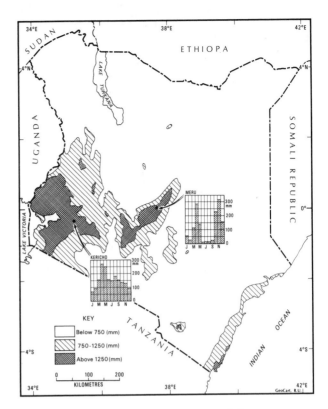

Figure 3. Mean annual rainfall map of Kenya. Typical seasonal rainfall distributions for eastern and western Kenya are shown.

associated with deep weathering leading to a reduction in rock strength (Thomas, 1974). Landslide potential will therefore be determined largely by the distribution of annual rainfall totals and intensities.

Rainfall patterns in Kenya are governed both by its equatorial location, its position on the east coast of Africa and locally by the rift valley axis. As can be seen from Figure 3 high rainfall areas are limited in extent; 75 per cent of the country is semi-arid to arid.

Mean annual rainfall totals in excess of 1250 mm are restricted to western Kenya and to high altitude areas, especially the rift valley margins and associated volcanic peaks. These are the areas in which rainfall regularly exceeds potential evaporation, leading to a soil moisture surplus for several months of the year (Table 1). Of particular note are Kericho on the western dip slope of the rift valley escarpment and South Kinangop and Kimakia in the Aberdares, localities at which soils are saturated on average for 8 or 9 months of the year.

High variability is a characteristic of annual rainfall in Kenya, particularly in the drier areas (Rowntree, 1988; Rowntree, 1989) where the more normal dry years are interspersed with years of heavy rainfall conducive to flooding. In areas with moderate mean annual rainfall the risk of landsliding will be greatly reduced relative to high rainfall areas, but it will not be removed altogether.

The seasonal distribution of rainfall is significantly different east and west of the rift valley as depicted in Fig. 3. To the east two marked rainy seasons, centred on the months of April and November, are related to the movement of the inter-tropical convergence zone, whereas to the west there is an extended rainy season from March through to August and a less marked dry season in the remaining months. As a consequence the concentration of rainfall in a shorter time period results in higher intensities to the east of the rift valley as can be

Table 1 Rainfall characteristics of selected
locations in Kenya

	Altitude (m)	Mean annual rainfall (mm)	Maximum 24 hour rainfall (mm)	Month of maximum rainfall	Months with surplus moisture*
Western Kenya:					
Koru	1560	1746	148.5	March	6
Kisii	1707	1957	108.4	March	0
Kapenguria	2134	1200	71.1	May	4
Kitale	1890	1191	99.5	April	1
Eldoret	2084	1124	100.0	April	1
Equator	2762	1219	73.7	Nov	1
Molo	2477	1177	84.1	April	3
Kericho	2134	2081	79.3	April	9
Eastern Kenya:					
Rumuruti	1768	776	111.8	Nov	0
South Kinangop	2591	1417	165.1	Feb	8
Kimakia	2439	2288	104.9	March	9
Ruiru	1608	1065	177.5	April	1
Muguga	2096	995	215.6	April	1
Kitui	1088	1162	1499.4	April	2
Voi	560	549	254.0	Dec	0
Malindi	91	1229	156.2	July	0

*Average number of months per year

seen from the maximum 24 hr rainfall totals
listed in Table 1.   It can also be seen that high
daily rainfall totals can be received at stations
with a low mean annual rainfall, for example 215.6 mm
at Muguga on the eastern edge of the rift valley
escarpment (M.A.P. 995 mm) and 254 mm at Voi on the
eastern plateau (M.A.P. 549 mm). Hence although the
potential for landslides is obviously higher in the
high rainfall areas, infrequent intense storms could
also trigger off landslides in low rainfall areas if
other conditions were suitable. Both to the west and
east the highest rainfall intensities tend to occur
at the start of the wet season in March or April so
that the landslide risk would be highest then.

5 PREVALENCE AND LOCATION OF STEEP SLOPES

Slope gradient is obviously a critical factor
controlling the distribution of landslides as failure
will only occur on slopes exceeding the critical
angle for the material in question. Reference has
been made above to Thomas's (1974) observation that
in tropical areas mass movements are generally
confined to slopes of between 30° to 60°.
Observations by the author in the Aberdares foothills
indicate a high frequency of valley side slopes of
20°, suggesting that this is close to the stable
angle for the deeply weathered Tertiary volcanics
which form the substrate. This observation is
substantiated by the work of Kamau (1981) reported
below. For the purposes of this paper it was assumed
that slopes of 20° and over are potentially unstable.
Their distribution is shown on Fig. 4, based on an
analysis of 1:250,000 topographic maps. The map
shows those areas where the macro-scale slopes,
generally with a relief of 300 m or more, are in
excess of 20°. It excludes others which may have
short steep valley side slopes, but in which the
regional gradient is more subdued.
  Comparison of Fig. 2 with Fig. 4 shows that areas
of steep slopes are located in the main along the
flanks of the rift valley, especially to the west in
the Elgeyo-Cherangani-Kapenguria area and to the east
in the Aberdares range. Steep slopes are also
associated with inselberg type hills of the
Precambrian rocks: in the Mathews and Ndoto ranges to

the north of Mount Kenya, the Machakos and Kitui
Hills to the south of Mount Kenya and the Taita Hills
near the southern border.

6 LANDSLIDE PROCESSES: AN ASSESMENT OF LANDSLIDE POTENTIAL

The geographic extent of landslide processes can be
related to the three factors of geology, rainfall and
slope gradient as outlined above. Figure 4 presents
an assessment of landslide potential on steep slopes
for various combinations of geology and rainfall.
  The areas predicted as having the highest landslide
potential are those with a combination of steep
slopes, high rainfall and deeply weathered volcanic
rocks. Fig. 4 shows that this combination occurs in
the Aberdares and the Tugen Hills. Areas such as
Mount Kenya (shaded on Fig. 4) satisfy two of the
criteria but lack the steep regional slope gradients.
Locally stream dissection and oversteepening as along
road cuttings will promote landslide activity here.
  High rainfall and steep slopes coincide with
basement rocks in the Cherangani and Taita Hills.
Evidence for landslide activity in humid areas
underlain by basement rocks is given by Temple and
Rapp (1972) for the Uluguru Mountains in Tanzania.
The risk of landsliding will depend on the depth of
soil development over the resistant parent material
and the degree of fracturing of the rocks as along
fault lines.
  As the mean annual rainfall diminishes so does the
potential for landsliding. However, the great
variability of annual rainfall at a station and the
high intensity of storm rainfall means that in the
sub-humid areas wetter periods occur with sufficient
frequency to make landsliding a significant hazard.
Hence the Elgeyo escarpment, with excessively steep
fault scarp slopes, but only a moderate rainfall,
will still have a significant landslide potential.
  In the semi-arid areas, as for example in the
Mathews and Ndoto ranges, debris avalanches rather
than landslides may be expected as a result of
infrequent intense or prolonged rainfall events.
Coarse debris produced by mechanical weathering would
collect in steep ravines and gullies during the drier
periods, to be set in motion during high rainfall
events. Numerous such debris avalanches were
observed by the author in 1985 on steep hillslides
within the rift valley floor near Lake Baringo (mean
annual rainfall 650 mm) following exceptionally heavy
rainfall in the first half of the year.

7 KENYAN LANDSLIDE RESEARCH

From Fig. 4 it can be inferred that the risk of
landsliding is high in the eastern footslopes of the
Aberdares range due to a combination of steep
slopes, deeply weatherd soils and high rainfall
distributed in two short rainy seasons. This
inference is borne out by a study by Kamau (1981)
reported by Barber (1983).
  Kamau (1981) reported 40 mass movements within a
300km² area draining the Aberdares backslope. 35 of
these were associated with humic andosols on slopes
of 22-36°, cultivated with tea, coffee, pyrethrum,
maize and vegetables. Andosols occur at high
altitude, are strongly leached and acidic. They have
a high water sorption capacity compared to the more
stable nitosols found at lower altitudes and behave
as an extra sensitive clay at failure. Movement took
the form of landslides and earth flows.
  Kamau estimated that the 40 mass movements had
mobilised about 1 mill. m³ of soil. Many of the
slides (80 per cent) were triggered by heavy rain in
May 1981 when more than 300 mm of rain fell in 8
consecutive days. There is evidence to suggest that
mass movements in this area have coincided with
annual rainfall totals of at least 3000 mm and a
recurrence interval of about 10 years.

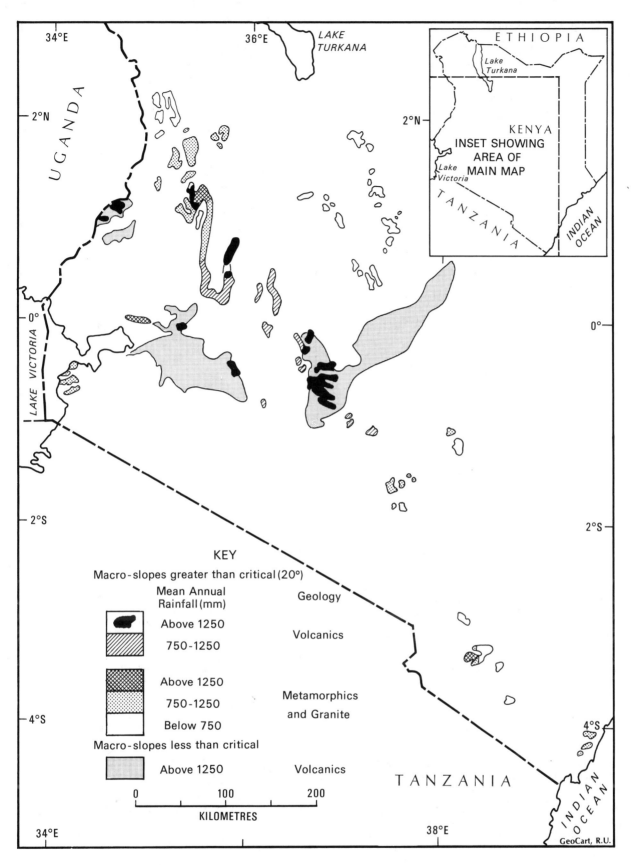

Figure 4. Distribution of landslide factors in Kenya. No occurrence of these factor combinations is found outside the area shown on the main map.

Two factors may exacerbate landsliding in this area. One is the clearance of forest cover as cultivation is extended in response to population increases, the other is the poor design of terraces aimed at soil conservation. The author has observed sliding caused by the collection of water in the upslope cutoff drains and the consequent saturation of the soil profile.

Kamau's study demonstrates the reality of landslide activity in one area of Kenya, but stands alone as the one example of landslide research. Substantiation of the inferences of landslide risk made earlier in this paper requires an extended survey of the rest of the country.

8 SOCIAL AND ECONOMIC IMPACTS

In 1979 Kenya had a total population of 15,327,061 (Republic of Kenya, 1981), increasing at 4 per cent per annum. Approximately 83 per cent of Kenya's population live in rural areas and depend on agriculture for their livelihood (Fox, 1988). There is a strong relationship between rural population density and rainfall (Morgan and Shaffer, 1966), the density increasing markedly above 1000mm and reaching a peak between 1625-1750m. It is thus apparent that the rural population is concentrated in those areas most prone to landsliding on account of high rainfall. In contrast, major urban centres tend to be located at the margins of the high rainfall areas and away from the localities at risk to landsliding.

Table 2. Population densities of areas with a high landslide risk

Location	mean density/km^2
Eastern Aberdares:	
forest edge locations	360
Tugen Hills	150
Taita Hills	168
Elgeyo Escarpment	84
Cherangani Hills	47

Table 2 shows the 1979 population density for those areas assessed above as having the highest landslide risk. Of particular note are the very high densities in the forest edge locations of the eastern Aberdares, shown by Kamau's (1981) study to be a landslide prone area.

The main economic impact of landslides will be felt through loss of productive farmland, damage to roads, and reservoir sedimentation. Statistics on landslide related deaths and damage to buildings are unavailable but numbers and costs are probably negligible.

The greatest social and economic impact of landslides is likely to be in the loss of farmland. Kenya's agricultural potential is largely determined by rainfall; high rainfall areas are often coincident with steep slopes. It is therefore inevitable that cultivation takes place on steep slopes, sometimes as high as 36° (Njoroge, 1983), and will be at risk to landsliding. A comparison of Fig. 5 with Fig. 4 shows that the main crop producing area encloses that of high landslide risk, whilst the coincidence of high rural population densities with a high landslide risk has already been demonstrated. These high population densities are associated with small farm size so that damage from a single slide could represent a grave economic loss to the individual farmer.

As long as population densities remained low, the steepest slopes remained under indigenous forest cover and examination of Figure 5 shows that many of the areas of highest landslide risk are designated as forest reserves. As the population numbers grow,

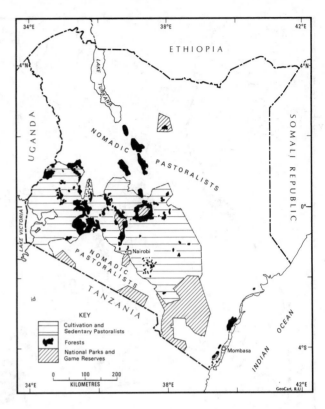

Figure 5. Major land-use patterns in Kenya

however, the steep forested land comes under increasing pressure from cultivators, grazers and wood gathers and the protection of forests has become a major concern. Research from elsewhere has demonstrated that the result of forest clearance is often an increase in landslide activity (Crozier, 1986; Selby, 1982) and there is no reason to believe that it would be otherwise in the Kenyan highlands.

A number of recently constructed roads are at risk to landsliding, in particular the Marigat-Kabarnet road up the Tugen escarpment and the Thuci-Ncubu road around the east side of Mount Kenya where a combination of high rainfall and deeply weathered volcanic soils is likely to render the steep road cuttings unstable.

A third economic impact of landslides arises from their contribution to river sediment loads. No estimates of the contribution made by mass movements to catchment sediment yields is available for Kenya (Barber, 1982), but it may well be significant in areas such as the Mathioya and Maragua catchments studied by Kamau. These catchments have an estimated sediment yield of 886 and 1356 t/km^2/yr (Dunne and Ongweny, 1976), but the separate contributions from surface wash erosion and mass erosion are not known. High river sediment loads is a serious problem due to siltation of reservoirs which store water for hydro-electric power generation and for irrigation projects downstream (Wain, 1983; Nyambok and Ongweny, 1979).

9 CONCLUSIONS

Because of the lack of relevant published research, the discussion in this paper has been based on deductive inference rather than on evidence of landslides per se. Although largely unrecognised to date by scientists and planners, it has been shown that there is a significant potential for landslide activity in certain areas of the country, many of which are associated with a high and increasing population density.

If population pressure forces agriculture to expand at the expense of forest onto steeper slopes, the present geomorphic risk of landslides will be converted to a social and economic hazard. At the same time financial investment into rural development projects such as roads and reservoirs will increase the economic cost of landslide damage. As a result it is likely that awareness of the significance of landslides in Kenya will grow. In anticipation of future planning needs, more detailed research is warranted in order to test the inferences made in this paper.

REFERENCES

Baker, B.H., P.A. Mohr and L.A.J. Williams 1972. Geology of the Eastern Rift System of Africa. The Geological Society of America, Special Paper 136.

Barber, R.G. 1983. The magnitude and sources of soil erosion in some humid and semi-arid parts of Kenya, and the significance of soil loss tolerance values in soil conservation in Kenya. In D.B. Thomas and W.M. Senga (eds.), Soil and Water Conservation in Kenya. Occasional Paper no. 42. p. 20-46. Institute for Development Studies and Faculty of Agriculture, University of Nairobi.

Cahen, L., N.J. Snelling, J. Delhal and J.R. Vail 1984. The Geochronology and Evolution of Africa. Oxford: Clarendon Press.

Crozier, M.J. 1986. Landslides: causes, consequences and environment. London: Croom Helm.

Dunne, T. 1979. Sediment yield and land use in tropical catchments. J. of Hydrol. 42: 281-300.

Dunne, T. and G.S. Ongweny 1976. A new estimate of sedimentation rates on the upper Tana River. Kenyan Geographer, 2: 20-38.

Fox, R.C. 1988. Environmental problems and the political economy of Kenya: an appraisal. Applied Geography, 8.4:

Kamau, N.R. 1981. A study of mass movements in Kangema area, Muranga District, Kenya. Project report submitted in partial fulfilment of the requirements for the Postgraduate Diploma in Soil Conservation, University of Nairobi.

Morgan, W.T.W. and N.M. Shaffer 1966. Population of Kenya: density and distribution. London: Oxford University Press.

Nilsen, T.H. 1986. Relative slope-stability mapping and land-use planning in the San Fransisco Bay region, California. In A.D. Abrahams (ed.), Hillslope Processes, p.389-413. The Binghampton Symposium in Geomorphology: International Series, No. 16. Boston: Allen and Unwin.

Njoroge, S.N.J. 1983. Soil and water conservation extension: the Kenyan experience. In D.B. Thomas and W.M.Senga (eds.), Soil and Water Conservation in Kenya, Occasional Paper no. 42. p.174-184. Institute for Development Studies and Faculty of Agriculture, University of Nairobi.

Nyambok, I.O. and G.S. Ongweny 1979. Geology, hydrology, soil erosion and sedimentation. In R.S. Odingo (ed.), An African Dam. Ecological Bulletins, 29: 36-46.

Pallister, J.W. 1956. Slope development in Buganda Geog. J. 112: 80-87.

Republic of Kenya 1981. Kenya Population Census 1979. Vol. I Nairobi: Central Bureau of Statistics.

Rowntree, K.M. 1985. The Geology of the area west of Lake Baringo, Kenya. Baringo Erosion Project Working Paper no 2. Nairobi: Department of Geography, Kenyatta University College.

Rowntree, K.M. 1988. Storm rainfall on the Njemps Flats, Baringo District, Kenya. J. Climatology, 8: 297-309.

Rowntree, K.M. 1989. Rainfall characteristics, rainfall reliability and the definition of drought, Baringo District, Kenya. S. African Geog. J.

Ruxton, B.P. and L. Berry 1961. Weathering profiles and geomorphic position on granite in two tropical regions, Rev. Geomorph. Dyn., 12: 16-31.

Temple, P.H. and A. Rapp 1972. Landslides in the Mgeta area, western Uluguru Mountains, Tanzania, Geograf. Ann., 54A: 154-93.

Thomas, M.F. 1974. Tropical Geomorphology: a study of weathering and landform development in warm climates. London: Macmillan.

Wain, A. 1983. Athi river sediment yields and significance for water resource development. In D.B. Thomas and W.M.Senga (eds.), Soil and Water Conservation in Kenya, Occasional Paper no. 42. p. 274-293. Institute for Development Studies and Faculty of Agriculture, University of Nairobi.

Landslides: Extent and Economic Significance, Brabb & Harrod (eds)
© 1989 Balkema, Rotterdam. ISBN 90 6191 876 6

Landslides: Extent and economic significance in southern Africa

P.Paige-Green
Division of Roads and Transport Technology, CSIR, Pretoria, RSA

ABSTRACT: Southern Africa is a region of contrasts with mainly flat, arid areas in the west and rugged, mountainous regions with a relatively high rainfall in the east. Landslides are a significant problem only in the eastern and southern coastal areas of South Africa, the western areas of Swaziland and in the rugged hills surrounding Lesotho. The rest of South Africa, Botswana and Namibia comprises limited areas with moderate landslide risk but in most areas instability is unlikely. It is estimated that the annual cost of landslide prevention, rehabilitation and associated expenses in southern Africa is about twenty million dollars (United States), incurred mostly in the eastern and southern areas of the region. The extent and processes of landslides in southern Africa are well understood and the expertise exists to deal with the design and rehabilitation of problem slopes, subject to financial constraints.

1 INTRODUCTION

Situated at the foot of the continent of Africa, southern Africa separates the Indian Ocean (with the warm Benguela current) from the Atlantic Ocean (with the cold Agulhas current). The region thus has unique climatic conditions appropriate to the geographic location (between 22 and 35°S latitude) and the meeting of the two currents, the extreme variation being of particular consequence. The geological conditions are also extremely variable and together with the climate have resulted in a complex geomorphology.

Southern Africa can best be described as a "Second World" region with a number of highly-developed, large metropolitan areas and vast, poorly-developed rural areas. The high population growth and consequent rapid urbanisation presently being experienced, has necessitated rapid improvements in the infrastructure. As is the case in many countries, land previously considered unfavorable for development is now being developed. Much of this land is in the higher rainfall areas, which are generally more mountainous, and landslides are often a consequence of the interference by man in the natural status quo. Land—slides are, however, also associated with development in some of the rural areas, especially housing developments and transport routes.

This paper attempts to give an overview of the extent and economic significance of landslides in southern Africa. The physiographic attributes of southern Africa are reviewed, the state of the art and knowledge of technology regarding landslides is assessed and the extent and economic significance of landslides in southern Africa is discussed in relation to the social and economic impact. The discussion is restricted to natural slopes and excavations in them and excludes embankments and fills associated with construction, industry or mining.

2 GEOGRAPHY

2.1 Location

For the purpose of this paper southern Africa is taken to include the Republic of South Africa, Transkei, Bophuthatswana, Venda, Ciskei, Lesotho and Swaziland and those portions of Namibia (formerly South West Africa) and Botswana south of the 22°S parallel. The geographical location of these countries and independent states is shown in Figure 1. Although these countries are all independent, they are generally interdependent in terms of their transport infrastructures and labour resources and are thus collectively grouped as southern Africa for the purpose of this paper. The area under discussion stretches for some 1 300 kilometres from north to south and 1 500 km from east to west.

2.2 Rainfall

The area is sub-tropical with a typical temperate climate. Geographically the rainfall is extremely variable, with long-term annual means varying between about 1 750 mm along the eastern escarpment to less than 25 mm in the western areas of the Namib desert, only some 1 500 kilometres distant. Figure 2 shows a simplified map of the annual rainfall. There is a marked seasonal distribution of rainfall with most of the area receiving more than 80 per cent of its rainfall between October and March (the southern African summer). The extreme south-westerly areas have a Mediterranean type of climate with a marked winter rainfall while the south coastal regions have rainfall almost evenly distributed throughout the year. Snowfalls occur on the mountains of the eastern and southern areas during winter, but are irregular and are typified by limited duration and extent.

The precipitation over most of southern Africa is of convective origin, typically occurring as late afternoon thunderstorms, often with between 75 and 100 mm falling within a few hours. Along the eastern and southern seaboards, the rainfall is typically orographic with prolonged periods of steady rainfall. As much as 600 mm has been recorded within a 24 hour period in the eastern coastal areas. Tropical cyclones originating in the Indian Ocean occasion— ally reach the north-eastern parts of southern Africa resulting in periods of extremely intensive rainfall.

Studies have shown that the rainfall follows a wet/dry oscillation lasting about 18 years (Tyson, 1986). During the wet periods, extremely heavy rainfall often occurs for extended periods and it is generally during these times that slope instability manifests itself, together with localised or regional

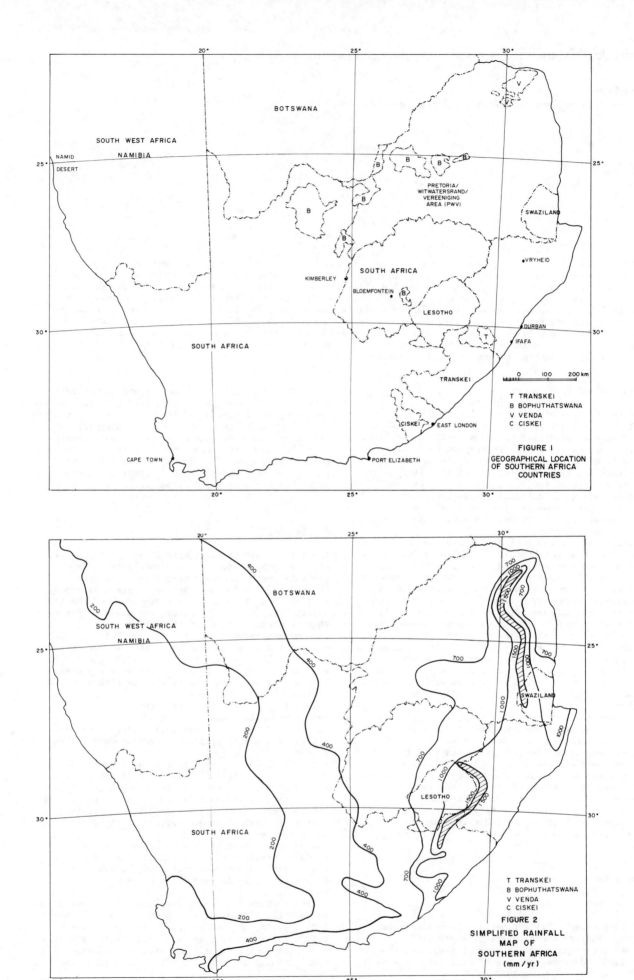

FIGURE I
GEOGRAPHICAL LOCATION
OF SOUTHERN AFRICA
COUNTRIES

T TRANSKEI
B BOPHUTHATSWANA
V VENDA
C CISKEI

FIGURE 2
SIMPLIFIED RAINFALL
MAP OF
SOUTHERN AFRICA
(mm/yr)

flooding. During the summer of 1987/1988 extensive flooding affected much of southern Africa, probably the beginning of a new wet cycle after a number of years of serious drought. The estimated return period of the rainfall was about 1:1000 years. Most of the cost figures referred to in the paper were obtained for the 1987 financial year and are thus more representative of the wet cycle than the dry one.

Figure 3 shows the areas of southern Africa in which there is a surplus in the soil moisture balance (Schulze, 1958). The shaded areas indicate those areas where the annual precipitation exceeds the potential evapotranspiration and an accumulation of soil moisture occurs. Less than five per cent of the significant landslides recorded in southern Africa occur outside this area.

2.3 Area and population

The region discussed in this paper covers about 1.9 million square kilometres with a road density of 0.16 km/km^2, one of the highest in Africa (International Road Federation, 1986).

The population of southern Africa is estimated at about 38 million, with a relatively low population density in the rural areas (less than one person per square kilometre in Namibia) and strong concentrations of the population in eight major metropolitan areas (including the Pretoria-Witwatersrand-Vereeniging (PWV) area and the four major cities along the eastern and southern coastal areas. It is estimated that more than six million of the total population of South Africa reside, or are employed, in the PWV area.

Apart from the PWV and the Bloemfontein-Kimberley areas most of the interior consists of rural agricultural or small mining communities with the bulk of the remaining population concentrated within about 100 km of the eastern and southern coasts.

3 LANDSLIDE LITERATURE AVAILABLE

Literature pertaining solely to southern African landslides consists mainly of case studies presented at local and regional conferences and published in local journals. These have not been compiled into a compendium nor summarised into a single volume. Many reports on landslide problems are in the form of confidential consultants' reports, but are often available for reference and research purposes from the authority for whom the investigation was carried out.

However, general reference works such as Schuster and Krizek (1978), Hoek and Bray (1974) and Zaruba and Mencl (1982) are all relevant to southern African conditions. The proceedings of the ISSMFE Regional Conferences for Africa usually contain information pertinent to local conditions.

4 GEOLOGY AND GEOMORPHOLOGY

The geology of southern Africa is extremely complex (too complex for a simplified small-scale map) for the relatively small area under discussion. Besides being the main source of many strategic minerals in the western world and having vast resources of other minerals (over 60 minerals and metals are economically mined), the rocks of southern Africa are some of the oldest ever dated (in excess of 3.5 billion years).

In brief, the geology of southern Africa consists of vast areas of relatively flat-lying sediments and volcanics of the Karoo Supergroup overlying older sedimentary rocks of the Cape Supergroup in the southern and eastern areas. These latter rocks have been heavily folded in the southern areas, resulting in a very rugged, steep topography. Beneath the sediments of the Cape Supergroup lies a complex succession consisting of numerous intruded and interlayered igneous, sedimentary and metamorphic materials all in excess of one billion years old. The geology is fully described in text-books such as Haughton (1969) and Truswell (1970).

The important materials regarding landslides are primarily the Cape and Karoo sedimentary rocks, some of the Archaeozoic granites and gneisses in the east (older than 3 billion years) and smaller outcrops of younger Cambrian granites in the south, which have been exposed by erosion of the overlying sediments. Most of the other materials occur in the more arid areas of southern Africa, and other than landslides induced by the oversteepening of inclined sedimentary sequences during excavation, seldom give rise to instability. Some of the shales and diabases of the Vaalian Erathem (about 2.2 billion years old) cropping out in the wet, eastern escarpment areas are also susceptible to instability.

Large areas of the geology in the north-western parts of South Africa, Botswana and Namibia are overlain by recent wind-blown Kalahari sands. Some of the highest sand-dunes in the world (> 400 m) occur in the western Namib desert areas (Readers Digest, 1978). These sands are often quite stable owing to a continuous skin which forms at the surface and cements the sand grains together. If this surface skin is broken, for example during construction, the dunes start moving with the wind, often necessitating continual maintenance of the adjacent roads.

Natural erosion processes have produced a relatively youthful topography in the eastern coastal areas with deep valleys cutting through to the Archaean basement granites and gneisses. In the southern Cape Fold Belt of South Africa, southward flowing rivers have incised deep valleys through the west-east trending mountain ranges, these being the most economic potential alignments for transportation routes. The topography over the remainder of the region can generally be classified as mature to old (King, 1963) with definite peneplanes.

The physiography (geomorphology (Figure 4 and Table 1)) of southern Africa is characterised by a series of escarpments around the coastline rising to between 1 000 and 1 500 metres generally within about 100 kilometres of the coast. The escarpment down the eastern part of the region (Province 11) is more pronounced than the rest, rising to almost 3 500 metres (the highest point in southern Africa) in the Lesotho Highlands (Province 2). To the west of this escarpment, the Highveld (Province 1) lies at an average altitude of between 1 000 and 1 500 metres. The Kalahari (Province 8) is situated mostly over Botswana and Namibia at an altitude of about 1 000 m with the inland Okavango Delta in northern Botswana lying at an altitude somewhat less than this.

Unlike the northern hemisphere, no recent glaciation has occurred in southern Africa (the last major glaciation being recorded some 300 million years ago). This has resulted in substantial thicknesses of residual weathered rock, the formation of extensive areas of pedocretes (mainly calcrete (caliche) and ferricrete) and often thick colluvium on hillslopes and valley bottoms.

Along the eastern seaboard, monoclinal folding towards the sea has resulted in all the strata dipping seawards at an angle of about 15 degrees. As nearly all the affected strata are sedimentary materials (e.g. tillites, mudrocks and sandstones) much of the local construction results in deep cuttings parallel to the strike of the material and consequent sliding into the excavation.

In general southern Africa is seismically stable with relatively few earthquakes of significant magnitude. It is geologically inactive with minimal fault movement and no volcanic or geyser activity.

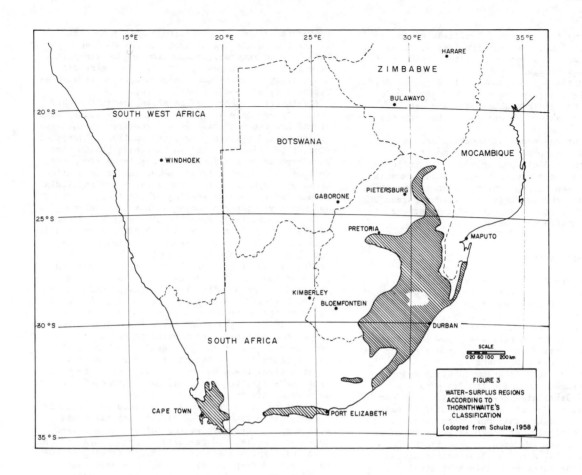

FIGURE 3

WATER-SURPLUS REGIONS
ACCORDING TO
THORNTHWAITE'S
CLASSIFICATION

(adapted from Schulze, 1958)

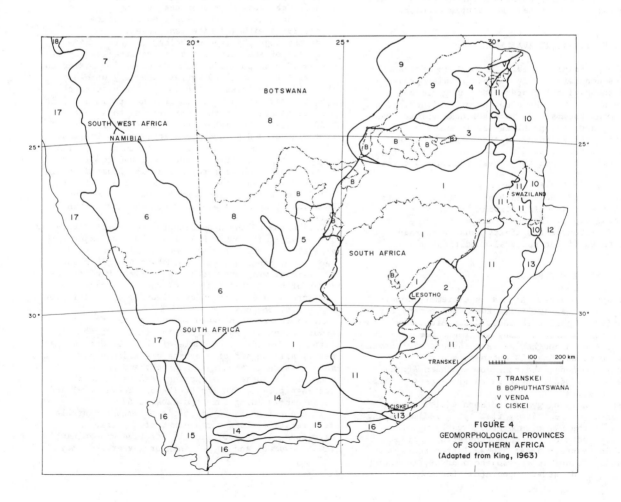

T TRANSKEI
B BOPHUTHATSWANA
V VENDA
C CISKEI

FIGURE 4
GEOMORPHOLOGICAL PROVINCES
OF SOUTHERN AFRICA
(Adapted from King, 1963)

Table 1. Geomorphologic provinces of southern Africa (mostly from King, 1963)

Province	Brief Description
1. Highveld	1200-1800 m, broad valleys, low hills
2. Lesotho Highlands	2500-3200 m, deep valleys, high cliffs
3. Central Transvaal Basin	900-1200 m, low ridges and valleys
4. Northern Transvaal	900-1400 m, plateaux and rugged hills
5. Kaap Plateau	1200-1800 m, plains and low hills
6. Cape Middle Veld	500-1000 m, plains, plateaux
7. Damaraland	1000-2000 m, plains, rugged hills
8. Kalahari	$\pm$ 1000 m, sand plains with small dunes
9. Limpopo Valley	< 1000 m, wide mature valley
10. Lowveld	200-600 m, plains with rolling hills
11. Eastern Uplands	600-3200 m, deep valleys, escarpments
12. Zululand Coastal Plain	< 100 m, flat plains, low hills
13. Eastern Coastal Belt	< 1000 m, deep valleys, steep slopes
14. Karoo	350-1200 m, rolling plains
15. Cape Folded Belt	500-2300 m, rugged hills, deep rivers
16. Southern Coast	< 1000 m, rolling hills deep valleys
17. Namib	< 1000 m, sand plains, inselbergs
18. Kaokoveld	1200-2000 m, plateaux, rugged hills

5 LANDSLIDE PROCESSES

5.1 Geographic extent of landslide processes

Figure 5 shows a landslide susceptibility map compiled on the basis of rainfall, water surplus, geomorphologic provinces, geology, published reports and personal observations by the author (Paige-Green, 1985). Identified on the map are those areas highly susceptible to instability (unfavourable geological, moisture and topographic conditions), not susceptible to instability (not more than one of the geological, moisture or topographic conditions is unfavourable) and those areas where instability may occur (not more than two of the conditions is unfavourable). Included on the map (in the instability susceptible areas) are regions such as the west coast where moving dunes often cause maintenance problems on roads. Some mountain ranges in the drier areas are classified as possible problem areas. Should slopes be oversteepened without adequate analysis, joint controlled translational or wedge failures may result. In the areas highly susceptible to landslides, high pore-water pressures are typically the cause of landslides.

It is of significance that much of the area classified as highly susceptible to landslides occurs in the more heavily populated eastern and southern coastal areas. Most of the major landslides therefore have an impact of some consequence on the population of southern Africa.

Botswana, Namibia and much of the western portion of southern Africa are almost totally free of major landslides.

5.2 Type and distribution of landslide processes

The majority of landslides are induced by human activities. Naturally occurring landslides of any significance are seldom recorded and those that do occur have minimal social or economic impact. However, once the soil and surface rock cover has been disturbed by excavation or loading, the prevalence of landslides is often evident with significant economic implications. It must be noted that since the increased use of engineering geological mapping and geotechnical investigations in the last decade or so, the incidence of major landslides has decreased significantly.

Water is by far the major cause of instability (Paige-Green, 1981; 1982; 1984), with nearly all of the reported major landslides occurring during or shortly after periods of prolonged rainfall in the water surplus areas of the region (Figure 3). Back-analysis of failures invariably indicates the presence of excessive pore-water pressures prior to failure. The failure plane is usually situated along gouge-filled joints or bedding planes or at the residual material/ rock interface. The failure in the former instance is usually caused by insufficient shear strength of the joint or bedding plane filling which has generally been reduced to its residual shear strength value by repeated desiccation/ soaking cycles. Failures at the interface of the soil/rock are typically caused by excessive pore-water pressures at the interface because of the differential permeability between the two materials. Failures of this type are usually restored to a stable condition by a suitable drainage system.

Oversteepening of excavations often results in three-dimensional structurally-controlled failures (joints, bedding planes or faults) which are independent of moisture and occur in both wet and arid areas. These can often be directly attributed to poor engineering judgement or practice, but may in some cases be associated with the acceptance of a higher risk of failure.

Very few of the soils in southern Africa are isotropic, homogeneous or infinite, most being heavily overconsolidated, heterogeneous and finite, thus invalidating most of the theories of classical slip circle failure. The mechanism of sliding is usually complex (Varnes, 1978) with the failure plane passing through residual soil and weathered rock and along structural planes of weakness.

The failure of some deep excavations in southern Africa may have resulted from a decrease in the shear strength of joints owing to stress relief on excavation of the overburden and laterally confining materials.

6 ASSESSMENT OF STATE-OF-KNOWLEDGE ON EXTENT OF LANDSLIDE PROCESSES

Existing knowledge on the extent of landslide processes is considered to be very high. Southern Africa is small enough for good liaison to exist between the road and railway authorities, property developers, consulting geotechnical specialists and the research bodies involved in instability problems. Nearly all major landslides are therefore investigated fully, the cause and process being identified prior to rehabilitation. Many of the reports are, however, of a confidential nature, as the findings may often be sensitive or financially or politically embarrassing. Greater effort should certainly be

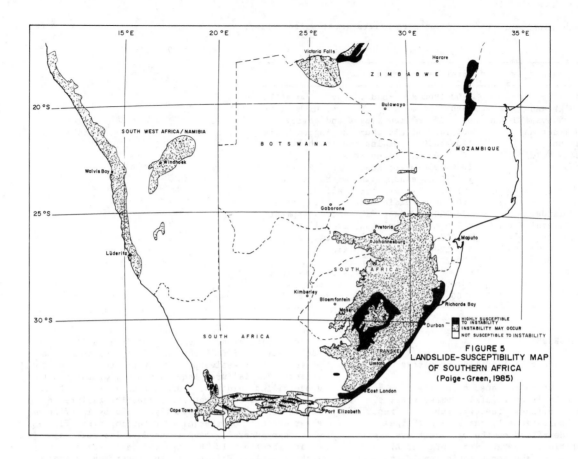

FIGURE 5
LANDSLIDE-SUSCEPTIBILITY MAP
OF SOUTHERN AFRICA
(Paige-Green, 1985)

■ HIGHLY SUSCEPTIBLE TO INSTABILITY
▨ INSTABILITY MAY OCCUR
□ NOT SUSCEPTIBLE TO INSTABILITY

made to compile a compendium of case studies for general release.

7 ASSESSMENT OF STATE-OF-THE-ART OF LANDSLIDE PREVENTION AND CONTROL

The quality of the engineers and geotechnical staff in southern Africa is as high as that in any developed area in the world. At least seven southern African universities have civil engineering faculties and many southern African engineers have obtained higher degrees in Europe and the United States. Although the number of practising engineers is low in terms of the population, there is adequate communication between them through technical journals, conferences and symposia to exchange new technologies and information.

Some of the most sophisticated mining techniques in the world (for both surface and underground mining) have been developed in southern Africa. The analysis techniques developed for large open-cast mines (finite- and boundary- element and probability analyses) are available to the geotechnical engineer for the analysis of large cuts where high risks (in terms of economic or social consequences) are likely.

Computerisation of slope analysis and design procedures is available throughout the region with most consulting engineers having access to programs varying from simple Bishop (1955) and Morgenstern and Price (1967) analyses to sophisticated packages incorporating the most recent models and developments in stability analysis.

High quality geological, geotechnical and soil maps and aerial photographs are available for most of the region, or are produced when necessary for inadequately mapped areas. Geotechnical sampling and testing too have achieved high levels of expertise in southern Africa. Sophisticated equipment for material characterisation and rock stabilisation is available, providing accurate data for use in stability analyses.

A draft recommendation on the investigation, design, construction and maintenance of cuttings for roads (NITRR, 1987) was released at a symposium attended by some 250 delegates during 1987. This is the state-of-the-art guide for non-specialist engineers and, besides giving the basics of good practice, has been designed to indicate for which structures, and at which stage in the investigation, specialist input should be obtained.

The state-of-the-art of landslide prevention and control is thus assessed as being very good.

8 SOME MAJOR LANDSLIDES: CASE HISTORIES

Many landslides have been investigated fully with reports submitted by the geotechnical consultants to the clients. Most of these are, however, confidential. Three published case-histories involving different modes of failure are summarised below to indicate problems typically encountered in southern Africa.

8.1 Failure at the residual soil/rock interface

In the Durban area several major landslides have occurred with the shear plane being at or near the interface between residual soil and the parent shale. A similar situation developed near Vryheid (Figure 1) during 1974 and has been investigated in great detail (Smedley and Nowlan, 1978).

An excavation 400 m long and 20 m deep at a batter of 1:2 showed signs of instability and developed cracks some six months after the end of construction. The slope was battered back to 1:3 but a bridge over the cutting indicated movement, which was soon so severe that it was decided to demolish one abutment and extend the bridge. With the onset of the summer rains further tension cracks developed, the bridge piers began moving and the railway line at the bottom of the cutting began to rise vertically.

The geological investigation indicated more than 10 metres of transported and residual material overlying a weathered mudstone with the soil/bedrock contact dipping at between 7 and 9°. The failure plane was found to follow a thin silty-clay layer at the contact, this having a cohesion (C') of 13.5 kPa and friction angle (ϕ') of 31°. Back-analysis, however, indicated a C' of 0 and ϕ' of 13.5° together with high pore-water pressures resulted in a Factor of Safety of less than unity. A dewatering system consisting of five 250 mm diameter wells bored to depths of between 25 and 40 m and fitted with submersible pumps was thus installed. This stabilised the slope. Eucalyptus trees were planted with the long term objective of replacing the pumps to remove the water.

8.2 Failure along gouge-filled bedding planes

An excavation 24 m deep, at an angle of 1:1.5 was cut through a thick, slightly-weathered tillite at Ifafa, south of Durban. It failed during construction (Knight et al, 1977) and was cut back further to 1:2 after which it failed again. Further flattening of the slope to 1:3 resulted in stabilisation. The failure plane was identified as a bedding plane containing a saturated clay gouge, starting at a tension crack and dipping into the excavation at an angle of 11°. Laboratory shear-box tests showed the peak and residual angles of friction to be 28 to 31° and 9.3° respectively. Ring shear tests on the other hand resulted in a cohesion of 2.26 kPa and ϕ of 5.4°.

It was concluded that although the angle of dip of the strata was greater than the residual angle of friction, an apparent cohesion in the non-saturated condition kept the slopes stable. Any increase in ground-water resulted in wetting up and sliding. It was also concluded that the material strength was larger than the laboratory-measured residual strengths or else the slope would have had to be reduced to the angle of dip.

8.3 Three-dimensional joint-controlled wedge failure

A typical joint-controlled wedge failure was reported by Tluczek (1980) from near East London (Figure 1). A road cutting 25 m deep through mainly unweathered sandstone and greywacke was excavated at an angle of 1:1/4, but as excavation proceeded signs of instability were noted. The cut was then flattened to 1:3/4 at which point stability was achieved. However, flattening of the slope was not possible in the vicinity of the abutments of an existing arch footbridge over the cutting. A three dimensional analysis of the intersecting fault and joint planes indicated a potential wedge failure (some 1870 tonnes) formed by two planes (170°/E85° and 060°/S50° (strike/dip)).

The back-analysis of a nearby plane failure provided shear strength parameters of C = 0 and ϕ = 53° or C = 22 kPa and ϕ = 35°. An analysis indicated that with a pore-water pressure ratio of 0.2, three forty tonne anchors would not maintain a Factor of Safety of 1.5. This was shown in practice with up to 6 mm of movement along the line of intersection of the planes before a further five 40t anchors were installed. This stopped all movement.

9 SOCIAL AND ECONOMIC IMPACT

Prior to analysing the social and economic impact of landslides in southern Africa, the prevalence of landslides in terms of the individual countries is discussed. The prevalence of landslides in Botswana and Namibia is negligible, with minor costs arising from the clearing of sand from the roads and railway lines. Lesotho generally has a very rugged topography and the incidence of instability is relatively high. However, few major excavations exist in Lesotho and

landslides are generally of minor consequence. The western areas of Swaziland are hilly with a very high rainfall and are susceptible to landslides. Like Lesotho, the population and road density in these areas is low and landsliding is of more nuisance value than significant economic consequence.

9.1 Injuries and fatalities

Although numerous landslides have occurred in the last decade, to the author's knowledge, no loss of life or major injury from landslides has been reported. The only social impact of consequence is thus the trauma associated with the loss of property.

9.2 Estimate of annual landslide damage in US dollars

The economic impact of landslides is extremely significant. This can be traced to a number of sources:
- cost of prevention
- cost of rehabilitation
- cost of closures
- cost of damage to associated structures

The quantification of the annual value of each of these is extremely difficult and is an educated guess at best. The annual cost of each identified source varies from year to year depending on the number of construction contracts, the quantity and intensity of the rainfall, the amount spent on the site investigation and testing, and the location of any failures. Many of the smaller landslides are rehabilitated by the district or regional maintenance crews, who remove the slide debris, restore or flatten the slopes, install gabion retaining walls, plant trees to assist with lowering ground-water or construct rock-traps. All these require a relatively low level of expertise and are carried out as a part of the routine maintenance programme. Although the cost may often be significant, perhaps running into tens of thousands of dollars, it is not accounted for separately in the overall maintenance budget. (Although the official exchange rate of about two South African Rand to one US dollar has been used in this paper, it should be noted that in terms of buying power the two currencies are almost equivalent.)

9.3 Cost of prevention

The cost of prevention is usually incorporated into the total design and construction contract cost unless indications of instability are noted during construction, in which case a separate contract for the prevention of failure is usually let. It is not very common, however, to design slopes with reinforcement in southern Africa; the usual practice is to decrease the batter where possible. The treatment of slopes with shotcrete or wire mesh is, however, often designed as part of the structure for road and rail cuttings, the cost of which probably does not exceed $100 000 per annum.

9.4 Cost of rehabilitation

The cost of rehabilitation of slope instability, especially when it affects roads and other structures is substantial. Estimates of the cost of rehabilitation of landslides affecting provincial and national roads and railway lines for the financial year 1987 indicate a total cost of between five and ten million dollars. Landslides in residential areas are seldom rehabilitated; the affected areas are usually declared unsuitable for further housing development.

9.5 Cost of closures

Most South African roads have alternative routes and major landslides which cause the complete closure of roads are nuisance rather than economically disastrous. Slight increases in travelling times and fuel consumption result, the economic value of these being almost impossible to determine. During recent flooding in the east coast areas, instability caused the closure of a road for five months. The alternative paved road involved a deviation of about 150 km and an unpaved alternative a deviation of 5 km. The closed road normally carried about 400 vehicles per day, a large proportion being timber transporters. The increased road user cost in this instance is estimated at about $50 000, with a substantial increase in the maintenance cost of the unpaved road under the increased traffic.

In the less developed areas such as Lesotho and Swaziland, the closure of roads could have significant economic consequences, as there are seldom viable alternative routes. The available alternative routes are either of extremely low standard or necessitate excessive deviations. Even in the rural areas of South Africa numerous unpaved roads are affected by landslides for limited periods. The expense involved in repairing unpaved roads is, however, fairly small as they can usually be easily and quickly cleared by bull-dozer.

Although the costs cannot be quantified accurately, significant losses are suffered by road closures.

9.6 Cost of damage to associated structures

The cost of damage to structures such as houses, roads and railway tracks by landslides may be significant. During the summer storms of 1987 some 100 houses were affected by landsliding in the greater Durban area alone (mainly in saturated residual sandstones) at an estimated cost of about $2.5 million. Heavy rainfall in February 1988 caused further damage, as movement in the local shales was activated and two railway lines were closed for more than 5 months at an estimated cost for the rehabilitation of the railway lines of between $300 and 400 thousand dollars.

9.7 Summary

It is estimated that the total social and economic impact of landslides during the 1987 financial year (a particularly wet period) was close to 20 million dollars. The "wet" period was probably the start of a new "wet" cycle likely to last for about nine years and so it is probable that similar costs will be incurred over this period. Taken in conjunction with the nine "dry" years the average annual cost of landslides is probably of the order of ten million dollars. Risks have been minimised where loss of life or injury may be caused by a landslide and this is manifested by the fatality-free record over the last decade. It is, however, a debatable point whether the risks associated with excavations on transportation routes in less heavily populated areas are acceptable or not.

10 PROFESSIONAL PERSONNEL AVAILABLE FOR LANDSLIDE RESEARCH AND APPLICATION

The number of professional personnel available for landslide research is probably inadequate. Research institutions such as the Council for Scientific and Industrial Research have a limited staff capacity for landslide research (3 or 4 geotechnical specialists) but the universities have a number of geotechnical specialists and post-graduate students available for research. Most of the larger civil engineering consulting firms, the roads authorities and local councils have adequate staff for the application of modern principles in their routine work programmes and limited research capabilities, but inadequate staff for large research projects.

11 THE QUESTION OF RISK

With the erratic climatic conditions, and problem geology and topography, the question of whether every slope should be designed for minimum or acceptable risk arises. The existing situation where the occurrence of major landslides is negligible during the dry periods of the wet/dry oscillation, but often a problem during the wet periods, must be considered in relation to the overall investments. The total annual costs of less than $20 million incurred during the wet periods should be viewed in the light of the total infrastructure investment, running into many billions of dollars. As a percentage of the total investment the annual cost is negligible although it is still a drain on the finances of the country as a whole, especially during periods of economic hardship. It is, however, the author's opinion that the existing situation could be improved slightly, especially regarding investigations for township developments, but the rural investigations are probably close to the optimum cost benefit ratio. To recover the information to attain zero risk would result in unacceptably high costs, with only a slightly improved probability of identifying those thin clay-gouge layers and temporary perched water tables which result in failure.

12 CONCLUSIONS

Southern Africa has a vast range of climatological, geological, topographical and geomorphological characteristics. Landslides are, however, restricted mainly to the wetter, more mountainous eastern areas, which are also the more densely populated parts of the sub-continent. The level of expertise on landslide problems in southern Africa and the application of this expertise is generally very high. However, the acceptance of a certain level of risk is common and a number of landslides occur during periods of excessive rain, which is cyclically based.

The cost of the rehabilitation of the landslides is substantial, (estimated at about twenty million US dollars per annum during wet periods), but the risks taken are such that no loss of life or major injuries have been incurred to date.

Since the importance of geological and specialist geotechnical input in development studies and planning was recognised, the incidence of landslides has been significantly reduced.

ACKNOWLEDGEMENTS

This paper was prepared as part of the research programme of the Division of Roads and Transport Technology (formerly the National Institute of Road and Transport Research) and is published with the permission of the Director. The assistance of a number of Road Authorities and Consultants with the provision of data used to estimate the costs and damage incurred is gratefully acknowledged.

REFERENCES

Bishop, A.W. 1955. The use of the slip circle in the stability analysis of slopes. Geotechnique, 5 (1):7-19.
Haughton, S.H. 1969. Geological history of southern Africa. Cape Town: Geological Society of South Africa.
Hoek, E. and Bray, J. 1974. Rock slope engineering.

London: Institution of Mining and Metallurgy.

International Road Federation (IRF). 1986. World road statistics. Washington, DC; IRF.

King, L.C. 1963. South African Scenery. Edinburgh; Oliver and Boyd.

Knight, K, Sugden, M.B. and Everitt, P.R. 1977. Stability of shale slopes in the Natal coastal belt. Proc 5th South East Asian Conf Soil Engng, Bangkok, p.201-212.

Morgenstern, N. and Price, V.E. 1975. The analysis of the stability of general slip surfaces. Geotechnique, 15 (1):79-93.

National Institute for Transport and Road Research (NITRR). 1987. The investigation, design, construction and maintenance of road cuttings. Draft Technical Recommendation for Highways, TRH18, Pretoria, CSIR.

Paige-Green, P. 1981. Current techniques in groundwater control applied to cut slopes. Trans Geol Soc South Africa, 84:161-167.

Paige-Green, P. 1982. The design and construction of cut slopes for roads in South Africa: A review of techniques. Proc Annual Transptn Convention, 3, H(ii), Pretoria.

Paige-Green, P. 1984. Investigation for cut slopes for roads in South Africa: A review of techniques. Proc 8th Reg Conf Africa on Soil Mech and Foundn Engng, Harare, Zimbabwe, p.475-479.

Paige-Green, P. 1985. The development of a regional landslide susceptibility map for southern Africa. Proc Annual Transptn Convention, Vol FB, Paper FB4, Pretoria.

Readers Digest. 1978. Illustrated guide to southern Africa. Cape Town: Readers Digest.

Schulze, B.R. 1958. The climate of South Africa according to Thornthwaite's rational classification. S Afr Geogr Jnl, 40:31-53.

Schuster, R.L. and Krizek, R.J.(Eds). 1978. Landslides: Analysis and control. Trans Res Board Spec Rep 176, Washington, DC.

Smedley, M.I. and Nowlan, P.H. 1978. A geotechnical investigation of a landslide on the Broodsnyers—plaas to Richards Bay coal line. The Civil Engineer in South Africa, July:169-173.

Tluczek, H.J. 1980. Limit equilibrium methods. Seminar on applications of numerical techniques in rock engineering, Pretoria, Int Soc Rock Mech.

Truswell, J.F. 1970. An introduction to the historical geology of South Africa. Cape Town: Purnell.

Tyson, P.D. 1986. Climatic change and variability in southern Africa, Cape Town: Oxford University Press.

Varnes, D.J. 1978. Slope movement types and processes. In Landslides: analysis and control, Schuster, R.l. and Krizek, R.J. (Eds), Trans Res Board Spec Rep 176, Washington, DC, p.34-80.

Zaruba, Q. and Mencl, V. 1969. Landslides and their control. New York: Elsevier.

Landslides: Extent and economic significance in China

Li Tianchi
Chengdu Institute of Mountain Disasters and Environment, Chinese Academy of Sciences, Chengdu, Sichuan, China

ABSTRACT   This paper examines the landslide situation in China, which has 15 provinces with landslide problems caused mainly by heavy rainfall and earthquakes. Landslides are a significant problem in southwest China, the Loess Plateau area, and Taiwan, including the provinces of Sichuan, Yunnan, Guizhou, Tibet, Shanxi, Shaanxi, Gansu, Qinghai, Huhei, Hunan, Fujian and Taiwan.

The annual economic losses due to landslides have been estimated to be about 0.5 billion US dollars, and the number of landslide-related fatalities per year exceeds 100. Organizations concerned with landslide study offer guidance on the possible approaches to prevention and control of landsliding.

1 INTRODUCTION

In China, about 75 per cent of the total land area is mountainous. Geological, geomorphological, and climatic conditions vary enormously, but landslides are a common problem. Landslide prevention and control are new disciplines, although China has the oldest records of landslide hazards in the world. During the 1950's, because knowledge about landslide identification and prevention was scanty, excavations on ancient landslides reactivated them with disastrous results. For instance, 136 big and small landslides occurred during 1954 through 1957 within 348 kilometers of the Baocheng railway from Baoji to Shangxiba. Railway service was interrupted several times during that period, and the cost of repair reached as much as 8,200 million yuan (2,200 million US dollars). A few studies of landslide identification and control were initiated after these incidents. More extensive studies, including regional landslide distribution, mechanisms of failure, and prevention and control techniques began in the early 1960's. The results of research on landsliding have been published mainly since 1975.

During the past 25 years, the author has had the opportunity to take part in fieldwork throughout China, including a survey of railway landslides in southeast China, highway landslide investigations in southwest, northwest and central China, field surveys of earthquake-induced landslides and rainfall-induced landslides, and a regional landslide study related to a hydroelectic project.

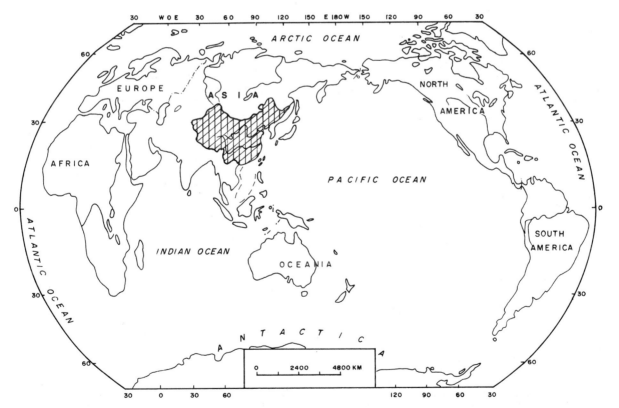

Figure 1. Location of China in the world.

This paper consists of three parts. The first part introduces the natural environment of China, with emphasis on topography, geology, and rainfall as elements of landsliding. The second part consists of the main triggering mechanisms and the geography of the different regions. The third part describes economic losses and fatalities due to landslides, and the control measures that are applicable in China.

2 NATURAL ENVIRONMENT

China is an large country located at the east coast of the largest continent (Eurasia) and on the western margin of the largest ocean (Pacific) (Fig. 1). It has a land area of about 9.6 million sq km which is about 6.5 percent of the total area of the earth. The population is more than 1 billion, which is approximately 23 percent of the total world population. In addition, China is one of the oldest countries in the world and has the oldest civilization.

2.1 Topography

China's topography looks like a staircase descending step by step from west to east, from the Qinghai-Xizang (Tibet) Plateau to the coastal area. The Qinghai-Xizang Plateau, the first step of the staircase, covers 2.2 million sq km with a mean elevation of 4000 m above sea level, the highest and largest plateau on earth. From the eastern and northern margins of the Qinghai-Xizang Plateau, eastward to the Dahinggan-Taihang-Wushan Mountains, the terrain drops abruptly to between 2,000 and 1,000 metres above sea level, forming the second step of the staircase. This area is composed mainly of plateaus and basins, such as the Loess and Yunnan-Guizhou Plateaus and Sichuan Basin.

From the eastern margin of the second step eastward to the coast, the land drops to less than 500 metres above sea level to form the third step, the largest plains of China: the northeast China plain, the north China plain and the middle and lower Changjiang Plain, with hilly country at the borders. East of the third step are the shallows, which extend from the coast into the sea. The average depth of the water is less than 200 meters.

Two large transitional zones from the first to the second step and from the second to the third step, have steep slopes, deep valleys, and many landslides.

2.2 Geology

China is very complex geologically. Rocks from Precambrian to Holocene age have been deformed repeatedly since Paleozoic time, resulting in extremely complicated geologic structures. Major active fault zones are shown in Fig. 2.

The rocks and sediments of China have, for the purposes of this paper, been grouped into 4 types related to their strength and landslide stability: loose, unconsolidated deposits; non-carbonate sedimentary rocks, carbonate rocks, and magmatic and metamorphic rocks (Fig. 3).

2.2.1 Loose, unconsolidated deposits of Quaternary age

These deposits are the chief surficial materials of China's plains, where most of the cities, agricultural fields, and industrial districts have been developed. Few landslides form in these deposits, except where slopes have been artificially steepened. Ground subsidence is a problem in several areas, such as in the Shanghai district.

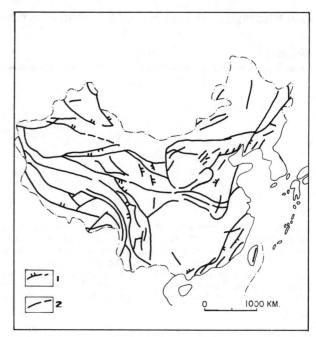

Figure 2. Active faults in China; 1, active thrust fault; 2, active normal fault and fault of indefinite nature.

Quaternary loess and loess-like deposits are widely distributed on plateaus, mountain slopes, intermontane basins, and piedmont plains in northern China, comprising a total area of more than 600,000 sq km. The deposits vary in thickness from a few meters to as much as 200 m. Two types of loess have been classified according to their origin and internal structure, plateau loess and secondary loess. Landslides are widespread in both types of loess, but large, lateral slides only occur in the first one.

2.2.2 Sedimentary rocks of Mesozoic and Cenozoic age

Continental clastic rocks of Mesozoic and Cenozoic age are widely distributed in central and south China. They consist of reddish conglomerate, sandstone and claystone, with a total thickness of several thousand meters. The Sichuan Basin, composed chiefly of Jurassic and Cretaceous reddish sandstone and claystone, is the largest basin with red hills in China. Hundreds of similar basins with red hills occur in central and south China, where the landslides are mainly debris creep, shallow debris slides, and consequent bedrock slides.

2.2.3 Carbonate rocks of Paleozoic age

These rocks are widely distributed in China, especially in south China and southwest China. They underlie a total area of about 1.3 million sq km, forming extensive karst topography. Few landslides develop in this terrain except where the carbonate rocks are associated with coal.

2.2.4 Magmatic and metamorphic rocks of different ages

Granite is widely distributed, especially in south China where the hot, wet climate promotes extensive chemical weathering and weakening of the granite to substantial depths. Slumps and other shallow landslides are widespread in these weathered

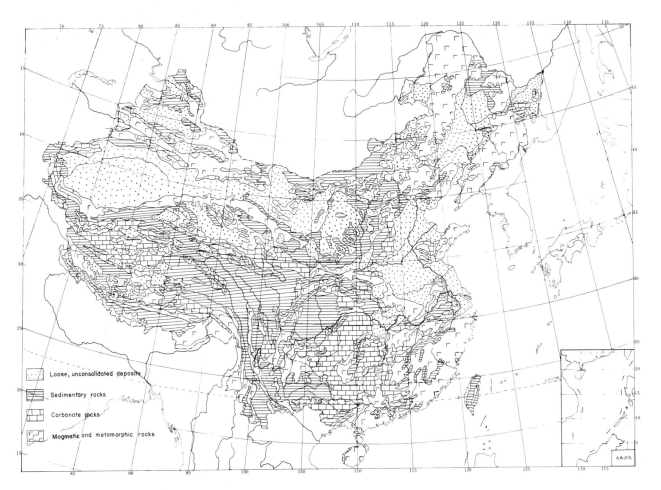

Figure 3. Map of China showing geology.

Loose, unconsolidated deposits

Sedimentary rocks

Carbonate rocks

Magmatic and metomorphic rocks

granites.

Cenozoic basalt is common in southwest and northeast China and on the eastern Nei Monggol Plateau. Few landslides develop in the basalt of northeast China, but many large landslides form in basalt in southwest China, such as in the northeast Yunnan and southwest Sichuan provinces. The landslides form where basalt cap rocks are underlain by Mesozoic sedimentary rocks.

2.3 Rainfall

The rainy season starts in early April in south China, early June in central China, and early July in north and northeast China. In southwest China, the southwestern monsoon dominates; it bursts northward in late May when the rainy season in Yunnan and West Sichuan begins. It does not stop until October when the southwestern monsoon retreats rapidly southward.

The distribution of annual precipitation in China is generally determined by distance from the sea; the farther away from the sea, the less abundant is the precipitation (Fig. 4).

Seasonal distribution of precipitation is also uneven. Most areas have their annual precipitation concentrated in summer when the warm, moist maritime monsoon dominates. Precipitation in one month usually accounts for more than one quarter or even one half of the total annual precipitation. In central China, because of the earlier arrival of the maritime monsoon, spring is the most important rainy season, followed by summer. In north and northeast China, however, summer rain comprises more than one half of the total annual precipitation, and there is a pronounced spring drought. In southwest China and

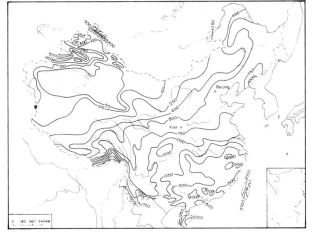

Figure 4. Distribution of annual precipitation in China. Isohyets in mm.

the southern Qinghai- Xizang Plateau, where the southwestern maritime monsoon from the Indian Ocean is the chief moisture source, there are clear-cut rainy and dry seasons. The rainy season begins in May and ends in October, accounting for 80 to 90 percent of the total precipitation.

According to records from 1950 to 1977, there were 263 heavy rainstorms each exceeding 400 mm/24 hrs in China, an average of 9 events per year. These rainstorms occur almost all over the country even in the inland area excepting the westernmost part. As shown in Table 1, ten rainstorms exceeded

273

Table 1. The Maximum Point of the 10 Major Rainfalls in China, from 1950 to 1977.

No.	Location			Rainfall in mm				Date
				6hrs	24hrs	3days	7days	
1.	Jiangxi	Lu Shan	Zhiwuyuan		900	1073		Aug. 17, 1953
2.	Guangdon	Taishan	Zhenhai	386	851	949	972	July 12, 1955
3.	Guangdon	Dianbai	Lidong	281	858	1030	1234	May 20, 1957
4.	Jiangsu	Rudong	Chaoqiao		822	934		Aug. 4, 1960
5.	Hebei	Neiqiu	Zhang	426	950	1456	2051	Aug. 4, 1963
6.	Taiwan	Taoyuan	Peishih		1248	1794		Sept. 10-11, 1963
7.	Taiwan	Yilan	Hsinliao		1672	2749		Oct. 17, 1967
8.	Henan	Biyang	Linzhuang	830	1060	1065	1631	Aug. 7, 1975
9.	Guangdong	Haifeng	Baishimen	460	884	1222	1513	May 30, 1977
10.	Nei Monggol	Uxin Qi	Muduo		1400			Aug. 1, 1977

Table 2. Maximum rainfall in China for different time periods from 1950 to 1981.

Duration	Amount(mm)	Date	Location		
5 min.	53	July 1, 1971	Shanxi	Taiyuan	Mei Dong Gou
20 min.	79	Sept. 10, 1964	Zhejiang	Dequing	Li Xi
60 min.	245	June 10, 1979	Guangdong	Chenghai	Do Xi Kou
65 min.	267	June 20, 1981	Shaanxi		Do Shi Cao
3 hrs.	495	Aug. 7, 1975	Henan	Biyang	Li Zhuang
6 hrs.	830	Aug. 7, 1975	Henan	Biyang	Li Zhuang
12 hrs.	954	Aug. 7, 1976	Henan	Biyang	Li Zhuang
24 hrs.	1672	Oct. 17, 1967	Taiwan	Yilan	Hsin Liao
2 days	2259	Oct. 17-18, 1967	Taiwan	Yilan	Hsin Liao
3 days	2749	Oct. 17-19, 1967	Taiwan	Yilan	Hsin Liao

800 mm in 24 hours, and four of these even exceeded 1,000 mm (two occurred in mainland China and the others in Taiwan). Maximum rainfall in China for different time periods is shown in Table 2. Some of these rains have even exceeded the maximum recorded in the same latitude in America (Zhao, 1986).

3. POPULATION

The total population of China on July 1, 1982, was 1,031,882,511, 22.6 percent of the world's total population and the most populous country. There are now 30 first-level administrative units (provinces, autonomous regions and national municipalities) in China (Fig. 5). Among the provinces, central and coastal provinces are the most densely populated and inland areas are more sparsely populated (Table 3.) For example, the population density of 7 coastal provinces, 3 municipalities and one autonomous region is 320 per square kilometer. The population density of sparsely populated Xizang, Qinghai, Xinjiang, Gansu, Ningxia, and Nei Monggol is only 11.8 per square kilometer (Qi Wen, 1984). Unfortunately, landslides are concentratred in densely populated provinces.

4. GEOGRAPHIC EXTENT OF LANDSLIDE PROCESSES

4.1 Main Triggers of Landslide Processes

Landslides are the result of the comprehensive action of geologic and geographical environments. Earthquakes and rainstorms constitute two of the most important landslide-inducing agents.

4.1.1 Earthquake-triggered landslides

China is a country with a lot of earthquakes: at least 656 M≥0 earthquakes happened from 780 B.C. to 1976 A.D. Earthquake-induced landslides were triggered in 35.5 percent of the total shocks except in marine areas, Xizang, and Taiwan (Feng and Guo, 1986).

One of the most seismically active regions is western Sichuan and Yunnan provinces. Earthquakes have caused many large-scale rockslides and rockfalls, some of them blocking rivers and forming lakes. Landslides and avalanches were induced widely during the earthquakes of Luhuo (Feb. 6, 1973, M=7.9) in Sichuan province, Zhaotong (May 11, 1974, M=7.1) in Yunnan province, Longling (May 29,

Table 3. Population of China by province, autonomous region (Ningxia, Nei Monggol, Xinjiang, Guangxi and Xizang) and municipality (Beijing, Tianjin and Shanghai).

Province	Area sq km (thousands)	Population (thousands)	Province	Area sq km (thousands)	Population (thousands)
Heilongjiang	46	3,267	Guangdong	22	5,930
Jilin	18	2,256	Guangxi	23	3.642
Liaoning	15	3,572	Sichuan	56	9,971
Shanxi	15	2,529	Yunnan	17	2,855
Hebei	19	5,301	Guizhou	38	3,255
Shandong	15	7,442	Nei Monggol	110	1,924
Jiangsu	10	6,052	Gansu	39	1,960
Anhui	13	4,967	Qinghai	72	389
Zheijiang	10	3,888	Ningxia	6.6	389
Fujang	12	2,593	Xinjiang	160	1,308
Taiwan	3.6	1,827	Xizang	122	189
Henan	16	7,442	Shaanxi	19	2,890
Hubei	18	4,780	Beijing	1.68	923
Hunan	21	5,401	Tianjin	1.1	776
Jinagxi	33	3,318	Shanghai	5.8	1,186

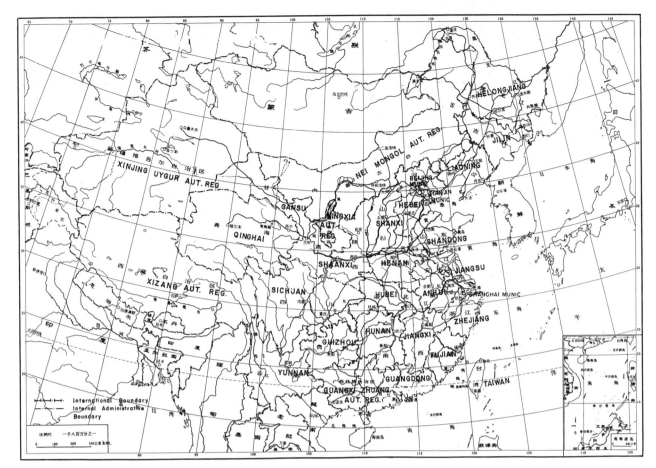

Figure 5. Map of China showing the first-level administrative units.

1976, M=7.4) in Yunnan province and Songpan–Pingwu (Aug. 16, 1976, M=7.2) in Sichuan province.

The earthquake of Kangding–Louding (Oct. 10, 1786, M=7.5) in Sichuan province caused many landslides, among them a landslide in Momianshan, Louding, which dammed the Dadu river for 10 days. When the landslide dam was overtopped and the dam failed, a great flood extended 1,400 km downstream and drowned as many as 100,000 people. The earthquake killed only 400–500 people.

The earthquake centered near Diexi in northwestern Sichuan province (Aug. 25, 1933, M = 7.5) induced many landslides and killed 6800 people. In the town of Diexi, all but one of the 577 residents were buried by a huge landslide. The landslide formed a dam 250 m high across the valley in the upper reaches of the Minjiang River, and created a lake. The dam was overtopped 45 days later and a flood of water rushed down the valley to a distance of 250 km. At least 2500 people were killed by this great flood (Li and others, 1986).

The Loess Plateau is another seismically very active region where thousands of earthquakes have been recorded since ancient times. Many of these caused extensive and disastrous landslides, including the following:

An earthquake in the Hongdong area in Shanxi province (Sept. 17, 1303, M = 8.0) induced landslides at Xunbao Mountain. The longest landslide was about 1600m long and 1400m wide.

The great earthquake of Tianshui (July 21, 1654, M = 8.0) in Gansu province triggered 59 huge loess landslides. The Luojiabao landslide was the largest with a length of 4.5 km and a width of 2 km.

The great earthquake of Tongwei (June 19, 1718, M = 7.5) in Gansu province induced 337 large landslides. One of the largest slides, at Yongning, was 8 km long, 3 km wide, and an area of 17 sq km. The slide buried about 2,000 families in the town.

During this century, the earthquake of Haiyuan (Dec. 16, 1920, M = 8.5) triggered 675 large loess landslides. The Dongjiaca landslide near Xiji moved 2,300 km northward and then 950 m southeastward where it dammed a river amd formed a lake 5 km long and 380 m wide. In the Xiji–Jingning region, more than 40 lakes were created by the landslides and 27 of them still exist.

A study of landslides triggered by earthquakes from 1973 to 1978 indicates a direct correlation between earthquake magnitude (M) and the total area(s) where seismogenic landslides and ground failures might develop. The equation of the correlation derived by the standard least–square regression method is:

$$\text{(1)} \qquad \log S \text{ (in km}^2\text{)} = 0.9246 \, M - 3.1089$$
$$\text{(2)} \qquad \log S \text{ (in km}^2\text{)} = 1.0719 \, M - 3.5899$$

Equation (1) is applied in southwest China, and equation (2) in north China (Li, 1979).

4.1.2 Rainfall–triggered landslides

More than 60,000 landslides were triggered in July, 1981 when heavy rainfall flooded Sichuan Province. A similar storm in July, 1982, triggered more than 80,000 landslides in eastern Sichuan Province. In August 1984, more than 1,000 landslides occurred in 8 districts in south Gansu.

Rainfall from 15 to 30 July, 1982 in eastern Sichuan Province has been studied in relation to landsliding. Average cumulative precipitation from that single storm was 632 mm, an amount that exceeded the mean monthly precipitation by 340 percent. The area devastated by the heavy downpour covered about 21,000 km², including the districts of Fengjie, Kaixian, Liangping, Wanxian, Zhongxian,

Table 4. Relation between rainfall and landslides in the eastern Sichuan basin, July 1982.

Location	Rainfall up to time a landslide occurred		Rainfall up to time landslide occurred in large numbers	
	Cummulative rainfall (mm)	Daily rainfall (mm)	Cummulative rainfall (mm)	Daily rainfall (mm)
Zhongxian	139.0	139.0	289.7	138.2
Yunuang	-	-	277.7	205.6
Kaixian	53.4	51.4	280.8	153.8
Lingping	177.0	177.0	279.3	102.3
Fengdu	99.0	88.0	-	-
Fengjie	113.7	47.9	218.9	10.1

Table 5. Typical landslides induced by rainfall in eastern Sichuan basin July, 1982.

Name of landslide	Location	Time Month, Day	Volume million (m^3)	Lithology	Cummulative rainfall (mm)	Daily rain (mm)
Nanzhuba	Fengdu	7/16	0.7	Mudstone	90.0	88.0
Shankou	Zhougxian	7/17	18.0	Mudstone	310.8	171.8
Yijian	Zhougxian	7/17	2.8	Mudstone	310.8	171.8
Jipazi	Yunyang	7/18	13	Debris	331.0	164.1
Tianbo	Yunyang	7/17	6.2	Mudstone	283.0	101.6
Geling	Yunyang	7/17	9.5	Mudstone	345.7	94.9
Baigou	Fengjie	7/16	1.2	Debris	138.3	58.5
Guadouzai	Liangping	7/28	5.62	Debris	210.5	83.2

Shizhu, and Fengdu. Farmlands, houses, roads, canals, and power stations were destroyed, and many people were killed by landslides (Li and Li, 1985).

Rainfall thresholds necessary to trigger landslides are shown in Table 4. Typical landslides induced by this rainfall are listed on Table 5.

Tables 4 and 5 and field investigations indicate that: 1) If cumulative precipitation of an area is 50 to 100 mm and daily precipitation is more than 50 mm, somewhat small-scale and shallow debris slides will occur; 2) when cumulative precipitation is 100 to 200 mm and daily precipitation is about 100 mm, the numbers of landslides have a tendency to increase with precipitation; and 3) when cumulative precipitation within 2 days of downpour exceeds 250 mm, a large number of landslides and a great disaster will occur.

Heavy rainfall also triggers many landslides in the Loess Plateau. In 1978, many landslides were triggered by heavy rainfall in Tianshui, Guansu Province. In the 1984 rainy season, more than 1,000 landslides occurred in the Wudu and Tianshui district, south Guansu. Some figures of precipitation triggering landslides in this province are in Table 6.

Table 6. Some figures for precipitation-triggered landslides in Gansu Province.

Location	Time day, month, year	Lithology	Rainfall in the early 10 days (mm)	Daily rain- fall (mm)
Tianshi	7/21/78	Loess	82.9	200.00
Huixian	8/21/84	Mudstone	294.0	120.00
Tianshui	8/3/84	Loess	63.1	52.7

The data in Tables 4, 5 and 6 indicate that the relationship between landslide incidence and precipitation is very complex. The relationship not only depends on precipitation intensity, duration, and cumulation but also on geological and topographical conditions. Based on the data mentioned above and on other statistical data, the amount of precipitation needed to induce landslide for different materials is given in Table 7.

Table 7. Rainfall needed to trigger landslide for different materials.

Type of landslide	Rainfall intensity (mm/h)	Daily rainfall (mm)	Cummulative rainfall (mm)
Small landslide of debris and loess	6.0	50.0	50.0-100.0
Moderate land- slide of debris, loess, and fratured rocks	10.0	120.0	150.0-200.0
Huge landslide of debris and bedrock	15.0	150.0	250.0

4.2 Landslide situation in natural regions of China

Approximately two-thirds of China is mountains where various landslide processes have a widespread distribution and frequent occurrence. The total area and location of these landslides has not been determined, but the general extent can be described from published literature and unpublished notes. Inasmuch as environmental factors influence landslide processes, landsliding in China is discussed in terms of various natural regions. The natural regions (Fig. 6) used are those completed by Zhao (1986) with some important modification. Rock types, structure, landform, precipitation, landslide type, and incidence are mentioned in the brief description of each natural region.

The four landslide categories on Figure 7 are arbitrary and subjective, owing to the lack of precise landslide statistics for much of the country, especially for very sparsely populated areas in the far western parts of China. Taiwan should be considered as having many small areas of high landslide incidence, but they are not shown because of the small scale of the map.

4.2.1 Da Hinggan Mountains

The Da Hinggan Mountains occupy the northwest corner of Heilongjiang province. These generally rolling mountains average about 1000 m in altitude, with a

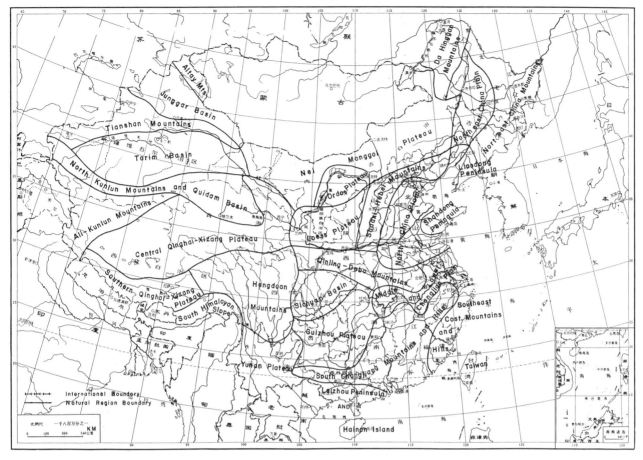

Figure 6. Map of China showing the natural regions. Modified from Zhao (1986).

few peaks as high as 1500 m. The mountains consist of folded and faulted Paleozoic and Mesozoic sedimentary rocks and large masses of granite. The climate is humid with annual precipitation of about 450 mm in the eastern part to 350 mm in the west.

There are a few small landslides locally where sedimentary rocks form steep fault scarps, or along the Heilongjiang, Ergun, and Nenjiang Rivers where the rocks undergo freezing and thawing and erosion by the river.

4.2.2 Northeast China Mountains

This region is located in eastern Heilongjiang and Jilin provinces, and northeastern Liaoning Province. It includes the Xiao Hinggan Mountains, the Sanjiang Plain, and the Changbai Mountains. The region is characterized by a series of northeast-trending mountains, mostly 500 to 1500 m in elevation, alternating with broad intermontane basins and valleys. The mountains consist of Mesozoic sedimentary rocks and large masses of granite and basalt. Climatically, the region is typically humid and temperate. Annual rainfall totals more than 800 mm in the southeast, about 500 mm in the central part, and only 450 mm in the Xiao Hinggan Mountains and the Sanjiang Plain.

There are a few landslides in the Baiah-Nahe area along the upper reach of the Mudanjiang River where basalt covers Cretaceous and Tertiary deposits.

4.2.3 Northeast China Plain

This region is sandwiched between two humps, the Da Hinggan Mountains and the northeast China Mountains. It is a piedmont flood plain with

Quaternary deposits about 30 to 50 m in thickness. Because of its relatively low topographic location, precipitation is less abundant than in the surrounding mountains, being about 600 mm in the southeast and decreasing to about 400 mm in the northwest.

Black soil characteristic of the region is composed mostly of Flood Plain deposits and a thin veneer of Quaternary loessic material. These materials are susceptible to slump, creep, and earthflow. Many small landslides occur in the tributaries of the Wuyur and Hulan Rivers. Slope failures of irrigation channels are very common in spring.

4.2.4 Liaodong and Shandong Peninsulas

This region includes the Liaodong Peninsula, the Shandong Peninsula, and the central Shandong Mountains and hills. The region consists mostly of undulating mountainous and hills, with a few peaks towering above 1000 m, and narrow strips of coastal and intermontane plains. The principal bedrock consists of Precambrian metamorphic rocks, granite, andesite, basalt, and limestone of Cambrian and Ordovician age. The climate is mild and humid. Annual precipitation is 600 to 900 mm, decreasing inland from the coast.

Liaodong Peninsula has the northeast-trending Qianshan Mountains with an average elevation of 500 m. These mountains consist mostly of strongly weathered and fractured Precambrian metamorphic rocks. Debris flows, slumps, and rockfalls occurred frequently in these rocks in recent years. Much of the area is highly developed, so that many landslides have been activated by works of man.

The Shandong Peninsula is composed mainly of

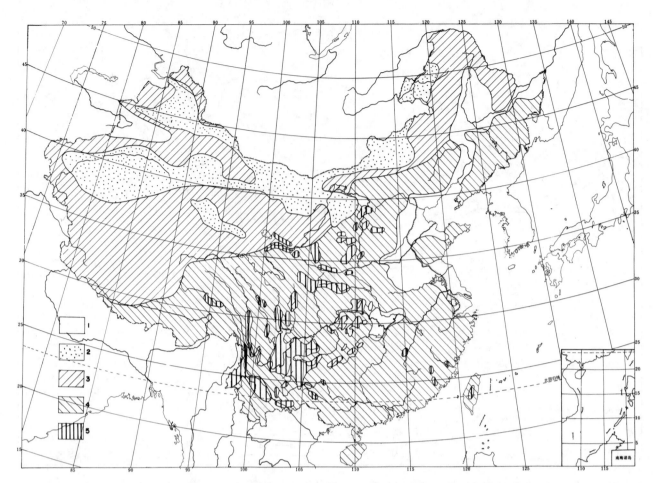

Figure 7. Distribution and intensity of landslide processes in China; 1, areas with hardly any landslides. Includes the plains in eastern and northeastern China composed of unconsolidated, cohesive soils which are susceptible to landsliding but lack relief to fail; 2, areas with hardly any landslides. Includes the Gobi and other desert areas in northern and northwestern China where rock properties, morphological and climatical conditions are not favorable for landsliding; 3, areas with few landslides. Climate not favorable for landsliding; 4, area with a moderate incidence of landsliding; 5, areas with a high incidence of landsliding. Rock properties, morphology, and climate are favorable for landsliding.

undulating hills, mostly under 300 m elevation, with a few higher rock mountains of more than 1000 m. The principal geological unit in the northern part is granite and metamorphic rocks of Precambrian age. Jurassic and Cretaceous sedimentary rocks occupy the southern part. Small debris slumps and slides are common in the upper reaches of the Daguhe River. The coastal area of the Shandong Peninsula is a rocky terrain with an irregular shoreline, a cover of Quaternary clay, sand, and gravel. Small slumps and falls form where the cliffs are undercut by waves.

The central Shandong Mountains and hills are composed mainly of horsts and grabens with many peaks above 1000 m including sacred Mount Taishan, the highest. Most of the rocks in this area consist of Precambrian metamorphic rocks and limestones of Cambrian and Ordovician age. Landslides are a common but not dominant part of the landscape.

4.2.5 North China Plain

This region is composed essentially of flat alluvial plains under 50 m (above sea level) in elevation, including the lower Liaohe Plain, the Haihe Plain, the Huanghe (Yellow) River flood plain, and the North Huaihe Plain. The climate is essentially subhumid characterized by a warm, rainy summer and a cold, dry winter. Annual precipitation is 500 to 800 mm.

Landslides are normally absent in this region of

low relief, though the final alluvial deposits are susceptible to landslide.

4.2.6 Shanxi-Hebei Mountains

This region consists of a series of parallel, folded and faulted mountains and intermontane graben basins and structural valleys. The region extends from the Taihang Mountains in the east to the Luliang Mountain in the west and from the Yanshan Mountain in the north to the western Henan Mountains in the south. The mountains are underlain by Precambrian metamorphic rocks, Cambrian and Ordovician limestone and shale, and Carboniferous and Permian sandstone and shale. Loess covers the basins and lower slopes of the mountains. Climatically, the region is transitional between subhumid and semiarid. Annual precipitation is 400 to 700 mm, decreasing from southeast to northwest.

In general landslide are scarce in the rocky mountains of this region. Landslides are abundant along the Huanghe River valley, where Tertiary sedimentary rocks and overlying loess are exposed on steep slopes, and in the seismically active Fenhe River valley, composed chiefly of alluvial and lacustrine deposits where small slumps, slides, and earthfalls are well developed. Many landslides along the Fenhe River valley have been triggered by earthquakes.

4.2.7 Loess Plateau

The Loess Plateau is limited by the Luliang Mountains in the east, by the Helan Mountains in the west, by the Qinling Mountains in the south, and by the Great Wall in the north. Administratively, it includes a part of Shaanxi, Shanxi, Gansu, and Ningxia provinces. The Loess Plateau has an elevation of 1200 to 1600 m. Geologically, it is characterized by loess deposits 30 to 60 m thick on the average and underlying sedimentary rocks of Jurassic to Tertiary age. Geomorphologically, the Loess Plateau is subdivided into three parts: (1) the broad Weihe River valley in the south, (2) high plains underlain by loess in the middle, (3) hilly areas underlain by loess in the north. Annual precipitation is 350 to 650 mm, of which 90 percent is concentrated in June to September.

Regional field investigations by the Chengdu Institute of Mountain Disasters and Environment and other institutions indicate that the geographic extent of landslides in the Loess Plateau is very uneven. Most landslides occur in the southern region crossed by the Weihe, Jinhe, and Taohe Rivers. The slopes of these river basins consist mainly of reddish claystone of Tertiary age overlain by comparatively thick Quaternary loess. Famous landslides in this area include the large loess landslide of Saleshan and a bedrock landslide near Putouyan. Near Baoji City in the middle reach of the Weihe River, there are 43 large loess landslides occupying nearly 90 percent of the total length of the bank (Zha and Zhang , 1987). More than 1,000 landslides have occurred recently in the three watershed areas, causing great damage (Li and Don, 1985). The landslides are induced mainly by rainfall, earthquakes, and artifical irrigation.

In the central and northern Loess Plateau, landslides are scarce except in a few small drainages, such as those near Qinan and Tongwei in Guansu Province and Xiji in the Ningxia-Hui Autonomous Region (Jin, 1987).

4.2.8 Middle and Lower Changjiang (Yangtze River) Plain

The middle and lower Changjiang Plain is essentially a region of low plains and wetlands underlain by fine-grained alluvial deposits susceptible to landsliding. Small slumps and earthflows are common along the river banks but are generally absent elsewhere in this region of low relief.

4.2.9 Southeast Coast Mountains and Hills

This region consists of three northeast-trending mountain ranges 500 to 1500 m high paralleling the southern coast of China. The mountains are underlain mainly by Mesozoic granite and igneous rocks. Within the mountains, are many small, graben-structured basins containing Cretaceous, Tertiary, and Quaternary deposits. Annual precipitation is 1,100 to 2,000 mm. Typhoons are frequent from July to September.

Many of the rocks in the region are jointed, faulted, and strongly weathered. Landslides, particularly debris slides, are frequent in the valleys of the Minjiang and Jiulongjiang Rivers.

4.2.10 Taiwan Island

Two-thirds of Taiwan is made up of mountains and hills, mostly in the middle and eastern parts of the island. The mountains are composed chiefly of sedimentary rocks, some of which have been subject to different grades of metamorphism. The mountains are the youngest in China and are still actively undergoing tectonic movement. Topographically, they are composed mainly of four north-south parallel mountain ranges, with 30 peaks over 3,330 m. Between these four mountain ranges are large, active faults, associated with major earthquakes and many landslides. The island is extremely wet and rather hot all year. Mean annual precipitation for the whole island is about 2,600 mm and is more than 5000 mm for the mountainous area south of Jilong (Zhao, 1986).

The rugged mountain topography, abundant and high intensity rainfall, and common earthquakes contribute to a severe landslide problem. According to Chou (1960), a landslide survey in six main watersheds in 1935 by N. Yoshei documented a total area of 13,870 hectares of landslides (Table 8).

Table 8. Landslide area in six watershed of Taiwan (Chou, 1960).

Location	Name of watershed	Area of landslide (Hectares)
Central Taiwan	Choshui River (upper reach)	3,980
	Chen-yu-lan Creek	3,390
	Wu River	1,570
	Ta-an Creek	990
Southern Taiwan	Tsengwen Creek (upper reach)	770
	Hsia-tan-shui Creek (upper reach)	3,170
Total		13,870

An island-wide land use and forest resources survey in 1956 indicated 117,400 hectares as "denuded and unplantable". Sheng (1966) analyzed the survey and estimated that active landslides cover about 40,000 ha in the mountainous watershed of Taiwan. The remaining 77,400 ha could be influenced or produced by landslides.

4.2.11 South Changjiang Hills and Basins

Hills and mountains in this region with 300 to 600 m in elevation are composed mostly of pre-Devonian metamorphic rocks and granite. Most of the highest peaks are composed of rhyolite, granite, and other hard igneous rocks. Between these hills and low mountains are scattered basins with sedimentary rock formations of Cretaceous and Tertiary age. Loose, landslide-prone red soil covers the foothills. Annual precipitation is 1400 to 1700 mm. Most of the landslides in the red basins and foothills are triggered by heavy rainfall rather than by earthquakes. In June 1982, for instance, a storm with 626.5 mm of rain during 7 days induced 9579 landslides in three counties in central Jiangxi (Yan, 1985).

4.2.12 Qinling-Daba Mountains

This region stretches from the lofty Qinhai-Xizang Plateau eastward almost to the Pacific Ocean, separating the country into north and south parts. It consists of a series of east-west folded mountains 1000 to 3000 m in elevation and intermontane basins. Administratively, this region includes south Gansu, north Sichuan, south Shaanxi, southwest Henan, north Hubei and west Anhui Provinces.

The geology of the region is very complex. The principal bedrock consists of metamorphic rocks of Precambrian age, granite and other igneous rocks of different geological age, limestone of Cambrian, Ordovician and Permian age, shale of Silurian age, and red beds of Cenozoic age. Annual precipitation varies from 1550 to 700 mm, decreasing from east to west and from south to north.

The Precambrian metamorphic rocks and Silurian shale are shown on Fig. 6 as being more susceptible to landsliding. The Beilongjiang River basin in west Qinglin and the Hanzhou Basin are two particularly slide-prone areas. In the upper reaches of the Jialing River, landslides are also relatively abundant.

The Beilongjiang River basin is along an east-west tectonic fracture zone several hundred kilometers long. Strong earthquakes are frequent in this area. More than 2000 landslides occur along the both banks of the river, especially in fractured metamorphic rocks. Some of these landslides such as Xiliupe, Suoertao, and Bailinping cover an area of nearly 4 sq km.

4.2.13 Sichuan Basin

The Sichuan Basin, with an area of about 260,000 sq km, is one of the largest inland basin in China. Administratively, it occupies 45 percent of Sichuan province and has a population of nearly 100 million, one of the most populous areas in the world.

The basin is occupied mostly by rolling hills and low mountains 450-700 m in elevation. These hills and mountains are underlain mainly by reddish sandstone and purple shale of Jurassic and Cretaceous age. The basin is subdivided into the Chengdu Plain in the west, the central hills, and the eastern ranges and valleys. Annual precipitation ranges from 900 to 1300 mm in the basin, and from 1500 to 1800 mm in the surrounding mountains.

Chengdu Clay distributed in the Chengdu Plain is shown on as susceptible to landsliding. Many earth slumps and slides occurred in this clay on banks of Dujiang Dam irrigation ditches. The reddish sandstone, mudstone and purple shale of Jurassic and Cretaceous age are also very susceptible to landsliding during the rainy season. Thousands of landslides occurred in these units during heavy rainfall in 1981 and 1982. In the valley of the Changjiang River, from Wanxian to Fengjie, where the rocks dip toward the valley and have been undercut by erosion, at least 15 huge landslides with volumes in excess of 10 million m³ have formed. Some of these landslides dammed the Changjiang River in historic time (Li and Liu, 1987). Landslides are also numerous along the famous Three Gorges of the Changjiang River from Fengjie to Yizhang, where the slopes expose Silurian shale and limestone and dolomite of Permian and Ordovician age.

4.2.14 Guizhou Plateau

Guizhou Plateau at an elevation mostly between 1000 to 2000 m is located at the second great topographic step in China. Annual precipitation is 1200 to 1400 mm.

Carbonate rock, mainly Paleozoic limestone, occupies more than 75 per cent of the total area. Karst topography is well developed and extensively distributed. Landslides have been sparse in the central part of the plateau, but are very common in the western margin of the plateau where bedrock is mostly limestone, sandstone, coal, and basalt of Permian age. Landslides are also common in the deep valleys of the northern and northeastern margin.

4.2.15 Yunnan Plateau

The Yunnan Plateau can be divided roughly into three geologic areas. The eastern part of the plateau is mostly carbonate rocks with karst. The central part is mainly red sandstone and shale of Jurassic and Cretaceous age. The western part of the plateau consists mainly of volcanic and igneous rocks.

Weather in the Yunnan Plateau is dominated by the southwestern (Indian) monsoon in the western part and the southeastern (Pacific) monsoon in the eastern part. Annual precipitation is 1000 to 1200 mm, decreasing both from the southeast and southwest to the middle north.

Landslides have been sparse in the eastern and central part of the plateau, and common in the drainage areas of the Xiaojiang River in the northeastern part of the plateau. In these drainage areas, many different kinds of rocks are exposed. An active fault system extends 300 km from north to south through this region. Metamorphic rocks of the Kunvang Group (Proterozoic) and tectonic melange are so strongly sheared and jointed along this fault that they crumble to powder in many places. They are shown on Fig. 6 as being very susceptible to landsliding. So many different kinds of landslides, including block glides, slumps, shallow debris slides, and debris flows have formed in the drainage area of Xiaojiang River that the region deserves the title of "museum of landslide and debris flow."

4.2.16 Hengduan Mountains

This region of about 40,000 sq. km is located in the transition from the Qinghai-Xizang Plateau, Sichuan Basin, and Yunnan Plateau. It includes west Sichuan, northwest Yunnan, and east Xizang Provinces. The region consists of a series of nearly north-south-trending, lofty mountain ridges and river gorges with a relative relief of more than 3,000 m.

Geologically, this region has a complex geological structure. Many active fault systems trending north to south generate earthquakes. The rocks are mainly sedimentary rocks of Mesozoic age and some intrusive and metamorphic rocks. The rocks have been folded, faulted and, in places, intensely sheared. Lacustrine deposits of Tertiary and Quaternary age occupy the valleys of the Jinsha River and its main tributaries. The region is under the influence of monsoons, both from the southeast and southwest. Annual precipitation is 400 to 1,000 mm, decreasing northwestward. The combination of steep slopes, sheared rocks, frequent earthquakes and heavy rainstorms and snowmelt combine to make this region one of the most landslide-prone areas of China. The highway from Sichuan to Xizang and the railway from Chengdu to Kunming have serious troubles from landslides nearly every year.

Landslides are characterized by large scale, frequent occurrence, and wide distribution. In historic time, the Jinsha, Yalongjiang, Dadu, and Minjiang Rivers were dammed by landslides several times, causing great floods downstream when the dam broke.

The Xigeda Formation, an early Quaternary lacustrine deposit, is one of the principal slide-prone units. During the construction of Dukou City from 1960 to 1987, more than 50 landslides occurred in this formation.

4.2.17 Leizhou Peninsula and Hainan Island

This region consists of the Leizhou Peninsula, Hainan Island, and the narrow coast of Guangdong and Guangxi Provinces. The topography is hilly, with an elevation generally below 150 m. Geologically, it is composed mainly of basalt and associated igneous rocks erupted during the Quaternary period in different stages. The region has a humid tropical climate and annual precipitation from 900 to 2500 mm. Typhoons occur frequently in summer and autumn.

Chemical weathering and leaching are intensive in the humid tropical climate. Basalt and igneous rocks form a red soil that is very susceptible to landsliding when it is saturated. Most of the landslides are earth slides and falls.

4.2.18 Nei Monggol Plateau

This region is located in the northernmost part of
China. It is composed mainly of an immense and
undulating plateau and a surrounding area with
elevations between 1,000 to 1,500 m. Geologically,
the plateau is made up of large masses of basalt
erupted during the Cretaceous, Tertiary, and
Quaternary periods, associated igneous rocks, and
sedimentary rocks of Tertiary age. The climate
varies from semiarid to arid with annual
precipitation varying from 400 mm in the east to 150
mm in the west. Landslides are scarce in this area
of low relief and aridity, but sedimentary rocks of
Tertiary age are locally susceptible to landsliding.

4.2.19 Ordos Plateau

The Ordos Plateau is located between the Nei Monggol
Plateau and the Loess Plateau. It consists
essentially of an uneven plateau surrounded by low
mountains and high plains. The climate is
transitional from semiarid to arid. Annual
precipitation is 300 mm in the eastern part
decreasing to 200 mm in the western part. The
western part of the plateau is underlain by
Cretaceous sandstone, conglomerate and shale, and
red sandstone and sandy clay of Tertiary age. The
landslide situation is similar to that of the Nei
Monggol Plateau.

4.2.20 Altay Mountains

The Altay Mountains are in the northern part of the
Xinjiang Uygur Autonomous Region and have an average
height of 3000 m. The mountains extend
southeastward to the People's Republic of Mongolia
and northwestward to the Soviet Union. Ancient
crystalline rocks and granite are the two main
geological units. These rocks have been deeply
weathered to red soil in some places. Climate is
sub-humid to arid, with annual precipitation from
250 mm in the east to 500 mm in the west.

A few areas of moderate landslide incidence lie on
the south flank of the Altay Mountains between
elevations of 1500 and 3200 m. Rockfalls in that
area are caused mostly by ice wedging and freeze-
thaw action.

4.2.21 Junggar Basin

The Junggar Basin is sandwiched between the Altay
Mountains and the Tianshan Mountains, with an
elevation from 500 to 1,000 m. It is composed
mainly of depositional gobi (lenticular sedimentary
deposits) around marginal piedmont plains and an
immense desert in the basin center. The climate is
semiarid to arid with annual precipitation from 100
to 300 mm, decreasing from west to east. Few
landslides are in this region.

4.2.22 Tianshan Mountains

The Tianshan Mountains running across the middle of
Xinjiang are composed of more than 20 parallel
ranges with elevations from 3,000 to 5,000 m,
interspersed with graben-structured and rhomboid-
shaped intermontane basins. Rocks in this region
are Precambrian schists and shale and Phanerozoic
sedimentary units intruded by igneous rocks and cut
by numerous faults. The Tianchan Mountains have a
sub-humid to arid climate, with precipitation
decreasing from 800 mm in the west to 100 mm in the

east.
In general, landslides are not common, although
some rockfalls have occurred on high peaks and
canyon walls and a few debris flows occurred on
easily-eroded glaciofluvial deposits. A few ancient
landslides might have occurred after glaciers
retreated in late Pleistocene time.

4.2.23 Tarim Basin

The Tarim Basin occupies the southern part of
Xinjiang and the northwestern part of Gansu
Provinces. It consists of the Tarim Basin proper,
the Turpan-Hami intermontane basin, and western Hexi
corridor. The basin topography dips generally from
west to east as well as from south to north with an
elevation from 1400 to 700 m. The climate is
extremely dry, with annual precipitation less than
90 mm over-all and even less than 10 mm in the
central part.

The ground surface of the basin consists of
immense desert, and denudational stony gobi, and
diluvial gravel gobi. In general, landslides are
rare owing to the arid climate and low relief.

4.2.24 Qaidam Basin and Northern Kunlun Mountains

This region is transitional from the Qinghai-Xizang
Plateau to the desert region of northwest China. It
includes the Qaidam Basin at an elevation from 2600
to 3000 m, the western Qilian Mountains, the Atlun
Mountains, and the northern flanks of the Kunlun
Mountain with elevations from 3,000 to 5,000 m.

The Qaidam Basin is underlain by loosely
consolidated Tertiary sediment in the northwestern
part, and by sandy loam or clay in the central
part. The mountains are composed mainly of schist,
gneiss, and dolomite of Precambrian age, and
sedimentary rocks of Paleozoic and Mesozoic age. The
rocks and sediment are cut by a series of large
fault systems trending nearly west. Climatically,
the region is the driest in the Qinghai-Xizang
Plateau, with annual precipitation decreasing from
200 mm in the east to 10-20 mm in the west.

Landslides are absent in the Qaidam Basin owing to
the low relief and very arid climate. Landslides do
occur in the mountains of Qilian, Altum and northern
Kunlun. Rockfalls and rock glaciers occur in the
high mountains.

4.2.25 Ali-Kunlun Mountains

The Ali-Kunlun Mountains region includes the west
Ali area and the southern flanks of the middle-
western section of the Kunlun Mountains. It
occupies the northwestern part of the Qinghai-Xizang
Plateau, and it borders on Kashmir in the west. The
Ali area is composed of the upper reaches of the
Indus River and the wide valley of Bangong Lake,
with elevations between 3800 and 4500 m. The crests
of the Kunlun Mountains are mostly above 6,000 m.

The climate of the region is arid and cold. The
annual precipitation varies from 20 to 100 mm in the
Kunlun Mountains and from less than 50 to 150 mm in
the Ali area. The region is underlain by Paleozoic
and Mesozoic sedimentary rocks, Cretaceous and
Tertiary shale, andesite and tuff, and Quaternary
sediment.

Shale and tuff in the region are very susceptible
to landsliding, but landslide incidence is low owing
to the very arid climate in the Ali area. Landslide
characteristics in the Kunlun Mountains are similar
to those in the northern Kunlun Mountains described
above.

4.2.26 Central Qinghai-Xizang Plateau

This region includes the central part of the Qinghai-Xizang Plateau, and the southern and eastern Qinghai-Qilian mountains, stretching from southwest to northeast. The central part of the plateau, with an elevation from 4500 to 4800 m, is characterised by an immense intermontane basin sandwiched between the east Kunlun Mountains in the north and the Gangdise-Nyainqentanglha Mountains in the south. The area is underlain mainly by extensive nonmarine sedimentary rocks of Paleozoic and Mesozoic age. The climate is cold and dry, with annual precipitation from 100 to 400 mm. There are a few small landslides on roadcuts, but landslides are generally absent in this region of dry and cold climate and low relief.

The eastern Qilian Mountains, which rise to nearly 5000 m above the Hexi Corridor, consist of parallel ranges and intermontane basins and valleys. The climate is semiarid to arid, with annual precipitation from 200 to 500 mm. Precambrian metamorphic rocks and Paleozoic and Mesozoic sedimentary rocks are the principal bedrock. Semi-consolidated and interbedded sandstone and claystone of Tertiary age are involved in areas of slump and debris flow along the Huanghe River and its main tributaries in east Qinghai.

4.2.27 Southern Qinghai-Xizang Plateau

The southern Qinghai-Xizang Plateau lies between the Gangdise-Nyainqentanglha mountains in the north and the Himalayas to the south. The Yarlung Zangbo and the Pumqu Rivers drain from the plateau.

The region is underlain by Precambrian metamorphic rocks and Phanerozoic sedimentary rocks and, in the Gangdise Mountains, large masses of granite. The climate is subhumid to arid, with annual precipitation from 500 mm in the east to 200 mm in the west.

Landslides are few in the western part of the region. Landslide and debris flows are common where fine-grained terrace material of glaciofluvial and glaciolacustrine origin occur in the northern foothills of the Himalayas and in the middle reaches of Yarlung Zanghbo. Floods caused by failure of glacial lake dams (jokulhlaups) at higher altitude brought great disasters to people in Diangi, Jilong and Jiangzi districts. Rockslides, rockfalls, and debris flows are major hazards to road transport-ation in the region. The highway from Xizang to Nepal has been destroyed several times by landslides and floods.

4.2.28 Southern Himalayan Slopes

This region includes the southern slopes of the eastern Himalayas with crests to 7000 m, and the Kangrigarbo Mountain with deep gorges and high ridges. It is bordered by the Yunnan Plateau on the southeast and by the Hengduan Mountains on the east.

The mountains are composed chiefly of Precambrian metamorphic rocks in the south, and metamorphosed Mesozoic sedimentary rocks in the north. The climate is warm and humid, which differs entirely from the dry and cold climate on the Qinghai-Xizang Plateau. Mean annual precipitation varies from 1000 to 4000 mm in the belt from 2500 to 3000 m elevation.

Landslides associated with glaciation, earth-quakes, and rainfall are common in this region.

5. ECONOMIC SIGNIFICANCE OF LANDSLIDES IN CHINA

China has suffered more fatalities and property loss from landslides than any other nation. The recorded history of landslideing goes back nearly 4,000 years to B.C. 1789. The landslide of Wudu in central China which killed 760 people in B.C. 186, is probably the oldest record of such a disaster in the world before Christ.

5.1 Fatalities due to landslides

The large number of landslide deaths in China is related to earthquakes, very heavy rainfall, and flooding from failure of landslide dams. Before the 18th century, the greatest loss of life in China in a single landslide was at Zhigui, western Hubei, in 1310, when 3466 people were killed. Similarly, more than 1000 people were killed in 1561 by a landslide on the north bank of the Changjiang River near Xitan, Zhigui district, western Hubei Province.

In the 18th century, China suffered two of the greatest losses to life and property from landslides. The first catastrophe was loess landslides induced by the Tongwei earthquake (M = 7.5) in 1718 in Gansu Province. At least 40,000 people were buried in these landslides. The second was in 1786 when a flood resulted from a landslide-dam failure in Momianshan, middle Dadu River, western Sichuan, where more than 100,000 people were drowned by the flood as far as 1400 km downstream (Qiu and Liu, 1985).

In the 20th century, the greatest number of deaths due to landslides occurred in 1920, when a magnitude 8.5 earthquake in Haiyuan, Ningxia, triggered a series of massive loess slides that killed at least 100,000 people, half the 200,000 deaths in this event (Close and McCormik, 1922). In recent years, most of deaths have been caused by major individual landslides. In November 1965, a large collapsing landslide in Permian basalt and tuff, occurred at Luguan district in the northern part of Yunnan Province. The moving mass had a volume of 450 million m^3, and it traveled at high speed over a path of 6 km long and 2 km wide. The landslide buried 4 villages and killed 444 people. Another landslide in the Nanjiang district, Sichuan Province, occurred on Sept. 13, 1975 on a slope of Triassic carbonate rocks. A moving mass with a volume of 7 million m^3, killed 195 people (Fig.8).

On July 3, 1980, a collapsing rock slide occurred in the upper reaches of the Yanohi River in the Yuanan district, Hubei Province (Fig. 9). The rock mass of about 700 thousand m^3 fell 200m, destroyed all buildings of the Yanchihe Phosphorite Mine in the valley, and claimed 284 lives.

A famous loess landslide occurred at Sale Shan on March 7, 1988 in the Dongxiang district of Gansu Province. The total volume of the moving mass was about 35 million m^3, covering an area of 2.7 km^2. The landslide buried 4 villages, filled two reservoirs and killed 277 people (Fig. 10).

During the rainy season of 1984, more than 1000 landslides were induced in Wudu and Tianshui prefectures, south Gansu. In Wudu prefecture only, 570 landslides threatened 14,245 families with a total of 70,049 people. For 231 of these landslides, 6,019 families had to make an urgent move. In the same year, over 300 landslides took place in the Xiheli district, south Gansu, resulting in collapse of 15,000 rooms and the loss of 128 lives.

Unusual amounts of rain in 1984 caused many landslides in the Loess Plateau area in Shaanxi Province. More than 99 people were buried by landslides in Yulin and Tongchuan Prefectures, although forecasts of landslides and rescue work saved many lives (Han and others, 1985).

Although there have been some very large and catastrophic slope failures in the Qinghai-Xizang Plateau, most have occurred in the high mountains where few people live. However, there have been two notable exceptions recently. The first was in July

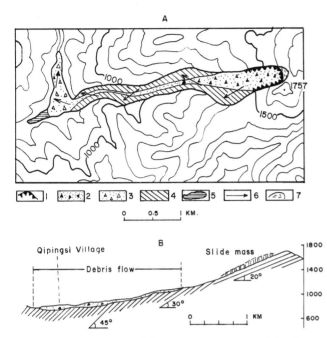

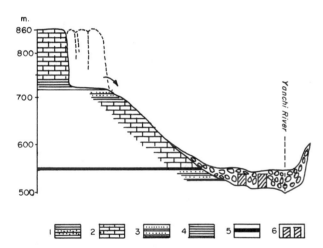

Figure 8. Sketch map (A) and cross-section (B) of Nanjian landslide, north Sichuan. 1, head scarp area, 2, unmobilized landslide deposits, 3, debris-flow deposits, 4, multiple-stroke dive and upthrust area, 5, slide-dammed lake, 6, movement direction, 7, potential slide area.

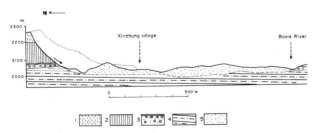

Figure 9. Slide and fall of dolomite in the valley of the Yanchi River in Yuanan district, Hubei Province. 1, silty shale, 2, thick dolomite, 3, mudstone and silty shale, 4, thin platy dolomite, 5, phosphorite deposits, 6, buildings of Yanchi River phosphorite mine.

Figure 10. Slide of loess at Sale Shan in Dongxiang district, Gansu Province. 1, loess, 2, older loess, 3, gravel and loam, 4, mudstone, 5, slide deposits.

1954, when a flood resulted from a glacier-dam failure at Sewang in Jiangzhi district, south Xizang. At least 450 people were swept away by the flood. The second was a catastrophic rock avalanche in the Karakorun Mountains, northwest Xizang, during construction of the Karakorun Highway in the 1960's. All soldiers of one company (about 150) were buried by the avalanche.

Landslides are frequently disastrous in densely populated Taiwan. In 1948, the Tungmen Powerplant with a capacity of 29,000 kw was buried by sediment created by landslides at the headwaters of Mukua Creek in east Taiwan (Chu, 1961). In May 1951, several days of intense rainfall led to the overtopping and failure of Tsao-Ling rockslide dam on the Chin-Shui Chi River, which was created by the earthquake of December 1941. In the subsequent flood, 154 people were killed and 564 homes and 3,116 ha of croplands were damaged. The total loss was estimated at about $0.4 million (Sheng, 1966, Chang, 1984).

Some of the best documented major catastrophes in historical and technical records and from the author's experiences are presented in Table 9. Landslide disasters since B.C. 186 are listed, each of which claimed more than 100 deaths. According to incomplete and rough statistics from technical records of eight provinces known to be most severely affected, deaths due to all kinds of landslides totaled more than 3,800 for the 36-year period from 1951 to 1987. An average of more than 100 people per year were killed by landslides. In fact, deaths due to landslide was much higher than the figures mentioned above, because there are some reported events not known to the author and more events have never been recorded or reported. Conservatively speaking, more than 140 people were killed by landslides each year in China for the past 36 years.

5.2 Economic losses due to landslides

In many provinces of China, economic losses due to landslides are great and are apparently growing. Landslides destroy or damage residential and industrial developments, agricultural and forest land, railways, and highways. The landslides also have a negative impact on the quality of water in rivers and streams.

The author was unable to obtain significant data from all provinces of China, but somewhat more accurate cost information can be estimated for landslide disasters in smaller geographic areas. Some large cities of China, such as Chongqing, Wanxian, Dukou, Lanzhou and Baoji are located in landslide-prone areas and dangerous debris flow and mudflow areas. Urban area landslides have been particularly costly. For example, in July 1964, a large mudflow in Lanzhou, the capital of Gansu Province, killed 137 people, buried 3 km of railway track, disrupted railway traffic for 34 hours, and inflicted considerable damage on the city. In August 1968, Lanzhou suffered another large loss of life from mudflow and $1.1 million in damages. During the past 35 years, 335 people were killed by landslides and mudflows in Lanzhou City.

The cost of landslide damages in Dukou City, southwestern Sichuan, were estimated to be about $0.3 million annually for the period 1965 to 1985. In Wanxian City, eastern Sichuan, landslide damage costs for the five year period from 1970 to 1975 were estimated to be about $0.2 million annually. Another $1 million was spent to stabilize a single landslide in the central city, where a department store, cinema, book store, and several hundred private houses were destroyed in 1972 and 1973.

In Chongqing City, one of the largest cities in China, more than 30 landslides were documented in the past 30 years. In the 1960's landslides damaged a big transformer station, two workshops of the

Table 9. Landslides in China that have killed at least 100 people.

Year	Province	Affected Area	Type of slope failure	Number of deaths
BC 186	Gansu	Wudu	Rock and debris avalanche	760
100	Hubei	Zhigui	Rock slide and avalanche	over 100
1310	Hubei	Zhigui	Rock slide and avalanche	3,446
1558	Hubei	Zhigui	Rock slide and avalanche	over 300
1561	Hubei	Zhigui	Rock slide and avalanche	over 1,000
1718	Gansu	Tongwei	Earthquake-induced landslide	40,000
1786	Sichuan	Luding	Flood resulting from landslide-dam-failure	100,000
1847	Qinghai	Beichuan	Loess and rockslide	Hundreds of deaths
1856	Sichuan	Qianjiang	Rockslide induced by earthquake	over 1,000
1870	Sichuan	Batang	Rockslide induced by earthquake	over 2,000
1897	Gansu	Ningyuan	Loess and rockslide	over 100
1917	Yunnan	Daguan	Rockslide	1,800
1920	Ningxia	Haiyuan	Loess landslides induced by earthquake	100,000
1933	Sichuan	Maowen	Flood resulting from landslide-dam-failure	2,429
1935	Sichuan	Huili	Rock and debris slide	250
1943	Qinghai	Gonghe	Loess and mudstone slide	123
1951	Taiwan	Tsao-Ling	Flood caused by landslide-dam-failure	154
1954	Xizang	Jiangzhi	Flood caused by glacier dam failure	450
1964	Gansu	Lanzhou	Landslide and debris flow	137
1965	Yunnan	Luguan	Rock landslide	444
1966	Gansu	Lanzhou	Landslide and debris flow	134
1972	Sichuan	Lugu	Debris flow	123
1974	Sichuan	Nanjiang	Landslide	195
1975	Gansu	Zhuanglong	Loess slide caused flooding along the shore of the reservoir and downstream	over 500
1979	Sichuan	Yaan	Debris flow	114
1980	Hubei	Yuanan	Rockslide and avalanche	284
1983	Gansu	Dong Xiang	Loess landslide	277
1984	Yunnan	Yinmin	Debris flow	121
1984	Sichuan	Guanlue	Debris flow	over 300
1987	Sichuan	Wushan	Rock avalanche	102

Chongqing Steel Plant, and other public properties. In recent years, two landslides occurred in the central part of the city, damaging about 300 houses, and causing about $1.2 million in damages.

Landslide disasters are serious in the coal fields of Guizhou, Sichuan, Shaanxi and Liaoning Provinces. For instance, the direct cost of 19 landslides were estimated to be $1.25 million, and indirect losses were about $13 million in the past 15 years in the Liupanshui coal mining area, west Guizhou. In the Weibei coal fields of Shaanxi, 160 landslides have been induced since coal mining began in 1950. A large number of buildings were destroyed and 54 people were killed (Wang, 1987). More recently, the Xiangshan landslide located in the Hancheng coal mining area caused substantial losses to the economy of China. The damages, including deformation of the subgrade and upper structure of large Hancheng Power Plant, ventilation shaft and railway tunnel, totaled $3 million. The cost of stabilization was estimated to be about $25 million, one of the costliest in China.

In many mountainous areas of China, damages to transportation facilities, mainly railways and highways, constitute a significant part of total landslide costs. The railway lines of Baoji-Chengdu, Chengdu-Kunming, Baoji-Tianshui, Xiangfan-Chongqing and Yingtan-Xiamen have been damaged more seriously than other railway lines.

During the period from 1954 to 1957, 2136 slope failures occurred within a section 848 km long of the Baoji-Chengdu railway from Baoji-Shangxiba. The renovation costs were estimated to be about $2200 million (Ju, 1987). During the past 30 years, landslides within a section 154 km in length of the Baoji to Tianshui railway interrupted the train traffic a total of 4,679 hours. The repair cost was $675 million.

Table 10 shows the cost of 8 individual landslide stabilizations. The average cost of a landslide stabilization was about $1.75 million. According to incomplete and rough statistics from 1974 to 1976,

there were more than 1000 landslides of middle and large scale along China's railway lines (Fu, 1982). It can be roughly calculated that the total costs of more than 1000 landslide stabilizations is nearly $2 billion.

Table 10. The cost of stabilizing 8 individual landslides.

Name of landslide	Name of railway	Cost of stabilization ($ Million)
Xiongjiahe	Baoji-Chengdu	2.2
Junshimiao	Baoji-Chengdu	0.9
Shizishan	Chengdu-Kunming	0.4
Tiexi	Chengdu-Kunming	6.6
K 118	Sichuan-Guizhou	1.7
Dazhongxi	Xiangfan-Chongqing	0.23
Zhaojiatang	Xiangfan-Chongqing	1.53
K 163	Yingtan-Xiamen	0.44

In 1980 alone, there were 963 cases of slope failure which damaged railways in southwestern China and interrupted railway transportation for a total of 1656 hours, according to the statistics made by the Chengdu Railway Administration Bureau. Direct losses were at least $6 million, the cost of interruption of transportation was $3 million, and the costs of repair were estimated to be about $9 million.

In 1985 to 1986 the Dongchuan railway was closed for more than 240 days due to landslide and debris flow damages. The total losses were more than $10 million.

The cost of landslides on highways in China are, perhaps, greater than those for railways. According to 1983 statistics, highways and roads extend 910,000 km in China, 11 times more than railways. In many mountainous areas, highways and roads are closed by landslides in the rainy season. In the rainy season of 1982, for example, 894 landslides formed a long 848 km of roads and highways in the

Table 11. Chronological table indicating the impact of landslides on navigation of the Changjiang River (Li and Liu, 1987).

Year	Location		Volume of landslide down to river ($10^4 \times m^3$)	Impact on navigation
377	Hubei	Xitan area	2600*	Dammed the river, forming a lake about 50 km long. Landslide obstructed navigation for many years.
1026	Hubei	Xitan area	1500*	Dammed the river and obstructed navigation in a 15 km section of the river for about 22 years.
1542	Hubei	Xitan area	860*	Blockage 2 km long obstructed navigation for 82 years and prevented navigation during the dry season.
1896	Sichuan	Xinglongtan	600*	Formed large rapids, obstructed navigation for 10 years, and prevented navigation in dry season.
1982	Sichuan	Jipazi	180	Formed low rapids and obstructed navigation for 4 years.
1985	Hubei	Xitan	200	None

Fenjia district, Sichuan Province; total losses due to landslides was $1.2 million. In 1984, about 3,244 landslides and debris flows damaged the main highways in Wudu and Tianshui prefectures and destroyed 457 bridges and culverts. In this same area, large debris flows from 1964 to 1978, killed 1,142 people, buried 17,544 rooms, and destroyed 2,266 ha of farmland.

The Sichuan-Xizang Highway, which starts at Chengdu in the east and extends 2,413 westward to Lhasa has suffered seriously from landslides, avalanches and debris flows year after year. In 1984, avalanches and debris flows in Belong destroyed 3 steel bridges, 5 suspension bridges, 10 km of highway, and many trucks. The traffic disruption lasted for about 7 months. Direct cost was estimated to be about $3 million.

Rivers are also adversely affected by landslides. The Changjiang is the longest and busiest river in China. At least five landslides have obstructed its navigation seriously during the last 2,000 years (Table 11). In 1982, a huge landslide with a volume of 15 million m^3 occurred on a slope underlain by sandstone and claystone of Late Jurassic age in Yunyang district, East Sichuan (Fig. 11). It destroyed about 775 acres of farm land, a refrigerator storehouse, a hospital, and many other buildings. The direct economic loss totaled $1.25 million. The front part of the slide with a volume of 1.8 million m^3 slid down to the Changjiang River,

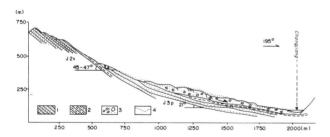

Figure 11. Jipazi slide of Jurassic sandstone and mudstone on the north bank of the Changjian River in the Yunyang district of Sichuan, which obstructed the navigation of the river; 1, sandstone, 2, mudstone, 3, slide deposits, 4, slope surface before sliding.

forming a navigation barrier. The costs of landslide stabilization and dredging were $32 million from 1982 to 1986.

Within mainland China, more information on cost of landsliding has been obtained for Sichuan, Gansu, Hubei, Yunnan, Guizhou, Shaanxi and Xizang Provinces than any other area. For instance, in Sichuan province, about $5 million was spent annually for

landslide disasters in the 15 years from 1970 to 1985. During the Sixth Five year plan period (1981-1985), the Central Government of China spent $210 million per year for disaster relief to compensate for losses caused by natural catastrophes, including floods, landslides, typhoons, and earthquakes. During the Seventh Five Year plan (1986-1990), $320 million is being spent for disaster relief annually.

Using the above information and previously unpublished data, the author estimates roughly that the annual losses due to landslide in China, including direct and indirect losses to public and private property, exceeded $0.5 billion per year for the past 36 years, from 1951 to 1987.

6. PREVENTION AND CONTROL OF LANDSLIDES

6.1 Brief History of Prevention and Control

A long time ago, the terms "mountain avalanche damming river", "slide sinking village" and "earth moving" were found often in Chinese literature. Landslide prevention and control in China seem to date back sometime before the Christian era. Before that, measures for preventing and controling landslides involved nothing except evacuations of local residents from hazardous areas of predictable landslides.

In the early 1950's, because knowledge about landslide identification and prevention was largly lacking, some public works were built on old landslide deposits. Excavation for other public works reactivated landslides such as those at the Xipo, Tanjiazhuang, Basihuijiang and Lueyang railway stations along the Baoji-Chengdu railway line. These landslides interrupted the railway several times and made geologists and civil engineers aware of the magnitude of the problem. Systematic study of landslide control was started after these incidents.

During the late 1950's, the methods of landslide control adopted involved surface drains, ground drains and retaining walls. A trench in the headwall and a buttress fill at the toe was the main measure used to treat landslides. The trench unloads the upper part of the slide and provides drainage, and the buttress supports the sliding mass.

From the 1960's to 1970's, based on previous experience, the old, large-scale landslides and areas where landslides are concentrated were avoided as much as possible in designing mountain railways, highways and other public works. For example, during the selection of the Chengdu-Kunming railway line, about 100 large-scale landslides were avoided. Some large landslides could not be avoided, however, so control measures were used to stabilize them before construction. During this period, staff geologists and engineers in the

Ministry of Railways and the Ministry of Water Conservancy studied extensively various techniques for controlling landslides. The techniques and understanding developed rapidly.

Since the 1960's, concrete piles have been used for some landslide control. Most of these are concrete-filled holes with square or rectanglar sections 1x1 m² to 2x3 m² in diameter. In recent years, these piles have been adopted extensively at landslides in general, because their anti-sliding capacity is very strong, the amount of concrete needed to stabilize the slide is less than other methods, construction is convenient, tools needed to dig the holes are operated easily, and the holes provide useful geologic information (Lin, 1984 and Wang, 1985). Piles are especially suitable in controlling active landslides. The dug pile is similar to the so-called shaft works, deep foundation pipes of large diameter, used in Japan (Taniguchi, 1985).

Recently, a vertical-anchor retaining wall has been tested and used to supersede gravity retaining walls. It can reduce masonry about 20 percent and is especially suitable where the slide outlet is higher than the slope foot.

Horizontal bore holes have been used to drain groundwater at some landslides. The lack of a factory in China to produce the special horizontal drillers limits the use of this method.

Chemical grouting to strengthen soil mass for controlling landslides is being tested. Lime pile and lime-sand pile have also been used in recent years for stabilizing soil embankments.

6.2 Prevention and Control Works of Landslides

The prevention and control works actually carried out in landslide areas are based on the following principles: The saving of human life has primary importance, and the safety of public structures, buildings, and road traffic, has secondary importance. If a slide dams a river the prevention of flooding becomes a major concern.

For convenience sake, landslides can be arbitrarily divided into two groups: those occurring on artificial slopes and those on natural slopes. The term "artifical slope" here implies not only the human or constructed slope but also the slope that has been partly excavated or filled by human activities. It includes, therefore, levees, dams, reservoir banks, road slopes and mining spoil. Landslides on natural slopes in wildland or forested watersheds receive less attention from the public. Those landslides are not usually subject to treatment unless they endanger railway, road, buildings, reservoirs or other important installations below. The control works, therefore, for these two kinds landslides are somewhat different. Of course, some control works can certainly be applied to both types. But in general, artificial slopes receive more intense treatment than natural slopes. The control works of landslides in China are summarized as follows:

List of landslide control works in China:
A. For artifical slopes
1. Avoidance; relocation, bridging, tunnel (or open-out tunnel)
2. Surface drainage; channel or ditch, prevention of water leakage
3. Subsurface drainage; tunnel, blind trench, stabilization trench vertical-drill drainage holes, horizontal bore holes, slope-seepage ditch, drainage well of ferroconcrete, drainage well with liner plates.
4. Support structures; retaining wall, anchor retaining wall, cribwork, gabion stabilization trench piling works

5. Excavation; removal, flattening, and benching
6. River structure work; erosion control dam, consolidated dam, reventment, groin, spur dikes
7. Other methods; planting vegetation, blasting and hardening
B. For natural slopes
1. River structure work; check dam, consolidated dam revetment, groin, spur dikes
2. Benching and diversions
3. Revegetation; grass reseeding, reforestation.

7. ORGANIZATION AND PUBLICATIONS

7.1 Organizations

There is no national landslide society in China as yet. Recently, several landslide societies or committees have been established in the provinces which have suffered from landsliding. The Landslide Committee of the Geographical Society of Sichuan was established in 1982, the Gansu Society of Landslides and Debris Flows in 1984, the Landslide and Debris Flow Committee of the Shaanxi Geology Society in 1985, the Landslide Control Committee of Shaanxi Civil Engineering in 1985, the Landslide Society of East China in 1987, and the Landslide Society of Hubei in 1987. These societies have, altogether, more than 1,000 members, composed mainly of researchers and engineers specializing in geology, geomorphology, topography, geophysics, civil engineering, erosion control, forestry, agricultural civil engineering and other fields concerned with landslides, from research institutes, universities and colleges, public organizations, consultants, and governmental agencies. They hold national or provincial symposia and seminars together or separately for information exchange on the results of studies on landslide processes and control methods and publish symposium proceedings. They also have international symposia such as the China-Japan Field Workshop on Landslide in 1987.

The Chengdu Institute of Mountain Disasters and Environment of the Chinese Academy of Sciences and the Northwest Institute of China Academy of Railway are two special institutions concerned with landslide study and control.

7.2 Publications

Since the 1970's, many monographs about landslides in China have been published, such as "Landslide," "Land-slides and their control," "Earthquakes and landslides," "The law and control of landslides," "Collected works on landslides" (No 1-5), "Landslide analysis and control." Additionally, a great number of articles and reports can be found in the journals "Mountain Research", "Soil and Water Conservation", "Engineering Geology" and "Civil Engineering."

REFERENCES

Chang, S.C. 1984. Tsao-Ling landslide and its effect on a reservoir project. Proc. IVth Intern'l Symp. on Landslides, Toronto, Sept. 16-21, 1984, 1:469-473.
Chou, H. 1960. Soil and water conservation. Taichung (Taiwan): Taiwan Provincial Agricultural College Press.
Chu, D.S.L. 1961. Improvement of the hydro-electric project in Taiwan. China Min. Econ. Affairs, Water Resrces Planning Comm., Taipei, Taiwan: 57.
Close, U. & E. McCormick 1922. Where the mountains walked. Nat'l Geographic Mag. 41, 5:445-464.

Feng, X. & A. Guo 1985. Earthquake landslide in China. Proc. IVth Intern'l Conf. & Fld Workshop on Landslides, Tokyo: 339-344.

Fu, Z. 1982. The seven classes and seven models of landslides classification and the law of landslide's distribution in Chinese railways. Proc. IVth Cong. Intern'l Assoc. Eng. Geol., New Delhi, 3:161-168.

Han, H., M. Hu and L. Han 1985. A brief report on gravitational erosion of collapse and landslide in Loess Plateau. Bulletin Soil and Water Conservatlon, 4:29-33.

Jin, Z. 1987. On the distribution of landslides in Loess Plateau, China. Proc. China-Japan Fld. Wrkshp on Landslide 1987, Xian-Lanzhou, China, 9-10.

Ju, H. & R. Gu 1987. Prognosis for landslide and studying of environmental engineering geologic zonation. Proc. Intern'l Symp. on Eng. Geol. Environment in Mountainous Areas, Beijing, China: 527-542.

Li, H. and Y. Don 1985. A general description on landslide in Gansu Province, China. Proc. IV Intern'l Conf. & Fld. Wrkshp. on Landslides. Tokyo: 457-461.

Li, M. and T. Li 1985. An investigation of landslide induced by heavy rainfall during July, 1982, in the eastern part of Sichuan Province, China. Proc. PRC-US-JPN Trilateral Symp. Eng. for multiple natural hazard mitigation, Jan. 1985, Beijing, China. L-5-1-L-5-14.

Li, T. 1979. A study on the relationship between earthquake and landslide and the prediction of seismogenic landslide area. In Collected Works of Landslides No. 2., Beijing. Railway Press: 127-132.

Li, T. and M. Li 1985. A preliminary study on landslide triggered by heavy rainfall. Proc. Intern'l Symp. on Erosion, Debris Flow and Disaster Prevention, Sept. 3-5, 1985, Tsukuba, Japan: 317-320.

Li, T., R.L. Schuster & J. Wu 1986. Landslide dams in south central China. In R.L. Schuster (ed.), Landslide Dams--processes, risk and mitigation, Amer. Soc. Civil Eng. Geotech. Spec. Pub. 8:146-162.

Li, T., S. Zhang and Z. Kang 1984. Sliding debris flow. In Memoirs of Lanzhou Inst. of Glaciology and Cryopedology, Chinese Acad. Sci., Beijing. Science Press 4:171-177.

Li, T. and X. Liu 1987. Impacts of landslides on Three Gorge Project of Yangtze River and treatments. In studies on the Impacts of Three Gorge Project on Ecology and Environment and Treatments. Beijing. Science Press: 632-655.

Lin, Z. 1984. Some characteristics and correction methods of landslides in Xigeda-group stratum in Dukou. In Landslide Committee of Geography Society of Sichuan and Chengdu Institute of Geography (ed.), Landslide Analysis and Control. Chonqing Branch Sci. & Techn. Literature Press: 106-111.

Qi, W. 1984. China, Beijing. Foreign Languages Press.

Qiu, J. and X. Liu 1985. A summary on damage of extraordinary collapse, landslide and debris flow in mountain areas of Southwest China. Bulletin of Soil and Water Conservation, 4:43-47.

Sheng, T.C. 1966. Landslide classification and studies of Taiwan (Chinese-American Joint Commission on Rural Reconstruction, Forestry Series 10. Taipei, Taiwan: 96.

Taniquchi, T. 1985. History of prevention and control of landslide in Japan.

Wang, G. 1985. The measures for controlling landslide on railway in China. Proc. IVth Intern'l Conf. & Fld. Wrkshp. on Landslides, Tokyo: 107-111.

Wang, T. 1987. Study on the character of landslide in coal field areas of Shaanxi Province. Proc.

the China-Japan Fld. Wrkshp. on Landslide, 1987, Xian-Lanzhou, China: 37-40.

Yan, G. 1985. The damages of collapse, landslide and debris flow and their control in Jiangxi Province. Bulletin of Soil & Water Conservation 4:38-42.

Zha, X. and J. Zhang 1987. Analysis of the stability of landslides in Baoji City. Proc. the China-Japan Fld. Wrkshp. on Landslide, 1987, Xian-Lanzhou, China: 45-48.

Zhao, S. 1986. Physical Geography of China. Beijing and New York: Science Press & John Willey & Sons: 209.

Landslides: Extent and Economic Significance, Brabb & Harrod (eds)
© 1989 Balkema, Rotterdam. ISBN 90 6191 876 6

Geological and economic extent of landslides in Japan and Korea

N.Oyagi
National Research Center for Disaster Prevention, Ibaraki-ken, Japan

ABSTRACT: Ratio of casualties by landslides recently reaches 90 % though the total ones have been gradually decreasing. Annual economic loss by landslides is estimated to be 30 to 6000 billion yen. For mitigation of those landslide disasters, prediction of location and characteristics of landslides is necessary as well as time prediction. Geological zonation for landslides seems valuable for this purpose. Japanese Islands can be divided into 15 zones for landslides. Each zone shows characteristic distribution of landslides and landslide properties. Problems on landuse and necessary scientific studies are also discussed.

1 INTRODUCTION

Extent of landslides includes very wide contents as that of other geological phenomena in space and time. For mitigation of landslide disasters, it is considered to be important for us to understand spatial distribution of landslide occurrence and its meaning with respect to geologic zones and geomorphological situation and to know cost due to landslide disasters and for countermeasures against landslide disasters.

Therefore, we shall firstly review geographical distribution of landslides and then consider geological zonation for landslides in relation to their distribution and characteristics of landslides mainly in Japan and partially in Korea. Secondly, economic extent will be discussed on the damage cost caused by landslides and cost for rescuing, restoration, control works and monitoring. Thirdly, future landslide problems and possibility for mitigation of landslide disasters will be considered with respect to landuse. Finally, important studies for the mitigation of landslide disasters will be discussed.

2 OCCURRENCE, GEOGRAPHICAL EXTENT OF LANDSLIDES AND ITS GEOLOGICAL SITUATION

2.1 Distribution and zonation of landslides

Landslides and landslide landforms distribute throughout Japanese islands. However, their distribution is not uniform but uneven making zones of landslide concentration. It is also inferred that a certain zone will have mainly a similar type of landslides. If we choose appropriate criteria and classify zones relevantly for a wide region, zonal classification will be helpful for study and practical prediction of sites and characteristics of landslides to occur. Those characteristics can be considered to be highly related to lithology or engineering property of the zones. Lithologic character in a part of a certain geological zone is usually similar to that of another part of the zone, owing to similar process in formation of the zones, that is similar original material in the same or similar sedimentation basins and similar type of diagenesis, metamorphism and tectonics affected. From this reason, geological zonation based on lithology will explain the setting of landslides or mass movements in regional point of view.

In this section, we review landslide distribution in several grade of recognition in social and scientific approaches. Then, zonation of landslides is discussed from geological and geomorphological viewpoints.

2.1.1 Designated landslide areas for control works

Distribution of designated landslides for control works based on the Landslide Law is shown in Figure 1. The Landslide Law was enacted in 1958 for mitigation of landslide damage through control measures and regulation of human activity on landslide areas. Those landslides are mainly "jisuberi" in Japanese which correspond to creeps, creep slides or very slow slides in type. The designated landslides reached 5539 in number and 2531 km² in total area (0.7 % of the whole area of Japanese Islands) in 1986 in Japan (Ministry of Construction). Landslide prone areas though inactive in the present reach 10288 in total number and 40 21 km² in total area (1.2 % of the whole area of Japanese Islands). The area of the designated landslides

Figure 1. Designated landslides for control works in Japan. (Japan Landslide society 1988).

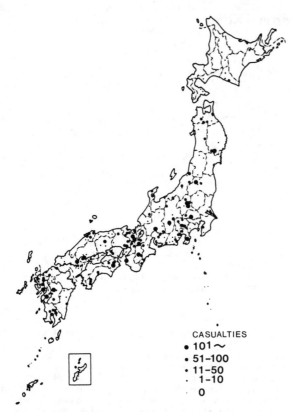

CASUALTIES
• 101〜
• 51-100
• 11-50
· 1-10
  0

Figure 2. Distribution of disasters caused by landslide disasters from 1945 to 1981 in Japan. (Land
Agency).

and the landslide prone areas occupies that of 1.9 %
of the Japanese Islands. However, if we include the
designated area for small-scale landslides of rapid
type on steep slopes and that for "Sabo" (erosion control) works including debris flows, the total area of
mass movements reaches 14663 km² and occupies 4.2 % of
the whole area of Japanese Islands.
    The landslide prone areas are omitted in Figure 1 as
they have similar distribution to the designated landslides. Densely concentrated areas are found in the
coastal region along the Sea of Japan, which includes
Shikoku. Less dense and sporadical distribution can be
seen in the northern part of Honshu, central and western parts of Honshu and Kyushu. Those areas of landslide distribution show clearly zonal concentration as
suggested above. This will be discussed with geological zonation later.
    Distribution of disasters caused by landslides in
wide category including rapid landslides, debris flows

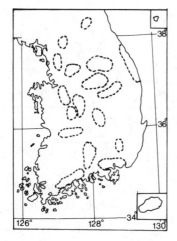

Figure 3. Heavily damaged area by landslide disasters
in Korea. (Woo 1984).
Dashed areas show the disaster areas.

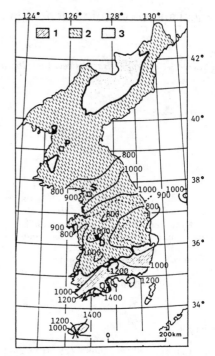

Figure 4. Annual precipitation (in mm) in Korea. (Central Meteorological Office 1982).
1: Red soil. 2: Forest brown soil. 3: Mountain forest
podosol. P : Pyongyang. S: Seoul. D: Daegeon.

and avalanches is shown in Figure 2 (Land Agency 1985)
in which a solid circle means the center of a disaster
caused not only by a landslide but also by many landslides induced by a single proximate inducing factor,
"trigger", a heavy rainfall or a strong earthquake
etc. The magnitude of disasters is indicated by the
different sizes of solid circles corresponding to the
number of casualties. Those locations distribute
mainly in central and western part of Japan and not
necessarily coincide with the densely concentrated
zones of designated landslides of Figure 1. This difference between those two figures suggests different

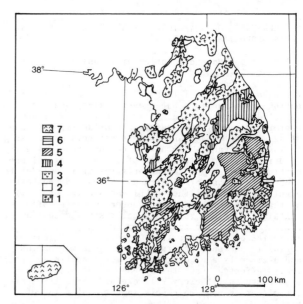

Figure 5. Symplified geological map of Korea. (Modi-
ied from Lee 1987.
1: Granitic rocks. 2: Pre-Cambrian rocks including
gneiss. 3: Cretaceous to Paleogene terrestrial volcanic rocks. 4: Paleozoic to Middle Mesozoic sediments.
5: Cretaceous sediments. 6: Tertiary sediments. 7:
Neogene volcanic rocks.

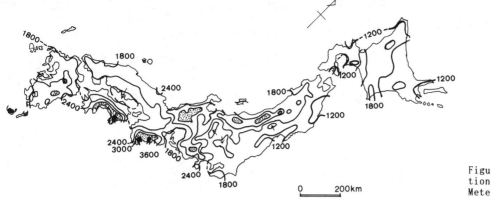

Figure 6. Annual precipitation in Japan. (Japanese
Meteorological Agency 1984).

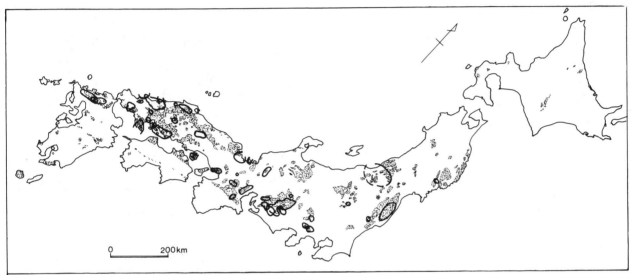

Figure 7. Landslide disasters in granitic rock areas.
(Oyagi 1976).
Solid lines show disaster areas from 1945 to 1988.
Areas underlain by grnitic rocks are shown by dots.

mechanisms between two types of landslide disasters
which will be discussed later.
    In Korea, heavily damaged areas by landslide disas-
ters are not restricted to special zones but rather
evenly distribute throughout the country (Figure 3,
Woo 1984). On this point, their distribution (Figure 3
) is fairly corresponded to the disaster distribution
caused by mass movements in Japan (Figure 2). The tendency that the heavily damaged areas are rather sporadic along the western coastal area of Korea facing the
Yellow Sea suggests geomorphic condition in this area
of low relief and gentle slopes as precipitation dif-

ference is not significant between the coastal area
and central area (Figure 4, Central Meteorological Office of Korea 1984) and also as geologic belts are
trending obliquely from southwest to northeast, that
the both areas are underlain by the same geology (Figure 5, Lee 1987).
    In case of Japan, one of the reasons of sporadic
occurrence of landslide disasters in northern part of
Honshu and Hokkaido (Figure 2) is due to smaller
amount of precipitation except snowfall (Figure 6, Japan Meteorological Agency 1984). However, another
reason is geology composed of semi-consolidated sediments as discussed later.
    On the other hand, landslide disasters which cause
large number of casualties are found in hard rock regions, especially in granitic rock regions. The disasters from 1945 to 1976 in granitic rock regions (Oyagi
1976) are shown in Figure 7. The most cases of landslide disasters occurred in the granitic rock regions.

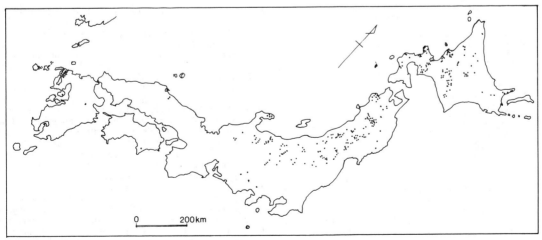

Figure 8. Distribution of large-scale landslides ( ) larger than 1 km² in area in Japan. (Machida et al. 1986).

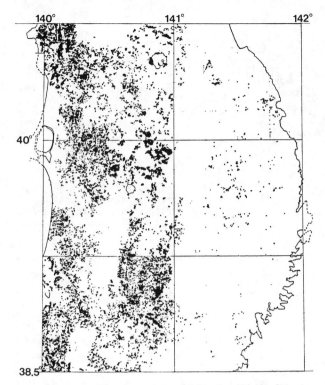

Figure 9. Distribution of landslide landforms (black coloured areas) in Tohoku District from air-photograph interpretation. (Oyagi et al. 1982).

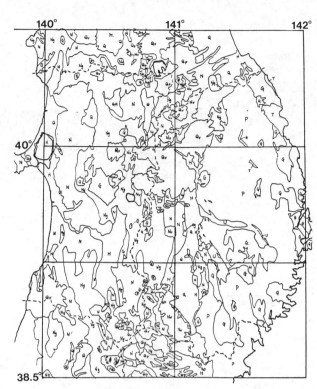

Figure 11. Geologic zoning for the distribution of landslide landforms in Tohoku District. (Oyagi et al. 1982).
Letters show geologic zones same to the Figure 12.

All the cases in granitic rock regions were caused by heavy rainfall related to Bai-u front, typhoons and localized thunderstorms. Comparison between distribution of designated landslides (Figure 1) and landslide disasters in granitic rock regions (Figure 7) suggests that geological zoning for landslides will effective to understand landslide disasters in wide regions.

2.1.2 Distribution of large-scale landslides

Figure 8 shows distribution of large-scale landslides with areas larger than 1 km² mapped by Machida et al. (1986). However, type of landslides is not discriminated into slow slides or rapid slide-flows. The distribution seams partially concordant with that of the designated landslides in the northern and central part of Japan though distribution density is similar. While in the southwestern part of Japan, large-scale landslides are hardly seen except in the northwestern corner of Kyushu. This meaning can only be understood with geological zoning which will be discussed later.

2.1.3 Landslide landforms in Tohoku District

Landslides and landslide landforms are mapped through air-photograph interpretation in Tohoku district of northern part of Japan. Their distribution clearly shows distinct zonation through different landslide concentration and magnitude (Figure 9, Oyagi et al. 1982). Landslides are very rare in the Kitakami Mountains of the eastern half of the Tohoku district, where is underlain by Palaeozoic to Mesozoic sedimentary rocks. They are very highly concentrated in zones running north to south in the Oh-u and Dewa Mountains of the western part of the Tohoku district, where is underlain by Miocene to Pliocene clastic sediments and volcanic rocks. Very large-scale landslides are found mainly on Quaternary volcanic bodies in the latter part.

Large-scale landslides of which source areas are larger than 1 km in width show more restricted distribution. They are mainly found in areas of maximum uplift in the Tohoku District during Quaternary Period and in those of Quaternary volcanoes which are shown

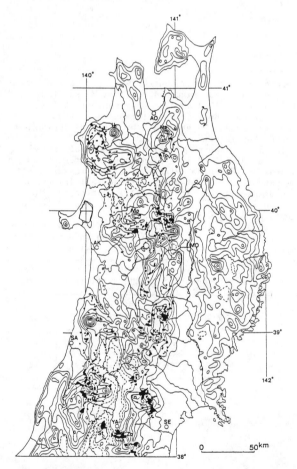

Figure 10. Distribution of large-scale landslides in Tohoku District with respet to Giphelflur. (Shimizu 1985).
Black areas show large scale landslides. Contors are of 200 m in interval and broken ones indicate depression.

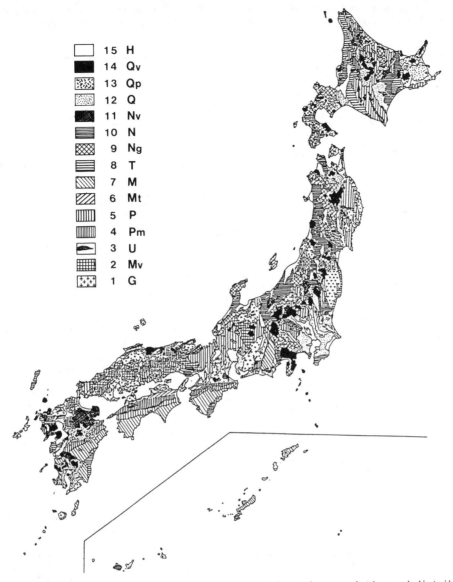

	15	H
	14	Qv
	13	Qp
	12	Q
	11	Nv
	10	N
	9	Ng
	8	T
	7	M
	6	Mt
	5	P
	4	Pm
	3	U
	2	Mv
	1	G

Figure 12. Geologic zoning for landslides in Japanese Islands. (Kuroda 1986).
Numerals show the zone numbers discussed in the text.

by concentric contours in Figure 10 (Shimizu 1985). As discussed above, very large-scale landslides are usually observed in volcanoes.

Through those discussions, possibility of geological zoning for landslides is highly suggested.

2.2 Relation between zonal distribution of landslides and geologic belts and geographic provinces

Geological zoning for landslides is discussed in this section based on various kind of landslide distribution maps and geologic maps. It will be also considered from geomorphological point of view.

2.2.1 Geological zoning for landslides

As reviewed in the preceding sections, landslides occur on specific areas or zones and some types of landslides are exclusively found on specific zones of geology in a region of almost similar climatic condition. Therefore, appropriate geological zoning will provide regional criteria on the characteristics of landslides needed for their prediction, countermeasures and better landuse. We have two possible approaches to make geological zoning for landslides. An ideal approach should have a vector of study originat-

ing from characteristics and distribution of landslides in a certain region concerned and then directing to geology or geologic sciences. This approach needs landslide distribution maps of high quality showing exact location, area and surface form or structure related to characteristics of landslides. The Tohoku district in Japan is well mapped appropriately for this purpose as shown in Figure 9 which is compiled from Landslide Maps of 1/50000 in scale (Shimizu, Oyagi and Inokuchi 1982,1984,and 1985, Shimizu and Oyagi 1986,1987 and 1988). Such trial is shown partially in Figure 11 in the same district. However, it will take long period for mapping throughout Japanese Islands. At this situation, another approach is still applicable. It is based on geologic belt classification. The classification by Koide (1955) has been famous and used for long time in Japan. It consists of three geologic zones for landslides; (1) Tertiary formation (2) Sheared zone and (3) Hot spring area. His classification is simple and comprehensive. However, many imcomplete points heve been pointed out for his classification by many researchers (e.g., Yatsu 1965) through recent studies on landslides. The Committee on geological classification of landslides of the Japan Landslide Society initiated investigation for description method of landslides for engineering purposes and their geological zonation in 1974 (Fujita 1985). Through discussion in the committee, Kuroda tried to classify the Japanese Islands into 15 engineering geological zones. His zonation is based on lithology of geologic zones and engineering properties for landslides. Procedure on mapping is essentially reinterpretation of geological maps of usual type.

Table 1. Geological, geomorphological and landslide characteristics of the fifteen zones classified for landslide studies and landslide disaster prevention in Japanese Islands. (Partially modified from the appendix talbe of Oyagi 1984).

GEOLOGIC ZONES FOR SLOPE DISAS.		LITHOLOGY			TOPOGRAPHICAL SITUATION	MAJOR SLOPE
Geol. zones	G. provin.	bedrocks	Decomposition	Slope deposits		Type and scale
15 H Late Pleistocene-Holocene clastics		clay,silt,sand & gravel.	is generally rare.	alluvial cones, debris,artificial fill.	alluvial areas, dunes,fans,cones.	lateral spreading by liquifaction,slides in debris & fills.
14 Qv Quaternary volcanic rocks	E-Japan volcanic belt W-Japan volcanic belt	lava & pyroclastic deposits(basalt-rhyolite)	is often severe due to hydrothermal alteration & local	is often thick and composed of clay to rubbly materials.	volcanoes.	very large-scale rock slides triggered by eruption. altered rock & debris slides, and flows. surficial slide & debris flows
13 QP Quaternary pyroclastic rocks & sediments	ditto	unconsol. pyroclastic flow deposits shirasu, loam.	Weathering is less in "shirasu" and more in loam.	is usually thin	dissected platform, hills & mountains.	surficial,unconsolidated material slides falls & flows. Small in scale.
12 Q Quaternary terrestrial or deltaic sediments	Boso Penin. Kanto Plane	unconsolidated gravel, sand & mud.	Materials near surface easily lose its cohesion.	is thick & developed.	hills,terraces & sand dunes.	surficial & debris slides.
11 Nv Pliocene-Early Pleistocene terrestrial volcanic rocks	Hohi volcanic zone S-Kyushu,	lava & pyroclastic rocks.	severely altered in some areas.	is composed of thick debris.	mesa,hills,low relief mountains.	surficial,debris & rock slides; debris flows.
10 N Miocene-Pliocene marine clastic deposits	Green-tuff region A1.NE-Japan A2.SW-Japan	semi-consolid. clastic rocks; rich in montmorillonite.	easily softened by weathering in mudstone.	is composed of thick landslide deposits.	hilly areas,cuesta slopes,landslide topography.	small to large mud-debris creeps & slides ;mud flows. Secondary slides in ls deposits.
	Non-Green-tuff region B1.Setouchi	consl. or semiconsl.clastic rocks(+ vol.rk).	easily weakened by weathering .	ditto	hilly area,low relief mountain area	debris & rock creeps & slides. Small in scale but rarely large.
	B2.Kanto, Boso & Oi	semi-consl.clst rocks,serpentinite blocks.	Clay in sheared zone along serp. blocks.	ditto	ditto	debris & rock creeps & slides;small in scale.
9 Ng Miocene submarine volcanic rocks	Green-tuff region	lavas & pyroclastic rocks altered to greenish,with thin clst.rocks.	acidic tuff,mudsonte & sheared zones are weathered into weak zones.	Thick debris & landslide deposits are observed.	mountain ranges of low to high relief ;cuesta.	large rock slides & medium debris slides in the Upper horizon; surficial slides & debris flows.
8 T Paleogene-Early Neogene non-marine to neritic deposits	Central Hokkaido,Kuji, Joban, NW & W-Kyushu	clastic rocks coal seams,covered by basalts in NW-Kyushu.	less weathered.	is sometimes thick.	low to intermediate relief mountain areas.	rock creeps and slides large and middle. Surficial slides and debris flows.
7 M Late Mesozoic-Paleogene eugeosyncline & flysh type dep.	Shimanto z. Mineoka z.	consl.& alternated beds of sandstone & shale.	fractured but less weathered than 10;altered along serpent.	is composed of thick debris.	medium to high relief mountain areas.	rapid & large rock slides, debris flows & creeps.
6 Mt Cretaceous marine deposits of turbidite f.	Kii Mts. Sanuki Mts.	consl.clastic rocks;folded.	less weathered but brecciated.	similar to 3.	ditto	debris creeps & slides on gentle slopes; rock slides.
5 P Paleozoic-Early Mesozoic eugeosyn. deposits non-meta.facies	Tan-ba zone Chichibu z. Ashio z.	consl.clastic rocks;floded.	less weathered but severely fractured along tectonic lines.	similar to 3.	ditto	rock & surf.slides & debris flows on steeps; debris creeps & slides on gentle slopes.
4 Pm Paleozoic-Early Mesozoic eugeo. deposits of metamorphic f.	Sanbagawa z. Sangun z. Kamuikotan	schists.	is intensive in pelitic schist; chlorite schist turns into clay.	is composed of thick debris & ls deposits.	high to medium relief mountain with specific type of piedmont slopes.	debris creeps and slides on piedmont slopes;rare in rock slides.
3 U Ultramafic & mafic intrusive rocks	Mikabu z. Mineoka z. Kamuikotan z.	(meta)gabbro, peridotite, serpentinite etc.	Rocks are highly fractured & clayey along sheared zones.	develops as thick debris on gentle slopes.	ditto	debris creep or slides & earth flows.(Some are deep sheeted).
2 Mv Late Mesozoic-Paleogene terrestrial volcanic rocks	SW-Japan	andesite-dacite lavas & pyrclst. rocks;shale & sandstone	residual weathered crusts in low relief areas;less in steep slopes.	consists of thick debris & soils.	mountains in low to high relief; usually stable.	surficial slides & debris flows. Rarely rock slides & earth flows.
1 G Acidic intrusive rocks & gneiss	NE-Japan Hidaka z. Kitakami z. Abukuma z. SW-Janan Ryoke z. Chugoku batholith	granite,adamelite,granodiorite,diorite, gneisses.	is severe by weathering:red paleosol on gentle slopes; young weathering zones on steep slopes.	is mainly composed of sandy material; generally thin (1 m) but in some areas thick.	low relief hilly areas with high valley density; high relief horst mountains.	Surficial slides in high density(450/km^2), weathered rock slides & debris flows are caused by heavy rainfalls.

Notes: G. prov. for geological province; f. for formation; z. for zone; P for prefecture. In the column of "Examples ..." I, II and III mean that the retardation of the landslide occurrence from the peak of triggers is short (< 1 Hour), intermediate (< 1 day) and long (> 1 day). Letters, r and e indicate rainfall and earthquake, respectively. In the column of "Human damage", L, M, and S mean number of casualties, larger than 50 persons, between 50 to 5, and fewer than 5 or 0.

MOVEMENTS AND DISASTERS — Examples in terms of retardation from peak of triggers	Damage Human	Damage Econ.	PREDICTION — in site & scale	PREDICTION — in time	CONTROL WORKS
Ie:Niigata(Niigata Eq.1964)	S	L	is probable from struc- ture,lithology & property	is probable from strain monitoring	Drainage,slope protection, piling,anchoring are effec-
III:Hisasue(Kawasaki,1965)	M	S	of deposits.	except earthquakes.	tive for fills.
Ie:Midorigaoka,Sendai(1978)	N	M			
Iv:Mt.Bandai(1888)	L	L	is probable from geologic	depends on prediction	are generally imposible.
Iv:Mt.Mayuyama(1792)	LL	L	structure & ground deform.	of volcanic activity.	
Iv:Sounzan(1958),Hakone volc.	M	M	is probable from mapping	is probable from usu-	Usual landslide control works
Ir:Mt.Akagi(1947).Ir:Mt.Usu(1947).Ir:Mt.Yakedake(often).	S-M	M	of altered areas.	al ls monitoring.	steam discharge.
	N	S	is probable from geologic	is probable from moni-	Check dams & drainage works
Ir:Mt.Sakurajima(very often)	N	M	& geomorphic survey.	toring of rainfall &	are effective.
Ir:Mt.Myoko(1978)	M	L		strain.	
Ir:Kagoshima P(1969)	M	M	is probable:slides occur	ditto.	Slope protection & erosion
Ir:Yokohama(1966)	L	M	on steep slopes but on		protection works are effec-
Ie:Hachnohe-Gonohe,Aomori(1968	M	M	gentle slopes(10°-28°) by		tive. Selection of sites is
Ie:Izu Peninsula(1978)	M	M	earthquakes.		important.
Ir:Boso Peninsula(1966)	L	M	ditto.	ditto.	ditto.
Ie:Dune sand slide at Funasaka (Fukui Eq.1948)	S	S			
II:Lss. in the Uonuma Formtion	M	M			
Ir:Isahaya(1957),Nagasaki P.	L	L	For surf.slides by similar	ditto & occasionally	ditto. For large landslides,
Ir:or II Masaki ls.(1972)	N	M	method to 1. For w. rock	by smell as in Masaki	structure is very important.
Ir:Nagasaki(1982)	L	L	slides by structure & to-	landslide.	
III:Komoro ls.(1982-,Nagano P)	N	M	pography.		
III:Yachi ls.(Akita P)	N	M	is usually probable from	is probable from mo-	Surface and subsurface drain-
III:Hiramaru,Matsunoyama,Mu- shigame lss.(Niigata P)	N	M	landslide topography.	nitoring of movement & strain of land-	nage, piling, soil removal & counter weight,etc.
III:Chausuyama ls.(Nagano P)	N	M		slides.	
III:Narao ls.(1976-,Nagano P)	N	M			
III:Lss in Sanda Basin(Hyogo Prefecture)	N	S	ditto.	ditto.	Full scale control works have been done in Kamenose landslide.
III:Kamenose ls.(Osaka P)	N	L			
III:Shorinsan ls.(Gunma P)	N	M	ditto.	ditto.	Surface & subsurface drain-
III:Abekura ls.(Kanagawa P)	N	M			age and piling works.
III:Kurumi ls.(Toyama P)	N	L	ditto.	ditto.	ditto.
Ir:Nen-ba & Saiko(1966,Yama- nashi P)	L	L	is similar to 11.	is probable from mo- nitoring of rainfall and strain.	Slope protection & erosion protection works & check dams.
III:Washio,Hirayama,Nagatashi- ro & other lss.(Nagasaki)	N-M	M-L	ditto.	ditto.	ditto.
Ir:Amakusa Kamijima (1972,Ku- mamoto P)	L	L	is similar to 11.	is probable from mo- nitoring.	Slope protection works & check dams.
Ir,II:Tozugawa(1853,Nara P)	L	L	is probable from rock	ditto but no success-	are difficult in rapid &
Ir,II:Aritagawa(1953,Wakayama)	L	L	creep topography.	full experience.	large slides but drainage is
Ir:Umegashima(1966,Shizuoka p)	M	M			effective in debris creeps.
III:Soro ls.(Chiba P)	N	M			
III:Sanuki Mts.(Tokushima &	N	M	is similar to 3 but some-	is similar to 3.	is similar to 3.
Ir:Amakusa Kamijima(1972,Ku- mamoto p)	L	L	times difficult in rock slides.		
III:Choja ls.(Kochi p)	N	M	is similar to 3.	is similar to 3.	is similar to 3.
IIr:Shigeto ls:(1972,Kochi P)	L	M			
Ir:Niyodogawa area(1975,Kochi)	L	L			
Ir:Fukui & Gifu P(1965)	M	L			
III:Zentoku,Morito,Kito lss.(Tokushima P)	N	S-M	is probable from topogra- phy of landslides.	is possible from strain monitoring.	Surface & subsurface drainage works are effective.
Ir:Nagasaki P(1982),West part of Shimane P(1983)	L	L	is similar to 11.	is probable from mo- nitoring.	Slope protection works & check dams.
III:Nishikawa,Minamidaio lss. (Kochi P)	N	M	is similar to 3 but larg- er in scale in Mikabu z.	is possible from strain monitoring.	Drainage works are effective ;piling is sometimes less ef-
III:Sawatari ls.(Ehime P)	N	S			fective because of deep
III:Soro ls.(Chiba P)	N	M			sheeted shear zones.
Ir:Aioi(1976,Hyogo P)	M	M	is similar to 11.	is probable from mo-	ditto & soil removal.
Ir:Hidagawa debris flow(1968, Gifu P)	L	M		nitoring rainfall & cracks on slopes.	
Ir:Nukeyama ls.(1976,Hyogo P)	S	L			
Ir:Rokko Mts.(1938,1961,1967)	L	L	is progressed for surfi-	is somewhat possible	Limited number of slopes are
Ir:Kure(1945,1967,Hiroshima P)	L	L	cial slides. Essential	in a wide region from	protected by retaining walls,
Ir:Inadani(1961,Nagano P)	L	L	factors are slope angle,	accumulated amount &	slope cribworks,anchoring &
Ir:East part of Shimane(1964)	L	L	profile pattern,depth of	intensity of rainfall	drainage etc.
Ir:Uetsu(1967,Niigata P)	L	L	slope layer,catchment	and is possible on	
Ir:Aich & Gifu P(1972)	L	L	area etc.	selected slopes by	
Ir:Shodoshima(1974,1976,Kaga- wa Prefecture)	L	L		strain monitoring.	

Fortunately, engineering properties of sedimentary
rocks are fairly related with geologic age in Japan.
Therefore, zoning includes lithologic character and
geologic age. The Japanese Islands can be divided into
fifteen zones on the lithological point of view for
landslides (Figure 12, Kuroda 1980, Kuroda et al. 1982
, Kuroda 1986, Japan Landslide Society 1980 and 1988)
as follows:

1. Granites and gneisses. (G for the abbreviation
for the zone).
2. Late Mesozoic to Paleogene Tertiary terrestrial
volcanic rocks and interculated shallow sea or non-
marine sedimentary rocks. (Mv).
3. Serpentine and associated basic rocks. (U).
4. Paleozoic to Middle Mesozoic sedimentary sequences, crystalline schist facies. (Pm).
5. Paleozoic to Middle Mesozoic sedimentary sequences, non-metamorphosed facies. (P).
6. Late Mesozoic turbidite facies sedimentary rocks.
(Mt).
7. Late Mesozoic to Paleogene Tertiary flysh-type
alternations. (M).
8. Paleogene to Early Neogene non-marine and neritic
sediments. (T).
9. Miocene submarine volcanic rocks. (Ng).
10. Miocene to Pliocene marine clastic sediments. (N).
11. Late Miocene to Early Pleistocene terrestrial
volcanic rocks. (Nv).
12. Quaternary terrestrial and neritic sediments of
clastic material. (Q).
13. Quaternary pyroclastic rocks and sediments. (Qp).
14. Quaternary volcanic rocks composing volcanic bodies and rock avalanche or mud flow deposits. (Qv).
15. Late Pleistocene to Holocene sediments filling
valleys of Wurm-maximum age and inter-mountain basins.
(H).

As some zones show more or less similarity each
other in landslide characteristics, lithology or geology, those zones can be grouped up into seven super
zones. Characteristics of each zone is discussed in
the following and its main description is shown in
Table 1 (Oyagi, 1984 & 1987).

Hard igneous acidic rocks and gneisses

The zone 1 consists of granitic rocks including granodiorite, adamerite and quartz diorite etc. and
gneiss. It covers about 11 % of the area of Japanese
Islands. Red or reddish brown residual soil and highly
weathered crusts are often preserved on low relief
areas at higher altitude horizons than the present
stream beds. However, valley side slopes are covered

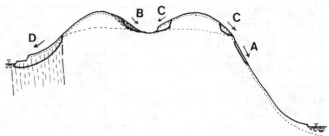

Figure 13. Schematic profile showing four types landslides in granite areas and their topographical condi-
ion. Letters show the types discussed in the text.

by very thin weathering crusts or surficial soil (Oyagi 1984). Landslide disasters in the zone 1 have been
extremely often and large in damage as shown in Table
1 and Figure 7. Those are caused by concentrated occurrence of small slides and debris flows. Four types
of slope movements (Figure 13) can be seen in this
zone:

A. Small surficial slides of 40 to 80 cm deep of
loose sandy soil on weakly weathered bedrock at steep
slopes.
B. Debris slides of 1 to 5 m deep in debris or colluvium accumulated at concave slopes in antient valleys.
C. Weathered material slides of 0.5 to 5 m deep in
residual weathered crusts.
D. "Bedrock" slides of 5 to 20 m deep in fractured
zones.

Characteristic and serious landslide disasters in
the zone 1 are caused by the surficial slides of the
type A and debris flows which are often related to
surficial slides and debris slides. Those slides are
induced by heavy rainfall and their occurrence density
is usually very high, for example, more than 450
slides/km^2 (Okuda & Yokoyama 1977) in Aichi Prefecture
in 1972. Material of surficial slides has been formed
on relatively steep slopes just below the break of
slopes of hills or low relief mountains and composed
of sandy soil with small value of cohesion less than
0.1 kgf/cm^2 and rather large internal friction angle
between 30 to 40 degrees (Matsukura & Tanaka 1983).
That of debris slides of the type B is slightly compacted sandy soil deposited on small and shallow val-

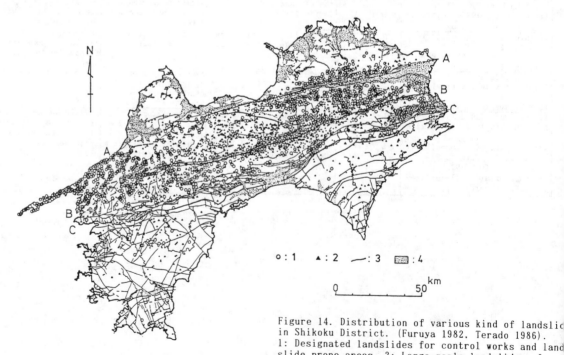

Figure 14. Distribution of various kind of landslic
in Shikoku District. (Furuya 1982, Terado 1986).
1: Designated landslides for control works and land
slide prone areas. 2: Large-scale landslides of rap
type. 3: Faults. 4: Alluvial areas.

ley heads or concave break of slope mainly during
Glacial Age in the Late Pleistocene time. Those large
slides often turn into debris flows and cause serious
disasters. The slides of the type C occur in residual
weathered crusts which are found on gentle slopes near
crests of low relief areas composed of clayey and
sandy material remaining original rock texture with
red, reddish brown and yellowish gray colour (Oyagi 19
68). Their mechanical properties show larger cohesion,
<0.4 kgf/cm², and smaller internal friction angle, 17-
30 degrees (Matsukura 1980). Rotational slides or
block glides in small scale are common and damage
caused by them are usually small.

Landslides in the zone 1 were generally caused when
shear stress exceeded shear strength of slope material, that seems to correspond to peak slides of Sassa
(1985).

The zone 2 is mainly composed of welded pyroclastic
rocks of Late Mesozoic to Early Paleogene in age. It
is characterized by surficial slides which are induced
by heavy rainfall as in the zone 1 but less in landslide occurrence density because of smaller valley
density and thinner surficial soil on steep slopes.
However, a few large-scale landslides were caused at
fractured zones or residual soil layers also by heavy
rainfall but the failure usually occurs several hours
after the peak of a rainfall. Debris flows from channels of high relief mountain cause big damage as in
the case of Hidagawa 1968 where a large debris flow
pushed down two coaches to the Hidagawa River and
killed 109 persons (Hatano & Oyagi 1988).

Consolidatred rocks of Paleozoic to Mesozoic sediments and related rocks

This super-zone consits of three zones and is characterized by high concentration of landslides specifical
ly in Shikoku (Furuya 1982, Terado 1986).

The zone 3 is generally found as a narrow zones along
fractured areas and its bedrocks are serpentinite, peridotite and basic rocks. Slope material consists of
clayey debris and is originated from fractured serpentinite and host sedimentary rocks. Landslides are
mainly creeps or slow slides with medium scale, 50-200
m wide and 200-1000 m long and 5-20 m deep. Slides
will be able to occur at residual shear strength. They
are fairly similar to landslides in the zone 10.

The zone 4 is composed of crystalline schists metamorphosed in the condition of low temperature and high
pressure, in which the most important example is found
in the Sambagawa Metamorphic zone. Slope movements in
the zone 4 are mainly debris creeps and debris slides.
Bedrock slides and debris flows are caused by heavy
rainfall but less common in occurrence. Slope material
involved in the slope movements is mainly debris or
colluvium deposits on gentle slopes which distribute
at several levels higher than river beds in front of
backward steep slopes (Furuya 1982). Those deposits
were thought to be originated from precedent large
bedrock slides (Terado 1986). Density of landslide
distribution is highly related to lithologic type of
bedrocks. It is the highest, 68 %, in pelitic schist,
28 % in psamitic schist and very low in mafic schist
(Fujita 1978). Materials of slide surfaces are usually
clayey and have large internal friction angle of 30 to
40 degrees with almost zero cohesion (Enoki et al. 198
7). Slope instability is affected by pore water pressure, reflecting the soil mechanical property of relatively high permiability and large amount of annual
precipitation. The movement in clayey debris seems to
be related to its residual strength but is also considered that underground erosion by groundwater is res
ponsible (Sassa & Takei 1977).

The zone 5 consists of non- or very weakly metamorphosed sedimentary rocks of Paleozoic to Mesozoic in
age. Slope movements in this zone are almost the same
with those in the zone 4. Debris creeps or slides of
large-scale are found on gentle slopes geomorphologically similar situation (50 to 100 m higher than river
bed) to slopes in the zone 4. Some of large-scale
debris creeps or slow slides occur along fault-sheared
zones in Shikoku. Soil mechanical property in the zone

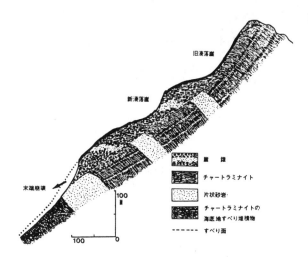

Figure 15. Schematic profile showing landslides in
chert-laminite areas. (Iwamatsu and Shimokawa 1986).

is also similar to that of the zone 4. However, large
debris slides and/or debris flows are often induced
by heavy rainfall. Those cause large disasters, because each slope movement has large volume and high
velocity in spite of lower occurrence density of slope
movements than that in the zone 1 (granitic rocks).

The zone 6 crops out inside (northern side) of the
zone 4 and is composed of alternation of shale and
sandstone which are well consolidated by diagenesis.
Slope movements and landforms are similar to those in
the zone 4. However, large bedrock slides caused by
heavy rainfall are reported (Terado 1975) and bedrock
creeps are found (S.Yokoyama, personal communication).

The zone 7 is underlain by the layers of thin alternation of sandstone and shale of flysh type deposits.
Those rocks include chert-laminite rocks (Yoshida 1981
) and are severely fractured. Slope movements are
characterized by large-scale landslides of extremely
rapid type and debris flows some of which are originated from large volume of deposits along valley
floors, for example at the upper part of the Abe River
(Machida 1966 & 1984). Bedrock creeps in large scale
are found in Central Honshu and South Kyushu (Figure
15, Iwamatsu & Shimokawa 1986). Those creeps are considered to be developed into bedrock slides.

Semi-consolidated Neogene deposits

The zones 8 to 10 constitute the important super-zone
for landslides in Japanese Islands. Rocks in the super
-zone are weakly or moderately consolidated , but are
generally softened through weathering at the layers
near slope surfaces. Creeps and slow slides in unconsolidated slope material (clayey debris and weathered
material from mudsotne) are common in the super-zone.

The zone 8 consists of Paleogene to Neogene deposits
of cyclothem type in which one cyclothem unit shows
sediments from coarse sandstone at the base to mud
stone at the top through fine sandstone, thin alternation of sandstone and mudstone. Coal seams are often
interculated in the upper part of mudstone of the unit
cyclothem. Slide surfaces of landslides are generally
found in thin tuff layers interculated in the coal
seams (Oyagi et al. 1970). Therefore, the landslides
move characteristically along the slide surfaces and
are translational in type. However, rotational slides
can be also developed on anti-dip slopes. Basalt lava
flows covering the cyclothem sediments make caprock
structure for landslides. This structure protects usual erosion or small-scale slides but rather produces large-scale landslides.

The zone 9 is mainly composed of lava flows and pyroclastic rocks originated from submarine volcanic activity, which are more or less altered to greenish
rocks rich in chlorite and zeolites and interculated

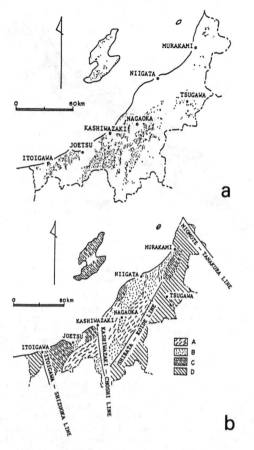

Figure 16. Distribution of landslides (a) and the type
of the late Cenozoic crustal movements (b) in Niigata
Prefecture. (Uemura 1986).
A: Area folded since Middle Pleistocene and mainly
composed of mudstone. B: Area subsided since Middle
Pleistocene and composed of conglomerate. C: Area
folded since Middle Pliocene and mainly composed of
mudstone. D: Area differencially uplifted since Middle
Pleistocene and composed of several type of geology.

with clastic rocks. Medium to large-scale bedrock and
unconsolidated material slides are common in type of
landslides in the areas underlain by rocks with layers
of highly altered tuff. Heavy rainstorms often induce
surficial or shallow debris slides and debris flows
which cause large disasters.

The zone 10 is the most prominent of the super zone of
the semi-consolidated Neogene deposits and consists of
Miocene to Early Pleistocene mudstone and sandstone
with subordinate amount of pyroclastic rocks. Important areas in the zone are found along the Sea of Japan, that is Niigata, Yamagata, Akita, Nagano and To-
yama prefectures. Landforms in the zone show hilly
terrains with high valley density and show irregular
concave slopes formed by landslides especialy in areas
underlain by mudstone.
Typical slope movements are creeps or slides of
clayey debris, weathered mudstone and hard shale.
They are relatively small to large in scale with 50-
500 m wide, 200-1000 m long and 5-40 m deep. Those
slope movements often develop into debris flows or mud
flows. Therefore, the rate of movements is extremely
slow (1 cm/year) to rapid (30 cm/min) in the term of
Varnes (1958 and 1978). Clays near slide surfaces includes high percentages (30-70 %) of smectite clay minerals, especially those derived from tuffaceous mudstone or siltstone. The clays have small value of cohesion (0.01-0.9 kg/cm²) and that of residual-internal
friction angle (0°-20°) (Iwanaga 1986). Those landslides move in a snowmelt season from March to early
May and in other rainy seasons from June to July or
November. Effect of tectonics is also important in
this zone. Niigata Prefecture can be divided into four

areas, A to D, on tectonic point of view as shown in
Figure 16 (Uemura 1982). The areas A and C are the
most prominent in landslide distribution and occurrence. Those belong to the zone 10 and are suffered
folding since Middle Pleistocene or Middle Pliocene.
Landslides concentrate near crests of anticlines or
upper parts of wings.

The zone 11 is composed of Late Miocene to Early
Pleistocene terrestrial volcanic rocks which are less
altered than those of the zone 9. Pyroclastic rocks in
the zone 11 are usually highly weathered to leave red
residual soil in hilly areas where debris slides and
weathered rock slides are caused by heavy rainfall,
for example the 1978 Nagasaki rainfall induced landslide disaster. Basalt lava flows with interculating
thin layers of weathered tuff are underlain by Tertiary cyclothem sediments of the zone 8 in the northwest
part of Kyushu. In this case, large-scale landslides
are common because of the cap-rock structure from the
rocks with large contrast in ductility (Uemura 1975)
between lava flows and the sediments.

Weakly consolidated sediments

The zone 12 is composed of weakly consolidated clastic
material from gravel to clay and constituting table
lands or hills. Typical slope movements are surficial
slides and unconsolidated material slides on the
shoulder part of steep slopes facing valleys. Those
slides are caused by heavy rainfall and earthquakes
when the peak shear strength is exceeded by shear
stress. They easily change into debris flows and often
damage housing areas.
Those layers lose their shear strength in a short
term, for example, from 1.5 to 0.2 t/m² in cohesion
and from 30 to 16 degrees in internal friction angle
(Akutagawa et al. 1977).

Quaternary volcanic rocks

Quaternary volcanoes provide highly hazardous super-
zone through volcanic activity and material from them.

The zone 13 is underlain by pyroclastic flow deposits
or ash fall deposits originated from gigantic eruptions which produced large calderas, for example Aira
and Aso calderas in Kyushu and Hakone caldera in Kanto
District. The areas of the zone are characterized by
table lands of low relief (20-100 m). Typical landslides in the zone are surficial slides and debris
slides on shoulder parts of steep slopes. They are induced by heavy rainfall and earthquakes.

The zone 14 consists of Quaternary volcanic bodies and
is one of the most hazardous zones in Japan. Gigantic
landslides are induced at an upper portion of a stra-
tovolcano by volcanic activities. It caused a catastrophic disaster as in the case of the 1980 eruption
of Mt.St.Helens. Similar cases occurred at Mt.Bandai
in 1888 and Mt.Mayuyama in 1792. The most important
inducing factors for those gigantic landslides are
outward force of a rising lava plug, volcanic gas
(steam) pressure, pore pressure of thermal water and
earthquakes.
Large-scale ladnslides smaller than gigantic ones
are often caused by earthquakes on slopes of upper
part of a volcano, specifically on those of radial
ridges cut at their basal portions through valley erosion, as in the case of Mt.Ontake in 1984. Some landslides are related to hydrothermal alteration of lava
flows and pyroclastic rocks, for example Sounzan Landslide in 1953. Other types of important slope movements are small to medium surficial slides and debris
slides which usually change to debris flows as in the
case of the debris flow disaster at Mt.Myoko in 1978.
Those are often caused by heavy rainfall on Mt.Sakura-
jima and Mt.Yakedake.

Non-consolidated deposits

Non-consolidated deposits should included two types of
deposits; 1) deposits on flat lands and 2) deposits on
slopes.
The zone 15 represents the deposits of the type 1,
which consists of clastic material of Holocene in age.

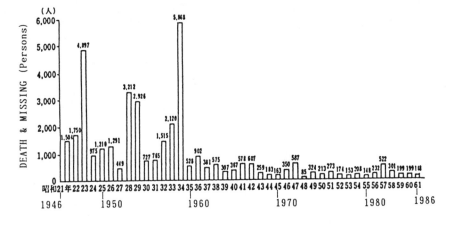

Figure 17. Casualties by natural disasters in Japan from 1946 to 1986. (Land Agency 1988).

They filled the valleys dissected during the Last Glacial Age. Sand dunes and artificial fills are also included in this deposits. Earthquakes mainly induce ground failures due to liquefaction.

The deposits of the type 2 are mainly surficial soil layers formed on slopes, debris, colluvium and/or landslide deposits formed through eolian fall, particle fall or transport, soil creep, soil flow, debris flow etc. during the Late Pleistocene and Holocene. Volcanic ash falls mostly contribute to formation of the surficial layers in Kyushu (Takeshita 1986) and probably most part of Japan. Eolian fall of loess from the northwestern part of China also seems to have important role to their formation during the Latest Pleistocene to the Early Holocene. Particle fall and soil creep becomes important in the Late Holocene, especially after occurrence of surficial slides in historic times.

Those deposits are important for landslides. However, they are not shown in Figure 12 because those deposits cover almost the whole slopes throughout Japanese Islands without nescessity to show in a small -scale map.

3 ECONOMIC EXTENT OF LANDSLIDES

3.1 Damage by landslides

Large damage of human lives and properties has been suffered in Japan by natural disasters for long times. Especially in one and a half decades after the World War II, we lost more than one thousand persons every year by the disasters caused by big typhoons and large earthquakes. Land devastation through clear cutting, digging out pine tree roots and wide cultivation on steep slopes during the War was thought to be related to those disasters. Large disasters had occurred until 1959 when the Isewan Typhoon caused the worst damage killing more than 5800 persons. However, casualties by natural disasters have been gradually decreasing from 1960 (Figure 17, Land Agency 1988). This decreasing in damage is dependent on very less attacks of big typhoons and large earthquakes and due to advancement of preparedness for those natural disasters including river bank reconstruction, improvement of meteorological observation system and development for antiquake structures.

However, ratio of death due to landslides and related phenomena has been rather relatively increasing and often very large among the whole casualties by natural disasters (Figure 18). The ratio kept above 50 % from 1967 to 1987. It reached to 83 % in 1984. Similar tendency can be also seen in an individual rainstorm disaster. For example, the 1982 Nagasaki heavy rainfall disaster killed 299 persons in which 89 % were due to landslides and debris flows (Oyagi et al. 1984).

Figure 19 shows annual damage cost caused by natural disasters from 1962 to 1986 in Japan. The cost includes several kinds of damages on facilities, institutions and establishments (Land Agency 1988). Its maximum value was 1800 billion yen (12 billion U.S. dollars) recorded in 1982, in which the Nagasaki disaster took great share. If the cost includes private

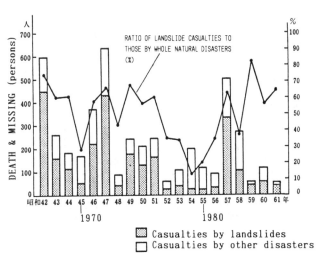

Figure 18. Rate of casualties by landslide disasters to those by whole natural disasters. (Land Agency 1988).

houses it is estimated about 1900 billion yen. Though it had been above 1000 billion yen from 1980 to 1983, it decreased below 900 billion yen since 1984. Its ratio to GNP (gross national product) had been gradually decreasing from 1.3 % to 0.25 % with large fluctuations from 1962 to 1986 (Figure 19). This decrease in the ratio is not necessarily apparent one, because it is roughly parallel to the tendency of decreasing in casualties discussed above.

The ratio of cost by landslide disasters is between 3 % to 10 % of water-related disasters from 1973 to 1984 (Ministry of construction 1987). During this term, annual damage cost by landslide disasters including debris flows in charge of the Ministry of Construction was about 14 billion yen and the maximum was 43 billion yen. However, lack of economic statistical data for landslide disasters prevents detail discussion on economic situation for large amount of budget of landslide control works in Japan. The ratio of disaster by debris flows among the whole landslide disasters was the largest from 20 % to 80 %. While in creeps or slow slides, that ratio was 2 % to 32 % and, in rapid slides, it took rather larger value 15 % to 50 %.

3.2 Cost for rescuing and restoration

3.2.1 Cost for rescuing

Outcome for rescuing has not been published because of complicated items in the cost. In national budget, 0.95 billion yen are appropriated for the rescuing cost in 1987 and 1.0 billion yen in 1988. Those consist of the cost for extra works of the Self Defence Force and

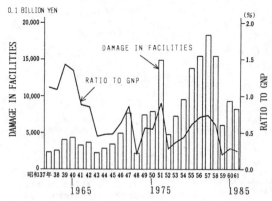

Figure 19. Annual damage cost of facilities for all kind of civilization from 1962 to 1986 in Japan. (Land Agency).

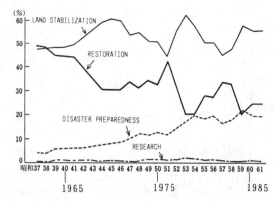

Figure 20. Ratio of restoration cost and other costs to the whole cost for disaster prevention. (Land Agency 1988).

the Ministry of Health and Welfare. Costs in police, fire fighting stations and other related organizations are not clear.

3.2.2 Cost for restoration

Soon after a big disaster restoration activities are usually carried out, for example reconstruction of roads, railways, water lines, river banks, bridges, river bed works, check dams, buildings, paddy fields etc. Those costs are proposed from various sections and/or divisions of city and prefecture offices, and payed from the reserve fund of the national and prefectural budget. The national government will pay in the high ratio the restoration cost of public facilities damaged by a severe disaster ("gekijin saigai" in Japanese) and a large disaster. The ratio ranges from 50 to 90 % and its criterion for identification is based on the relation between the total damage and the annual income of the local government which is concerned to the disaster.

The restoration cost appropriated for disasters is 44.7 billion yen in the national budget of the 1988 fiscal year. It was 39.3 billion yen in that of the 1987. The ratio of the restoration cost in the total cost for natural disasters has been gradually decreasing from 50 to 22 % with somewhat large fluctuations since 1962 (Figure 20). The ration of the latter , the total cost for the disasters, has been also decreasing from 8 to 4.6 % (Figure 21).

For recent landslide disasters, it reached 25 billion yen in 1982 when Nagasaki city was suffered the severe disaster caused by the heavy rainfall. It was 15 billion yen in 1985 when the Jizukiyama Landslide occurred in Nagano city, which killed 26 aged persons in a hospital.

3.3 Cost for control works

For designated landslides of the slow movement type (Jisuberi), control works have been done since 1958 when the Landslide Law was enacted. The total cost for those control works in the 1988 fiscal year is 92.64 billion yen (712.6 million US dollars), in which the cost of about 45 % is payed by the national government and the rest is payed from prefectural governments. Jurisdiction for those landslides is divided into three offices, the Ministry of Cnostruction , the Ministry of Agriculture and the Forestry Agency. Twenty landslides are directry treated by those offices of the national government. The cost of the control works for those landslides is 10.05 billion yen of which 60 % is payed out of the national treasury. The most other designated landslides are treated by offices of prefectural governments and the payment from the national treasury is 30 to 50 % of the cost of the control works.

The cost for a landslide ranges from 0.5 million yen to 8 billion yen for one year. Its maximum is 18.5 billion yen from 1962 to 1986 at Kamenose Landslide in Osaka Prefecture.

For rapid type landslides on steep slopes, the total cost of the control works for 2265 sites in 1987 was 59.4 billion yen. It reaches 26.2 million yen per landslide. The budget for the total cost in the 1988 fiscal year is about 69.8 billion yen. The great increase is due to the first year of the second five years' project. The half of those cost is payed by the national government. The rest of the cost should pay beneficiaries. This mechanism began from 1969 when the Law for prevention of disaster by steep slope failures was enacted. However, many people disliked designating the slopes behind their houses for control works because they should pay the half of the cost of the works and because they feared falling their land values. Recently, importance of disaster preparedness as been gradually understood by most of the people.

For debris flows and landslides of rapid type, two kinds of works are appropriated, that is "Sabo" (erosion control) and "Chisan" (forest and mountain slope stabilization). The budget of the former work was 233.7 billion yen in the 1987 fiscal year and that of the latter was 211.9 billion yen. The national government payed 61 % for the former and 63 % for the latter. Some part of the budget have been used for the control works for debris flows or rapid landslides. However, we have no detail data differentiating those costs.

3.4 Cost for warning and monitoring

Monitoring is usually done at a landslide for three purposes:
  1. To identify a moving zone on a wide slope.
  2. To confirm the effect of control works done at the landslide.
  3. To prepare warning for an unexpected occurrence of failure on the treated slope.
  Monitoring for identification of a moving zone is

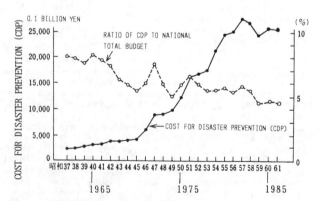

Figure 21. Ratio of the total cost for the prevention against natural disasters to the whole national budget. (Land Agency 1988).

300

generally done before designing the control works at a landslide or a landslide prone slope to know the geometry of the landslide exactly. Equipments for monitoring are mainly extensometers, tiltmeters of bubbled level type, pipe-straingages in a bore hole and/or inclinometers in a borehole.

For confirmation of effect of control works, several kinds of monitoring are carried out other than the observation using the equipments used for identification of a moving zone mentioned above, for example, triangulation, distance measurement by light wave survey, groundwater level observation, measurement of runout volume from drainage system, precipitation etc.

For warning of unexpected failures at the treated slope, observation of surface movement using extensometers is usually done. Those equipments have a simple mechanism to make alarm output from presetting the velocity at some value, for example 5 mm/hour. Recently, however, they are connected with a personal computer with a program to calculate time to the final failure at the slope based on the Saito's time prediction method for landsliding.

This type of monitoring can be applicable for any other hazardous slopes. Slide prone slopes along railways and important highways are monitored by this method. However, other slopes near houses are very rarely monitored.

Cost for monitoring ranges from only 80 thound yen of an extensometer, if a man try to set and observe it by himself at a small slope, to 40 million yen at an important landslide treated by several control works. As a special case, the Jizukiyama landslide is monitored by the cost of 127 million yen which include monitoring for landslide movements by 49 extensometers, 14 tiltmeters, 59 pipe-straingages etc., monitoring for deformation of such control work structures as cast-in-pipe foundations, anchors, steel piles etc., and autommatic data analysis by computers (N.Yamaura, personal communication).

4 DISCUSSION

How will be the landslide disasters in future? Will they increase or decrease? How will they change in characteristics? The occurrence of landslide phenomena has been thought perfectly natural phenomena as a kind of erosion. It is partially true but it has been often related to human activity. The 18 cases of 25 landslides which caused large disaster killing more than 10 persons from 1947 to 1984 were related more or less to such human activities as cutting at the lower part of slopes, filling at the upper part of the slopes, underground mining just beneath the slopes, clear cut of natural forests etc.(Hatano and Oyagi 1988). This fact means that occurrence of landslides will increase through development on slopes in several countries if stabilization measures will not be appropriate in site condition and for long term stability. Control works using man made materials, for example steel, concrete, plastics etc. have limited life time. Instability of a slope will be accelerated by an action of a large destabilizing input above some threshold. Weakening of slope material will develop faster on a cut slope than a natural slope (Okuzono 1983). Opportunity to develop slopes will increase in several decades in many other countries, especially in mountainous countries with high population density as those along island arc zones similar to Japan. This forcast tells increase of landslides and their disasters in those countries. The small-scale landslides will have important meaning for disaster producers as well as large-scale landslides.

Can we mitigate the damage caused by the disasters? And can we perfectly prevent occurrence of landslides? It is difficult to stabilize all landslides perfectly. However, we are able to mitigate landslide disasters through various measures including control works, site and time prediction, warning and evacuation system, information networks and selection of better landuse. Those measures should be applied with good timing effectively. For new development, planning of landuse seems the most important.

For the better choice of mitigation measures, it is desired to accelerate the development of landslide sciences and technology. Mechanisms or organizations for application of several measures are also needed. Lows to support those mitigation measures for landslide disasters are necessary as "Landslide Law" in Japan.

The most important problem for landslide scientists will be what directions landslide sciences must be developed. Those are considered as follows:

1. To know exactly distribution and characteristics of landslides and related phenomena and disasters caused by them.
2. To understand those distribution and characteristics from the geological and geomorphological point of view.
3. To clarify origin, cause and mechanism on landslides and related phenomena.
4. To estimate short and long term effect of human activity on slopes.
5. To estimate state of instability or stability of whole slopes in the area concerned.
6. To find out relevant landuse in the area from the viewpoint of mitigation of landslide disasters.

For items 1 and 2, discussions in the early part in this paper will provide a good example. Geological zonation (based on lithology) for landslides will be used for prediction of sites and characteristics of landslides and landslide disasters. Similar zonation will be applicable for other island arc zones. In alpine zones or continental areas, zonation will be fairly different from that of the island arc zones but methodology will be still similarly applicable.

The items 3 to 5 are beyond the scope of this paper. However, those problems are not only essential for scientific interests but also important to take reliable countermeasures for ladnslide disasters. Studies should be intensified on these problems.

Landuse, the item 6, is not necessarily simple. Though it is thought to be purely social problem, but it is largely affected by natural condition, that is landform, soil condition, geologic constitution, climate and biological environment, and material movement including landslides, floods and other phenomena. Improper landuse has been often forced to pay much cost to restorate landslide disasters. For example at the Jizukiyama landslide and its surrounding areas, a toll road was constructed and a hospital for aged men and many houses were built. Unfortunately, the developer organization had little knowledge on the antient landslide landform at the site for development. As a result, those were destroyed by the 1985 landslide disaster and the restoration cost to stabilize the landslide and to reconstruct the housing area should be payed about 15 billion yen from 1985 to 1988. It tells importance of careful planning precedent to new development or change in landuse.

5 CONCLUSION

Landslides and landslide landforms distribute throughout Japanese Islands. They cause very often large disasters. The ratio of casualties by landslide disasters among the whole natural disasters often reaches 90 % recently in Japan. Economic loss by landslide disasters has been also very large. It is estimated from 30 to 6000 billion yen per year for the recent decade.

The ratio of casualties by landslide disasters to those by natural disasters has been rather increasing for the recent two decades, though the latter have been decreased. This fact suggests difficulty in effective prevention measures for landslide disasters in comparison with such other disasters as flood, wind and earthquake disasters (in this case, those by earthquake-induced landslides are included in landslide disasters). The reason for the difficulty is dependent on several conditions as follows:

1. Instable slopes are infinitive in number.
2. Factors related to slope instability are various
3. Shreshold level of each factor is various in strength and working duration and in combination of other factors acting the slope concerned.
4. Landuse on slopes is various.
5. Effect by control works on a slope is relatively low in cost-benefit point of view.

To advance mitigation measures for landslide disas-

ters, studies pointed out in the precedent chapter are
important. One of the basic approaches for those problems is geological zonation as discussed in the chapter 2. Similar type of geological zonation to Japanese
Islands will be applicable in other island arc zones.
Or similar approach will be useful in alpine or continental areas.

Acknowledgement: I would like to express my sincere
thanks to Dr. Earl E. Brabb for encouragement to me
writing this article and also to Dr. Y.S. Shin for
collecting and giving me data of landslide disasters
in Korea. I am grateful to the members of "the Commit-
ee on Terminology in Geology and Geomorphology for
Landslides" of the Japan Landslide Society for discussion and especially to Dr. Kazuo Kuroda for his
courtesy to use his original map.

REFERENCES

Akutagawa, M. & H.Kazama 1977. Formation of potential
failure surface on the slope of the Narita Formation
, 14th Nat.Dis.Sci.Symp., p.305-308.*
Central Meteorological Office (Korea) 1984. Annual
climatological report. C.M.O.,Seoul, Korea.
Depertment of Erosion Control, Ministry of Construc-
tion 1987. Sabo Binran (Notes for Erosion Control).
Enoki, M., N.Yagi, R.Yatabe & K.Kunitomi 1987. Stabi-
lity analysis on landslides in "Fractured zone".
Proc.26th Conf.Japan Landslide Soc. p.224-227.*
Furuya, T. 1982. On some characters of the so-called
fractured zone type landslides. J.Japan Landslide
Soc. 18,4:54-58.
Fujita, T. 1978. Geological characteristics of land-
slides of the crystalline schist type in Southwest
Japan. Proc.IIIrd Int.Cong.IAEG. Sec.1,1:278-288.
Fujita, T. 1985. Geological zonation of landslides in
Japan, with reference to the landslide regions in
the Kinki District. Proc.IVth ICFL. p.79-84.
Hatano, S. & N.Oyagi 1988. Photo-geomorphic clue to
detection of ladnslide-prone slopes, drawn from no-
table disasters in postwar Japan. 16th ISPRS, Kyoto,
Comt.VII, W.G.5, T26, Brief Paper for Poster Session
Iwamatsu, A. & E.Shimokawa 1986. Creep-type large-
scale ladnlsides of well-cleaved argillaveous rocks.
Mem.Geol.Soc.Japan, 28:67-76.
Japan Landslide Society 1980 & 1988. Landslides in
Japan.
Japan Meteorological Agency 1984. Climatic atlas of
Japan.
Koide, H. 1955. Landslides in Japan. Tokyo Keizai-
shimposha.*
Kuroda, K. 1980. Landslide zonation in Japan. Oral
presentation at the IInd ICFL, Tokyo.
Kuroda, K., N.Oyagi & H.Yoshimatsu 1982. Geological zo-
nation of landslides in Japan. J.Japan Landslide
Soc. 18,4:17-24.
Kuroda, K. 1986. Geological zonation of landslides in
Japanese Islands, in Mem.Geol.Soc.Japan. 28:13-19.
Land Agency 1985 & 1988. Bosai Hakusho (Disaster Pre
vention, White Paper).
Lee, D.S. (ed) 1987. Geology of Korea. Geological
Society of Korea, Kyohak-Sa.
Machida, H. 1966. Rapid erosional development of moun-
tain slopes and valleys caused by large landslides
in Japan. Geogr.Rev.Tokyo Metropol.Univ., 1:55-78.
Machida, H., T.Furuya, S.Nakamura & I.Moriya 1986.
Large-scale landslides in Japan. In Scale, type, oc-
currence frequency of landslide disasters and re-
lated groundwater movement, S.Shindo (ed). 165-184.*
Matsukura, Y. 1980. On the physical and dynamic properties of some soils distributed in the vicinity of
Mt.Tsukuba. Geogr.Rev.Japan, 53,10:54-61.
Matsukura, Y. & Y.Tanaka 1983. Stability analysis for
soil slips of two gruss-slopes in Southern Abukuma
Mountains, Japan. Tran.Japanese Geomor.Union, 4,2:
229-239.
Ministry of Construction 1987. Sabo-binran (Records
on erosion control works).
Okuda, S. & K.Yokoyama 1977. Stream systems and landslides. In S.Tanaka (ed) Studies on the relation
between landslides and geologic and geomorphic
structures. p.16-22.*
Okuzono, M. 1983. Design and maintainace of cut

slopes. Kashima-shuppan Co.*
Oyagi, N. 1968. Weathering-zone structure and land
slides of the area of granitic rocks in Kamo-Daito.
Shimane Prefecture. Rep.Coop.Res.Dist.Prev., 14:113-
127.
Oyagi, N., M.Oishi & T.Uchida 1970. Structural factors
of the Washiodake landslide in the Hokusho Region,
Northwest Japan. Rep.Coop.Res.Dist.Prev.,22:115-140.
Oyagi, N. 1976. Landslide disasters in granitic rock
regions. Proc.13th Nat.Dis.Sci.Sym.,p.231-232, and
supplement maps.*
Oyagi, N., F.Shimizu & T.Inokuchi 1982. Characteristic
distribution of landslides and their relation to
geology in Tohoku District, Northern Japan. J.Japan
Landslide Soc., 18,4:34-38.
Oyagi, N., K.Nakane & T.Fukuzono 1984. Report of in-
vestigation of the disasters caused by the July,1982
rainstorm in Nagasaki City and its suburbs. Natural
Disaster Research Studies, 21.*
Oyagi, N. 1984. Landslides in weathered rocks and re-
sidual soils in Japan and surrounding areas. Proc.IV
ISL. Toronto, 3:1-31.
Oyagi, N. 1987. Geological zoning for landslides.
Proc.US-Asia Conf.Eng.Mitigating Nat.Haz.Damage.
Bangkok,p.C9-1 - C9-14.
Sassa, K. & A.Takei 1977. Consider vertical subsidence
in slope unstabilization II. J.Japan landslide Soc.
14,3:7-14.
Sassa, K. 1985. The geotechnical classification of
landslides. Proc.IVth ICFL,p.31-40.
Shimizu, F. 1985. Disstribution characteristics of
large-scale landslides in the Tohoku District North-
east Japan. Proc.IVth ICFL, Tokyo. p.489-492.
Shimizu, F., N.Oyagi & T.Inokuchi 1982. Landslide maps
, Part 1. Shinjo area, 16 maps.
Shimizu, F., N.Oyagi & T.Inokuchi 1984. Landslide maps
, Part 2, Akita, Oga & Sakata Area, 23 maps.
Shimizu, F., N.Oyagi & T.Inokuchi 1985. Landslide maps
, Part 3, Hirosaki & Fukaura areas, 18 maps.
Shimizu, F. & N.Oyagi 1986. Landslide maps, Part 4,
Murakami area, 11 maps.
Shimizu, F. & N.Oyagi 1987. Landslide maps, Part 5,
Aomori & Sendai. 28 maps.
Shimizu, F. & N.Oyagi 1988. Landslide maps, Part 6,
Fukushima, Aikawa and Nagaoka (Sadogashima) areas,
22 maps.
Takeshita, K. 1985. Processes of slope and soil forma-
tions on steep mountain slopes. Tran.Japan Geomor.
Union,6,4:317-332.
Terado, T. 1975. Large landslides in the eastern part
of Shikoku. Rep.Anan Tech.Col., 11:91-100.
Terado, T. 1986. The distribution of landforms caused
by large-scale mass movement on Shikoku Island and
their regional characteristics. Mem.Geol.Soc.Japan,
No.28,p.221-232.
Uemura, T. 1975. Classification and prediction on
landslides. Rep.Nat.Dis.Spc.Res., A-50-6, p.3-12.*
Uemura, T. 1982. Geologic diagnosis of landslides in
Niigata Prefecture. J.Japan Landslide Soc., 18,4:39-
43.
Varnes, D.J. 1958. Landslide types and processes. In
Landslides and engineering preactice. E.B.Eckel (ed)
.. H.R.B. Spc.Rep., 29, p.20-47.
Varnes, D.J. 1978. Slope movement types and process-
es. In Landslides: analysis and control. R.L.Schus-
ter & R.J.Krizek (ed), T.R.B..N.A.S. Spc.Rep., 176
p.11-33.
Woo, Bo Myeong 1984. Landslide disaster countermeas-
sures in Korea. J.Korean Forestry Soc. 63:51-60.
Yatsu, E. 1965. Sur la classification des glissements
de terrain. Earth Science, 76:34-37.
Yoshida, S. 1981. Chert-laminite:Its petrographical
description and occurrence in Japanese geosynclines.
J.Geol.Soc.Japan,87,3:131-141.

The asterisk * means the title of the paper is translated by the present author from Japanese tittle without English one.

Landslides: Extent and Economic Significance, Brabb & Harrod (eds)
© 1989 Balkema, Rotterdam. ISBN 90 6191 876 6

Occurrence and significance of landslides in Southeast Asia

E.W.Brand
Geotechnical Control Office, Hong Kong

ABSTRACT: This paper reviews the landslide situation in the Southeast Asian countries of Hong Kong, Indochina (Cambodia, Laos, Vietnam), Indonesia, Malaysia, the Philippines, Singapore, Sri Lanka, Taiwan and Thailand, all of which have common problems with failures caused by heavy tropical rainfall in steep residual and colluvial terrain. Additionally, Indonesia suffers from 'lahars' on the slopes of active volcanoes. Landslides are of appreciable significance in all the countries except Singapore, and they are of major importance in Hong Kong, Indonesia and Taiwan. The state-of-the-art of landslide mitigation, however, is assessed as being very high only in Hong Kong, although it is improving steadily in Malaysia and Taiwan in step with economic development.

1 INTRODUCTION

This paper reviews the situation with regard to landslides and their significance in the countries of Southeast Asia. It is based heavily on earlier state-of-the-art reports by the Author (Brand, 1984, 1985a), and it may be regarded as an update of those reports to December 1988.

For historical reasons associated with the foundation of the Southeast Asian Geotechnical Society, Southeast Asia is defined here as comprising Hong Kong, Indochina (Cambodia, Laos, Vietnam), Indonesia, Malaysia, the Philippines, Singapore, Sri Lanka, Taiwan and Thailand. The map in Figure 1 shows the geographical locations of these countries. The total land area of these countries, which is almost 4 million sq. km, straddles the equator and is contained almost entirely within the tropics. Apart from Indochina, Thailand and West Malaysia, the countries together comprise many thousands of islands which stretch for more than 6 000 km from east to west and more than 3 500 km from north to south. The total population of the area is about 400 million.

The Southeast Asian region is physiographically and geologically as complex as any area of the world (Bemmelen, 1949). For this reason, combined with the vast size of the land area and the paucity of relevant published literature, a landslide review of the region can only be superficial. Notwithstanding this fact, it is possible to provide a unified review of a few major aspects of the landslide situation of the whole area. With the notable exceptions of some large deltaic plains, the majority of the terrain of the total land area of Southeast Asia is hilly or mountainous. The mainly warm, wet climatic conditions have resulted in varying depths of weathering of a wide range of igneous,

metamorphic and indurated sedimentary rocks, to give profiles which grade from residual soils at the surface through to unweathered bedrock at depth. The footslopes of the hills are commonly carpeted with layers of colluvium. The landslide problems in the region are therefore confined virtually entirely to failures in residual profiles and colluvium, with the notable exception of the 'lahars' which occur on volcanic slopes in Indonesia.

The region experiences high seasonal rainfalls, sometimes in excess of 5 500 mm annually, with intensities that can exceed 150 mm per hour, and rainfall is the direct cause of the majority of landslides that occur. In addition, a large part of the Southeast Asian archipelago is in the Pacific earthquake belt, which runs through much of Indonesia and the Philippines and skirts the tip of Taiwan (Figure 2). Landslides in those three countries are therefore sometimes associated with seismic activity. In addition, Indonesia is plagued by active volcanoes which play a major part in the country's severe landslide problems.

The term 'landslide' is used in this paper to describe any type of slope failure of significance. The types of failure which occur in Southeast Asia vary greatly. They include failures of natural slopes and of man-made cuttings and embankments, and they range in size from a few tens of cubic metres to millions of cubic metres of earth materials.

The significance of landslides in the Southeast Asian countries varies greatly in terms of human casualties, economic loss and disruption of communications. Unfortunately, no reliable statistics are available as to the casualties or damage caused by landslides in each of

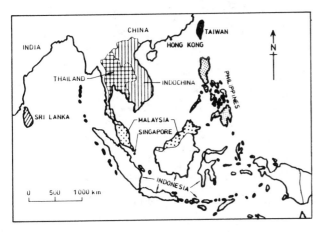

Figure 1. Map showing Southeast Asian countries

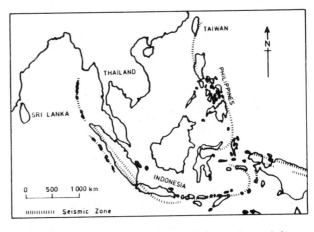

Figure 2. Pacific earthquake belt in Southeast Asia

the countries. UNESCO publishes an annual report which purports to give these statistics, but the data from this source are far from reliable. There are three main reasons for this deficiency. Governments rarely keep accurate central records of casualties or damage from natural disasters, and when they do, 'landslides' is commonly contained within some broader disaster category such as 'typhoons' or 'earthquakes'. In addition, political and social factors sometimes result in distortions in reported casualty and damage figures. It is certain, however, that a few thousand people are killed and injured annually by landslides in Southeast Asia, and very extensive damage is regularly done to property and communications.

The earlier papers by the Author (Brand, 1984, 1985a) are the only publications which have attempted to cover the scope of the subject matter dealt with in this review. The three volume treatise by Bemmelen (1949), however, is worthy of mention as an outstanding work of reference on the geography and geology of the Southeast Asian archipelago, which also gives a small amount of information on landslides caused by rainfall, earthquakes and volcanic activity in the region. Also worth mentioning is the bibliography on landslides in Asia published recently by AGE (1987).

Each of the Southeast Asian countries is reviewed separately in this paper, except that Cambodia, Laos and Vietnam are dealt with collectively under the heading 'Indochina' because of the tiny amount of information available on these three countries. In addition, some general guidance is given on the published engineering methods that are applicable to the tropically weathered rock profiles of the region.

2 LANDSLIDES IN HONG KONG

The Territory of Hong Kong has a land area of only 1 033 sq. km and a population of nearly six million. It consists of Hong Kong Island, the Kowloon peninsula and the 'New Territories' (Figure 3). The combination of hilly terrain and high seasonal rainfall gives rise to severe landslide problems which, in such an intensely urban environment, are unique in Southeast Asia and almost unparalleled anywhere. There is now a very extensive body of published literature relevant to the landslide problems of Hong Kong, as evidenced by a recent comprehensive bibliography (Brand, 1988a). The papers by Lumb (1975, 1979) summarised the landslide situation in Hong Kong as it existed in the early 1970s, and the paper by Brand (1985d) gave a fuller and more recent picture of the state-of-the-art of landslide mitigation in the Territory.

The natural terrain of the majority of the Territory is very hilly. On Hong Kong Island and Kowloon, virtually no naturally flat land exists. The land rises steeply on Hong Kong Island to over 550 m in a distance of about 1.5 km from the sea; the narrow strip of flat land along the north shoreline has almost all been reclaimed from the sea. The once hilly peninsula of Kowloon has now mostly been levelled to provide material for reclamation, but isolated hills of up to 100 m in height still exist. Even in the New Territories, there is little low-lying land, peaks of more than 400 m being common. Natural slopes throughout the Territory are steep, more than 60% of the land area being steeper than 15° and about 40% being steeper than 30°.

The geology of Hong Kong, which is summarised in Figure 3, has been described by Ruxton (1960) and Allen & Stephens (1971), and the engineering geology has been summarised by Lumb (1975), Brand & Phillipson (1984) and McFeat-Smith et al (1989). The main rock types are granite and acid volcanic rocks, which together cover the major portion of the Territory and are by far the most important from an engineering point of view. The small amounts of sedimentary and metamorphic rocks are of much less importance, although some specific landslide problems have been associated with these. Granite predominates in those areas of the Territory where building development is densest. The granite is extensively weathered almost everywhere, depths of up to 60 m of silty-sandy residual

soil (grades IV to VI) being common, with large corestones in the matrix or exposed on the surface (Lumb, 1975, 1983). The volcanic rocks consist mainly of tuffs and rhyolite, with a residual soil mantle of up to 20 m thick. About 20% of the land area is carpeted with colluvium, up to 30 m thick in places; this is prone to the formation of 'pipes' or 'tunnels' as a result of internal erosion (Brand, Dale & Nash, 1986), and these features can be of major significance to the hydrogeology of an area (Leach & Herbert, 1982).

Excellent rainfall data is available in Hong Kong, hourly records having been kept at the Royal Observatory on the Kowloon peninsula since 1884, except for the period 1940 to 1946, and over a number of years at many other locations. The average annual rainfall is 2 225 mm, with a minimum recorded figure since 1884 of 901 mm (in 1963) and a maximum of 3 248 mm (in 1982). Nearly 80% of the rain falls during the period May to September. Rainfall intensities can be high, with 24-hour rainfalls of more than 250 mm and one-hour rainfalls in excess of 50 mm occurring fairly frequently. The maximum 24-hour rainfall recorded at the Observatory is 697 mm (in 1889) and the maximum one-hour intensity recorded at any location is 157 mm (in 1966). A sophisticated system of automatic gauges has enabled excellent correlations to be obtained between rainfall and the occurrence of landslides (Brand, Premchitt & Phillipson, 1984; Brand, 1985b). This has shown that short-term rainfall intensity is the critical factor, with a 'trigger' value of 70 mm/hr, irrespective of antecedent rainfall.

On average, several hundred failures occur in Hong Kong each year (So, 1971, 1976), but most of them are small or occur in undeveloped terrain, and the consequences are therefore light. All too frequently in the past, however, there have been severe consequences in terms of casualties and damage. In fact, a significant landslide event, in which a large number of failures occur in one day causing considerable disruption and damage, can statistically be expected to take place in Hong Kong about once every two years (Lumb, 1975, 1979; Brand, 1985b). The whole range of slope 'features' is prone to landslides, including natural slopes, soil cut slopes, rock cut slopes, earth fill slopes, retaining walls and boulders. The majority of failures, and usually those with the most severe consequences, take place in man-made features or are triggered by man-made features, particularly cut slopes in soil (grades IV to VI). Case histories of failures have been presented in many publications (e.g. Lumb, 1975; Hencher & Martin, 1984; Hencher et al, 1984; Hudson & Hencher, 1984; Brand, 1985b, 1985d; Irfan et al, 1988).

Cut slope failures in soil (grades IV to VI), or in mixed

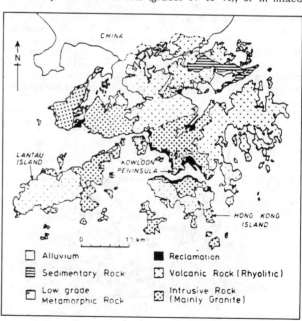

Figure 3.   Geological map of Hong Kong

soil and rock now constitute by far the most common form of landslide. The volcanic rock profiles are more susceptible to failure than the granite profiles, and colluvium is frequently involved. The failures nearly always occur suddenly during intense rain without prior warning, but there are a few reported cases of large failures taking place slowly (Lumb, 1975; Brand, 1985b, 1985d). Most slip surfaces are shallow, the thickness of the failed zone usually being less than 3 m. Relict joints and other discontinuities are often important in dictating the mode of failure in soil slopes (Hunt, 1982; Koo, 1982a, 1982b; Irfan & Woods, 1988). Cut slope failures in rock (grades I to III) occur much less frequently than in soil slopes, but such joint-controlled failures do occur, and they can involve fairly large volumes of material (Beattie & Lam, 1977; Brand, Hencher & Youdan, 1983). Some of the rock slopes in Hong Kong are particularly high and steep.

There are good historical records in Hong Kong which indicate that severe landslide events occurred as long ago as 1889, but the severity has generally increased with time as the Territory has become more densely populated and as the amount of hillside cutting has increased. The earliest well-documented landslide disaster is that which occurred at Po Hing Fong in the Mid-levels district of Hong Kong Island on 17 July 1925. A high masonry retaining wall on the steep slope collapsed onto a row of terraced multi-storey dwelling-houses, completely destroying five of them and killing 75 people. There had been heavy rain for three days prior to the failure, and 126 mm of rain was recorded at the Observatory in the two hours immediately before failure occurred. There was a great deal of flooding in the low-lying parts of Hong Kong Island and Kowloon, and a large number of cut slope failures occurred.

Since fairly intensive building development began on slopes in the 1950s, there have been some very severe and even disastrous landslide events. Those that occurred between 1950 and 1976 have been summarized by Lumb (1975, 1979), and this record has been extended recently by Brand, Premchitt & Phillipson (1984) and Brand (1985b). There were significant landslide events, often with high numbers of casualties, in 1951, 1952, 1955, 1957, 1960, 1964, 1966, 1968, 1971, 1972, 1976, 1978, 1982 and 1983. Of particular severity and significance were the events of 1966, 1972,

1976 and 1982, and these are worthy of special mention.

On 12 June 1966, a trough of low pressure brought rains which caused more damage and disruption than any rainfall event in living memory. After several days of continuous rain, 401 mm was recorded at the Observatory in a 24-hour period from the morning of 11 June, of which 108 mm fell in one hour on the moring of 12 June (Chen, 1969). At other locations, the comparative figures reached 525 mm and 157 mm. There were hundreds of landslides (So, 1971), 180 of which occurred on Hong Kong Island alone, and there was extensive flooding in coastal areas. Although there was no single large landslide that caused many deaths, a total of 64 people were killed. Virtually every main road was blocked, and many buildings and other facilities were badly damaged.

The most disastrous landslide event in Hong Kong's history occurred on 18 June 1972. More than 650 mm of rain fell at the Observatory during the three days 16 to 18 June, 280 mm being recorded in one 24-hour period; as much as 560 mm was measured in 24 hours elsewhere. There were many landslides, mainly on Hong Kong Island, which caused a total of 250 casualties (Government of Hong Kong, 1972a, 1972b). The majority of the dead and injured, however, were involved in the two major landslides at Po Shan Road in the Mid-levels district of Hong Kong Island, and at Sau Mau Ping in northeast Kowloon. On the steep natural colluvial hillside above Po Shan Road, a 120 m long and 67 m broad failure occurred on a surface about 10 m deep (Government of Hong Kong, 1972b; Vail, 1984). This demolished a four-storey building and a 13-storey apartment block, killing 67 people (Figure 4). At the Sau Mau Ping housing estate, a 35 m high fill slope liquefied under the intense rain and inundated the single-storey dwellings in the village area at the foot of the slope (Figure 5), killing 71 people (Government of Hong Kong, 1972a, 1972b; Vail, 1984; Vail & Beattie, 1985).

Hundreds of landslides also occurred on 25 August 1976 after a record 416 mm of rain had fallen at the Observatory. 57 people were killed or injured. Yet another fill slope failure occurred at Sau Mau Ping, this time killing 18 people (Morgenstern, 1978, Vail & Beattie, 1985). The resulting report into the disaster (Government of Hong Kong, 1977) recommended the establishment of a government organization to rectify existing unsafe slopes and to control the designs of new ones. As a result, the Geotechnical Control Office was established in July 1977.

The rainfall for 1982 (3 248 mm) was the highest recorded since records began in 1884. Two severe landslide events took place on 29 May and 16 August, of which the first

Figure 4.   Po Shan Road disaster of June 1972

Figure 5.   Sau Mau Ping disaster of June 1972

was by far the worst (Brand, Premchitt & Phillipson, 1984; Brand, 1985d). About 1 500 failures were observed from aerial photographs after the event, which occurred during a period in which 394 mm of rain fell in 24 hours at the Observatory and up to 440 mm elsewhere, with a maximum intensity of 111 mm in one hour. The vast majority of the failures were fortunately small and of little consequence. There was no major single disaster, but 48 casualties resulted from many small landslides in squatter villages on steep slopes.

Slope design practice in Hong Kong has improved radically over the past decade. The engineering methods applied to slopes are now governed largely by the Geotechnical Manual for Slopes (Geotechnical Control Office, 1984). Although this is intended to be a guidance document only, it has considerably more influence than might be supposed, because it forms the basis on which slope designs are checked for adequacy by the Geotechnical Control Office. The recommended approach to the stability assessment of soil slopes is that of 'classical' limit equilibrium analysis by means of one of the proven methods for non-circular surfaces (e.g. Janbu, 1954, 1973; Morgenstern & Price, 1965), together with an appropriate factor of safety. Because of the steep natural terrain of much of the land area, it would be unrealistic and enormously costly for high factors of safety to be used for slope design purposes. These are therefore kept to the minimum possible values with due regard for the risk involved in any particular situation. The Geotechnical Manual specifies the appropriate factor of safety in relation to the likely loss of life or economic loss in the event of a failure; it ranges in value from only 1.4 for high risk situations to 1.2 for low risk ones.

Hong Kong has a very effective system of geotechnical control over slope safety, the responsibility for which is vested in the Geotechnical Control Office (GCO). The GCO is concerned with a wide range of geotechnical engineering activities related to the safe and economic utilization and development of land, with particular emphasis on the stability of existing and future slopes associated with both buildings and engineering works. It is responsible for investigating the stability of existing slopes, for designing and executing landslip preventive works to public slopes and for making recommendations on the need for preventive works to private slopes. It exercises geotechnical control over public and private developments by checking the geotechnical aspects of the designs of works and the standards of site supervision.

Designs of new slopes and site formation works are checked by the GCO on the basis of the recommendations made in the Geotechnical Manual for Slopes (Geotechnical Control Office, 1984), and the same checking processes and high standards are applied to the public and private sectors alike.

Several years ago, the GCO completed a major exercise to catalogue the almost 9 000 existing slopes and retaining structures in Hong Kong, and to place them into ranked order of priority for the purposes of investigation, with a view to establishing the need for preventive works in each case. A computer program was devised to sort and rank the slopes and walls on the basis of a number of simple parameters, such as height, slope angle, geology, proximity and use of the nearest building, and general slope condition (Brand, 1988b; Koirala & Watkins, 1988). This ranked list is now used as the basis for the Office's ongoing programme of geotechnical studies and landslide preventive works, on which about US$10 million is spent annually.

A particularly important geotechnical study was carried out during 1979 to 1981 to assess the overall stability of the Mid-levels area, which has a long history of slope failures (including the Po Shan Road disaster) and retaining wall failures (including the Po Hing Fong disaster). Pending the outcome of the in-depth study, a moratorium was imposed on building developments in the area during the period May 1979 to June 1982, at which time new legislation was enacted to enable more stringent geotechnical controls to be placed on all future site development and foundation designs and works in the area, particularly to prevent excessively deep excavations in the hillsides. The basic data obtained from the study in respect of geology, hydrology and soil properties has been published in

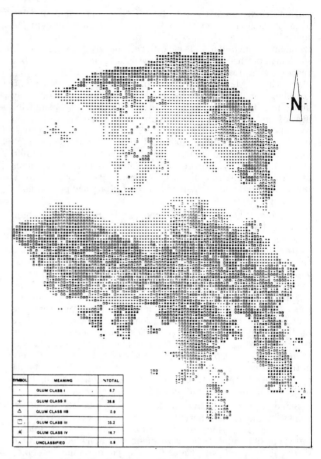

SYMBOL	MEANING	% TOTAL
.	GLUM CLASS I	8.7
+	GLUM CLASS II	38.8
△	GLUM CLASS IIB	0.0
□.	GLUM CLASS III	35.2
✳	GLUM CLASS IV	16.7
^	UNCLASSIFIED	0.8

Figure 6.   Computerised Geotechnical Land Use Map

summarized form by the Geotechnical Control Office (1982b) and by Rodin et al (1982).

A great deal of effort is expended by the GCO in providing adequate geotechnical input to the process of land use planning (Burnett & Styles, 1982; Burnett et al, 1986). For this purpose, a sophisticated system of terrain evaluation has been applied to the whole Territory, as described in detail by Brand, Styles & Burnett (1982), Styles et al (1984) and Brand (1988b). As part of this, a computerised data base (GEOTECS) has been devised (Styles et al, 1986), from which derivative maps can be readily produced (Figure 6). This information is available to the public in a series of twelve comprehensive published reports (Styles & Hansen, 1989). A number of guidance documents for geotechnical engineering practice are also available (Geotechnical Control Office, 1982a, 1982b, 1984, 1987, 1988).

The whole range of engineering methods is employed for constructon in Hong Kong's steep terrain, where the design of the site formation works is often the most challenging part of a project. The site formation design practices adopted, and the difficulties that these incur, have been discussed by Flintoff & Cowland (1982). All possible types of slope remedial measures have also been adopted. The cheapest one for a particular job is of course favoured, but this generally involves cutting back the slope, which is often not possible because of the restricted site area or because of the loss of valuable land. A large number of retaining walls are therefore used, and drainage measures are also common. Sub-surface horizontal drains up to 100 m long have been applied to a few situations (Craig & Gray, 1985), as have drainage tunnels.

Although not complacent about the landslide situation, Hong Kong has made huge strides over the past ten years in alleviating the consequences of the extreme rainfall events to which the Territory is subjected about once every two years. The standards of engineering practice have improved radically, and fairly stringent control measures are employed to ensure that the probability of failure of a newly designed slope or retaining structure is very low.

In addition, the government's longterm programme of landslip preventive measures, on which about US$100 million has been spent since 1977, is gradually rectifying unsatisfactory slopes and walls constructed in the past.

The standards of engineering practice in Hong Kong are thought to be very high, and significant progress has been made towards an understanding of the mechanisms of rain-induced landslides, although there still remain significant areas of technical doubt that will take many years to resolve. However, on the basis of what has been achieved, the high level of expertise available, the general awareness of landslide problems, and the effort expended on their alleviation, the state-of-the-art of landslide prevention and control in Hong Kong can be claimed to be very high relative to most countries of the world.

3 LANDSLIDES IN INDOCHINA

'Indochina' is the English form of the name used by the French for their former colony between the eastern border of Thailand and the South China Sea (Figure 1). It comprises the three countries of Cambodia (formerly Kampuchea), Laos and Vietnam, which together have a land area of more than 752 000 sq. km and a population of about 71 million, the majority of whom occupy the coastal areas of Vietnam. The southern portion of Indochina (including all of Cambodia) consists largely of the huge deltaic plain of the Mekong River, on which the capital of Cambodia (Phnom Penh) and the former capital of South Vietnam (Saigon) are situated. The northeastern portion, which abuts the Gulf of Tongking, consists of the large deltaic plain of the Red River, on which the capital of Vietnam (Hanoi) is sited. A mountain chain runs from northwest to southeast, with elevations of up to 2 000 m, and covers the majority of the land area of Laos and much of Vietnam. The capital of land-locked Laos (Vientiane) sits on a plain next to the Mekong River on the northeast border of Thailand. There is also a small range of mountains (Candomones) along the southwest coast of Cambodia that continues into Thailand. Sedimentary rocks, mainly sandstones, limestones and shales, predominate throughout Indochina. Rainfall varies greatly between the limits of about 1 900 mm per year on the east coast to more than 5 000 mm in some mountainous regions, and the vast majority of this falls between May and September; no data is available on rainfall intensity.

A great deal of published information is available on the geology of Indochina, mostly from the French colonial period. Excellent summaries of the geology have been published by Workman (1977) and Fontaine & Workman (1978), and a conference on the geology of Indochina was held fairly recently (Department of Geology, Vietnam, 1986). Unfortunately, however, the Author's enquiries have led to not a single publication that deals in any way with landslides in any of the three countries, and the Author can therefore only attempt to surmise in the vaguest way about the occurrence of landslides and their economic significance.

The vast majority of the people of Indochina live on the flat lands of the Mekong and Red Rivers and along the eastern coastal strip. The mountainous areas are very sparsely populated, with less than 4.5 million people occupying the whole of the 247 000 sq. km area of Laos. Rural areas in mountainous regions are very poorly developed in all the countries, and roads and railways are few. In the wet season, it is certain that some landslides occur in natural slopes in mountainous areas as part of the normal geological evolution. In addition, there will be the customary failures of cuttings on rural roads. To the Author's knowledge, there have been no news reports of landslide disasters, and it seems unlikely that any occur. Therefore, it must be concluded that landslides are of little significance in Indochina with its present low level of economic development. In time, however, the significance of landslides will increase considerably with the construction of more roads through hilly terrain and the improvement in facilities which always accompanies economic development.

4 LANDSLIDES IN INDONESIA

Indonesia is a vast country which stretches in an island chain that sits astride the equator for more than 5 000 km from east to west (Figure 1). The main land masses of Sumatra, Java, Kalimantan, Sulawesi and Irian Jaya constitute the major portion of the total land area of more than 1.9 million sq. km, but there are thousands of smaller islands in addition. The total population of more than 163 million is very unevenly spread throughout the country, in that about 98 million people inhabit the single island of Java, which has an area of only 132 000 sq. km, making it one of the most densely populated areas in the world. In contrast, the neighbouring island of Sumatra has only 29 million people occupying an area of about 474 000 sq. km.

Java and Sumatra (Figures 7 & 8) are by far the most important parts of Indonesia from an economic point of view, and they together contain the vast majority of the population and virtually all the main cities. In this respect, Java is absolutely dominant, containing the capital city of Jakarta and the other major cities of Bandung and Surabaya. It is perhaps for these reasons that Java and Sumatra are the two islands of Indonesia that have received by far the most attention in the landslide literature. It is also because of this fact that this paper will confine itself solely to considerations of landslides in Java and, to a much lesser extent, Sumatra. However, because so little published information is available, it is difficult for the Author to present anything but a very inadequate review of the landslide situation on these two major islands.

It is worth mentioning in passing that the interesting but brief booklet produced by the Geological Survey (1981) outlines (in the Indonesian language) the landslide problems of the whole of the country, although it describes actual failures only for Java and Sumatra. The paper by Panjaitan et al (1981) reviews some problems of 'erosion' on the island of Timor, which is east of Java. Other than these two documents, no other publications are known to the Author which deal with the landslide problems of any part of Indonesia other than Java and Sumatra.

The islands of Java and Sumatra are geologically similar (Figures 7 & 8). Uplifted sediments (mainly sandstone, limestone and shale) are in many places overlain by volcanic material, much of which is of recent origin. The islands are geologically young, the dominant feature being the chain of volcanoes, many of which are still active, that runs longitudinally through each of them. As a result of frequent volcanic and seismic activity, much of the area is intensively folded and faulted, to leave mixtures of volcanic material and sediments which have been weathered to clayey soils at the surface. Recent volcanic ash and breccia deposits are spread widely over many slopes to give a covering of very weak material of high permeability.

The topographies of the two islands are also similar, dictated as they are by the chains of volcanoes. The broad north coastal strip of low relief contrasts sharply in each case with the hilly and mountainous terrain that comprises the majority of the land area. Elevations along the southern sides are typically between 700 and 1 000 m, but dramatic rises occur to over 2 000 m in many places and to more than 3 000 m in some. Steep natural slopes are commonplace. Rainfall averages more than 2 500 mm annually, varying from about 1 800 mm along the north coast of Java to more than 6 000 mm in the mountainous regions of the islands, and it is spread throughout the year.

There are three important agencies of landslides in Indonesia, either singularly or in combination - rainfall, volcanic activity and earthquakes. Of these, rainfall is undoubtedly the one most frequently responsible for landslides, but volcanic eruptions have historically accounted for the highest casualties. Although the country is seismically very active because of its justa-position with the Pacific earthquake belt (Figure 2), it is not clear whether earthquakes are a major primary cause of landslides or whether they merely constitute a secondary influence, although the indications are that it is the latter. Regional seismic zoning has been undertaken on the basis of observations made from the beginning of the century

Figure 7.   Geological map of Java

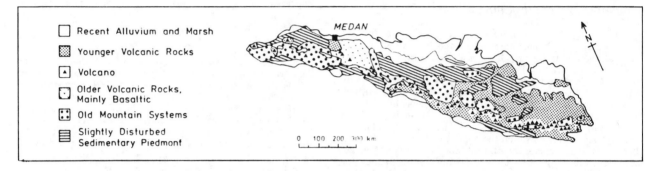

Figure 8.   Geological map of Sumatra

(Soetadi, 1962), but little appears to have been done with regard to engineering seismology (Purbo-Hadiwidjojo, 1971). Many hundreds of earthquake shocks are felt annually in Java and Sumatra, but only a few of these are strong shocks (Priyantono et al, 1980; Effendi et al, 1981). The seismic activity of the area cannot, of course, be dissociated from the volcanic activity.

Active volcanism is a very special cause of landslides in Indonesia, which is the most volcanically active area in the world, with almost 150 centres of activity (Bemmelen, 1949). Since 1800, a volcanic calamity has occurred statistically about once every three years, causing a total of approximately 135 000 deaths, and destroying hundreds of towns and villages and thousands of hectares of farmland. The best known of these is the violent eruption in 1883 which completely destroyed the small island of Krakatoa between Java and Sumatra, causing massive destruction and thousands of deaths. There have, however, been other major volcanic catastrophies (Furuya, 1978) which are not generally known about outside of Indonesia, and in some of these the main agency of death and destruction has been landslides.

Landslides on the slopes of active volcanoes take two main forms - lava avalanches and mud-flows. Some of the volcanoes on Java (e.g. Kelud, Merapi, Semeru and Galunggung) are recognised as being particularly dangerous in this respect, and permanent observation posts were established many years ago to provide early warnings to the inhabitants of the surrounding areas of impending serious conditions (Bemmelen, 1949; Sumaryono & Kondo, 1985). The tens of millions of cubic metres of water in the crater lake of the Kelud Volcano have been ejected by eruptions on many occasions, causing devastating mudflows ('lahars') which travelled long distances and killed many people. The Kelud lahar of 1919 alone killed 5 110 people and destroyed 104 towns and villages (Sigit, 1965). Lava flows from active volcanoes can also be a major landslide hazard (Geological Survey, 1981). Some idea of the extent of this hazard can be gained from the fact that the records of the Volcanological Survey (1958) show that more than 8 000 separate lava 'avalanches' occurred on the slopes of the Merapi Volcano during the last six months of 1953. Over the years, considerable economic resources have been expended in an attempt to mitigate such disasters by large-scale civil engineering works to drain water from the craters of major volcanoes and to control the flow of debris (Takanashi, 1981; Legowo, 1981, 1985; Sumaryono

& Kondo, 1985; Suryo & Clarke, 1985). The paper by Suryo & Clarke (1985) gives an excellent account of these measures. On the slopes of all the major volcanoes, the lahar flow paths have been mapped, and land development is now controlled in these areas, although not always effectively.

Volcanic activity also results in the accumulation of debris and ashes on the slopes of volcanoes, and landslides in the form of mudflows ('cold lahars') are then often caused by high intensity rainfall. In this regard, it is of particular interest that Bemmelen (1949) suggested that a mudflow of recent ash was likely to occur if 70 mm of rain fell in 35 minutes. Landslides of this nature are very prevalent in Java and Sumatra, and they have caused many casualties in some densely populated areas of Java (Geological Survey, 1965; Furuya, 1978; Sumaryono & Kondo, 1985; Suryo & Clarke, 1985).

A fairly recent rain-induced mud-flow catastrophy has been described by Tjojudo & Sulastoro (1982). On 26 December 1980, after two days of heavy rain, a total of 87 separate landslides occurred in a 30 minute period on the southeastern slope of the 3 078 m high Cirema Volcano about 100 km east of Bandung on Java Island. Several of the failures combined together to form large mudflows that travelled extremely rapidly over very long distances downhill, killing 160 people and injuring another 48, destroying 59 houses and blocking over 5 km of road. Investigations after the failures showed that the angles of the majority of the slopes on which failures occurred were from 30° to 60° and that the thickness of the layer of failed material varied between only 0.5 and 1.5 m. The further eruptions of Galunggung in 1982/83 and their disastrous effects have been described by Sudradjat & Tilling (1984).

Landslides in natural slopes occur frequently in the mountainous parts of Indonesia. On occasion, the consequences are disastrous when large masses of debris fall onto rural villages, which are commonly close to steep slopes where there are springs. Such landslides result in up to 2 000 casualties per year. In February 1984, there were reports in Indonesian newspapers of a number of landslide catastrophies in Java and Sumatra. Heavy rains at the end of the month brought about severe flooding in Java which caused a large failure 125 km southeast of Bandung that completely buried a school, a mosque and 46 houses, and damaged almost 400 others, but with no casualties. Within a few days of this, the village of

308

Simulungun on the slopes of a volcano 180 km south of Medan in Sumatra was engulfed by a mud-flow that killed 25 people and caused a great deal of damage. A similar landslide in 1982 had overwhelmed the nearby village of Sihobuk and killed 62 people. In May 1987, a major catastrophy occurred in west Sumatra when a large natural slope, with limestone quarry workings at the toe, collapsed during heavy rain and killed 132 people in the village below (Figure 9).

As in the other countries of Southeast Asia, the cutting of slopes has triggered innumerable failures in Indonesia. These have been of particular significance for roads and irrigation canals (Purbo-Hadiwidjojo, 1971). An inventory made by the Indonesian Highways Department early this decade showed that the number of slope failures that had occurred on roads in Java averaged 1.4/km (Rachlan, 1982); these were classified as 'slides', 'flows' and 'falls'. Although there are rarely casualties involved, the frequent road cutting failures are serious in terms of economic loss and social consequences. The problem is increasing as more roads are built and as some of these are designed to improving standards in terms of width and alignment.

Engineering methods of cut slope design have not been used in Indonesia, the trial-and-error technique being employed, usually without adequate engineering geological input. However, a limited amount of terrain evaluation has been used to good effect in some areas (Dowling, 1983; Saroso et al, 1983). As might be expected, the large majority of failures associated with cut slopes involve various types of volcanic material, but some failures also occur in slopes composed of shales, limestones and alluvial deposits. Landslides associated with cuttings are most commonly of the shallow translational type, and they frequently extend a long way back into the natural slopes above. The failure surface is often the interface between the highly weathered material (grades V & VI) and the more stable parent rock below. Also common is for rain infiltration to cause a thin layer of high permeability volcanic debris to slide off the surface of low permeability shale below.

In 1981, the Indonesian Road Research Institute initiated a research project jointly with the British Transport & Road Research Laboratory to improve the geotechnical engineering aspects of road location and design in Indonesia (Dowling, 1983; Heath & Saroso, 1988). If pursued rigorously over a number of years, this work should yield a great deal of useful information that will enable the number of slope failures to be reduced in future from its present high figure. Some initial results have already been published by Heath & Saroso (1988).

Indonesia obviously suffers severely from landslide problems, the major causes of which are volcanic activity and rainfall, often in combination. The human casualties and economic loss are undoubtedly enormous even by the standards of other countries in Southeast Asia. Over many years, appreciable resources have been expended on mitigating the effects of volcanic eruptions and associated landslides, and the state-of-the-art in this regard is undoubtedly quite high. Little attention, however, has been given to the landslides which affect rural villages and highways, and the available information suggests that the state-of-the-art in this regard is low.

5 LANDSLIDES IN MALAYSIA

Malaysia is a country which is physically divided into two distinct pieces (Figure 1). West Malaysia (formerly named Malaya) is attached to the peninsular portion of Thailand; it has an area of about 130 000 sq. km and a population of about 12 million. East Malaysia comprises the two states of Sarawak and Sabah on the northern part of the island of Borneo; it has an area of more than 200 000 sq. km and a population of only about 4 million. The capital of the country is Kuala Lumpur in the southwest of West Malaysia.

The tiny population of East Malaysia is found almost entirely along the relatively flat land of the north coast, and there are few building developments or roads into the hills to the south. There is virtually nothing published on the landslides on the natural slopes of the interior. The Author knows of only one mention of landslide problems near a centre of population in Sabah (Hunt, 1971), although Cook & Younger (1987) have given useful information on the influence of geology on the empirical design of road cuts in Sarawak. This section must therefore confine itself entirely to considerations of landslides in West Malaysia, on which adequate information exists.

The topography of West Malaysia (Figure 10) is a natural continuation of that along peninsular Thailand. The low-lying coastal areas on the east and west are separated by a central 'spine' of hills generally less than 500 m high, but rising in thin ribbons to 1 500 m and even reaching more than 2 000 m in places. The geology (Gobbett & Hutchinson, 1973) is complex, but it can be summarised as shown in Figure 10. The two large granitic intrusions into the sedimentary rocks essentially form the high relief. The rocks are almost everywhere very deeply weathered, considerable thicknesses of residual soil covering the terrain.

The physical characteristics of the weathered rocks of West Malaysia have been described by Bulman (1967). The igneous rocks, the majority of which are granites, weather to firm sandy clay soils, which often contain fine gravel size quartz particles, overlying grade I to III rocks. The engineering properties of this residual soil have been investigated by Ting & Ooi (1976). The sedimentary rocks range from firm quartzite and sandstone to soft phyllites and shales, both of which weather deeply to give a grade VI consistency of firm clayey silt that becomes cohesionless on drying; cut slopes formed in this material are less stable than those in other sedimentary rocks. The climate of West Malaysia has been described by Dale (1959/60). The peninsula is hot and wet all the year round, with an average annual rainfall of about 3 100 mm on the east coast, decreasing to about 2 500 mm on the west coast. Rainfall intensities can exceptionally reach more than 100 mm/hour

Figure 9.   Landslide disaster in West Sumatra in May 1987

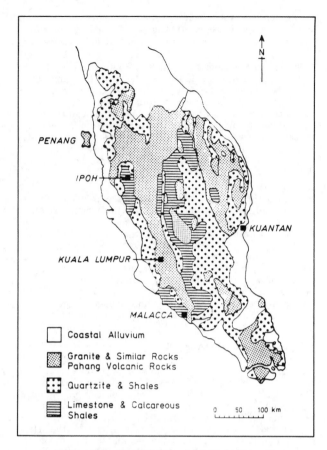

Figure 10. **Geological map of West Malaysia**

Legend:
- ☐ Coastal Alluvium
- ▨ Granite & Similar Rocks Pahang Volcanic Rocks
- ⊞ Quartzite & Shales
- ☰ Limestone & Calcareous Shales

0   50   100 km

in some places (Ministry of Agriculture, 1977).

Although relevant literature is in very short supply, there are a few good reference papers on landslides in West Malaysia. Tan (1984) and Ting (1984) have given brief reviews of all forms of slope failures in the country, and Narayanan & Hengchaovanich (1986), Moh et al (1987) and Tan (1986) have given full accounts of failures on major highways. In addition, Tan (1986) has described rock slope failures brought about by site formations for building construction. The spate of landslides on relatively new highways in Malaysia led the Public Works Department (1984) to organize a Seminar, the Proceedings of which are of considerable interest.

In the mountainous regions of West Malaysia, landslides in natural slopes are fairly common (Ting, 1984). These usually take the form of shallow slides, 3 or 4 m deep, in the residual soil (grades V & VI) mantles to give failure surfaces almost parallel to the slope face. It has been found that steep natural terrain is particularly prone to landslides when the natural vegetation is removed. Failures of this kind have caused problems on mountain roads and in hill stations, but there are no records of any disastrous consequences.

The rapid economic development in Malaysia over the last three decades has resulted in the construction of many new roads. Older roads tended to follow the ground contours as far as possible to minimise the difficulties and costs of construction. The newer roads have been built to much higher standards of alignment, and this has meant that cut slopes have become higher and that it has been necessary to excavate in a wide range of geotechnical profiles. The variations in the residual materials from the geological and the weathering points of view are such that it is not generally possible to apply satisfactory slope design procedures, engineering judgement and precedent being relied upon for the determination of cut slope angles. In addition, it seems that embankments have often not been properly compacted or adequately drained. As a result, failures in road cut and fill slopes are fairly common, and they represent a continuing heavy maintenance commitment and a substantial economic nuisance, although casualties

are almost unknown from these failures.

The Kuala Lumpur to Karak highway, running west to east across the mountains, has been particularly badly affected by landslides (Tan, 1987; Moh et al, 1987). The East-West Highway across the mountains in the north of the country has also suffered badly; details of failures in cut slopes and fill slopes, and of remedial measures, have been given by Narayanan & Hengchaovanich (1986).

A survey of road cuttings in West Malaysia was conducted by Bulman (1967) in an attempt to produce empirical guidelines for the common geological profiles encountered. The recommendation made by him for mean design angles varied from 60° to 70° in hard rock, at the one extreme, to 35° to 40° for 'soft' shales, schists and phyllites, at the other. An amount of terrain evaluation has also been completed for West Malaysia which provided some valuable basic planning and land use data (Beaven et al, 1972; Lawrance, 1972, 1978), but this work highlighted the difficulties of applying the technique to densely forested terrain. For cut slopes on the Kuala Lumpur to Karak highway, Moh & Woo (1986) and Moh et al (1987) devised an interesting slope 'risk' categorization system based on both assessed instability and on consequences of failure to form the basis of priorities for preventive works. This used five categories of 'instability' and four of 'consequence'. The system is in some ways similar to that used in Hong Kong (Brand, 1988b; Koirala & Watkins, 1988), and it might have application elsewhere.

Failures in cut slopes are very often initiated by the occurrence of surface erosion, which sometimes takes place rapidly immediately after a cut slope is formed (Tan, 1984; Ting, 1984). A serious example of this is shown in Figure 11. Gullying and tunnelling are generally initiated along joints and other discontinuities. The most seriously affected residual materials have been found to be those with low clay contents and low clay/silt ratios, or with high dissolved sodium cation contents (Yuen, 1980). The importance of slope protection to minimize erosion is obvious, and turf, sprayed bitumen and cement or lime stabilized soil have been suggested for this purpose (Bulman, 1967).

Rockfalls are not common in Malaysia, because unweathered outcrops are rare, but they do occur (Tan, 1986). It is ironical, therefore, that a large number of fatalities have been suffered by rockfalls at Gunong Cheroh on the outskirts of Ipoh. Gunong Cheroh is one of many closely jointed limestone hills outcropping along the Kinta Valley which are bounded by sheer cliffs. Several known rockfalls of significant size have occurred over the years from the near vertical faces of this particular outcrop, which occupies a land area of only 2.7 hectares. In early 1927, several people were killed when a Hindu temple beneath a cliff was badly damaged, and further falls in the 1950s caused the authorities to close permanently the main road beneath the southern cliffs. On 18 October 1973, during prolonged rain, a major disaster occurred when an estimated 9 000 cu. m slab of rock dropped from the eastern cliffs onto a large wooden dwelling house and other

Figure 11. **Surface erosion in 30 m high granite cut slope**

flimsy structures below. More than 40 people were killed (Shu & Lai, 1988).

The most serious landslides in Malaysia have been of unusual cause, in that they have been associated with the open excavations used to pursue tin mining in the northwestern part of the country. The alluvial overburden to the limestone is removed by hydraulic jetting with gravel pumps, and the resulting alluvial slopes, typically 10 m high, have been known to collapse catastrophically killing workers in the mine below and sometimes causing casualties in buildings at the surface near the rim of the excavation. Between 1960 and 1980, 392 accidents in tin mines resulted in 404 deaths and considerable property damage; slope failures alone caused 246 deaths and 25 injuries. The most devastating kind of failure encountered is of the flow-slide type, in which the debris liquifies and flows rapdily as a viscous fluid over a long distance.

Because of its bad record of casualties, the tin mining industry in Malaysia has recently begun to replace its previous ad hoc methods of slope construction by properly engineered techniques which include the use of stability analyses, and geotechnical engineering methods are increasingly being employed generally throughout Malaysia to prevent and control all forms of potential landslide situations. In addition, a whole range of remedial measures is applied to failures, although the more sophisticated methods (e.g. drainage caissons, long horizontal drains) are employed only rarely. There is, however, an increasing all round application of engineering methods to the ground engineering aspects of projects, especially major ones. The present state-of-the-art of landslide prevention and control in Malaysia can therefore be assumed to be moderate, and there are clear indications that the situation is improving in keeping with the country's economic development.

6 LANDSLIDES IN THE PHILIPPINES

The Philippines consists of more than 7 000 islands arranged approximately in a 1 700 km north-south archipelago (Figure 12), the largest islands being Luzon in the north and Mindanao in the south. The total land area of almost 300 000 sq. km is occupied by about 54 million people. The capital, Quezon City, is adjacent to the principal city of Manila, and together they constitute by far the largest and most important centre of population, cities like Cebu and Davao being much smaller and of comparatively little commercial importance. A high proportion of the terrain of the islands is hilly to mountainous with a narrow coastal strip of low relief. The geology is very complex, with sandstones, shales and volcanic rocks predominating. These are everywhere deeply weathered and thick colluvial deposits are widespread. Although always warm, the Philippines has large variatons in the pattern and quantity of annual rainfall. In general terms, the south and east parts of the country experience rainfall spread fairly evenly throughout the year, whereas the western parts of the central and northern islands have a pronounced wet season from May to October, the rest of the year being dry. The annual rainfall ranges from about 1 000 mm to more than 4 000 mm in some places.

The Philippines is traversed from north to south by the Pacific earthquake belt (Figure 2), and earthquakes of magnitude 5.0 or above are fairly common, especially along the eastern island chain. Higher magnitudes of 5.3 to 6.9 are also experienced, particularly on east Mindanao and north Luzon, the latter having been known to experience magnitudes in excess of 7.0. Considerable damage and casualties have occurred on some occasions, perhaps the worst being those from the earthquake that shook Manila in July 1968.

It is very difficult to review the landslide situation in the Philippines in a meaningful way because of the negligible amount of published literature on the subject. Extensive enquiries by the Author have produced only three relevant papers. The excellent review by Santos (1985) of the geotechnical engineering practice in the country's residual soils provides basic geological and geotechnical information for the Philippines, including geological, soil and slope

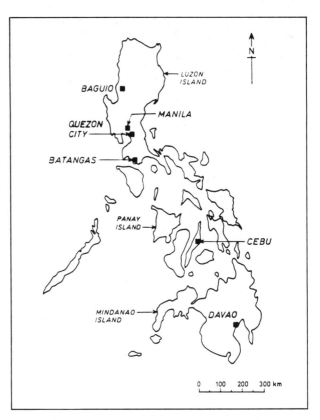

Figure 12. Map of the Philippines

maps. Paderes & Contreras (1969) earlier described some landslides near Baguio, and Belloni (1984) discussed failures in weathered tuff at a pumped storage scheme east of Manila.

It is probable that landslides occur regularly throughout the mountainous parts of the Philippines during times of heavy rain. However, the majority are of relatively minor economic consequence, and virtually no information is available about these except for a few cases of failures in the hilly area of the country in the vicinity of Manila. Information on these has been obtained privately by the Author and from the paper by Paderes & Contreras (1969).

The most important part of Luzon which is prone to landslides is the area around Baguio, 200 km north of Manila. The town is situated on a small plateau-like area at an elevation of about 1 500 m surrounded by hills that rise in a short distance to more than 2 500 m. The rainfall in this area is one of the highest in the Philippines, averaging about 3 400 mm annually, with a pronounced wet season from May to October. From June to September, the monthly rainfall averages 676 mm, with a peak average of 818 mm in August. Exceptional daily rainfalls can equal the average monthly figures; the highest daily rainfall recorded since records began in 1909 is 979 mm, which fell on 17 October 1967. The geology of the Baguio area is predominantly of volcanic rocks, mainly basaltic in composition, which consist largely of agglomerate and tuffs. There are quartz diorite intrusions in places. Weathering is fairly deep, the silty clay soil mantle having a montmorillonite-halloysite mineralogy.

Landslides in the Baguio City area itself do not occur too frequently, mainly because the relatively flat terrain has generally not necessitated much cutting for site formation purposes. Those failures that have occurred have not resulted in casualties, but a considerable amount of damage has been done. At the Fort Del Pilar parade ground on the edge of the town, an uncompacted fill platform formed in a streambed failed soon after construction, and again in June 1963 after the construction of a large toe retaining wall. In January 1968, a 50 m high natural slope at Magasaysay Avenue failed slowly, damaging several houses and necessitating debris removal and the installation of an extensive surface drainage system.

A few kilometres south of Baguio on the Kennon Road, a spectacularly large landslide occurred in November 1968 in an area of 50° natural slopes that are many hundreds of metres high (Paderes & Contreras, 1969). The deep-seated failure of about 800 000 cu. m of material appears to have occurred during heavy rain along two large faults orientated to form a wedge-shaped mass of volcanic rock. There were no casualties, but the Bued River was completely blocked, and a road bridge across the river was destroyed. In addition, the exploratory workings of a nearby underground mine were flooded for some time. It is thought that the blasting for the mining operations might have been partially responsible for the landslide.

Two significant failures occurred in June 1973 at Camp Allen and at the Baguio General Hospital. The residential Camp Allen was damaged when excavations for the foundations of a building caused a 50 m high, 30° natural slope to fail after initial tension cracks had formed. At the back of the hospital, where a road runs through the middle of the 90 m high, 30° natural slope, slow movements of a 400 m length of the soil slope took place above the road over a number of years during the wet season, before remedial drainage measures carried out in 1975 arrested the movements.

Failures in cut slopes on highways are very common, particularly on highways constructed to high standards of alignment in the last 20 years. The Pan-Philippine Highway in central Luzon has been particularly badly affected, with almost 100 failures having taken place to-date. Roads are often impassable for days or weeks in some areas, and the economic disruption is therefore considerable.

A disastrous landslide took place during August 1978 in Zambales Province, 130 km northwest of Manila, where the terrain is very hilly. Records from the meteorological station in the area show that the average annual rainfall is 3 685 mm and that more than 1 000 mm of this falls in the month of August. There had been 228 mm of continuous rain over a three-day period which caused a creek to overflow onto a steep natural slope above the cutting for the Olongapo-Sybic National Highway. The thin soil mantle of weathered gabbro slid off the fresh rock, completely destroying a section of the highway and killing 30 people who lived below the level of the road. This is thought to have been one of the worst landslides, in terms of casualties, ever to have occurred in the Philippines. More recently, 23 people were reported to have died in a landslide that occurred in May 1988 near Davao in the southern island of Mindanao.

The most catastrophic landslides in the Philippines in terms of human casualties have, ironically, been caused by uncontrolled digging for gold. Near the east coast of Mindanao, several 'gold rush' areas have experienced uncontrolled mining activities, involving haphazard surface excavations, primitive tunnels and vertical shafts, and the uncontrolled heaping of spoil. Several hundred people have been killed by resulting landslides during heavy rains at Diwalwal, Boringot and Baleleng. The worst incidents occurred in October 1985 and April 1988.

The small amount of available information makes it difficult for the Author to judge meaningfully the level of the state-of-the-art of landslide prevention and control in the Philippines. The indications are, however, that engineering methods are not applied to slope stability assessments, and that remedial measures are confined to the very simplest kind. The general state-of-the-art is therefore almost certainly low.

7 LANDSLIDES IN SINGAPORE

Singapore is a small island of only 580 sq. km that sits virtually on the equator at the southern tip of West Malaysia (Figure 1). It has a population of less than three million people spread fairly evenly throughout the country. The main business and commercial centre (Singapore City) is located around the mouth of the Kallang River at the southeast corner of the island (Figure 13).

A geological survey report and good geological maps

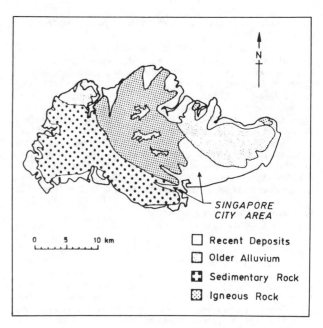

Figure 13. Geological map of Singapore

at a scale of 1:25 000 are available for Singapore (PWD Singapore, 1976). The geology is summarized in Figure 13. The sedimentary rocks (Jurong Formation) are mainly sandstone, shale and conglomerate, and the igneous rocks belong largely to the granite suite. The bedrocks are everywhere fairly deeply weathered, and on the eastern side of the island they are covered by alluvial and marine deposits that can be more than 50 m thick. There is a small amount of reclaimed land on the southern coast. The topography of Singapore is not rugged but is generally of moderately low relief of gently rolling hills, 20 to 40 m high, with some steep natural slopes that rise to about 80 m in a number of places. The highest point (176 m) is at Bukit Tamah near the centre of the island. The climate is hot and humid all the year round, with an average annual rainfall ranging from 1 600 mm in the southwest to 2 500 mm in the central region. The majority of the rain occurs in the monsoon period from October to January, and short-term intensities can be high. In December 1968 and December 1978, 431 mm and 512 mm of rain respectively were recorded in a 24-hour period.

There is a reasonable amount of published information on landslides in Singapore, most of which has become available since 1980. The papers by Ramaswamy (1975), Ramaswamy & Aziz (1977) and Pitts (1984a, 1984b) provide general background on the geology and engineering geology of the country, and they all touch upon slope stability problems. The occurrence of landslides has been briefly reviewed by Ramaswamy & Aziz (1980), and Pitts (1983) has given details of the form and causes of many small slope failures that occurred in west Singapore during heavy rain in November 1982. A recent paper by Tan et al (1987) has provided a fairly comprehensive review of the problem.

The majority of landslides in Singapore are cutting failures, but they sometimes extend back into the natural slopes above roads or site formations of building developments. Failures, which are always associated with heavy rainfall, usually involve only soils, although rock slope failures occur very occasionally (Pitts, 1988). Failure surfaces are invariably very shallow, and the volumes of material involved are generally not large. Pre-existing discontinuities have been found to control the form of many of the failures (Pitts & Andrievich, 1984; Pitts, 1985). An example of a failure in a road cutting is shown in Figure 14.

Casualties and serious damage from landslides are virtually unknown in Singapore. Road blocks, which are quickly cleared, and the disruption of construction activities are generally the most serious consequences, although some damage to private housing has been sustained on occasion. It would seem, however, that landsliding and

Figure 14. Typical road cutting failure in Singapore

its effects are becoming more significant as the land area of the small island state is increasingly developed. Some significant cut slope failures were brought about by heavy rains at the beginning of 1984 (Pitts & Andrievich, 1984; Broms et al, 1986; Tan et al, 1987) and again at the beginning of 1987.

The comprehensive range of remedial measures adopted for landslides has been fully described in the papers by Ramaswamy et al (1981) and Tan et al (1987). Failed slopes are most commonly regraded and provided with surface drainage. Where space is not available for this, retaining walls are usually employed, which may take the form of reinforced concrete walls, crib walls, tied-back sheet piles or contiguous bored piles. Ground anchors and soil nails are also occasionally used where appropriate, as are subsurface drainage measures. Geofabric has been adopted for remedial works on a few slopes (Broms & Wong, 1985).

By comparison with many other Southeast Asian countries, landslides are of minor consequence in Singapore, although their impact is becoming increasingly significant. No radical control measures have been adopted by the government in the way that they have in Hong Kong, but the Public Works Department does exercise limited control over the design of cut slopes and site formations. Unfortunately, no overall terrain evaluation of the country is available to guide land use planning or to highlight areas of potential instability, and cut slope 'design' is based largely on precedent and experience. The last few years, however, have seen some research activity into the mechanisms of slope failures (Pitts & Andrievich, 1984; Pitts, 1985; Pitts & Cy, 1987), which will eventually enhance engineering practice in slope design. At the present time, therefore, the state-of-the-art in Singapore with regard to landslide mitigation can best be described as 'moderate'.

8 LANDSLIDES IN SRI LANKA

Sri Lanka (formerly Ceylon) is an island of about 66 000 sq. km and 16 million people situated at the southeastern tip of India. The capital, Colombo, is on the southwest coast, and other significant centres of population include Galle, Jaffna and Kandy (Figure 15). The physiography is characterised by three plains of erosion (peneplains) which constitute the Hill Country at the centre of the island. The outer land area rises gradually from the sea to a maximum elevation of only about 150 m. Inside this is the first peneplain, with an elevation generally between 200 and 300 m. There is a steep step to the second peneplain at an average elevation of about 700 m. There is a further dramatic rise to the third peneplain, which consists of a complex group of hills, plateaus and basins at elevations between 1 500 m and 1 800 m, but there are a few peaks that are over 2 000 m.

The book by Cooray (1967) describes fully the geology of Sri Lanka. The main rock types are igneous and metamorphic (particularly gneiss), and these are deeply weathered in the Hill Country to give considerable

thicknesses of residual soil as well as carpets of colluvium on footslopes.

Sri Lanka's climate is generally hot and wet all the year round. The annual rainfall ranges from about 1 000 mm at the coast to more than 5 000 mm in parts of the Hill Country (Senanayake, 1986a). No data have been published on very short-term rainfall intensities, but Senanayake quotes extreme two-day rainfalls in parts of the Hill Country of 750 mm in 1974 and 550 mm in 1984, as well as a five-day figure of 1 050 mm in 1986.

Unitil very recently, information on landslides in Sri Lanka was available only from five publications. The landslide problems of the Hill Country were touched upon by Pattiaratchi (1956) and Dissanayake (1973). Cooray (1958) described 15 landslides that occurred in December 1957 in the Kandy area, and Balasubramaniam et al (1975) reviewed the types of landslide that are prevalent in the Hill Country on the basis of thirteen major failures that had occurred in recent years, particularly during the heavy rains of November 1972. The paper by Balasubramaniam et al (1977) describes failures and remedial works that occurred in cut slopes during excavations for penstocks at the Ukuwela Power House near Kandy.

The thirteen landslides described by Balasubramaniam et al (1975)(Figure 15), were classified by them according to the categories of Skempton & Hutchinson (1969) as being: three rock falls, three flows, four translational slides, and three complex failures. The majority of these were associated with road cuttings, but they also involved large portions of natural slopes in some cases. All were triggered by heavy rainfall. There were significant casualties in some instances (including 40 deaths from two landslides in 1972), and the economic consequences were considerable.

Two major landslides that occurred in the Hill Country during heavy rain in January 1971 are worth special mention. On the Rattota-Gammaduwa Road, a slump-mudflow completely destroyed an 85 m length of the highway (slide no. 11 in Figure 15). Movement took place over a period of several days on a failure surface which extended about

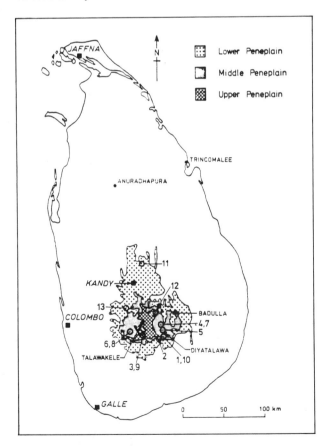

Figure 15. Physiographic map of Sri Lanka showing land-slides studied by Balasubramaniam et al (1979)

313

560 m up the natural slope; the sliding mass was estimated to be about 200 000 cu. m. Another slump-mudflow failure occurred in a few minutes at Ragala (slide no. 12), where a similar quantity of bouldery colluvium slid into a rapidly flowing stream. The destruction by the liquid debris of buildings downstream resulted in 19 deaths and a number of injuries.

In recent years, there have been some major landslide events in Sri Lanka which have focussed attention on the problem. In May 1984, four separate large landslides occurred in the southwest of the country on rolling hills inland from Galle, with a loss of more than 40 lives. In January 1986, many landslides occurred in the Hill Country in a weeks continuous rain, the death toll exceeding 50, with many thousands rendered homeless and extensive damage to roads and property. Most of the casualties and damage were centred on the plantation district of Nuwara Eliya in the Upper peneplain (Figure 15).

The May 1984 landslide event has received some attention in the literature, with brief descriptions of the four failures being provided by M. Jayawardene (1986) and Mampitiyaarachchi (1986). These were all failures of cut slopes less than about 20 m high in soil and soft gneiss rock. The rainfall for the two-day period preceding the failures was about 500 mm, and all four took place some time between 2:00 and 2:30 p.m. (Jayawardene, 1986), which seems to suggest that the trigger was the short-term intensity of rainfall in much the same way as in Hong Kong (Brand, Premchitt & Phillipson, 1984).

Largely as a result of the January 1986 disasters, a Seminar on Landslides was held in Kandy in June 1986 (Institute of Fundamental Studies, 1986), at which some useful papers were presented. D. Jayawardene (1986) briefly summarized the occurrence of landslides and some of the contributing factors, and Dimantha (1986) discussed the role of the land use pattern. Priyasekapa (1986) presented the daily rainfall data for many meteorological stations in the Hill Country and postulated that a two-day rainfall in excess of 450 mm was necessary to trigger landslides. The paper by Madduma Bandara (1986) is of great interest as a historical summary of landsliding in Sri Lanka and its relation to land use management. Dahanayake (1986) and Wilbert Kehelpannala (1986) gave outline case histories of some of the 1986 landslides, one of which occurred after more than 400 mm of rain had fallen in the area in five hours. The volume of material involved was in each case in excess of 100 000 cu. m. It seems that failures were in many cases associated with the extensive deforestation of hillsides.

The 1984 and 1986 disasters in Sri Lanka led to proposals for programmes of landslide mitigation (Madduma Bandara, 1986; Senanayake, 1986a, 1986b). The lengthy paper by Senanayake (1986a) is a particularly important document, in that it sets down an outline programme of landslide mitigation in Sri Lanka and describes the decisions already taken in this direction by the government. Central to the mitigation strategy is landslide hazard mapping, which could be used for the purposes of long-term land use planning and management. The National Building Research Organization (NBRO) has embarked upon a long-term programme of landslide studies to form the basis of the landslide mitigation measures. This programme includes the use of terrain evaluation techniques to produce hazard maps for the landslide-prone areas. In addition, the NBRO programme is attempting to devise some form of early warning system for landslides by means of survey techniques and the installation of simple instrumentation. The Author is sceptical, however, that any practicable warning system will be able to be devised, since the occurrence of landslides in the Hill Country is almost certainly related closely to short-term rainfall intensity.

Despite the long history of landsliding in parts of Sri Lanka, it is clear that no serious thought has been given to landslide mitigation until very recently. It must therefore be concluded that the overall state-of-the-art is low. Recent disasters, however, have created a political atmosphere in which some measure of governmental support is now being given to the development of mitigation procedures, and some improvements in the situation should result over the next decade.

9 LANDSLIDES IN TAIWAN

Taiwan is an island off the southeast coast of China, with a land area of almost 36 000 sq. km and a population of about 19 million. The capital is Taipei, and other important cities include Taichung and Kaohsiung (Figure 16). Physiographically, the island is divided from north to south by the Nitaka mountain range, which leaves a broad lowland coastal strip on the west side and a much narrower one on the east side. Only about one-third of the land area is relatively flat, with an elevation below 100 m, and the rest is hilly or mountainous. In the central mountain chain, elevations of more than 3 000 m are not uncommon, and almost 4 000 m is reached east of Chiayi.

Geologically, Taiwan is composed largely of sedimentary and metamorphic rock formations which occur in long narrow belts roughly parallel to the longitudinal axis of the island (Figure 16). The predominant sedimentary rocks are sandstone and shale in thin beds which dip steeply and which are often in contact with mudstone. In most places, the rocks are deeply weathered, with considerable depths of residual soil (grades V & VI) occurring at the ground surface. The fractured state of the rock and the many faults and folds are evidence of the tectonic disturbance to which the formations have been subjected.

Because Taiwan is situated on the Pacific earthquake belt (Figure 2), earthquakes occur frequently in the island. Records kept since 1907 indicate that an average of 269 shocks are felt annually (Hsu, 1971). The highest frequency of occurrence is in the vicinity of Hualien on the east coast, where there is an average of 116 shocks annually. The average annual rainfall is 2 430 mm, concentrated in the period from May to October. This varies locally from less than 2 000 mm along the west coast to more than 3 000 mm in the mountains. The maximum figure of almost 6 000 mm falls in the mountains due east of Taipei. More than 1 000 mm has been known to fall in one day, and intensities greater than 150 mm/hour have been recorded.

The combination of the generally steep terrain, the high seasonal rainfall and the frequent seismic activity, brings about large numbers of landslides in Taiwan. These affect some urban areas, but they pose the biggest problem for

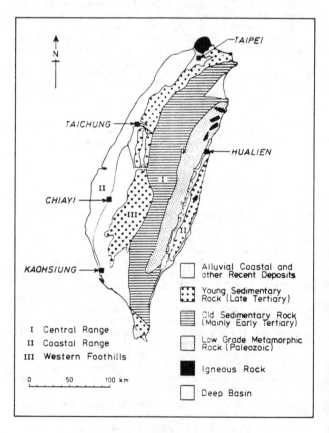

Figure 16. Geological map of Taiwan

highways, railways, reservoirs, transmission lines and mining operations. The frequency and economic consequences of landslides have increased steadily over the years as the land use pattern has changed with the pace of economic development, such that more building and engineering projects are being constructed on increasingly steep terrain. Although the landslide problem is of very significant economic consequence, it is of note that there are few reports of human casualties.

Unlike for most other countries of Southeast Asia, Taiwan boasts a good volume of published literature that is directly relevant to the widespread landslide problems. Good summary papers about landslides in natural terrain have been written by Lee (1981) and Hung (1981), and specific aspects of landslides, especially in cut slopes, have been reviewed by Hung (1977), Moh (1977), Woo et al (1981), Chen & Wang (1985), Moh et al (1987), Chang (1988b) and Hung (1988); all of these present some excellent case histories. Other useful review publications in the English language are those by Sheng (1966) and Fang (1977), and there are also some in the Chinese language (Hung, 1979; Lin et al, 1979; Woo, 1979; Shieh, 1984). Failures initiated by seismic activity have been described by Hsu (1971) and Yen & Wang (1977).

Landslides in natural terrain are commonplace in Taiwan. Some examples have been described by Hung (1981, 1988), Shieh & Chen (1986) and Chang (1986b, 1987). From aerial photographs, Lee (1981) catalogued 7 810 such failures between 1965 and 1977 in the catchment areas of 22 streams and rivers. These were estimated to amount to a total failed area of more than 104 sq. km and a total mass movement of almost 151 million cu. m of material; these figures suggest that the average landslide is very shallow indeed. More than 95% of all the failures occurred in slopes steeper than 30°. Scouring by streams was thought to be the single most important landslide triggering agency, followed by runoff erosion and geological factors. Interbedded sandstone and shale was found to be the geological condition which was by far the most prone to landslides.

One very spectacular landslide event in natural terrain is worthy of special mention. On 17 December 1941, a strong earthquake struck Taiwan, with its epicentre 10 km southeast of Chiayi (Hsu, 1971; Hung, 1977). Many landslides occurred in the vicinity, and about 350 people were killed. At Tsaoling, 30 km northeast of Chiayi, a landslide involving more than 100 million cu. m of sandstone and shale resulted in the formation of a natural dam 130 m high in the Chingshui River (Hsu & Leung, 1977; Hung, 1980; Chang, 1984). On 10 August 1942, heavy rain caused another 150 million cu. m of material to slide from the same slope, to leave the natural earth dam in the river with dimensions of about 1 200 m long, 170 m high, 4 800 m wide at the base, and 100 to 300 m wide at the top. This remains intact to the present day (Hung, 1977, 1980), except for a failure in the dam on 18 May 1951 which resulted in 154 deaths and a great deal of damage to homes and farmland (Chang, 1984).

Another landslide of large proportions is that which occurred in the natural slope above the Mukua River at Wuchia in the mountains of east central Taiwan (Moh, 1977). This landslide, which consists of many millions of cubic metres of heavily jointed schist, has been active for at least fifty years. A hydro-electric power station on the right bank of the river was buried by debris in August 1944 only three years after its completion. The landslide, which covers five hectares and is greatly influenced by a fault that runs through the area, continues to move during heavy rainstorms. Remedial measures are considered economically not feasible.

At Pitan in the centre of the island, a slow-moving lanslide once threatened the East-West Highway and the spillway and intake structures of a large reservoir on the Tachia River (Moh, 1977). Slope failure had occurred several times during the reservoir's construction, and some remedial measures had then been undertaken in the form of concrete buttresses at the toe of the 35° to 40° slope that threatened the spillway. Slope indicators installed at that time later showed that the slope had started to move again on a shear plane about 30 m deep after the East-West Highway was

cut through it in the early 1970s. Further remedial works were therefore carried out (completed 1977) which consisted of a number of 50 m deep drainage shafts driven normal to the slope surface and connected by a longitudinal outlet shaft.

Also on the East-West Highway, a sudden failure of more than one million cu. m of material occurred in October 1972 during heavy rain and blocked the highway for about one month (Moh, 1977). In 1963, numerous landslides took place in the catchment area of the Tuchan River, about 60 km northeast of Chiayi (Moh, 1977) after 1 136 mm of rain had fallen in three days. More than 33 hectares of forestry land were affected by the landslides, and a village was buried by debris that was carried downstream. The remedial works undertaken over a four-year period essentially comprised a comprehensive soil/water conservation scheme. A similar remedial scheme has been employed for the slow-moving Pinzaunair landslide that was observed over several years on a slope of the 4 000 m high Alishan Mountain (Moh, 1977).

The importance of landslides in cut slopes or associated with cut slopes on highways, particularly in Taiwan, is well illustrated in the published literature by a number of case histories of major failures. Hung (1977, 1981) has described several classical rock slides in cuts, including a wedge failure in slate and a failure along a bedding plane between limestone and mudstone that killed 20 people and destroyed a railway line.

The Fushan landslide (Frisch et al, 1975; Moh 1977) took place on the outskirts of Taipei in 1972 after excavations in the interbedded sandstone and shale had commenced at the toe of the slope for the purposes of road widening. Several buildings on the slope were damaged by the sudden, but not large, downhill movements, and the main road and railway line beneath the slope were both threatened. Back-analysis was performed on the slope to enable remedial measures to be designed, which included the provision of a gravity retaining wall at the toe and the installation of some subsurface drainage.

The important North-South Freeway runs for almost 340 km along the west side of Taiwan from Taipei to Kaohsiung. Construction commenced in 1973 and took more than four years to complete. A number of very significant failures have occurred in cut slopes on the northern half of the road where it passes through the young sedimentary rocks and recent deposits (Moh, 1977; Lin et al, 1979; Woo et al, 1981; Moh et al, 1987). During construction in May 1973, a major landslide took place during a heavy rainstorm at station 62 K, which delayed construction for several months. At this location, a 1 to 2 m thick layer of firm lateritic soil covered a 10 m thick layer of poorly cemented lateritic gravel and boulders, beneath which was siltstone and mudstone. Excavation of the 700 m long cut resulted in heavy seepage from the water-table in the gravel layer, and a large part of the excavation collapsed. A horizontal drainage tunnel was installed as a remedial measure, and this proved to be satisfactory.

Since its completion, the North-South Freeway has suffered many landslides during the rainy season (Tsai & Hwang, 1985; Moh et al, 1987). In the vicinity of station 17 K, a 45° cut slope had been formed in a 25° natural slope of sandstone and shale. The 130 000 cu. m landslide that occurred suddenly on 23 September 1977 involved a 285 m long section of the cutting and extended more than 100 m up the slope. The failure surface was seen to be a bedding plane between the sandstone and thin interbedded shale layers that was almost parallel to the surface of the natural slope (Figure 17). The successful remedial measures consisted of short horizontal drains normal to the slope, together with prestressed ground anchors bearing on 0.6 x 0.6 m concrete cross-beams placed on the rock surface; in addition, the surface was protected with mesh-reinforced shotcrete (Moh et al, 1987).

At station 128 K on the North-South Freeway, a wedge-shaped landslide of about 25 000 cu. m occurred in 1978 in a 85 m high sandstone and shale cutting (Woo et al, 1981). At station 21 K, movements in a 50 m high slope cut at only 26° have been a continual problem since the highway construction stage, when some failures

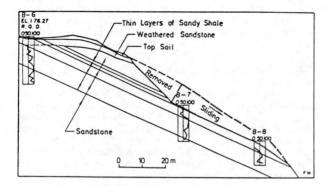

Figure 17. Cross-section of landslide on Taiwan North-South Freeway (Moh, 1977)

necessitated the provision of concrete surface drains together with a concrete grid system of surface protection. By September 1980, however, continuing creep movements of the slope had caused considerable damage to the surface protection and the drainage system, and subsequent extensive remedial works were undertaken to stabilise the slope. These involved the removal of a large volume of material to flatten the slope, and improved surface drainage.

In recent years, positive steps have been taken to develop a scientific framework for studies leading to the establishment of landslide mitigation procedures in Taiwan. Engineering geological zoning for the country is now fairly well advanced (Chang, 1986, 1988), and attempts have been made to develop models of a geomorphological kind to quantify the likelihood of failure of specific slopes (Chen, 1988; Hung, 1988). In parallel with these efforts, modern methods are increasingly being employed for remedial works, including caisson retaining walls (Guo et al, 1987; Hsiung et al, 1987), various anchor types (Kao et al, 1986), and subsurface drainage by wells and horizontal drains (Chen, 1985).

It is abundantly clear from the examples that have been quoted that landslides are of high economic consequence in Taiwan. Although casualty figures do not generally feature in the published literature, it is difficult to believe that the frequency and size of landslides do not result in significant numbers of human casualties. Whereas the available publications demonstrate that a considerable amount of engineering expertise is employed in remedying failures, they also indicate that geotechnical engineering methods are not extensively employed at the earthworks design stage. Overall, therefore, the state-of-the-art in Taiwan with respect to landslide prevention and control is assessed as being moderate.

10 LANDSLIDES IN THAILAND

Thailand consists of a continuous land mass of almost 514 000 sq. km which stretches 1 600 km from Burma in the north to West Malaysia in the south (Figures 1 & 18). The mainly rural population of 51 million is spread fairly evenly throughout the country, except that appreciably higher densities exist on the Central Plain, on the North East Plateau and along the south coast. The capital city of Bangkok, with a population of more than 7 million, completely dominates the economic and commercial life of the country, the other noteworthy towns (e.g. Chiang Mai, Korat, Khon Kaen, Songkhla) being small and unimportant by comparison.

The six physiographic regions of the country can be broadly characterised by three types of topography. The Central Plain is flat and featureless over its entire area, the elevation being everywhere less than 10 m above sea level. The North East Plateau is an area of low or moderate relief, with an average elevation of about 200 m. The rest of the country is hilly to mountainous, peaks of over 2 000 m occurring in the northwest and of up to 1 800 m on the peninsula.

The geology of Thailand is outlined in Figure 18. Detailed geological maps are not available for the majority of the country, and the best overall geological guide available in the English language is probably that published by Landplan I (1982), but comprehensive geological bibliographies have been produced by Nutalaya et al (1982) and Workman (1982). In the mountainous areas along the western border with Burma, the geology is very complex, granites and volcanic rocks being interbedded with shale and limestone in many places, and gneiss and schist appearing at the surface in others. Outcrops are everywhere deeply weathered.

The climate is warm and wet for most of the year, but there are appreciable variations in rainfall over the country. In the mountainous areas, this ranges from 1 200 mm annually in the Chiang Mai area to more than 2 600 mm on the peninsula. Over most of the country, the rains fall during the period from May to October, but they are spread throughout the year in the peninsula. No detailed data on rainfall intensities are available.

Concentrations of population are nowhere sited on hilly terrain in Thailand, and no urban landslide problems therefore exist. Failures do occur fairly frequently in the wet season, however, in road cuttings in the North and West Continental Highlands; landslides elsewhere are rare. Although only a few isolated casualties have been caused by the vast majority of landslides, the failures constitute a continuous economic burden, and they sometimes cause rural communities to be isolated for long periods while roads are cleared of debris.

Slope failures are usually shallow and involve deeply weathered material on or above cuttings formed to steep slope angles determined on the basis of local experience, usually without the provision of drainage other than roadside ditches. Granitic soil mantles (weathering grades V & VI) and closely jointed gneiss and schist (grades II to IV) are the materials most commonly involved. The volume of landslide debris is typically several hundred cubic metres, but large landslides can consist of up to 30 000 cu. m or more.

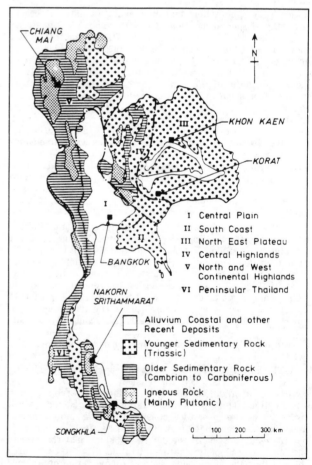

Figure 18. Geological map of Thailand

The landslide remedial measures employed in Thailand consist simply of debris removal, modification of slope geometry and, occasionally, surface drainage provisions. Where a large landslide occurs, it is often more economical to reroute a section of the road rather than to carry out work to rectify the failed slope.

There is very little source information on landslides in Thailand. Ruenkrairergsa (1978) briefly discussed the occurrence of landslides in road cuttings in the northern part of the country and suggested a few simple 'design' rules. Ruenkrairergsa & Chinpongsanond (1980) elaborated on the earlier paper and gave some information about local micro-seismicity which, it was claimed, could be a secondary effect in triggering landslides in that part of the country because of its proximity to the Sumatra-Burma earthquake belt. A report by Poopath (1978) described some costly failures to embankments of compacted lateritic soil built almost normal to the surface of a natural slope of up to 35°.

Very recently (November 1988), a major landslide catastrophy has been reported to have occurred in the extreme south of Thailand near Nakorn Srithammarat (Figure 18). Although the details were unclear at the time this paper was written, the available press reports indicated that as many as 700 people were killed when prolonged heavy rain caused large flowslides to inundate whole villages. Massive damage appears to have been done to roads, railway lines and power and telephone lines over a large area. It will doubtless be some time before the magnitude and causes of the disaster are fully assessed, but deforestation of the hillsides in the area almost certainly played a major part.

A few worthwhile studies were carried out on landslides in northwestern Thailand some time ago by students from the Asian Institute of Technology (Deeswasmongkol, 1976; Guo, 1976; Sutcharit, 1977; Tanomtin, 1979), and there were associated studies on the engineering properties of the weathered granite (Bergado, 1976; Brenner et al, 1978), but no overall survey of the landslide situation appears to have been attempted even for the badly affected areas. It is possible that significant progress could be made in reducing the frequent landslides in Thailand by the adoption of the terrain evaluation approach to initial design, which hitherto has not been employed.

With the increasing number of roads being constructed in the mountainous areas of Thailand, and the continuing deforestation, the number of landslides can be expected to increase. The Department of Highways of the Ministry of Communications is continually reviewing the landslide problem in an attempt to reduce the number of failures and their effects on roads, but only limited resources are available for this work, and progress is slow. The present overall state-of-the-art with regard to landslide mitigation is therefore low. However, the November 1988 catastrophy in the south of the country might result in more attention being paid to the landslide problem.

11. CONCLUSIONS

This paper has attempted to review the landslide situation in the countries of Southeast Asia and to assess the state-of-the-art with regard to landslide prevention and control in each of them. Considerable emphasis has been placed on the present engineering practices employed in Hong Kong, becasue of its pre-eminence in this field.

A high proportion of the land area of Southeast Asia is hilly or mountainous and, because of its position in the tropics, it suffers high seasonal rainfalls. Rain-induced landslides in the residual and colluvial terrain are therefore common in all the countries. Indonesia additionally suffers from hot and cold 'lahars' from its many mountain volcanoes. Seismic activity also plays some part in landslide events in Indonesia, the Philippines and Taiwan, but not elsewhere in the region.

Table 1 provides some basic information on the eleven countries dealt with in this review, and it summarizes the assessed significance of landslides in each country and the present state-of-the-art with regard to landslide mitigation.

Landslides have significant human and economic consequences in all the Southeast Asian countries except Singapore, and possibly Indochina, their effects being particularly severe in Indonesia and Hong Kong, and to a lesser extent in Taiwan and Sri Lanka. Despite this, the state-of-the-art of landslide mitigation is very high only in Hong Kong. In Indonesia, the extreme hazards from lahars are now under reasonable control, and the state-of-the-art in this regard can best be described as 'moderate'; that which applies to other forms of landslides in Indonesia, however, is at a regrettably low level. The state-of-the-art is also low in both the Philippines and Sri Lanka, although recent landslide events in Sri Lanka appear to have initiated longterm improvements. Although geotechnical engineering methods are generally only applied in cases of remedial measures to failed slopes in Malaysia and Taiwan, there is increasing awareness of the need to apply these to new site formation works, and the state-of-the-art in these countries is considered to be moderate and improving.

With the exception of Indonesia, the majority of landslides in the region are triggered by man's activities, including slope cutting, earth fill construction, the denudation of hillside vegetation, and the destruction or alteration of natural drainage regimes. There is also plenty of evidence to suggest that totally inadequate consideration is given to the overall geomorphological features of the terrain at the time that new projects are planned or designed. The more extensive use of terrain evaluation techniques would probably go a considerable way towards improving this situation.

Cut slopes are notably the most hazardous from a landslide point of view. The vast majority of these in Southeast Asian countries were never designed but were simply formed on the basis of experience or empirical rules; this is especially true of road cuttings. In the majority of situations, this approach is still found to be economically more satisfactory than the adoption of an engineering design approach, because the total cost of remedial measures to failed slopes is always likely to be appreciably less than the additional cost involved in ensuring that all cut slopes incorporate an engineering factor of safety. The trial-and-error approach is not acceptable, however, where a landslide can cause loss of life or significant economic loss or social disruption. Many of the countries of Southeast Asia are now having to face landslide problems which come into this category, and engineering methods are therefore beginning to be increasingly used. The urban environment of Hong Kong has necessitated the use of advanced geotechnical engineering methods for slope design for some years, and this largely explains its present very high state-of-the-art.

It is extremely difficult to quantify in any meaningful way the real economic consequences of landslides, and the Author has not attempted to do so for the Southeast Asian countries. Such quantifications would be highly subjective and could only be judged relative to the overall wealth of each country. Although attempts have been made to quantify the economic consequences of landslides in some countries (e.g. Smith, 1958; Schuster, 1978), such attempts have of necessity had to confine themselves to assessing the direct costs attributable to landslide damage and remedial works, which probably represents only a small proportion of the true cost of landslide mitigation measures. For example, the Government of Hong Kong spends only about US$10 million per year directly on landslide preventive works, but the real cost of the Territory's mitigation measures is probably closer to US$100 million, in that large resources are expended in both the public and private sectors on ensuring the integrity and safety of new building and civil engineering works constructed in hilly terrain.

One very obvious conclusion can be drawn from the information summarized in Table 1. As would be expected, the state-of-the-art of landslide mitigation bears some relationship to the human and economic significance of landslides in a particular country. However, it is also strongly related to the wealth of the country, as expressed in terms of the per capita Gross Domestic Product. It is not surprising that poorer countries do not commit large resources to landslide mitigation, and conversely, it is

Table 1. Southeast Asian countries with assessments of landslide significance and state-of-the-art of landslide mitigation

Country	Area (sq. km)	Population * (millions)	Population Density * (people/sq. km)	Per Capita GDP * (US$)	Volume of Relevant Literature	Human & Economic Significance of Landslides	State-of-the-Art of Landslide Mitigation
Hong Kong	1 033	6	5 421	6 268	Very High	Very High	Very High
Indochina	752 164	71	95	144	Nil ?	Low ?	Low ?
Cambodia	181 035	7	40	108			
Laos	236 798	4	17	105			
Vietnam	334 331	60	179	151			
Indonesia	1 919 263	163	85	529	Moderate (Lahars)	Very High	Moderate (Lahars)
Java	132 164	98	742				
Sumatra	473 960	29	61		Very Low (Others)		Low (Others)
Remainder	1 313 139	35	27				
Malaysia	330 669	16	47	2 002	Moderate	Moderate	Moderate
West	131 235	12	91				
East	199 434	4	20				
Philippines	299 765	54	180	603	Very Low	Moderate	Low
Singapore	580	3	4 414	6 827	Moderate	Low	Moderate
Sri Lanka	65 610	16	241	372	Low	High	Low
Taiwan	35 980	19	531	3 097	High	High	Moderate
Thailand	513 517	51	100	752	Low	Moderate	Low

* 1985 figures obtained from "The World in Figures", Economist Publications, London, 1987 edition.

to be expected that the resources expended will increase, and the state-of-the-art will improve, as the economic affluence of a country increases. Good examples of this are Hong Kong, Malaysia and Taiwan.

By any standards in the world, the Southeast Asian countries are plagued by landslide problems that are responsible for a continual stream of human casualties and which impose a perenial economic burden. Ironically, it can be expected that the severity of the situation will increase as economic development brings about more engineered facilities, particularly high quality roads. Considerable improvements will therefore be necessary in the levels of landslide mitigation now practised in many of the countries of the region. There are hopeful signs that these are beginning to occur, but some Southeast Asian countries have still to begin to tackle the landslide problem in a meaningful way.

ACKNOWLEDGEMENTS

Invaluable assistance was given to the Author by a number of individuals who responded generously in providing information about landslides in Southeast Asian countries which formed the basis of the major part of this paper. The help is gratefuly acknowledged of Mr W. Heath (Indonesia), Dr Ting Wen Hui (Malaysia), Mr Jose Rolando R. Santos and Dr Guillermo R. Balce (Philippines), Dr J. Pitts and Mr Tan Siong Leng (Singapore), Prof. A. Thurairajah (Sri Lanka), and Dr Z.C. Moh and Dr S.M. Woo (Taiwan). Others who provided some information or copies of publications included Dr H. Fontaine and Dr D.R. Workman (Indochina), Ir Aziz Jayaputra and Mr J.S. Younger (Indonesia), Mr Ronaldo A. Almero and Dr Arhtur Saldivar-Sali (Philippines), Prof. J.J. Hung (Taiwan), and Prof. A.S. Balasubramaniam, Prof. Prinya Nutalaya and Dr D.T. Bergado (Thailand).

This paper is published with the permission of the Director of Civil Engineering Services of the Hong Kong Government.

REFERENCES

References which do not relate specifically to a particular country are listed under 'General'. All other references are listed under the heading of the country to which they pertain. The countries are listed alphabetically after 'General'.

General

AGE (1987). Bibliography on Landslides in Asia. Asian Infcrmation Center for Geotechnical Engineering, Bangkok, 95 p.

Bemmelen, R.W. van (1949). The Geology of Indonesia. Govt Printing Office, The Hague, Netherlands, 3 vols, 1027 p. & 16 plates & 47 maps.

Brand, E.W. (1982). Analysis and design in residual soils. Proc. ASCE Spec. Conf. Engineering and Construction in Tropical & Residual Soils, Honolulu, 89-143.

Brand, E.W. (1984). Landslides in Southeast Asia: A state-of-the-art report. Proc. 4th Int. Symp. Landslides, Toronto, (1), 17-59. (Addendum, (3), 105-106).

Brand, E.W. (1985a). Landslides and their control in Southeast Asia. Geotechnical Engineering in Southeast Asia, ed. A.S. Balasubramaniam et al, 167-183. A.A. Balkema, Rotterdam.

Brand, E.W. (1985b). Predicting the performance of residual soil slopes. (Theme Lecture). Proc. 11th Int. Conf. SMFE, San Francisco, (5), 2541-2578.

Brand, E.W. (1985c). Geotechnical engineering in tropical residual soils. (Special Lecture). Proc. 1st Int. Conf. Geomechanics in Tropical Lateritic & Saprolitic Soils, Brasilia, (3), 23-91. (Discussion, (3), 92-99).

De Mello, V.F.B. (1972). Thoughts on soil engineering applicable to residual soils. Proc. 3rd S.E. Asian Conf. Soil Eng., Hong Kong, 5-34.

Hoek, E. & Bray, J.W. (1977). Rock Slope Engineering (2nd

edition). Inst. Mining & Met., London, 402 p.

Houghton, D.A. (1987). A review of slope safety in SE Asian quarries and mines. Contractor (Hong Kong), October 1987, 7-13.

Janbu, N. (1954). Application of composite slip surface for stability analysis. Proc. Europ. Conf. Stability of Earth Slopes, Stockholm, (3), 43-49.

Janbu, N. (1973). Slope stability computations. Embankment Dam Engineering (Casagrande Volume), ed. R.C. Hirschfeld & S.J. Poulos, 47-107. Wiley, New York.

Morgenstern, N.R. & Price, V.E. (1965). The analysis of the stability of general slip surfaces. Geotechnique, (15), 79-83.

Schuster, R.L. (1978). Introduction. Landslide Analysis and Control, ed. R.L. Schuster & R.J. Krizek. Transp. Res. Brd Spec. Rep. 176, 1-10.

Skempton, A.W. & Hutchinson, J.N. (1969). Stability of natural slopes and embankment foundations. Proc. 7th Int. Conf. SMFE, Mexico City, state-of-the-art vol., 291-340.

Smith, R. (1958). Economic and legal aspects (of landslides). Landslides and Engineering Practice. Highway Res. Brd Spec. Rep. 29, 6-19.

Hong Kong

Allen, P.M. & Stephens, E.A. (1971). Report on the Geological Survey of Hong Kong. Hong Kong Govt Press, 116 p. & 2 maps.

Anderson, M.G. & Richards, K.S. (ed.)(1987). Slope Stability: Geotechnical Engineering and Geomorphology. Wiley, Chichester, UK, 654 p.

Beattie, A.A. & Attewill, L.J.S. (1977). A landslide study in the Hong Kong residual soils. Proc. 5th S.E. Asian Conf. Soil Eng., Bangkok, 117-188.

Beattie, A.A. & Chau, E.P.Y. (1976). The assessment of landslide potential with recommendations for future research. Hong Kong Engineer, (4), no. 1, 27-44. (Discussion, (4), no. 2, 55-62).

Beattie, A.A. & Lam, C.L. (1977). Rock slope failures their prediction and prevention. Hong Kong Engineer, (5), no. 7, 27-40. (Discussion, (5), no. 9, 27-29).

Brand, E.W. (1985b). Predicting the performance of residual soil slopes. (Theme Lecture). Proc. 11th Int. Conf. SMFE, San Francisco, (5), 2541-2578.

Brand, E.W. (1985d). Landslides in Hong Kong. Proc. 8th S.E. Asian Conf. Geotech. Eng., Kuala Lumpur, (2), 107-122. (Discussion, 122-123).

Brand, E.W. (1987). Some aspects of field measurements for slopes in residual soils. Proc. 2nd Int. Symp. Field Measurements in Geomechanics, Kobe, Japan, (1), 531-545.

Brand, E.W. (1988a). Bibliography on the Geology and Geotechnical Engineering of Hong Kong to December 1987. Geotech. Control Office, Hong Kong, 151 p.

Brand, E.W. (1988b). Landslide risk assessment in Hong Kong. Proc. 5th Int. Symp. Landslides, Lausanne, (2), 1059-1074.

Brand, E.W., Dale, M.J. & Nash, J.M. (1986). Soil pipes and slope stability in Hong Kong. Qtrly Jour. Eng. Geol., (19), 301-303.

Brand, E.W., Hencher, S.R. & Youdan, D.G. (1983). Rock slope engineering in Hong Kong. Proc. 5th Int. Rock Mech. Cong., Melbourne, (C), 17-24.

Brand, E.W. & Hudson, R.R. (1982). CHASE - An empirical approach to the design of cut slopes in Hong Kong soils. Proc. 7th S.E. Asian Geotech. Conf., Hong Kong, (1), 1-16. (Discussion, (2), 61-72 and 77-79).

Brand, E.W. & Phillipson, H.B. (1984). Site investigation and geotechnical engineering practice in Hong Kong. Geotech. Eng., (15), 97-153.

Brand, E.W., Premchitt, J. & Phillipson, H.B. (1984). Relationship between rainfall and landslides in Hong Kong. Proc. 4th Int. Symp. Landslides, Toronto, (1), 377-384.

Brand, E.W., Styles, K.A. & Burnett, A.D. (1982). Geotech-

nical land-use maps for planning in Hong Kong. Proc. 4th Cong. IAEG, New Delhi, (1), 145-153.

Burnett, A.D., Koirala, N.P. & Hee, A. (1986). Engineering geology and town planning in Hong Kong. Proc. Landplan III, Hong Kong, 25-42.

Burnett, A.D. & Styles, K.A. (1982). An approach to urban engineering geological mapping as used in Hong Kong. Proc. 4th Cong. IAEG, New Delhi, (1), 167-176.

Chen, T.Y. (1969). The Severe Rainstorms in Hong Kong during June 1966. Supplement to Meteorological Results, 1966. Royal Observatory, Hong Kong, 82 p.

Craig, D.J. & Gray, I. (1985). Groundwater Lowering by Horizontal Drains. Geotech. Control Office, Hong Kong, 123 p.

Flintoff, W.T. & Cowland, J.W. (1982). Excavation design in residual soil slopes. Proc. ASCE Spec. Conf. Engineering and Construction in Tropical & Residual Soils, Honolulu, 539-556.

Geological Society (1983). Proceedings of the Meeting on the Surficial Deposits of Hong Kong, Hong Kong, September 1983. Geol. Soc. Hong Kong, 183 p.

Geological Society (1985). Proceedings of the Conference on Geological Aspects of Site Investigation, Hong Kong, December 1984. Geol. Soc. Hong Kong, 241 p.

Geological Society (1987). Proceedings of the Symposium on the Role of Geology in Urban Development (Landplan III), Hong Kong, December 1986. Geol. Soc. Hong Kong, 601 p.

Geotechnical Control Office (1982a). Guide to Retaining Wall Design. Geotech. Control Office, Hong Kong, 154 p.

Geotechnical Control Office (1982b). Mid-levels Study: Report on Geology, Hydrology and Soil Properties. Geotech. Control Office, Hong Kong, 2 vols, 265 p. & 54 drgs.

Geotechnical Control Office (1984). Geotechnical Manual for Slopes (2nd edition). Geotech. Control Office, Hong Kong, 295 p.

Geotechnical Control Office (1987). Guide to Site Investigation. Geotech. Control Office, Hong Kong, 353 p.

Geotechnical Control Office (1988). Guide to Rock and Soil Descriptions. Geotech. Control Office, Hong Kong, 189 p.

Government of Hong Kong (1972a). Interim Report of the Commission of Inquiry into the Rainstorm Disasters, 1972. Hong Kong Govt Printer, 22 p.

Government of Hong Kong (1972b). Final Report of the Commission of Inquiry into the Rainstorm Disasters, 1972. Hong Kong Govt Printer, 94 p.

Government of Hong Kong (1977). Report on the Slope Failure at Sau Mau Ping, August 1976. Hong Kong Govt Printer, 105 p. & 8 drgs.

Government of Hong Kong (1983). Buildings Ordinance (and Building Regulations). Laws of Hong Kong, Chapter 123, revised edition 1983. Hong Kong Govt Printer, 304 p. (With continual amendments).

Hencher, S.R. & Martin, R.P. (1984). The failure of a cut slope on the Tuen Mun Road in Hong Kong. Proc. Int. Conf. Case Histories in Geotech. Eng., St Louis, Missouri, (2), 683-688.

Hencher, S.R., Massey, J.B. & Brand, E.W. (1984). Application of back analysis to some Hong Kong landslides. Proc. 4th Int. Symp. Landslides, Toronto, (1), 631-638.

Hudson, R.R. & Hencher, S.R. (1984). The delayed failure of a large cut slope in Hong Kong. Proc. Int. Conf. Case Histories in Geotech. Eng., St Louis, Missouri, (2), 679-682.

Hunt, T. (1982). Slope failures in colluvium overlying weak residual soils in Hong Kong. Proc. ASCE Spec. Conf. Engineering and Construction in Tropical & Residual Soils, Honolulu, 443-462.

Irfan, T.Y., Koirala, N.P. & Tang, K.Y. (1987). A complex slope failure in a highly weathered rock mass. Proc. 6th Int. Cong. Rock Mech., Montreal, (1), 397-402.

Irfan, T.Y. & Woods, N.W. (1988). The influence of relict

discontinuities on slope stability in saprolitic soils. Proc. 2nd Int. Conf. Geomechanics in Tropical Soils, Singapore, in press.

Koirala, N.P. & Watkins, A.T. (1988). Bulk appraisal of landslide risk in Hong Kong. Proc. 5th Int. Symp. Landslides, Lausanne, (2), 1181-1186.

Koo, Y.C. (1982a). Relict joints in completely decomposed volcanics in Hong Kong. Canad. Geotech. Jour., (19), 117-123.

Koo, Y.C. (1982b). The mass strength of jointed residual soils. Canad. Geotech. Jour., (19), 225-231.

Leach, B. & Herbert, R. (1982). The genesis of a numerical model for the study of the hydrogeology of a steep hillside in Hong Kong. Qtrly Jour. Eng. Geol., (15), 243-259.

Lumb, P. (1965). The residual soils of Hong Kong. Geotechnique, (15), 180-194. (Discussion, (16), 1966, 78-81 and 359-360).

Lumb, P. (1972). Landslides in Hong Kong. Proc. 1st Int. Symp. Landslide Control, Kyoto, 91-93.

Lumb, P. (1975). Slope failures in Hong Kong. Qtrly Jour. Eng. Geol. (8), 31-65.

Lumb, P. (1979). Statistics of natural disasters in Hong Kong, 1884-1976. Proc. 3rd Int. Conf. Applications of Statistics and Probability to Soil and Struc. Eng., Sydney, (1), 9-22.

Lumb, P. (1983). Engineering properties of fresh and decomposed igneous rocks from Hong Kong. Eng. Geol., (19), 81-94.

Massey, J.B. & Pang, P.L.R. (1988). Stability of slopes and excavations in tropical soils. (General Report). Proc. 2nd Int. Conf. Geomechanics in Tropical Soils, Singapore, in press.

McFeat-Smith, I., Burnett, A.D., Workman, D.R. & Chow, E.C. (1989). Geology of Hong Kong. Bull. Assoc. Eng. Geol. (USA), in press.

McNicholl, D.P. & Cho, G.W.F. (1985). Surveillance of pore water conditions in large urban slopes. Groundwater in Engineering Geology, ed. J.C. Cripps et al, 403-415. (Discussion, 423). Geol. Soc., London.

Morgenstern, N.R. (1978). Mobile soil and rock flows. Geotech. Eng., (9), 123-141.

Premchitt, J., Brand, E.W. & Phillipson, H.B. (1985). Landslides caused by rapid groundwater changes. Groundwater in Engineeing Geology, ed. J.C. Cripps et al, 87-94. Geol. Soc., London.

Rodin, S., Henkel, D.J. & Brown, R.L. (1982). Geotechnical study of a large hillside area in Hong Kong. Hong Kong Engineer, (10), no. 5, 37-45.

Ruxton, B.P. (1960). The geology of Hong Kong. Qtrly Jour. Eng. Geol., (115), 233-260 (plus 2 plates & 1 map).

So, C.L. (1971). Mass movements associated with the rainstorm of June 1966 in Hong Kong. Trans. Inst. British Geographers, no. 53, 55-65.

So, C.L. (1976). Some problems of slopes in Hong Kong. Geografiska Annaler, (58a), no. 3, 149-154.

Styles, K.A. & Hansen, A. (1989). Geotechnical Area Studies Programme: Territory of Hong Kong. Geotech. Control Office, Hong Kong, GASP Rep. no. XII, in press.

Styles, K.A., Hansen, A. & Burnett, A.D. (1986). Use of a computer-based land inventory for delineation of terrain which is geotechnically suitable for development. Proc. 5th Int. Cong. IAEG, Buenos Aires, (3), 1841-1848.

Styles, K.A., Hansen, A., Dale, M.J. & Burnett, A.D. (1984). Terrain classification methods for development planning and geotechnical appraisal: a Hong Kong case. Proc. 4th Int. Symp. Landslides, Toronto, (2), 561-568.

Vail, A.J. (1984). Two landslide disasters in Hong Kong. Proc. 4th Int. Symp. Landslides, Toronto, (1), 717-722.

Vail, A.J. & Attewill, L.J.S. (1976). The remedial works at Po Shan Road. Hong Kong Engineer, (4), no. 1, 19-27.

Vail, A.J. & Beattie, A.A. (1985). Earthworks in Hong Kong - their failure and stabilization. Proc. Symp. Failures in Earthworks, London, 15-28.

Indochina

Department of Geology, Vietnam (1986). Proceedings of the First Conference on the Geology of Indochina, Ho Chi Minh City, December 1986, 3 vols.

Fontaine, H. & Workman, D.R. (1978). Review of the geology and mineral resources of Kampuchea, Laos and Vietnam. Proc. 3rd Reg. Conf. Geology and Mineral Resources of Southeast Asia, Bangkok, 539-603.

Workman, D.R. (1977). Geology of Laos, Cambodia, South Vietnam and the eastern part of Thailand. Inst. Geol. Sciences, UK, Overseas Geol. & Min. Resources no. 50, 33 p.

Indonesia

Bemmelen, R.W. van (1949). The Geology of Indonesia. Govt Printing Office, The Hague, Netherlands, 3 vols, 1027 p. & 16 plates & 47 maps.

Blong, R.J. (1984). Volcanic Hazards: A Sourcebook on the Effects of Eruptions. Academic Press, London, 427 p.

Dowling, J.W.F. (1983). Terrain evaluation for road engineering in Indonesia. Proc. 4th Conf. Road Eng. Assoc. Asia & Australasia, Jakarta.

Effendi, I., Priantono, T., Tjokrosapoetro, S. & Budiono, K. (1981). Overview of disastrous tectonic earthquakes in the period between April 1979 - April 1980. Bull. Geol. Res. & Dev. Centre, Indonesia, no. 4, 18-20.

Elifas, J.D. (1987). Landslides in Indonesia, its occurrences and the effort made to overcome the problems. Proc. US-Asia Conf. Engineering for Mitigating Natural Hazards Damage, Bangkok, C5.1-C5.15.

Furuya, T. (1978). Preliminary report on some volcanic disasters in Indonesia. S.E. Asian Studies, (15), no. 4, 591-597.

Geological Survey of Indonesia (1981). Gerakantanah di Indonesia (Landslides in Indonesia). Geol. Survey of Indonesia, Bandung, 50 p. & 2 maps.

Heath, W. & Saroso, B.S. (1988). Natural slope problems related to roads in Java, Indonesia. Proc. 2nd Int. Conf. Geomechanics in Tropical Soils, Singapore, in press.

Legowo, D. (1981). Volcanoe debris control applied in Indonesia. Jour. Hydrology, New Zealand, 71-79.

Legowo, D. (1985). On the recent technical development of Sabo works in Indonesia. Proc. Int. Symp. Erosion, Debris Flow and Disaster Prevention, Tsukuba, Japan, 439-443.

Maru, K. & Shaw, R.D. (1984). Ground instability during open excavation at the Saguling Hydroelectric Project, West Java. Proc. 4th Int. Symp. Landslides, Toronto, (2), 137-142.

Pandjaitan, B.T.D., Udihartono, Amron, M., Sunandi & Suprapto, P. (1981). Erosion potential and its damage in Timor. Proc. S.E. Asian Reg. Symp. Problems of Soil Erosion and Sedimentation, Bangkok, 129-141.

Priyantono, T., Effendi, I. & Budiono, K. (1980). The earthquakes of 2 November, 1979 and 16 April, 1980 in the Garut and Tasikmalaya areas, West Java. Bull. Geol. Res. & Dev. Centre, Indonesia, no. 3, 13-18.

Purbo-Hadiwidjojo, M.U. (1971). The status of engineering geology in Indonesia: 1970. Bull. IAEG, no. 4, 33-41.

Rachlan, A. (1982). Penanggulangan lonsoren jalan di Indonesia (Preventive and corrective works for road landslides in Indonesia). Proc. 2nd Indonesian Geotech. Conf., Jakarta, (3), 22 p. [In Indonesian].

Saroso, B.S., Dowling, J.W.F. & Heath, W. (1983). Terrain evaluation for road engineering in Indonesia. Proc. 4th Conf. Road Engineering Assoc. Asia & Australasia, Jakarta, (5), 575-599.

Sigit, S. (1965). The rehabilitation of the Kelut tunnel works. Bull. Geol. Survey Indonesia, (2), no. 1, 3-11.

Soetadi, R. (1962). Seismic zones in Indonesia. Met. & Geophys. Inst., Geophysical Notes no. 2.

Sudradjat, A. & Tilling, R. (1984). Volcanic hazards in Indonesia: The 1982-83 eruption of Galunggung. Episodes, (7), no. 2, 13-19.

Sumaryono, A. & Kondo, K. (1985). Development programme of volcanic mud-flow forecasting and warning system in Indonesia. Proc. Int. Symp. Erosion, Drbris Flow and Disaster Prevention, Tsukuba, Japan, 481-486.

Surjo, I. (1965). Casualties of the latest activity of the Agung volcano. Bull. Geol. Survey Indonesia, (2), no. 1, 22-26.

Suryo, I. & Clarke, M.C.G. (1985). The occurrence and mitigation of volcanic hazards in Indonesia as exemplified at the Mount Merapi, Mount Kelut and Mount Gelunggung volcanoes. Quart. Jour. Eng. Geol., (18), 79-98.

Takanashi, K. (1981). Basic concepts for debris control work in Mt Kelud, East Java, Indonesia. Proc. Symp. Problems of Soil Erosion & Sedimentation, Bangkok, 477-489.

Tjojudo, S. & Sulastoro (1982). Hipotesa makanisme longsoran di sekitar gunung gegerhalang kabupaten majalengka ditinjau dari segi geologi teknik (Hypothesis regarding the mechanism of landslides on mountainsides from a geotechnical viewpoint). Proc. 2nd Indonesian Geotech. Conf., Jakarta, (2), 22 p. [In Indonesian].

Volcanological Survey of Indonesia (1958). Bulletin for the Period 1950-1957. Volc. Survey Indonesia, Bandung.

Walton, G. & Gunawan, R. (1979). Some observations on surface instability associated with coal deposits in south Sumatra. Qtrly Jour. Eng. Geol., (12), 41-50.

Wesley, L.D. (1977). Shear strength properties of halloysite and allophane clays in Java, Indonesia. Geotechnique, (27), 125-136.

Wirasuganda, S. (1983). Landslide hazard investigation and mitigation in Indonesia. Bull. Directorate of Environmental Geol., Indonesia, no. 1.

Malaysia

Beaven, P.J., Lawrence, C.J. & Newill, D. (1972). A study of terrain evaluation in West Malaysia for road location and design. Proc. 4th Asian Reg. Conf. SMFE, Bangkok, (1), 411-416.

Bulman, J.N. (1967). A survey of road cuttings in western Malaysia. Proc. S.E. Asian Reg. Conf. Soil Eng., Bangkok, 289-300, (Discussion, 590-591).

Cheang, K.K. (1986). Landslide-related geotechnical engineering problems in Malaysia. Proc. Landplan III, Hong Kong, 335-346.

Cook, J.R. & Younger, J.S. (1987). The engineering geology of road projects in North Borneo. Planning and Engineering Geology, ed. M. Culshaw et al, 419-428. Geol. Soc., London.

Dale, W.L. (1959/60). The climate of Malaya. Jour. Trop. Geog., (13), 23-27 and (4), 11-28.

Endicott, L.J. (1982). Analysis of piezometer data and rainfall records to determine groundwater conditions. Hong Kong Engineer, (10), no. 9, 53-56.

Gobbett, D.J. & Hutchinson, C.S. (ed.)(1973). Geology of the Malay Peninsula. Wiley, Toronto, 459 p.

Hengchaovanich, D. (1984). Practical design and construction of roads in mountainous terrain. Proc. Sem. Design & Construction of Roads in Mountainous Terrain in Malaysia, Kota Bharu, Malaysia, 14 p.

Hunt, T. (1971). Some engineering characteristics of soils in the vicinity of Kota Kinabalu, Sabah, Malaysia. Geotech. Eng., (2), 1-20.

Jacobsen, G. & Hunt, T. (1969). The engineering geology of Kota Kinabalu. Bull. Malaysian. Geol. Survey, no. 11.

Lawrance, C.J. (1972). Terrain evaluation in West Malaysia. Part 1 - Terrain classification and survey methods. Transp. & Road Res. Lab., UK, Rep. LR506, 35 p.

Lawrance, C.J. (1978). Terrain evaluation in West Malaysia. Part 2 - Land systems of South West Malaysia.

Transp. & Road Res. Lab., UK, Supp. Rep. 378, 167 p. & 2 maps.

Lawson, C.R. & Hengchaovanich, D. (1984). Use of subsurface drainage as part of the remedial measures carried out on the East-West Highway project. Proc. Sem. Design & Construction of Roads in Mountainous Terrain in Malaysia, Kota Bharu, Malaysia, 29 p.

Ministry of Agriculture (1977). Urban drainage design standards and procedures for peninsular Malaysia. Min. Agric. & Rural Develop., Malaysia, Drainage & Irrigation Div., Proc. no. 1.

Moh, Z.C., Guo, W.S. & Huang, C.T. (1987). Ground failures in Southeast Asian countries. Proc. US-Asia Conf. Engineering for Mitigating Natural Hazards Damage, Bangkok, C2.1-C2.24.

Moh, Z.C. & Woo, S.M. (1986). Geotechnical considerations in the planning and design of highways in mountainous terrain. Proc. 1986 S.E. Asian Cong. Roads, Highways & Bridges, Kuala Lumpur, B1-B27.

Narayanan, A. & Hengchaovanich, D. (1986). Slope stabilisation and protection for roads in mountainous terrain with high rainfall. Proc. 13th Australian Road Research Board Conf., Adelaide, (13), part 3, 152-161.

Nossin, J.J. (1964). Geomorphology of the surroundings of Kuantan (Eastern Malaya). Geol. Mijnbou., (43), 157-182.

Public Works Department (1984). Proceedings of the Seminar on Design and Construction of Roads in Mountainous Terrain in Malaysia, Kota Bharu, Malaysia, July 1984. Public Works Dept, Kuala Lumpur, 74 p.

Shu, Y.K. & Lai, K.H. (1980). Rockfall at Gunung Cheroh, Ipoh. Geol. Survey Malaysia, Geol. papers, (3), 1-9.

Tan, B.K. (1984). Landslides and their remedial measures in Malaysia. Proc. 4th Int. Symp. Landslides, Toronto, (1), 705-709.

Tan, B.K. (1986). Landslides and hillside development - recent case studies in Kuala Lumpur, Malaysia. Proc. Landplan III, Hong Kong, 373-382.

Tan, B.K. (1987). Engineering geological studies of landslides along the Kuala Lumpur - Karak highway, Malaysia. Proc. Int. Symp. Engineering Geological Environment in Mountainous Areas, Beijing, (1), 347-356.

Ting, W.H. (1984). Stability of slopes in Malaysia. Proc. Symp. Geotechnical Aspects of Mass & Material Transportation, Bangkok, 119-128.

Ting, W.H. & Ooi, T.A. (1976). Behaviour of a Malaysian residual granite as a sand-silt-clay composite soil. Geotech. Eng., (7), 67-79.

Ting, W.H., Toh, C.T. & Chee, S.K. (1987). Movement and stability of slopes in residual soil fills. Proc. 9th Asian Reg. Conf. SMFE, Kyoto, (1), 503-507.

West, G. & Dumbleton, M.J. (1970). The mineralogy of tropical weathering illustrated by some West Malaysian soils. Qtrly Jour. Eng. Geol. (3), 25-40.

Philippines

Belloni, L. (1984). Slope failures in surficial weathered tuff at the Kalayaan pumped storage power plant, Laguna, Philippines. Philippine Geologist, April-June 1984, 36-54.

FAPE (1973). The Philippine Atlas - A Historical, Economic and Educational Profile of the Philippines. Kyodo Print. Co., Tokyo, 3 vols.

Obra, B.Q., Miranda, F.E. & Santiago, N.G. (1984). Case study: Baguio landslides. Paper presented at Reg. Sem./ Workshop Preparedness for Geologic Disaster in Southeast Asia and the Pacific Region, Manila, 25 p.

Paderes, A.E. & Contreras, P.P. (1969). Landslide problems along Kennon Road, Benguet Province. Jour. Geol. Soc. Philippines, (23), no. 3, 121-138.

Santos, J.R.R. (1985). Sampling and testing of residual soils in the Philippines. Sampling and Testing of Residual Soils: A Review of International Practice, ed. E.W. Brand & H.B. Phillipson, 139-152. Scorpion Press, Hong Kong.

Singapore

Broms, B.B. & Wong, I.H. (1985). Stabilization of slopes with geofabric. Proc. 3rd Int. Geotech. Sem., Singapore, 75-83.

Broms, B.B., Wong, I.H. & Pitts, J. (1986). Landslide investigations using the weight penetrometer. Proc. ASCE Conf. Use of Insitu Testing in Geotechnical Engineering, Blacksburg, Virginia.

Gasim, M.B. & Brunotte, D.A. (1987). Structural behaviour of Crocker Formation and its implications to landslides. Proc. 9th S.E. Asian Geotechnical Conf., Bangkok, (1), 1.57-1.68.

Lo, K.W., Leung, C.P., Hayata, K. & Lee, S.L. (1988). Stability of excavated slopes in the weathered Jurong Formation of Singapore. Proc. 2nd Int. Conf. Tropical Soils, Singapore, (1), in press.

Pitts, J. (1983). The form and causes of slope failures in an area of west Singapore Island. Singapore Jour. Trop. Geog., (12), 162-168.

Pitts, J. (1984a). A review of geology and engineering geology in Singapore. Qtrly Jour. Eng. Geol., (17), 93-101. (Discussion, (18), 291-293).

Pitts, J. (1984b). A survey of engineering geology in Singapore. Geotech. Eng., (15), 1-20.

Pitts, J. (1985). Structural control of landslides in soils in Singapore. Proc. 4th Int. Conf. & Field Workshop on Landslides, Japan, 251-256.

Pitts, J. (1988). Stability of a rock slope at Bukit Batok new town, Singapore. Proc. 2nd Int. Conf. Case Histories in Geotechnical Engineering, St Louis, Missouri, (2), 115-121.

Pitts, J. & Andrievich, E. (1984). Appropriate investigation and analysis methods for slope stability in Singapore. Proc. Int. Symp. Geotechnical Aspects of Mass & Material Transportation, Bangkok, 333-343.

Pitts, J. & Cy, S. (1987). In situ soil suction measurements in relation to slope stability investigations in Singapore. Proc. 9th Europ. Conf. SMFE, Dublin, (1), 79-82.

PWD Singapore (1976). Geology of the Republic of Singapore. Public Works Dept, Singapore, 94 p. & 10 maps (scale 1:25 000).

Ramaswamy, S.D. (1975). Some geological problems of Singapore as applied to civil engineering. Proc. 2nd Reg. Conf. Geology and Mineral Resources of Southeast Asia, Jakarta, 213-221.

Ramaswamy, S.D. & Aziz, M.A. (1977). Identification and excavability of weak rocks of Singapore. Jour. Inst. Engrs Singapore, (17), no. 2, 51-67.

Ramaswamy, S.D. & Aziz, M.A. (1980). Rain induced landslides of Singapore and their control. Proc. Int. Symp. Landslides, New Delhi, (1), 403-406.

Ramaswamy, S.D., Aziz, M.A. & Narayanan, N. (1981). Some methods of control of erosion of natural slopes. Proc. S.E. Asian Reg. Symp. Problems of Soil Erosion and Sedimentation, Bangkok, 327-339.

Tan, S.B., Tan, S.L., Lim, T.L. & Yang, K.S. (1987). Landslide problems and their control in Singapore. Proc. 9th S.E. Asian Geotech. Conf., Bangkok, (1), 1.25-1.36.

Sri Lanka

Balasubramaniam, A.S., Dissanayake, J.B. & Karunaratne, G.P. (1975). Some aspects of landslides in Sri Lanka. Proc. 4th S.E. Asian Conf. Soil Eng., Kuala Lumpur, 5.1-5.8.

Balasubramaniam, A.S., Munasinghe, N.T.K., Tennakoon, L.L. & Karunaratne, G.P. (1977). Stability of cut slopes for installation of penstocks. Proc. Int. Symp. Geotechnics of Structurally Complex Formations, Capri, (1), 29-39.

Cooray, P.G. (1958). Earthslips and related phenomena in the Kandy District, Ceylon. Ceylon Geographer, (12), 75-90.

Cooray, P.G. (1967). An Introduction to the Geology of Ceylon. National Museums of Ceylon Publication, Colombo, 324 p.

Dahanayake, K. (1986a). Case studies of some landslides in the Central Highlands of Sri Lanka. Proc. Multi-disciplinary Sem. Landslides, Kandy, Sri Lanka, 8-14.

Dahanayake, K. (1986b). Causes of landslides and some short term remedies to minimize damages. Proc. 42nd Session Sri Lankan Assoc. Advancement of Science, Colombo.

Dimantha, S. (1986). Influence of land use on landslides. Proc. Multi-disciplinary Sem. Landslides, Kandy, Sri Lanka, 15-23.

Dissanayake, J.B. (1973). Landslides in Sri Lanka. Jour. Inst. Engrs Sri Lanka, (1), no. 1, 26-29.

Institute of Fundamental Studies (1986). Proceedings of the Multi-disciplinary Seminar on Landslides, Kandy, Sri Lanka, June 1986, 92 p.

Jayawardene, D. (1986). Landslides - Geological aspects. Proc. Multi-disciplinary Sem. Landslides, Kandy, Sri Lanka, 3-5.

Jayawardene, M.P.J. (1986). Landslides in Pasdun Koralaya, Kalutara District, Sri Lanka, May 22, 1984. Proc. Asian Regional Symp. Geotechnical Problems and Practices in Foundation Engineering, Colombo, 77-81.

Madduma Bandara, C.M. (1986). Recent earthslips in the hill country - Human implications. Proc. Multi-disciplinary Sem. Landslides, Kandy, Sri Lanka, 58-70.

Mampitiyaarachchi, D.K. (1986). Earth movements in natural slopes. Proc. Asian Regional Symp. Geotechnical Problems and Practices in Foundation Engineering, Colombo, 83-90.

Pattiaratchi, D.B. (1956). Some aspects of engineering geology in Ceylon. Proc. 12th Session Ceylon Assoc. Advancement of Science, part 2, 141-175.

Priyasekapa, G.D. (1986). Impact of rainfall on landslide areas of Nuwara Eliya and Badulla. Proc. Multi-disciplinary Sem. Landslides, Kandy, Sri Lanka, 24-29.

Senanayake, K.S. (1986a). Natural disaster mitigation planning and some aspects of landslide disasters in Sri Lanka. Paper prepared for Int. Sem. Planning for Crisis Relief, Nagoya, Japan, 28 p.

Senanayake, K.S. (1986b). Geotechnical problems and geotechnical engineers' role in the development of low lying lands. Proc. 42nd Session Sri Lankan Assoc. Advancement of Science, Colombo.

Wilbert Kehelpannala, K.V. (1986). A preliminary study of some recent, major landslides in Sri Lanka. Proc. 42nd Session Sri Lankan Assoc. Advancement of Science, Colombo.

Taiwan

Chang, S.C. (1984). Tsao-Ling landslide and its effects on a reservoir project. Proc. 4th Int. Symp. Landslides, Toronto, (1), 469-473.

Chang, S.C. (1986a). A preliminary report on the engineering geological zoning and types of highway slope failure of Taiwan. Jour. Eng. Environment (Taiwan), no. 7, 89-105.

Chang, S.C. (1986b). A case study on the disastrous rockfall in Taichi Gorge on May 25, 1986. Jour. Eng. Environment (Taiwan), no. 1, 5-19. [In Chinese with English abstract].

Chang, S.C. (1987). Studies on an earthflow and its related phenomena in Yang-Ming-Shan National Park, Taiwan. Jour. Eng. Environment (Taiwan), no. 8, 21-39. [In Chinese with English abstract].

Chang, S.C. (1988a). The engineering geological division of Taiwan related to landslide types. Proc. 5th Int. Symp. Landslides, Lausanne, (1), 95-101.

Chang, S.C. (1988b). Landslides and their social impact assessment in northern Taiwan. Proc. CCNAA-AIT Joint

Sem. Research and Application for Multiple Hazards Mitigation, Taipei, Taiwan, 559-571.

Chen, H.W. (1985). Drainage for landslide stabilization. Sino-Geotechnics, no. 12, 35-52. [In Chinese].

Chen, H.W. (1988). The assessment of failure vulnerability on potentially unstable slopes. Geol. Soc. China (Taiwan), Memoir no. 9, 10 p. [In Chinese].

Chen, R.H. & Wang, C.L. (1985). The failure mechanism of lateritic soil slopes. Proc. ROC-Japan Sem. Multiple Hazards Mitigation, Taipei, Taiwan, 1093-1105.

Cheng, Y. & Chang, S.Y. (1979). Caissons used for landslide stabilization. Proc. 6th Asian Reg. Conf. SMFE, Singapore, (1), 213-215.

Guo, W.S., Wong, L.W. & Hsiung, K.I. (1987). Caisson for slope stabilization. Proc. 9th S.E. Asian Geotech. Conf., Bangkok, (1), 1.45-1.56.

Hsiung, K.Y., Chiou, H.Y. & Wong, L.W. (1987). Design of caisson retaining wall for slope stabilization. Sino-Geotechnics, no. 17, 32-44. (In Chinese).

Hsu, H.L. (1982). Study of the mechanical characteristics of the land sliding under natural dynamics occurred in Tsao-Ling area. Jour. Eng. Environment (Taiwan), no. 8, 41-53. [In Chinese with English abstract].

Hsu, M.T. (1971). Seismicity of Taiwan and some related problems. Bull. Int. Inst. Seismology and Earthquake Eng., (8).

Hsu, T.L. & Leung, H.P. (1977). Mass movements in the Tsaoling area, Yulin-Hsien, Taiwan. Proc. Geol. Soc. China (Taiwan), no. 20, 114-118 (plus 2 plates).

Hung, J.J. (1977). General report on rockslides in Taiwan. Proc. Advisory Meeting Earthquake Eng. and Landslides, Taipei, Taiwan, 126-142.

Hung, J.J. (1979). Geotechnical processes in the prevention and control of landslides in Taiwan. Proc. Sem. Slope Stability and Landslides, Taipei, Taiwan, 147-172. (In Chinese).

Hung, J.J. (1980). A study of Tsaoling rockslides, Taiwan. Jour. Eng. Environment (Taiwan), (1), 29-39. [In Chinese with English abstract].

Hung, J.J. (1981). The role of environmental factors in landslides. Jour. Eng. Environment (Taiwan), no. 2, 63-72. [In Chinese with English abstract].

Hung, J.J. (1984). Slope stability problems and scientific researches on slope lands in Taiwan. Sino-Geotechnics, no. 7, 4-6. [In Chinese].

Hung, J.J. (1988). Landslides and related researches in Taiwan. Geol. Soc. China (Taiwan), Memoir no. 9, 20 p.

Interior Ministry (1983). Regulations for Building Developments on Hillsides. National Building Regulations, 69-79. Interior Ministry, Taiwan.

Kao, T.C., Guo, W.S. & Wang, C.H. (1986). Stabilization of a legendary hanging cliff. Proc. 13th Australian Road Research Board Conf., Adelaide, (13), part 3, 107-114.

Lee, S.W. (1981). Landslides in Taiwan. Proc. S.E. Asian Reg. Symp. Problems of Soil Erosion and Sedimentation, Bangkok, 195-206.

Lee, S.W. (1984a). On landslides problems in Taiwan. Sino-Geotechnics, no. 7, 43-49. (In Chinese).

Lee, S.W. (1984b). A preliminary study of Feng-Ping Chi landslide in Hualien. Jour. Eng. Environment (Taiwan), no. 5, 25-35. [In Chinese with English abstract].

Lee, S.W. (1985). Review of landslide investigation and research in Taiwan and prospects for the future. Proc. Symp. Geotechnical Engineering, Taipei, Taiwan, 241-253. [In Chinese].

Lin, A.S.N., Lin, C.S., Chou, I.C. & Hsu, H.L. (1986). Study on the characteristics of the land sliding occurred in Er-jen Mountain on Taiwan. Jour. Chinese Soil & Water Cons. (Taiwan), (17), no. 2, 183-199. [In Chinese with English abstract].

Lin, M.W., Toh, F.L., Chung, T.C. & Lui, C.S. (1985). Investigation of a landslide at Kaishow secondary school and monitoring of remedial works. Proc. Symp. Geotechnical Engineering, Taipei, Taiwan, 279-313.

[In Chinese].

Lin, S.C. & Hsu, H.L. (1986). Study on the soil mechanical characteristics of the land-slide occurred in Chin-Qua-Shih District. Jour. Eng. Environment (Taiwan), no. 7, 81-87. [In Chinese with English abstract].

Lin, Y.T., Chen, P.T. & Chen, H.S. (1979). Landslide problems of Taiwan highways. Proc. Sem. Slope Stability and Landslides, Taipei, Taiwan, 83-116. [In Chinese].

Moh, Z.C. (1977). Landslides in Taiwan - some case reports. Proc. Advisory Meeting Earthquake Eng. and Landslides, Taipei, Taiwan, 199-218.

Moh, Z.C., Guo, W.S. & Huang, C.T. (1987). Ground failures in Southeast Asian countries. Proc. US-Asia Conf. Engineering for Mitigating Natural Hazards Damage, Bangok, C2.1-C2.24.

Sheng, T.C. (1966). Landslide classification and studies in Taiwan. Joint Commission on Rural Reconstruction, Taiwan, Forest Series, no. 10, 97 p.

Shieh, C.I. (1984). Geotechnical experiences regarding slope failures in Lishan-Tehchi area. Sino-Geotechnics, no. 7, 50-61. [In Chinese].

Shieh, T.J. & Chen, C.H. (1986). Study on the Shih-Lung landslide. Jour. Chinese Soil & Water Cons. (Taiwan), (17), 107-126. [In Chinese with English abstract].

Tsai, K.J. & Hwang, P.J. (1985). Study of the side-slope stability analysis for the Central Cross-Island Highway. Proc. ROC-Japan Joint Sem. Multiple Hazards Mitigation, Taipei, Taiwan, 1107-1128.

Wong, L.W. (1985). Analysis of piezometer data of colluvial slopes. Sino-Geotechnics, no. 12, 26-34. [In Chinese].

Woo, S.M. (1979). Slope problems in sandstone and shale formation. Proc. Sem. Slope Stability and Landslides, Taipei, Taiwan, 117-145. [In Chiense].

Woo, S.M., Guo, W.S., Yu, K. & Moh, Z.C. (1984). Engineering problems of gravel deposits in Taiwan. Proc. Sem. Design & Construction of Roads in Mountainous Terrain in Malaysia, Kota Bharu, Malaysia, 19 p.

Woo, S.M., Moh, Z.C., Yu, K. & Guo, W.S. (1981). A study on the causes of some rock slope failures along highways in Taiwan. Proc. 3rd Conf. Road Assoc. Asia & Australasia, Taipei, Taiwan, 761-777.

Yen, B.C. & Wang, W.L. (1977). Seismically induced shallow hillside slope failures. Proc. 6th World Conf. Earthquake Eng., New Delhi, (7), 43-48.

Thailand

Bergado, D.T. (1976). Engineering Properties of Granite Residual Soil along the Hod-Mae Sariang Highway. MEng Thesis, Asian Inst. Tech., Bangkok, 178 p.

Brenner, R.P., Nutalaya, P. & Bergado, D.T. (1978). Weathering effects on some engineering properties of a granite residual soil in northern Thailand. Proc. 3rd Cong. IAEG, Madrid, session 2, (1), 23-36.

Deeswasmongkol, N. (1976). Engineering Geology of Some Rockslides in the Northern Part of Thailand. MSc Thesis, Asian Inst. Tech., Bangkok, 153 p.

Guo, W.S. (1976). Engineering Geology of the Ban Mae Hoh Landslides, Northern Thailand. MSc Thesis, Asian Inst. Tech., Bangkok, 146 p.

Landplan I. (1982). Guidebook for Post Symposium Excursion: Bangkok-Kamphaengphet-Sukothai-Tak-Lampang-Chiangmai-Bangkok. Proc. Landplan I, Bangkok, 103 p.

Nutalaya, P., Petchvarun, V. & Saengsuwan, C. (1977). Bibliography of the geology and mineral resources of Thailand (Thai languauge). Proc. 3rd Reg. Conf. Geology and Mineral Resources of Southeast Asia, Bangkok, 823-843.

Poopath, V. (1977). An investigation of slope failures in the northern part of Thailand. Proc. 8th Int. Road Fed. Conf., Tokyo.

Ruenkrairergsa, T. (1977). Some problems of road construction and maintenance in weathering rocks in northern

Thailand. Proc. 8th Int. Road Fed. Conf., Tokyo.

Ruenkreirergsa, T. & Chinpongsanond, P. (1980). Geological and seismological aspects of landslides. Proc. Int. Symp. Landslides, New Delhi, (1), 85-88.

Sutcharit, R. (1977). Engineering Geology of Some Landslides on the Tak - Mae Sot Highway, Northern Thailand. MSc Thesis, Asian Inst. Tech., Bangkok, 227 p.

Tanomtin, C. (1979). Investigation of Landslides along Doi Inthanon Highway. MSc Thesis, Asian Inst. Tech., Bangkok, 210 p.

Workman, D.R. (1978). Bibliography of the geology and mineral resources of Thailand, 1951-1977 (English, French and German Languages). Proc. 3rd Reg. Conf. Geology and Mineral Resources of Southeast Asia, Bangkok, 823-843.

Landslides: Extent and Economic Significance, Brabb & Harrod (eds)
© 1989 Balkema, Rotterdam. ISBN 90 6191 876 6

Slope failure: Extent and economic significance in Afghanistan and Pakistan

J.F.Shroder, Jr.
University of Nebraska, Omaha, Nebr., USA

ABSTRACT: The complex geology of Afghanistan and Pakistan is a mosaic of sutured and uplifted crustal fragments whose widely varying rock types, great relief and high seismicity cause many slope failures. Rock falls and rock slides are among the most common failure types in the Hindu Kush and western Himalayas; one large sackung failure has also been recognized. Debris falls and debris slides from the many deposits of rubble and regolith are common; catastrophic debris flows are frequent during monsoon rains. Earth falls, earth slides and earth flows are least common but do occur in loess and clay-rich saprolites. Large complex slump-and-flow landslips occur in unstable sedimentary rock in a few places in northern Afghanistan. Many of these different landslips have repeatedly destroyed highways, homes and water works, as well as damming rivers and causing later break-out floods. Slope-failure hazards are an important constraint on development in the two countries, although little data on monetary or human losses are collected as yet.

1 INTRODUCTION

The lands of Afghanistan and Pakistan are comprised of a wide variety of rock types and relief associated with the Cenozoic and ongoing collision between the Eurasian and Indian plates. Pre-Cenozoic plate fragments are also incorporated into the mix of tectonic zones. This mosaic provides ideal conditions for many kinds of slope failure, inasmuch as high relief, widely varying rock types, shattered rocks, high seismicity and monsoon rains contribute to significant instability, although aridity does reduce hazard in some places.

1.1 Regional geology and slope failure

Slope-failure hazard zonation in Afghanistan and Pakistan (Fig. 1 and Tables 1 & 2) is presented herein on the basis of a plate-tectonic framework first developed from the work of others to assist in analysis of metallogeny (Shroder, 1984). Use of this base map for natural-hazard zonation represents a convenient scheme of subdivision derived from the regional geology and tectonics so that spatial relationships can be discussed more easily. The map is keyed to Tables 1 and 2, which list the basic geology of the region as mapped by Shareq et al. (1977) and Kazmi and Rana (1982). Hazard subdivision into high, moderate and low categories is based generally upon relief, rock type, rock structure, seismic potential, and precipitation.

Northern Afghanistan is characterized by the pre-collisional Eurasian plate margin and basin zones ("Turan epi-Hercynian platform"). This is a zone of moderate potential for slope failure that consists largely of sediments and sedimentary rock slightly to moderately deformed along monoclinal flexures and faults (Table 1). Similarly, much of southern and eastern Pakistan is the pre-collisional Indo-Pakistan plate margin platform and basin zones. Most of this area is covered with relatively undeformed and stable sediments and sedimentary rocks over-

lying a few knobs of Precambrian crystalline rock. The entire northern and western margin of this zone, however, is bounded by the uplifted Himalayan, Suleiman and Kirthar fold belts that are variously faulted and have moderate failure potential (Table 2). Much of the remaining rock and topography of the two countries is the product of intense folding and faulting produced by suturing along the plate boundaries. Failure potential in these areas is moderate to high.

The collision in northern Pakistan and northeastern Afghanistan was the most direct part of the plate impact, and the intense deformation extensively fractured and metamorphosed wide areas of bedrock, as well as uplifting some of the highest mountains in the world. The oblique collision zones, extending from the Makran coast in southwest Pakistan and running north and northeast into Afghanistan, are narrow bands of intense deformation along interplate marginal geosuture structures, as well as interplate marginal obduction zones (Tables 1 & 2; Fig. 1).

Closely associated with the collision structures are the structures and sediments of the interplate marginal convergence boundaries. These consist mainly of folded flysch basins and some volcanics. In central and western Afghanistan also occur the Iran-Afghanistan micro-continental plate fragments ("central Afghan mass of Baikal consolidation"). This area is a series of fault-bounded blocks that collided with the Eurasian plate prior to the main Himalayan orogeny. At present these blocks are being forced to move southwest between the two strike-slip dominated, interplate-marginal geosuture structures that bound it.

In the intermontane basins between fault blocks and folds of the lower ranges of both countries, and in downwarped basins piggy-backed upon thrust sheets, occur a variety of Cenozoic sediments and sedimentary rock. Lacustrine and braid plain deposits, some of which have unstable slopes, are the most common types, with extensive stable alluvial fan deposits along basin edges. In the Bal-

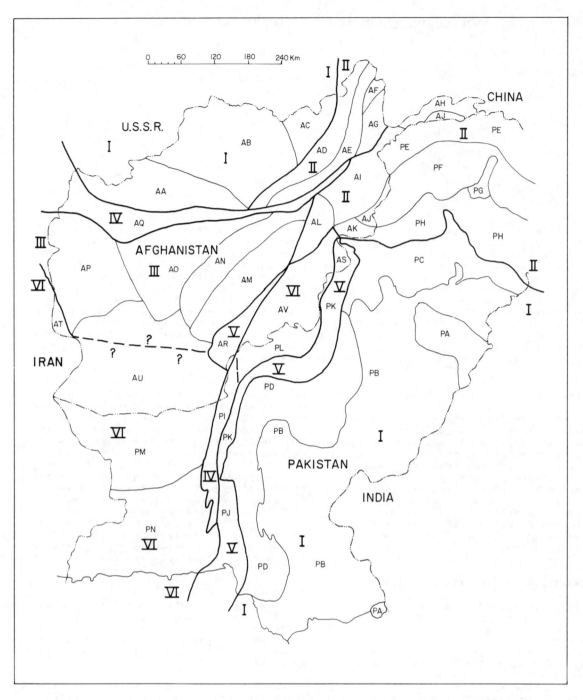

Figure 1. Index map of Afghanistan and Pakistan based on plate-tectonic divisions (Shroder, 1984). Divisions are keyed by Roman numeral and letter to Tables 1 and 2.

uchistan area of southwestern Pakistan, a wide variety of potentially failure-prone marine sedimentary rocks have been folded and thrust into rugged basins and ranges, but the high aridity and minimum human disturbance reduce landslip hazards. In a few places thick weathered saprolites developed across preexisting rocks, and in some of these areas slope failures have occurred.

1.2 General geomorphology and slope failure

The Quaternary history and geomorphology of these regions are a product of complex interaction between the exceptionally active and accelerating Cenozoic tectonism suffic-

ient to uplift some mountain massifs as much as 0.5 mm/yr (Zeitler, 1985). The steepness of the terrain and its maximum relief are among the greatest on Earth, with a 5.94 km rise in 11 km at Rakaposhi Peak (7790 m) and 7 km in 21 km at Nanga Parbat (8125 m) in northern Pakistan. These high values are further reflected in rapid denudation rates produced by vigorous geomorphic processes (Goudie et al., 1984; Ferguson, 1984; Shroder, in press; Shroder et al., in press). This geomorphic variability is controlled by great vertical climatic zonation affected by regional climatic change, together with strong topographic variation, rain shadows, and the like. The result of all these factors is an environment in which slope fail-

326

ure is a highly active process and one which is of considerable annual economic loss to both countries.

The high magnitude of relief, the overall steepness of slopes and the large amount of

debris accumulation in the mountains of Afghanistan and Pakistan show much evidence of past catastrophic events and potential for further slope instability. The high potential energy, however, is underutilized in the lower valleys because of the extreme aridity in many places. Slope failure grows in both absolute and relative importance

Table 1. Geologic zones in Afghanistan, arranged according to plate tectonic theory after data by Shareq et al. (1977). Roman numerals and letters are keyed to Figure 1. Slope-failure hazard is identified as high, moderate and low.

I.   Pre-collisional Eurasian plate margin - platform and basin zones.
      AA Murghab block - Moderate.
      AB Balkh block - Moderate.
      AC Fore-Badakshan molasse - Moderate.
II.   Interplate marginal collisional zone.
      AD Surkhab-Jaway block complex - High.
      AE Western Badakshan "   " - High.
      AF Shewa-Nakchirpar   "   " - High.
      AG Eastern Badakshan "   " - High.
      AH Wakhan block - High.
      AI Nurestan block - High.
      AJ Konar block - High.
      AK Spin-Gar block - High.
      AL Kabul block syntaxis - Moderate.
III. Iran-Afghanistan micro-continent plate.
      AM Arghandab-Tirin block - Moderate.
      AN Helmand block - Moderate.
      AO Harutrod block - Moderate.
      AP Shindand-Kishmaran block - Low.
IV.   Interplate marginal geosuture structures.
      AQ Hari Rod - Panjsher fault system - Moderate to High.
V.   Interplate marginal obduction zones.
      AR Tarnak block - Moderate to High.
      AS Khost block - Moderate to High.
VI.   Interplate marginal convergence boundary.
      AT Asparan block - Low.
      AU Chagai volcanic arc and magmatic belt - Low.
      AV Katawaz flysch basin - Low to Moderate.

Table 2. Tectonic zones in Pakistan (taken in part from Kazmi and Rana; 1982). Zonation deemphasizes areas of low relief and emphasizes border areas contiguous with Afghanistan. Slope-failure hazard is defined as high, moderate and low.

I.   Pre-collisional Indo-Pakistan plate margin - platform and basin zones.
      PA Shield block - Low.
      PB Foreland belt of platform slopes, downwarp and upwarp zones, and foredeeps - Low.
      PC Himalayan fold belt - Moderate.
      PD Suleiman and Kirthar fold belts - Moderate.
II.   Interplate marginal collision zone.
      PE Karakoram (Tethyan) fold belt - High.
      PF Kohistan volcanic and calc-akaline magmatic belt - High.
      PG Nanga Parbat-Haramosh massif - High.
      PH Himalayan crystalline schuppen zone - High.
IV.   Interplate marginal geosuture structures.
      PI Chaman-Nal Ornach fault system (with flysch) - Moderate.
V.   Interplate marginal obduction zone.
      PJ Bela Ophiolite belt - Moderate.
      PK Balla Dhor-Zhob-Kurram ophiolite belt and schuppen zone - Moderate.
VI.   Interplate marginal convergence boundary.
      PL Kararkhorsan flysch basin - Low.
      PM Chagai volcanic arc and calc-alkaline magmatic belt - Low.
      PN Makran flysch basin - Low.

Figure 2. South side of Salang Pass in Afghanistan. The long shed in the left center leads to a tunnel entrance through the pass. This and other sheds protect against snow avalanches, as well as against the many rock falls from the well fractured and frost-shattered granite and other crystalline rocks of the area.

with altitude as precipitation and weathering activity increase. Evidence of mass movement is ubiquitous, as would be expected in mountainous terrain anywhere, but activity is not uniform and varies in character and geomorphological significance with respect to altitude, lithology and relief.

Goudie et al., (1984) noted in the Hunza Karakoram that snow avalanching is the most important mechanism at high elevations (>4500 m) and during late winter and spring may extend down to 2500 m. Rock falls and rock slides occur virtually anywhere there are steep rocky cliffs and slopes, a quite common situation. Debris falls and slides are important at lower elevations in the zones of large magnitude valley-fill sediment storage in the mountains. Earth falls, slides and flows are not common and the few of these occur only where there are thick saprolites and loess. Complex types of failure involving extensive movements of large landslide blocks lying above unstable sedimentary rocks occur but are not common. They are known mostly only in central Afghanistan and in the Himalayan foothills of Pakistan.

In spite of economic losses and because of the pressures of the war in Afghanistan and the overwhelming desire for development of mineral resources in both countries (Shroder, 1984), little attention has been paid to slope-failure hazards in either country. Instead effort has been directed to produce basic bedrock maps, together with a few minor reports on such surficial phenomena as ground water or soil types. No slope-failure hazard maps have been made to date, and no records are known to be kept of problem areas. Along the impressive Karakoram Highway to China in the valleys of the Indus, Gilgit and Hunza Rivers, for example, the Pakistani army must continually repair damage from large and small slope failures. Little information about this problem is available, in part because of the sensitive strategic nature of the highway. In addition, neither general monetary loss nor the toll in human lives are known for any slope failures in these regions. Thus the following discussion of slope failure is highly generalized and offers only a minimum view of some landslips observed by the writer or extracted from a small literature. Furthermore, slightly greater information is available concerning slope failure in the Karakoram and western Himalayas of Pakistan than from Afghanistan because of the longer tradition in India and Pakistan of record keeping and access by western explorers and scientists. Perhaps in future years as the importance of data collection on slope failures becomes better known to officials of the two countries, greater information will become available.

2 SLOPE FAILURE TYPES

The term "landslide" is avoided in many places herein in order to reduce usage of nomenclature in which sliding is not involved. The terms "landslip" and "slope failure" are preferred. Thus rock slides and debris slides are discussed as landslides, but falls, flows, topples, and sackung failures are not. Complex forms involve a variety of types of material and movement and are discussed as well. The classification of Varnes (1978) is used throughout.

2.1 Rock falls, rock slides and sackung

Rock falls and slides from the cliffs of Afghanistan occur widely but are a special problem only on the few highways that have been constructed in the past three decades. Two areas are known for problems - the Salang Pass road from the northern border south to Kabul, and the highway from Kabul to Jalalabad through the gorge of the Kabul River. The Salang Pass, eroded by glaciers and rivers through the heavily fractured granites and other crystalline rocks of the Hindu Kush, is constantly threatened by snow avalanches and slope failures. The top of the pass at about 4000 m altitude has a long tunnel protected by sheds at both ends. Several other protective concrete sheds on the south side deflect mass-movement events and avalanches (Fig. 2). The road through the Kabul gorge has a much lower incidence of slope failures because of its lower and more arid location but the sheer cliffs produce considerable hazard none-the-less (Fig. 3).

Figure 3. Sheared and deformed crystalline rocks of the Kabul gorge on the main highway between Kabul and Jalalabad cities. The highway is cut into the cliff at left center and extends from the building in the center and across to left.

Elsewhere in Afghanistan, the many deep narrow canyons are dammed in some places by massive slope failures. Most of the resulting lakes are small, linear in orientation, and some are rather short-lived as they are readily susceptible to both infilling as well as to downcutting of the dam by the parent river. Many of these lakes are valuable sources of water in arid Afghanistan.

In Badakshan Province in the northeast, for example, 2.5 km long Lake Skazer in Jangale Sare Hawdz occurs on the upper Kokcha River directly below its confluence with the Munjan and Anjuman Rivers (Bruckl, 1935; Mirzad, 1970). Several other lakes also occur a few tens of kilometers downstream, between Skazer and the famous lapis lazuli mines, and appear to be also the result of rock slides. Desio (1975) mapped several old lake deposits and drained landslip lakes on the upper Kokcha and its tributaries, including the large (20 km long, 2 km wide) Khash valley between Faizabad and Jarm (Fig. 4). The unusual flat-floored valley was thought to be the result of slope failure at both ends, but the exact sites of the presumed failures were not located. There was no evidence either for glacial moraines which could have formed a dam, so the open-ended valley remains enigmatic. However,

just above Ardar, at about 5 km from the confluence of the Warduj valley in the Baharak basin, a rock slide clearly dammed the valley. The failure came from a rounded scar on the right (north) bank and left a heap of granitoid gneiss blocks across the valley. The blocks are mingled with abundant clay and rest upon a terrace also rich in clay. Desio (1975) thought that the fine clastics were related to lacustrine deposition following failure of the slope and blockage of the valley. Similarly, at the headwaters of the Kokcha River about 8 km northeast down from the Panshir valley and Anjuman Pass, occurs a 700 m long lake that was dammed by a rock slide (Grotzbach and Rathjens, 1969; photograph by Dunsheath and Baillie, 1961).

In central Afghanistan, in the Ajar valley and in several other nearby canyons (lat 35° 15'-30'; long 67° 15'-30'), several small lakes have been dammed by rock falls. Weippert (1964) mapped eight such lakes there, and Shank, Petocz and Habibi (1977) noted that a major earthquake in the early 1960's caused near vertical canyon sides of Ajar valley to collapse and produce a natural dam behind which 1.5 km long Lake Chiltan laps against sheer rock walls rising 350 m above.

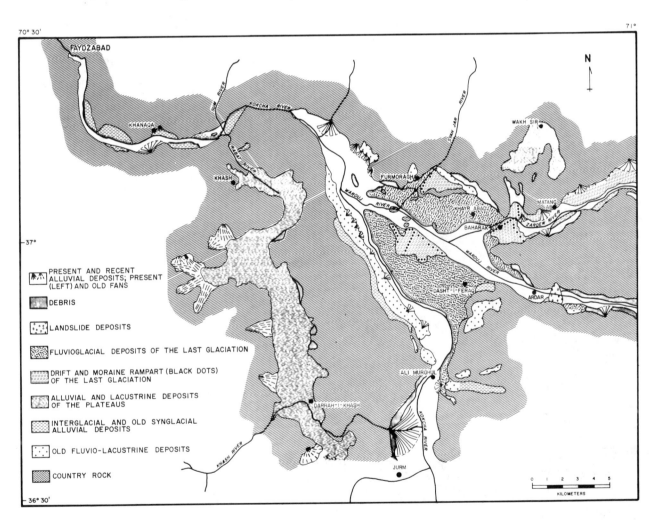

Figure 4. Quaternary deposits in the Baharak and Khash valley areas of Afghanistan (after Desio, 1975).

Because the relief is so much greater in Pakistan, rock slides and rock falls into canyons and onto roads are more common than in Afghanistan Several of these failures are well-known historical events because they blocked rivers and later caused floods; some of the failures are less well known to the average citizen but are notorious to road crews because of the continued danger.

Two major rock slides blocked rivers in the western Himalaya in the 19th century. The most famous of these happened in January 1841 when an earthquake brought down the Hatu Pir spur of Nanga Parbat mountain into the gorge of the Indus River (Fig. 5). The site was initially unstable because a rock spur was thrust over unconsolidated sediments along the Raikot fault; in fact Butler and Prior (1988) misidentified this rock slide and located it at nearby Lichar because of the confusing relationships of bedrock thrust above Quaternary sediments there. L. Owen (oral commun., 1988) believes that the thick debris-flow sediment behind and overlying the fault blocks at Lichar argues for a greater antiquity of the site than only 150 yr of accumulation since 1841. In any case, the resulting dam created a lake 150 m deep and 30 km upstream past Bunji and the confluence of the Hunza and Gilgit rivers. As the impounded Indus rose to its full height on the terrace plateau by the Bunji fort, the rock slope opposite failed as well (Fig. 6) and made a giant wave in the lake (Drew, 1875). A note on birch bark was sent downriver to warn of the hazard but no one at the mountain front seems to have understood the implications. In the early days of June 1841, the dam failed and a wall of water, mud and rocks roared down the Indus gorge and out onto the plains below. A Sikh army encamped near Tarbela was overwhelmed, with at least 500 soldiers perishing immediately, and the plain was "sown with barren sand" (Abbot, 1848; Becher, 1859; Mason, 1929).

The location of the 1841 landslip was relocated and mapped as a part of this study (Fig. 7). The vertical drop of the rock slide was about 1.9 km and the horizontal distance of transport was about 4 km. The rock slide overrode earlier diamicts of unknown origin as well as one or more glacial tills. As the slide was in motion, it first moved west and produced typical hummocky topography on the east side of the Indus. Further along it was deflected by bedrock and other diamict hills on the west bank of the Indus and was then forced to move northwest. The surviving interior portion of the landslide at this location has several longitudinal ridges and furrows of about a meter amplitude, as well as considerable open-matrix mega-porosity characteristic of some rock slides. The distal end of the landslide is marked by hummocks of jumbled rubble that are about 2-4 m high.

When the landslide dam broke, the flood waters produced scour marks and four mega-ripples with an average wavelength of 110 m and an amplitude of 2.5-6 m just upstream of the dam. Flood scour scars also occur a few kilometers downstream on both sides of the Indus between the landslide and the Raikot Bridge (Fig. 5). Also upstream at Bunji the Indus was diverted around the new Bunji rock slide, undercut the plateau opposite, and part of it failed in a debris fall or debris slide as well. At present the Indus cuts through the toe of the Bunji rock slide, a position it probably attained by a subsequent river diversion event caused by one of the common debris flows that periodically sweep down the Bunji fan.

The other major catastrophic failure of some note occurred at Pungurh in the Hunza River gorge. There a rock fall on the left or south bank of the river blocked the gorge for six months until the dam burst in August. Yet another catastrophic flood wave swept down the Indus, raising the water level at Attock by some 9 m in less than 10 hours. The discharge may have been well over 1 million m^3/sec (Goudie et al., 1984).

Figure 5. Photograph of the Indus valley below Nanga Parbat peak. The 1841 rock slide is in the background below the long fresh talus slopes. The Lichar fault blocks occur above the Raikot fault and unconsolidated sediments in the center.

Figure 6. Bunji rock slide of 1841 that is now traversed by the Karakoram Highway. This massive slide moved into the lake that rose behind the earthquake-generated rock slide of 1841 further downstream. The Bunji rock slide created a giant wave in the lake which swept across to Bunji town and caused further debris fall or slide collapse of the steep banks there.

Many other large slope failures have been recognized elsewhere in the middle and upper Indus, Gilgit and Hunza River valleys (Burgisser et al., 1982; Mason, 1935), but as with the 1841 rock slide, many of the largest are along the Raikot fault zone (Fig. 8). One example is the Sumari slide above Sassi village which overrode the Hurban fault branch of the Raikot system (Shroder et al., in press). Sumari village is located in the graben at the head of the slide, which is filled with till from the last glaciation, indicating that the slope failure predated at least part of the last glaciation (Figs. 8 & 9).

At Khalola River, 10 km south of Sassi, 67-85° dips on foliation planes into the Indus gorge have produced the massive Burumdoin rock slide (Figs. 8 & 10). The landslide originated in Iskar gneiss and is about 1.3 km wide and a similar amount in length. The vertical drop from the crown to the river was about 1 km. The thickness of the rock moved, based on the height of the main and side scarps, was about 300 m. Most of the rock slide went into the Indus River and was subsequently eroded away. Possible landslide debris from the original failure is plastered against terrace sediments on the opposite side of the river. At the present time, intermittant movement of the pulverized remains of the Iskar gneiss force rock debris at the level of the highway out over the bedrock lip and into the Indus 30-100 m below (Figs. 8 & 11). The post landslide downcutting by the river is not yet sufficient to detach another rock slide below the highway along a new shear plane at the river level, but this could happen at a future time.

The event sequence leading to failure at the site is as follows: (1) glaciation down the Indus valley oversteepened the side slopes; (2) glaciation from nearby Haramosh Peak issued from Khalola River canyon into

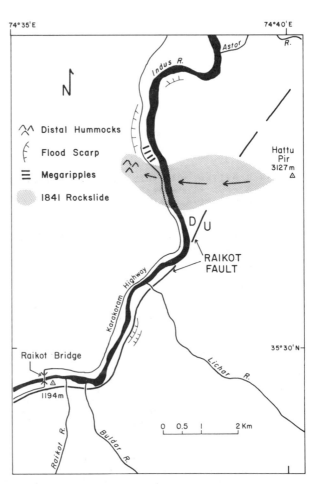

Figure 7. Sketch map of the rock slide that blocked the Indus River in 1841. The failure from Hattu Pir was produced by an earthquake, presumably from movement on the Raikot fault that crosses beneath the slide.

331

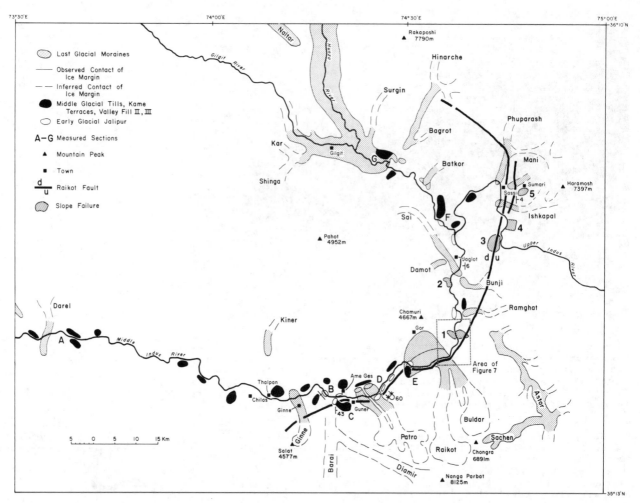

Figure 8. Index map of ice-related deposits of Pleistocene age and recent slope failures in the Indus valley and its tributaries. 1 - Main rock slide of 1841; 2 - Bunji rock slide of 1841; 3 - Sarkun Lake and area of <u>sackung</u> failure; 4 - Burumdoin rock slide at Khalola; 5 - Sumari slope failure.

the Indus valley; (3) initial failure of the Burumdoin rock slide occurred; (4) post-failure downcutting into bedrock by the Indus River left the landslide several hundred meters above the river; (5) subsequent minor movements of the landslide periodically have closed the highway to Skardu.

Across the gorge to the west on the narrow ridge cut by the Raikot fault between Haramosh and Nanga Parbat peaks, a large number of antislope scarps show characteristics of massive <u>sackung</u> failure (Zischinsky, 1966; Bovis, 1982). Directly on top of the ridge, topographically anomalous Sarkun lake occurs in a narrow, graben-like structure. The ridge is about 2600 m high and 7 km wide and occurs inside a tight loop where the Indus River is offset by the fault. This steep topography and the fault produces a situation ideal for the formation of <u>sackung</u> in which the entire ridge is fracturing and settling downward (Fig. 8).

Extensive rock-fall deposits on glaciers in the Karakoram Himalaya provide evidence of what appears to be a common but rarely observed process there. Hewitt (1988) discovered that three catastrophic slope failures in July 1986 deposited about 20×10^6 m^3 of rock fragments over 4.1 km^2 on the Bualtar Glacier in Hunza. A small amount of carbonate appeared to have been calcined by frictional heating, presumably at the base

of the initial sliding mass. The glacier subsequently accelerated its movement in and below the landslide mass several months after the rock slide occurred.

2.2 Debris falls, slides and flows

Movements of debris from the many cliffs and slopes of till and other rubble deposits are exceptionally common in Afghanistan and Pakistan. Two causes of movement that are most prevalent are seismicity and monsoon rains. During several large earthquakes in Pakistan in 1984, for example, debris falls and slides occurred all along the Karakoram Highway and some temporarily blocked it. Small failures went on for several days afterwards and provided exciting driving as material rattled down onto vehicles and highway. During the summer monsoons, the water load on the slopes produces similar effects.

In several places in the foothills of the Himalayas, large debris slides have developed from regolith overlying metamorphic bedrock or from unstable sedimentary rock. Near the Thakot Bridge across the Indus, for example, the Karakoram Highway is repeatedly severed by a debris slide that moves down steeply-dipping foliation planes (Fig. 12). In general wherever the river valleys paral-

332

lel foliation, the highways on the slopes above are menaced by such phenomena. In the valley of the Jehlum River near the Pir Panjal Range of eastern Pakistan, the red shales and clastics of the Muree Formation are unstable and periodically remove the highways as the slopes fail. In some places river undercutting of thick valley-fill sediments has produced large debris slides. The Karakoram Highway near the Astor River crosses over such a failure (Fig. 13).

Probably the most common catastrophic mass-movement mechanism in the Himalayas and Hindu Kush, however, is the debris flow in which mixtures of fine and coarse clastics, soils, and organic matter are mobilized by rain and snowmelt. Waves or walls of this slurry rush down gullies, build up fans, and devastate fields and bridges. Small storms on peaks can generate many flows that wash into the arid lower valleys in the mountain rainshadows. Larger, more destructive debris flows have blocked some of the larger rivers and overwhelmed major structures as a result of intense storms or break-out floods from water bodies blocked behind ice or rock glaciers (Goudie, et al., 1984).

Major rainstorms in Afghanistan and Pakistan that generate debris flows are largely the result of penetration of the summer monsoon rains deep into mountainous areas (Fig. 14). Such areas generate large debris flows for two main reasons; (1) debris accumulates over the years in mountain valleys through freeze and thaw and other slope processes; and (2) when the rare rains do occur, the greater aridity in the deep valleys and the lack of vegetation allow free erosion of the debris masses. The result commonly is a striking display of the catastrophic energy of the debris flow.

Wasson (1978) described a debris flow in August 1975 at Reshun in the Hindu Kush of Pakistan. The flow was initiated by a landslip that was triggered by rainfall. Boulders 1-2 m long were carried in the flow, which had a maximum surface velocity of 3 m/sec and lasted for about 65 minutes.

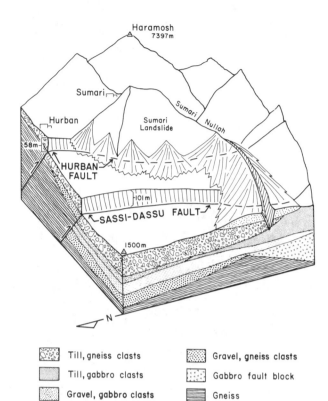

	Till, gneiss clasts		Gravel, gneiss clasts
	Till, gabbro clasts		Gabbro fault block
	Gravel, gabbro clasts		Gneiss
	Lacustrine silt		Gravel, diverse clasts

Figure 9. Schematic block diagram showing stratigraphy and landforms along the Hurban and Sassi-Dassu faults of the Raikot fault system. The Sumari rock slide is associated with the Hurban fault and Sumari village is located in the pull-away graben at the head.

Figure 10. Burumdoin rock slide at Khalola below Haramosh Peak on the road to Skardu. Foliation planes of the Iskar gneiss dip into the Indus River below and make an unstable situation.

Total volume of sediment transported was about $10^5 m^3$. Debris flows of this magnitude appear to have a return interval at this locality of >30 yr, but smaller ones occurred in the same location in <10 yr. The large debris flow left deposits of boulders and finer sediment in embayments along the sides of the stream, thinly veneered the stream banks, and deposited lobes on the apex of the Reshun fan.

Goudie (et al., 1984) pointed out that evidence for debris-flow activity in the Hunza Karakoram and elsewhere in the western Himalaya is widespread and includes: (1) striped talus slopes; (2) debris-flow fans veined by debris-flow tracks; (3) debris-flow infillings in gullies; (4) large depositional lobes in valley bottoms; and extensive accumulations of matrix-supported sediments within the Hunza Valley infills.

Debris flows (mud-rock flows in the Chinese literature) in the vicinity of Batura Glacier in 1974 were plentiful, large and damaging (Cai et al., 1980; Wang et al., 1984). For example, on 12 April 1974 a giant debris flow burst out of Balte Bare

Figure 11. Bedrock lip at base of Burumdoin rock slide. The white obelisk in the upper right is a "martyrs monument" to the men killed in construction and maintenance of the road across the rock slide.

Figure 12. Debris slide of regolith down foliation planes dipping parallel to the hillslope near Thakot Bridge on the Karakoram Highway.

Figure 13. Large debris slide crossed by the Karakoram Highway near Astor and Ramghat valleys. This landslide is, however, too small to include on Figure 8. The talus above the 1841 rock slide that dammed the Indus is in the background to the south.

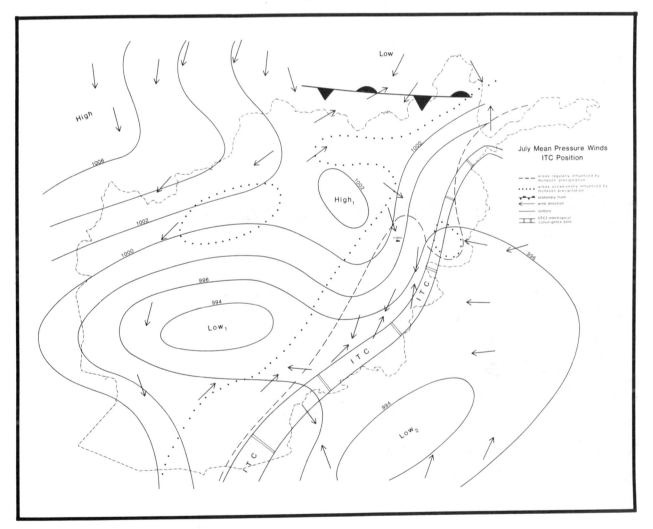

Figure 14. Map of the summer monsoon climate of Afghanistan showing zones where monsoon rains penetrate only occasionaly (after Sival, 1977). Such areas tend to be locations of catastrophic debris flows.

canyon with a front as high as 20-30 m. At a maximum discharge of 6300 m^3/sec, it poured out 5 million m^3 of earth, debris and rock. Soon a huge fan was formed 300-400 m wide, over 150 m long and 80-100 m high. The Hunza River discharge was only 40-50 m^3/sec at the time and was quickly blocked to form a lake 8 km long. The innundation overwhelmed one of the new 120-m long Friendship Bridges built along the Karakoram Highway by the Chinese, and it destroyed many agricultural fields as well. Although the exact cause of the catastrophic flow was not known for certain, it later was discovered that the Balt Bare Glacier was a surging type, and it was therefore suspected that the flow was related to rapid release of surface or basal meltwater, as is common with such types of glaciers.

Also in the same area near Batura Glacier, heavy rains on 4 August 1974 caused several other debris flows. One came down with a front 5 m high and a maximum discharge of 250 m^3/sec. In only ten minutes a big fan spread out at the mouth of the gulley to about 220 m wide, 300 m long and 4 m thick. The 40-m wide Hunza River was dammed in spite of a high summer discharge of 250 m^3/sec. The flood rose rapidly upstream but within an hour the rushing torrent broke through the dam and drained the lake.

A decade ago in the Ghorband Valley in the Hindu Kush of central Afghanistan, I observed several large debris flows in motion (Fig. 15). On 6 July 1978, after several days of increasing clouds and thunder, heavy monsoon rains occurred on all the peaks and lighter rain fell in the valleys. Flash floods were common. After being turned back by a 100-m wide debris flow in the Seli-Gauand tributary to the Ghorband River, our party nearly was overwhelmed by a debris flow as we crossed the Senjedak River valley a short distance away. Suddenly over the noise of the truck engine, we heard a roar as a wall of mud and rocks about 3-m high

and moving at an estimated 5 m/sec burst over the bridge in front of us. Velocity varied in a series of surges which occurred as rocks and debris formed temporary dams. Viscosity varied in the flow, with periodic boulderless mud lobes emerging and slopping up on the valley sides. As we stood beside the flow to take pictures, we noticed that the ground was shaking and the surrounding canyon walls were releasing small rocks which rolled down into the flowing mass. The flow went on for over an hour and when it seemed nearly finished I drove the truck across through wet and viscous mud. The mud stayed wet for several days afterward, according to Afghans who attempted foot crossings. Several weeks later when I returned, however, it had dried into a compact, indurated mass.

The Senejedak drainage basin measures less than 10 km in length and heads on the Koh-i-Usturgarden ridge at about 3300 m, so it is not large. Apparently the rare penetration of the monsoon into this region (Fig. 14) triggered the events.

Similarly in the Gilgit Valley on 21 July 1984, I observed a sudden cloudburst in the Kohistan ranges directly south of Bagrot Valley (Fig. 8). This caused a large debris flow to descend a small gulley and overwhelm another Chinese-built bridge of the Karakoram Highway. About half of the concrete structure was broken away by the 1-2-m long boulders that struck the upstream side (Fig. 16). This flow was highly conspicuous from the mountain slopes several km away because of the dust clouds from infalling debris caused by the ground shaking (fig. 17), and because of the roaring noise that echoed through the Gilgit Valley. The Gilgit River was dammed for less than an hour before the river broke through.

Figure 15. Debris flow in motion in the Senjedak River valley, Afghanistan in 1978.

Figure 16. Partly destroyed bridge on the Karakoram Highway
damaged by a debris flow in 1984. The bridge was built too low
in the gulley for the size of the drainage basin above it.

Figure 17. Dust in a gulley (Fig. 16) from a debris flow moving
down a fan in the Gilgit valley in 1984. The Gilgit River at the
lower left was dammed for about an hour.

2.3 Earth falls, slides and flows

Because so much of Afghanistan and Pakistan are arid and underlain by well indurated
rock, earth falls, slides and flows are not
common. The steep loess hills of northern
Afghanistan do fail in this fashion in some
places, and the Cretaceous shales of central
Afghanistan also behave similarly from time
to time. In general, however, these kinds
of landslips are rare in Afghanistan.

On the other hand, the well-watered foothills of the Himalayas in Pakistan do support earthen types of slope movements. On
the Chattar Plain, for example, deeply
weathered, clay-rich saprolites slide and
flow slowly downhill (Lawrence and Shroder,
1985). These features compare favorably
with similar African examples in which deep
weathering of granitic rocks in the tropical
highlands provides plentiful clays that subsequently fail in a variety of earth flows

and slides. Failure and consequent dissection is repetitive at the same sites, and provides exposure of the deeper coarse clastics for their later incorporation into the local transport and depositional systems (Shroder, 1976). At the Chattar Plain sites, 4-10 m of residual soils have developed over corestones and parent bedrock below. Failure of these materials is well advanced in the Chattar uplands, especially where accelerated by modern rapid deforestation (Fig. 18). The Karakoram Highway in this southern region is repeatedly disturbed by large slow landslips that locally choke rivers with debris.

2.4 Complex and undifferentited failures

In Central Afghanistan at the edge of the pre-collisional Eurasian plate margin, but in the variably deformed southwest end of the Surkhab-Jaway block (Fig. 1), occurs a

Figure 18. Earth flow near Chattar Plain in the foothills of the Himalayas. Light-colored scarps along the ridge line mark the right flank of the failure, with most of the remander of the slope in the foreground moving slowly (<1 m/yr) to the left. The highway is disrupted periodically by the movement. This area was covered with a dense coniferous forest only a few decades ago.

Figure 19. Large mass movement of Cretaceous shales and clastics at the downstream end of the lakes of Band-i-Amir in central Afghanistan. The failure came from the south side (left) of the escarpment and produced the large hummocky lobe in the center.

Figure 20. Complex slump and flow in Cretaceous shales and clastics on Koh-i-Jedachel at the upstream end of the Band-i-Amir lakes in central Afghanistan.

Figure 21. Complex slump and flow in sedimentary rocks of Cretaceous age on the southeast-facing slope above Darya-i-Bum valley between Bamiyan and Band-i-Amir in central Afghanistan.

variety of carbonates, conglomerates and shales that are subject to complex slope failure. Most of these are of the slump and flow type, and some are associated topographically, and perhaps genetically, with unusual carbonate-dammed lakes (Lapparent, 1966; Lang and Lucas, 1970; Jux and Kempf, 1971; Albul et al., 1975). Agitation of the carbonate-saturated water is known to cause some of the carbonate precipitation at the Band-i-Amir and other lakes; such water movement over landslip debris may be par-

tially responsible for the close association of the two phenomena.

In any case, downstream a few hundred meters from the lowermost of the eight lakes of Band-i-Amir, a large landslip has filled half the valley with rock rubble from the cliffs of shale and conglomerate above (Fig. 19). At the southeast upper end of the lake complex, Koh-i-Jedachel has on its slopes a variety of slump blocks and flowed shale on it (Fig. 20).

Between the Band-i-Amir lakes and the

town of Bamiyan to the east occur several flat-floored valleys, some of which have been dammed by carbonate precipitation and perhaps landslips as well. For example, both Koh-i-Zardkamar above the Bariki River valley and the southeast-facing slopes of Koh-i-Tawa above the Darya-i-Bum valley have large landslips (Fig. 21).

3 CONCLUSION

Delineation of type, cause, magnitude and time of occurrence of major slope-failure hazards in Afghanistan and Pakistan is important as development progresses in these widespread fragile environments with their local heavy population pressures. Numerous field excursions, large- and small-scale field mapping, photography, remote sensing and a literature search were used in this study to locate sites of prior or active instability. The basic passive conditions favoring high-magnitude slope failure events in Afghanistan and Pakistan include well-jointed and well-fractured metamorphic rocks, unstable fine clastics and plentiful deposits of rubble, exceptional maximum relief produced by especially active uplift and downcutting, and oversteepened slopes from faulting, glaciation and river under-cutting. Personal observation of failure induction by seismicity and torrential rains added understanding of some of the high-frequency events such as rock falls and slides, debris falls and slides, and the rapid wet debris flows. Road building and deforestation are the main human causes of failure hazards. The main highways through the Himalayas and Hindu Kush are cut repeatedly by large and small slope failures, especially during monsoon season. Constant maintenance is required in all seasons and most is provided by the armies of both Afghanistan and Pakistan.

Slope-failure hazards to highways, settlements and water works in these countries limit development in some areas. This study provides preliminary data for analytical and predictive capabilities that can be used to establish later protective and remedial measures. Such hazard mitigation is a vital necessity according to the recent call by the U.S. National Research Council for an international Decade for Hazard Reduction. Continued, more intensive analysis of unstable slopes in Afghanistan and Pakistan is important in developing remote areas in the face of the expanding and migrating human populations in these countries.

REFERENCES

Abbott, J. 1848. Report on the cataclysm of the Indus taken from the lips of an eyewitness. J. Asiatic Soc. Bengal. 17:231-233.

Albul, S.P., A.A. Makhorin & V.M. Chmyriov 1975. Mineral carbon dioxide water in central Afghanistan (in Russian). Izvestia Vysshihk Uchebnykh Zavedenii Geologia I Razvedka, Moscow. 41:103-107, pt. 2.

Becher, J. 1859. Letter addressed to R.H. Davies Esquire, Secretary to the Government of the Punjab and its dependencies. J. Asiatic Soc. Bengal. 28:219-228.

Bovis, M.J. 1982. Uphill-facing (antislope) scarps in the coast mountains, southwest British Columbia. Geol. Soc. America Bul. 93:804-812.

Bruckl, K. 1935. Uber die Geologie von Badakshan und Kataghan (Afghanistan). N.J. Miner. Geol. u. Palaont. Abhandl., Beit. 74:360-401, Abt. 13.

Burgisser, H.M., A. Gansser & J. Pika 1982. Late glacial lake sediments of the Indus valley area, northwestern Himalayas. Ecologae geol. Helvetica. 75:51-63.

Butler, R.W. & D.J. Prior 1988. Tectonic controls on the uplift of the Nanga Parbat massif, Pakistan Himalayas. Nature. 333:247-250.

Cai Xiangxing, Li Nienjie & Li Jan 1980. The mud-rock flows in the vicinity of the Batura Glacier (in Chinese with English abstract). In Batura Glacier Investigation Group, Exploration and Research on the Batura Glacier, Science Press, Beijing. 146-152.

Desio, A. 1975. Geology of central Badakshan (north-east Afghanistan). Italian Expeditions to the Karakoram (K2) and Hindu Kush, Scientific Repts., III Geology-Petrology. Leiden, Brill.

Drew, F. 1875. The Jummo and Kashmir territories; a geographical account. London, E. Stanford.

Dunsheath, J. & E. Baillie 1961. Afghan quest: the story of their Abinger Afghanistan expedition: 1960. London, Harrap.

Ferguson, R.I. 1984. Sediment load of the Hunza River. In K.J. Miller (ed.), The International Karakoram Project. 2:456-495. England, Cambridge Univ. Press.

Goudie, A.S., D. Brundsen, D.N. Collins, E. Derbyshire, R.I. Ferguson, Z. Hashmet, D.K.C. Jones, F.A. Perrott, M. Said, R.S. Waters, W.B. Whalley 1984. The geomorphology of the Hunza Valley, Karakoram Mountains, Pakistan. In K.J. Miller (ed.), The International Karakoram Project. 2:359-410. England Cambridge Univ. Press.

Grotzbach, E. & K. Rathjens 1969. Die heutige und jungpleistozane Vergletscherung des Afghanischen Hindukusch. Z. f. Geomorphologie, Supplementbd. 8:58-75.

Hewitt, K. 1988. Catastrophic landslide deposits in the Karakoram Himalaya. Science. 242:64-67.

Jux, U. & E. Kempf 1971. Stauseen durch Travertinabsatz im zentral afghanischen Hochgebirge. Z.F. Geomorphologie, Supplementbd. 12:107-137.

Kazmi, A.H. & R.A. Rana 1982. Tectonic map of Pakistan. Geol. Sur. Pakistan.

Lang, J. & G. Lucas 1970. Contribution a l'etude de biohermes continentaux: barrages des lacs de Band-e-Amir (Afghanistan central). Bul. Soc. geol. France (7), n. 5, XII:834-842.

Lapparent, A.F. de 1966. Les depots de travertin des montagnes Afghanes a l'ouest de Kaboul. Rev. Geog. Phys. & Geol. dynam. (2), fasc. 5 VIII:351-357.

Lawrence, R.D. & J.F. Shroder, Jr. 1985. Tectonic geomorphology between Thakot and Manshera, northern Pakistan. Geol. Bul. Univ. Peshawar 18:153-161.

Mason, K. 1929. Indus floods and Shyok glaciers. Himalayan J. 1:10-29.

Mason, K. 1935. The study of threatening glaciers. Geog. J. 85:24-41.

Mirzad, A.G. 1970. On the lakes of Afghan-
    istan and their origin (in Russian with
    English abstract). Moscow University
    Seriia Geografiia, n. 5. 15:106-110.
Shank, C.C., R.G. Petocz & K. Habibi 1977. A
    preliminary management plan for the Ajar
    Valley wildlife reserve. UNDP FAO & Afghan
    Ministry of Ag. Dept. Forests & Range
    Report.
Shareq, A., W.M. Chmyriov, K.F. Stazhilo
    Alekseev, V.I. Dronov, P.J. Gannon, B.K.
    Lubemov, A.Kh. Kafarskiy, E.P. Malyarov,
    L.N. Rossovskiy 1977. Mineral resources of
    Afghanistan. UNDP Project AFG/74/012.
Shroder, J.F.,Jr. 1976. Mass movement on
    Nyika Plateau, Malawi. Z.f. Geomorphol-
    ogie, 2:56-77.
Shroder, J.F.,Jr. 1984. Comparison of
    tectonic and metallogenic provinces of
    Afghanistan to Pakistan. Geol. Bul. Univ.
    Peshawar 17:87-100.
Shroder, J.F.,Jr. in press. Landforms and
    hazards in the Nanga Parbat Himalaya.
    American Scientist.
Shroder, J.F.,Jr., M. Saqib Khan, R.D.
    Lawrence, I.P. Madin, S.M. Higgins in
    press. Quaternary glacial chronology and
    deformation in the Himalaya of northern
    Pakistan. In L. Malinconico & R.J. Lillie
    (eds.), Geophysics and tectonics of the
    western Himalaya. Geol. Soc America Spec.
    Paper.
Sivall, T.R. 1977. Synoptic-climatological
    study of the Asian summer monsoon in
    Afghanistan. Geograf. Annaler 59A:67-87.
Varnes, D.J. 1978. Slope movement types and
    processes. In R.S. Schuster & R.J. Krizek
    (eds.), Landslides; analysis and control.
    p. 11-33. Washington, DC, Transport Res.
    Brd. Comm. Sociotechnical Sys. Nat. Res.
    Council Nat. Acad. Sci. Rept. 176.
Wang Wenying, Huang Maohuan, Chen Jianming
    1984. A surging advance of Balt Bare
    glacier, Karakoram mountains, 1:76-83. In
    K.J. Miller (ed.), The International
    Karakoram Project. England, Cambridge
    Univ. Press.
Wasson, R.J. 1978. A debris flow at Reshun,
    Pakistan Hindu Kush. Geografiska Annaler
    60A:151-159.
Weippert, D. 1964. Zur Geologie des Gebeites
    Doab-Saighan-Hajar (Nord-Afghanistan).
    Beih. geol. Jb. 70:153-184.
Zeitler, P.K. 1985. Cooling history of the
    NW Himalaya, Pakistan. Tectonics 4:127-
    151.
Zischinsky, U. 1966. On the deformation of
    high slopes. Intern. Soc. Rock Mechanics,
    Proc. First Con. Lisboa, 6/11 2:179-185.

Landslides: Extent and Economic Significance, Brabb & Harrod (eds)
© 1989 Balkema, Rotterdam. ISBN 90 6191 876 6

Landslides: Extent and economic significance in Australia, New Zealand and Papua New Guinea

R.J.Blong
Macquarie University, New South Wales, Australia

G.O.Eyles
Soil Conservation Centre, Aokautere, New Zealand

ABSTRACT: This paper examines the style, extent, and major factors influencing the distribution of landslides in three countries. Most attention is focussed on the economic impacts of the landslides. In Australia, an arid continent distant from tectonic plate margins and characterised by old, hard rocks, landslides are generally small scale, relatively rare, and have only limited economic impact. In New Zealand and Papua New Guinea, located on active margins of the Indo-Australian plate and characterised by young, mobile rocks and high orographic rainfalls, landslides are frequent, sometimes large, and often produce considerable economic damage to several sectors of the economy. Loss of life and damage increase in the order Australia, New Zealand, Papua New Guinea.

1 INTRODUCTION

This paper examines the style, extent, and economic significance of landslides in Australia, New Zealand, and Papua New Guinea (Figure 1). *Landslides* include slumps, slides, flows, and avalanches, with the terminology generally following that of Varnes (1978). Discussion is based on a review of published and unpublished works and the experience of the authors and is organized into sections dealing with each country separately.

2 AUSTRALIA

2.1 The setting

The Australian continent lies near the centre of the Indo-Australian tectonic plate which was produced by the break-up of Gondwanaland and separation from Antarctica about 95 Ma ago and the coalescence of the Indian and Australian plates about 44 Ma ago (Veevers, 1984). Much of the margin of the continent is characterised by steep escarpments with pronounced stream incision, considerable relief, and a variety of active geomorphic processes. On the other hand, most of the interior is low-lying with gentle slopes formed on flat-lying, old (pre-Cenozoic) hard rocks on which the imprint of current geomorphic processes may be difficult to discern.

The contrast between the continental interior and the continental margins is further drawn by rainfall distribution which marks the interior as arid or semi-arid (less than 500 mm annual precipitation). The 16 million people of Australia live almost entirely around the coastal margin. With only a few towns with populations more than 10,000, the population density in the interior averages less than 0.3 persons/square km; Land use is mainly extensive pastoralism although there are significant areas of mining activity.

2.2 Landslide distribution

Given the distribution of landforms, precipitation, and population, it is not surprising that landslides are concentrated largely around the margins of the continent. It is also not surprising that many landslides are the consequence of human activities.

No national survey of landslide distribution has yet been undertaken, although Ingles (1974) has provided some details and Joyce (1979) has summarised landslide sites and characteristics in Victoria. However, it seems likely that the major area where landslides produce economic consequences are in Tasmania, Victoria, New South Wales, and Queensland

- along the eastern seaboard of the country. Major areas of landsliding are shown on Figure 2, in some cases with an indication of periods when landslides have occurred.

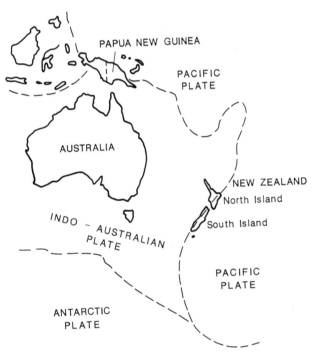

Figure 1. Australia, New Zealand, and Papua New Guinea in relation to the boundaries of tectonic plates.

2.3 Landslide lithologies and slopes

In Tasmania, particularly in the Tamar Valley, slides, slumps and mudflows, occur on overconsolidated and fissured Tertiary clays and sandy clays where interbedded gravels and basalts provide seepage. In the northwest of the state landslides occur on Tertiary basalts and on Permian mudstones; again, the occurrence of springs is important. On the Tertiary lake sediments of the Tamar Valley the threshold angle for slope stability is about 7°, but on weathered basalt soils the critical angle is $12-14^{\circ}$. Some of the large slides are reactivated ancient block slides and slump-earthflows. Certainly, many modern landslides are located on large ancient failures (Stevenson, 1971; Donaldson, 1980).

In Victoria, landslide problem areas occur in the Strzelecki and Otway Ranges and in urban areas, including the Dandenong Ranges close to Melbourne. Most failures occur in Tertiary sediments (sandstones and mudstones) and basalts or on weathered Mesozoic sediments (arkose and mudstone), particularly on strongly undulating terrain. Many of the larger landslides are of the slump-earthflow type. Large failures (>20 ha) appear to be inactive, but small slides (<2 ha) occur within the larger inactive ones. These shallow translational movements are also common (Danvers-Powers, 1892; Barlett, 1966; Joyce, 1979; Cooney, 1980; Brumley, 1983).

Numerous earthflows also occur in the Wodonga area of northern Victoria, particularly on south-facing slopes of more than 12-15° (Clutterbuck, unpub.). Some slides in Miocene sediments in the Baccus Marsh area west of Melbourne have had important economic consequences (Harding, 1952).

In New South Wales, four main areas of landsliding have been recognised: Illawarra, Campbelltown-Picton, Gosford-Wyong (respectively, south, southwest, and north of Sydney) and Lismore (Figure 2). The first three areas occur in the essentially flat-lying Permo-Triassic sandstones and shales of the Sydney Basin. Other minor areas of slope failures also occur in Sydney's northern and northwestern suburbs. The interbedded shales, lateral water movement, and minor bedding dips are probably important controls on landslide distribution in these areas. Slides, slump-earthflows, and debris avalanches are all common.

Critical slope angles on shales are about 11° in the Picton area but fall to as low as 6° in other areas on the same shale units. Large rockfalls have also occurred from steep sandstone cliffs in the Sydney Basin, particularly where subsurface coal mining continues (Chesnut, 1980; Blong and Dunkerley, 1976; Pells et al., 1987; Cunningham, 1988).

A number of landslide areas occur around Lismore in the north of New South Wales on krasnozem, chocolate, and yellow podsolic soils, particularly on Jurassic shales (Melville, 1976). Landslides are also common in some suburbs of Newcastle ; some of these are related to mining-induced subsidence. Other numerous small areas of landslides occur in the coastal foothills and along the central highlands (Chesnut, 1980).

In Queensland landslides occur mainly in rural areas but there is an increasing landslide hazard in the suburbs of Cairns and Toowoomba. Slope failures occur on thick colluvium and deep soils on Paleozoic metamorphics in the Cairns area, on Tertiary basalts around Toowoomba and in the Gold Coast hinterland, on poorly consolidated Tertiary sediments in the Brisbane and Buderim (100 km to the north) areas, and on weathered Mesozoic argillites in the Lockyer Valley. More than 1000 landslides have been reported in the latter area. In the Mount Tambourine area south of Brisbane, slopes on basalts steeper than about 8° are zoned as hazardous (Hofmann, 1980; Findlay, 1980).

In the Northern Territory there are numerous areas of natural rockfalls along escarpment areas and planar slides occur in shales and greywackes between the Adelaide River and Katherine. In the Bullocky Point area of Darwin, coastal slumping has continued for at least 40 years. However, most stability problems are minor and man-made (Weber, 1980).

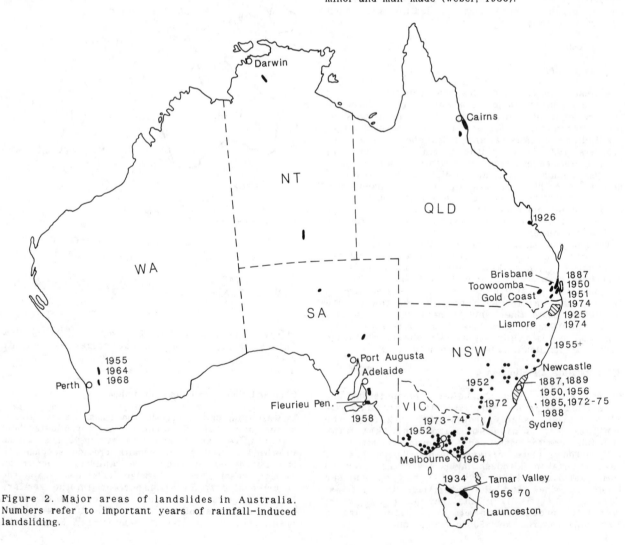

Figure 2. Major areas of landslides in Australia. Numbers refer to important years of rainfall-induced landsliding.

Few landslides have been described from Western Australia. Landslides occur on chalk and greensand near Gingin to the northeast of Perth. Earthflows also occur in colluvium on slopes of more than 15^0 in the Chittering Valley, also northeast of Perth. Minor slumping along river banks away from the faultline was produced by the M=6.9 1968 Meckering earthquake (Pilgrim and Conacher, 1974; Biggs and Mather, 1980; Everingham et al., 1982).

In South Australia earthflows occur on Permian glacigene sediments on the Fleurieu Peninsula on slopes of $8-20^0$ and rockslides and slab and wedge failures are found in the Flinders and Mount Lofty Ranges. At Oodnadatta, west of Port Augusta and in the Leigh Creek coalfield, large-scale slumps with volumes of more than 0.5 million m^3 occur in rocks. In the Adelaide Hills, small debris slides are quite common in road batters particularly on southwest-facing slopes in winter (Bourman, 1976; Selby, 1980a, b).

2.4 Landslides and climate

Although few studies have been made in Australia of the relationship between landslide occurrence and precipitation characteristics, there can be little doubt that, at least for Eastern Australia, many failures are induced by heavy rainfalls. In the Illawarra area of New South Wales rainfalls of more than 430 mm/month almost invariably produce landslides. With rainfalls of more than 350 mm/month landslides are common but Young (1978) has noted that landslides have also occurred in 40% of the months receiving 250-375 mm. In the Lismore area landslides tend to occur only when 30-day precipitation totals exceed 450 mm (Melville, 1976).

In many areas landslides seem to be associated more with long-saturating rains rather than the shorter duration high intensity falls. This generalisation applies particularly to areas where earthflows are common and it is possible to recognise very wet years when landslides have been widespread; for example, in the Picton area landslides occurred in the 1880's, 1896-1900, 1905, 1915, 1922, 1950, 1956, 1969, 1974 and 1988. However, in at least some of these years, high intensity rains produced small slides, often on areas of old earthflows.

Some of the wet years noted above, for example, 1950, 1956, 1974, and 1988, also produced major landslides in other parts of eastern Australia, notably in the Illawarra area, Wodonga (northern Victoria), Queensland, and in northern Tasmania. Some of the other years in which rain-induced failures have been common are indicated on Figure 2.

Human interference in the form of clearing vegetation and the consequent reduction in evapotranspiration, and moisture inputs from blocked natural drainage lines, garden hoses, downpipes, leaking pipes and septic systems have exacerbated stability problems in urban areas in wet years. Other human interference, particularly the construction of cut slopes for house foundations and access roads, have created many of the slope stability problems evident in Eastern Australia.

2.5 Economic consequences

Very few people have died or suffered physical injury as a result of landslides in Australia; for the last 100 years deaths average one every four or five years.

Although economic consequences of landslides are widespread in urban areas in Australia, the most severe consequences have been experienced in Tasmania as several large landslides in this state have been active for long periods of time in urban areas. Between 1956 and 1980 more than 50 houses were destroyed or became uninhabitable and there were also considerable effects on transport routes and grazing and agricultural land.

The Lawrence Vale landslide (near Launceston) was active between 1956 and 1970 and resulted in the destruction of 30 houses. Houses riding passively on the main lobes were little damaged but those near the margins suffered severely (Knights, 1977). A further 5 houses were destroyed in an adjacent slide. Both slides may be part of an older, deep-seated, failure.

At Beauty Point, also in the Tamar Valley, 12 houses have been destroyed and a further 16 have suffered structural damage. This earthflow has been active for at least 70 years (Stevenson, 1975).

In the Penguin-Ulverstone area the wet winter of 1931 produced a movement of 3.8 m over a width of 200 m in interbedded basalts, gravels and clays, seriously disrupting road and rail links. To stabilise this slide a total of 550 m of tunnels were constructed in the 1930's and converted to French drains at a cost of less than US$10,000 in 1936 dollars (Balsille, 1936). The tunnels drained up to 80,000 litres/day in 1935.

Rockfalls have occasionally blocked roads and railways and damaged buildings. In northwest Tasmania the small lumber port of Stanley, located at the foot of 100 m high dolerite cliffs, has suffered damage to the lumber mill and the railway line, the latter providing access to the wharf. In 1979 it was reported that a loose mass of about 500 tonnes, 50 m above the base of the cliff offered an unacceptable risk. It was recommended "that the area under the cliff be evacuated and closed to the public. The problem then became one with political implications, because this action would have meant closure of the railway and timber company operations. This would, in turn, close the port facilities and ultimately spell economic doom to the township of Stanley" (Donaldson, 1980). Subsequently, a rock fall fence was constructed from used materials to provide protection at a cost (1980 dollars) of less than US$25,000.

The consequences of the Lawrence Vale and Beauty Point landslides also included political action. The Lawrence Vale Landslip Act of the Tasmanian Parliament (1961), for example, "provided that the land which moved and destroyed houses should be acquired by the City of Launceston and the previous owners compensated to the extent of 75% of the valuation of the land and houses. The Act only applied to dwelling houses which were permanently owner occupied. No other property was considered" (Stevenson, 1975).

In the Strzelecki Ranges of Victoria annual expenditure on roads damaged by landslides exceeds US$80,000, with roads narrowed by slumping, covered by earthflows and displaced laterally by creeping colluvium. Widening of roads creates further potential for slope failures. As Brumley (1983) notes:

"Landslides also degrade the quality of farm pastures, damage crops, destroy dams, add to the sediment load of streams and occasionally damage buildings, fences and powerlines. The recent trend toward more extensive land use by subdividing large farms into smaller hobby farms and residential allotments increases the potential for landslide damage ...".

In the Bacchus Marsh area of Victoria a 1952 slide carried away more than 100 m of concrete irrigation channel and disrupted the water supply for 2000 residents, necessitating pumping of an additional 4.5 million liters per day for at least several months (Harding, 1952). In the Dandenong Ranges close to Melbourne minor failures are common with consequent damage to roads and batters. In 1892 a debris avalanche with a volume of more than 30,000 m^3 destroyed a house, carried away fences and buried many hectares of good land under debris (Danvers-Powers, 1892).

In New South Wales landslides have been described as the second most costly hazard if maintenance costs are included (Chesnut, 1980). Apart from costs associated with road and railway links most of the economic consequences of landslides have been in damage to houses in the Illawarra area, Gosford-Wyong, Newcastle, Lismore and in the suburbs of Sydney. Most of these failures can be attributed in part to the lack of statutory controls on the development of potentially unstable areas (Chesnut, 1980).

The Wollongong area (Illawarra) has been one of the most important in Australia in terms of the economic and other consequences of landslides (Figure 3). As Young (1979) notes:

"heavy rains in 1973-74 initiated failures that destroyed at least 11 houses, very severely damaged at least 12 others, and threatened many more. Yet demand for housing on the naturally unstable slopes of the 400 m high escarpment which rises behind the city continues and, as is the norm for this and other natural hazards ..., many homeowners, homebuyers, and public authorities do not perceive or respond constructively to the threat. This is so despite the fact that no insurance cover is available against the possibility, that no compensation is available at all from either the N.S.W. or Australian governments, and that the only relief is purchase of disrupted land by council for a nominal sum to release the affected party from rates payment. The owner of a slipping house thus lives under constant emotional and financial stress as his home is disrupted a little more after each heavy rain, and cooperative efforts to solve the problem with neighbours are ... discouraged by uncertainties of responsibility and multiple ownership".

Young and Johnson (1977) estimated that the cost of the 1974 landslides to private individuals was at least US$2 million (1988 dollars). In addition, there was substantial expenditure of public monies on repairs to roads, water mains, and other public services.

Figure 3. Houses demolished in Buttenshaw Drive, Wollongong, September 1974. (Photo: A.R.M. Young).

Until the 1970's the highway through Picton was the main Sydney-Melbourne route. In 1950 this road was closed for two months when there were 17 landslides onto the road (mainly earthflows) and a major failure in a road fill. Binns (1950) described the disruption to traffic in July 1950 as the most serious road problem facing the New South Wales Department of Main Roads. In November 1969 the road was again closed for 20 days, mainly by slides in fills and cut slopes. Although the road is no longer the main highway, despite considerable expenditure on horizontal drainage and other stabilising measures, failures in fills continue.

Significant economic consequences have also occurred in rural areas. Melville (1976) reports a variety of failures in banana plantations on the far North Coast of New South Wales in 1973-74. Roads were closed and one property lost 20% of its productive land. At Toonumbar Dam 5% of the foreshore was regarded as slip-prone, with one

landslide extending across 13 ha. Farm roads and fences also suffer frequently in the Picton area. One 1950 earthflow changed the course of a river, effectively halving an area of arable land and making it uneconomic for further cultivation.

In Queensland, although landslides occur mainly in rural areas, considerable damage has been done to residential property in Brisbane, Ipswich, Toowoomba, and Buderim. In Oxley, a Brisbane suburb, high pore fluid pressures have produced problems since 1971. In 1974 rapid drawdown in the Brisbane River following severe flooding damaged properties in Coronation Drive. In the Corinda area ten houses on Tertiary sediments were destroyed or had to be relocated after sliding during the same floods (Findlay, 1980). In Toowoomba minor damage has occurred to houses and street alignments as a result of small recent slides and slumps in basaltic clays. In Buderim, of more than 20 houses built on landslide colluvium, three have been condemned and abandoned after being damaged by a 1-2 m movement on a 3-5 m deep shearplane (Barker, 1980). Landslides have also affected major roads in this area. Hofmann (1980) estimated average annual losses due to landsliding at about US$2.5 million (1988 dollars), with most of these losses occurring in rural areas.

Indirect costs from landslides have not been assessed but total to quite substantial amounts. Such costs include, in addition to those already mentioned: reduced real estate values; loss of tax revenues as a result of devaluation; loss of revenues to developers; and, loss of agricultural, forestry, and/or industrial productivity (Donaldson, 1980). Willmott (1984) has noted that adverse effects (i.e. indirect costs) in rural areas include erosion of disrupted soil, germination of noxious weeds, limitations on machinery access to disrupted areas, bogging hazards for stock in ponded areas, loss of stock in crevices, and blocking of streams with slide debris.

3 NEW ZEALAND

3.1 The setting

The New Zealand land mass lies across the boundary between the Indo-Australian and Pacific plates (Figure 1). Relative movement of these plates has rarely been simple, with oblique compression dominating over much of the landmass (Lewis, 1980). As a consequence of this compression, block faulting and uplift of both basement and Tertiary cover rocks have proceeded at a rapid rate.

Both lateral and vertical tectonic movements associated with plate interaction create a dominantly steepland environment. Active volcanism adds a further complexity to the structure and stratigraphy.

The stratigraphy of New Zealand is dominated by Mesozoic and Cenozoic rocks with local occurrences of Paleozoic rocks. The typically highly deformed, and indurated Mesozoic rocks – greywackes, argillites, schists, gneisses and plutonic rocks – form the bulk of the mountain lands. Adjacent areas are usually weaker Mesozoic and sometimes lower Tertiary argillites and mudstones, the latter typically less indurated and deformed.

Relief is predominantly steep with 65% of the landmass having slope angles in excess of 15°.

These factors, together with a cyclonic weather pattern and frequent high intensity rainfall cells, create an environment in which landslide erosion predominates.

3.2 Landslide distribution

The New Zealand Land Resource Inventory – NZLRI – (NWASCO, 1975, 1979), a national, computerised land resource survey at 1:63,360 scale, provides the basis for analysing the extent and potential for erosion in New Zealand. The use of landform units to represent erosion has been necessary due to the extremely large number of erosion scars covering the landscape. Figures 4, 5, and 6 are distribution maps based on these data and illustrate the distribution of landform units containing flows, debris slides, debris avalanches, and slumps (Eyles, 1983).

The distribution patterns indicate causal factors are not uniform, but are related to a variety of influences, the interrelations of which are often not well understood. The following sections briefly review the main factors involved in landslide erosion in New Zealand.

Figure 6. The distribution of map units containing slump erosion.

Figure 4. The distribution of map units containing earthflow erosion.

Figure 5. The distribution of map units containing slips (soil and earth) and debris avalanche erosion.

3.3 Vegetation

Vegetation cover is recognised as having a dominant influence on landslide occurrence. While it is accepted that landslides occur naturally under undisturbed indigenous forest, conversion to pasture increases the probability of landsliding. Selby (1976) estimated that in the North Island conversion of greywacke hill country from forest to pasture increased the probability of a landslide of given magnitude by a factor of 3. O'Loughlin and Pearce (1976) showed that clear felling of beech-podocarp-hardwood forests increased the landslide rate from 1 /km^2 to 20 /km^2. They also illustrated that a similar trend occurs with exotic forest removal, with root strength reducing rapidly following felling. For instance, the root strength of *Pinus radiata* reduces to a minimum within 5 to 8 years of felling. Landslide susceptibility increases during that time. The converse also applies, with stability increasing following the planting of forest trees with new plantings taking 5 to 10 years to effectively reduce landslide potential.

Whether animal damage to vegetation increases the potential for landsliding is less well understood. Some writers (e.g. McKelvey, 1959; Cunningham and Stribling, 1978) attributed increased rates of landsliding and surficial erosion in areas of indigenous mountain vegetation to its ill-thrift or destruction by introduced browsing and grazing animals. Other writers point to the difficulty of proving such assumptions (e.g. Veblen and Stewart, 1968) and to the lack of correlation between landsliding and the destruction of animal-susceptible species. Some recent studies (Jane and Green, 1983) show that while vegetation mortality is thought to be the most important factor predisposing an area to erosion, little evidence can be found to relate this mortality to animal damage.

3.4 Earthquakes

Earthquakes are recognized for their ability to initiate landslides in New Zealand. In the South Island mountains Whitehouse and Griffiths (1983) have shown an observed frequency of landslides with volumes >1 x 10^{6}m^3 of one per 244 years over the last 10,000 years. They considered this figure was biased by the masking effect of more recent erosion. A count of all large landslides over the last 1700 years, a period in which they believe other erosion does not mask the high magnitude events, indicated a frequency of one per 94 years. Whitehouse and Griffiths considered most of these large landslides were triggered by earthquakes with only a few produced by large storms. In the North Island there are also examples of large landslides, the causes of which are unknown, but which are likely to have been triggered during earthquakes (e.g. Stephens, 1975).

The effects of earthquakes on the generation of smaller landslides was illustrated by Pearce et al. (1985) in a study of the Murchison earthquake (M=7.6) in 1929. This event initiated more than 1850 landslides with areas >0.25 ha within a 1200 km^2 area close to the epicenter. These landslides were mainly on well-bedded and jointed calcareous mudstones and fine sandstones in steep mountainlands. However, in 1968 the Inangahua earthquake (M=7.1) in the same area produced only 21 new landslides. The 1855 Wellington earthquake (M=8.0) caused "a great many large slips" (Robbins, 1958) in the Rimutaka Range adjacent to Wellington.

These New Zealand examples suggest that significant numbers of landslides occur only when earthquake magnitudes are greater than 7.0.

3.5 Climate

Climate plays a dominate role in initiating landslides. Grant (1985) identified five erosional phases in New Zealand's recent history, four occurring before the impact of introduced animals and three of these before the impact of European settlement. He attributes these phases to climatic changes and also considers that there has been an increase in storminess since the 1950's but this is difficult to confirm due to the incomplete records of erosion-causing storms.

Figure 7. The location of areas experiencing serious damage due to landslide erosion in New Zealand since 1970 (Modified from Eyles and Eyles, 1983). Shaded areas have occurred since 1983.

Figure 7 illustrates the distribution of significant erosion-causing storms since 1970 (Eyles and Eyles, 1982). Rainfall intensities causing these events are of two types. The first is high intensity events with rainfalls ranging as high as 900 mm over 3 days; the second is long periods of near field capacity soil moisture levels as a result of prolonged, light rainfall with additional relatively small triggering storms. Accurate measurements of rainfall intensity during these storms, however, are difficult to obtain due to the cellular nature of the high intensity rainstorms and the very dispersed rainfall measuring network in rural areas. It has, therefore, been difficult to predict landslide-triggering storm rainfalls. That it can be done has been shown by Crozier and Eyles (1980) who developed an antecedent excess rainfall model and applied this to Wellington City, using the city's rainfall records and known landslide damage events.

A standard measure of storm severity has been to measure the landslide density or the number of scars per ha (Table 1). However, data in the table are not strictly comparable in some cases. For instance, Crozier and Eyles (1980) excluded areas of low relief from analysis and although the Otoi results indicate 2.25% bare ground, actual values for different landforms ranged from 0.1% to 8.5%. Furthermore, in some cases, only landslide scars have been included whereas others also include debris tails.

Table 1: Comparison of landslide densities from selected storm events in New Zealand.

	LOCALITY	EROSION DENSITY (scars/km^2)	% AREA ERODED
1	Tangoio (Eyles, 1971)	31	1
2	Pakaraka (Crozier & Eyles, 1980)	478	9.7
3	Wairarapa (Crozier & Eyles, 1980)	98	
4	West Coast (O'Loughlin & Pearce, 1976)	19	
5	Thames - Te Aroha (Salter et al., 1983)	86	
6	Otoi Harmsworth et al., 1987)	157	2.25

3.6 Lithology

There is generally a close relationship between surface lithology and landslide type and landslide potential. Table 2 indicates some general relationships taken from the NZLRI for map units which have a landslide erosion severity of "severe" or greater (Eyles, 1983).

Table 2: The dominant rock types in which landslide erosion occurs with a severe or greater erosion ranking (after Eyles, 1983)

EROSION TYPE	DOMINANT ROCK TYPE	
Earthflow	Mudstones	-fractured -bedded -bentonitic
	Argillite	-crushed
Slump	Mudstones	-fractured -bedded -massive
	Argillite	-crushed
	Unconsolidated sediments	
Debris avalanche	Controlled by slope angle	
Slip (debris slide or flow)	Mudstone	-bedded -massive -fractured

Earthquake-induced landslide frequency can also be related to rock type. Pearce et al. (1985), for instance, found the Paleozoic sedimentary, metamorphic and some granite rocks had significantly less landsliding than on the Tertiary sediments despite being closer to the epicenter of the 1929 Murchison earthquake.

3.7 Economic costs

The economic cost of landslides in New Zealand is difficult to analyse as in major storms landslides generally occur in association with flooding. It is the flooding which has the immediate economic impact, with sedimentation on the flats having a longer term, but equally obvious, impact. Whether the sediment results from landslide erosion or from fluvial erosion cannot be distinguished accurately enough to allow analysis. Hawley (1984), making an "informed guess", estimated the average cost of landslides in the pastoral hill country is in excess of US$12 million per year. The immediate costs can be large, for instance, following Cyclone Bola, a three day cyclone in March 1988 during which up to 900 mm of rainfall fell and which devastated the eastern region of the North Island (Figure 8), the government allocated US$30 million to compensate farmers for losses in production but the resulting damage to land, housing, communications and services is estimated to have cost more than US$72 million (Stephens et al., 1988).

Figure 8. Landslides caused by Cyclone Bola in March 1988, when over 600 mm of rain fell in a three day period at Waerangaokuri, 20 km west of Gisborne, New Zealand. The large central landslide caused a dam which flooded the country road for three weeks before the dam wall was artificially opened. (Photo: N. Trustrum).

In addition to these immediate costs, there are longer term costs associated with losses in pastoral productivity from landslide scars. Recent hill country pasture productivity studies by Trustrum and Hawley (1986), indicate pasture production levels will not recover to within 20% of the original, uneroded production levels. Reduction in volume of the plant rooting medium, causes lower fertility and a reduced moisture holding capacity on the slip scar soil. As profits are made on the top 20-30% of productivity, such production losses have a significant effect on the long term viability of hill country farming enterprises.

Landslides in urban environments are common in cut and fill situations, with houses and roads being undercut or inundated with debris. The most significant urban landslide in New Zealand's history occurred in 1979 when the Abbotsford block slide, covering 18 ha, caused the destruction or removal of 64 houses in the Green Island Borough of Dunedin (Figure 9). The main movement comprised a 50 m slide at a rate of about one meter per minute and occurred within an hour, sliding along a bedding plane failure surface dipping at 7 [o]. Initial movement is thought to have begun as early as 1968 (Gallen, 1980). Cost is difficult to assess. While claims under the Earthquake and War Damage Act amounted to US$1.5 million it is likely that the total cost of the one landslide to New Zealand was US$6 to 9 million.

Figure 9. The Abbotsford landslip in Dunedin, New Zealand's most disastrous urban landslide which covered 18 ha and caused the destruction or removal of 64 houses. (Photo: L. Homer).

While the economic cost of landslides is considerable in New Zealand, the loss of human life attributable to landslides has been low, with a national average of less than one per year (Hawley, 1984).

4 PAPUA NEW GUINEA

4.1 The setting

Papua New Guinea lies at the collision margin of the Indo-Australian and Pacific plates (Figure 1). Minor plates have formed at the margin, probably as a result of shearing, to produce a complex pattern of highly seismic zones and steep mountain ranges.

Along some sections of the the north coast of the mainland, uplift rates exceed 1mm/y. Massive limestones as young as Eocene in age rise to more than 3000 m above present sea level. Detachment tectonics have played an important part in the evolution of the relief of the highlands (Jenkins, 1974). Parts of the country are amongst the most seismic places on earth. With this setting, landslide activity is profoundly influenced by seismicity.

In many parts of the highlands the massive limestones are underlain by weak shales and mudstones of Cretaceous age. Exposures of these and other weak rocks provide preferential sites for landslide occurrence. In many areas the weaker rocks form slopes of lower angles, but most of the landscape is one of steep slopes with V-shaped valleys and narrow ridge crests. Some of the clay shales have amongst the highest recorded erosion rates (Blong and Humphreys, 1982).

The Pleistocene and Recent history of Papua New Guinea has been punctuated by explosive eruptions of numerous volcanoes along the Bismarck Arc (Figure 1). Volcanic ash (tephra) is widespread across large areas of Papua New Guinea; for example, Blong and Pain (1978) indicated that more than 50,000 km² of the central highlands has received a tephra mantle of more than 2 m in thickness. The steep slopes of the volcanoes encourage the development of lahars after eruptions and highlight the possibility of volcano sector collapse to form massive debris avalanches.

With a landmass area of about 460,000 km² and a population of 3.6 million, Papua New Guinea is not densely populated but some of the intermontane valleys of the central highlands are densely settled, leading to soil erosion and slope stability problems. However, most of the country is covered in thick tropical vegetation. Rainfall throughout the country is high and often seasonal. In many areas annual rainfall exceeds 3000 mm; some areas of the highlands receive, on average, more than 6000 mm/y. At Tabubil the average annual rainfall (1970-1977) is 7800 mm (Byrne et al. 1978).

4.2 Landslide distribution

Little information is available about the distribution of landslides in Papua New Guinea. The data on Figure 10 represent a compilation of landslides described in published work and the authors' experience rather than an accurate portrayal of actual landslide occurrence. Nonetheless, it seems likely that the two main factors influencing landslide distribution are weak lithologies and earthquake energy.

4.3 Landslides and lithology

Mapping of tephra occurrence in order to assess the long term stability of hillslopes (Blong and Pain, 1978) suggests the following order of decreasing stability of lithologies:

1. 25-35°
   Quaternary volcanics
   Limestones

2. 12°+
   Tertiary volcanics
   Greywackes, indurated sandstones and siltstones
   Quaternary fan sediments

3. c.6°
   Clay shales/friable mudstones
   Quaternary lake sediments (clays and silts)

The greywackes are represented across large areas of the country by the sandier facies of the Chim Formation (Bain, McKenzie & Ryburn, 1975). Although variable in strength, with some slopes up to about 24° retaining much of their upper Pleistocene tephra mantle, the mean threshold slope indicating stability appears to be about 12°. On the other hand, the montmorillonite-illite clay shales in the Chim Formation, here informally named Chim shale, are rarely stable on slopes of 10°. Lithologies equivalent to Chim shale (Cretaceous to lower Tertiary in age) are very widespread and are probably the most landslide-prone materials in Papua New Guinea.

Landslides on Chim shale are commonly of the mudslide type (cf. Brunsden, 1984). They tend to be large, often tens of hectares in area and up to several kilometers in length. The Tarpa River slide near Kundiawa began moving in 1979, evidently triggered by an earthquake. This mudslide was at least 70 ha in area with a volume of between 2.5 and 7.5 million m³. Almost all of the failed mass was composed of colluvial material (i.e. the product of earlier landslides), reworked blocks of Chim shale, and limestone boulders (Blong, 1986a). Another example, the Maiapa slide near Porgera, has been active for at least 30 years, with portions of the mass moving on different occasions. The total area of this slide is nearly 8 km².

Chim shale mudslides are very common in some areas. Near Kundiawa at least 12 mudslides occur along a 16 km stretch of the Highlands Highway. Near Porgera, the 5-6 km access road to the mine site crosses at least 12 mudslides although only 4 of these have been active in the last decade (Blong, 1985).

4.4 Landslides and earthquakes

Numerous earthquakes this century have produced thousands of landslides, but it is only in the last 20 years that reasonably intensive investigations of these slope failures have been made.

The 1985 Bialla earthquake (M=7.1) produced an

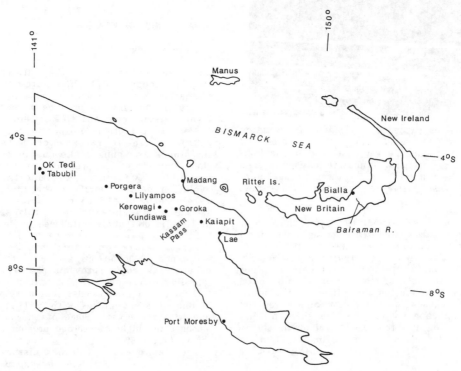

Figure 10. Papua New Guinea, indicating places mentioned in the text.

enormous number of slope failures in the central mountains of New Britain. King and Loveday (1985) indicated that two "very large" debris avalanches occurred in the Bairaman River. The larger failure had a mass of about 180 million m^3 and involved the failure of a 200 m thick limestone (Figure 11). Debris filled 3 km of the valley floor up to 200 m thick and dammed a lake more than 3 km long and over 100 m deep. The lake was breached using explosives in September 1986 (King, 1987a).

In the New Ireland earthquake of 1985 (M=7.2), landslides stripped the soil and vegetation cover from up to 50% of the surface of small headwater catchments (Buleka, 1985). The 1970 Madang earthquake (M=7.0) produced 60 km^2 of shallow debris avalanches in an area of 240 km^2, with up to 60% stripped in some small headwater areas (Pain, 1972). Pain and Bowler (1973) suggested that earthquake-initiated landsliding occurs about once in 200 years on average in the mountains along the north coast of mainland Papua New Guinea.

4.5 Landslide potential

The large failures associated with earthquakes, particularly the 1985 Bialla event, raises the question of potential in other parts of the country for very large failures. A debris avalanche with an estimated volume of 5-7 million m^3 occurred from the Hindenburg wall in 1977, and field investigations in the Porgera area indicate that several debris avalanches with minimum volumes of more than 10 million m^3 have occurred there in the last 20-30,000 years. It seems likely that all areas of the country where massive limestone cliffs are underlain by clay shales and similar weak rocks have the potential to produce large-scale failures. The average frequency of such events, on a nationwide basis, is probably more than one failure every 10 years (cf. Byrne et al. 1978).

In 1888 the cone of Ritter volcano collapsed into the sea; the debris avalanche generated may have had a volume of 4-5 km^3. Earlier large debris avalanche

Figure 11. The large debris avalanche generated by the May 10, 1985 Bialla earthquake (M=7.0) in New Britain. The Bairaman River (foreground) was dammed – see text. (Photo. P.L. Lowenstein).

scars have been identified on a number of volcanoes in the highlands and along the Bismarck Sea (Blong, 1986b; Johnson, 1987). Future collapses are certainly a possibility.

4.6 Economic consequences

Numerous lives have been lost in landslides in Papua New Guinea in recent years. The annual average number of deaths from landsliding is probably between 5 and 10, but it seems likely that many deaths are not recorded. In September 1988 a landslide (probably a debris avalanche) in the Kaiapit area killed 75 people and buried three villages under more than 50 meters of debris.

Other economic costs resulting from landsliding are also high despite the low population densities. Road closures are common; the highlands highway, the only road link from Lae on the coast to the intermontane valleys of the interior has occasionally been closed for a week or more by landslides, particularly in the Kassam Pass area. Damage to bridges also occurs (Ghartey and Anderson, 1983). Upgrading of roads has frequently also meant higher maintenance costs on larger cuts and fills. Many of the deaths from landsliding also occur along roads.

In other areas, villages have been abandoned because of slope stability problems. Potential landslides near the Hindenburg wall have forced the movement of villages (e.g. Anderson, 1980). Landslides also occur in the highlands town of Goroka (Warren, 1988). At Kerowagi two landslides in 1978 and 1979 with a volume of 6000 m^3 caused the abandonment for a year of 27 new houses constructed in the Police Barracks and the expenditure of more than US$60,000 on remedial works (Goldsmith, 1979; Blong, 1986a). The Tarpa River slide also destroyed several houses and large areas of food gardens. Landslides in almost every province in the country damage gardens and houses built of local materials almost every year. A typical example, the Lilyampos debris avalanche in 1987 killed one man and seven pigs and destroyed 10 houses, 22 coffee gardens, and a number of vegetable gardens (King, 1987b).

Individual landslides have also caused considerable damage to modern engineering structures and communication networks. A failure in a slope during the construction of a tailings dam at Ok Tedi forced the redesign of part of the gold-copper project and litigation in the courts, not yet finalised, but possibly involving more than US$100 million. The Tarpa River slide destroyed an electricity tower on the main leg of the highlands network. While replacement costs were probably less than US$20,000, large parts of the network were without power, or reliant on backup generators with insufficient capacity, for a week or more. Shops in Kundiawa and elsewhere lost freezer stocks and a range of interesting social effects stemmed from the power failure (see Blong, 1986a). Submarine landslides in 1966 and 1968 damaged telephone cable connections; the latter break in the SEACOM cable cost more than US$150,000 (1969 values) to repair (Denham, 1969; Krause et al., 1970).

5 CONCLUSIONS

Landslides in the three countries discussed here present a number of contrasts. In Australia, landslides are concentrated near the continental margins, in the moister more populated parts of the country where geomorphic processes are more active. Most landslides have volumes of less than 1-10,000 m^3. Earthflows are common in pasture and other relatively little-disturbed areas, and commonly move following long saturating rains. Houses, roads and railways bear the brunt of the damage produced by landslides, particularly in the south east of the country. Deaths are rare. As Hollingsworth (1982) notes, landslide risks in urban areas are a relatively recent phenomenon.

In New Zealand landslides are much more common,

including some large spectacular failures which are triggered by earthquakes and/or cyclonic rains. Most failures are less than 100-1000 m^3, but landslides with volumes greater than 100,000 m^3 are not uncommon. Individual landslides have damaged tens of houses and other structures, but deaths resulting from landslides are relatively infrequent. Landslides have been shown to have profound long-term impacts on agricultural productivity.

In Papua New Guinea, landslides occur on an even more spectacular scale with very large failures occurring every few years. Earthquakes and weak lithologies, coupled with high annual rainfalls appear to be the main driving forces. Landslides with volumes >1 million m^3 are not uncommon. Most landslides are natural rather than man-induced. The annual average death rate from landslides may approach 10, about an order of magnitude higher than in New Zealand. Food gardens are commonly destroyed although the wide dispersal of the gardens belonging to an community suggest that these effects are unlikely to be very severe. Houses and roads suffer the most damage, though rarely in urban areas. However, modern well-engineered structures have also been severely affected and produce a significant damage bill and other social impacts which result from the scarcity of modern resources.

It is not unreasonable to relate these broad variations in landslide style and frequency to differences in tectonic setting. At one extreme, the Australian continent lies distant from the plate margins and has landforms of generally low relief developed on old and hard rocks. Much of the continent is arid. At the other extreme, Papua New Guinea has formed at an active collision margin from recently uplifted rocks including massive limestones and weak shales. Rapid uplift and incision have produced steep slopes in an active seismic environment. Rainfalls are an order of magnitude higher than in Australia.

Economic and other costs of landsliding cannot be assessed with any accuracy for the three countries. However, it seems likely that costs increase from Australia to New Zealand, to Papua New Guinea. In each step it may be that costs increase by an order of magnitude.

6 REFERENCES

Anderson. G.R. 1980. Review of landslide at Bolovip Mission. Geol. Surv. Papua New Guinea Prof. Opinion 21-80.

Bain, J.H.C., D.E. Mackenzie & R.J. Ryburn 1975. Geology of the Kubor Anticline, Central Highlands of Papua New Guinea. Aust. Bur. Min. Resources Bulletin 155 (PNG 9).

Barker, R.M. 1980. Buderim landslide investigation. Geol. Surv. Qld. Record. 1980/37: 209-216.

Bartlett, A.H. 1966. Investigation of landslides affecting some Victorian roads. Proc. Aust. Road Res. Board. 3(2): 732-746.

Balsille, G.D. 1936. Large land slip near Ulverstone. Tasmania, arrested by deep drainage. Trans. Inst. Eng. Aust. 8(1): 365-371.

Biggs, E.R. 1980. Geological hazards in Western Australia. Geol. Surv. Qld. Record. 1980/37: 175-196.

Biggs, E.R. & R.P. Mather 1980. Environmental and engineering mapping techniques. Geol. Surv. Qld. Record. 1980/37: 269-274.

Binns, C.S. 1950. Report on slips and landslides on State Highway No 2, Hume Highway, Shire of Wollondilly between Camden and Picton on 26 July, 1950. Dept. Main Roads N.S.W. Report 39M.976.

Blong, R.J. 1985. Mudslides in the Papua New Guinea Highlands. Proc. 4th Int. Conf. and Field Workshop on Landslides. Tokyo. 277-282.

Blong, R.J. 1986a. Natural hazards in the Papua New Guinea highlands. Mountain Research and Development. 6(3): 233-246.

Blong, R.J. 1986b. Pleistocene volcanic debris avalanche from Mount Hagen, Papua New Guinea. Aust. Jnl. Earth Sci. 33: 287-294.

Blong, R.J. & D.L. Dunkerley 1976. Landslides in the Razorback area, New South Wales, Australia. Geografiska Annaler. 58A(3): 139-147.

Blong, R.J. & G.S. Humphreys 1982. Erosion of road batters in Chim shale, Papua New Guinea. Civil Eng. Trans., Inst. Eng. Aust. CE24(1): 62-68.

Blong, R.J. and C.F. Pain, 1978. Slope stability and tephra mantles in the Papua New Guinea highlands. Geotechnique. 28(2): 206-210.

Bourman, R. 1976. Environmental geomorphology: examples from the area south of Adelaide. Royal Geog. Soc. Aust. S.A. Branch. 75: 1-23.

Brumley, J. 1983. Slope stability in the Strzelecki Ranges, Victoria. Collected case studies in Engineering Geology, Hydrogeology, and Environmental Geology. M.J. Knight, E.J. Minty & R.B. Smith (eds.). Geological Society Australia Special Publication 11:127-147.

Brunsden, D. 1984. Mudslides. In D. Brunsden & D.B. Prior (eds.). Slope instability. Wiley, Chichester. 363-418.

Buleka, J. 1985. Geological report on the effects of the earthquake of 3 July 1985 centred in Southern New Ireland. Geol. Surv. Papua New Guinea Report 85/20.

Byrne, G.M., M.M. Ghiyandiwe & P.M. James 1978. Ok Tedi landslide study. Geol. Surv. Papua New Guinea Report 78/6.

Chesnut, W.S. 1980. The extent and severity of geological hazards in New South Wales. Geol. Surv. Qld. Record 1980/37: 4-65.

Cooney, A.M. 1980. Urban geological hazards in Victoria. Geol. Surv. Qld. Record. 1980/37: 74-96.

Crozier, M.J. & R.J. Eyles 1980. Assessing the probability of rapid mass movement. Third Australia - New Zealand Conference on Geomechanics. Wellington. 2:47-53.

Cunningham, A. & P.W. Stribling 1978. The Ruahine Range. Water and Soil Technical Publication No 13. Wellington.

Cunningham, D. 1988. A rockfall avalanche in a sandstone landscape, Nattai North, New South Wales. Aust. Geographer 19(2): 221-229.

Danvers-Powers, F. 1892. Notes on the late landslip in the Dandenong Ranges, Victoria. Aust. Ass. Advancement Sci. 4: 337-340.

Denham, D. 1969. Recent damaging earthquakes in New Guinea. Earthquake engineering symp. 16 October, 1969, Inst. Eng. Aust. Vic. Div.

Donaldson, R.C. 1980. Geological hazards in Tasmania. Geol. Surv. Qld. Record. 1980/37: 97-118.

Everingham, I.B., A.J. McEwin, & D. Denham 1982. Atlas of isoseismal maps of Australian earthquakes. Bureau Mineral Resources Bull. 214.

Eyles, R.J. 1971. Mass movement in Tangoio Conservation Reserve, northern Hawkes Bay. Earth Science Journal 5(2): 79-91.

Eyles, G.O. 1983. The distribution and severity of present erosion in New Zealand. New Zealand Geographer 39(1): 12-28.

Eyles, R.J. & G.O. Eyles 1982. Recognition of storm damage events. Proc. 11th New Zealand Geog. Conf. Wellington, 1981: 118-123.

Findlay, J.K. 1980. Landslide investigation techniques applicable in Southeast Queensland. Geol. Surv. Qld. Record. 1980/37: 201-208.

Gallen, R.G. 1980. The commission of enquiry into the Abbotsford disaster. N.Z. Govt. Printer, Wellington, 196p.

Ghartey E. & G. Anderson 1983. Engineering geological assessment of a landslide at the Ankura River bridge, Mendi-Hagen highway, Southern Highlands Province. Geol. Surv. Papua New Guinea Tech. Note 35-83.

Goldsmith, R.C.M. 1979. Landslide investigation at Kerowagi Police Barracks, Chimbu Province. Geol. Surv. Papua New Guinea Prof. Opinion 15/79.

Grant, P.J. 1985. Major periods of erosion and alluvial sedimentation in New Zealand during the late Holocene. Jnl. Royal Soc. New Zealand 15(1): 7-121.

Harding, H.E. 1952. Extensive landslide at Bacchus Marsh. Aqua. 4:10-15.

Harmsworth, G.R., G.D.Hope, M.J.Page, & P.A. Manson 1987. An assessment of storm damage at Otoi in northern Hawke's Bay, Soil Conservation Centre, Aokoutere. Pub. No. 10, 76p.

Hawley, J.G. 1984. Slope instability in New Zealand. In Natural Hazards in New Zealand. I. Spedon & M.J. Crozier (eds.). N.Z. National Commission for UNESCO, Wellington: 88-133.

Hofmann, G.W. 1980. Geological hazards in Queensland. Geol. Surv. Qld. Record. 1980/37: 119-147.

Hollingsworth, P.C. 1982. Landslides and residential development - a review of the social impact and methods of mitigation. Geol. Soc. Aust. Eng. Geol. Group. Landslide hazards in hillside development. 1-17.

Ingles, O.G. 1974. Unstable landforms in Australia. Water Res. Foundation Aust. Report 42.

Jane, G.T. & T.G.A. Green 1983. Biotic influences on landslide occurrence in the Kaimai Range. N.Z. Jnl. Geology & Geophysics. 26: 381-393.

Jenkins, D.A.L. 1974. Detachment tectonics in western Papua New Guinea. Geol. Soc. Am. Bull. 85: 533-548.

Johnson, R.W. 1987 Large-scale volcanic cone collapse: the 1888 slope failure of Ritter volcano, and other examples from Papua New Guinea. Bull. Volc. 49: 669-679.

Joyce, E.B. 1979. Landslide hazards in Victoria. In R.L. Heathcote and B.G. Thom. Natural hazards in Australia. Aust. Academy Sci. Canberra. 234-247.

King, J. 1987a. The breaching of the Bairaman Dam, East New Britain Province. Geol. Surv. Papua New Guinea Report 87/9.

King, J. 1987b. Preliminary report on Lilyampos landslide and debris flow, Enga Province. Geol. Surv. Papua New Guinea Tech. Note 14-87.

King, J. & I. Loveday 1985. Preliminary geological report on the effects of the earthquake of 11th May 1985 centred near Bialla, West New Britain. Geol. Surv. Papua New Guinea Report 85/12.

Knights, C.J. 1977. Investigation of the Lawrence Vale landslip. Tasmania Dept. Mines Rept. 1977/53.

Krause, D.C., W.C. White, D.J.W. Piper & B.C. Heezen 1970. Turbidity currents and cable breaks in the Western New Britain trench. Geol. Soc. Am. Bull. 81: 2153-2160.

Lewis, K.B. 1980. Quaternary sedimentation on the Hikurangi oblique subduction and transform margin, New Zealand. In P.F. Ballance & H.G. Reading (eds.). Sedimentation in oblique-slip mobile zones. Spec. Publ. Intern. Assoc. Sedimentologists. 4: 171-189.

McKelvey, P.J. 1959. Animal damage in North Island protection forests. N. Z. Science Review. 28-34.

Melville, A.R. 1976. Landslides on the Far North Coast. Soil Conservation Jnl. N.S.W.: 180-186.

NWASCO, 1979. Our land Resources. National Water and Soil Conservation Authority. Wellington, 79p.

NWASCO, 1975-1979. New Zealand Land Resource Inventory Worksheets, 1:63,360. National Water and Soil Conservation Authority. Wellington.

O'Loughlin, C.L. & A.J. Pearce 1976. Influence of Cenozoic geology on mass movement and sediment response to forest removal, north Westland, New Zealand. Bull. Int. Assoc. Eng. Geol. 14: 41-46.

Pain, C.F. 1972. Characteristics and geomorphic effects of earthquake-initiated landslides in the Adelbert Range, Papua New Guinea. Eng. Geol. 6: 261-274.

Pain, C.F. & J.M. Bowler 1973. Denudation following the November 1970 earthquake at Madang, Papua New Guinea. Z. Geomorph. Suppl. Bd. 18: 92-104.

Pearce, A.J., C.L. O'Loughlin, & A.J. Watson 1985. Medium-term effects of landsliding and related sedimentation evaluated fifty years after an M7.7 earthquake. Proc. Int. Symp. Erosion, debris flow and disaster prevention. Tsukuba, Japan. 291-296.

Pells, P.J.N., J.C. Braybrooke, G.P. Kotze, & J. Mong 1987. Cliff line collapse associated with mining activities. Soil slope instability and stabilisation, B. Walker & R. Fell (eds.), 359-385. Rotterdam, Balkema.

Pilgrim A.T. & A.J. Conacher 1974. Causes of earthflows in the southern Chittering Valley, Western Australia. Aust. Geog. Studies, 12: 38-56.

Robbins, R.G. 1958. Direct effect of the 1855 earthquake on the vegetation of the Orongorongo Valley, Wellington. Trans. Roy. Soc. N.Z. 85(2): 205-212.

Salter, R.T., T.F. Crippen, & K.E. Noble 1983. Storm damage assessment of the Thames – Te Aroha area following the storm of April 1981. Soil Conservation Centre, Aokoutere. Pub. No.1. 54p.

Selby, J. 1980a. Notes on geological hazards in South Australia. Geol. Surv. Qld. Record 1980/37: 66-73.

Selby, J. 1980b. Landslides in South Australia. Third Australia – New Zealand Conference on Geomechanics. Wellington. 2:69-72.

Selby, M.J. 1976. Slope erosion due to extreme rainfall: a case study from New Zealand. Geog. Ann. 58(A): 131-138.

Smith, W.D. & K.R. Berryman 1986. Earthquake hazard in New Zealand: inference from seismology and geology. Bull. Roy. Soc. N.Z. 24: 223-242.

Stephens, P.R. 1975. Determination of procedures to establish priorities for erosion control as determined in the southern Ruahine Range, North Island, New Zealand. Unpub. MAgSci thesis, Massey Univ.

Stephens, P.R., C.M. Trotter, R.C. De Rose, P.F. Newsome, & K.S. Carr 1988. Use of SPOT satellite data to map landslides. Paper presented to the 9th Asian Conf. on Remote Sensing. Bangkok.

Stevenson, P.C. 1971. A mudspring and a landslip at Deviot. Tasmania Dept. Mines Tech. Rept. 14:79-82.

Stevenson, P.C. 1975. A predictive landslip survey and its social impact. Second Australia-New Zealand Conference on Geomechanics. Institution of Engineers Australia, Brisbane: 10- 15.

Trustrum, N.A. & J.G. Hawley, 1986. Conversion of forest land use to grazing – a New Zealand perspective on the effects of landslide erosion on hill country productivity. In A.J. Pearce & L.S. Hamilton (eds.). Proc. Seminar on Land Use Planning in a Watershed Context, Gympie, Australia, FAO 1986/3: 73-93.

Varnes, D.J. 1978. Slope movement types and processes. In Landslides: Analysis and control. (ed. R.L. Schuster & R.J. Krizek). Nat. Acad. Sci. Spec. Report 176: 11-33.

Veblen, T.T. & G.H. Stewart 1968. The effects of introduced wild animals on New Zealand forests. Annals Assoc. Am. Geographers. 72(3): 372-397.

Veevers, J.J. (ed.) 1984. Phanerozoic earth history of Australia. Clarendon Press, Oxford.

Warren, E. 1988. Land stability zoning of Goroka and environs. Geol. Surv. Papua New Guinea Report 88/1.

Weber, B. 1980. Geological hazards in the Northern Territory. Geol. Surv. Qld. Record. 1980/37: 148-174.

Whitehouse, I.E. 1983. Distribution of large rock avalanche deposits in the central Southern Alps, New Zealand. N.Z. Jnl. Geol. & Geophys. 26: 272-279.

Whitehouse, I.E. & G.A. Griffiths 1983. Frequency and hazard of large rock avalanches in the central Southern Alps, New Zealand. Geology. 11: 331-334.

Willmott, W.F. 1984. Forest clearing and lanslides on the basalt plateaux of south east Queensland. Qld. Agric. Jnl. Jan-Feb. 15-20.

Young, A.R.M. 1978. The influence of debris mantles and local climatic variations on slope stability near Wollongong, Australia. Catena. 5: 95-107.

Young, A.R.M. 1979. Landslip in Wollongong: a case history. 227- 233.

Young, R.W. & A.R.M. Johnson 1977. The physical setting: environmental hazards and urban planning. In R. Robinson (ed.). Urban Illawarra. Sorrett, Melbourne. 51-55.

Landslides: Extent and Economic Significance, Brabb & Harrod (eds)
© 1989 Balkema, Rotterdam. ISBN 90 6191 876 6

Landslide hazard in the Pacific Islands

M.J.Crozier
Victoria University of Wellington, New Zealand

ABSTRACT: The landslide hazard for the tropical islands of the western Pacific is assessed indirectly by reference to their volcanicity, seismic activity, landform, location and social and administrative factors. Ongoing tectonic activity accompanied by rapid weathering and erosion has produced islands with high relief, steep slopes and restricted areas for habitation within the lower catchments. These inherently hazardous conditions are from time to time subjected to high intensity, tropical cyclone rainfalls which are known to produce devastating landslide episodes. In parts of the region sensitivity of the terrain to landslides has been enhanced by mining activity and deforestation. The small size, isolation and recent political reorganization of many island groups present special problems in hazard mitigation within the region.

1 INTRODUCTION

The islands included in this section of the world survey lie within the western Pacific in a region extending north of the equator almost to Japan, east to include Samoa, south as far as Tonga and west towards the coasts of Papua New Guinea and the Philippine Islands. The islands surveyed are listed in Table 1 and located in Figure 1.

Table 1. General information of Pacific Islands surveyed

Island groups[1,2]	Area (km²)	Population (thousands)	Population density (per km²)	Rainfall mm/y
Caroline Islands[3]	702	88[b]	125	2667
Ponape (district)	375	27[c]	72	4859
Truk (district)	126	45[c]	357	3556
Fiji	18776	705[a]	39	2974
Kiribati	690	66[a]	91	1250-3000
Tarawa	9	22[b]	2444	1977
Loyalty Islands	1981	16[b]	8	1778
Mariana Islands (district)	475	19[a]	40	1483
Tinian	101	0.9[c]	9	1483
Marshall Islands	171	36[a]	199	2000-4000
Bikini	<10	0.1[c]	10	2000
Eniwetak	<10	0[c]	0	2000
Kwajalein	<10	5[c]	500	2718
Midway Island	5	0.5[b]	87	1069
New Caledonia Island	16372	128[b]	8	1000-3000
Palau Islands	460	12[a]	25	2540-3810
Samoa	3028	192	63	2000-5000
Western Samoa	2831	161[a]	57	2897
American Samoa	197	32[b]	162	4850
Solomon Islands	29785	280[a]	10	2000-3000
Tonga	699	137[a]	9	1920
Volcano Islands	32	<2[b]	63	1200
Iwo Jima	22	<0.5[b]	23	1200
Wake Island	7	0.3[b]	43	991

Notes
1 This list contains all those Pacific Islands allocated to the Pacific Island sector of the world survey. There are, however, some notable omissions including: Cook Islands, Guam, Nauru, Niue and Tuvalu.
2 The primary grouping is based on geographic affinity rather than constitutional integrity.
3 Treated as equivalent to the Federated States of Micronesia.
a Economic and Social Commission for Asia and Pacific, Bangkok, 1986.
b Statesman's Yearbook (1987-88).
c Office of High Commissioner of Trust Territories.

Parts of the region, geologically and meteorologically, are among the most violent on earth. Most of the inhabitants derive their subsistence directly from the land and the livelihood of all is closely dependent upon the vagaries of the natural environment. Yet little systematic work has been done on the natural hazards of the region and even less on landslides and their social and economic importance. Blong and Johnson (1986) in a useful review of the geological hazards in the much larger and more populous south east Asian and south west Pacific region, admit they have been unable to offer a comprehensive treatment of the problem largely because of the 'current unavailability of appropriately compiled data.' The problem is much more pronounced in the Pacific Islands region, where resident scientific personnel are scarce and where attention to landslides has been largely devoted to clearing up the mess. Even in a review of the more ubiquitous and economically pervasive subject of soil erosion in the South Pacific, Eyles (1987) expresses the same concern when he concludes with two revealing counterpoints:

Surface erosion processes of sheet, rill and gully erosion predominate on volcanic islands, with landslide erosion having a significant sediment generation effect,

and further:

On no South Pacific Island has the extent, severity and rate of soil erosion been mapped or adequately assessed.

The Pacific Islands covered by this sector of the world survey of landslides all have three characteristics in common; isolation, small land areas, and relatively high and erratic rainfalls. The first two factors contribute enormously to island vulnerability. A small island nation can be completely engulfed by a hazardous event of even limited physical magnitude in world terms, leaving few if any local resources to cope with the situation. These resources are already limited by an incomplete transition towards a fully developed commercial economy which has seen a decline in traditional mechanisms for handling disasters without adequate replacement by modern hazard mitigation strategies. Compounding the problem of size is the factor of isolation. Communication and transportation links over vast distances of ocean can be tenuous under normal conditions but under stress they can lead to long delays in obtaining relief. In recent years economies of scale have resulted in ocean-going vessels becoming larger, thus reducing the number and frequency of shipping contacts with the small island states and increasing their isolation. The third factor, high and erratic rainfall, reflects the turbulent atmospheric conditions within the region which, in combination with certain geological and geomorphic conditions, can produce devastating landslides. Fortunately inherently unstable terrain does not occur throughout the entire Pacific but is located in clearly identifiable zones.

Natural hazards other than landslides are prevalent within the region. The entire area is subject to tsunamis, low islands can suffer prolonged drought, a large part of the region experiences wind, rain

SOUTH PACIFIC

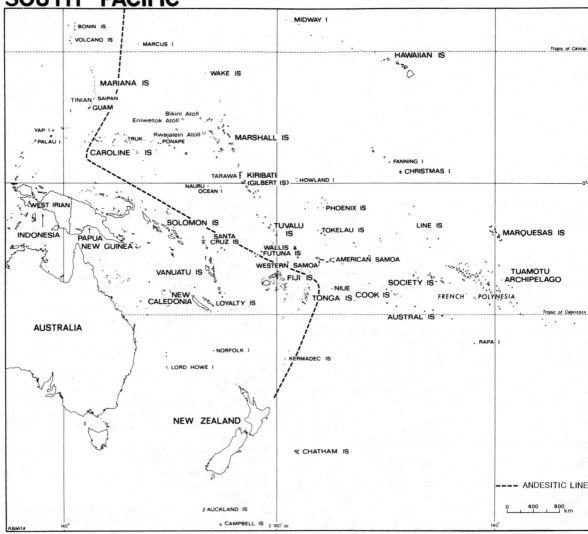

Figure 1. Location of island groups.

and storm surge damage commonly associated with
cyclones and parts of the region are threatened by
volcanic and seismic hazard.

2 GEOLOGIC SETTING

The region under consideration is found within the
central to western part of one of earth's oldest and
largest features, the Pacific Basin. Most of the
world's active volcanoes, seismic activity and recent
uplift, occur in the Pacific 'rim of fire' which
surrounds this basin, and passes through the region
on its western margin.

There are two geological divisions within the
region. One consists of Pacific crustal plate
material and occupies most of the region and the
other is made up of parts of the Australian, Philippine
and Eurasian plates and covers the western areas of
the region. In effect the divisions are separated by
the andesitic line (Figure 1) which has its physical
expression on the ocean floor in the form of deep sea
trenches. From south to north these are the Kermadec-
Tonga trench running northeast of New Zealand and
north of Fiji, the Vityaz trench north east of Vanuatu,
the Manus trench north of the Solomon Islands and
Papua New Guinea, the Palau, Yap and Mariana trenches
east of their respective island groups, and the continuation of the Mariana trench into the Bonin trench
east of Iwo Jima which continues east of Japan to
complete the western boundary of the Pacific Basin.
The location of island groups with respect to these

two geological divisions is given in Table 2.

Table 2. Geological affinities of Pacific Island
groups

Eastern Division	Western Division
Kiribati	Fiji
Marshall Islands	Loyalty Islands
Midway Island	Mariana Islands
Samoa	New Caledonia
Wake Island	Palau
Caroline Island	Solomon
(excluding Yap)	Vanuatu
	Volcano Islands
	Caroline Islands: Yap
	Tonga

In the eastern of these two divisions only low
intensity seismic shaking is experienced, the area
is relatively stable tectonically and there is little
active volcanism except outside the region in the
Hawaiian and Galapagos Islands. Volcanic rocks are
of the effusive rather than eruptive variety consisting largely of basalt.

The geology of the western division exhibits
'continental' and continental plate boundary, subduction features. In contrast to the eastern

358

division, volcanic rocks are of the acid and more eruptive variety with the exception of ultramafic rocks in New Caledonia. The larger islands contain a range of metamorphic and sedimentary rocks often strongly folded and faulted. Many large earthquakes originate in the division and there is active volcanism (Table 3). The violent explosive activity from volcanoes within the western division results from rapid vesiculation of basalt, andesite, dacite and rhyolite magmas. The volatiles producing these reactions are thought to derive from groundwater and from dehydration processes in the subducting slabs of lithosphere beneath the volcanoes.

Table 3. Islands volcanically active in historic times

Mariana Islands
  Uracas (Farallon de Pajaros)
  Guguan
  Pagan
  Asuncion
  Alamagan
  Esmeralda Bank
  Ruby
Solomon Islands
  Bougainville (Bagana)
  Savo
  Cook Submarine
  Tinakula
  Mendana
  Simbo
Vanuatu
  Vanua Lava
  Mt Ambrym
  Mt Benbow
  Lopevi
  Yasur (Tana)
  Hunter (Fearn)
  between Epi and Tongoa
  between Traitori Head and High Rock
Samoa
  Savai'i
  Matavanu
  submarine near Olosenga
Tonga
  Niuafo'ou
  Fonualei
  Late Island
  Metis shoal
  Tofua
  Falcon Island
  many submarine eruptions
Volcano Is
  Iwo Jima
  Shin Iwo Jima

Blong and Johnson (1986) have discussed some of the historic hazard events associated with the tectonic instability of this division. These include lava flows from Ambrym (Vanuatu) on 29 June 1929 which destroyed three mission stations and food supplies for about 300 villagers. Tephra falls occurred in the region in 1950 and 1951 causing the evacuation of 700 people, 48 of whom were tragically killed by a tropical cyclone a few weeks later. An earthquake near San Christobal (Solomon Islands) in October 1931 produced a tsunami which killed perhaps 20-25 people and destroyed 18 villages. Other damaging earthquakes have taken place in the Solomon Islands including one of Richter magnitude 7.5 which occurred in February 1984. Damaging earthquakes have also been recorded in Taveuni (220 km north of Suva, Fiji) in 1979 and Viti Levu (Fiji) 1953. Ground subsidence caused by a Richter magnitude 7.2 earthquake in June 1977 on Tongatapu (Tonga) and followed four other earthquakes of similar magnitude in the same area recorded this century in the years 1913, 1917, 1943

and 1948 (Campbell et al. 1977).

The mineral ores that occur in the western division - notably nickel and chromium (New Caledonia), phosphate (Palau, Nauru and other islands), manganese (Solomons), gold (Fiji) - are in some areas extensively exploited by open cast methods causing steep surrounding terrain to become susceptible to mass movement.

3 LANDFORM

Using the definition proposed by Wiens (1962), the islands of the region can be classified as 'high' or 'low' islands. Low islands are those under about 10 metres above sea level and are mostly but not exclusively coral atoll or reef islands. High islands are all those with altitudes greater than those of low islands and are largely volcanic in origin commonly with uplifted reef limestone margins. Some of the large islands in this survey have formations with 'continental' affinities such as New Caledonia (metamorphic rocks) and Viti Levu of Fiji (metamorphic rocks and granite). The high islands in this region are all associated with contemporary reefs (however, not all Pacific high islands have contemporary reefs). The Pacific islands surveyed belong to eight geomorphic types, Figures 2 and 3. In these diagrams which are based on those of Cumberland (1954) and British Naval Intelligence (1945) the term 'motu' refers to a small island of coral rock and 'makatea' refers to a 'raised' former lagoon floor consisting of dry cavernous limestone.

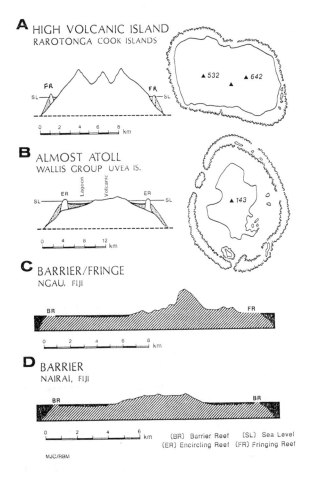

Figure 2 HIGH ISLAND TYPES

A HIGH VOLCANIC ISLAND
RAROTONGA COOK ISLANDS

B ALMOST ATOLL
WALLIS GROUP UVEA IS.

C BARRIER/FRINGE
NGAU, FIJI

D BARRIER
NAIRAI, FIJI

(BR) Barrier Reef   (SL) Sea Level
(ER) Encircling Reef (FR) Fringing Reef

MJC/RBM

Figure 2. High island types (after Cumberland, 1954; Naval Intelligence Division, 1945).

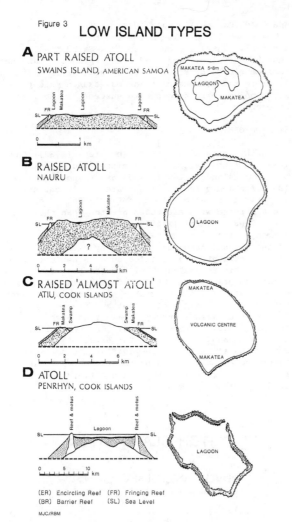

Figure 3

LOW ISLAND TYPES

A PART RAISED ATOLL
SWAINS ISLAND, AMERICAN SAMOA

B RAISED ATOLL
NAURU

C RAISED 'ALMOST ATOLL'
ATIU, COOK ISLANDS

D ATOLL
PENRHYN, COOK ISLANDS

(ER) Encircling Reef (FR) Fringing Reef
(BR) Barrier Reef   (SL) Sea Level

MJC/RBM

Figure 3. Low island types (after Cumberland, 1954).

4 TROPICAL CYCLONES

The most important landslide triggering agent
within the region is high intensity rainfall associated
with tropical cyclones. Tropical cyclones known also
as hurricanes and typhoons develop over tropical
waters with temperatures of at least 26°C where
differential heating from isolated land areas causes
pockets of extremely low pressure air to develop.
Air circulates around the relatively calm centre of
the cyclone (clockwise in Southern hemisphere and
anticlockwise in the Northern hemisphere) generating
winds often greatly in excess of 63 k/h. Energy to
develop and sustain the cyclone is supplied from warm
ocean waters and moisture laden air spirals upward,
eventually cooling to release latent heat and intense
rainfall. This rainfall producing mechanism can be
enhanced by the orographic effect of high islands.

In both hemispheres, topical cyclones develop in
summer in the inter-tropical convergence zone about
10-15° either side of the equator and move westwards
at rates of 15-25 k/h gathering in intensity as they
proceed. As they approach the Asian and Australian
waters they track polewards following erratic paths,
peaking in intensity just before leaving the tropics.
However tropical cyclones have on occasions been
sufficiently strong to produce landslide triggering
rainstorms as far south as Dunedin, New Zealand
(45° S.Lat.).

Distinct tropical cyclone seasons exist (Table 4)
centred on the months of February in the southern
Pacific and August and September in the northern
Pacific. The annual frequency of these events in any
one location can vary dramatically but records

Table 4. Numbers of tropical cyclones in the southwest Pacific[1] for
the 40 years from November 1939 to April 1979 with relative monthly
percentages in brackets.

Decade	Oct/Nov	Dec	Jan	Feb	Mar	Apr/May/Jun	Total
1939/49	1(2)	8(12)	17(26)	18(28)	16(25)	5(7)	58
1949/59	2(3)	8(11)	21(29)	20(27)	15(20)	7(10)	64
1959/69	6(7)	13(15)	16(19)	23(27)	19(22)	8(10)	72
1969/79	6(6)	13(13)	23(23)	25(25)	17(17)	16(16)	91
40 years	15	42	77	86	67	36	285

indicate that certain localities are much more likely
to experience potentially destructive cyclones than
other areas (Table 5).

Table 5. Occurrence of potentially damaging tropical
cyclones/typhoons

Northern Pacific Islands[1]	Annual frequency
Volcano	1.00
Midway	0
Wake	0
Mariana	2.25
Caroline	
Yap	1.50
Truk	1.50
Ponape	0.75
Marshall	0.50
Palau	0.25
Kiribati	0
Southern Pacific Islands[2]	
Solomon	0.40
Samoa	0.20
Vanuatu	1.10
Fiji	0.50
Tonga	0.60
Loyalty	0.40
New Caledonia	0.80

Notes
1 Data from interpretation of map by Wiens (1962)
   based on four year survey of storms with winds
   greater than 64 knots.
2 Data from interpretation of chart by Revell (1981)
   based on ten year survey of storms with winds
   greater than 48 knots.
3 Refers to the maximum number of storm tracks that
   crossed any part of island group or district waters.

Britton (1986) has estimated that of the 80 tropical
cyclones which occur each year, on average about 30
of these will form in the Pacific. Even though most
of these do not track directly over inhabited areas,
associated high tides caused by storm surges, wave-
induced coastal erosion and strong winds can be
experienced by islands in their vicinity. Between
1980 and 1985 cyclone damage occurred on 11 occasions
in Fiji, three in Vanuatu, two in Tonga and one in
the Solomons. Probably only about one in 10 tropical
cyclones produce landsliding. Since 1985 there have
been two extremely severe landslide-producing
tropical cyclones: Namu, May 1986 in the Solomons
(Figure 4) and Bola, East Cape, North Island, New
Zealand, in March 1988 where thousands of hectares
of pasture land experienced up to 30% loss of land
area from mass movement. The economic losses resulting from recent tropical cyclone activity have been
outlined by Britton (1986) as follows:

The average annual cost of tropical cyclones in
the Pacific region, and the Indian Ocean (including
the Bay of Bengal and the Arabian Sea) was over
US$2,500 million in the years 1965-78. Over this
whole fourteen year period, tropical cyclone-
induced damage totalled US$36,000 million. More
specifically, cyclone damage to insured property in
Fiji over the last seven years was in excess of
F$140 (US$124) million, causing considerable
anxiety for the re-insurance industry. The passage

360

Figure 4. Winds up to 100 knots and rainfall over a four day period of 2330 mm were brought to the Solomon Islands by cyclone Namu in May 1986 (Photo: Peter Stephens).

of TC "Isaac" over Tonga in 1983 resulted in damage to property and crops in excess of T$18 (US$11) million. Whilst these cost estimates may be small when compared to tropical cyclone damage in other locations (for instance, the Insurance Council of Australia has assessed the annual level of payable claims from northern Australian tropical cyclone damage to be A$25 (US$15) million; TC "Tracy" in 1974 cost insurers A$198 (US$120) million), one has to keep in mind that the relative scale of damage and costs incurred by tropical cyclones in small nations can be very great to those affected, even though the actual figures may be small by comparison to events in large countries which have larger populations and possess much greater material resource bases.

Since Britton's report, UNDRO (1988) has recorded the following cyclone impacts in the region during 1987. The Tokelau Islands (north of Samoa) were hit by cyclone Tusi on 15 January causing tens of thousands of US dollars damage including damage to 61% of the residences in Nukunonu and uprooting 5000 coconut and breadfruit trees. Cyclone Uma struck Vanuatu on 7 February killing 45 people, destroying more than 90% of the homes in Port Vila and producing damage amounting to US$150 million. On 1 April the Cook Islands were hit by a cyclone which destroyed or damaged 80% of the buildings in the main town of Avarua, disrupted water supplies, blocked the harbour and caused in all US$25.4 million damage. In Truk and Moen Islands (Caroline Islands) tropical storm Nina killed five people on 21 October and caused US$6 million damage. Many other tropical cyclones occurred in areas bordering the region and a landslide killing ten people was reported on 17 April from Huahine, French Polynesia.

5 ENVIRONMENTAL MODIFICATION

5.1 Deforestation

Even with high relief and intensive tropical cyclone rainfall some of the terrain in the region has been sufficiently stable tectonically for long enough to allow equilibrium slopes to develop. Slope angle and height appear to be below the critical limits for mass movement on forested, unpopulated slopes. The main island of New Caledonia which lies to the west of the Pacific/Australian plate boundary appears to have developed this level of stability on unmodified slopes and natural landslides are uncommon. Very extreme events may produce landslides on such slopes but they have a correspondingly low frequency. For example, in the area affected by landslides during cyclone Wally (1980) in Viti Levu (Fiji), oral history suggested that landslides were previously unknown and there were few geomorphic indications of previous

mass movement (Crozier et al. 1981). Cyclone Wally was a very extreme event in terms of rainfall intensity and even relatively stable natural slopes were affected (Figure 5). However, more frequent

Figure 5. Heavily forested slopes with no previous landslide history failed during cyclone Wally rainstorm of 900 mm in 24 hours in Fiji, 1980.

lower intensity tropical cyclone rainstorms do produce landslides on modified slopes as was demonstrated by the preferential confinement of landslides to garden clearings (Figure 6) during cyclone rainfall on Viti Levu (Howorth and Penn, 1979).

Figure 6. A shallow debris flow confined to a cassava garden, Korovou Village, Fiji.

On most of the high islands in the region 'slash and burn' shifting agriculture has been practised for centuries. Throughout the region population and economic pressures have shortened the rotation period of these plots thus decreasing the net stabilizing influence of forest vegetation in a given area. Baines (reported in Eyles, 1987) observes that in the western Solomons the rotation period has dropped from 25 years to 3 or 4 years. In western Viti Levu, Whitehead (1952) observed that following forest fires the soil 'in a large mass, avalanches down to choke creeks, block roads, railways and smother rich farmlands'.

As noted by Crozier et al. (1981) the stabilizing effect of tropical rainforest depends to a large extent on storm intensity and the nature of the substrate. In very extreme rainstorms slopes fail irrespective of the type of vegetation. What is more the failure plane of landslides on deeply weathered volcanic rock is generally many meters below the root zone (Figure 7). However antecedent moisture levels may be lower on forested slopes than on equivalent deforested slopes, and, in shallow regolith, root binding will be effective. The stabilizing effect of forest vegetation is most

Figure 7. Deep-seated rotational landslide in weathered volcanic rock showing tree roots terminating well above the shear plane, Fiji.

noticeable during medium magnitude triggering rainstorms. The effect of vegetation on sheet, rill and gully erosion is much more pronounced and this is discussed for the South Pacific by Eyles (1987).

5.2 Mining

Governments and mining companies are today more mindful of the need to mitigate and prevent harmful environmental modification, than they were when many of the Pacific's large scale mineral exploitation activities began. Unfortunately expeditious disposal methods have left vast quantities of landslide-prone rock waste and tailings precariously dumped on hillslopes, ravines and water courses (Figure 8).

Figure 8. The effects of overburden disposal from open cast nickel mining, New Caledonia.

Disposal of overburden and side cast from exploration and operation phases for quarries, staging areas and access roads have by mass movement destroyed slope vegetation, choked river channels and accelerated coastal progradation.

Within the survey region, New Caledonia's main island has suffered greatly from the effects of open cast mining for nickel (particularly during the nickel boom of 1968-1973), chromium, iron, cobalt, manganese, and coal. Bird et al. (1984) observe for this island that:

  ...the existing vegetation cover has been disrupted by landslides, gullying and associated fan deposition in areas where erosion was initiated or accelerated by such activities as road making, mineral prospecting and depletion of natural vegetation by clearing, grazing and burning...the onset of widespread erosion during the past century and a half has been largely the outcome of human activities in an

environment where the relatively steep hillsides mantled by deep weathering materials evidently depended on the persistence of a retentive natural vegetation.

In the high islands of the Pacific, settlement is generally confined to coastal strips and garden plots are never far from villages or roads or communication pathways. Mining activity as well as forestry operations, however, have commonly occurred in previously unmodified or long abandoned land often within steep headwaters of the catchment. Adjustments to the terrain through erosional and depositional activity can therefore work their way throughout entire catchments, eventually affecting coastal settlements (Figure 9).

Figure 9. Downstream debris accumulation from debris flow and flooding during cyclone Namu, Solomon Islands, May 1986 (Photo: Peter Stephens).

6 ADMINISTRATION AND HAZARDS

The Pacific Island region has been one of the last parts of the world to undergo decolonization. Governments and constitutional arrangements for administration are relatively new and responsibilities for land management have only recently been addressed. There are few indigenous institutional mechanisms in place for such activities as resource and hazard assessment. Indeed at the time of this survey the constitutional administrations for New Caledonia (including Loyalty Islands), Palau and Fiji are still in a state of flux while Vanuatu continues to face the political uncertainties of recent independence.

The small size, geographic isolation and previous dependence on larger nations means that the islands of this region tend to look outside for the solution to natural hazards, often through an international mouthpiece of collective island nation groupings such as the South Pacific Bureau for Economic Cooperation (SPEC) based in Suva, Fiji. Bilateral aid from former neighbouring countries such as Australia and New Zealand is still considered essential.

The former Trust Territories of the Pacific are now virtually autonomous within a compact of free association with the United States of America. These include the Republic of Palau (still under a United Nations Trusteeship), the Federated States of Micronesia (including the states of Yap, Truk, Ponape and Kosrae), the Republic of the Marshall Islands and the Commonwealth of the Northern Mariana Islands which, since 1976, has been a de facto commonwealth of the United States of America. Midway Island and Wake Island are United States possessions; the former administered by the Navy for the Department of Defence and the latter along with the territory of American Samoa, administered ultimately by the Department of the Interior. Disaster planning in all these American affiliates is handled from the USA Office of Foreign Disaster Assistance (OFDA), the Federal Emergency Management Agency (FEMA) and related research is undertaken by the Pacific Island Development Programme

(PIDP) at the East/West Centre, Honolulu. The former Trust Territories established a local Disaster Control Office in 1971 which complemented other small government agencies with engineering, material testing and land resource survey capabilities.

New Caledonia including the Loyalty Islands are overseas territories of France and Noumea (the capital) is the site of the scientific laboratories of ORSTOM. The Volcano Islands are one of four groups of islands making up the Ogasawara Islands which constitutionally form a municipal unit of Tokyo. The remaining islands are constitutionally independent and are part of or associated with the British Commonwealth. The United Nations Disaster Relief Organization (UNDRO) has responded on a number of occasions in recent years to requests for assistance from mainly the independent island nations. This organization works closely with offices of indigenous governments including, for example, the well developed Emergency Services Committee (EMSEC), and the Department of Relief, Rehabilitation and Rural Housing of Fiji. Tonga has an interdepartmental committee for similar purposes, convened by the Minister of Works and moves have been made to set up a National Disaster office. Tonga Red Cross also has a Disaster Preparedness Committee. Vanuatu has had a Relief and Rehabilitation Committee and is currently setting up a National Disaster Plan. All of the independent nations of the Pacific come under the umbrella of the Economic and Social Commission for Asia and Pacific (ESCAP) and a number of other United Nations agencies.

The approach to the hazard situation and particularly to the landslide problem in the Pacific region is characterised by the development of a bureaucracy designed to expedite the acquisition and distribution of external aid after the event. There is clearly a need for the indigenous scientific programmes and external funding to be directed towards the gathering of data necessary for establishment of landslide hazard maps. The lack of emphasis on prevantative and control measures compared to post-event mitigation is reflected in the soil erosion problem where Eyles (1987) observed that 'No countries in the South Pacific, except Fiji, actively administer a soil conservation organisation.'

With the progressive shift of administrative responsibility from the village (Figure 10) to

Figure 10. In some communities villagers are still responsible for repairing landslide damage of common facilities such as water supply systems, Korovou Village, Fiji.

centralized government and the increasing dependency on international trade and relief, many of the local indigeneous mechanisms for coping with disaster have been eroded (Lewis, 1981). Indeed reliance on external support has meant that villages isolated by landslide-prone communication lines (Figure 11) have become even more vulnerable than in the past (Crozier, 1982). The scale and sometimes damaging consequences of relief dependency within the Pacific

have been discussed by Cijffers (1987).

Figure 11. Damage to Queen's Highway, Viti Levu, Fiji, during cyclone Wally, April 1980.

7 TYPES OF LANDSLIDES

Apart from the work of Howorth and Penn (1979) Crozier et al. (1981) and Stephens et al. (1987), there appears to be very few systematic studies on landslides for this region. Until the information base is improved resort must be made to tropical island analogues from other parts of the world. However, Crozier et al. (1979) indicate that, in Fiji, debris flows, some of which reached 10 m in depth, were the most commonly occurring form of rainstorm-induced landslide. Many of these were initiated as deep-seated, multiple rotational slumps off old surfaces of deeply weathered volcanic rock (Figure 12). The topography of eroded volcanic forms

Figure 12. Deeply weathered volcanic rock can produce rotational slides in excess of 10 meters depth, Fiji.

enhances the channelisation of debris flows to the extent that they commonly form fluid and rapidly flowing debris torrents which can travel considerable distances onto the populated coastal fringes (Figure 13).

The effects of cyclone Namu which struck the Solomon Islands 18-19 May 1986 killing over 100 people have been studied from aerial photographs by Stephens et al. (1987). Table 6 shows the nature of their results. The fluid and catchment-wide character of these landslides is again evident (Figure 14).

8 LANDSLIDE HAZARD POTENTIAL

The landslide hazard potential has been assessed in a subjective semi-quantitative manner with reference to six factors, volcanicity, seismicity, slope angle,

Figure 13. Debris torrents carrying large quanitites of vegetation are the most common form of cyclone-induced landslide, Korovou Village, Fiji.

Table 6. Summary data on landslides and deposition resulting from cyclone Namu, Solomon Islands, 18-19 May 1986. A section of survey results produced by Stephens et al (1987).

Watershed Name	Area (km^2)	Number of deep-seated landslides	Percent of watershed with shallow landslides (in 3 severity classes)[1]			Percent of watershed affected by deposition
			Slight	Moderate	Severe	
GUADALCANAL						
Aola	109	0	0	0	0	0
Kolovaghamela	88	2	23	0	0	0
Kombito	110	0	6	0	0	0
Lughumboko	216	5	54	0	0	0
Lungga	394	9	15	13	0	4
Mandonu	116	0	0	0	0	0
Matepono	202	4	41	10	3	31
Mbalisuna	235	42	7	23	28	0
Mberande	234	24	13	41	5	5
Mbokokimbo	385	15	28	14	0	1
Mbonehe	67	0	3	0	0	0
Ngalimbiu	262	37	16	12	33	18
Nggurambusa	205	9	16	15	0	0
Rere	109	3	33	7	0	4
Tavangaoa	173	7	32	7	0	7
Tenaru	170	0	1	0	0	17

Note:
1 'Slight' represents 1-5% eroded, 'moderate' 6-15% eroded and 'severe' greater than 15% eroded.

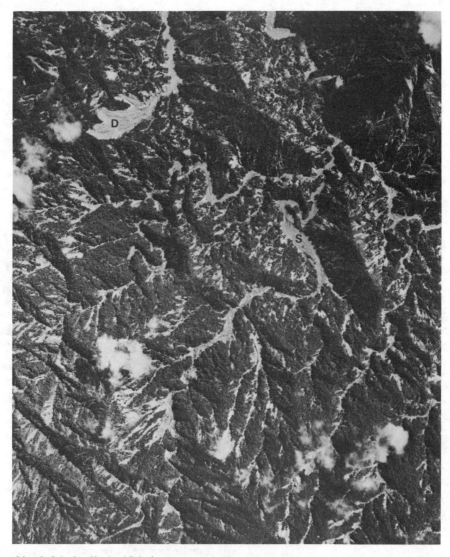

Figure 14. Channelized debris flows (debris torrents) characterized the mass movement during cyclone Namu, Solomon Islands, May 1986 (Photo: Stephens et al. 1988).

cyclone frequency, landslide reports in the literature and population density. The range of conditions represented by each factor has been divided into three classes: class one indicates little or no influence on the hazard; class two, moderate influence; class three high influence. Table 7 lists the specific cri-

teria for classifying each factor and Table 8 represents the landslide hazard of the island groups based on these criteria.

On the basis of this very generalized assessment it appears that all the high islands surveyed have a high landslide hazard potential. Those most at risk are the

Volcano Islands, Marianas especially the northern group, Solomons, Samoa, Vanuatu and Tonga. This group is closely followed by Fiji, and the high islands of Yap, Truk, Ponape, New Caledonia and Palau. The remaining islands surveyed have insufficient relief to qualify as constituting any landslide hazard.

Table 7. Classification of landslide hazard factors

	Classes		
	1	2	3
Volcanicity (V)	no volcanoes or highly eroded	dormant volcanoes	active volcanoes
Seismicity (S)	severe earthquakes unknown	occasional severe earthquakes	severe earthquakes common
Slope angle (A)	flat-gentle	moderate	steep greater than 20°
Cyclone frequency (C)	none or rare	0.2 to 0.49 per year	greater than 0.49 per year
Landslide reports (L)	no reference	reference	significant reference
Population (P) density	up to 49 per km²	50-99 per km²	over 100 per km²

Table 8. Relative importance of landslide hazard factors (see Table 7 for Key)

		1	2	3
Volcano Islands (Kazan Retto) Volcanic islands of Kito-Iwo shima Iwo Shima Minami-Iwo Shima Tuffs agglomerates and older sedimentary	V			X
	S			X
	A			X
	C			X
	L		X	
	P		X	
Midway Atoll	V	X		
	S	X		
	A	X		
	C	X		
	L	X		
	P		X	
Wake Atoll	V	X		
	S	X		
	A	X		
	C	X		
	L	X		
	P	X		
Mariana (northern) Anatahan, Sariguan, Guguan, Alamagan, Pagan, Agrihan, Asuncion, Maug, Farallon de Pajaros Andesitic volcanics	V			X
	S			X
	A			X
	C			X
	L			X
	P	X		
Mariana (southern) Saipan, Tinian, Agiguan, Rota, Farallon de Medinilla, Volcanic cores and raised limestone	V		X	
	S			X
	A		X	
	C			X
	L			X
	P	X		
Yap (Caroline) Yap proper is volcanic + 15 coral atolls islands	V		X	
	S		X	
	A		X	
	C			X
	L	X		
	P			X
Vanuatu 12 major and 16 lesser islands all high volcanic and raised limestone + 40 small islands	V			X
	S			X
	A			X
	C			X
	L	X		
	P	X		
Fiji Two large and 260 small islands. Mainly high but also low coral islands. Volcanic granitic and sedimentary rocks	V	X		
	S		X	
	A			X
	C			X
	L			X
	P	X		
Tonga 150 islands, high and low, volcanic and limestone	V			X
	S			X
	A			X
	C			X
	L	X		
	P			X
Loyalty Three main islands mainly coral	V	X		
	S	X		
	A	X		
	C		X	
	L	X		
	P	X		
New Caledonia (Main Island) High ultramafic metamorphic and sedimentary rocks	V	X		
	S	X		
	A			X
	C			X
	L		X	
	P	X		
Truk (Caroline) About 100 islands of which 14 are volcanic, rest low islands	V	X		
	S	X		
	A			X
	C			X
	L	X		
	P			X
Ponape (Caroline) and Kusaie Two volcanic + 8 coral atolls	V	X		
	S	X		
	A			X
	C			X
	L	X		
	P		X	
Palau 200 high volcanic and limestone islands + 4 atolls	V	X		
	S			X
	A		X	
	C		X	
	L	X		
	P	X		
Marshall 29 Atolls + 5 low coral islands	V	X		
	S	X		
	A	X		
	C			X
	L	X		
	P			X
Kiribati All low: 11 atolls and 5 reef islands	V	X		
	S	X		
	A	X		
	C	X		
	L	X		
	P		X	
Solomon 12 major high volcanic islands + 30 small islands	V			X
	S			X
	A			X
	C		X	
	L			X
	P	X		
Samoa (Western) 2 large and 7 small high islands	V			X
	S			X
	A			X
	C		X	
	L			X
	P		X	
Samoa (American) 4 high + 1 atoll	V		X	
	S			X
	A			X
	C		X	
	L	X		
	P			X

REFERENCES

Blong, R.J. & R.W. Johnson 1986. Geological hazards
   in the southwest Pacific and southeast Asian
   region: identification, assessment and impact.
   BMR Journal of Australian Geology and Geophysics
   10:1-15.
Bird, E.C.F., J-P. Dubois & J.A. Iltis 1984. The
   impacts of opencast mining on the rivers and
   coasts of New Caledonia. NRTS-25/UNUP-505 The
   United Nations University, Tokyo.
Britton, N. 1986. Cyclone preparedness in the South
   Pacific. United Nations Disaster Relief
   Organization News. Sept/Oct: 18-21.
Cijffers, K.M. 1987. Disaster relief: doing things
   badly. Pacific Viewpoint 28(2):95-118.
Crozier, M.J. 1982. The character of natural
   hazards in different physical and social settings.
   Proceedings of the Eleventh New Zealand Geography
   Conference, Wellington, 1981:106-8.
Crozier, M.J. 1986. Landslides: causes, consequences
   and environment. London: Croom Helm.
Crozier, M.J., R. Howorth & I.J. Grant 1981. Land-
   slide activity during cyclone Wally, Fiji: a case
   study of the Wainitubatolu catchment. Pacific
   Viewpoint 22(1):69-80.
Cumberland, K.B. 1954. Southwest Pacific. New
   Zealand: Whitcombe and Tombs Ltd.
Eyles, G.O. 1987. Soil erosion in the south Pacific.
   Environmental Studies Report, 37. University of
   the South Pacific: Institute of Natural Resources.
Franco, A.B., M.P. Hamnett & J. Makasiale, 1982.
   Disaster preparedness and disaster experience in
   the south Pacific. Pacific Islands Development
   Programme, East-West Center, Honolulu.
Howorth, R. & N. Penn 1979. Survey of Viti Levu
   storm of 5 May 1979: interim report. University
   of the South Pacific Information Bulletin 12: 3-10.
Lewis, J. 1981. Some perspectives on natural
   disaster vulnerability in Tonga. Pacific Viewpoint
   22: 145-162.
Naval Intelligence Division 1945. Pacific islands
   volumes I to IV. Great Britain.
Revell, C.G. 1981. Tropical cyclones in the south-
   west Pacific. New Zealand Meteorological Service,
   Misc. Pub. 170.
Stephens, P.R., N.A. Trustrum & J.R. Fletcher 1987.
   Natural hazard mapping - Solomon Islands. Stream-
   land 51. National Water and Soil Conservation
   Authority Wellington, New Zealand.
Stephens, P.R., N.A. Trustrum, J.R. Fletcher &
   S. Danitofea 1988. Reconnaisance mapping of
   erosion caused by Cyclone Namu, Solomon Islands.
   Asian-Pacific Remote Sensing Journal 1(1):57-64.
UNDRO 1988. Disasternews in brief 1 January-31 Dec-
   ember 1987. Geneva: Office of the United Nations
   Disaster Relief Coordinator.
Whitehead, C.E. 1954. Soil erosion and soil con-
   servation in Fiji. Fiji Agricultural Journal
   25(1-2).
Wiens, H.J. 1962. Atoll Environment and Ecology.
   New Haven: Yale University Press.

Landslides: Extent and Economic Significance, Brabb & Harrod (eds)
© 1989 Balkema, Rotterdam. ISBN 90 6191 876 6

Undersea landslides: Extent and significance in the Pacific Ocean

Homa J.Lee
US Geological Survey, Menlo Park, Calif., USA

ABSTRACT: Submarine slope failures have been identified in the Pacific Ocean by observing a) coastal evidence of shoreline disruption, b) damage to offshore structures such as communications cables, c) observed changes in sea-floor morphology over a short time span, and d) distinctive bottom and subbottom morphologic features. A review of twenty case studies of submarine landslides shows the range of scale and morphology of these landslides, mechanisms of failure initiation, and methods used to observe them. The floors of fjords, with steep slopes and rapidly deposited glacio-fluvial sediment, are the most susceptible to failure. Next most susceptible are active river deltas on the continental shelf, in which underconsolidation, gas charging and storm or earthquake loading can lead to instability. Other sites exist within canyon-fan systems and seemingly all areas of the open continental slope. Finally, some of the largest landslide deposits ever observed occur off Hawaii and involve major parts of the islands.

1 INTRODUCTION

The Pacific Ocean and its surrounding seas cover roughly one-third of the earth's surface. Latitudes range from 70°S to 65°N and include the ice shelfs of Antarctica, the glaciated coasts of Alaska, Canada, Chile and New Zealand, coral atolls, volcanic islands, the mouths of four of the world's ten largest rivers in terms of sediment discharge, the coasts of five continents, most of the deep-sea trenches and all of the submarine sediment types. In addition, an almost continuous ring of seismically active zones surrounds the ocean, large winter storms rage in the high latitudes of the Pacific, and hurricanes and typhoons strike the lower latitudes. Within this basin, a myriad of submarine landslides is found, ranging from small failures in fjords, submarine canyons, basins, and continental shelves and slopes to some of the largest landslide deposits ever observed, involving major parts of some of the Hawaiian Islands.

This paper can only touch the surface of the subject of submarine landsliding in the Pacific Ocean. For one reason, the environment is too large and too diverse to be covered adequately in one paper. The number of landslides present in such a large area must also be large. Second, there have not been many of the types of surveys that can be used to discern undersea landslides. Almost certainly, most of the undersea landslide deposits have not been discovered. In spite of these limitations, however, the extent and form of landsliding in the Pacific can be speculated upon based on the few well-documented investigations. Accordingly, we can surmise that undersea landsliding is an important hazard to offshore and coastal development, as well as being a major geologic agent for transporting sediment to deeper water.

This paper introduces undersea landslides of the Pacific through a series of examples that bracket the range of failure types and identification methods. Next, the extent and morphology of Pacific Ocean landslides are discussed within a framework that groups the features by their environment: fjords and other bays and inlets, the open continental shelf, canyons on the continental slope (including deep-sea fan deposits), and the open continental slope. Features are grouped by environment rather than geography or nearest country to show the similarities of features from various locations. The time of occurrence of these landslides frequently cannot be identified. However, the cause can often be estimated by

considering the impact of environmental parameters, e.g., seismicity, storm climate, deposition rate, or gas charging, and the engineering behavior of the sediment. Finally, the potential impact of undersea landslides upon man's activities is considered, in terms of both present development and potential future exploitation, as we attempt to use more and more of the ocean's resources.

2 THE NATURE OF SUBMARINE LANDSLIDES IN THE PACIFIC AND METHODS AVAILABLE FOR FINDING THEM (EXAMPLES)

Undersea landslides typically occur without making their presence known to human observers, and many present on the sea-floor occurred before recorded history. As a result, we must often use sophisticated remote sensing equipment to locate these features and interpret their causes, times of occurrence, and potential hazard. Even the accurate identification of a bottom feature as a landslide deposit can be difficult, given that other submarine processes can form similar features. The following discussion begins with landslides that caused disruption of the shoreline and whose times and conditions of occurrence are well known. Next considered are events that damaged offshore installations or caused observable changes in bathymetry and are thought to have been caused by landslides. Discussed last are sea-floor features that have been surveyed using remote acoustic sensing methods. The discussions that follow are grouped around well-investigated examples that bracket the range of both undersea landslide types and methods of investigating them.

2.1 Landslides that caused disruption of the shoreline

1. Howe Sound, British Colubia: failure in a coarse-grained fan delta.

In an early paper on the varieties of submarine slope failures, Terzaghi (1956) described a disasterous failure of the front of a sand and gravel fan delta on the northwest shore of Howe Sound, British Columbia. On August 22, 1955, a few minutes after low tide, part of a warehouse collapsed and fell into the sound. Twenty-five minutes later, piles began to disappear from beneath a dock and reappear 35 m away. Finally the entire dock, along with parts of nearby buildings, were destroyed. A post-failure investigation indicated that a fairly limited, perched

body of silty sediment, located at the top of a 28°
offshore slope, was lost to landsliding and in the
process destroyed the onshore structures. The low
tide caused the failure through a rapid drawdown
mechanism. Such a mechanism often occurs when the
level of water against a partially submerged slope
drops so rapidly that pore water pressures corresponding to the original water level cannot dissipate.

Later sidescan-sonar studies by Prior and others
(1981) showed offshore chutes, scarps, blocks and
steps that supported Terzaghi's interpretation of the
cause of the coastal damage. However, the complexities of the observed offshore features indicate that
slope instability is more widespread in the area than
previously thought and is a major component of development of the fan delta.

2. Kitimat Arm, British Columbia: failure in muddy
sediment at a fjord head delta front.

Kitimat Arm is located on the northern part of the
coast of British Columbia and has a history of underwater landslides. The most significant recent failure
occurred on April 27, 1975 shortly after low tide. A
piled jetty 1500 m^2 in area disappeared under water,
and waves up to 8.2 m in height flooded adjacent low-
lying areas (Murty, 1979; Prior and others, 1982a).
In one area, the underwater landslide expanded to
include movement of subaerial slopes. The deposits
remaining after the undersea landslide were mapped
using sidescan-sonar and a 3.5 kHz subbottom profiler
(Prior and others, 1982b). A sidescan-sonar mosaic
showed a complex pattern of features near the margin
of the slide including pressure ridges, scarps, blocky
areas and outrunner blocks (illustrated conceptually
in Fig. 1). Core samples showed the failed sediment
to be highly charged with biogenic gas. The immediate
cause of failure was rapid drawdown brought about by
an extreme low tide. Contributing to failure were a
high rate of sedimentation on a steep (up to 8°)

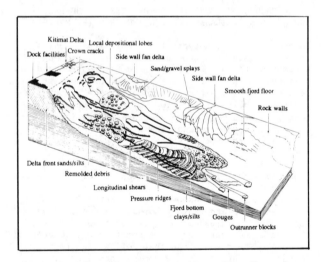

Figure 1. Schematic representation of seafloor features at the head of Kitimat Arm, showing components
of the 1975 submarine landslide (from Prior and
others, 1982a, printed with permission of Springer-
Verlag NY, Inc.).

delta-front slope, gas charging of sediment, and
recent man-made construction activity (installation of
a crib wharf and a breakwater that loaded and disturbed the sediment).

3. Port Valdez, Alaska: catastrophic earthquake-
induced failure of a submarine slope.

The Alaska earthquake of March 27, 1964, caused
severe damage and fatalities in a number of south-
central Alaskan communities. Much of the damage and
perhaps most of the deaths could be attributed to
undersea landslides and resulting violent local waves
(Plafker and others, 1969). The largest number of

deaths (31) in any one community occurred in Valdez,
which was situated on the seaward edge of a large
glacial outwash delta composed of saturated silty sand
and gravel. The earthquake triggered a massive
submarine failure in the delta deposits involving up
to 100 million cubic meters of sediment. The failure
(Fig. 2) destroyed the harbor facilities and nearshore
installations. Slide-generated waves and seiches

Figure 2. Submerged slide area at Valdez, Alaska,
following the 1964 Alaska earthquake. Dashed lines
indicate dock area destroyed by the landslide (from
Coulter and Migliaccio, 1966).

further damaged the downtown area. Removal of support
from the face of the delta by the submarine landslide
allowed parts of the delta still remaining to move
seaward and caused parts of the shore area to subside
below high-tide level. Following the disaster a study
was made of the stability of the remaining delta on
which the remnants of the town still stood. The study
showed that the waterfront was still unstable under
seismic loading conditions. To prevent another tragedy, the entire town of Valdez was relocated to a more
stable location 5 km to the west (Coulter and
Migliaccio, 1966). The cause of the undersea landslide at Valdez was intense earthquake loading and
resulting strength loss in cohesionless sediment on a
steep (average of 16°) delta-front slope. Low tide
and drawdown during large waves may have further
aggravated the situation.

4. Hamana Lake, Japan: the earthquake-induced
Imagiri-Guchi failure.

Hamana Lake was the second largest fresh water lake
in Japan before 1498 when an earthquake with an
estimated magnitude of 8.6 occurred 80 km away.
During the earthquake, failure occurred in the sand
dune that separated the lake from the open sea. The
lake was opened to the sea through the newly created
passage termed "Imagiri-Guchi (Mouth Failed Just
Now)." Recent geotechnical evaluations and bathymetric surveys have shown that the failure was likely

caused by liquefaction of the dune sand and undersea sediment. Continuing failure of these materials during major earthquakes can be expected (Okusa and others, 1988).

5. Awashima Island, Japan: failure induced by storm waves.

In March, 1974, a winter low pressure system passed through the Japan Sea. Storm waves of 4 m height, 110 m wavelength, and 8.5 s period and winds of up to 20 m/s attacked the southeast coast of Awashima Island off the western coast of northeast Japan. The most intense part of the storm occurred between morning and midnight of March 22. At 5 PM a 150 m long breakwater began to tilt into the sea. Near midnight the office for the village on the island began to tilt and the coastal side of the village kindergarden sank. The kindergarden and office eventually collapsed into the sea and a second breakwater was lost. In all, 500 m of the coast failed and the sea bottom was covered with silt and fine sand intercalated with gravelly sand; one-million cubic meters of sediment were displaced. Bathymetric surveys before and after the failure showed major changes in bottom topography. Failure occurred by storm wave loading of the seafloor and was possibly aggravated by recent uplift that occurred ten years before when a 7.5 magnitude earthquake uplifted the eastern side of the island by 0.8 to 1.5 m (Okusa and others, 1988).

6. San Francisco Bay: failure of an excavated trench slope.

In August, 1970, construction was proceeding on a new shipping terminal for the Port of San Francisco. As part of the construction an underwater trench was being cut into San Francisco Bay mud. The trench would ultimately be filled with sand, which would provide a better foundation for a pile-supported dock than would the bay mud. On August 20, after a section of the trench about 150-m-long had been excavated, the dredge operator found that the clamshell bucket could not be lowered to the depth from which mud had been excavated hours before. Sidescan sonagraphs showed that a failure had occurred involving a 75 m section of the trench. Within the next four months, another failure occurred involving an additional 60 m of the trench, and a debris dike placed near the trench collapsed. The submarine slope failure was used as a well-documented case study to test methods for measuring the strength properties of soft cohesive sediment (Duncan and Buchignani, 1973). The failure was caused by oversteepening of a sea-floor slope as a result of construction activities.

2.2 Landslides that disrupt undersea facilities

The case histories discussed above occurred under conditions favorable for documentation in that each landslide had a visible effect on onshore activities and the conditions at the time of failure are well known. Nearly as favorable in terms of facilitating documentation are landslides that disrupt undersea facilities.

1. Western New Britain Trench: cable breaks indicate frequent submarine landslides and turbidity currents.

The Solomon Sea region has an extremely high level of seismic activity. Twenty-three earthquakes with magnitudes greater than 6.9 occurred in the region between 1900 and 1974 (Everingham, 1974). In addition, the Markham River, which drains eastern New Guinea is heavily sediment laden and rapidly builds river delta lobes at the head of the western New Britain Trench. A submarine telecommunications cable is frequently broken at the point where it crosses the trench floor about 230 km to the east of the Markham River mouth. For example, the cable was broken in both 1966 and 1968 following magnitude 6.0 and 6.7 earthquakes, respectively. Krause and others (1970) speculated that the cause of the cable breaks was earthquake-induced undersea landsliding in the Markham River delta that transformed into high-speed (30 to 50 km/hr) turbidity currents that flowed out onto the

trench floor. Submarine cable breaks are a common occurrence throughout the world and often follow major seismic events. They may be one of the most expensive manifestations of undersea landslides; for example, the cost of repairing a broken cable near Madang, New Guinea, following a magnitude 7.1 earthquake in 1970 was A$500,000 (Everingham, 1974).

2. Off southwest Oahu: current sensors displaced by turbidity currents generated during a hurricane.

A suite of current sensors was deployed down the submerged slope off the southwest coast of the Hawaiian island of Oahu as part of an environmental study to investigate the feasibility of generating usable power from ocean thermal energy conversion. The top of the slope is a 60° to 80° escarpment that slopes down to a debris apron of 5° to 10°. The slope is covered by a sediment that is nearly pure calcareous ooze. During the passage of Hurricane Iwa in November of 1982, four of the sensors were displaced downslope, increasing their water depths by as much as 220 m. Downslope movement of up to 2.4 km occurred at speeds up to 300 cm/s. Because measured currents were not in themselves sufficient to move the anchors for the current sensors and because the temperatures at the sensors increased during the event, the movement was most likely caused by a turbidity current. Submersible observations disclosed small (roughly 2 m) scarps and ridges within 100 m of locations where smooth bottom was observed before the turbidity current episode began. Dengler and others (1984) postulated that the hurricane induced slope failures above the current sensors. Four of the slope failures generated turbidity currents carrying a mixture of sediment and water at frontal speeds of 300 cm/s or greater. The turbidity currents moved the current sensors and also broke or buried four communications cables that had been located farther down the slope (Fig. 3).

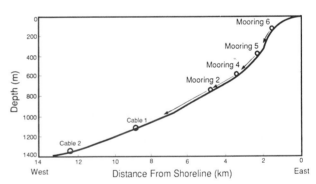

Figure 3. Profile off Oahu, Hawaii, of depth vs. distance from shoreline with positions of two cables and moorings affected by seafloor landslides induced by Hurricane Iwa in 1982. Arrows next to open circles indicate the total downslope motion of moorings (after Dengler and others, 1984).

2.3 Landslides identified by changes in bathymetry

Some offshore landslides have been identified by major changes in bathymetry or relief. The following three examples of landslides did not noticeably influence the coast or offshore installations at their time of occurrence. However, subsequent investigations showed that the failures did occur and the time of occurrence could be associated with specific events.

1. Klamath River delta, California: earthquake-induced failures on a 0.25° slope.

On November 8, 1980, a large magnitude (6.5 M_L to 7.2 M_L) earthquake occurred offshore about 75 km northwest of Eureka, California (Lajoie and Keefer, 1981). Minimal onshore effects were limited to coastal rockfalls, sand boils on beaches, minor damage to structures, minor local settlement and one bridge collapse. Immediately following the earthquake, commercial fishermen reported the appearance of ridges and scarps on the sea-floor off the mouth of the Klamath River, roughly 60 km from the epicenter of the earthquake. Surveys were conducted of the area in December, 1980, and May and October, 1981, to map the changes and determine their causes. Using high-resolution seismic-reflection profiling and sidescan sonography, a failure zone measuring 20 km in length and 1 to 5 km in width was delineated. These data could be compared with previous survey information to precisely document the changes that occurred during the earthquake (Field and others, 1982). Sonographs (Fig. 4) of the failure zone showed a toe ridge, pressure ridges, sand boils, collapse features, sediment flows and gas seeps (Field and Jennings, 1987) all occurring on a slope of about 0.25°. The elongated failure zone lies roughly parallel to the coastline and is 10 to 15 km offshore in about 60 m of water. The seaward boundary of the failure coincides with the contact between muddy sand and sandy clayey silt. The muddy sand apparently failed during the earthquake through a process of liquefaction, lateral spreading, and sediment collapse and spreading (Field

Figure 4. Sidescan sonograph of the seaward edge of a failure zone in the Klamath River prodelta, northern California, following the earthquake of November, 1980 (after Field and others, 1982).

and others, 1982).

2. Sagami Bay, Japan: earthquake-induced landslides cause sea valleys to deepen.

Following the great earthquake of 1923, the Japanese resurveyed onshore and offshore areas in which the earthquake had been most destructive. In contrast with only a few meters of elevation change on land, the offshore surveys showed up to 400 m of vertical modification to the sea-floor. These surveys were conducted in the days when depths were measured by lowering a weight to the sea-floor, and navigation was much inferior to what it is today. Accordingly, the soundings cannot give an accurate picture of the sea-floor changes that occurred. However, by carefully replotting all the data, Shepard (1933) surmised that there was substantial evidence that real changes did occur. Submarine valleys appeared to be deepened and shoaling occurred in other limited areas. Submarine fault displacement cannot account for the changes because displacements greatly exceed any ever recorded on land. Shepard (1933) concluded that the changes must have been caused by massive submarine landslides, one of the first suggestions of this mechanism. Locations where shoaling occurred likely received landslide material from elsewhere.

3. Scripps Submarine Canyon, California: storm-induced slump opens an unknown tributary.

During the last week in May, 1975, a graduate student at the Scripps Institution of Oceanography was conducting an underwater inspection of his research area offshore of La Jolla, California, and observed significant changes in water depth relative to a similar inspection several days earlier. Between the two inspections, a major storm had occurred. Subsequent surveys by Marshall (1978) disclosed a new

abrupt scarp of 1.8 m in water depths of 9 m to 11 m. Below the scarp, the sediment had a pock-marked appearance, possibly reflecting escape of pore water. A bathymetric survey disclosed a new irregular valley below the scarp that was about 400 m long, up to 100 m wide and 8 m to 30 m deep. Marshall (1978) believed the valley to have been caused by a failure that moved about 100,000 m^3 of sediment into Scripps Canyon which lies directly below the new valley. Furthermore, exposed bedrock in the lower portion of the valley showed that the slump formed in a small bedrock tributary to Scripps Canyon that had been previously filled with sediment. Apparently the storm caused liquefaction and flow of sediment in the headward and central part of the slump valley (Marshall, 1978).

2.4 Landslides identified by morphology

Most reported undersea landslides have been observed using sub-bottom profiling and sidescan-sonar after their times of occurrence. The landslides are recognized by their distinctive morphologies and occasionally dated by assessing the amount of sediment deposited on scarps and scars after the failures occurred. The causes of the failures cannot be known exactly although quantitative analysis can lead us to suspect some mechanisms more than others (Lee and Edwards, 1986, Schwab and Lee, 1988). The following examples show the variety of landslides of this type that have been reported from locations around the Pacific Ocean.

1. Alsek Prodelta, Alaska: mobilized flows in a deltaic setting on the continental shelf.

A detailed sidescan-sonar mosaic of the sea-floor was obtained over a 10 km by 2 km area within the Alsek River prodelta about 4 km off the mouth of Dry Bay, northeast Gulf of Alaska (Molnia, 1982). Analysis of this mosaic (Fig. 5), along with related bathymetric and seismic reflection data and sediment samples, allowed delineation of sediment slides and flows of varying morphology. Nearshore sands contain isolated sediment failures and depressions, but these cover less than 10% of the area. Deposits of clayey silt somewhat farther offshore contain slumps, slides and flows of several types over most of the mosaic area. The failures vary from fairly small features (most with lateral dimensions less than 50 m) in the west to well-developed linear troughs in the east. In a geotechnical analysis, Schwab and Lee (1983) showed that the clayey silt is particularly susceptible to strength loss during cyclic loading events and that storm waves likely caused the failures in the Alsek prodelta. The mobility demonstrated by the prodelta failures likely resulted from incorporation of water into the clayey silt during lengthy storm wave loading to the point that the sediment strength was below the shearing stress exerted by gravity on the 0.5° to 1.3° shelf (Schwab and Lee, 1988). Nearby slumps on the shelves off Yakutat and off the Malaspina Glacier occurred in somewhat deeper water and show fairly limited mobility. These slumps likely resulted from earthquake loading. Insufficient time was available during the loading events to allow much incorporation of water into the sediment fabric (Lee and Edwards, 1986, Schwab and Lee, 1988).

2. Kayak Trough, Gulf of Alaska: large submarine slide of unknown origin.

High-resolution seismic profiles from the continental shelf of the northeastern Gulf of Alaska show a large submarine slide at the eastern edge of the Copper River Prodelta. The slide (Fig. 6) has moved down a slope of about 1° to the bottom of Kayak Trough and is 18-km-long, up to 15-km-wide and 115-m-thick (Molnia and others, 1977). The slide has a very irregular surface morphology and disrupted internal acoustic reflectors as well as pull-apart scarps with reliefs of 5 m to 10 m. Sidescan sonographs show the scarp region to contain a series of small steplike scarps, each marking a separate failure plane. The

NORTH

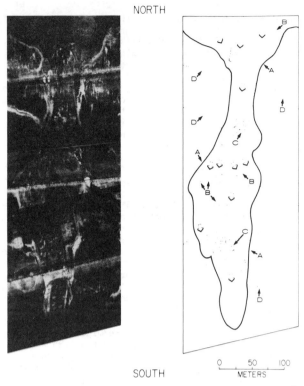

SOUTH

```
0    50   100
    METERS
```

Figure 5. Side-scan sonograph mosaic and interpretive
sketch of an area of the Alsek prodelta (northeast
Gulf of Alaska) showing: A) boundaries of an elongate
sediment gravity-flow deposit (large arrows indicate
the flow direction), B) coalescing sediment-gravity
flows, C) collapsed, blocky sea floor or "flake-type"
slides, and D) headwall escarpments of slope failures
that predate the elongate sediment-gravity-flow
deposit (from Schwab and Lee, 1988).

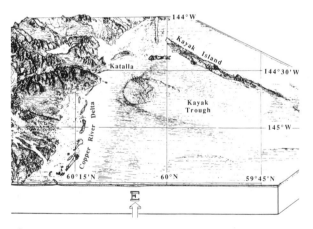

Figure 6. Physiographic diagram of Kayak Trough,
Alaska, showing a massive submarine slide (after
Molnia and others, 1977).

scarps are in water depths of 40 m to 100 m and the
toe is at a depth of over 200 m. Because surface
features of the slide are well-preserved and the
sedimentation rate is high, the slide is probably
recent, perhaps caused by the 1964 Alaska earthquake
or the 1899-1900 Yakutat earthquakes (Hampton and
others, 1978). The slide moved with enough momentum
to carry the toe of the slide past the thalweg of the
trough (Carlson and Molnia, 1977).

3. Ulleung Basin, Sea of Japan: zoned landslide
deposits on basin slope, rise, and floor.

Using high-resolution subbottom profiling (3.5 kHz),
Chough and others (1985) mapped extensive landslide

deposits all along the southern margin of the Ulleung
Basin, a backarc basin lying between Korea and
Honshu. Within this margin, the continental slope
lacks submarine canyons and has an average declivity
of about 3.5°. Landslide deposits on the slope were
classified (generally following Varnes, 1958) according to their appearance on acoustic reflection profiles into rockfalls, slides and slumps. Rockfalls
are highly reflective, hummocky, and lacking in
internal reflectors. Slides lie downslope from glide
planes and contain continuous internal reflectors
within each slide block. Slumps also show a glide
plane and lie below scars, but in addition show
distortion of internal structure. Chough and others
(1985) indicated that virtually the entire southern
basin slope, along a 200 km sector, was involved in
one or more of these types of failure. However, the
ability to resolve the extent of these failures is
limited by the 25 km trackline spacing of the subbottom acoustic reflection lines. In any case, the
extent of failure on the slope is significant. The
basin rise, lying below the base-of-slope, is occupied
by wedge- or lense-shaped deposits that are acoustically transparent. Chough and others (1985) speculated that these deposits were caused by debris flows
that mobilized from the slides and slumps on the basin
slope above. The basin floor contains acoustically
reflective and evenly layered sediment interpreted to
be thin-bedded turbidites. Chough and others (1985)
speculated that these deposits originated when debris
flows became turbulent and gave rise to turbidity
currents. Failure in this environment was likely more
extensive during glacial time when a large volume of
terrigenous sediment was transported to the shelfbreak
and could be acted upon by storm waves. However, a
geotechnical analysis (Chough and Lee, 1987) showed
that moderate earthquake activity could cause slopes
to fail under present interglacial conditions.

4. Modern Huanghe (Yellow River) subaqueous delta:
slope failure processes on slopes of 0.1° to 0.4°.

The modern Huange Delta was investigated using high-
resolution subbottom profiling and sidescan sonar
(Prior and others, 1986a). The surveys disclosed
deposits characteristic of widespread slope instability and mass-movement processes. There were two
general bottom morphology types: heavily disturbed
areas displaying features with a sharp, "fresh"
appearance along the 30-km-long northeast delta face
and more subdued, moderately disturbed areas along the
southern delta face. The subdued appearance in this
area is apparently the result of burial of mass deformation features. Within the heavily disturbed areas,
there are two major types of bottom features. First,
silt-flow gullies are very wide (100 to 500 m), shallow and flat-floored features that can achieve lengths
of 3 to 4 km. The gullies are found on slopes of 0.3
to 0.4°. Second, there are enclosed collapse depressions (Fig. 7) that occur on slopes of 0.1 to 0.2° in
water depths of 10 to 13 m. The collapse depressions
are absent from the steeper delta slopes in shallower
water where the silt-flow gullies are more common.
Collapse depressions are bounded on all sides by low-
relief scarps that range in height from 0.5 m to 1.5
m. The features range from 10-m-diameter circles to
extensive irregular areas up to 500 m across. Prior
and others (1986a) speculated that failure and downslope flow occurred on slopes of 0.3° to 0.4° forming
the silt gullies. On more gentle slopes, failure
takes the form of sediment collapse and subsidence.
Extensive failure of the Huange delta occurs mainly
because of extremely rapid sedimentation. The modern
delta has occupied its present position only since
1976 and has prograded 15 km into the Gulf of Bohai
during that time. With such rapid sedimentation, most
of the sediment fabric is supported by excess pore
water pressures rather than intergranular or effective
stresses. The resulting shear strength of this highly
underconsolidated sediment is low and incapable of
resisting gravitational and storm-wave loading. Storm
wave heights in the Gulf of Bohai can reach 7 m and
are most intense along the northeast face of the delta

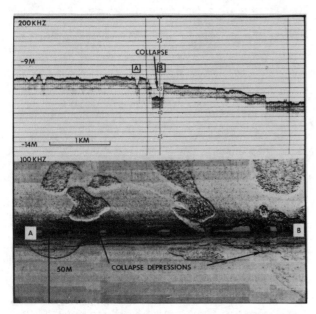

Figure 7. 200-kHz fathometer and 100-kHz side-scan sonar data across an area of irregular-shaped collapse features on the lower delta front of the modern Huanghe River delta, China (from Prior and others, 1986a, printed with permission of Springer-Verlag NY, Inc.).

where the level of bottom disturbance appeared most fresh. The sediment is derived in large part from the loess deposits of interior China and is likely highly sensitive to loss of shear strength during cyclic loading caused by storm waves. Loading from frequent nearby earthquakes of magnitudes of 7 to 7.5 could also cause failure in Huange Delta silts. Gas charging has not been observed in the delta and is not considered to be a major factor in causing failure (Prior and others, 1986a).

5. Kidnappers Slump, New Zealand: failure on a gentle open continental slope.

An early investigation of slumping of modern sediment on continental slopes was conducted by Lewis (1971). One of the failures investigated was the Kidnappers Slump which is about 15 km off the coast of North Island, New Zealand, near Napier. The landslide (Fig. 8) was identified using subbottom acoustic profiling and was found to extend 11 km down the slope and about 45 km along the slope. The head of the slump is in 250 m of water, and the toe is deeper than 500 m. Failure occurred on a glide plane that is roughly parallel to the sea-floor surface and slopes seaward at 1° to 4°. In the upper part of the failure, there are tensional depressions formed during separation; near the toe, beds are folded and thrusted due to compression. The landslide mass is 20-m- to 50-m-thick. The landslide involves sediment that is 20,000 years old and therefore occurred within the last 20,000 years. The landslide was probably caused by earthquake loading and strength degradation within sandy silt beds (Lewis, 1971).

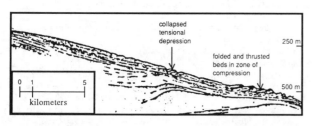

Figure 8. Tracing of continuous seismic profile across the Kidnappers Slump, east of North Island, New Zealand (after Lewis, 1971, printed with permission of Elsevier Publishing Co.).

6. Ranger Submarine Slide: failure on a 3° slope off the coast of Mexico.

The Ranger Submarine Slide lies northwest of Sebastian Vizcaino Bay, Baja California, Mexico. The failure was discovered using subbottom acoustic profiling equipment which disclosed an exposed slide surface (scar) that dips at 3°. In deeper water there is a large slide sheet. The scar extends from water depths of 1200 m to 1500 m, and the distorted remains of the slide sheet are in 1500 m to 1700 m of water. The scar covers nearly 125 km^2 and the slide sheet itself covers more than 300 km^2. Movement of the slide was apparently stopped when the sheet collided with a gentle, anticlinal fold of sediment. Failure probably occurred in the late Pleistocene because only a thin bed of sediment has been deposited on the scar. Failure was likely caused by seismic activity associated with displacement along faults that cut recent sedimentary units (Normark, 1974).

7. Gaviota mudflow, California: failure on a borderland basin slope.

A mudflow lies on the mainland slope of the Santa Barbara Basin of southern California (Nardin and others, 1979a; Lee and Edwards, 1986). The failure deposit is a 6- to 10-m-thick zone, about 2-km-long by 2-km-wide, that heads on a 4° slope in 395 m and toes in 510 m of water. The failure likely occurred within the past few hundred years, as determined by measuring the amount of newly deposited sediment on the slide scar and estimating an average sedimentation rate. The mudflow (Fig. 9) was discovered by high-resolution subbottom acoustic profiling.

Historical tectonic movement within the Santa Barbara Basin is indicated by minor to moderate seismic activity. For example, the recent moderate (M_L = 5.1) earthquake of 1978, located off Santa Barbara, California, caused local peak ground acceleration of 0.40 g, and an even larger earthquake shook

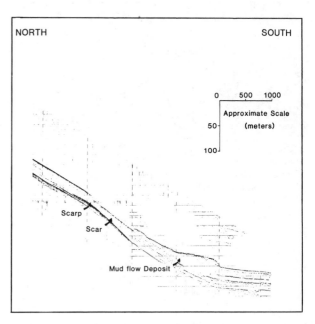

Figure 9. High resolution acoustic profile across a mudflow deposit on the Santa Barbara Basin slope near Gaviota, California (from Lee and Edwards, 1986).

the area in 1925. Lee and Edwards (1986) performed a geotechnical analysis on gravity cores from the mudflow and determined that the failure was likely seismically-induced. The specific location of the failure was caused by a combination of factors of which the seaward fining of sediment grain size and the location of the base-of-slope were the most important.

8. Humboldt slide zone: slumping on the continental
slope near Eureka, California:
   A slump about 150 km² in area lies off Eureka,
California, approximately 50 km north of the Mendocino
Ridge-Gorda Escarpment (Field and others, 1980; Lee
and Edwards, 1986; Field and others, 1987). The sediment failure (Fig. 10) was discovered using high-resolution subbottom profiling techniques and involves
late Quaternary sediment overlying the Eel Plateau.
The sediment failure heads in 460 m of water and

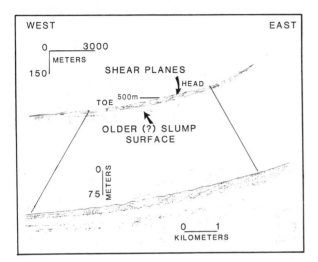

Figure 10. Line drawing and detail of high-resolution
acoustic profile across the Humboldt Slump near
Eureka, California (from Lee and Edwards, 1986).

extends more than 6 km downslope to a water depth of
580 m. The slump mass has a marked hummocky upper
surface due to rotation along numerous shear planes,
and individual blocks within the mass exhibit backro-
tated bedding. The failed mass is at least 30 and
possibly as much as 85 m thick. The slump lies on an
average 1.5° slope below a steeper (4°) slope. The
continental margin in this area is seismically active
and is characterized by large (M_L>7) and frequent
earthquakes.
   Lee and others (1981) performed a geotechnical
analysis of the slump and determined that it was
seismically induced. Sediment grain size decreases
offshore and plasticity increases. The sediment
composing the slump is somewhat more susceptible to
seismic loading than the other sediment in the area.
   9. Hawaiian Islands: giant submarine landslides.
   Recently collected evidence shows that some of the
world's largest landslides have occurred on the steep
submerged volcanic slopes of the Hawaiian Islands.
The possibility of such giant landslides was first
suggested by Moore (1964), who observed distinct
irregular topography northeast of Oahu and north of
Molokai (Fig. 11). Moore interpreted a series of
elongate blocky seamounts with tilted relatively flat
upper surfaces as representing debris that was emplaced during two major landslides. The larger landslide, northeast of the island of Oahu, is more than
160 km long and 50 km wide. The blocky seamounts that
represent slide debris are from 8 to 25 km long and 5
to 15 km wide. The second landslide extends northward
down the slope from the north side of the island of
Molokai and is 50 km wide and more than 80 km long.
These slides appear to represent failure of entire
half flanks of Hawaiian volcanoes, along their rift
zones and extending back toward their summit calderas
(Lipman and others, 1988).
   The best documented of these giant Hawaiian submarine landslides is the Alika debris slide (Lipman and
others, 1988), which was discovered off the west coast
of the island of Hawaii. The Alika slide was mapped
in detail using subbottom acoustic profiling and the

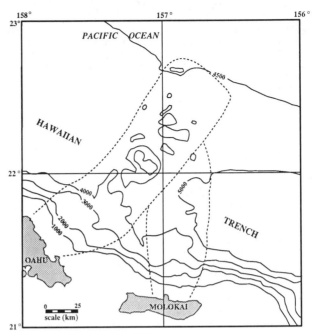

Figure 11. Irregular seafloor topography adjacent to
the islands of Oahu and Molokai, Hawaii, indicative of
two giant submarine landslides. Inferred boundaries
of landslides are shown by dotted lines (after Moore,
1964).

GLORIA (Geologic Long-Range Inclined Asdic) sidescan-
sonar system. The slide covers a total area of about
4000 km². The deposits consist of two debris avalanche lobes that were derived from the same source
area and probably formed in rapid succession. Failure
began near shore in a 10° to 15° breakaway area that
is characterized by large block slumps. The lower
part of the failure deposit has distinct hummocky
topography and marginal levees, lies on gentle (0.3°
to 0.6°) slopes and resembles many subaerial volcanic
debris avalanche deposits. The hummocky zone grades
into a fringing zone of isolated blocks that may be
several kilometers across. The smoothly sedimented
axis of the Hawaiian Deep adjacent to the Alika slide
has been interpreted to be mainly turbidite deposits. The longitudinal profiles of the Alika slide are
analogous to those observed at subaerial debris
avalanches such as Mount St. Helens and imply rapid
flow velocities, perhaps several hundred kilometers
per hour (Lipman and others, 1988). Although the
exact age of the slide is unknown, it is at least
13,000 years and is probably less than a few hundred
thousand years. The Hawaiian slides are unique in
that they mainly include failure of volcanic rock,
whereas most other reported undersea landslides have
developed in unconsolidated sediment.
   The possible future occurence of comparably large
landslides presents a potential hazard to parts of the
Hawaiian Islands, both because they might include
large subaerial parts of the islands and because they
might generate tsunamis. Gravel deposits on the island of Lanai suggest that a giant wave rode up as
much as 325 m above mean sea level and probably was
caused by a large submarine landlide south of Lanai
(Moore and Moore, 1984). Such failures have an
apparent low recurrence interval of perhaps only one
per 25,000 to 100,000 years (Lipman and others, 1988),
thus their risk to the Hawaiian populace is minimized.

3 ENVIRONMENTS OF PACIFIC OCEAN LANDSLIDES

The twenty examples given above bracket the range of
environments, causes, survey techniques and effects of
landslides in the Pacific Ocean. They do not in them-

selves, however, indicate the relative importance of each of the environments or causes, nor do they allow us to assess the extent of the landslide hazard on the Pacific sea-floor. In the following section, each of the environments in which landslides have been known to occur is discussed in terms of its relative significance.

3.1 Fjords

Of all the environments of the Pacific basin, fjords with high sedimentation rates are likely the most susceptible to failure, both in terms of the proportional areal extent of deposits that can become involved in mass movement and also in terms of the recurrence interval of failures at the same location. Fjords are glacially eroded valleys that have been inundated by the sea. They are often fed by sediment-laden rivers and streams that drain active glaciers. These factors lead to environmental conditions that are highly conducive to sediment failure. The submerged sides of glacial valleys are commonly extremely steep and may extend to great depths, perhaps 1000 m or more. There is typically a delta at the head of the fjord formed by streams that may drain the glacier that initially eroded the valley. These fjord-head deltas have foresets that dip at 5° to 30° between 10 and 50 m water depth. In greater depths, the prodelta dips at angles of 0.1° to 5° to the flat basin floor, typically at depths between 100 and 1000 m (Syvitski and others, 1987). The glacial streams feeding these deltas carry rock flour and coarse sediment that is highly susceptible to strength loss during cyclic loading from earthquake or storm-wave loading events. Also, the sediment may be deposited so rapidly that pore water pressures cannot dissipate completely as sedimentation continues. The resulting underconsolidated state of the sediment causes abnormally low static shear strengths. Organic matter brought down with the glacial debris may be abundant, and it can decay and produce bubble-phase gas that may also lead to elevated pore water pressures and low shear strength. These steep slopes with low-strength sediment that is susceptible to cyclic loading may fail seasonally or semi-continuously yielding numerous small-scale failures. Failures of a seasonal or simi-continuous nature have been reported in many fjords in British Columbia; for example, Bute Inlet (Syvitski and Farrow, 1983; Prior and others, 1986b), Knight Inlet (Syvitski and Farrow, 1983) and North Bentinck Arm (Kostaschuk and McCann, 1983). In some situations, however, fjord-head delta slopes may fail infrequently instead and produce catastrophic effects, such as occurred in Valdez (Coulter and Migliaccio, 1966), Seward (Lemke, 1967) and Whittier (Kachadoorian, 1965) in the 1964 Alaska earthquake or in Kitimat Arm (Prior and others, 1982a, 1982b).

The sidewall slopes of fjords can also be unstable. Normal suspension fallout on the steep (10° to greater than 90° overhangs) submerged valley sides can frequently lead to small failures (Farrow and others, 1983). Even more important are failures on side-entry deltas that build out rapidly onto the sidewall slopes. The failures in Howe Sound (Terzaghi, 1956; Prior and others, 1981) are of this type as are reported failures in Knight and Bute Inlets, British Columbia (Syvitski and Farrow, 1983).

Finally, the deep fjord basins, which tend to have slopes of less than 0.1°, commonly receive failed sediment masses and flows from the side walls and fjord-head deltas. If these landslides incorporate enough water during their movement, they can progress through viscous sediment gravity flows into turbulent turbidity currents. These fluidized failures can be fed into and across the basins by channels (Syvitski and others, 1987).

Failures in fjords can generate tsunamis and sieches that may cause major damage to coastal communities. During the 1964 Alaska earthquake, much of the damage and many of the fatalities in Valdez (Coulter and

Migliaccio, 1966), Whittier (Kachadoorian, 1965), Seward (Lemke, 1967) and perhaps Chenaga (Plafker and others, 1969) resulted from giant waves generated by submarine landslides in fjords. Waves of 8.2 m in height were generated during the major sea-floor failure at Kitimat Inlet, British Columbia in 1975 (Murty, 1979).

Because fjords are typically found in rugged mountainous terrain, the fjord-head deltas and side-entry deltas are frequently the only flat land available. These marginally stable to unstable locations become the sites of communities and other coastal developments. Not only do these developments become vulnerable to natural slope failure, which might occur frequently even if the developments were not present, but man's activities can also lead to additional failures. For example, a river channel stabilization program at Howe Sound (Terzaghi, 1956) caused rapid delta growth to be localized and probably contributed to the ultimate failure. At Rupert Inlet on Vancouver Island, British Columbia, mine tailings are dumped into a fjord and these produce turbidity currents with a frequency of two to five days (Hay and others, 1983).

Records of submarine slope failures in fjords of the Pacific basin have mainly been obtained from Alaska and Canada. There is no reason to believe, however, that similar failures do not also occur in fjords of the southern hemisphere. Topography, sedimentation conditions, and seismicity of New Zealand and Chilean fjords are comparable to those of Alaska and Canada. At least one example of a failure of a fjord delta front in New Zealand has been reported (Brodie, 1964).

3.2 Active river deltas on the continental shelf

Active river deltas are the next most likely sites for slope instability in the Pacific basin. Rivers contribute large quantities of sediment to relatively localized areas on the continental margins. Depending upon a variety of environmental factors, including wave and current activity and the configuration of the continental shelf and coastline, thick deltaic deposits can accumulate fairly rapidly. These sediment wedges and blankets can become the locations of sediment instability and landsliding partly because of their thickness. To create large, deep-seated landslides, a thick deposit of comparatively low strength sediment is needed. Because most of the continental shelves were subaerially exposed during the last glacial cycle, most pre-Holocene sediment on the shelves is overconsolidated. The shear strengths of these older deposits are commonly high enough to resist downslope gravitational stresses on the gentle shelves and all but the very greatest storm- and earthquake-induced stresses as well. Only in areas of active deposition will significant thicknesses (tens of meters) of younger sediment be found lying above the strong older sediment. These younger deposits tend to be normally- or underconsolidated. In addition, decaying organic matter can produce gas charging that can further reduce strength. As a result, thick deposits of normally- to highly underconsolidated and possibly gas-charged sediment characterize many active river deltas on the continental shelves. These locations may fail under gravitational loading or during storms or earthquakes.

According to data presented by Milliman and Meade (1983), about 7×10^9 metric tons of suspended sediment are discharged by rivers into the Pacific Ocean and its marginal seas each year. This value represents about one-half of the total for the world's oceans. The locations of the major sedimentary depocenters provide some information on where undersea landslides might be expected in deposits on the continental shelf. For example, the Huange (Yellow) River is the largest single source of suspended sediment and deposits about 10^9 tons per year. The numerous examples of sediment instability found in the modern delta and the large extent of the instability

374

field (Prior and others, 1986a) are not suprising given this tremendous sediment load that originates in the loess deposits of central China.

Also not surprising are the failures of active deltas on the continental shelf off southern Alaska. Glacially fed rivers debouching into the Gulf of Alaska or adjacent sounds and inlets contribute 0.45 x 10^9 tons of sediment per year, about one-half of which is carried by the Copper River (Milliman and Meade, 1983). Failures have been identified in Holocene sediment all along the margin including within the Kayak Trough (Molnia and others, 1977) and Alsek prodelta (Schwab and Lee, 1983; Schwab and others, 1988) failure areas discussed previously. Also included are major landslides in the Copper River prodelta (Reimnitz, 1972), off Icy Bay and the Malaspina Glacier (Carlson, 1978; Lee and Edwards, 1986) and off Yakutat (Schwab and Lee, 1983). The landslides are likely induced by either storm waves or earthquakes; the high incidence of landsliding arises because of the intensity of the environmental loading, the thickness of the Holocene sediment wedge and the tendency of the glaciomarine sediment to lose shear strength when cyclically loaded. Although sedimentation rates are high, underconsolidation does not appear to be a major factor on the Gulf of Alaska shelf because of the relatively high permeability and low compressibility of the sediment (Lee and Schwab, 1983).

The Yukon River is a major source of sediment (60 x 10^6 tons per year) in the northern Bering Sea. Although the prodelta is nearly flat, there are indications of localized liquefaction of the sandy silt (Clukey and others, 1980) during large storms. The liquefied sediment is particularly susceptible to resuspension and erosion. Another major sediment source is Canada's Fraser River which delivers 20 x 10^6 tons per year to the Strait of Georgia. Gullies indicative of sediment failure exist on the slope of the delta front and compressional ridges exist at the base of the slope (Luternauer and Finn, 1983). In the case of both the Yukon and Fraser Rivers, the fairly coarse-grained and low plasticity nature of sediment from northern and glacially dominated environments likely contributes to the sensitivity of the sea-floor to cyclic loading and failure.

Oceanic islands, including New Zealand, New Guinea, Borneo, Japan, Taiwan, Sumatra and Luzon contribute a suspended sediment load of 3 x 10^9 tons per year, over 20% of the world's total (Milliman and Meade, 1983). Such a large amount of sediment arises from the active uplift and volcanism of the islands, as well as heavy rainfall, steep slopes, intense human activity and small drainage basins. Although the stability of offshore deposits has not been investigated to any great extent for many of these islands, the potential for landslides is likely high in many areas. The frequent occurrence of cable breaks off the Markham River delta in New Guinea (Krause and others, 1970) is probably indicative of a severe landslide hazard around other large islands in the western Pacific. Other failures have been reported near the Abra River delta on Luzon (Shepard and others, 1977), on the continental shelf of Taiwan (Chien and Tian, 1982), near Suva, Fiji (Houtz, 1962) and on the continental shelf off South Island, New Zealand (Carter and Carter, 1985).

Excluding the Columbia and Colorado Rivers, the west coasts of the United States, Mexico and South America have relatively short, steep rivers with small drainage basins. The sediment yield of many of these rivers is poorly known but it may be high and extremely variable (Milliman and Meade, 1982). There are indications of slope instability in some of the deltas, for example, the Esmeralda in Ecuador (Shepard, 1932), the Rio Balsas in Mexico (Reimnitz and Gutierrez-Estrada, 1970) and the Klamath River in the United States (Field and others, 1982). Landslide potential in most of the other deltas is unknown.

Many areas receive little sediment from rivers and are covered by relict deposits. For example, most of the continent of Australia is arid and low-lying and is not a major sediment contributor. Similarly, the deserts of northern Chile, southern Peru and Baja California, Mexico, have few rivers and low sediment yield.

Deltas of some of the major rivers that deposit large amounts of sediment along the margins of the Pacific do not display geomorphic evidence of mass instability. Such lack of evidence does not automatically preclude previous failure in these deposits because waves and burial can erase the failure effects. However, the lack of failure features likely indicates that the occurrence of failure is less frequent or extensive in these deltas. Most notable is the delta of the Changjiang (Yangtze) River in China, which is the fourth largest contributor of suspended sediment in the world, delivering about 0.5 x 10^9 tons of sediment per year. Although there are significant deltaic deposits as well as pockmarks resulting from expulsion of biogenic gases, the seafloor is otherwise featureless (Butenko and others, 1985). Continental shelf deposits derived from China's Zhu Jiang (Pearl) River (Liang and Lu, 1986) and the Columbia River (Nittrouer and Sternberg, 1981) also appear to be featureless. Evidently, a combination of circumstances is needed for mass instability to occur on the continental shelf. High sedimentation rates are needed so that a sufficiently thick bed of Holocene sediment can develop. Relatively low permeability (fine grained or plastic sediment) can allow high pore water pressures to be retained and produce underconsolidation. Also needed, however, are environmental factors, such as storms or earthquakes, that can generate shear stresses in the sediment that exceed the shear strength. These can be augmented by shear strength degradation during cyclic loading, such as seems to be particularly common with glacially derived sediment. Finally, the configuration of the continental shelf, including its slope, can influence the stability of deltas. The interaction of all of these factors, rather than one factor alone, ultimately determines whether mass instability will occur.

3.3 Submarine canyon-fan systems

Submarine canyon-fan systems serve as conduits for passing large amounts of sediment from near the continental shelf to the deep sea. In the Pacific, the presence of extensive, thick sediment fans and abyssal plains off the west coasts of the United States and Canada and within the Bering and other marginal seas testifies to the importance of mass movement mechanisms associated with these systems. These mechanisms are capable of bringing sand-size and even coarser particles to locations hundreds of kilometers from shore. Landsliding appears to be one element in the system that allows massive deposition on submarine fans to occur. According to one model (Hampton, 1972), sediment accumulations in canyon heads begin to move as coherent landslide blocks following some trigger, such as an earthquake or storm. As the blocks move downslope, the resulting jostling and agitation causes disintegration and subsequent incorporation of surface water. The debris flow that is produced displays increasingly fluid-like behavior. As the debris flow continues on its path, further dilution by surrounding water occurs, particularly as sediment is eroded from the front of the flow. Ultimately, a dilute turbulent cloud is created that has a density below 1.1 g/cm^3. The resulting turbidity current can flow for long distances (up to hundreds of kilometers) at moderate to high velocities (1 to 8 or more m/s, Shepard, 1963; Bowen and others, 1984; Reynolds, 1987). When the turbidity current enters a depositional phase, it leaves deposits that have distinctive textural characteristics (Bouma, 1962; Middleton and Hampton, 1976).

Landsliding, particularly within submarine canyons, appears to be an important, if not essential, part of the process of building deep-sea fans, which are among

the most extensive sedimentary features of the earth's
surface. However, the circumstances surrounding these
slope failures and their subsequent conversion into
turbidity currents are poorly understood. Storms
cause sediment movement in canyons, perhaps by inducing slumping near the canyon heads (Marshall, 1978) or
perhaps by introducing or resuspending enough sediment
to form a density current directly (Shepard and
Marshall, 1973; Reynolds, 1987). Earthquakes also
cause landslides in canyon heads and subsequent
turbidity current flow (Malouta and others, 1981;
Adams, 1984), but details of this process are lacking. Major earthquakes and other shocks do not always
cause canyon-head landslides (Dill, 1969). However,
landslides can occur under low sea, aseismic conditions (Shepard, 1951). Landsliding in canyon heads
and turbidity current mobilization were likely more
common during the glacial cycles of the Pleistocene
(Nelson, 1976; Barnard, 1978) because of increased
sediment supply and possibly increased storm wave
loading.

3.4 The open continental slope

A final common, but poorly understood, environment for
undersea landsliding in the Pacific Ocean is the
intercanyon area of the continental slope. Landslides
have been reported all around the Pacific margin along
slopes removed from submarine canyon-fan systems.
Included are slopes off Central America (Baltuck and
others, 1985), Mexico (Normark, 1974), southern
California (Buffington and Moore, 1962; Haner and
Gorsline, 1978; Field and Clarke, 1979; Nardin and
others, 1979a; Nardin and others, 1979b; Ploessel and
others, 1979; Field and Edwards, 1980; Field and
Richmond, 1980; Hein and Gorsline, 1981; and Thornton,
1986), central and northern California (Field and
others, 1980; Richmond and Burdick, 1981), Alaska
(Marlow and others, 1970; Hampton and Bouma, 1977;
Carlson and others, 1980), Japan and Korea (Jacobi and
Mrozowski, 1979; Chough and others, 1985; Okusa and
others, 1988), Taiwan (Chen and Tian, 1982), and New
Zealand (Lewis, 1971, 1985; Herzer, 1975, 1979). The
failures are found near river mouths and far removed
from them, as well as in both arid and humid climates. Ages of the failures are seldom known, so we
cannot determine whether they occurred under glacial
or interglacial conditions. Most were probably
seismically induced because the typical continental
slopes of 5° or less would be expected to be statical-
ly stable and storm wave loading is seldom a major
factor much below the shelf break (Lee and Edwards,
1986). Occurrence seems to correlate with sedimentation rate, slope declivity, and seismicity but the
relationship is complex (Field, 1981). An example of
the extent of landsliding that can occur on open
slopes within a particularly well investigated region,
the Southern California Continental Borderland, is
given in Fig. 12, which summarizes the known intermediate to large zones of mass failure in this region.
The continental slope appears to be an area of
extensive mass wasting. However, the recurrence
interval for failures in this environment cannot be
estimated with any degree of accuracy nor can the
likelihood of failure at any specific location. These
landslides are among the least understood in the
oceans. Because they are likely an important sediment
transport mechanism and will become hazards to
offshore development as deeper water construction
sites become important, these landslides are a clear
target for future research.

4 ECONOMIC IMPORTANCE

Pacific undersea landslides have a great potential for
destroying both onshore and offshore structures and
for taking human life. This potential was realized
during the 1964 Alaska earthquake, when $24 million
(equivalent to $80 million in 1988) in damage loss

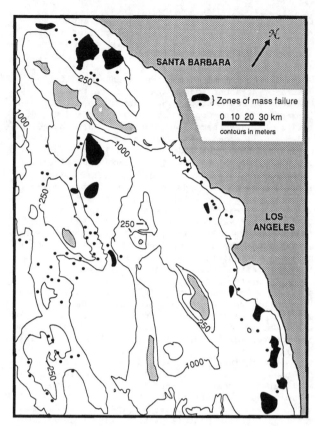

Figure 12. Known submarine landslide deposits on basin
slopes in the southern California borderland (after
Field and Edwards, 1980).

occurred to the port and harbor facilities in Seward,
Valdez, and Whittier (Hansen and others, 1966). Most
of this damage could be attributed either directly to
undersea landslides or to sea waves and seiches caused
by them. Likewise, most of the 57 deaths in Seward,
Valdez, and Whittier were related to undersea landsliding.
On a smaller scale, the failures in Kitimat and Howe
Sound caused damage losses totalling $1.3 million and
$3.5 million, respectively (1984 dollars, written
communication, Brian Bornhold, Geological Survey of
Canada). Failures causing damage of this magnitude
are probably fairly common and can be expected at
least each decade, if not more often.
Occurring most frequently are submarine cable
breaks. These appear to occur in the Pacific on an
average of about once every two years and require
roughly $150,000 to repair (personal communication,
Carl Jeffcoat, AT&T Bell Laboratories, Whippany,
NJ). If lost service is considered, the cost of these
breaks is much greater.
Damage to offshore oil facilities from undersea
landslides has not been as extensive in the Pacific as
it has been in the Gulf of Mexico and elsewhere.
However, the potential exists for landslide-induced
damage to these facilities, with associated loss of
life, and is commonly considered during the planning
and design stages. The added expenses involved in
siting and constructing these offshore structures to
minimize losses from undersea landslides can be
significant.

5 SUMMARY AND CONCLUSIONS

Landslides are common all along the margins of the
Pacific Ocean and on the slopes of Pacific islands.
Frequently, they intersect the shoreline and damage or
destroy coastal or other subaerial structures. They
can cause serious loss of life, as they did in the

1964 Alaska earthquake. The landslides are often very large, exceeding in scale the largest landslides reported on land. Three giant Hawaiian landslide deposits have areas ranging from 4000 to 8000 km^2. Four of the other example landslides have areas between 100 and 400 km^2. A landslide off the Malaspina Glacier in the Gulf of Alaska has an area of 1000 km^2 (Carlson, 1978). Although many of these submarine landslides occur on slopes comparable to those of landslides on shore (10° to 30°), some of the features develop on nearly flat surfaces. Two of the example landslides reported above (Huanghe and Klamath deltas) occurred on slopes less than 0.5°. Five other example failures were on slopes of 4° or less.

The occurrence of some Pacific Ocean submarine landslides was readily apparent to human observers as coastal communities were damaged or flooded and offshore installations were destroyed. More commonly the landslides occur unobserved. Only by comparing survey information from before and after the events can the occurrence of some failures be determined. Where relevant pre-failure survey information is lacking, the landslide features can only be recognized by the morphology apparent on geophysical records. The cause and time of failure can only be guessed. Remote sensing, using acoustic profiling and side-scanning, plays a major role in locating and mapping undersea landslide deposits.

The floors of fjords are the most likely locations for landslides along the Pacific margin. Within the fjords, fjord-head and side-entry deltas are perhaps the most susceptible and have been the location of most of the landslides that have damaged coastal structures and caused loss of life. These failures have typically been induced by either low tides or earthquakes and have been aggravated by coastal development. The next most likely locations for submarine landslides are active deltas on the continental shelf. Certain deltas, characterized by intense seismic or storm wave loading, rapid sedimentation, and sediment that has a low permeability or loses strength easily to cyclic loading show elaborate mass deformation features or are the sites of frequent submarine cable breaks. Other deltas, including those of some of the largest rivers of the Pacific basin, contain apparently stable sediment. The occurrence or non-occurrence of failure appears to be related to a combination of environmental factors, rather than any one factor.

Pacific undersea landslides are also a major factor in the system of canyons and fans through which nearshore sediment is transported to the deep sea. According to most models, landslides in the heads of canyons commonly initiate the flow of sediment down these systems. These landslides can be a direct hazard to communications cables and other offshore structures. Also, their record, preserved in channel and fan deposits, can be used to estimate the recurrence of extreme loading events, such as major earthquakes.

A final environment for Pacific undersea landslides is the open continental slope. Landslides have been mapped on the slope all around the Pacific margin. Knowledge of causes is sketchy but quantitative analysis indicates that most were induced by seismic loading. There is a correlation between the failure location and relative sediment supply, but the details of the relation are still unknown.

ACKNOWLEDGMENTS

The author thanks B. Bornhold (Canada), S. Okusa (Japan), H.J. Lee (Korea), and K.B. Lewis (New Zealand) for information on undersea landslides near each of their countries. Helpful reviews of this manuscript were provided by M.E. Field, D.K. Keefer, and W.C. Schwab.

REFERENCES

Adams, John 1984. Active deformation of the Pacific northwest continental margin. Tectonics. 3: 449-472.

Baltuck, M., E. Taylor, and K. McDougall 1985. Mass movement along the inner wall of the Middle America Trench, Costa Rica. In R. von Huene, J. Aubouin, and others, Initial Reports of the Deep Sea Drilling Project, v. 84, p. 551-165. Washington: U.S. Government Printing Office.

Barnard, W.D. 1978. The Washington continental slope: Quaternary tectonics and sedimentation. Mar. Geol. 27: 79-114.

Bouma, A.H. 1962. Sedimentology of some flysch deposits. Amsterdam: Elsevier.

Bowen, A.J., W.R. Normark, and D.J.W. Piper 1984. Modeling of turbidity currents on Navy Submarine Fan, California Continental Borderland. Sedimentology. 31: 169-185.

Brodie, J.W. 1964. The fiordland shelf and Milford Sound. In T.M. Skerman (ed.) Studies of a southern fiord, p. 15-23. New Zealand Oceanographic Inst. Mem. 17.

Buffington, E.C. and D.G. Moore 1962. Geophysical evidence on the origin of gullied submarine slopes, San Clemente, California. J. of Geol. 71: 356-370.

Butenko, J, J.D. Milliman and Y.-c. Ye 1985. Geomorphology, shallow structure, and geological hazards on the East China Sea. Cont. Shelf Res. 4: 121-142.

Carlson, P.R. 1978. Holocene slump on continental shelf off Malaspina Glacier, Gulf of Alaska. Am. Assoc. of Petroleum Geologists Bull. 62: 2412-2426.

Carlson, P.R. and B.F. Molnia 1977. Submarine Faults and slides on the continental shelf, northern Gulf of Alaska. Mar. Geotech. 2: 275-290.

Carlson, P.R., B.F. Molnia, and M.C. Wheeler 1980. Seafloor geologic hazards in OCS Lease Area 55, eastern Gulf of Alaska. Proc. of the 12th Offshore Tech. Conf., Houston. 1: 593-603.

Carter, L. and R.M. Carter 1985. Current modification of a mass failure deposit on the continental shelf, North Canterbury, New Zealand. Mar. Geol. 62: 193-211.

Chen, M.-P. and W.-M. Tian 1982. Marine geotechnical properties and stability of the continental margin deposits off Hua-Lien, northeast of Taiwan. Acta Oceanographica Taiwanica. 13: 23-68.

Chough, S.K., K.S. Jeong and E. Honza 1985. Zoned facies of mass-flow deposits in the Ulleung (Tsushima) Basin, East Sea (Sea of Japan). Mar. Geol. 65: 113-125.

Chough, S.K. and H.J. Lee 1987. Stability of sediments on the Ulleung Basin slope. Mar. Geotech. 7: 123-132.

Clukey, E., D.A. Cacchione and C.H. Nelson 1980. Liquefaction potential of the Yukon prodelta, Bering Sea. Proc. of the 12th Offshore Tech. Conf., Houston. 315-325.

Coulter, H.W. and R.R. Migliaccio 1966. Effects of the earthquake of March 27, 1964 at Valdez, Alaska. U.S. Geol. Survey Professional Paper 542-C.

Dengler, A.T., P. Wilde, E.K. Noda, and W.R. Normark 1984. Turbidity currents generated by Hurricane Iwa. Geo-Mar. Letters. 4: 5-11.

Dill, R.F. 1969. Earthquake effects on fill of Scripps Submarine Canyon. Geol. Soc. of Am. Bull. 80: 321-328.

Duncan, J.M. and A.L. Buchignani 1973. Failure of underwater slope in San Francisco Bay. J. of the Soil Mech. and Found. Div., ASCE. 79: 687-703.

Everingham, I.B. 1974. The major Papua New Guinean Earthquakes near Madang (1970) and beneath the north Solomon Sea (1971). Proc., Fifth World Conf. on Earthquake Eng. 1: 3-6.

Farrow, G.E., J.P.M. Syvitski and V. Tunnicliffe 1983. Suspended particulate loading on the macrobenthos in a highly turbid fjord: Knight Inlet, British Columbia. Can. J. of Fisheries and Aquatic Sci. 40(suppl. 1): 273-288.

Field, M.E. 1981. Sediment mass-transport in basins: controls and patterns. In R.G. Douglas, E.P. Colburn, and D.S. Gorsline, Short Course Notes Pacific Section, SEPM, San Francisco, p. 61-83.

Field, M.E. and S.H. Clarke 1979. Small-scale slumps and slides and their significance for basin slope processes, Southern California Borderland. SEPM Spec. Pub. No. 27, p. 223-230.

Field, M.E., S.H. Clarke, and M.E. White 1980. Geology and Geologic Hazards of Offshore Eel River Basin, Northern California Continental Margin. U.S. Geol. Survey Open-File Rep. 80-1080.

Field, M.E. and B.D. Edwards 1980. Slopes of the southern California borderland: a regime of mass transport. In M.E. Gield, A.H. Bouma, I.P. Colburn, R.G. Douglas, and J.C. Ingle (eds.), Proc. of the Quat. depositional envir. of the Pacific Coast: Pacific Coast Paleogeog, Symp. No. 4, SEPM, Pacific Section, p. 169-184.

Field, M.E., J.V. Gardner, A.E. Jennings, and B.D. Edwards 1982. Earthquake-induced sediment failures on a 0.25° slope, Klamath River delta, Cal. Geol. 10: 542-546.

Field, M.E., J.V. Gardner, D.E. Drake, and D.A. Cacchione 1987. Tectonic morphology of offshore Eel River Basin, California. In H. Schymiczek and R. Suchsland (eds.), Tectonics, Sedimentation, and Evolution of the Eel River and Other Coastal Basins of Northern California, San Juaquin Geological Society Misc. Publication No. 37, p. 41-48.

Field, M.E. and A.E. Jennings, 1987. Seafloor gas seeps triggered by a northern California earthquake. Mar. Geol. 77: 39-51.

Field, M.E. and W.C. Richmond 1980. Sedimentary and structural patterns on the northern Santa Rosa-Cortes Ridge, southern California. Mar. Geol. 34: 79-98.

Hampton, M.A., 1972. The role of subaqueous debris flow in generating turbidity currents. J. of Sed. Pet. 42: 775-793.

Hampton, M.A. and A.H. Bouma 1977. Slope instability near the shelf break, western Gulf of Alaska. Mar. Geotech. 2: 309-331.

Hampton, M.A., A.H. Bouma, P.R. Carlson, B.F. Molnia, E.C. Clukey, and D.A. Sangrey 1978. Quantitative study of slope instability in the Gulf of Alaska. Proc. of the 10th Offshore Tech. Conf. 2307-2318.

Haner, B.E. and D.S. Gorsline 1978. Processes and morphology of continental slope between Santa Monica and Dume submarine canyons, Southern California. Mar. Geol. 28: 77-87.

Hay, A.E., J.W. Murray and R.W. Burling 1983. Submarine channels in Rupert Inlet, British Columbia: I. Morphology. Sed. Geol. 36: 289-315.

Hein, F.J. and D.S. Gorsline 1981. Geotechnical aspects of fine-grained mass flow deposits: California Continental Borderland. Geomar. Letters. 1: 1-5.

Herzer, R.H. 1975. Uneven submarine topography south of Mernoo Gap-the result of volcanism and submarine sliding (note). New Zealand J. of Geol. and Geophys. 18: 183-188.

Herzer, R.H. 1979. Submarine slides and submarine canyons on the continental slope off Canterbury, New Zealand. New Zealand J. of Geol. and Geophys. 22: 391-406.

Houtz, R.E. 1962. The 1953 Suva earthquake and tsunami. Bull. of the Seis. Soc. of Am. 52: 1-12.

Jacobi, R.D. and C.L. Mrozowski 1979. Sediment slides and sediment waves in the Bonin Trough, Western Pacific. Mar. Geol. 29: M1-M9.

Kachadoorian, R. 1965. Effects of the earthquake of March 27, 1964 at Whittier, Alaska. U.S. Geol. Survey Prof. Paper 542-B.

Kostaschuk, R.A. and S.B. McCann 1983. Observations on delta-forming processes in a fjord-head delta, British Columbia. Sed. Geol. 36: 269-288.

Krause, D.C., W.C. White, D.J.W. Piper, and B.C. Heezen 1970. Turbidity currents and cable breaks in the western New Britain Trench. Geol. Soc. of Am. Bull. 81: 2153-2160.

Lajoie, K.R. and D.K. Keefer 1981. Investigations of the November 1980 earthquake in Humboldt County, California. U.S. Geol. Survey Open-File Rep. 81-397.

Lee, H.J. and B.D. Edwards 1986. Regional method to assess offshore slope stability. J. of Geotech. Eng. 112: 489-509.

Lee, H.J., B.D. Edwards, and M.E. Field 1981. Geotechnical analysis of a submarine slump, Eureka, California. Proc. of the 13th Annual Offshore Tech. Conf. 53-65.

Lee, H.J. and W.C. Schwab 1983. Geotechnical framework, northeastern Gulf of Alaska. U.S. Geol. Survey Open-File Rep. 83-499.

Lemke, R.W. 1967. Effects of the earthquake of March 27, 1964, at Seward, Alaska. U.S. Geol. Survey Prof. Paper 542-E.

Lewis, K.B. 1971. Slumping on a continental slope inclined at 1°-4°. Sedimentology. 16: 97-110.

Lewis, K.B. (compiler) 1985. New seismic profiles, cores, and dated rocks from the Hikurangi Margin, New Zealand. NZOI Oceanographic Field Report 22.

Liang, Yuan-bo and Bo Lu 1986. Acoustic environment and physico-mechanical properties of the shelf seabed off the Pearl River mouth. Mar. Geotech. 6: 377-392.

Lipman, P.W., W.R. Normark, J.G. Moore, J.B. Wilson and C.E. Gutmacher 1988. The giant Alika debris slide, Mauna Loa, Hawaii. J. of Geophys. Res. 93: 4239-4299.

Luternauer, J.L. and W.D.L. Finn 1983. Stability of the Fraser River delta front. Can. Geotech. J. 20: 603-616.

Malouta, D.N., D.S. Gorsline, and S.E. Thornton 1981. Processes and rates of recent (Holocene) basin filling in at active transform margin: Santa Monica Basin, California Continental Borderland. J. of Sed. Pet. 51: 1077-1095.

Marlow, M.S., D.W. Scholl, E.C. Buffington, R.E. Boyce, T.R. Alpha, P.J. Smith and C.J. Shipek 1970. Buldir Depression-a late Tertiary graben on the Aleutian Ridge, Alaska. Mar. Geol. 8: 85-108.

Marshall, N.F. 1978. Large storm-induced sediment slump reopens an unknown Scripps Submarine Canyon tributary. In D.J. Stanley and G. Kelling (eds.), Sedimentation in Submarine Canyons, Fans, and Trenches, p. 73-84. Stroudsburg, PA: Dowden, Hutchinson and Ross, Inc.

Middleton, G.V. and M.A. Hampton 1976. Subaqueous sediment transport and deposition by sediment gravity flows. In D.J. Stanley and D.J.P. Swift (eds.), Marine Sediment Transport and Environmental Management, p. 197-218. New York: Wiley and Sons.

Milliman, J.D. and R.H. Meade 1983. World-wide delivery of river sediment to the oceans. The J. of Geol. 91: 1-21.

Molnia, B.F., P.R. Carlson and T.R. Bruns 1977. Large submarine slide in Kayak Trough, Gulf of Alaska. Geol. Soc. of Am. Reviews in Eng. Geol. 3: 137-148.

Molnia, B.F., 1982. Geology of the Alsek sediment instability area. U.S. Dept. of Commerce, NOAA, OCSEAP Final Report 47, p 40-55.

Moore, J.G. 1964. Giant submarine landslides on the Hawaiian Ridge. U.S. Geol. Survey Prof. Paper 501-D.

Moore, J.G. and G.W. Moore 1984. Deposit from a giant wave on the island of Lanai, Hawaii. Science. 226: 1312-1315.

Murty, T.S. 1979. Submarine slide-generated water waves in Kitimat Inlet, British Columbia. J. of Geophys. Res. 84: 7777-7779.

Nardin, T.R., F.J. Hein, D.S. Gorsline, and B.D. Edwards 1979a. A review of mass movement processes, sediment and acoustic characteristics and contrasts in slope and base-of-slope systems versus canyon-fan-basin floor systems. SEPM Special Publication No. 27, p. 61-74.

Nardin, T.R., B.D. Edwards and D.S. Gorsline 1979b. Santa Cruz Basin, California Borderland: dominance of slope processes in basin sedimentation. SEPM Special Publication No. 27, p. 209-221.

378

Nelson, C.H., 1976. Late Pleistocene and Holocene depositional trends, processes, and history of Astoria Deep-Sea Fan, northeast Pacific. Mar. Geol. 20: 129-173.

Nittrouer, C.A. and R.W. Sternberg 1981. The formation of sedimentary strata in an allochthonous shelf environment: the Washington continental shelf. Mar. Geol. 42: 201-232.

Normark, W.R. 1974. Ranger submarine slide, northern Sebastian Vizcaino Bay, Baja California, Mexico. Geol. Soc. of Am. Bull. 85: 781-784.

Okusa, S., K. Nemoto, T. Nakamura and S. Aoki 1988. Submarine landslides in the Cenozoic orogenic island arc. Proc. of the Fifth Inter. Symp. on Landslides.

Plafker, G., R. Kachadoorian, E.B. Eckel, and L.R. Mayo 1969. Effects of the earthquake of March 27, 1964 on various communities. U.S. Geol. Survey Prof. Paper 542-G.

Ploessel, M.R., S.C. Crissman, J.H. Rudat, R. Son, C.F. Lee, R.G. Randall, and M.P. Norton 1979. Summary of potential hazards and engineering constraints, proposed OCS Lease Sale No. 48, offshore Southern California. Proc. of the 11th Offshore Tech. Conf. 355-363

Prior, D.B., B.D. Bornhold, J.M. Coleman, and W.R. Bryant 1982a. Morphology of a submarine slide, Kitimat Arm, British Columbia. Geol. 10: 588-592.

Prior, D.B., B.D. Bornhold and M.W. Johns 1986b. Active sand transport along a fjord-bottom channel, Bute Inlet, British Columbia. Geol. 14: 581-584.

Prior, D.B., J.M. Coleman, and B.D. Bornhold 1982b. Results of a known seafloor instability event. Geo-Mar. Letters. 2: 117-122.

Prior, D.B., W.J. Wiseman, and R. Gilbert 1981. Submarine slope processes on a fan delta, Howe Sound, British Columbia. Geo-Mar. Letters. 1: 85-90.

Prior, D.B., Z.-S. Yang, B.D. Bornhold, G.H. Keller, N.Z. Lu, W.J. Wiseman, L.D. Wright, and J. Zhang 1986a. Active slope failure, sediment collapse, and silt flows on the modern subaqueous Huanghe (Yellow River) delta. Geo-Mar. Letters. 6: 85-95.

Reimnitz, E. 1972. Effects in the Copper River delta. In The Great Alaska Earthquake of 1964, vol. 6: oceanography and Coastal Engineering, p. 290-302. Washington: National Research Council, National Academy of Sciences.

Reimnitz, E. and M. Gutierrez-Estrada 1970. Rapid changes in the head of the Rio Balsas submarine canyon system, Mexico. Mar. Geol. 8: 245-258.

Reynolds, Suzanne, 1987. A recent turbidity current event, Hueneme Fan, California: reconstruction of flow properties. Sedimentology. 34: 129-137.

Richmond, W.C. and D.J. Burdick 1981. Geologic hazards and constraints of offshore norhern and central California. Proc. of the 13th Annual Offshore Tech. Conf. 4: 9-17.

Schwab, W.C. and H.J. Lee 1983. Geotechnical analyses of submarine landslides in glacial marine sediment, northeast Gulf of Alaska. In B.F. Molnia, Glacial-Marine Sedimentation, p. 145-184. New York: Plenum Press.

Schwab, W.C. and H.J. Lee 1988. Causes of two slope-failure types in continental-shelf sediment, north-eastern Gulf of Alaska. J. of Sed. Pet. 58: 1-11.

Schwab, W.C., H.J. Lee, and B.F. Molnia 1988. Causes of varied sediment gravity flow types on the Alsek prodelta, northeast Gulf of Alaska. Mar. Geotech. 7: 317-342.

Shepard, F.P. 1932. Landslide-modifications of submarine valleys. Trans., Am. Geophys. Union. 13: 226-230.

Shepard, F.P. 1933. Depth changes in Sagami Bay during the great Japanese earthquake. The J. of Geol. 41: 527-536.

Shepard, F.P. 1951. Mass movements in submarine canyon heads. Trans., Am. Geophys. Union. 32: 405-418.

Shepard, F.P. 1963. Submarine canyons. In M. Hill (ed.), The Sea, p. 480-506. New York: John Wiley.

Shepard, F.P. and N.F. Marshall 1973. Storm-generated current in La Jolla Submarine Canyon, California. Mar. Geol. 15: M19-M24.

Shepard, F.P., P.A. McLoughlin, N.F. Marshall, and G.G. Sullivan 1977. Current-meter readings of low-speed turbidity currents. Geol. 5: 297-301.

Syvitski, J.P.M., D.C. Burrell and J.M. Skei 1987. Fjords: Processes and Products. New York: Springer-Verlag.

Syvitski, J.P.M. and G.E. Farrow 1983. Structures and processes in bayhead deltas: Knight and Bute Inlet, British Columbia. Sed. Geol. 36: 217-244.

Terzaghi, Karl 1956. Varieties of submarine slope failures. Proc. of the 8th Tex. Conf. on Soil Mech. and Found. Eng, p. 1-41.

Thornton, S.E. 1986. Origin of mass flow sedimentary structures in hemipelagic basin deposits: Santa Barbara Basin, California Borderland. Geo-Mar. Letters. 6: 15-19.

Varnes, D.J. 1958. Landslide types and processes. In E.B. Eckel (Ed.), Landslides and Engineering Practice. Highway Res. Board Spec. Rep. 29, p. 20-47.

Landslides: Extent and Economic Significance, Brabb & Harrod (eds)
© 1989 Balkema, Rotterdam. ISBN 90 6191 876 6

Landslides in southeastern Europe: Extent and economic significance

P.Anagnosti
Energoprojekt, Belgrade, Yugoslavia

S.Cavounides
Greece

G.Petrides
Cyprus

ABSTRACT: This review of geological and geotechnical conditions relevant to landsliding from northwestern Yugoslavia to Greece and Cyprus has been made to establish and emphasize the most common situations where landslides occur. The area under consideration is complex geologically with very different mechanisms and size of landslide events.

Activities of various governmental and professional bodies in Yugoslavia, Greece, and Cyprus are discussed briefly to establish a background for planning landuse and adequate legislation. No figures are available for the extent or financial value of landslide problems, but the damage is substantial.

1. INTRODUCTION

This report provides an overview of the landslide problems in a 1,600 km-wide area of southeastern Europe, comprising Yugoslavia, Greece, and Cyprus. The location of these countries in relation to other European countries is shown on Fig. 1.

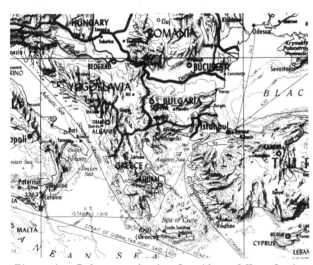

Figure 1. Index map showing location of Yugoslavia, Greece, and Cyprus.

The size of the area and obvious differences in climatic conditions, ranging from central European continental climate at the northern frontier of Yugoslavia to the warm, Mediterranean climate in southern Greece and Cyprus, present considerable difficulties for the attempt to generalize landslide problems.

On the other hand, the geological conditions, such as types of rocks, tectonic features, and stratigraphy are fairly similar in spite of the size of the area. This similarity may be attributed to the pronounced effects of the Alpine orogeny and the development of the southeastern Alps extending along the Adriatic coast. The principal tectonic elements are shown on Fig. 2. The oldest rocks are schists of early Paleozoic age. These are overlain by innumerable intrusions and outpouring of lava. The main sedimentary formations are carbonate rocks and flysch of Late Carboniferous age. These are overlain by Pliocene and Pleistocene fanglomerates.

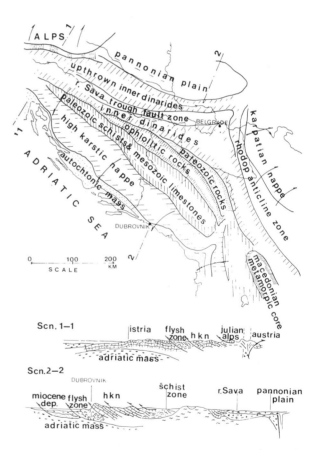

Figure 2. Tectonic elements of Yugoslavia (after K. Petkovic).

The Alpine orogeny produced many low-angled faults with dips ranging between 15° and 40°. The thrust faults are zones of weakness that help promote landsliding. The principal faults and earthquake epicenters are shown on Fig. 3.

In northern Yugoslavia, the average temperature during winter ranges from -4° to -8° and during summer from 15° to 22°C. The annual average precipitation varies from 800 to 1,000 mm. In southern Greece and Cyprus, the average winter

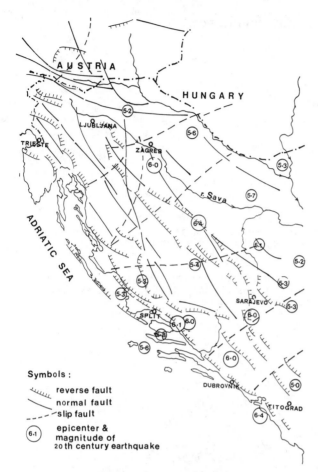

Figure 3. Principal faults, epicenter, and magnitude of twentieth century earthquakes.

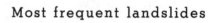

zone

| Most intensive seismic activity |
| Most frequent landslides |

Figure 4. Relation between landslides and seismic activity in southeastern Europe.

temperature is approximately 8 to 10°C, and the average summer temperature is 30 to 32°C. The annual average precipitation is as much as 400 mm. However, some mountainous areas have much more precipitation, from 1,000 mm to 1,800 mm and even extremes as much as 5,000 mm, and much lower temperatures.

The average length of the main rivers is relatively small, averaging 300 to 350 km. Many rivers discharging directly into the sea are even shorter, 50 to 100 km in length. These rivers have high gradients and high erosion potential. When they are associated with flood water, they trigger large landslides and rockfalls and create deep gullies in weathered rocks and unfavorably-oriented bedded rocks. Numerous landslides and rockfalls are also triggered by earthquakes. The relation between landslides and seismic activity is shown on Fig. 4.

2. MAIN CONDITIONS FOR GROUND INSTABILITY

Materials susceptible to slope failure are:
(a) Hard rocks (limestone, dolomite, volcanic rocks, etc.); discontinuities caused by structural forms, and their dips.
(b) Mixed hard and soft rocks (flysch, diabase and chert, etc.); unfavorably located and oriented structural forms, and pronounced weathering in soft rocks.
(c) Soft rocks (claystone, marl, etc.); zones of extensive weathering and cracks.
(d) Loess; structural instability when saturated.
(e) Deposits of soil and rock fragments overlying weathered zones of weak rocks.
(f) Residual clay, clayey soil deposits and fine-grained soil accumulations on hillside slopes.

2.1 Landslides in Yugoslavia

Government agencies for prevention and/or mitigation of landslides are established in each administrative district, but their activity is mostly oriented around safeguarding people living in endangered areas. The initial signs of hillslope instability that might endanger people and their property are rarely recognized and appreciated. Thus posterior activities are more or less common practice for a great majority of the damaging landslides. Expert advice is usually requested for post-landslide stabilization work. Road construction, hydro-electric plants, and dams have appropriate attention paid during the design stage to secure the area against instability.

Due to the appreciable complexity of landslides in Yugoslavia, the common procedure is for specialists in soil mechanics, rock mechanics and engineering geology to study landslide problems. Solutions are also subject to competition between construction firms which may propose their alternatives based on experience and available facilities.

All works required for stabilization of landslides and other remedial measures (such as the relocation of people, houses, industrial plants, etc.) are financed either by governmental funds for natural disasters or by investors who constructed in the landslide-prone areas. In some cases, insurance provides the funds for stabilization or remedial measures. No published official figures are available for the financial value of landslide problems.

Town planning regulations and land-use legislation are not sufficiently specific in relation to land-slide problems. These regulations also do not consider natural and other differences in the various parts of the Yugoslav territory.

Highly populated areas and areas with high-cost structures, plants, and other facilities have more carefully written regulations that require consider-ing evaluation of land stability during the land planning stage, but the extent of the insight and its quality have not been specifically determined.

382

2.2 Landslides in Greece

The limited information obtained by the author does not permit the extensive review of all landslide situations which have undoubtedly happened in the long history of Greece. Similarities in geological structure with Yugoslavia and information from available publications and other communications indicate that the landslide problem in Greece is significant. The landslides appear in two basic forms:

1. As phenomena created during geological time with gradual changes in environmental conditions which are bringing the area to the state of potential instability or very slow creep movements. Extreme seasonal changes in precipitation or ground water, with or without other contributive factors, such as surface erosion or earthquakes, result in active landslides.

2. As phenomena created by various human activities in areas prone to the development of landslides. These human activities may consist of improper use of land, change in vegetation, change in ground water, change in ground shape by excavations and fills, etc. Such changes may be associated with the seasonal changes in natural factors mentioned above.

The significance of landslides in Greece is also related to the density of population and the general use of land for industrial development, for other building activities, for tourism, etc. The consequences of landslides are felt mostly in densely populated areas and in areas associated with construction activities.

Several landslide articles have been published in Greece, particularly during the last 20 to 30 years when contemporary concepts of soil mechanics, rock mechanics, and engineering geology became introduced into design studies. National associations and governmental institutions, such as the Ministry of Public Works, Energy, etc., supported studies and investigations related to landslide problems. Most of the valuable publications, particularly case studies, were published in Greek, so that their usefulness to the world community is limited.

As the result of these investigations, Greece can be roughly divided into zones of less frequent and more frequent landslides, depending on the basic geological situation and amount of slope. Areas underlain by clay and flysch are the most susceptible to landsliding.

2.3 Landslides in Cyprus

Cyprus is also susceptible to landslides. Some areas were formed as the result of large-scale slope instability during the Pleistocene when precipitation was much higher and erosion of the ground more intensive than at present. The presence of highly-weathered volcanic and metamorphic rocks, tuff, bentonite, and remnants of ancient landslides provide landslide-prone materials and a permanent threat for human lives and construction activities. Landslides in these materials commonly cause damage to villages, local roads, and pipelines. In particular, complex landslides are extensive in the Troodos massif area which occupies about 40 percent of Cyprus.

Systematic investigations of landslides in some parts of the country were carried out recently by the National Geological Survey Department, some of them in association with the British Geological Department. Stabilization works for individual landslides were carried out by the Public Works Department and the Water Development Department of the Cyprus Government Ministries and different municipalities.

3. CASE HISTORIES

3.1 Landslide of Gradot Hill

A landslide at Gradot Hill, southeastern Yugoslavia, on Sept. 5, 1956 was 800 m long and 70 m deep. It formed a dam and created a lake (Fig. 5). Vertical

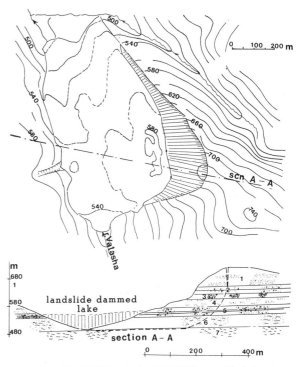

Figure 5. Landslide at Gradot Hill, southeastern Yugoslavia. 1, tuff; 2, cemented sand and gravel; 3, conglomerate; 4, weathered tuff; 5, weakly-cemented gravel; 6, sandy clay; 7, clay.

movement of the soil mass at the top of the slide was about 130 m, whereas the total height of the hill-side involved in the landslide was 230 m. The hill is underlain by horizontally-bedded tuff, conglomer-ate, and clay of Cenozoic age. The beds at the bottom of the valley consist of clay with peak shear strength of about $0'_p = 24^\circ$ and residual strength of about $0'_r = 15^\circ$. Gradual cutting of the hill by river erosion and long term shear strain in the bed of clay caused the gradual decrease in shear strength and eventual displacements, which began as large vertical cracks in the more rigid overlying strata (tuff and conglomerate).

A few earthquakes probably contributed to the development of cracks and a decrease of overall strength. The landslide involved an area of about 200,000 m^2 and a volume of approximately 20 million m^3 of material.

Treatment of the landslide consisted of constructing a regulated outflow channel for safe release of accumulated water in the reservoir. The channel has a rectangular shape consisting of articulated sections placed over the landslide dam.

3.2 Zavoj Slide in Visotchitza River Valley

This landslide occurred from 23 to 27 February, 1963 during intensive melting of snow. During several days of intensive displacements approximately 4 million m^3 of clayey soil slid over a surface 1,500 m long. About 1.5 million m^3 of soil reached the bottom of the Visotchitza River valley and formed a dam 35 m heigh and 500 m long along the river channel (Fig. 6). A lake approximately 10 km long with 30 million m^3 of water was formed and created a threat for heavily populated areas downstream.

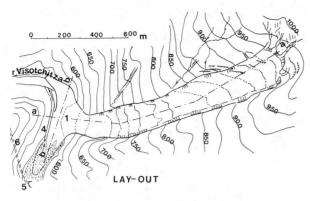

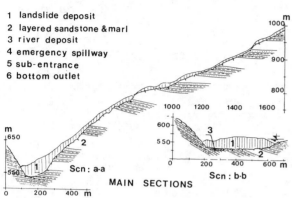

1 landslide deposit
2 layered sandstone & marl
3 river deposit
4 emergency spillway
5 sub-entrance
6 bottom outlet

MAIN SECTIONS

Figure 6. Landslide near Zavoj in the Visotchitza River valley.

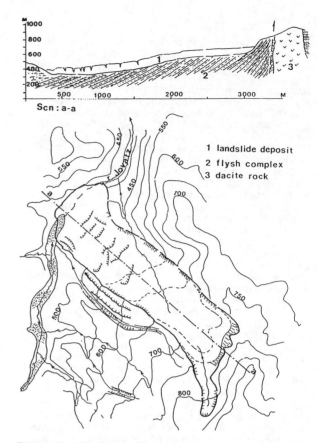

1 landslide deposit
2 flysh complex
3 dacite rock

Figure 7. Landslide in the Jovatz River valley, southeastern Yugoslavia.

One feature of this landslide is the very long distance over which the soil mass traveled from its original location to the Visotchitza River. The slope angle is about 17° where the slide began and about 50° just above the river bed. This change in slope accelerated soil movement and increased pressure in the central part of the landslide near the river channel.

The flow of water into the landslide-dammed reservoir was about 25 m³ per second, a high rate requiring a temporary spillway. The permanent release of accumulated water was accomplished by a 2.5 m diameter tunnel excavated into the left bank at river bed elevation. The entrance of the tunnel was provided by blasting the last 10 m of rock and reshaping this cavity after discharge of the accumulated water. From 1985 to 1988, the upstream part of the embankment created by the slide was used as the downstream part of an earth-rock fill dam 80 m above the river bed.

The Zavoj landslide destroyed approximately 100 houses and other structures. About 150,000 m² of the village could not be rehabilitated and was abandoned. Approximately 15 km² of arable land along the Visotchitza River was flooded and could not be used until the reservoir was drained two years later.

3.3 Landslide in Jovatz River Valley

A landslide on Feb. 15, 1977 involved an area of approximately 3 km² (Fig. 7). The initial stage of fast movement of about 70 m/day lasted several days. Ground displacements stopped in 20 days. The landslide dammed the Jovatz River creating a lake with about 500,000 m³ of water.

Bedrock in this area consists of alternating sandstone, siltstone, marl and claystone of late Eocene age, popularly referred to as flysch. These rocks are overlain by thick colluvium. The ground surface is cut by many deep gullies formed by

surface waters. During the rainy season, particularly during spring, the area is water-saturated. Water is particularly abundant in the upper parts of the hill where it flows from strongly fissured volcanic rocks.

Drainage of the water from the reservoir in order to reduce or eliminate the threat to the heavily populated Morava River valley with a railway line connecting Yugoslavia to Greece was considered but not implemented. Instead, an existing erosion control structure situated 3 km downstream was reconstructed and reshaped to serve as an overflow weir with large bottom outlets. A retaining structure 10 m high was built to provide a 5 million m³ reservoir to accommodate the inflow of soil and water from the slide area. Fortunately, the weather stayed dry for several months after the landslide so that overtopping of the landslide dam was only about 0.3 to 0.5 cu. m/sec. Equilibrium has been established between stream velocity and the shape and grade of the new stream bed over the landslide dam.

In the Jovatz River valley, therefore, no attempts were made to provide safe discharge of the accumulated waters but, instead, a downstream control structure was constructed.

The Jovatz landslide destroyed 20 farms and covered about 7 km² of arable land with impounded water. The farms were rebuilt and the land used again about two years after the landslide occurred.

3.4 Landslides in Greece

Case histories from Greece cordially made available by Mr. S.Cavounidis (Athens, Greece) consist of landslides in marly clays associated with overlying conglomerates and underlying limestone (Figs. 8, 9 and 10). A common feature of these landslides is complex geological structure, high ground water, extensive weathering, and weak material. Other

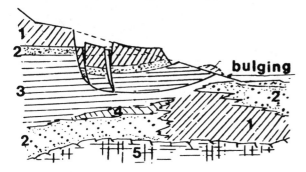

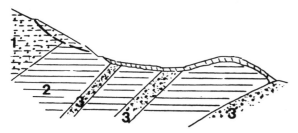

Figure 8. Tsaousi landslide, Greece: 1, marly clay; 2, sand, gravel and conglomerate; 3, blue marly clay; 4, red clay; 5, limestone.

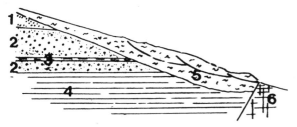

Figure 9. Kati landslide, Greece: 1, sand, gravel and clay; 2, marly clay; 3, sand, gravel and conglomerate.

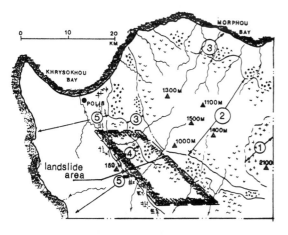

Figure 10 Kiafa landslide, Greece: 1, sand; 2, sand, gravel and conglomerate; 3, sandy marly clay; 4, marl; 5, soft organic clay; 6, limestone.

landslides in Greece have been described by Koukis (1982 and 1988).

3.5 Landslides in Cyprus

A case history from Cyprus, communicated by Mr. G.Petrides (Nicosia), is related to a larger area with complex geological conditions where various types of landslides have formed (Fig. 11). The landslides are related to low shear strength of bentonitic clay capped by limestone, conglomerate and sandstone, and high ground water.

4. CONCLUSIONS

Landslides are a substantial problem in Yugoslavia, Greece, and Cyprus, but the geographic extent of the problem and the magnitude of the damages have not been documented.

REFERENCES

Anagnosti, P. 1988. Instability phenomena in the zone of the Alpine Arc. ISLS Lausanne.
Cavounidis, S. & others 1979. A report on slide on slopes of marly clay. 6th ARC, 1, Singapore.
Cavounidis, S. & others 1987. Kiafa slide. IXth Int. CSMFE, Dublin.
Koukis, G. 1982. Mass movements in the Greek territory: a critical factor for environmental evaluation and development. Proc. 3rd Int. Cong. IAEG 3:233-243, New Delhi.
Koukis, G. 1988. Slope deformation phenomena related to the engineering geological conditions in Greece. ISLS Lausanne.
Makropoulos, R.C. & P.W.Burton 1984. Greek tectonics and seismicity. Tectonophysics 106:274-304. Elsevier Sc.P.
Northmore, Hobbs, Charalambous & Petrides 1988. Complex landslide in the Kannavion, Melange and Mamonia formations of southwest Cyprus. ISLS Lausanne.

Figure 11. Landslide area in Cyprus. 1, Troodos gabbro; 2, diabase; 3, pillow lava/andesite; 4, Mamonia melange; 5, chalk, limestone, and marl.

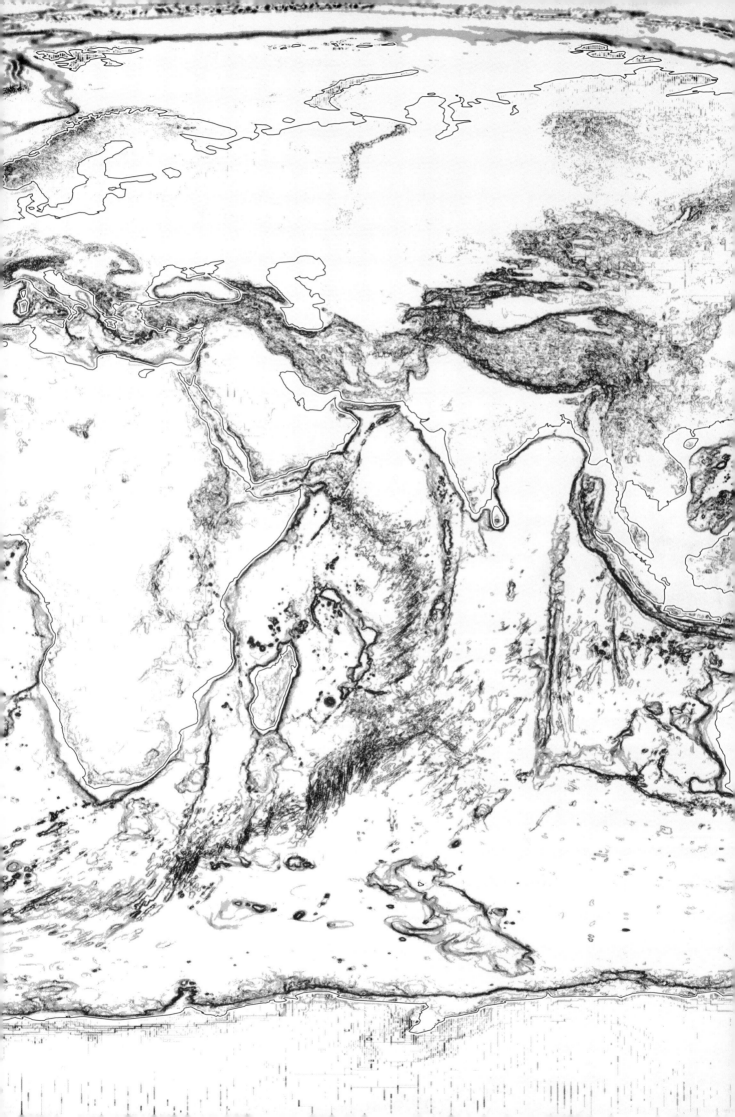